WAVE PROPAGATION

Edited by **Andrey Petrin**

Wave Propagation

Edited by Andrey Petrin

Published by InTech

Janeza Trdine 9, 51000 Rijeka, Croatia

Publishing Process Manager Ana Nikolic

Technical Editor Teodora Smiljanic

Cover Designer InTech Design Team

Additional hard copies can be obtained from orders@intechopen.com

Wave Propagation, Edited by Andrey Petrin
p. cm.
ISBN 978-953-307-275-3

For EU product safety concerns:
IN TECH d.o.o., Prolaz Marije Krucifikse Kozulić 3, 51000 Rijeka, Croatia,
info@intechopen.com or visit our website at intechopen.com.

Meet the editor

Andrey Petrin graduated and received the Ph.D. degree in physics and mathematics from the Moscow Institute of Physics and Technology, Moscow, Russia, in 1983 and 1988, respectively. In 1983, he joined the Space Corporation "Energia," U.S.S.R. In 1990, he joined the Moscow State University of Forestry as a Lecturer of Physics. Since 1996, he has been with the Theoretical Department of the Joint Institute for High Temperatures, Russian Academy of Sciences, Moscow, as a leading scientific researcher. He specialized in theory of electromagnetic waves propagation in complex media, nonlinear interaction of microwave and magnetized plasma of gas discharges, engineering applications of plasma sources, electrically stimulated heat and mass transfer in liquids, heat transfer in nanoscales, nanofocusing of light. He is the author of more than 40 papers in referenced scientific journals and the inventor of 4 U.S. patents.

Contents

Preface

In the recent decades, there has been a growing interest in telecommunication and wave propagation in complex systems, micro- and nanotechnology. The advances in these engineering directions give rise to new applications and new types of materials with unique electromagnetic and mechanical properties. This book is devoted to the modern methods in electrodynamics which have been developed to describe wave propagation in these modern materials, systems and nanodevices.

The book collects original and innovative research studies of the experienced and actively working scientists in the field of wave propagation which produced new methods in this area of research and obtained new and important results.

Every chapter of this book is the result of the authors achieved in the particular field of research. The themes of the studies are varied from investigation on modern applications such as metamaterials, photonic crystals and nanofocusing of light to the traditional engineering applications of electrodynamics such as antennas, waveguides and radar investigations.

The book contains 26 chapters on the following themes:

- Wave Propagation in Metamaterials, Micro/nanostructures;
- Light Wave Propagation and Nanofocusing;
- Antennas and Waveguides;
- Wave Propagation in Plasmas;
- Electromagnetic Waves Absorption and No Reflection Phenomena;
- Nonlinear Phenomena and Electromagnetic Wave Generation;
- Radar Investigations.

It is necessary to emphasise that this book is not a textbook. It is important that the results combined in it are taken "from the desks of researchers". Therefore, I am sure that in this book the interested and actively working readers (scientists, engineers and students) will find many interesting results and new ideas.

Andrey Petrin
Joint Institute for High Temperatures of Russian Academy of Science, Russia

a_petrin@mail.ru

Part 1

Wave Propagation in Metamaterials, Micro/nanostructures

1

Wave Propagation Inside a Cylindrical, Left Handed, Chiral World

Pierre Hillion
Institut Henri Poincaré
France

1. Introduction

Isotropic metamaterials, with negative permittivity and permeability, also called left-handed [Veselago, 1958] [Pendry, 2000], [Sihvola 2007] have opened the way to new physical properties, different from those obtained with a conventional material. So, it is natural to inquire how these properties transform when these metamaterials are also chiral because of the importance of chirality in Nature. Some left-handed materials [Pendry 2006] and metamaterials [Grbic & Elefthertades, 2006] have been recently manufactured. We consider here a metachiral circular cylindrical medium with negative permittivity and permeability endowed with the Post constitutive relations [Post, 1962]. Using the cylindrical coordinates $r > 0$, θ, z and, assuming fields that do not depend on θ, we analyze the propagation of harmonic Bessel beams inside this medium. Two different modes exist charac-terized by negative refractive indices, function of permittivity and permeability but also of chirality which may be positive or negative with consequences on the Poynting vector carefully analyzed. The more difficult problem of wave propagation in a spherical, left handed, chiral world is succinctly discussed in an appendix.

2. Cylindrical Maxwell's equations and Post's constitutive relations

With the cylindrical coordinates $r > 0$, θ, z, the Maxwell equations in a circular cylindrical medium are for fields that do not depend on θ

$$\begin{aligned} &-\partial_z E_\theta + 1/c\, \partial_t B_r = 0, \\ &\partial_z E_r - \partial_r E_z + 1/c\, \partial_t B_\theta = 0, \\ &(\partial_r + 1/r)E_\theta + 1/c\, \partial_t B_z = 0 \end{aligned} \tag{1a}$$

$$\begin{aligned} &\partial_z H_\theta + 1/c\, \partial_t D_r = 0, \\ &\partial_z H_r - \partial_r H_z - 1/c\, \partial_t D_\theta = 0, \\ &(\partial_r + 1/r)H_\theta - 1/c\, \partial_t D_z = 0, \end{aligned} \tag{1b}$$

$$(\partial_r + 1/r)B_r + \partial_z B_z = 0\,,$$
$$(\partial_r + 1/r)D_r + \partial_z D_z = 0 \tag{2}$$

We write $-|\varepsilon|$, $-|\mu|$ the negative permittivity and permeability [Pendry, 2006] in the meta-chiral cylindrical medium and the Post constitutive relations are [Post, 1962]

$$\mathbf{D} = -|\varepsilon|\ \mathbf{E} + i\xi\,\mathbf{B}\,, \qquad \mathbf{H} = -\mathbf{B}/|\mu|\ + i\xi\,\mathbf{E}\ ,\ i = \sqrt{-1} \tag{4}$$

in which ξ is the chirality parameter assumed to be real.
From (2), (3), we get at once the divergence equation satisfied by the electric field

$$(\partial_r + 1/r)E_r + \partial_z E_z = 0 \tag{4}$$

3. Cylindrical harmonic Bessel modes

3.1 The Bessel solutions of Maxwell's equations

Substituting (3) into (1b) gives

$$\partial_z(-B_\theta/|\mu| + i\xi\,E_\theta) + 1/c\partial_t(-|\varepsilon|\ E_r + i\xi\,B_r) = 0$$
$$\partial_z(-B_r/|\mu| + i\xi\,E_r) - \partial_r(-B_z/|\mu| + i\xi\,E_z) - 1/c\partial_t(-|\varepsilon|\ E_\theta + i\xi\,B_\theta) = 0 \tag{5}$$
$$(\partial_r + 1/r)(-B_\theta/|\mu| + i\xi\,E_\theta) - 1/c\partial_t(-|\varepsilon|\ E_z + i\xi\,B_z) = 0$$

Taking into account (1a), Eqs.(5) become

$$-1/|\mu|\ \partial_z B_\theta - |\varepsilon|/c\ \partial_t E_r + 2i\xi/c\ \partial_t B_r = 0$$
$$-1/|\mu|\ \partial_z B_r + 1/|\mu|\ \partial_r B_z + |\varepsilon|/c\ \partial_t E_\theta + 2i\xi/c\ \partial_t B_\theta = 0 \tag{5a}$$
$$-1/|\mu|\ (\partial_r + 1/r)B_\theta + |\varepsilon|/c\ \partial_t E_z + 2i\xi/c\ \partial_t B_z = 0$$

Applying the time derivative operator ∂_t to (5a) gives

$$1/c\ \partial_t\partial_z B_\theta + |\varepsilon|\,|\mu|/c^2\ \partial_t^2 E_r - 2i\xi|\mu|/c^2\ \partial_t^2 B_r = 0 \tag{6a}$$

$$1/c\ \partial_t\partial_z B_r - 1/|c\ \partial_t\partial_r B_z - |\varepsilon|\,|\mu|/c^2\ \partial_t^2 E_\theta + 2i\xi|\mu|/c^2\ \partial_t^2 B_\theta = 0 \tag{6b}$$

$$1/c\ \partial_t(\partial_r + 1/r)B_\theta - |\varepsilon|\,|\mu|/c^2\ \partial_t^2 E_z + 2i\xi|\mu|/c^2\ \partial_t^2 B_z = 0 \tag{6c}$$

Using (1a) and the divergence equation (4), we have in the first and last terms of (6a)

$$1/c\ \partial_t\partial_z B_\theta = \partial_r\partial_z E_z - \partial_z^2 E_r = -\ (\partial_r^2 + 1/r\partial_r - 1/r^2 + \partial_z^2)E_r$$

$$1/c^2\ \partial_t^2 B_r = 1/c\partial_t\partial_z E_\theta$$

Substituting these two relations into (6a) and introducing the wave operator [Morse & Feshbach, 1953]

$$\Delta_1 = \partial_r^2 + 1/r\partial_r - 1/r^2 + \partial_z^2 - |\varepsilon|\,|\mu|/c^2\,\partial_t^2 \tag{7}$$

we get the equation

$$\Delta_1 E_r + 2i\xi|\mu|/c\,\partial_z\partial_t E_\theta = 0 \tag{8a}$$

We have similarly in (6b) for the first two terms and for the last one taking into account (1a)

$$1/c\,\partial_t\partial_z B_r - 1/c\,\partial_t\partial_r B_z = (\partial_r^2 + 1/r\partial_r - 1/r^2 + \partial_z^2)E_\theta$$

$$1/c^2\,\partial_t^2 B_\theta = 1/c\partial_t\partial_r E_z - 1/c\partial_t\partial_z E_r$$

so that Eq.(6b) becomes

$$\Delta_1 E_\theta + 2i\xi|\mu|/c\,(\partial_r\partial_t E_z - \partial_z\partial_t E_r) = 0 \tag{8b}$$

Finally in (6c) , the first and third terms are according to (1a)

$$1/c\,\partial_t(\partial_r + 1/r)B_\theta = (\partial_r^2 + 1/r\partial_r + \partial_z^2)E_z$$

$$1/c^2\,\partial_t^2 B_z = -1/c(\partial_r + 1/r)\,\partial_t E_\theta$$

and, taking into account these two relations, we get

$$\Delta_0 E_z - 2i\xi|\mu|/c\,(\partial_r + 1/r)\,\partial_t E_\theta = 0 \tag{8c}$$

in which Δ_0 is the wave operator

$$\Delta_0 = \partial_r^2 + 1/r\partial_r + \partial_z^2 - |\varepsilon|\,|\mu|/c^2\,\partial_t^2 \tag{9}$$

We look for the solutions of Eqs.(8a,b,c) in the form inside the cylindrical medium:

$$\begin{aligned} E_r(r,z,t) &= E_r\,J_1(k_r r)\exp(i\omega t + ik_z z) \\ E_\theta(r,z,t) &= E_\theta\,J_1(k_r r)\exp(i\omega t + ik_z z) \\ E_z(r,z,t) &= E_z\,J_0(k_r r)\exp(i\omega t + ik_z z) \end{aligned} \tag{10}$$

in which J_0, J_1 are the Bessel functions of the first kind of order zero and one respectively while E_r, E_θ, E_z are arbitrary amplitudes. From now on, we use the two parameters

$$n^2 = |\varepsilon|\,|\mu|/c^2, \qquad \alpha = 2\xi|\mu|/c \tag{11}$$

Let us now substitute (10) into (8a,b,c). The Bessel functions with $k^2 = k_r^2 + k_z^2$ satisfy the following relations with the exponential factor $\exp(i\omega t + ik_z z)$ implicit

$$\Delta_1 J_1(k_r r) = -(k^2 - \omega^2 n^2)\,J_1(k_r r), \qquad \Delta_0 J_0(k_r r) = -(k^2 - \omega^2 n^2)\,J_0(k_r r) \tag{12a}$$

$$\partial_r J_0(k_r r) = -k_r J_1(k_r r), \qquad (\partial_r + 1/r)\,J_1(k_r r) = k_r J_0(k_r r) \tag{12a}$$

then, using (12a,b) we get the homogeneous system of equations on the amplitudes E_r, E_θ, E_z

$$(k^2-\omega^2n^2)E_r + i\omega\alpha k_z E_\theta = 0$$
$$(k^2-\omega^2n^2)E_\theta - \omega\alpha k_r E_z - i\omega\alpha k_z E_r = 0 \quad (13)$$
$$(k^2-\omega^2n^2)E_z - \omega\alpha k_r E_\theta = 0$$

This system has nontrivial solutions when its determinant is null and a simple calculation gives

$$(k^2-\omega^2n^2)\,[(k^2-\omega^2n^2)^2 - \alpha^2\omega^2k^2] = 0 \quad (14)$$

Leaving aside $k^2-\omega^2n^2= 0$, the equation (14) implies

$$k^2-\omega^2n^2 \pm\alpha\omega k = 0 \quad (15)$$

or in terms of refractive index $m = ck/\omega$: $m^2 \pm \alpha cm - cn^2 = 0$. These equations have four solutions, two positive and two negative But, it has been proved [Ziolkowski & Heyman, 2001] that in left handed materials, m must be taken negative: $m = -\,|\alpha c \pm (\alpha^2c^2 + 4n^2c^2)^{1/2}|$ so that introducing the $\gamma > I$ parameter

$$\gamma = (1+4n^2\alpha^{-2})^{1/2} = (1+|\varepsilon|\,|/|\mu|\,\xi^2)^{1/2} \quad (16)$$

the equation (15) has the two negative roots

$$k_1 = -\omega\,|\alpha|\,(1+\gamma)/2, \qquad k_2 = \omega\,|\alpha|\,(1-\gamma)/2 \quad (17)$$

So, there exist two modes with respective wave numbers k_1, k_2 able to propagate in the meta-chiral cylindrical medium, with two different negative indices of refraction $m_{1,2}. = ck_{1,2}/\omega$.

3.2 Amplitudes of harmonic Bessel beams

The **B**, **D**, **H** components of the electromagnetic field have the form (10), that is.

$$(B_r,D_r,H_r)(r,z,t) = (B_r, D_r, H_r)\,J_1(k_r r)\exp(i\omega t + ik_z z)$$
$$(B_\theta,D_\theta,H_\theta)(r,z,t) = (B_\theta, D_\theta, H_\theta)\,J_1(k_r r)\exp(i\omega t + ik_z z) \quad (18)$$
$$(B_z,D_z,H_z)(r,z,t) = (B_z, D_z, H_z)\,J_0(k_r r)\exp(i\omega t + ik_z z)$$

Then, in agreement with (15) and (17), we first assume $k_1^2- \omega^2n^2 = \alpha\omega k_1$. Deleting the exponential factor from (10), (18) and using (12a), we get at once from (13) in terms of $E_\theta \cong E_1$ with $k_r^2+k_z^2 = k_1^2$

$$E_r = -ik_zE_1/k_1, \qquad E_\theta = E_1, \qquad E_z = k_rE_1/k_1 \quad (19a)$$

Substituting (18) into (1a), taking into account (19a) and using (12a) give

$$B_r = ck_zE_1/\omega, \qquad B_\theta = ick_1E_1/\omega, \qquad B_z = ick_rE_1/\omega \quad (19b)$$

and, with (19a,b) substituted into the Post constitutive relations (3), we get

$$D_r = -ik_zD_1^{\dagger}E_1, \quad D_\theta = -k_1D_1^{\dagger}E_1, \quad D_z = -k_rD_1^{\dagger}E_1, \quad D_1^{\dagger} = |\varepsilon|/k_1 - c\xi/\omega \qquad (19c)$$

$$H_r = -k_zH_1^{\dagger}E_1, \quad H_\theta = -ik_1H_1^{\dagger}E_1, \quad H_z = -ik_r H_1^{\dagger}E_1, \quad H_1^{\dagger} = c/\omega|\mu| - \xi/k_1 \qquad (19d)$$

Similarly, with $k_2^2 - \omega^2n^2 = -\alpha\omega k_2$, $k_r^2+k_z^2 = k_2^2$, $E_\theta \cong E_2$, we get from (13 and (12a)

$$E_r = ik_zE_2/k_2, \qquad E_\theta = E_2, \qquad E_z = -k_rE_2/k_2 \qquad (20a)$$

$$B_r = ck_zE_2/\omega, \qquad B_\theta = -ick_2E_2/\omega, \qquad B_z = ick_rE_2/\omega \qquad (20b)$$

and, substituting (20a,b) into (3)

$$D_r = -ik_zD_2^{\dagger} E_2, \quad D_\theta = -k_2 D_2^{\dagger} E_2, \quad D_z = k_r D_2^{\dagger} E_2, \quad D_2^{\dagger} = |\varepsilon|/k_2 - c\xi/\omega \qquad (20c)$$

$$H_r = -k_zH_2^{\dagger}E_2, \quad H_\theta = ik_2 H_2^{\dagger}E_2, \quad H_z = -ik_r H_2^{\dagger}E_2, \quad H_2^{\dagger} = c/\omega|\mu| + \xi/k_2 \qquad (20d)$$

The expressions (19), (20) give in terms of E_1, E_2 the amplitudes of the two modes propagating in metachiral un-bounded cylindrical worlds.

3.3 Energy flow of Bessel waves

Using (10), (18) the Poynting vector $\mathbf{S} = c/8\pi\ (\mathbf{E}\wedge\mathbf{H}^*)$ where the asterisk denotes the complex conjugation, gives for the first mode

$$\begin{aligned} S_{1,r}(r,z,t) &= c/8\pi(E_\theta H_z^* - E_zH_\theta^*)(r,z,t) \\ &= 0 \\ S_{1,\theta}(r,z,t) &= c/8\pi(E_zH_r^* - E_rH_z^*)(r,z,t) \\ &= -ck_rk_z H_1^{\dagger}/4\pi k_1 J_0(k_rr) J_1(k_rr)\ |E_1]^2 \\ S_{1,z}(r,z,t) &= c/8\pi(E_rH_\theta^* - E_\theta H_r^*)(r,z,t) \\ &= ck_z H_1^{\dagger}/4\pi\ J_1^2(k_rr)\ |E_1]^2 \end{aligned} \qquad (21)$$

Now, according to (11) and (17):

$$\xi/k_1 = -2\xi[\omega|\alpha|(1+\gamma)]^{-1} = -c(\xi/|\xi|\ \omega|\mu|)\ (1+\gamma)^{-1} \qquad (22)$$

so that since according to (19d), $H_1^{\dagger} = c/\omega|\mu| - \xi/k_1$, we get

$$H_1^{\dagger} = c/\omega|\mu|\ [1+\xi/\ |\xi|(1+\gamma)^{-1}] \qquad (22a)$$

and $H_1^{\dagger} > 0$ whatever the sign of $\xi/|\xi|$ is. So for $k_z > 0$ (resp. $k_z < 0$) the z-component of the energy flow runs in the direction of the positive (resp.negative) z axis while according to (10) and (18), Bessel waves propagate in the opposite direction with the phase velocity $v_z = -\omega/k_z$. Consequently S_z and v_z are antiparallel, but, because $S_{1,\theta}$ is not null, the phase velocity is not strictly antiparallel to the energy flow.

A similar calculation for the second mode gives $S_{2,r}(r,z,t) = 0$ and

$$S_{2,\theta}(r,z,t) = ck_rk_z\, H_2\dagger/4\pi k_2\, J_0(k_rr)\, J_1(k_rr)\, |E_2]^2$$
$$S_{2,z}(r,z,t) = ck_z\, H_2\dagger/4\pi\, J_1^2(k_rr)\, |E_2]^2 \qquad (23)$$

Taking into account (11), (17), we have

$$\xi/k_2 = 2(\xi/\omega\, |\alpha|\,)\,(1-\gamma)^{-1} = c(\xi/|\xi|\,\omega\,|\mu|)(1-\gamma)^{-1} \qquad (24)$$

and, since according to (20d) $H_2\dagger = c/\omega|\mu| + \xi/k_2$, we get taking into account (24)

$$H_2\dagger = c/\omega|\mu|\,[1+ \xi/|\xi|\,(1-\gamma)^{-1}] \qquad (25)$$

$H_2\dagger$ is positive for $\xi/|\xi| = -1$ and for $\xi/|\xi| = 1$ with $\gamma > 2$ leading to the same conclusion as for the first mode while for $\xi/|\xi| = 1$ and $1 < \gamma < 2$ Bessel waves propagate in the same direction [Hu & Chui, 2002]. So, the harmonic Bessel waves may be considered as partially left-handed.

3.4 Evanescent waves

It is implicitly assumed in the previous sections that the wave numbers k_r, k_z are real which implies k_r^2, k_z^2 smaller than k_1^2, k_2^2 with $|k_2| < |k_1|$ according to (17). Suppose first $k_r^2 > k_1^2$, then

$$k_{1,z} = \pm\, i(k_r^2 - k_1^2)^{1/2}, \qquad k_{2,z} = \pm\, i(k_r^2 - k_2^2)^{1/2} \qquad (27)$$

with the plus (minus) sign in the $z > 0$ ($z< 0$) region to make $\exp(ik_zz)$ exponentially decreasing, the only solution physically acceptable. Both modes are evanescent but only the second mode if $k_1^2 > k_r^2 > k_2^2$.

Suppose now $k_z^2 > k_1^2$ then

$$k_{r(1,2)} = \pm ik_{s(1,2)}, \qquad k_{s(1,2)} = (k_z^2 - k_{1,2}^2)^{1/2} \qquad (28)$$

and

$$J_0(\pm ik_sr) = I_0\,(k_sr)\,, \qquad J_1(\pm ik_sr) = \pm I_1\,(k_sr) \qquad (28a)$$

in which I_0, I_1 are the Bessel functions of second kind of order zero, one respectively.These functions are expo-nentially growing with r and physically unacceptable in unbounded media.. Of course, if $k_1^2 > k_z^2 > k_2^2$ the first mode can exist.

4. Discussion

Wave propagation in chiral materials is made easy for media equipped with Post's constitutive relations because as electromagnetism, they are covariant under the Lorentz group. In a metachiral material, the refractive index m depends not only on ε, μ but also on the chirality ξ and in cylindrical geometry m may have four different expressions among which only the two negative ones are physically convenient. But, the Poynting vector **S** depends on the sign of ξ so that **S** and the phase velocity **v** may be parallel or antiparallel but not strictly because, as easily shown, the Poynting vector **S** is orthogonal to **E** but not to **H**, So that **E**, **H**, **S** do not form a cartesian frame. So, metachiral cylindrical media have some particular features. Wave propagationin uniaxially anisotropic left-handed materials is discussed in [Hu & Chui, 2002]. Incidentally, a cylindrical world has been envisged by Einstein [Eddington, 1957].

Appendix A: Wave propagation in spherical, left handed, chiral media.

1. Maxwell's equations in spherical metachiral media

With the spherical polar coordinates r, θ, ϕ, the Maxwell equations in a spherical medium are for fields that do not depend on ϕ

$$\begin{gathered}(1/r\sin\theta)\,\partial_\theta(E_\phi\sin\theta) + c^{-1}\partial_t B_r = 0\\ -1/r\partial_r(rE_\phi)] + c^{-1}\partial_t B_\theta = 0\\ 1/r\,[\partial_r(rE_\theta) - \partial_\theta E_r] + c^{-1}\partial_t B_\phi = 0\end{gathered} \tag{A.1}$$

And

$$\begin{gathered}(1/r\sin\theta)\,\partial_\theta(H_\phi\sin\theta) - c^{-1}\partial_t D_r = 0\\ 1/r\partial_r(rH_\phi)] + c^{-1}\partial_t D_\theta = 0\\ 1/r\,[\partial_r(rH_\theta) - \partial_\theta H_r] - c^{-1}\partial_t D_\phi = 0\end{gathered} \tag{A.2}$$

with the divergence equations

$$\begin{gathered}(1/r^2)\partial_r(r^2B_r) + (1/r\sin\theta)\,\partial_\theta(\sin\theta\,B_\theta) = 0,\\ (1/r^2)\partial_r(r^2D_r) + (1/r\sin\theta)\,\partial_\theta(\sin\theta\,D_\theta) = 0\end{gathered} \tag{A.3}$$

We look for the solutions of these equations in a metachiral material endowed with the constitutive relations (3) that is

$$\mathbf{D} = -|\varepsilon|\,\mathbf{E} + i\xi\,\mathbf{B}, \qquad \mathbf{H} = -\mathbf{B}/|\mu| + i\xi\,\mathbf{E}, \qquad i = \sqrt{-1} \tag{A.4}$$

Substituting (A.4) into (A.2) gives a set of equations depending only on **E** and **B**:

$$\begin{gathered}(1/r\sin\theta)\,\partial_\theta[\sin\theta\,(-B_\phi/|\mu| + i\xi E_\phi)] - c^{-1}\partial_t[-|\varepsilon|E_r + i\xi B_r] = 0\\ 1/r\partial_r[r(-B_\phi/|\mu| + i\xi E_\phi)] - c^{-1}\partial_t[-|\varepsilon|E_\theta + i\xi B_\theta] = 0\\ 1/r\partial_r[r(-B_\theta/|\mu| + i\xi E_\theta)] - 1/r\partial_\theta[-B_r/|\mu|) + i\xi E_r) - c^{-1}\partial_t[-|\varepsilon|E_\phi + i\xi B_\phi] = 0\end{gathered} \tag{A.5}$$

while, taking into account (A,3), (A.4), the divergence equation for **E** is

$$(1/r^2)\partial_r(r^2E_r) + (1/r\sin\theta)\,\partial_\theta(\sin\theta\,E_\theta) = 0 \tag{A.6}$$

Substituting (A.1) into (A.5), the Maxwell equations become

$$(-1/|\mu|r\sin\theta)\,\partial_\theta(\sin\theta\, B_\phi) + |\varepsilon|c^{-1}\partial_t\, |E_r - 2i\xi c^{-1}\partial_t\, B_r = 0$$
$$(-1/|\mu|r)\,\partial_r(rB_\phi) - |\varepsilon|\, c^{-1}\partial_t\, E_\theta + 2i\xi\, c^{-1}\partial_t B_\theta = 0 \qquad (A.7)$$
$$(-1/|\mu|r)\,[\partial_r(rB_\theta) - \partial_\theta B_r) + |\varepsilon|c^{-1}\partial_t|E_\phi - 2i\xi\, c^{-1}\partial_t B_\phi = 0$$

To look for the solutions of Eqs.(A.7) taking into account (A.1) is a challenge imposing simplifying assumptions, as for instance $B_\phi = 0$, which seems to be the most evident.

2. 2D-electromagnetic harmonic field

For a time harmonic field $\partial_t \Rightarrow i\omega$ and if $B_\phi = 0$, Eqs.(A.7) reduce to

$$|\varepsilon|\; E_r - 2i\xi B_r = 0\,, \qquad |\varepsilon|\; E_\theta - 2i\xi B_\theta = 0$$
$$-1/|\mu|r\,[\partial_r(rB_\theta) - \partial_\theta B_r) - i\omega|\varepsilon|\; c^{-1}E_\phi = 0 \qquad (A.8)$$

Now let $\mathbf{B}(r,\theta) = \nabla\Phi(r,\theta)$ be the gradient of a magnetic scalar potential Φ

$$B_r = \partial_r\Phi, \qquad B_\theta = 1/r\;\partial_\theta\Phi \qquad (A.9)$$

Substituting (A.9) into the third relation (A.8) gives $E_\phi = 0$ so that since $B_\phi = 0$, we have according to (A.4) $D_\phi = H_\phi = 0$. So, all the ϕ-components of the electromagnetic field are null and consequently, we have to deal with a 2D-field.

With the first two relations (A.8) substituted into (A.4), we get

$$\{H_r, H_\theta\} = -\{B_r, B_\theta)(1/|\mu| + 2\xi/|\varepsilon|), \qquad \{D_r, D_\theta\} = -|\varepsilon|/2\,\{E_r, E_\theta) \qquad (A.10)$$

So, according to (A.9), we have just to determine the potential Φ. Then, using the equations fulfilled by the spherical Bessel functions $j_n(kr)$ and by the Legendre polynomials $P_n(\theta)$ where n is a positive integer

$$\partial_r{}^2 j_n(kr) + 2/r\; j_n(kr) + [k^2 - n(n+1)/r^2]\; j_n(kr = 0$$
$$\partial_\theta{}^2 P_n(\theta) + \cos\theta/\sin\theta\;\partial_\theta P_n(\theta) + n(n+1)\,P_n(\theta) = 0 \qquad (A.8)$$

the divergence equation (A.3) is satisfied with $\phi_n(r,\theta) = j_n(kr)\,P_n(\theta)$ and $k^2 = \omega^2|\varepsilon|\;|\mu|c^{-2}$ since

$$(\Delta + k^2)\phi_n = 0, \qquad \Delta = 1/r^2\partial_r(r^2\partial_r) + 1/r^2\sin\theta\;\partial_\theta(\sin\theta\;\partial_\theta) \qquad (A.12)$$

So, the potential $\Phi(r,\theta,)$ with the complex amplitudes A_n) is

$$\Phi(r,\theta) = \textstyle\sum_0^\infty A_n\, j_n(kr)\, P_n(\theta) \qquad (A.13)$$

which achieves to determine the 2D-electromagnetic field.

In this situation where $E_\phi = H_\phi = 0$, the Poynting vector has only a non null component $S_r = S_\theta = 0$ and

$$S_\phi = c/8\pi \, \mathrm{Re}\,(E_r H_\theta^* - E_\theta H_r^*) \tag{A.14}$$

in which the asterisk denotes the complex conjugation. But, acording to (A.4) and (A.8):

$$\begin{aligned} E_r H_\theta^* &= -2i\,(\xi/|\varepsilon| \;\; |\mu|)\,(1+2\xi^2|\mu|/|\varepsilon|) B_r B_\theta^* \\ E_\theta H_r^* &= -2i\,(\xi/|\varepsilon| \;\; |\mu|)\,(1+2\xi^2|\mu|/|\varepsilon|) B_\theta B_r^* \end{aligned} \tag{A.15}$$

Substituting (A.15) into (A.14) gives

$$S_\phi = (c/8\pi/|\varepsilon| \;\; |\mu|)\,(1+2\xi^2|\mu|/|\varepsilon|)\,\mathrm{Im}\{B_r B_\theta^* - B_\theta B_r^*\} \tag{A.16}$$

Now, according to (A.4), (A.8), (A.10) we have with $B^2 = |B_r|^2 + |B_\theta|^2$

$$\mathbf{E.D^*} = -2\xi^2 B^2/|\varepsilon|, \qquad \mathbf{BH^*} = -B^2\,(1/|\mu| + 2\xi^2/|\varepsilon|) \tag{A.17}$$

So, the energy density $w = 1/8\pi\,(\mathbf{E.D^*} + \mathbf{BH^*})$ is

$$w = -1/8\pi\,((1/|\mu| + 4\xi^2/|\varepsilon|) \tag{A.18}$$

The energy density is negative which suggests that spherical left handed chiral media behave as a rotating plasma [Ilysionis et al] 2009], [Lashmore-Davies, 2005] and that the solutions obtained with the magnetic potential Φ are of the Alfven wave type. This result could also be a consequence of the approximations imposed here.

5. References

Eddington A.S. (1957) *The Mathematical Theory of Relativity*. Academic Press, Cambridge.

Grbic A. & Eleftheriades G.V. (2002) Experimental verification of backward-wave radiation from a negative refractive index material. *J.Appl.Phys.* Vol.92 (2002) 5930-5935.

Hu L. & Chui S.T.(2002) Characteristics of electromagnetic wave propagation in uniaxially aniotropic left-handed materials, *Phys.Rev.B* Vol.66, 08510.

Ilysionis V.I., Khalzov I.V. & Smolyakov A.I.(2009). Negative energy waves and MHD stability of rotating plasmas. *Nucl.Fusion* Vol. 49, 035008.

Lashmore-Davies C.N (2005). Negative energy waves. *J. of Plasma Phys.*Vol. 71, 101-10.

Morse P.M. &. Feshbach H.(1953). *Methods of Theoretical Physics*. Mac Graw Hill, New York.

Pendry J.B. (2000), Negative refraction makes a perfect lens. *Phys.Rev.Lett.* Vol. 85, 3966-3969.

Pendry J.B., Schurig D. & Smith D.R., (2006) Controlling electromagnetic fields, *Science* Vol. 312,1780-1782.

Post E.J., (1962) *Formal Structure of Electromagnetism*, North Holland, Amsterdam.

Sihvola A.A. (2007), Metamaterials in electromagnetism. *Metamaterials* Vol.1, 1-11.

Veselago V. (1968), The electrodynamic of substances with simultaneously negative ε and μ, *Sov.Phys.Usp. Vol.10. 509-514.*

Ziolkowski R.W. & Heyman E., (2001) Wave propagation in media having negative permittivity and permeability. *Phys.Rev.E* Vol.64, 056605.

2

Microwave Sensor Using Metamaterials

Ming Huang[1] and Jingjing Yang[2]
School of Information Science and Engineering,
Yunnan University, Kunming, Yunnan 650091
PR China

1. Introduction

It is well known that electromagnetic wave can be applied to wireless communication, radar, heating and power transmission. It has brought human society into industrial and information age. The mechanism behind these applications is electromagnetic radiation, i.e. energy transportation and conversion. Recently, research shows that evanescent wave, which is non-radiation wave associated with guided wave theory, can be amplified by metamaterials. Moreover, the sensitivity and resolution of sensor using metamaterials can be dramatically enhanced. The metamaterial-assisted sensors hope to fuel the revolution of sensing technology.

This chapter aims at the study of metamaterial-assisted sensors. Firstly, traditional microwave sensors are concisely reviewed, with emphasis on the method of microwave coupling, and the working principle of resonant microwave sensor. Then, the properties of metamaterials and its application are introduced. Enhancement of sensitivity in slab waveguide with TM mode and subwavelength resolution properties of perfect slab lens are demonstrated. The mechanisms of metamaterials for amplifying evanescent wave and for improving sensitivity and resolution are studied. Thirdly, the simulation results of the microwave sensor using epsilon near-zero materials (ENZ) and metamaterial particles are given. At last, the metamaterial-assisted open resonators are discussed.

2. Microwave sensor

Different from the famous radar and remote sensing technology, microwave sensor can measure properties of materials based on microwave interaction with matter, and it can be used to provide information about the moisture content, density, structure and shape of materials, and even chemical reaction. Microwave sensor offers many advantages over traditional sensor such as rapid measurement, nondestructive, precise, fully automated and it can be made in a laboratory or on-line. General speaking, there are two kinds of microwave sensor, i.e., nonresonant and resonant. A decided advantage of resonant sensor compared with the nonresonant sensor is that resonant sensor have higher sensitivity, simpler signal processing, and lower cost (Kupfer, 2000; Huang et al, 2007).

One of the key issues related to microwave sensor is the technique of coupling microwave to test materials. Dozens of techniques have been proposed in the past (Von Hippel, 1995). Generally, there are eight categories coupling mechanisms for microwave sensor as shown in Fig. 1, i.e., transmission-aperiodic-closed (TAC), reflection-aperiodic-closed (RAC), transmission-aperiodic-open (TAO), reflection-aperiodic-open (RAO), transmission-resonant-closed (TRC), reflection-resonant-closed (RRC), transmission-resonant-open (TRO), and reflection-resonant-open (RRO) (Kraszewski, 1991). The schematic diagram of TRC sensor is shown in Fig. 2. It consists of a resonant cavity, a microwave generator, an interface circuit, a detecting circuit and a computer. The computer is in charge of signal processing. The detecting circuit is composed of a linear detector, a low-pass filter and a high speed A/D converter. Measurement principle is the microwave cavity perturbation techniques. Let Q_0 and f_0 be the quality factor and resonance frequency of the cavity in the unperturbed condition, respectively, and Q and f the corresponding parameters of the cavity loaded with sample. When a small sample is introduced in a cavity, it causes a frequency shift $\Delta\omega$, and the Q factor is changed accordingly. These quantities are given by (Huang et al, 2007):

$$\Delta\omega/\omega = -\omega_0(\varepsilon_r^{'} - 1)\int_{ve} E_0^* \cdot E dv \,/\, 4W \tag{1}$$

$$1/Q - 1/Q_0 = 2\varepsilon_0\varepsilon_r^{''}\int_{ve} E_0^* \cdot E dv \,/\, 4W \tag{2}$$

where $W = \int_{ve}[(E_0^* \cdot D_0 + H_0^* \cdot B_0) + (E_o^* \cdot D_1 + H_0^* \cdot B_1)]dv$, $\Delta\omega = f - f_0$, $f_0 = \omega_0/2\pi$, E_0^*, D_0, B_0 and H_0 are the fields in the unperturbed cavity, D_1 and H_1 are the fields in the interior of the sample. V_c and V_e are the volumes of the cavity and the sample respectively, dv is the elemental volume. Obviously, the measurement of Q factor and resonant frequency is a key issue for TRC sensor, and the larger the Q factor of the cavity the higher the sensitivity.

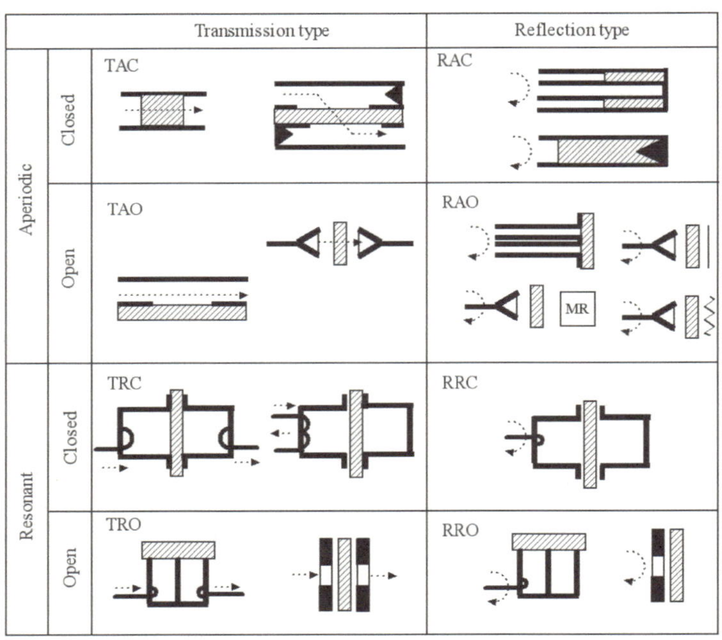

Fig. 1. Classification of eight generic types of microwave sensors for monitoring properties of industrial materials (Kraszewski, 1991)

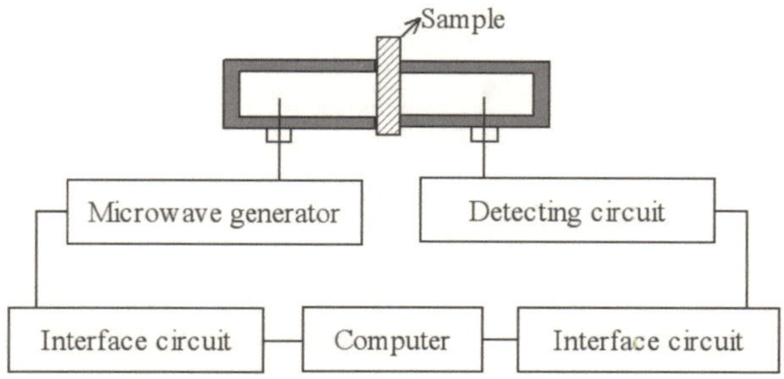

Fig. 2. The schemitic diagram of the microwave sensor

3. Metamaterials

3.1 Overview of metamaterials and its sensing applications

In the 1960s, Russian theoretical physicist Veselago realized that if materals could be properly engineered, their index of refraction could be negative (Veselago, 1968). Veselago's work implied that flat materials with simultaneously negative values of permittivity and permeability could act like lenses and produce other counterintuitive phenomena, such as a reverse Dopple effect and negative refraction. It took more than 30 years, but in 2001, researchers led by Pendry, a theoretical physicist at Imperial College London, and Smith, now at Duke University, made just such a material (Shelby et al, 2001). Since then, electromagnetic metamaterials, which is artificial materials, whose permittivity and/or permeability can be designed to continuously change from negative to positive values, attract great attention all over the world. Many novel applications are proposed based on metamaterials, such as cloak, concentrators, directive antenna, superscatterers, superabsorbers, transparent devices and so on. Now, physicists and materials scientists realize that the losses is a big problem for practical applications of metamaterials (Service, 2010).

However, the properties of metamaterials, usually regarded as detrimental, are either of no importance for sensor application or may even enhance it. For instance, the sharp resonant peaks in dispersion are convenient since they may mean an increased sensitivity of sensor and actually the sharper the peaks are, the better. Since the sensor can be made to operate at a single frequency, there is no need for broadband materials. One of the unavoidable drawbacks of the metamaterials, large absorption losses, is practically of no importance here (Zoran et al, 2007). Moreover, due to evanescent waves can be amplified by metamaterials, the sensor using metamaterials possesses the characteristics of sub-wavelength resolution (Pendry, 2000) and high sensitivity. Since sensitivity and resolution are two vital parameters of sensor, metamaterials will open up an avenue for improving the performance of sensors. The first revolutionary products using metamaterials might be the metamaterial-assisted sensors.

3.2 Enhancement of sensitivity in slab waveguide

Fig. 3 shows the model of integrated slab waveguide. In the case of time harmonic field and lossless, i.e., both ε and μ are all real numbers, suppose that magnetic field H is polarized along the y-axis, TM wave travels in the z-direction, and $H_y = H_y(x)e^{j(\omega t-\beta z)}$, then wave equation is simplified into the following well-known scalar Helmholtz equation:

$$\partial^2 H_y(x)/\partial z^2 + (\omega^2 \varepsilon(x)\mu(x) - \beta^2)H_y(x) = 0 \tag{3}$$

and

$$E_z(x) = (1/j\omega\varepsilon(x))(\partial H_y(x)/\partial x). \tag{4}$$

where ω is angular frequency of the field and β is the propagation constant in z-direction, which can be written as $\beta = k_0 N$, where k_0 is the free space wave number and N is the modal effective index. The structure of the model is illustrated as followings: in the non working region, a guiding layer is sandwiched between semi-infinite cladding and substrate, where d_1 is the thickness of the guiding layer, ε_c and μ_c, ε_g and μ_g, ε_s and μ_s are the permittivity and permeability of the cladding, guiding and substrate layer, respectively. In the working region, there is a layer of metamaterials with negative permittivity ε_m, negative permeability μ_m, and thickness d_2.

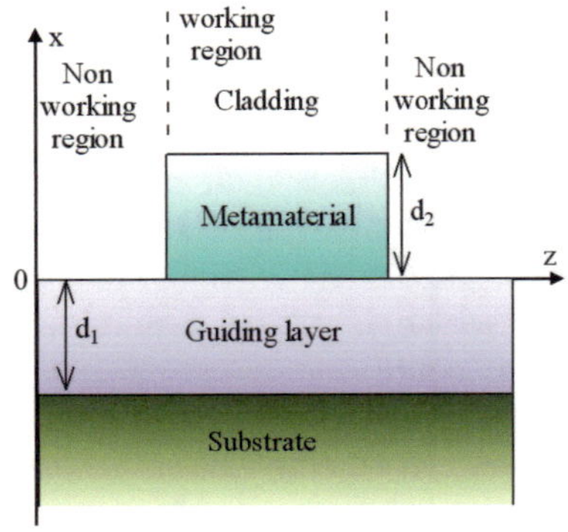

Fig. 3. Schematic structure of integrated slab waveguide sensor with a metamaterial layer

In the working region, the solution forms of Eq. (3) in each layer are shown as followings:

$$H_{yc}(x) = A e^{-\gamma_c (x-d_2)} e^{-j\beta z}, \qquad x > d_2, \tag{5}$$

$$H_{ym}(x) = \left(B_1 e^{-\gamma_m x} + B_2 e^{\gamma_m x}\right) e^{-j\beta z}, \qquad 0 < x < d_2, \tag{6}$$

$$H_{yg}(x) = (D_1 \cos(\gamma_g x) + D_2 \sin(\gamma_g x)) e^{-j\beta z}, \quad -d_1 < x < 0, \tag{7}$$

$$H_{ys}(x) = F e^{\gamma_s (x+d_1)} e^{-j\beta z}, \qquad x < -d_1, \tag{8}$$

where $\gamma_c = \sqrt{\beta^2 - \varepsilon_0 \varepsilon_c \mu_c \omega^2}$, $\gamma_m = \sqrt{\beta^2 - \varepsilon_0 \varepsilon_m \mu_m \omega^2}$, $\gamma_g = \sqrt{\varepsilon_0 \varepsilon_g \mu_g \omega^2 - \beta^2}$ and $\gamma_s = \sqrt{\beta^2 - \varepsilon_0 \varepsilon_s \mu_s \omega^2}$, respectively. Constants A, B_1, B_2, D_1, D_2 and F represent the amplitude of the waves in each layer, and need to be determined from the boundary conditions that require H_y and E_z to be continuous at $x=-d_1$, 0, and d_2, respectively. Substituting Eqs. (5)-(8) into Eq. (4), we get E_z for each layer, and also the boundary conditions and the corresponding constants as follows:

$$\tan(\gamma_g d_1)=(\gamma_g\varepsilon_g D_1-\gamma_g\varepsilon_g D_2)/(\gamma_g\varepsilon_g D_1+\gamma_g\varepsilon_g D_2) \tag{9}$$

$$B_1=0.5(D_1-\gamma_g\varepsilon_m D_2/\gamma_m\varepsilon_g), B_2=0.5(D_1+\gamma_g\varepsilon_m D_2/\gamma_m\varepsilon_g), \tag{10}$$

$$B_1=0.5(1+\gamma_c\varepsilon_m/\gamma_m\varepsilon_c)e^{\gamma_m d_2}A, \tag{11}$$

$$B_2=0.5(1-\gamma_c\varepsilon_m/\gamma_m\varepsilon_c)e^{-\gamma_m d_2}A, \tag{12}$$

After some algebraic manipulations the dispersion equation of the working region can be written as:

$$\gamma_g d_1=\arctan\left(\frac{\gamma_m\varepsilon_g}{\gamma_g\varepsilon_m}\left(\frac{(\gamma_m\varepsilon_c+\gamma_c\varepsilon_m)-(\gamma_m\varepsilon_c-\gamma_c\varepsilon_m)e^{-2\gamma_m d_2}}{(\gamma_m\varepsilon_c+\gamma_c\varepsilon_m)+(\gamma_m\varepsilon_c-\gamma_c\varepsilon_m)e^{-2\gamma_m d_2}}\right)\right)+\arctan\left(\frac{\gamma_s\varepsilon_g}{\gamma_g\varepsilon_s}\right)+m\pi, \tag{13}$$

where m=0, 1, 2,... is the mode order. When d_2 approaches zero, i.e., no metamaterial is available, Eq. (13) is reduced to the well-known dispersion equation for a three-layer slab linear waveguide given by

$$\gamma_g d_1=\arctan\left(\frac{\gamma_s\varepsilon_g}{\gamma_g\varepsilon_s}\right)+\arctan\left(\frac{\gamma_c\varepsilon_g}{\gamma_g\varepsilon_c}\right)+m\pi. \tag{14}$$

When $\gamma_c\varepsilon_m=-\gamma_m\varepsilon_c$, according to Eqs. (6), (7), (11) and (12), we get B_1=0, and $A=D_1e^{\gamma_m d_2}$. Compared with that in the nonworking region, where $A=D_1$, the amplitude of evanescent wave is enhanced by $e^{\gamma_m d_2}$ due to the existence of metamaterial layer. Eq. (13) can be reduced to Eq. (14). Therefore, the same dispersion equation holds for both the working region and the nonworking region, and consequently the propagation constants for the two regions are identical.

Fig. 4 shows the distribution of magnetic field $H_y(x)$. It can be seen that the evanescent wave in cladding layer is enhanced by metamaterials. It indicates that metamaterials could be used to amplify the evanescent wave in the cladding layer without affecting the propagation constant of the waveguide.

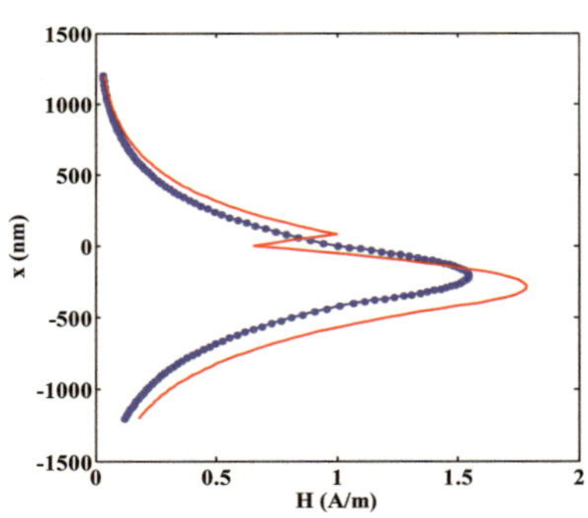

Fig. 4. Magnetic field $H_y(x)$ distribution in three-layer waveguide conventional sensor (blue dot line) and the proposed sensor with metamaterials (red line), d_2 is the thickness of the metamaterials

For the sake of simplicity in the evaluation of the sensitivity, it is assumed that $\gamma_c = k_0 q_c$, $\gamma_m = k_0 q_m$, $\gamma_g = k_0 q_g$ and $\gamma_s = k_0 q_s$, where $q_c = \sqrt{N^2 - \varepsilon_c}$, $q_m = \sqrt{N^2 - n\varepsilon_m}$, and n is a negative number, $q_g = \sqrt{\varepsilon_g - N^2}$ and $q_s = \sqrt{N^2 - \varepsilon_s}$. The sensitivity of the proposed waveguide sensor is obtained in a condensed form, three normalized effective indices X_s, X_c and X_m and three asymmetry parameters a_s, a_c and a_m are defined as $X_s = q_s/q_g$, $X_c = q_c/q_g$, $X_m = q_m/q_g$, $a_s = \varepsilon_s/\varepsilon_g$, $a_c = \varepsilon_c/\varepsilon_g$, and $a_m = \varepsilon_m/\varepsilon_g$. In the light of these assumptions, Eq. (13) can be written as

$$k_0 q_g d_1 = \arctan(X_s/a_s) + \arctan(X_m b_1/(a_m b_2)) + m\pi. \tag{15}$$

where $b_1 = (a_c X_m + a_m X_c)\text{-}(a_c X_m \text{-} a_m X_c)e^{-2k_0 X_m q_g d_2}$ and

$$b_2 = (a_c X_m + a_m X_c) + (a_c X_m \text{-} a_m X_c)e^{-2k_0 X_m q_g d_2}.$$

In the case of homogenous sensing, the sensitivity S is defined as the ratio of the modal effective index N change of the guided mode to the refractive index change of the cover n_c. The sensitivity $S_2 = (\partial N / \partial n_c)$ is calculated by differentiating Eq. (15) with respect to N and n_c.

After some algebraic manipulations, the sensitivity of proposed sensor can be written as

$$S_2 = \frac{\sqrt{a_c q} X_m \left[(2\frac{X_m}{\varepsilon_g} q_g^{\ 2} + \frac{a_m}{X_c})e^{-f}(b_1 + b_2) - (2\frac{X_m}{\varepsilon_g} q_g^{\ 2} - \frac{a_m}{X_c})(b_2 - b_1) \right]}{\left(A_{mTM} + \frac{a_s(1 + X_s^{\ 2})}{X_s(a_s^{\ 2} + X_s^{\ 2})} \right)\left(\frac{a_m^{\ 2} b_2^{\ 2} + X_m^{\ 2} b_1^{\ 2}}{a_m} \right) + b_1 b_2 \frac{C_1}{X_m} + G_1 + G_2}, \tag{16}$$

where $f = 2k_0 X_m q_g d_2$, $G_1 = (b_2 - b_1)(a_c C_1 + a_m X_m C_2/X_c)$,

$$G_2 = e^{-f}(b_2 + b_1)((a_c X_m - a_m X_c)f/X_m \text{-} a_c C_1 + C_2 a_m X_m/X_c),$$

$A_{mTM} = \arctan(X_s/a_s) + \arctan(X_m b_1/(a_m b_2)) + m\pi$, $C_1 = 1 + X_m^{\ 2}$, and $C_2 = 1 + X_c^{\ 2}$.

In a similar manner, Eq. (14) is differentiated to obtain the sensitivity S_1 of the three-layer waveguide conventional sensor without metamaterials. As a result, it can be obtained that

$$S_1 = \frac{2/\sqrt{a_c q} - \sqrt{a_c q}}{1 + X_c^{\ 2} + rF}. \tag{17}$$

where $q = (1 + X_s^{\ 2})/(a_s + X_s^{\ 2})$, $r = X_c(a_c^{\ 2} + X_c^{\ 2})/a_c$,
$F = \arctan(X_s/a_s) + \arctan(X_c/a_c) + m\pi + a_s(1 + X_s^{\ 2})/(X_s(a_s^{\ 2} + X_s^{\ 2}))$. The variation curves of sensitivity S_2 with the different values of thickness d_2 are shown in Figs. 5. It is can be noted that sensitivity goes up with the increasing thickness d_2 of the metamaterials. This is an unique property of metamaterials. It indicates that sensitivity can be enhanced by increasing the thickness of metamaterial layer.

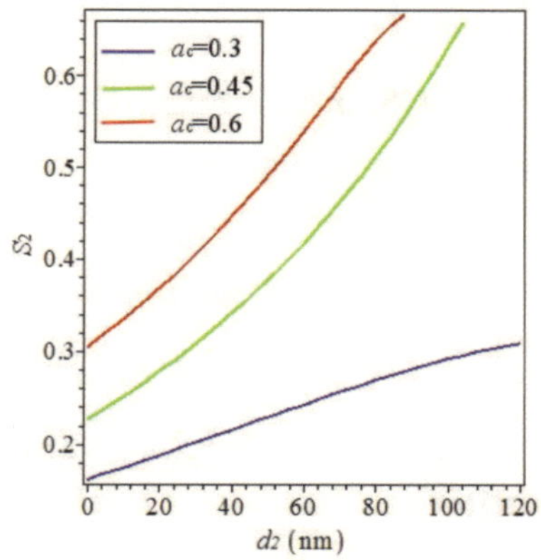

Fig. 5. Sensitivity S_2 versus the thickness of the metamaterials d_2 for different values of a_c, a_s=0.62, a_m=-0.5, n=-0.6, and d_1=400nm

We have verified that the metamaterials can amplify evanescent waves, and the sensitivity of sensors with TM mode can be dramatically enhanced compared with the conventional three-layer TM wave waveguide sensor without metamaterials. The other works about slab waveguide sensor can also be found in the work of Qing and Taya et al (Qing et al, 2004; Taya et al, 2009) However, to the best of our knowledge, there is no report about experimental results till now.

3.2 Enhancement of resolution based on metamaterials

Assume TE wave transmitting along z-axis in Fig. 6, when $\omega^2 c^{-2} > k_x^2 + k_y^2$, $k_z = \sqrt{\omega^2 c^{-2} - k_x^2 - k_y^2} \approx \omega c^{-1} = \beta = 2\pi/\lambda$, the electric field of radiation wave is defined as $E^i = E_y^i e^{-j(\omega t - \beta z)}$. Its transmission properties on the interface between two dielectrics can be obtained according to the equation of plane wave (Guru & Hiziroglu, 1998).

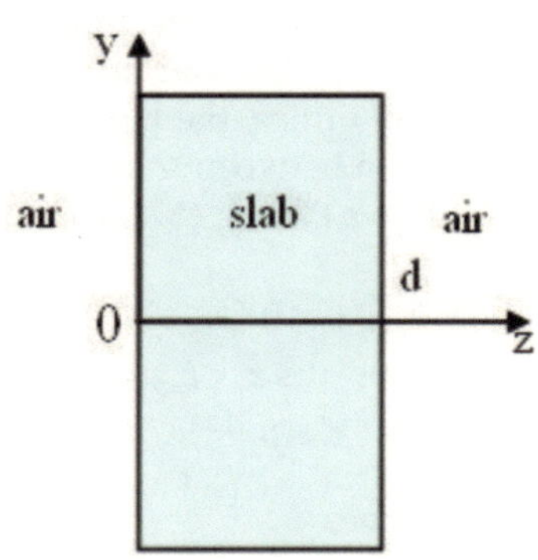

Fig. 6. Model of the metamaterial slab

When $\omega^2 c^{-2} < k_x^2 + k_y^2$, the electric field of evanescent wave is given by $E^i = E_y^i e^{j(k_z z + k_x x - \omega t)}$, where the wave vector $k_z = j\sqrt{k_x^2 + k_y^2 - \omega^2 c^{-2}}$, implies exponential decay. The reflected and transmitted wave are $E^r = E_y^r e^{j(-k_z z + k_x x - \omega t)}$, $E^t = E_y^t e^{j(k_z' z + k_x x - \omega t)}$, where $k_z' = j\sqrt{k_x^2 + k_y^2 - \varepsilon\mu\omega^2 c^{-2}}$, the corresponding magnetic field H_x can be obtained according to $\nabla \times \vec{E} = -\partial \vec{B}/\partial t$, and $H_x = 1/j\omega\mu\mu_0 (\partial E_y/\partial z)$. By matching boundary condition of E_y and H_x at $z = 0$, we can obtain the following equations $E_y^i + E_y^r = E_y^t$, $(k_z/\omega\mu_0)E_y^i - (k_z/\omega\mu_0)E_y^r = (k_z'/\omega\mu\mu_0)E_y^t$. After some algebraic manipulations the coefficient of transmission and reflection can be written as

$$t = E_y^t/E_y^i = 2\mu k_z/(\mu k_z + k_z') \text{ and } r = E_y^r/E_y^i = (\mu k_z - k_z)/(\mu k_z + k_z'). \tag{18}$$

Conversely a wave inside the medium incident on the interface with vacuum experiences transmission and reflection as follows:

$$t' = 2k_z'/(k_z' + \mu k_z) \text{ and } r' = (k_z' - \mu k_z)/(k_z' + \mu k_z). \tag{19}$$

To calculate transmission through both surfaces of the slab we must sum the multiple scattering events,

$$T = tt'e^{jk_z'd} + tt'r'^2 e^{3jk_z'd} + tt'r'^4 e^{5jk_z'd} + \ldots = tt'e^{jk_z'd}/(1 - r'^2 e^{2jk_z'd}) \tag{20}$$

By substituting from (19) and (20) and taking the limit,

$$\lim_{\substack{\mu \to -1 \\ \varepsilon \to -1}} T = e^{-jk_z'd} = e^{-jk_z d} \tag{21}$$

The reflection coefficient is given by

$$\lim_{\substack{\mu \to -1 \\ \varepsilon \to -1}} R = 0 \tag{22}$$

The Eqs.(21) and (22) was firstly derived by Pendry (Pendry, 2000). It indicates that both propagating wave and evanescent wave contribute to the resolution of the image. Therefore, there is no physical obstacle to perfect reconstruction of the image beyond practical limitations of apertures and perfection of the lens surface.

Based on the method of transformation optics, the performance of perfect slab lens can be simulated. The coordinate transformation between transformed space and the original space of the perfect slab lens is shown as follows (Wang et al, 2008).

$$x = \begin{cases} x' + b & x' < 0 \\ \delta x' + b & 0 \le x' < b, y = y', z = z' \\ x' + \delta b & x' > b \end{cases} \tag{23}$$

The corresponding material parameters are

$$\varepsilon_{x'} = \mu_{x'} = 1/\delta, \varepsilon_{y'} = \mu_{y'} = \delta, \varepsilon_{z'} = \mu_{z'} = \delta \tag{24}$$

where δ is a small number with positive value and the perfect slab lens can be obtained as δ goes to zero. Simulation results of the slab lens are shown in Fig. 7. From Fig.7 (a) and (b), it is seen that the propagating wave as well as the evanescent wave, are directionally guided along x-axis to the right side by the slab lens, and the object totally duplicated to the image plane. The same phenomenon can be observed when reduce the size of the object or diminish the thickness of the slab, as shown in Fig. 7(c) and (d). The image of the object can be enlarged with proper coordinate transmission.

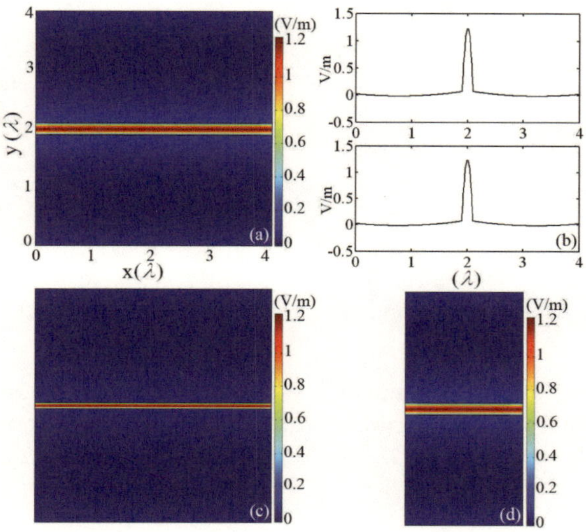

Fig. 7. Electric field (E_z) distribution in the planar perfect lens. (a) The thickness of the slab is $b = 4\lambda$, the size of the object is $t = 0.2\lambda$. (b) Electric field distribution on the object (up) and imaging (down) plane. (c) $b = 4\lambda$, $t = 0.1\lambda$. (d) $b = 2\lambda$, $t = 0.2\lambda$

Since the perfect lens was proposed by Pendry, imaging beyond diffraction limit have gained much attention. A series of perfect lensing structures including cylinders, spheres, corner perfect lens and superlens were developed. But the perfect resolution is difficult to realize for the inevitable losses of the realizable materials. At microwave frequency band, subwavelength detecting with resolution of 0.037λ has been realized by Shreiber et al (Shreiber et al, 2008).

4. Microwave sensor using ENZ metamaterials

The model of the microwave sensor based on ENZ metamaterials (Wu et al, 2008) is shown in Fig. 8(a). It is a 3D "dumbbell-shaped" coaxial metallic waveguide filled with ENZ metamaterials. Fig. 8(b) is a cross-sectional view, where the red region denotes the ENZ metamaterials, the grey region denotes the metallic conductor, and yellow region located at the center denotes the testing sample. The ENZ metamaterials is characterized by a Drude-type model with relative permittivity $\varepsilon = 1 - \omega_p^2 / \omega(\omega + j\Gamma)$, where ω_p is the plasma frequency and Γ is the collision frequency (rad/s). In the simulations, we have taken $\omega_p R / c = \pi / 2$. Note that at $f = f_p$, the permittivity of the ENZ metamaterials is given by $\varepsilon \approx j\Gamma / \omega_p \approx 0$. The incident wave is the fundamental TEM mode. The structure is uniform along the z direction. The simulation results of transmission coefficients for a variation of sample permittivity are shown in Fig. 9. It's seen that the transmission coefficients have two resonant peaks. The low-frequency peak is red shifted with the increase of sample permittivity, while the high-frequency peak keeps unchanged. Dielectric properties of the testing sample can be inversed by measuring the position of the low-frequency peak. When increasing the length L of the waveguide, results show that the high-frequency peak can be red shifted. It means that the high-frequency peak is mainly dependent on the Fabry-Perot resonance.

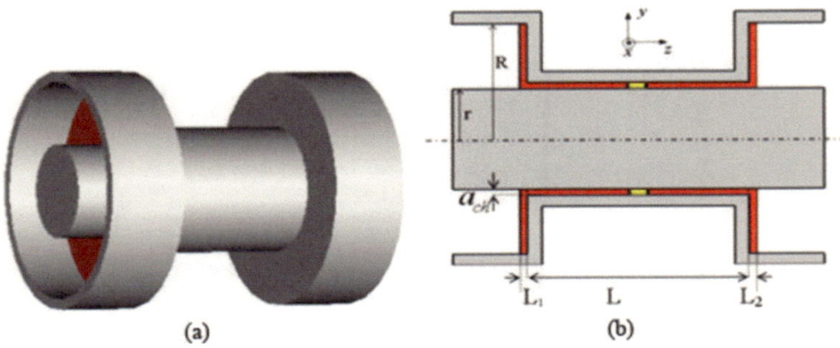

Fig. 8. (a) Model of the "dumbbell -shaped" coaxial metallic waveguide. (b) Cross-section of the model. Red region: ENZ metamaterials; Yellow region: testing sample; Grey region: metallic conductor ($R = 20mm$; $r = 8.7mm$; $L = 1.5R = 30mm$; $L_1 = L_2 = a_{ch} = R / 20 = 1mm$). [15] (Wu et al, 2008)

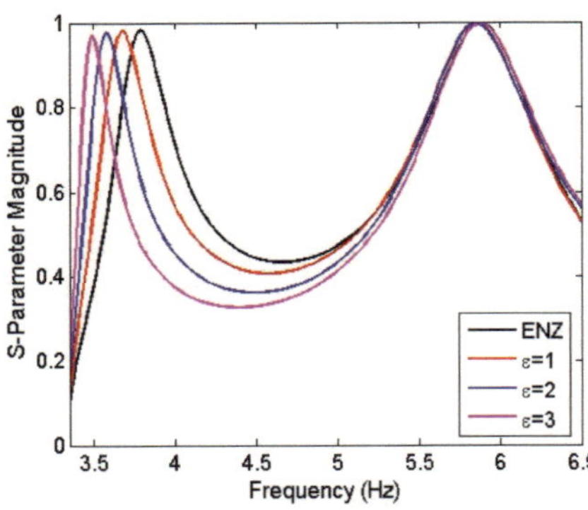

Fig. 9. Transmission coefficient for a variation of sample permittivity (Wu et al, 2008)

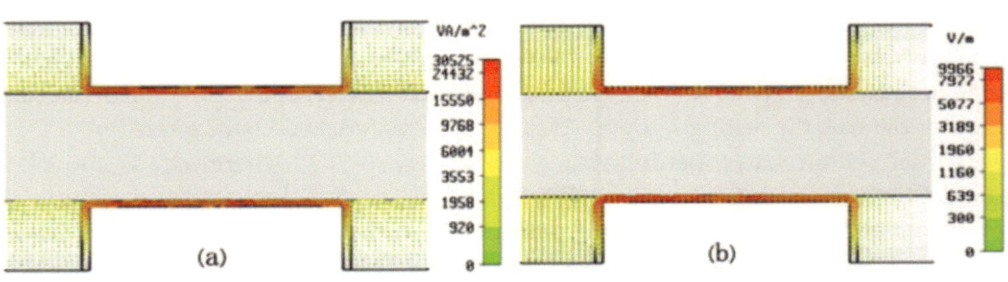

Fig. 10. The power flow distribution (a) and the electric field distribution (b) at the low-frequency transmission peak (Wu et al, 2008)

To investigate the working principle of the ENZ metamaterial-assisted microwave sensor, the power flow and the electric field distribution inside the waveguide are calculated as shown in Fig. 10. It is seen that the the power flow is squeezed through waveguide filled with ENZ metamaterials and perfectly recovered to original status at the output port. The electric field is confined into the detection zone, and as a consequence, this region will be rather sensitive to any slight change in sample permittivity.
The above simulation results show that the electromagnetic wave can be tunnelled and squeezed through 3D coaxial waveguide filled with ENZ metamaterials. The effects are similar to literature results of other guidewave structure (Silveirinha & Engheta, 2006; Alù & Engheta, 2008). The ENZ metamaterial-assisted wave guide have potential applications in microwave sensor.

5. Microwave sensor using metamaterial particles

5.1 Microwave sensor based on a single metamaterial particle

The model of the waveguide filled with a single metamaterial particle is shown in Fig. 11(a). The metamaterial particle with the thickness of 0.44mm is located at the center of the WR-14 rectangular waveguide. Since the topology structure of the particle influence the performance of the microwave sensor, the resonators with two different geometries are discussed, as shown in Fig. 11(b) and 11(c). Dimensions of the square asymmetric single split ring resonator (aSSR) are chosen as as g=1mm, w=0.8mm and L=6.68mm. The position of the gap is defined by the asymmetric parameter d_x. The resonator is designed onto a 0.127mm thick substrate with relative permittivity of 3. Fig. 11(c) shows the cross section of the rectangular aSSR resonator with the aspect radio of 1/2. Its perimeter and geometric parameters are the same as the square aSSR. The black region represents dielectric sample. The simulations were conducted with the commercial software CST. Perfect electric conductor boundaries were applied to all sides of the model to simulate the waveguide. The waveguide is terminated on both sides by a waveguide port through which the electromagnetic waves enter and exit the model. The incident wave is the fundamental TE_{10} mode with electric field polarized along the y-axis and it propagates along the z-axis.

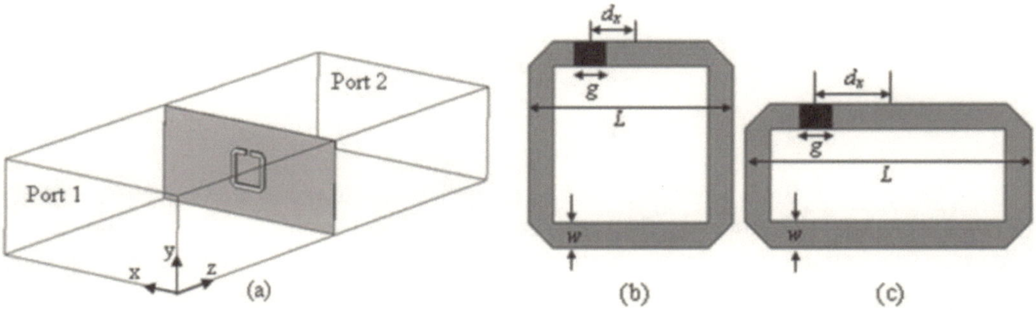

Fig. 11. (a) Simulation model of the asymmetric metamaterial particle-assisted microwave sensor. (b), and (c) show the cross section of the square aSSR and rectangular aSSR. Black region denotes the dielectric sample located in the gap (Yang et al, 2010)

The variation of Q factor as a function of asymmetric parameter is shown in Fig. 12. The position of the slit is successively moved from a close-to-symmetry position to the left resonator arm. From Fig. 12, it is seen that Q factor increases with d_x and reaches moderate values above 1400 for the square aSSR and above 1200 for the rectangular aSSR, at d_x =0.9mm and d_x =1.24mm, respectively. Therefore, the Q factor of the asymmetric metamaterial particle-assisted microwave sensor can be selected by varying the position of split or the aspect ratio of the resonator ring, which enables a high degree of design flexibility.

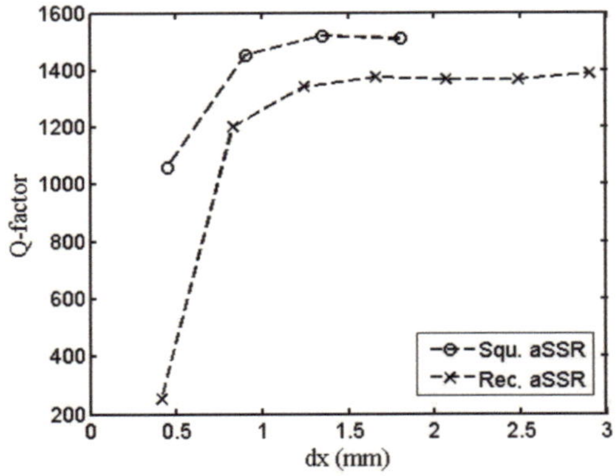

Fig. 12. The Q factor of the asymmetric metamaterial particle-assisted microwave sensor for a sweep of d_x (Yang et al, 2010)

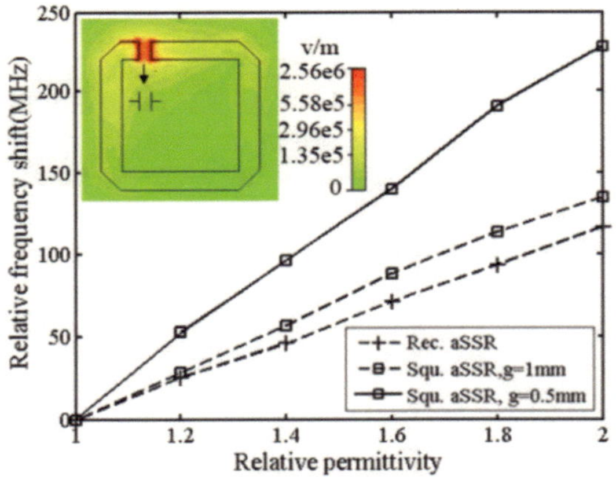

Fig. 13. Relative frequency shift of the spectral response as a function of relative permittivity of the sample located in the gap. The inset shows the normal electric field distribution for square aSSR when g=0.5mm, ε_r =1 (Yang et al, 2010)

To explore the sensitivity of the asymmetric metamaterial particle-assisted microwave sensor, dielectric material induced relative frequency shift was simulated. Results show that the peak position of transmission coefficient S_{21} is red shifted with the increase of relative permittivity of the sample located in the detecting zone. The relation between relative frequency (Δf) and sample permittivity is shown in Fig. 13. It indicates that Δf increases linearly with ε_r, and thus, both the square and the rectangular metamaterial resonators can be used in dielectric sensing. Dielectric properties of the testing sample can be inversed by detecting the relative frequency shift of the output signal of the microwave sensor. Furthermore, the microwave sensor based on the square aSSR will be more accurate in dielectric sensing, since it induces a larger Δf than the rectangular aSSR for detecting the same sample. The solid line in Fig. 13 indicates that when reducing the gap size of the square aSSR to g=0.5mm, a much larger frequency shift can be obtained. Therefore, the sensitivity of the designed microwave sensor can be further improved by diminishing the gap size of the metamaterial resonator ring. The gap of the asymmetric metamaterial resonator ring can be modelled by a capacitor as shown in the inset of Fig.13. Diminishing the gap size g results in an enhancement of the electric field, while diminishing the distance between the left and right plates of the capacitor leads to an enhancement of the energy reservation. Field enhancement is the main reason for the improved sensitivity of the asymmetric metamaterial particle-assisted microwave sensor.

The above simulation results show that the spectral response and Q factor of the sensor can be flexibly tailed to design requirement by varying the asymmetry parameter or the topological structure of the resonator. Two resonator arms of the asymmetric structure can strongly confine the field into the gap which is very sensitive in dielectric environment. Moreover, diminishing the gap size will improve the sensitivity to a great extent. Due to the spectral response of the asymmetric metamaterial resonator can be tailed to microwave, terahertz, infrared or optical band by miniaturizing the size of the structure, the asymmetric metamaterial particle-assisted sensor will have potential applications in a wide frequency band for sensing minute amounts of dielectric sample substance.

The excitation of trapped modes by asymmetrically split rings was firstly reported by Fedotov et al.(Fedotov, et al, 2007). Then, Al-Naib et al. (Al-Naib et al, 2008) realized experimentally the thin film sensor with high Q factor by locating one unit cell of double split ring resonator in the single mode rectangular waveguide. It is believed that more experimental results about the sensors will be reported in the future.

5.2 Microwave sensor based on coupled metamaterial particles

When a waveguide is filled or partially filled with metamaterials with simultaneously negative permittivity and permeability, the transmission can be enhanced (Marqués et al, 2002). In this section, a pair of coupled metamterial particles with Ω shape is located in the rectangular waveguide to form the microwave sensor. The simulation model is shown in Fig. 14(a). The geometric parameters for the Ω-shaped particles are chosen as w = 0.144 mm, h = 4.5 mm, d = 0.1 mm, R_1=1mm and R_2 = 1.4 mm (see Fig. 14(b)). The cross section of the rectangle waveguide is axb=15mmx7.5mm. The length is l=12mm. The incident wave is the fundamental TE_{10} mode propagates along the x-axis, and the incident power is 1W. The input port is at x=0, the output port is at x=12mm. Electric field and the magnetic field vectors are along the y and z-axis, respectively.

Fig. 15 shows the comparison of transmission coefficient (S21) between the rectangular waveguide filled by one pair of Ω-shaped particles and the hollow waveguide. It indicates that the cutoff frequency of the hollow waveguide is 10GHz, which is in good agreement with the theoretical results ($\lambda_c = 2a$). Interestingly, the transmission coefficient of the waveguide filled with coupled metamaterial particles has two enhancement peaks at the frequency of 8.424GHz and 9.864GHz. It is worth noting that the amplitude of the transmission coefficient S21 is greater than 1 at the two resonant frequencies.

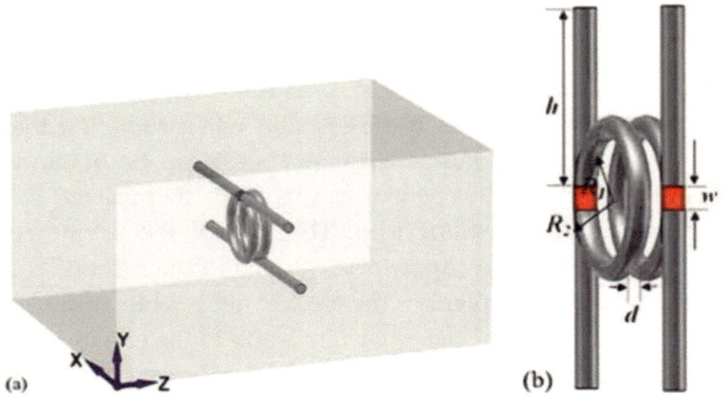

Fig. 14. (a) Perspective of the model of waveguide filled by one pair of Ω-shaped particles. (b) The Ω-shaped particles. Red region denotes the detecting zone, i.e., gap of the Ω-shaped particle (Huang et al, 2009)

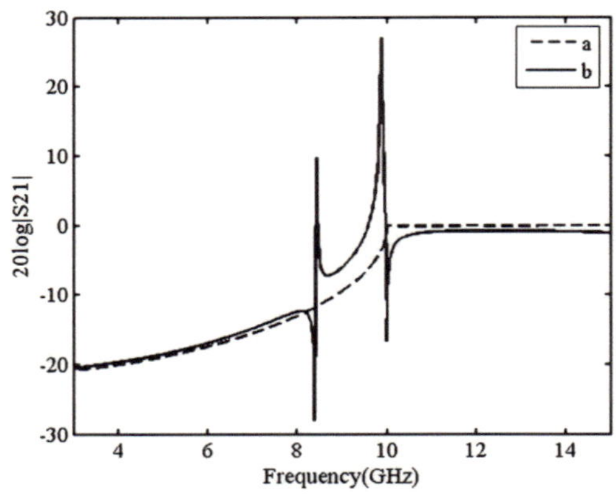

Fig. 15. Transmission coefficient (S21) as a function of frequency. Line a and b denote S21 of hollow waveguide and the waveguide filled with coupled metamaterial particles, respectively (Huang et al, 2009)

To explore the characteristic of evanescent mode in dielectric sensing, transmission spectrum was simulated for a variation of sample permittivity, as shown in Fig. 16. It is found that in the frequency range of 7GHz to 9GHz, the evanescent mode is red shifted with the increase of permittivity, while the peak position of the cutoff mode, in the frequency range of 9.7GHz to 10.1GHz, keeps unchanged. Therefore, different dielectric materials located the detecting zone can be inversed by measuring the peak frequency of the transmission coefficient in the evanescent mode. Besides, it should be noted that the cutoff mode is dependent on the Fabry-Perot resonance, since the peak position changes with the size of waveguide. Another interesting aspect, which is visualized in Fig. 16, is the fact that the transmission responses for different dielectric are progressively shifted towards left, but their peaks are not consistent with the dielectric constant increase. The results show that the peak frequency of transmission coefficient (S21) depends critically on the permittivity of the sample in the detecting zone, while the amplitude of S21 is a complicated function of the permittivity and topological structure of the coupled metamaterial particles.

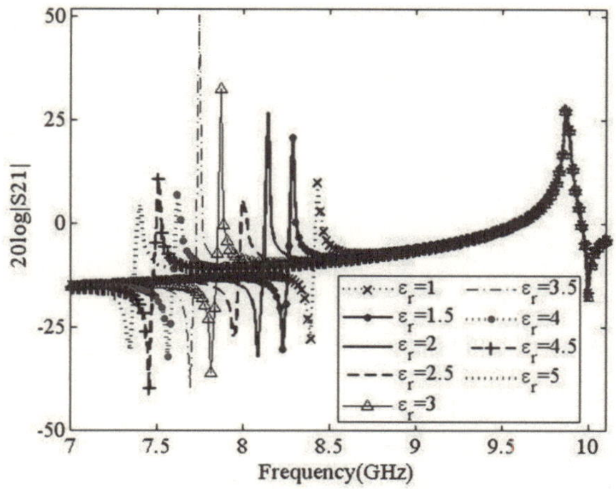

Fig. 16. Transmission coefficient (S21) versus frequency for different dielectric materials in the detecting zone (Huang et al, 2009)

To quantify the sensitivity of the evanescent mode for dielectric sensing, the performance of the metamaterial-assisted microwave sensor is compared with the traditional microwave cavity. We closed both ends of a hollow waveguide with metallic plates, which forms a conventional microwave cavity (axbxl=15x7.5x12mm^3), and computed the resonant frequency of the cavity located with dielectric sample. Table 1 shows a comparison between the relative frequency shift, i.e., $\Delta f_N = f_N(\varepsilon_1) - f_N(\varepsilon_r)$ of the waveguide filled with coupled metamaterial particles, and that of the conventional microwave cavity, i.e., $\Delta f_C = f_C(\varepsilon_1) - f_C(\varepsilon_r)$. Where, ε_1 and ε_r denotes the relative permittivity of the air and the dielectric sample, respectively. It indicates that minium (respectively maximum) frequency shift of the waveguide filled with Ω-shape coupled metamaterial particles is 360 times (respectively 450 times) that of the conventional microwave cavity. As a consequence, the waveguide filled with Ω-shape coupled metamaterial particles can be used as a novel microwave sensor to obtain interesting quantities, such as biological quantities, or for monitoring chemical process, etc. Sensitivity of the metamaterial-assisted microwave sensor is much higher than the conventional microwave resonant sensor.

ε_r	1.5	2	2.5	3	3.5	4	4.5	5
Δf_N	144	288	432	558	684	810	918	1026
Δf_C	0.4	0.7	1.1	1.3	1.6	1.8	2.2	2.5
$\Delta f_N/\Delta f_C$	360	411	393	429	428	450	417	410

Table 1. Comparison of the relative frequency shift (MHz) between the waveguide filled with coupled metamaterial particles and the conventional cavity

In addition, the microwave sensor can also be constructed by filling the other type of coupled metamaterial particles into the rectangular waveguide. For example, the meander line and split ring resonator coupled metamaterial particle (Fig. 17(a)); the metallic wire and split ring resonator (SRR) coupled metamaterial particle (Fig. 17(b)). The red regions shown in Fig. 17 denote the dielectric substances. Fig. 17(c) and (d) are the front view and the vertical view of (b).

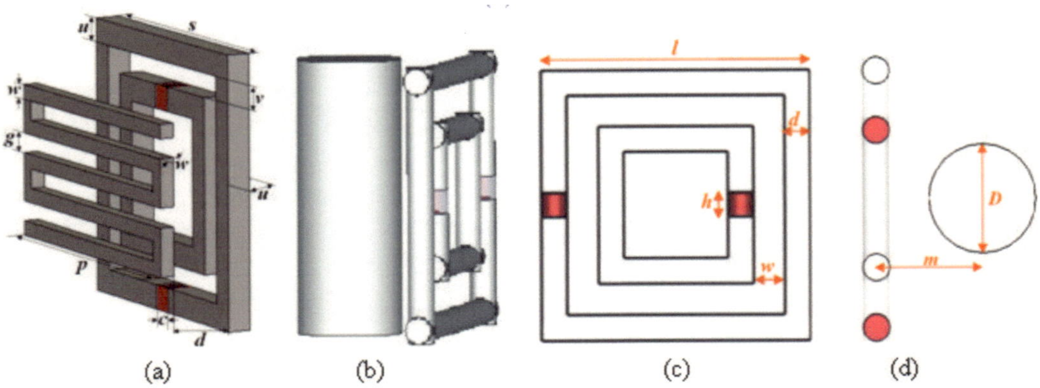

Fig. 17. (a) Configuration of the particle composed of meander line and SRR. w = 0.15mm, g = 0.2 mm, p = 2.92 mm, d=0.66mm, c=0.25mm, s=2.8mm, u=0.25mm, and v=0.25mm. (b) Configuration of the particle composed of metallic wire and SRR. (c) and (d) are the front view and the vertical view of (b). l=1.302mm, h=0.114mm, w=0.15mm, d=0.124mm, D=0.5mm, m=0.5mm

Transmission coefficient of the waveguide filled with any of the above two couple metamaterial particles also possesses the characteristic of two resonant peaks. When it is used in dielectric sensing, electromagnetic properties of sample can be obtained by measuring the resonant frequency of the low-frequency peak, as shown in Fig. 18.

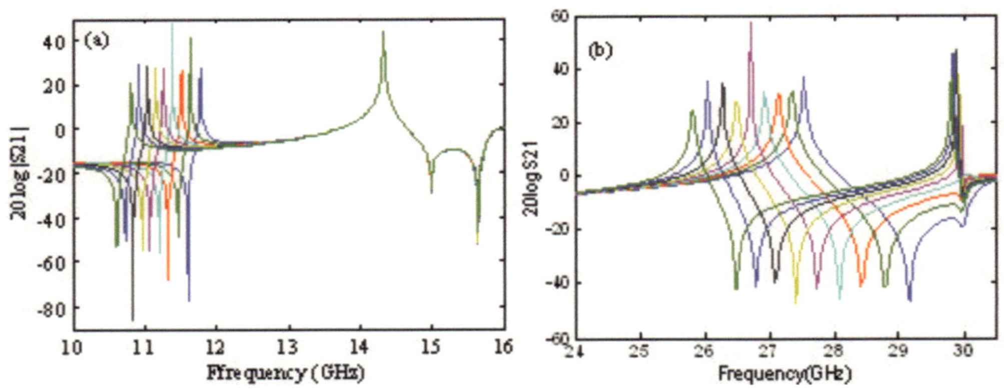

Fig. 18. Transmission coefficient (20log | S21 |) versus frequency for a variation of sample permittivity. (a) The wave guide is filled with coupled meander line and SRR. (b) The wave guide is filled with coupled metallic wire and SRR. From right to the left, the curves are corresponding to dielectric sample with permittivity of 1, 1.5, 2, 2.5, 3, 3.5, 4, 4.5, and 5, respectively

From the above simulation results, we can conclude that the evanescent wave in the waveguide filled with coupled metamaterial particles can be amplified. The evanescent mode is red shifted with the increase of sample permittivity. Therefore, the waveguide filled with couple metamaterial particles can be used as novel microwave sensor. Compared with the conventional microwave resonant sensor, the metamaterial-assisted microwave sensor allows for much higher sensitivity.

5.3 Microwave sensor based on stacked SRRs

Simulation model of the microwave sensor based on stacked SRRs is shown in Fig. 19. The size of the waveguide is axbxL=22.86x10.16x12.8mm, as shown in Fig. 19(a). Fig.19(b) is the front view of the SRR with thickness of 0.03mm. It is designed onto a 0.127mm thick substrate with relative permittivity of 4.6. The geometric parameters for the SRR are chosen as L=1.4mm, g=s=w=0.3mm, P=2mm, so that the sensor works at the frequency between 8-10.5GHz. Fig. 19(c) is the layout of the stacked SRRs, the distance between two unit cell is U=0.75.

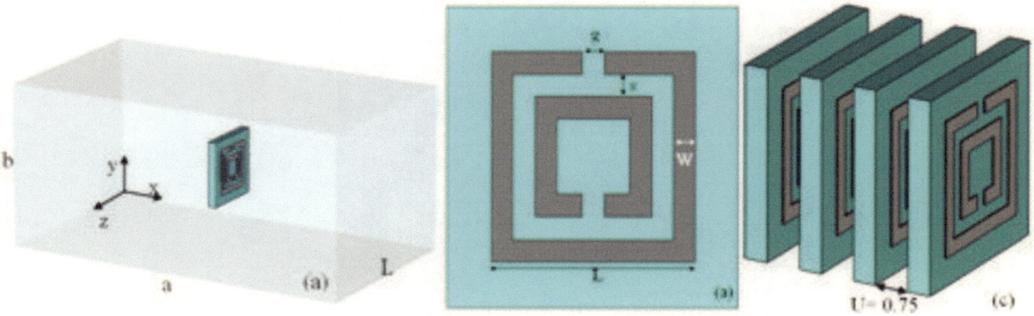

Fig. 19. (a) The microwave sensor based on stacked SRRs. (b) Front view of the SRR cell. (c) Layout of the stacked SRRs

Firstly, the effective permeability of the stacked SRRs is simulated using the method proposed by Smith et al (Smith et al, 2005). The simulation results are shown in Fig. 20. It is seen that the peak value increases with the number of SRR layer, and a stabilization is achieved when there are more than four SRR layers. Then, in what follows, the microwave sensor based on stacked SRRs with four layers is discussed in detail.

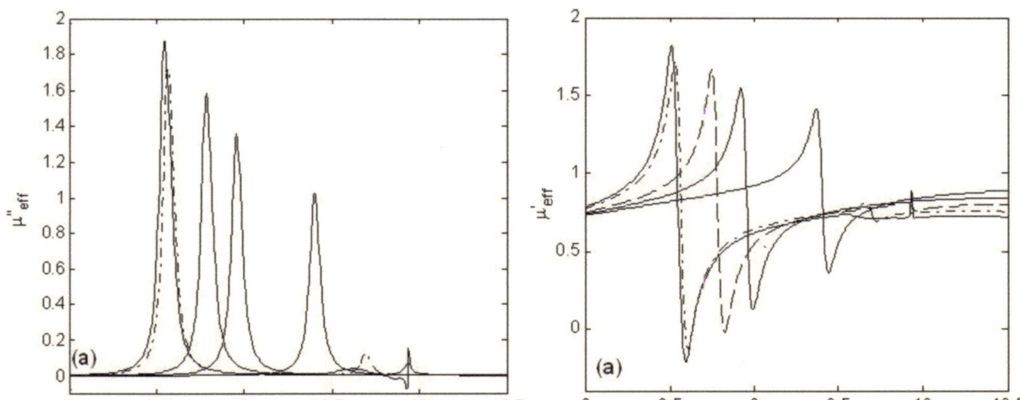

Fig. 20. Effective permeability of the stacked SRRs. (a) Real part. (b) Imaginary part. From right to left, the curves correspond to the simulation results of the stacked SRRs with one layer, two, three, four and five layers

Fig. 21 shows the electric field distribution in the vicinity of the SRR cells. It is seen that the strongest field amplitude is located in the upper slits of the SRRs, so that these areas become very sensitive to changes in the dielectric environment. Since the electric field distributions in the slits of the second and the third SRRs are much stronger than the others, to further investigate the potential application of the stacked SRRs in dielectric sensing, thickness of the SRRs is increased to 0.1mm, and testing samples are located in upper slits the second and the third SRRs. Simulation results of transmission coefficients for a variation of sample permittivity are shown in Fig. 22.

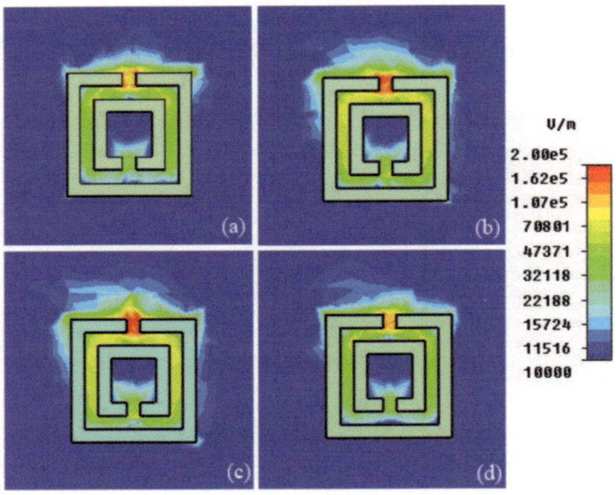

Fig. 21. Electric field distribuiton in the vicinity of the four SRRs. (a) The first SRR layer (x=-0.734 mm). (b) The second SRR layer (x=0.515 mm). (c) The third SRR layer (x=1.765 mm). (d) The fourth SRR layer (x=3.014 mm)

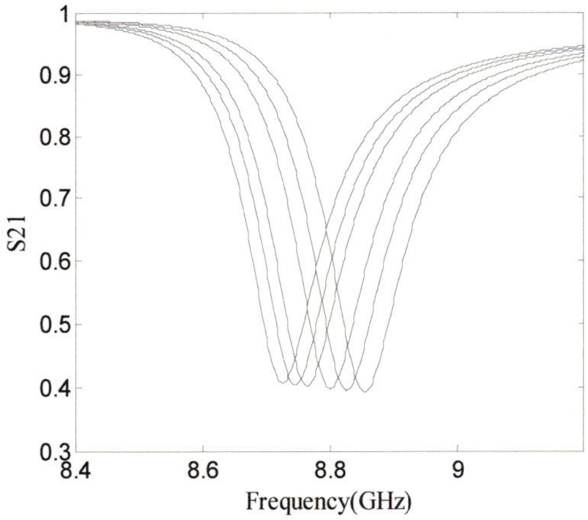

Fig. 22. Transmission coefficient as a function of frequency for a variation of sample permittivity. From right to the left, the curves are corresponding to dielectric sample with permittivity of 1, 1.5, 2, 2.5, 3 and 3.5, respectively

In conclusion, when the stacked SRRs are located in the waveguide, sample permittivity varies linearly with the frequency shift of the transmission coefficient. Although the periodic structures of SRRs (Lee et al, 2006; Melik et al, 2009; Papasimakis et al, 2010) have been used for biosensing and telemetric sensing of surface strains, etc. The above simulation results demonstrate that the stacked SRRs can also be used in dielectric sensing.

6. Open resonator using metamaterials

6.1 Open microwave resonator

For the model shown in Fig. 23, suppose the incident electric field is polarized perpendicular to the plane of incidence, that is, $\vec{E}^{(i)} = E^{(i)}\vec{e}_y$, then the incident, reflected, and refracted (transmitted) field can be obtained as

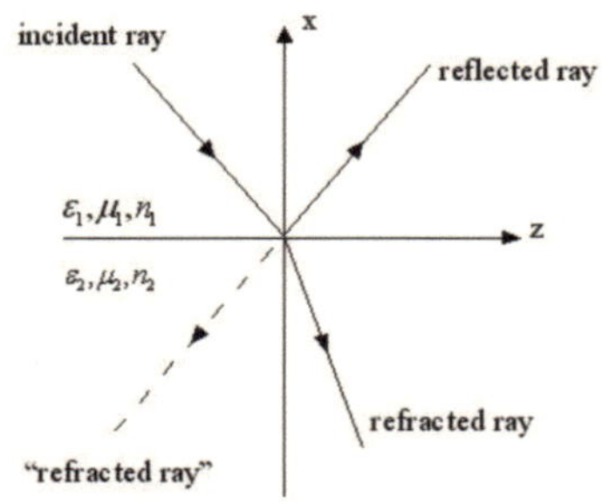

Fig. 23. Snell's law for $n_1 > 0$ and $n_2 > 0$ (real line). The dashed line for $n_1 > 0$ and $n_2 < 0$

$$\frac{E_0^{(t)}}{E_0^{(i)}} = \frac{2n_1 \cos\phi_1 / \mu_1}{n_1 \cos\phi_1 / \mu_1 + n_2(1 - n_1^2 \sin^2 \phi_1 / n_2^2)^{1/2} / \mu_2} \tag{25}$$

$$\frac{E_0^{(r)}}{E_0^{(i)}} = \frac{n_1 \cos\phi_1 / \mu_1 - n_2(1 - n_1^2 \sin^2 \phi_1 / n_2^2)^{1/2} / \mu_2}{n_1 \cos\phi_1 / \mu_1 + n_2(1 - n_1^2 \sin^2 \phi_1 / n_2^2)^{1/2} / \mu_2} \tag{26}$$

where $E_0^{(t)}$, $E_0^{(r)}$ and $E_0^{(i)}$ are the amplitudes of the transmitted, reflected, and the incident electric fields, respectively. Provided that $(n_1^2 / n_2^2)\sin^2 \phi_1 < 1$, the above formulas are valid for positive as well as negative index midia. For $(n_1^2 / n_2^2)\sin^2 \phi_1 > 1$, the expression

$$(1 - n_1^2 \sin^2 \phi_1 / n_2^2)^{1/2} = \pm j(n_1^2 \sin^2 \phi_1 / n_2^2 - 1)^{1/2}. \tag{27}$$

The $-$ sign is chosen because the transmitted field must not diverge at infinity for $n_2 > 0$. The + sign is chosen for $n_2 < 0$. If $n_1 > 0$ and $n_2 < 0$ and if $\varepsilon_2 = -\varepsilon_1$ and $\mu_2 = -\mu_1$, then $E_0^r = 0$. This means that there is no reflected field. Some interesting scenario shown in Fig. 24 can be envisioned. Fig. 24(a) illustrates the mirror-inverted imaging effect. Due to the exist of many closed optical paths running across the four interfaces, an open cavity is formed as shown in Fig. 24(b), although there is no reflecting wall surrounding the cavity.

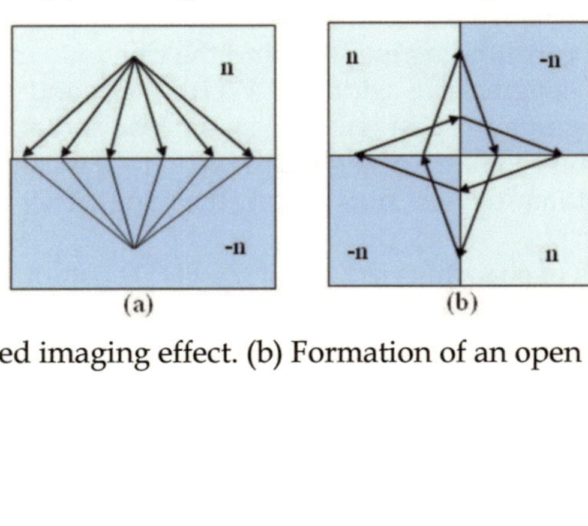

Fig. 24. (a)Mirror-inverted imaging effect. (b) Formation of an open cavity

As shown in Fig. 25(a), the open microwave resonator consists of two homogenous metamaterial squares in air. Its resonating modes are calculated using eigenfrequency model of the software COMSOL. Fig.25 (b) shows the mode around the frequency of 260MHz. It is in agreement with the even mode reported by He et al. (He et al, 2005). In the simulation, scattering boundary condition is added to the outer boundary to model the open resonating cavity. From Fig. 25(b), it is seen that electric field distribution is confined to the tip point of the two metamaterial squares. Therefore, it will be very sensitive in dielectric environment. The dependence of resonant frequency on the permittivity of dielectric environment is shown in Table 2. It is seen that when the permittivity changes from 1 to $1+10^{-8}$, the variation of resonate frequency is about 14KHz. The variation of resonant frequency can be easily detected using traditional measuring technique. Therefore, the open cavity based on metamaterials possesses high sensitivity, and it has potential application for biosensors.

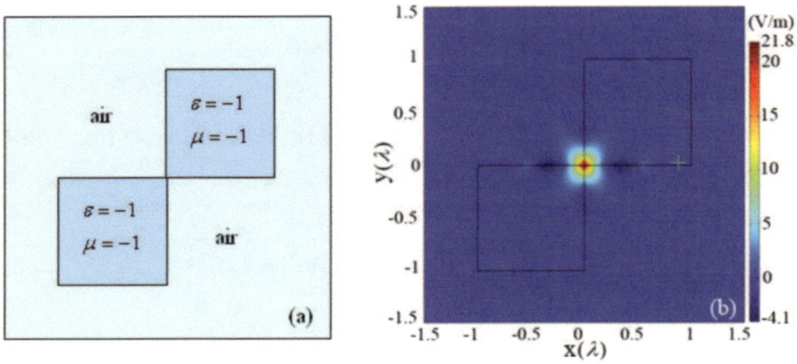

Fig. 25. (a) A subwavelength open resonator consisting of two homogenous metamaterial squares in air. (b) The electric field (Ez) distribution for (a)

Frequency(MHz)	260.481	260.467	260.336	259.794	255.372	240.485
Permittivity	1	$1+10^{-8}$	$1+10^{-7}$	$1+5\text{x}10^{-7}$	$1+10^{-6}$	$1+5\text{x}10^{-6}$

Table 2. The relation between resonate frequency and environment permittivity

The open resonator using metamaterials was first suggested and analyzed by Notomi (Notomi, 2000), which is based on the ray theory. Later, He et al. used the FDTD to calculate resonating modes of the open cavity.

6.2 Microcavity resonator

Fig. 26(a) shows a typical geometry of a microcavity ring resonator (Hagness et al, 1997). The two tangential straight waveguides serve as evanescent wave input and output couplers. The coupling efficiency between the waveguides and the ring is controlled by the size, g, of the air gap, the surrounding medium and the ring outer diameter, d, which affects the coupling interaction length. The width of WG1, WG2 and microring waveguide is 0.3μm. The straight waveguide support only one symmetric and one antisymmetric mode at $\lambda = 1.5$ μm. Fig. 26(b) is the geometry of the microcavity ring when a layer of metamaterials (the grey region) is added to the outside of the ring. The refractive index of the metamaterials is n=-1.

Fig. 27 is the visualization of snapshots in time of the FDTD computed field as the pulse first (t=10fs) couples into the microring cavity and completes one round trip(t=220fs). When refractive index of the surounding medium varies from 1 to 1.3, the spectra are calculated,

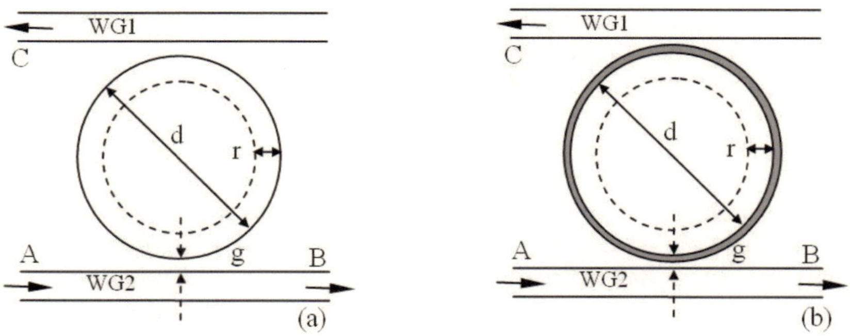

Fig. 26. (a) The schematic of a microcavity ring resonator coupled to two straight waveguides. (b) A metamaterial ring (the grey region) is added to the out side of the microring. d=5.0μm, g=0.23μm, r=0.3μm, the thickness of the metamaterials is r/3

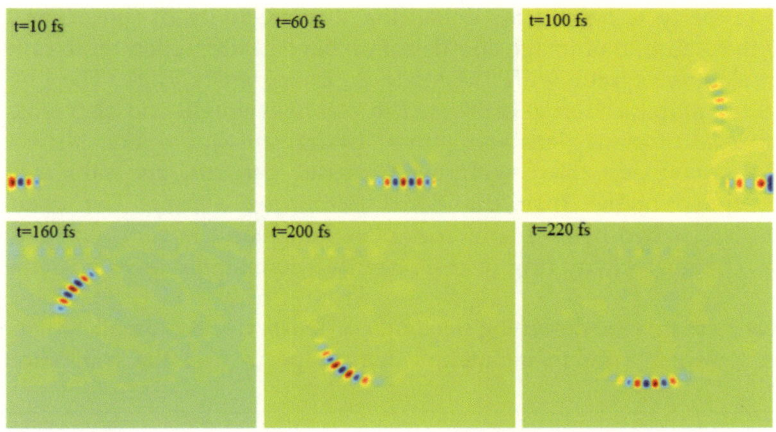

Fig. 27. Visualization of the initial coupling and circulation of the exciting pulse around the microring cavity resonators

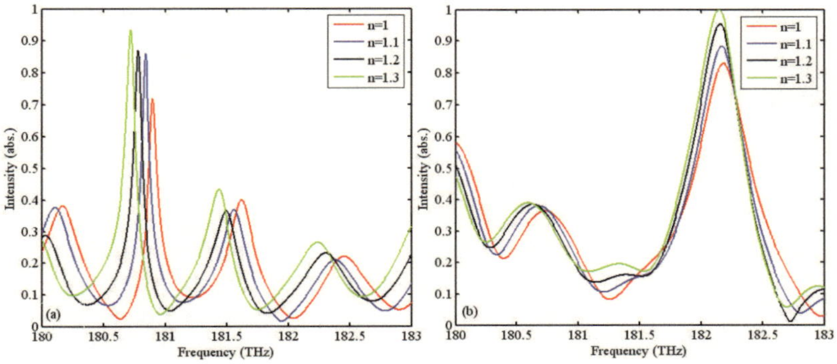

Fig. 28. Spectra for the surrounding medium with different refractive index. (a) Results for the microring cavity without metamaterial layer. (b)Results for the microring cavity with metamaterial layer

as shown in Fig. 28. From Fig. 28(a), it is seen that the resonance peak of the microring cavity without metamaterial layer is highly dependent on the refractive index of the surrounding medium, and it is red shifted with the increase of refractive index. From Fig. 28(b) we can clear observe that the resonance peaks are shifted to the high frequency side when metamaterial layer is added to the outside of the microring ring resonator. Meanwhile, the peak value increases with the increase of the refractive index of surrounding medium.

Due to its characteristics of high Q factor, wide free spectral-range, microcavity can be used in the field of identification and monitoring of proteins, DNA, peptides, toxin molecules, and nanoparticle, etc. It has attracted extensive attention world wide, and more details about microcavity can be found in the original work of Quan and Zhu et al (Quan et al, 2005; Zhu et al, 2009).

7. Conclusion

It has been demonstrated that the evanescent wave can be amplified by the metamaterials. This unique property is helpful for enhancing the sensitivity of sensor, and can realize subwavelength resolution of image and detection beyond diffraction limit. Enhancement of sensitivity in slab waveguide with TM mode is proved analytically. The phenomenon of evanescent wave amplification is confirmed in slab waveguide and slab lens. The perfect imaging properties of planar lens was proved by transmission optics. Microwave sensors based on the waveguide filled with metamaterial particles are simulated, and their sensitivity is much higher than traditional microwave sensor. The open microwave resonator consists of two homogenous metamaterial squares is very sensitive to dielectric environment. The microcavity ring resonator with metamaterial layer possesses some new properties.

Metamaterials increases the designing flexibility of sensors, and dramatically improves their performance. Sensors using metamaterials may hope to fuel the revolution of sensing technology.

8. Acknowledgement

This work was supported by the National Natural Science Foundation of China (grant no. 60861002), the Research Foundation from Ministry of Education of China (grant no. 208133), and the Natural Science Foundation of Yunnan Province (grant no.2007F005M).

9. References

Alù, A. & N. Engheta. (2008) Dielectric sensing in ε-near-zero narrow waveguide channels," *Phys. Rev. B,* Vol. 78, No. 4, 045102, ISSN: 1098-0121

Al-Naib, I. A. I.; Jansen, C. & Koch, M. (2008) Thin-film sensing with planar asymmetric metamaterial resonators. *Appl. Phys. Lett.,* Vol. 93, No. 8, 083507, ISSN: 0031-9007

Fedotov, V.A.; Rose, M.; Prosvirnin, S.L.; Papasimakis, N. & Zheludev, N. I. (2007) Sharp trapped-Mode resonances in planar metamaterials with a broken structural symmetry. *Phys. Rev. Lett.,* Vol. 99, No. 14, 147401, ISSN: 1079-7114

Guru, B. S. & Hiziroglu, H. R. (1998). Plane wave propagation, In: *Electromagnetic Field Theory Fundamentals,* Guru, B. S. & Hiziroglu, H. R. (Ed.), 305-360, Cambridge University Press, ISBN: 7-111-10622-9, Cambridge, UK, New York

Huang, M.; Yang, J. J.; Wang, J. Q. & Peng, J. H. (2007). Microwave sensor for measuring the properties of a liquid drop. *Meas. Sci. Technol.*, Vol. 18, No. 7, 1934–1938, ISSN: 0957-0233

Huang, M., Yang, J.J., Sun, J., Shi, J.H. & Peng, J.H. (2009) Modelling and analysis of Ω-shaped double negative material-assisted microwave sensor. *J. Infrared Milli. Terahz. Waves,* Vol. 30, No. 11, 1131-1138, ISSN: 1866-6892

He, S.; Jin Y.; Ruan, Z. C. & Kuang, J.G. (2005). On subwavelength and open resonators involving metamaterials of negative refraction index. *New J. Phys.,* Vol. 7, No. 210, ISSN: 1367-2630

Hagness, S. C.; Rafizadeh, D.; Ho, S. T. & Taflove, A.(1997). FDTD microcavity simulations: design and experimental realization of waveguide-coupled single-mode ring and whispering-gallery-mode disk resonators. *Journal of lightwave Technology,* Vol. 15, No. 11, 2154-2164, ISSN: 0733-8724

Kupfer, K. (2000). Microwave Moisture Sensor Systems and Their Applications, In: *Sensor Update,* Kupfer, K.; Kraszewski, A. & Knöchel, R, (Ed.), 343-376, WILEY-VCH, ISBN: 3-527-29821-5, Weinheim (Federal Republic of Germany)

Kraszewski, A. W. (1991). Microwave aquametry-needs and perspectives. *IEEE Trans. Microwave Theory Tech.,* Vol. 39, No. 5, 828-835, ISSN: 0018-9480

Lee, H. J. & Yook, J. G. (2008). Biosensing using split-ring resonators at microwave regime. *Appl. Phys. Lett.,* Vol. 92, No. 25, 254103, ISSN: 0003-6951

Marqués, R.; Martel, J.; Mesa, F. & Medina, F. (2002). Left-Handed-Media simulation and transmission of EM waves in subwavelength split-ring-resonator-loaded metallic waveguides. *Phys. Rev. Lett.,* 89, No.18, 183901, ISSN: 0031-9007

Melik, R.; Unal, E.; Perkgoz, N. K.; Puttlitz, C. & Demir, H. V. (2009). Metamaterial-based wireless strain sensors. *Appl. Phys. Lett.,* Vol. 95, No. 1, 011106, ISSN: 0003-6951

Notomi, M. (2000). Theory of light propagation in strongly modulated photonic crystals: Refractionlike behavior in the vicinity of the photonic band gap. *Phys. Rev. B,* Vol.,62, No. 16, 10696-10705, ISSN: 1098-0121

Pendry, J. B. (2000). Negative Refraction Makes a Perfect Lens. *Phys. Rev. Lett.,* Vol. 85, No. 18, 3966-3969, ISSN: 0031-9007

Papasimakis, Ni.; Luo, Z.Q.; Shen, Z.X.; Angelis, F. D.; Fabrizio, E. D.; Nikolaenko, A. E.; & Zheludev, N. I. (2010). Graphene in a photonic metamaterial. *Optics Express,* Vol. 18, No. 8, 8353-8359, ISSN: 1094-4087

Qing, D. K. & Chen, G. (2004). Enhancement of evanescent waves in waveguides using metamaterials of negative permittivity and permeability. *Appl. Phys. Lett.,* Vol. 84, No. 5, 669-671, ISSN: 0003-6951

Quan, H.Y.; & Guo, Z.X. (2005). Simulation of whispering-gallery-mode resonance shifts for optical miniature biosensors. Journal of Quantitative Spectroscopy & Radiative Transfer, Vol. 93, No. 1-3, 231–243, ISSN: 0022-4073

Shelby, R. A.; Smith, D. R. & Schultz, S. (2001). Experimental verification of a negative index of refraction. *Science,* Vol. 292, No. 5514, 77-79, ISSN: 0036-8075

Service, R. F. (2010). Next wave of metamaterials hopes to fuel the revolution. *Science,* Vol. 327, No. 5962, 138-139, ISSN: 0036-8075

Shreiber, D.; Gupta, M. & Cravey, R. (2008). Microwave nondestructive evaluation of dielectric materials with a metamaterial lens. *Sensors and Actuators,* Vol. 144, No.1, 48–55, ISSN：0924-4247

Silveirinha, M. & Engheta, N. (2006). Tunneling of Electromagnetic Energy through Subwavelength Channels and Bends using ε-Near-Zero Materials. *Phys. Rev. Lett.,* Vol. 97, No. 15, 157403, ISSN: 0031-9007

Smith, D. R.; Vier, D C; Koschny, Th. & Soukoulis, C. M. (2005). Electromagnetic parameter retrieval from inhomogeneous metamaterials. *Phys. Rev. E,* Vol. 71, No. 3, 036617, ISSN: 1539-3755

Taya, S. A.; Shabat, M. M. & Khalil, H. M.(2009). Enhancement of sensitivity in optical waveguide sensors using left-handed materials. *Optik,* Vol. 120, No.10, 504-508, ISSN: 0030-4026

Von Hippel A, (1995). Dielectric measuring techniques, In: *Dielectric Materials and Applications,* Hippel A. V., (Ed.) 47-146, Wiley/The Technology Press of MIT, ISBN: 0-89006-805-4, New York

Veselago, V. G. (1968). The electrodynamics of substances with simultaneously negative values of ε and μ. *Sov. Phys. Usp.,* Vol. 10, No. 4, (1968) 509-514, ISSN: 0038-5670

Wang,W.; Lin, L.; Yang, X. F. Cui, J. H.; Du, C. L. & Luo, X. G. (2008). Design of oblate cylindrical perfect lens using coordinate transformation. *Optics Express,* Vol. 16, No. 11, 8094-8105, ISSN: 1094-4087

Wu, Z.Y.; Huang, M.; Yang J. J.; Peng, J.H. & Zong, R. (2008). Electromagnetic wave tunnelling and squeezing effects through 3D coaxial waveguide channel filled with ENZ material, *Proceedings of ISAPE 2008,* pp. 752-755, ISBN: 978-1-4244-2192-3, Kunming, Yunnan, China, Nov. 2008, Institute of Electrical and Electronics Engineers, Inc., Beijing

Yang J. J.; Huang, M.; Xiao, Z. & Peng, J. H.(2010). Simulation and analysis of asymmetric metamaterial resonator-assisted microwave sensor. *Mod. Phys. Lett. B,* Vol. 24, No. 12, 1207–1215, ISSN: 0217-9849

Zoran, J.; Jakšić, O.; Djuric,Z. & Kment, C. (2007). A consideration of the use of metamaterials for sensing applications:field fluctuations and ultimate performance, *J. Opt. A: Pure Appl.,* Vol. 9, No. 9, S377–S384, ISSN: 1464-4258

Zhu, J.G.; Ozdemir, S. K.; Xiao, Y. F.; Li, L.; He, L.N. Chen, D.R. & Yang, L.(2009). On-chip single nanoparticle detection and sizing by mode splitting in an ultrahigh-Q microresonator. *Nature Photonics,* Vol. 4, No.1, 46-49, ISSN: 1749-4885

3

Electromagnetic Waves in Crystals with Metallized Boundaries

V.I. Alshits[1,2], V.N. Lyubimov[1], and A. Radowicz[3]
[1]*A.V. Shubnikov Institute of Crystallography, Russian Academy of Sciences, Moscow, 119333*
[2]*Polish-Japanese Institute of Information Technology, Warsaw, 02-008*
[3]*Kielce University of Technology, Kielce, 25-314*
[1]*Russia*
[2,3]*Poland*

1. Introduction

The metal coating deposited on the surface of a crystal is a screen that locks the electromagnetic fields in the crystal. Even for a real metal when its complex dielectric permittivity ε_m has large but finite absolute value, electromagnetic waves only slightly penetrate into such coating. For example, for copper in the wavelength range $\lambda = 10^{-5}$–10^{-3} cm, from the ultraviolet to the infrared, the penetration depth d changes within one order of magnitude: $d \approx 6 \times (10^{-8}$–$10^{-7})$ cm, remaining negligible compared to the wavelength, $d << \lambda$. In the case of a perfect metallization related to the formal limit $\varepsilon_m \to \infty$ the wave penetration into a coating completely vanishes, $d = 0$. The absence of accompanying fields in the adjacent space simplifies considerably the theory of electromagnetic waves in such media. It turned out that boundary metallization not only simplifies the description, but also changes significantly wave properties in the medium. For example, it leads to fundamental prohibition (Furs & Barkovsky, 1999) on the existence of surface electromagnetic waves in crystals with a positively defined permittivity tensor $\hat{\varepsilon}$. There is no such prohibition at the crystal–dielectric boundary (Marchevskii et al., 1984; D'yakonov, 1988; Alshits & Lyubimov, 2002a, 2002b)). On the other hand, localized polaritons may propagate along even perfectly metalized surface of the crystal when its dielectric tensor $\hat{\varepsilon}$ has strong frequency dispersion near certain resonant states so that one of its components is negative (Agranovich, 1975; Agranovich & Mills, 1982; Alshits et al., 2001; Alshits & Lyubimov, 2005). In particular, in the latter paper clear criteria were established for the existence of polaritons at the metalized boundary of a uniaxial crystal and compact exact expressions were derived for all their characteristics, including polarization, localization parameters, and dispersion relations.

In this chapter, we return to the theory of electromagnetic waves in uniaxial crystals with metallized surfaces. This time we will be concerned with the more common case of a crystal with a positively defined tensor $\hat{\varepsilon}$. Certainly, under a perfect metallization there is no localized eigenmodes in such a medium, but the reflection problem in its various aspects and such peculiar eigenmodes as the exceptional bulk (nonlocalized) polaritons that transfer energy parallel to the surface and satisfy the conditions at the metallized boundary remain.

We will begin with the theory for the reflection of plane waves from an arbitrarily oriented surface in the plane of incidence of the general position, where the reflection problem is solved by a three-partial superposition of waves: one incident and two reflected components belonging to different sheets of the refraction surface. However, one of the reflected waves may turn out to be localized near the surface. Two-partial reflections, including mode conversion and "pure" reflection, are also possible under certain conditions. The incident and reflected waves belong to different sheets of the refraction surface in the former case and to the same sheet of ordinary or extraordinary waves in the latter case. First, we will study the existence conditions and properties of pure (simple) reflections. Among the solutions for pure reflection, we will separate out a subclass in which the passage to the limit of the eigenmode of exceptional bulk polaritons is possible. Analysis of the corresponding dispersion equation will allow us to find all of the surface orientations and propagation directions that permit the existence of ordinary or extraordinary exceptional bulk waves. Subsequently, we will construct a theory of conversion reflections and find the configurations of the corresponding pointing surface for optically positive and negative crystals that specifies the refractive index of reflection for each orientation of the optical axis. The mentioned theory is related to the idealized condition of perfect metallization and needs an extension to the case of the metal with a finite electric permittivity ε_m. The transition to a real metal may be considered as a small perturbation of boundary condition. As was initially suggested by Leontovich (see Landau & Lifshitz, 1993), it may be done in terms of the so called surface impedance $\zeta = 1/\sqrt{\varepsilon_m}$ of metal. New important wave features arise in the medium with $\zeta \neq 0$. In particular, a strongly localized wave in the metal (a so-called plasmon) must now accompany a stationary wave field in the crystal. In a real metal such plasmon should dissipate energy. Therefore the wave in a crystal even with purely real tensor $\hat{\varepsilon}$ must also manifest damping. In addition, in this more general situation the exceptional bulk waves transform to localized modes in some sectors of existence (the non-existence theorem (Furs & Barkovsky, 1999) does not valid anymore).

We shall consider a reaction of the initial idealized physical picture of the two independent wave solutions, the exceptional bulk wave and the pure reflection in the other branch, on a "switching on" the impedance ζ combined with a small change of the wave geometry. It is clear without calculations that generally they should loss their independency. The former exceptional wave cannot anymore exist as a one-partial eigenmode and should be added by a couple of partial waves from the other sheet of the refraction surface. But taking into account that the supposed perturbation is small, this admixture should be expected with small amplitudes. Thus we arise at the specific reflection when a weak incident wave excites, apart from the reflected wave of comparable amplitude from the same branch, also a strong reflected wave from the other polarization branch. The latter strong reflected wave should propagate at a small angle to the surface being close in its parameters to the initial exceptional wave in the unperturbed situation.

Below we shall concretize the above consideration to an optically uniaxial crystal with a surface coated by a normal metal of the impedance ζ supposed to be small. The conditions will be found when the wave reflection from the metallized surface of the crystal is of resonance character being accompanied by the excitation of a strong polariton-plasmon. The peak of excitation will be studied in details and the optimized conditions for its observation will be established. Under certain angles of incidence, a conversion occurs in the resonance area: a pumping wave is completely transformed into a surface polariton--plasmon of much higher intensity than the incident wave. In this case, no reflected wave arises: the normal component of the incident energy flux is completely absorbed in the metal. The conversion solution represents an eigenmode opposite in its physical sense to customary leaky surface waves known in optics and acoustics. In contrast to a leaky eigenwave containing a weak «reflected» partial wave providing a leakage of energy from the surface, here we meet a pumped surface polariton-plasmon with the weak «incident» partial wave transporting energy to the interface for the compensation of energy dissipation in the metal.

2. Formulation of the problem and basic relations

Consider a semi-bounded, transparent optically uniaxial crystal with a metallized boundary and an arbitrarily oriented optical axis. Its dielectric tensor $\hat{\varepsilon}$ is conveniently expressed in the invariant form (Fedorov, 2004) as

$$\hat{\varepsilon} = \varepsilon_o \hat{I} + (\varepsilon_e - \varepsilon_o)\mathbf{c} \otimes \mathbf{c}, \tag{1}$$

where $\hat{I}$ is the identity matrix, **c** is a unit vector along the optical axis of the crystal, $\otimes$ is the symbol of dyadic product, ε_o and ε_e are positive components of the electric permittivity of the crystal. For convenience, we will use the system of units in which these components are dimensionless (in the SI system, they should be replaced by the ratios $\varepsilon_o/\varepsilon^0$ and $\varepsilon_e/\varepsilon^0$, where ε^0 is the permittivity of vacuum).

In uniaxial crystals, one distinguishes the branches of ordinary (with indices "o") and extraordinary (indices "e") electromagnetic waves. Below, along with the wave vectors $\mathbf{k}_\alpha$ ($\alpha = o, e$), we shall use dimensionless refraction vectors $\mathbf{n}_\alpha = \mathbf{k}_\alpha/k_0$ where $k_0 = \omega/c$, ω is the wave frequency and c is the light speed. These vectors satisfy the equations (Fedorov, 2004)

$$\mathbf{n}_o \cdot \mathbf{n}_o = \varepsilon_o, \qquad \mathbf{n}_e \cdot \hat{\varepsilon}\mathbf{n}_e = \varepsilon_o \varepsilon_e. \tag{2}$$

For real vectors $\mathbf{n}_o$ and $\mathbf{n}_e$, the ray velocities (the velocities of energy propagation) of the corresponding bulk waves are defined by

$$\mathbf{u}_o = \frac{c\mathbf{n}_o}{\varepsilon_o}, \qquad \mathbf{u}_e = \frac{c\hat{\varepsilon}\mathbf{n}_e}{\varepsilon_o\varepsilon_e} = \frac{c}{\varepsilon_o\varepsilon_e}\left[\varepsilon_o\mathbf{n}_e + (\varepsilon_e - \varepsilon_o)(\mathbf{n}_e \cdot \mathbf{c})\mathbf{c}\right]. \tag{3}$$

Formulas (3) show that, in the ordinary wave, energy is transported strictly along the refraction vector, whereas, in the extraordinary wave, generally not.

For our purposes, it is convenient to carry out the description in a coordinate system associated not with the crystal symmetry elements, but with the wave field parameters. Let us choose the x axis in the propagation direction **m** and the y axis along the inner normal **n** to the surface. In this case, the xy plane is the plane of incidence where all wave vectors of the incident and reflected waves lie, the xz plane coincides with the crystal boundary, and the optical axis is specified by an arbitrarily directed unit vector **c** (Fig. 1). The orientation of vector $\mathbf{c} = (c_1, c_2, c_3)$ in the chosen coordinate system can be specified by two angles, θ and φ. The angle θ defines the surface orientation and the angle φ on the surface defines the propagation direction of a stationary wave field.

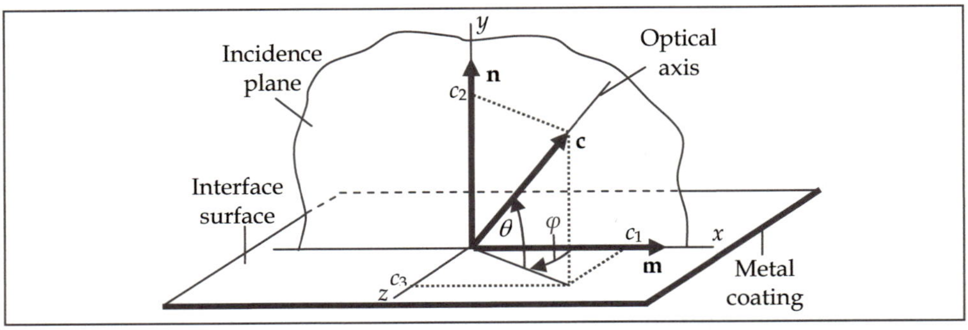

Fig. 1. The system of *xyz* coordinates and the orientation **c** of the crystal's optical axis

The stationary wave field under study can be expressed in the form:

$$\begin{pmatrix} \mathbf{E}(x,y,t) \\ \mathbf{H}(x,y,t) \end{pmatrix} = \begin{pmatrix} \mathbf{E}(y) \\ \mathbf{H}(y) \end{pmatrix} \exp[ik(x - vt)] \,. \tag{4}$$

The y dependence of this wave field is composed from a set of components. In the crystal ($y > 0$) there are four partial waves subdivided into incident (i) and reflected (r) ones from two branches, ordinary (o) and extraordinary (e):

$$\begin{pmatrix} \mathbf{E}(y) \\ \mathbf{H}(y) \end{pmatrix} = C_o^i \begin{pmatrix} \mathbf{E}_o^i(y) \\ \mathbf{H}_o^i(y) \end{pmatrix} + C_o^r \begin{pmatrix} \mathbf{E}_o^r(y) \\ \mathbf{H}_o^r(y) \end{pmatrix} + C_e^i \begin{pmatrix} \mathbf{E}_e^i(y) \\ \mathbf{H}_e^i(y) \end{pmatrix} + C_e^r \begin{pmatrix} \mathbf{E}_e^r(y) \\ \mathbf{H}_e^r(y) \end{pmatrix} . \tag{5}$$

Here the vector amplitudes are defined by

$$\begin{pmatrix} \mathbf{E}_o^{i,r}(y) \\ \mathbf{H}_o^{i,r}(y) \end{pmatrix} = \begin{pmatrix} \mathbf{e}_o^{i,r} \\ \mathbf{h}_o^{i,r} \end{pmatrix} \exp(\mp i p_o k y) \,, \tag{6}$$

$$\begin{pmatrix} \mathbf{E}_e^{i,r}(y) \\ \mathbf{H}_e^{i,r}(y) \end{pmatrix} = \begin{pmatrix} \mathbf{e}_e^{i,r} \\ \mathbf{h}_e^{i,r} \end{pmatrix} \exp[i(p \mp p_e) k y] \,. \tag{7}$$

In Eqs. (4)-(7), **E**, **e** and **H**, **h** are the electric and magnetic field strengths, k is the common x component of the wave vectors for the ordinary and extraordinary partial waves: $k = \mathbf{k}_o^{i,r} \cdot \mathbf{m} = \mathbf{k}_e^{i,r} \cdot \mathbf{m}$, $v = \omega/k$ is the tracing phase velocity of the wave, and $C_o^{i,r}$ and $C_e^{i,r}$ are the amplitude factors to be determined from the boundary conditions. The upper and lower signs in the terms correspond to the incident and reflected waves, respectively.

In the isotropic metal coating ($y < 0$) only two partial waves propagate differing from each other by their *TM* and *TE* polarizations:

$$\begin{pmatrix} \mathbf{E}(y) \\ \mathbf{H}(y) \end{pmatrix} = \left(C_m^{TM} \begin{pmatrix} \mathbf{e}_m^{TM} \\ \mathbf{h}_m^{TM} \end{pmatrix} + C_m^{TE} \begin{pmatrix} \mathbf{e}_m^{TE} \\ \mathbf{h}_m^{TE} \end{pmatrix} \right) \exp(-ikp_m y) \,. \tag{8}$$

By definition, the above polarization vectors are chosen so that the *TM* wave has the magnetic component orthogonal to the sagittal plane and the electric field is polarized in this plane, and for the *TE* wave, vice versa, the magnetic field is polarized in-plane and the electric field - out-plane:

$$\mathbf{h}_m^{TM} \,||\, (0,0,1)\,,\;\; \mathbf{e}_m^{TM} \,||\, \mathbf{n}_m \times \mathbf{h}_m^{TM}\,,\;\; \mathbf{e}_m^{TE} \,||\, (0,0,1)\,,\;\; \mathbf{h}_m^{TE} \,||\, \mathbf{n}_m \times \mathbf{e}_m^{TE}\,. \qquad 9)$$

The refraction vectors of the partial waves in the superpositions (5) and (8) are equal

$$\mathbf{n}_o^{i,r} = n(1,\; \mp p_o,\; 0)^T\,,\;\; \mathbf{n}_e^{i,r} = n(1,\; p \mp p_e,\; 0)^T\,,\;\; \mathbf{n}_m = n(1,\; -p_m,\; 0)^T\,. \qquad (10)$$

Here, the superscript T stands for transposition and $n = k/k_0 = c/v$ is the dimensionless wave slowness also called the refractive index. The parameters p_o, p_e, p and p_m that determine the dependences of the partial amplitudes on depth y can be represented as

$$p_o = \sqrt{s-1}\,,\quad p_e = \sqrt{\frac{\gamma}{A}\left(s - \frac{B}{A}\right)}\,,\quad p = (1-\gamma)\frac{c_1 c_2}{A}\,,\quad p_m = \frac{R}{\zeta n}\,, \qquad (11)$$

where we use the notation

$$s = \varepsilon_o / n^2\,,\;\; \gamma = \varepsilon_e / \varepsilon_o\,,\;\; A = 1 + c_2^2(\gamma - 1)\,,\; B = 1 - c_3^2(1 - 1/\gamma)\,, R = \sqrt{1 - (\zeta n)^2}\,. \qquad (12)$$

The orientation of the polarization vectors in (5), (6) is known from (Born & Wolf, 1986; Landau & Lifshitz, 1993) and can be specified by the relations

$$\mathbf{e}_o^{i,r} \,||\, \mathbf{n}_o^{i,r} \times \mathbf{c}, \qquad \mathbf{e}_e^{i,r} \,||\, \mathbf{n}_e^{i,r}(\mathbf{n}_e^{i,r}\cdot \mathbf{c}) - \boldsymbol{\varepsilon}_o \mathbf{c}, \qquad \mathbf{h}_a^{i,r} = \mathbf{n}_a^{i,r} \times \mathbf{e}_a^{i,r}\,, \qquad a = o, e\,. \qquad (13)$$

Substituting relations (10) into (9) and (13) one obtains

$$\begin{pmatrix} \mathbf{e}_o^{i,r} \\ \mathbf{h}_o^{i,r} \end{pmatrix} = N_o^{i,r} \begin{pmatrix} (\mp p_o c_3,\; -c_3,\; c_2 \pm p_o c_1)^T \\ -n[p_o(p_o c_1 \pm c_2),\;\; c_2 \pm p_o c_1,\;\; c_3 s]^T \end{pmatrix}, \qquad (14)$$

$$\begin{pmatrix} \mathbf{e}_e^{i,r} \\ \mathbf{h}_e^{i,r} \end{pmatrix} = N_e^{i,r} \begin{pmatrix} \{c_1 - [c_1 + (p \mp p_e)c_2]/s,\;\; c_2 - [c_1 + (p \mp p_e)c_2](p \mp p_e)/s,\;\; c_3\}^T \\ n[(p \mp p_e)c_3,\qquad -c_3,\qquad c_2 - (p \mp p_e)c_1]^T \end{pmatrix}, \qquad (15)$$

$$\begin{pmatrix} \mathbf{e}_m^{TM} \\ \mathbf{h}_m^{TM} \end{pmatrix} = \begin{pmatrix} \zeta(R,\; \zeta n,\; 0)^T \\ (0,\;\; 0,\;\; 1)^T \end{pmatrix}, \qquad \begin{pmatrix} \mathbf{e}_m^{TE} \\ \mathbf{h}_m^{TE} \end{pmatrix} = \begin{pmatrix} (0,\;\; 0,\; -\zeta)^T \\ (R,\; \zeta n,\; 0)^T \end{pmatrix}. \qquad (16)$$

The normalization in (14)-(16) was done from the conditions $|\mathbf{h}_{o,e}^{i,r}| = 1$ and $|\mathbf{h}_m^a| = 1$. It already presents in (16) and the factors $N_{o,e}$ in (14), (15) are specified by the equations

$$1/N_o^{i,r} = \sqrt{\varepsilon_o[(c_2 \pm c_1 p_o)^2 + c_3^2 s]}\,, \qquad 1/N_e^{i,r} = n\sqrt{1 + (p \mp p_e)^2 - [c_1 + c_2(p \mp p_e)]^2}\,. \qquad (17)$$

3. Boundary conditions and a reflection problem in general statement

The stationary wave field (4) at the interface should satisfy the standard continuity conditions for the tangential components of the fields (Landau & Lifshitz, 1993):

$$\mathbf{E}_t\,|_{y=+0} = \mathbf{E}_t\,|_{y=-0}, \qquad \mathbf{H}_t\,|_{y=+0} = \mathbf{H}_t\,|_{y=-0}\,. \tag{18}$$

When the crystal is coated with perfectly conducting metal, the electric field in the metal vanishes and the boundary conditions (18) reduces to

$$\mathbf{E}_t\,|_{y=+0} = 0\,. \tag{19}$$

When the perfectly conducting coating is replaced by normal metal with sufficiently small impedance $\zeta = \zeta' + i\zeta''$ ($\zeta' > 0$, $\zeta'' < 0$), it is convenient to apply more general (although also approximate) Leontovich boundary condition (Landau & Lifshitz, 1993) instead of (19):

$$(\mathbf{E}_t + \zeta\mathbf{H}_t \times \mathbf{n})_{y=+0} = 0\,. \tag{20}$$

Below in our considerations, the both approximations, (19) and (20), will be applied. However we shall start from the exact boundary condition (18).

3.1 Generalization of the Leontovich approximation

The conditions (18) after substitution there equations (5)-(8) and (16) take the explicit form

$$\begin{pmatrix} e^r_{ox} & e^r_{ex} & \zeta R & 0 \\ e^r_{oz} & e^r_{ez} & 0 & -\zeta \\ h^r_{ox} & h^r_{ex} & 0 & R \\ h^r_{oz} & h^r_{ez} & 1 & 0 \end{pmatrix} \begin{pmatrix} C^r_o \\ C^r_e \\ C^{TM}_m \\ C^{TE}_m \end{pmatrix} = -C^i_o \begin{pmatrix} e^i_{ox} \\ e^i_{oz} \\ h^i_{ox} \\ h^i_{oz} \end{pmatrix} - C^i_e \begin{pmatrix} e^i_{ex} \\ e^i_{ez} \\ h^i_{ex} \\ h^i_{ez} \end{pmatrix}. \tag{21}$$

Following to (Alshits & Lyubimov, 2009a) let us transform this system for obtaining an exact alternative to the Leontovich approximation (20). We eliminate the amplitudes C^{TM}_m and C^{TE}_m of the plasmon in metal from system (21) and reduce it to the system of two equations:

$$\left[\begin{pmatrix} e^r_{ox} & e^r_{ex} \\ e^r_{oz} & e^r_{ez} \end{pmatrix} + \zeta \begin{pmatrix} -Rh^r_{oz} & -Rh^r_{ez} \\ h^r_{ox}/R & h^r_{ex}/R \end{pmatrix}\right] \begin{pmatrix} C^r_o \\ C^r_e \end{pmatrix} + \left[\begin{pmatrix} e^i_{ox} & e^i_{ex} \\ e^i_{oz} & e^i_{ez} \end{pmatrix} + \zeta \begin{pmatrix} -Rh^i_{oz} & -Rh^i_{ez} \\ h^i_{ox}/R & h^i_{ex}/R \end{pmatrix}\right] \begin{pmatrix} C^i_o \\ C^i_e \end{pmatrix} = 0\,. \tag{22}$$

Taking into account the matrix identities

$$\begin{pmatrix} -Rh^{i,r}_{oz} & -Rh^{i,r}_{ez} \\ h^{i,r}_{ox}/R & h^{i,r}_{ex}/R \end{pmatrix} = \begin{pmatrix} -h^{i,r}_{oz} & -h^{i,r}_{ez} \\ h^{i,r}_{ox} & h^{i,r}_{ex} \end{pmatrix} + (1-R) \begin{pmatrix} h^{i,r}_{oz} & h^{i,r}_{ez} \\ h^{i,r}_{ox}/R & h^{i,r}_{ex}/R \end{pmatrix} \tag{23}$$

and the explicit form of two-dimensional vectors $\mathbf{E}_t = (E_{tx}, E_{tz})^T$ and $\mathbf{H}_t = (H_{tx}, H_{tz})^T$ residing in the xz plane, namely

$$\mathbf{E}_t = C^i_o \begin{pmatrix} e^i_{ox} \\ e^i_{oz} \end{pmatrix} + C^r_o \begin{pmatrix} e^r_{ox} \\ e^r_{oz} \end{pmatrix} + C^i_e \begin{pmatrix} e^i_{ex} \\ e^i_{ez} \end{pmatrix} + C^r_e \begin{pmatrix} e^r_{ex} \\ e^r_{ez} \end{pmatrix}, \tag{24}$$

$$\mathbf{H}_t = C^i_o \begin{pmatrix} h^i_{ox} \\ h^i_{oz} \end{pmatrix} + C^r_o \begin{pmatrix} h^r_{ox} \\ h^r_{oz} \end{pmatrix} + C^i_e \begin{pmatrix} h^i_{ex} \\ h^i_{ez} \end{pmatrix} + C^r_e \begin{pmatrix} h^r_{ex} \\ h^r_{ez} \end{pmatrix}, \tag{25}$$

system (22) reduces to the following equation

$$\{\mathbf{E}_t + \zeta \mathbf{H}_t \times \mathbf{n} + \zeta(1-R)\hat{N}\mathbf{H}_t\}_{y=+0} = 0\,, \tag{26}$$

where the function $R(\zeta n)$ was defined in (12), and $\hat{N}(\zeta n)$ is the 2×2 matrix:

$$\hat{N}(\zeta n) = \begin{pmatrix} 0 & 1 \\ 1/R(\zeta n) & 0 \end{pmatrix}. \tag{27}$$

Notice that equation (26) is equivalent to an initial set of conditions (18). Impedance ζ in Eq. (26) is not assumed to be small and this expression only includes crystal fields (5)-(7). Thus, equation (26) is the natural generalization of Leontovich boundary condition (18).
However, the impedance ζ of ordinary metals (like copper or aluminum) may be considered as a small parameter, especially in the infrared range of wavelengths. In this case, function $R(\zeta n)$ in equation (26) [see in (12)] can be expanded in powers of the small parameter $(\zeta n)^2$, holding an arbitrary number of terms and calculating the characteristics of the wave fields with any desired precision. This expansion comprises odd powers of the parameter ζ:

$$\mathbf{E}_t + \zeta \mathbf{H}_t \times \mathbf{n} + \frac{1}{2}\zeta^3 n^2 \left(\hat{N}_1 + \sum_{s=1}^{\infty} (\zeta)^{2s} \frac{(2s-1)!!}{2^s (s+1)!} \hat{N}_{2s+1} \right) \mathbf{H}_t = 0\,, \tag{28}$$

where the set of matrices $\hat{N}_m$ ($m = 2s + 1$) is defined by the expression

$$\hat{N}_m = \begin{pmatrix} 0 & 1 \\ m & 0 \end{pmatrix}. \tag{29}$$

In view of our considerations, from expansion (28) it follows that the discrepancy between Leontovich approximation (20) and the exact boundary condition starts from the cubic term $\sim\zeta^3$; hence the quadratic corrections $\sim\zeta^2$ to the wave fields are correct in this approach.

3.2 Exact solution of the reflection problem

Now let us return to the reflection problem, i.e. to the system (21), which, together with relations (14) and (15), determines the amplitudes of superpositions (5) and (8). The right-hand side of (21) is considered to be known. When the reflection problem is formulated, only one incident wave is commonly considered by assuming its amplitude to be known (while the other is set equal to zero). The refractive index n, which directly determines the angle of incidence, is also assumed to be known, while the amplitudes of the reflected waves in the crystal and those of the plasmon components in the metal are to be determined.
Being here interested only in wave fields in the crystal, we can start our analysis from the more simple system (22) of only two equations with two unknown quantities. Omitting bulky but straightforward calculations we just present their results in the form of the reflection coefficients for the cases of an ordinary incident wave,

$$r_{oo} = \frac{C_o^r}{C_o^i} = -\frac{D(-p_o, p_e) N_o^i}{D(p_o, p_e) N_o^r}, \qquad r_{eo} = \frac{C_e^r}{C_o^i} = -\frac{D_{eo} N_o^i}{D(p_o, p_e) N_e^r}, \tag{30}$$

and an extraordinary incident wave,

$$r_{ee} = \frac{C_e^r}{C_e^i} = -\frac{D(p_o, -p_e)N_e^i}{D(p_o, p_e)N_e^r}, \qquad r_{oe} = \frac{C_o^r}{C_e^i} = -\frac{D_{oe}N_e^i}{D(p_o, p_e)N_o^r}. \tag{31}$$

In the above equations the following notation is introduced

$$D(p_o, p_e) = (c_1 p_o - c_2)(1 + p_o \zeta n / R)\{c_1 p_o^2 - c_2(p + p_e) - \zeta nRs[c_2 - c_1(p + p_e)]\} \\ + c_3^2 s(p_o + \zeta nRs)[1 + (p + p_e)\zeta n / R], \tag{32}$$

$$D_{eo} = 2c_3 p_o s(1 - \varepsilon_o \zeta^2)(c_2 + c_1 \zeta n / R), \tag{33}$$

$$D_{oe} = 2c_3 p_e(1 - \varepsilon_o \zeta^2)(c_2 - c_1 \zeta n / R). \tag{34}$$

One can check that these expressions fit the known general equations (Fedorov & Filippov, 1976). Before beginning our analysis of Eqs. (30)-(34), recall that we consider only the crystals (and frequencies) that correspond to a positively defined permittivity tensor ($\varepsilon_o > 0$ and $\varepsilon_e > 0$). Depending on the relation between the components ε_o and ε_e , it is customary to distinguish the optically positive ($\varepsilon_e > \varepsilon_o$, i.e., $\gamma > 1$) and optically negative ($\varepsilon_e < \varepsilon_o$, i.e., $\gamma < 1$) crystals. Figure 2 shows the sections of the sheets of the refraction surface for these two types of crystals by the *xy* plane of incidence for arbitrary orientation of the boundary and propagation direction. Among the main reflection parameters shown in Fig. 2, the limiting values of the refractive indices $\hat{n}_o$ and $\hat{n}_e$ play a particularly important role:

$$\hat{n}_o = \sqrt{\varepsilon_o}, \qquad \hat{n}_e = \sqrt{\varepsilon_o A / B}. \tag{35}$$

These separate the regions of real and imaginary values of the parameters p_o and p_e:

$$p_o = \sqrt{(\hat{n}_o / n)^2 - 1}, \qquad p_e = \sqrt{[(\hat{n}_e / n)^2 - 1]\gamma B / A^2}. \tag{36}$$

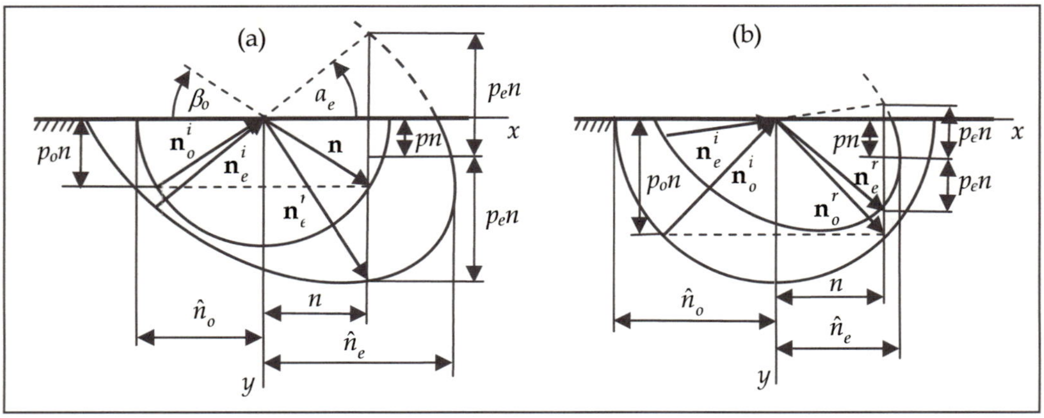

Fig. 2. Sections of the ordinary and extraordinary sheets of the refraction surface by the *xy* plane of incidence and main parameters of the reflection problem for optically positive (a) and optically negative (b) crystals

The parameter p_o remains real only in the region $0 \le n \le \hat{n}_o$, i.e., as long as the vertical straight line in Fig. 2 crosses the corresponding circular section of the spherical refraction sheet for ordinary waves (or touches it). Similarly, the parameter p_e remains real only in the region $0 \le n \le \hat{n}_e$. In both regions of real values, the refractive index $n = 0$ describes the reflection at normal incidence.

Thus, when the stationary wave motions along the surface are described, three regions of the refractive index n should be distinguished:

$$\text{(I) } 0 < n \le \min\{\hat{n}_o, \hat{n}_e\}, \text{ (II) } \min\{\hat{n}_o, \hat{n}_e\} < n \le \max\{\hat{n}_o, \hat{n}_e\}, \text{ (III) } n > \max\{\hat{n}_o, \hat{n}_e\}. \quad (37)$$

In the first region, both p_o and p_e are real — this is the region of reflections where all partial waves are bulk ones. This situation automatically arises in an optically positive crystal with an ordinary incident wave (Fig. 2a) or in an optically negative crystal with an extraordinary incident wave (Fig. 2b).

In region II, one of the parameters, p_o or p_e, is imaginary. This is p_o in an optically positive crystal and p_e in an optically negative one. Therefore, in the general solutions found below, one of the partial "reflected" waves may turn out to be localized near the surface. In particular, in this region, the amplitudes C_e^r (30) and C_o^r (31), respectively, in optically negative and positive crystals describe precisely these localized reflection components.

Finally, in region III, both parameters, p_o and p_e, are imaginary. In other words, in this region, a stationary wave field is possible in principle only in the form of surface electromagnetic eigenmodes — polaritons at fixed refractive index n specified by the poles of solutions (30) and (31), i.e., by the equation

$$D(n) = 0. \quad (38)$$

In this paper devoted mainly to the theory of reflection, only regions I and II (37) can be of interest to us. In principle, Eqs. (30) and (31) completely solve the reflection problem. In contrast to the problem of searching for eigenmodes, where the dispersion equation (38) specifying the admissible refractive indices n should be analyzed, the choice of n in the case of reflection only fixes the angle of incidence of the wave on the surface. In this case, the crystal cannot but react to the incident wave, while Eqs. (30) and (31) describe this reaction. However, the reflection has peculiar and sometimes qualitatively nontrivial features for certain angles of incidence. For example, the three-partial solution can degenerate into a two-partial one, so only one reflected wave belonging either to the same sheet of the refraction surface (simple reflection) or to the other sheet (mode conversion) remains instead of the two reflected waves. At the same time, when grazing incidence is approached, the total wave field either tends to zero or remains finite, forming a bulk polariton. Below, we will consider the mentioned features in more detail for the particular case of perfect metallization ($\zeta = 0$) when explicit analysis give visible results.

4. Specific features of wave reflection from the perfectly metallized boundary

The found above general expressions for reflection coefficients (30), (31) remain valid if to put into (32)-(34) $\zeta = 0$ and $R = 1$. As a result, we come to the much more compact functions

$$D(p_o, p_e) = (c_1 p_o - c_2)(c_1 g - c_2 p_e) + c_3^2 p_o \varepsilon_o / n^2, \quad (39)$$

$$D_{eo} = 2c_2c_3p_o\varepsilon_o / n^2, \tag{40}$$

$$D_{oe} = 2c_2c_3p_e, \tag{41}$$

where the new function $g(n)$ is introduced

$$g(n) = \frac{\varepsilon_o}{n^2} - \frac{1}{A} = p_o^2 - \frac{c_2}{c_1}p = \frac{A}{\gamma}\left(p_e^2 - \frac{c_3^2}{A^2}(\gamma - 1)\right). \tag{42}$$

With these simplifications we can proceed with our analysis basing on (Alshits et al., 2007).

4.1 Simple reflection

Let us consider the first type of two-partial reflections known as a pure (or simple) reflection. In this case the incident and reflected waves belong to the same refraction sheet, i.e., both components are either ordinary or extraordinary. It is obvious that such reflections take place when the amplitudes C_e^r in (30) or C_o^r in (31) become zero. This occurs when D_{oe} (40) or D_{eo} (41) vanishes, respectively. It is easily seen that both types of pure reflections are defined by the same criterion:

$$c_2\, c_3 = 0. \tag{43}$$

As follows from Eq. (43), the pure reflections of both ordinary and extraordinary waves in the crystals under consideration should exist independently of one another in the same two reflection geometries. This takes place only in those cases where the optical axis belongs either to the crystal surface ($c_2 = 0$) or to the plane of incidence ($c_3 = 0$). Since the optical axis in this case has a free orientation in these planes and since the angle of incidence is not limited by anything either, the pure reflections in three-dimensional space $\{n, \mathbf{c}\}$ occupy the surfaces defined as the set of two planes: $c_2 = 0$ and $c_3 = 0$. Let us consider in more detail the characteristics of pure reflections in these two geometries.

4.1.1 The optical axis parallel to the surface

In this case, $c_2 = 0$, i.e., $\theta = 0$, and the xy plane of incidence perpendicular to the surface makes an arbitrary angle φ with the direction of the optical axis. The main parameters for the independent reflections of ordinary and extraordinary waves take the form

$$\mathbf{n}_{o,e}^{i,r} = n\{1,\ \mp p_{o,e}(n),\ 0\}^T; \tag{44}$$

$$\begin{pmatrix} \mathbf{e}_o^{i,r} \\ \mathbf{h}_o^{i,r} \end{pmatrix} = \begin{pmatrix} (\pm c_3 p_o,\ c_3,\ \mp c_1 p_o)^T \\ n(c_1 p_o^2,\ \pm c_1 p_o,\ c_3\varepsilon_o / n^2)^T \end{pmatrix}, \quad \begin{pmatrix} \mathbf{e}_e^{i,r} \\ \mathbf{h}_e^{i,r} \end{pmatrix} = \begin{pmatrix} (c_1 p_o^2,\ \pm c_1 p_e,\ c_3\varepsilon_o / n^2)^T n / \varepsilon_o \\ (\mp c_3 p_e,\ -c_3,\ \pm c_1 p_e)^T \end{pmatrix}; \tag{45}$$

$$C_o^r = C_o^i, \qquad C_e^r = -C_e^i. \tag{46}$$

As above, the upper and lower signs in Eqs. (44) and (45) correspond to the incident (i) and reflected (r) waves, respectively. In Eq. (11) for $p_e(n)$, we should take into account the fact that $A = 1$ and $B = c_1^2 + c_3^2/\gamma$ in this case. The angles of incidence are defined by $n < \hat{n}_{o,e}$. For brevity, the normalizing factors in Eqs. (45) are included in the amplitudes $C_{o,e}^{i,r}$.

Given (44)-(46), the pure reflection of the electric component of an ordinary wave in this geometry can be described by the combination

$$\mathbf{E}_o(x,y,t) = C_o\left(\mathbf{e}_o^i \exp(-ip_o ky) + \mathbf{e}_o^r \exp(ip_o ky\right)\exp[ik(x-vt)]. \tag{47}$$

And the pure reflection of the magnetic component of an extraordinary wave is specified by a similar superposition:

$$\mathbf{H}_e(x,y,t) = C_e\left(\mathbf{e}_e^i \exp(-ip_e ky) - \mathbf{e}_e^r \exp(ip_e ky\right)\exp[ik(x-vt)]. \tag{48}$$

4.1.2 The optical axis parallel to the plane of incidence

In this case, $c_3 = 0$, i.e., the azimuth $\varphi = 0$, while the angle θ is arbitrary, which corresponds to arbitrarily oriented crystal surface and plane of incidence passing through the optical axis. The main parameters of the independently reflected waves are given by the formulas

$$\mathbf{n}_o^{i,r} = n\{1,\ \mp p_o(n),\ 0\}^T, \qquad \mathbf{n}_e^{i,r} = n\{1,\ p \mp p_e(n),\ 0\}^T; \tag{49}$$

$$\begin{pmatrix} e_o^{i,r} \\ h_o^{i,r} \end{pmatrix} = \begin{pmatrix} (0,\ 0,\ 1)^T \\ n(\mp p_o,\ -1,\ 0)^T \end{pmatrix}, \qquad \begin{pmatrix} \mathbf{e}_e^{i,r} \\ \mathbf{h}_e^{i,r} \end{pmatrix} = \begin{pmatrix} (\pm p_e,\ \gamma / A^2 \pm pp_e,\ 0)^T \\ (0,\ 0,\ 1)^T \varepsilon_e / An \end{pmatrix}; \tag{50}$$

$$C_o^r = -C_o^i, \qquad C_e^r = C_e^i. \tag{51}$$

At $c_3 = 0$ in Eq. (11) for $p_e(n)$, we have $B = 1$ and $A = c_1^2 + \gamma c_2^2$. Thus, according to Eqs. (50), the pure reflection of ordinary waves is described by the partial *TE* modes with the electric component orthogonal to the sagittal plane. Similarly, the partial components of the pure reflection of extraordinary waves are formed by the *TM* modes with the magnetic component perpendicular to the same plane. In the case under consideration, the analogues of Eqs. (47) and (48) are even simpler:

$$\mathbf{E}_o(x,y,t) = C_o(0,\ 0,\ 1)\sin(p_o ky)\exp[ik(x-vt)], \tag{52}$$

$$\mathbf{H}_e(x,y,t) = C_e(0,\ 0,\ 1)\cos(p_e ky)\exp[ik(x+py-vt)]. \tag{53}$$

4.2 Exceptional bulk polaritons

4.2.1 Simple reflections of ordinary waves at grazing incidence

As we see from Fig. 2, the grazing incidence of an ordinary wave is realized at $n = \hat{n}_o$, when, according to Eq. (36), $p_o = 0$. In this case, the simple reflection of an ordinary wave in the $c_2 = 0$ and $c_3 = 0$ planes behaves differently as grazing incidence is approached, $p_o \to 0$. As follows from Eqs. (44)-(46), in the former case where the optical axis is parallel to the surface ($c_2 = 0$), the incident and reflected partial waves at $p_o = 0$ are in phase and together form an ordinary exceptional bulk wave:

$$\begin{pmatrix} \mathbf{E}(x,t) \\ \mathbf{H}(x,t) \end{pmatrix} = C_o \begin{pmatrix} \mathbf{e}_o \\ \mathbf{h}_o \end{pmatrix} \exp\left(i\frac{\omega}{c}(\hat{n}_o x - ct) \right). \tag{54}$$

The refraction vector of the wave under consideration and its vector amplitude are

$$\mathbf{n}_o = (1,\ 0,\ 0)\hat{n}_o,\quad \hat{n}_o = \sqrt{\varepsilon_o}\,,\qquad \begin{pmatrix}\mathbf{e}_o\\ \mathbf{h}_o\end{pmatrix} = \begin{pmatrix}(0,\ 1,\ 0)\\ (0,\ 0,\ 1)\hat{n}_o\end{pmatrix}, \tag{55}$$

and the energy flux (Poynting vector) in this wave, $\mathbf{P}_o = \mathbf{E}_o \times \mathbf{H}_o$, lies at the intersection of the crystal surface with the sagittal surface, i.e., $\mathbf{P}_o \parallel x$ (Fig. 3a).

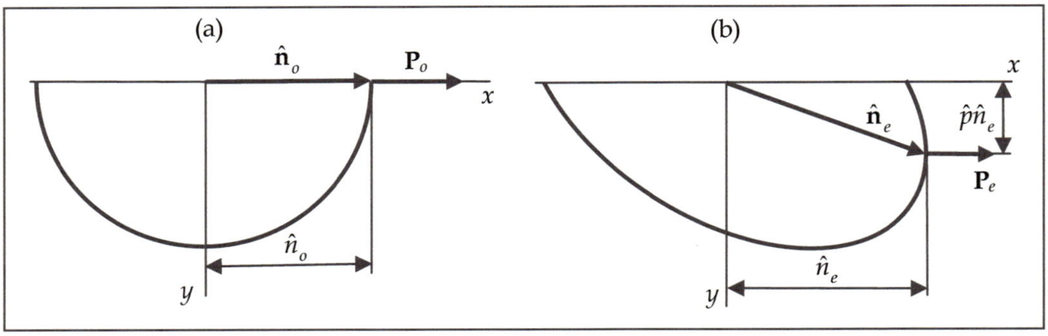

Fig. 3. Characteristics of the ordinary (a) and extraordinary (b) bulk polaritons that emerge at $c_2 = 0$ and $c_3 = 0$, respectively

On the other hand, the pure reflection of ordinary waves in the sagittal plane parallel to the optical axis ($c_3 = 0$, $c_2 \neq 0$) as grazing incidence is approached ($p_o \to 0$), according to Eqs. (52), gives antiphase incident and reflected partial waves that are mutually annihilated. In other words, no exceptional bulk polariton emerges on the branch of ordinary waves in this plane. The qualitative difference in the behavior of grazing incidence in the $c_2 = 0$ and $c_3 = 0$ planes that we found has a simple physical interpretation. For the limiting wave arising at grazing incidence to exist, its polarization $\mathbf{E}_o$, according to the boundary condition (19), must be orthogonal to the crystal surface, $\mathbf{E}_o \parallel \mathbf{n} \parallel y$. As we see from Eqs. (45), this is actually the case for an ordinary wave at $c_2 = 0$ and as $p_o \to 0$. However, the incident and reflected components in the sagittal plane, according to Eqs. (50) and (52), give polarization $\mathbf{E}_o$ that is not orthogonal, but parallel to the surface; hence the annihilation of these components.

4.2.2 Simple reflections of extraordinary waves at grazing incidence

The grazing incidence of extraordinary waves is considered similarly. It relates to $n \to \hat{n}_e$ and by (36), to $p_e \to 0$. In this case, the reverse is true: the incident and reflected waves are in antiphase (46) and, hence, are annihilated in the $c_2 = 0$ plane, and being in phase (51) when the optical axis is parallel to the sagittal ($c_3 = 0$) plane, which generates a bulk polariton:

$$\begin{pmatrix}\mathbf{E}(x,y,t)\\ \mathbf{H}(x,y,t)\end{pmatrix} = C_e \begin{pmatrix}\mathbf{e}_e\\ \mathbf{h}_e\end{pmatrix} \exp\left(i\frac{\omega}{c}[\hat{n}_e^0(x+py) - ct]\right), \tag{56}$$

where $\hat{n}_e^0$ is $\hat{n}_e$ taken at $c_3 = 0$ and $p_e = 0$. For the wave under consideration, we have

$$\mathbf{n}_e = (1,\ p,\ 0)\hat{n}_e^0,\qquad \begin{pmatrix}\mathbf{e}_e\\ \mathbf{h}_e\end{pmatrix} = \begin{pmatrix}(0,\ 1,\ 0)\\ (0,\ 0,\ 1)\hat{n}_e^0\end{pmatrix}, \tag{57}$$

$$\hat{n}_e^0 = \sqrt{\varepsilon_o A} = \sqrt{\varepsilon_o c_1^2 + \varepsilon_e c_2^2}, \quad p = (\varepsilon_o - \varepsilon_e) c_1 c_2 / (\hat{n}_e^0)^2 . \tag{58}$$

As we see from Eq. (57), the bulk polariton (56) is actually polarized in accordance with requirement (19): $\mathbf{e}_e \,||\, \mathbf{n}$. Note that the refraction vector $\mathbf{n}_e$ (57) is generally not parallel to the surface. But the Poynting vector of the wave $\mathbf{P}_e$ still lies at the intersection of the sagittal plane and the crystal surface, $\mathbf{P}_e \,||\, x$ (Fig. 3b). One can show that this is a general property of exceptional waves for $\zeta = 0$ holding even in biaxial crystals (Alshits & Lyubimov, 2009b).
In the special case of $c_2 = c_3 = 0$, which corresponds to the propagation direction x along the optical axis, the sheets of the ordinary and extraordinary waves of the refraction surface are in contact. As a result, degeneracy arises:

$$p_o = p_e = p = 0 , \qquad \hat{n}_o = \hat{n}_e = \sqrt{\varepsilon_o} , \qquad \mathbf{n}_o = \mathbf{n}_e = (1,\ 0,\ 0) , \tag{59}$$

and solutions (54) and (56) merge, degenerating into the corresponding *TM* wave. Since the uniaxial crystal in the case under consideration is transversally isotropic, the orientation of the xy coordinate plane is chosen arbitrarily: for any fixed boundary parallel to the optical axis, a bulk wave with a polarization vector $\mathbf{E}_c$ orthogonal to the surface and an energy flux $\mathbf{P}_c \,||\, x$ can always propagate along the latter.

4.2.3 Proving the absence of other solutions

Thus, exceptional *ordinary* bulk polaritons (54) emerge when the optical axis is parallel to the crystal surface. At the same time, similar *extraordinary* eigenmodes (56) exist if the optical axis is parallel to the sagittal plane. In both cases, *TM*-type one-partial solutions with an energy flux $\mathbf{P}_{o,e}$ parallel to both the crystal surface and the sagittal plane occur (Fig. 3).
Let us show that the dispersion equation (38), (39),

$$D = (c_1 p_o - c_2)(c_1 g - c_2 p_e) + c_3^2 p_o (p_o^2 + 1) = 0 \tag{60}$$

has no other eigensolutions. In principle, an exceptional bulk polariton does not need to belong to the family of simple reflections. It could also be a two-partial one, i.e., consist of the bulk component of one branch corresponding to outer refraction sheet and the admixing localized component of the other branch. Examples of such mixed solutions are known both for crystals with a metallized surface [in the special case of $\varepsilon_o = 0$ and $\varepsilon_e > 0$ (Alshits & Lyubimov, 2005)] and at the open boundary of a crystal with a positively defined tensor $\hat{\varepsilon}$ (Alshits & Lyubimov, 2002a and 2002b). However, it is clear that any such wave with or without an admixture of inhomogeneous components carries energy parallel to the surface, i.e., its bulk component should have a zero parameter p_o or p_e (Fig. 3).
Substituting into (60) Eqs. (11) and (42) for p_e and g taken at $p_o = 0$,

$$c_2(c_1 g - c_2 p_e) = 0 , \tag{61}$$

it is easy to see that at $\gamma < 1$ the parameter g is real, while the parameter p_e is imaginary and, apart from the already known solution $c_2 = 0$, Eq. (60) has no other solutions, since the localization parameter p_e (11) does not become zero at $c_2 \neq 0$. At $\gamma > 1$, when p_e is also real, Eq. (60) is equivalent to the requirement

$$c_2^2(c_2^2 + c_3^2) = 0 , \tag{62}$$

which again leads to the solution $c_2 = 0$.
At $p_e = 0$, Eq. (60) takes the form

$$c_3^2[Ap_o + (\gamma - 1)c_1c_2] = 0 , \tag{63}$$

where it is considered that $p_e = 0$ at $n = \hat{n}_e$ and, according to Eqs. (35) and (36),

$$p_o = \sqrt{(1-\gamma)(c_2^2 + c_3^2 / \gamma) / A} . \tag{64}$$

This time the complexity in the dispersion equation (63) arises at $\gamma > 1$; the purely imaginary parameter p_o at $c_3 \neq 0$ does not become zero, so $c_3 = 0$ is the only root of Eq. (63). At $\gamma < 1$, it is convenient to rewrite Eq. (63) as

$$c_3^2\{\gamma^2 c_2^2 + [1 - (1-\gamma)^2 c_2^2]\} = 0 . \tag{65}$$

Since the expression in braces is positive at any direction of the optical axis, $c_3 = 0$ again remains the only root of the dispersion equation.
Thus, there are no new solutions for exceptional bulk polaritons other than the one-partial eigenmodes (54) and (56) in crystals with a perfectly metallized boundary found above.

4.3 Mode conversion at reflection

Let us now turn to the other, less common type of two-partial reflections where the wave incident on the surface is converted into the reflected wave of the "conjugate" polarization branch (i.e., belonging to the other refraction sheet). We pose the following question: Under what conditions does the mode conversion take place at reflection and what place do the orientation configurations allowing a two-partial reflection with the change of the refraction sheet occupy in the three-dimensional space $\{n, \theta, \varphi\}$ of all reflections? To answer this question, let us turn to solutions (30), (31), (39). Conversion arises for the incident ordinary wave if we choose the angle of incidence (or n) in such a way that $C_o^r = 0$ in (30), which is equivalent to the requirement $D(-p_o, p_e) = 0$. At the same time, according to (31), the incident extraordinary wave will turn into an ordinary wave at reflection if $D(p_o, -p_e) = 0$.

4.3.1 The equation for the conversion surface and its analytical solution

Here, one remark should be made. Clearly, the two-partial conversion reflection is reversible if the reflections from left to right and from right to left are kept in mind. We mean that the simultaneous reversal of the signs of the refraction vectors for the incident and reflected waves automatically converts the reflected wave into the incident one and the incident wave into the reflected one. Certainly, this reversed reflection is mathematically equivalent to the original one — the so-called reciprocity principle (Landau & Lifshitz, 1993). Symbolically, this can be written in the form: $o \rightarrow e = o \leftarrow e$. It is much less obvious that two conversion reflections in one direction, $o \rightarrow e$ and $e \rightarrow o$ (see Fig. 2), also satisfy the boundary conditions for the same geometry of the problem (i.e., the set $\{n, \mathbf{c}\}$).
Thus, the form of the conversion wave superpositions is determined by the equations

$$D_o \equiv -D(-p_o, p_e) \equiv (c_1 p_o + c_2)(c_1 g - c_2 p_e) + c_3^2 p_o (p_o^2 + 1) = 0\,, \tag{66}$$

$$D_e \equiv D(p_o, -p_e) \equiv (c_1 p_o - c_2)(c_1 g + c_2 p_e) + c_3^2 p_o (p_o^2 + 1) = 0\,, \tag{67}$$

which are not identical and one would think that they should specify two different sheets, $n = n_{o\to e}(\mathbf{c})$ and $n = n_{e\to o}(\mathbf{c})$, of the mode conversion surface. However, this is not the case. As we will show below, these sheets merge into one common conversion surface $n = n(\mathbf{c})$.
It is convenient to represent the sought-for mode conversion surface as the locus of points located at distance n from the coordinate origin along each $\mathbf{c}$ corresponding to the roots of Eqs. (66) and (67). From physical considerations, this surface should not go beyond region I (37). Let us prove that the physical roots n of Eqs. (66) and (67) belonging to region I are common ones. After certain transformations, they can be reduced to

$$D_{o,e} = \frac{g - p_o p_e}{1-\gamma}\{f(z) \mp c_1 c_2 (\gamma - 1)\} = 0\,, \tag{68}$$

where the upper sign corresponds to the first subscript, the function $f(z)$ is defined by

$$f(z) = \frac{[A(1-c_2^2)p_o^2 + \gamma c_2^2 + c_3^2](g + p_o p_e)}{[c_1^2 g + c_3^2(p_o^2+1)]p_o + c_2^2 p_e}\,, \tag{69}$$

and p_o, p_e, and g are the known functions (11) and (42) of the variable $z = n/\sqrt{\varepsilon_o} \equiv 1/\sqrt{s}$. It can be shown that the positive function $f(z)$ decreases monotonically in the domains of its existence $0 < z < 1$ (at $\gamma > 1$) or $0 < z < A/B < 1$ (at $\gamma < 1$). Since it is larger in absolute value than the second term on the right-hand side of Eq. (68) in the upper limit, i.e., $f(z_{\max}) > |c_1 c_2(\gamma - 1)|$, the expression in braces in (68) has no physical roots. Thus, the two complex irrational equations (66) and (67) can be reduced to one simple equation:

$$g - p_o p_e = 0\,. \tag{70}$$

This implies that the mode conversion surface is actually a single-sheet one and the processes $o \to e$ and $e \to o$ are represented in space by the same pointing surface.
In the given domain of the variable z, one has $g + p_o p_e > 0$ and the following identity holds:

$$\frac{g - p_o p_e}{1-\gamma} = \frac{g^2 - p_o^2 p_e^2}{(1-\gamma)(g + p_o p_e)} = \frac{A^2[(c_1^2 + c_2^2)z^4 - (c_1^2 + A)z^2 + A(c_1^2 + c_3^2)]}{g + p_o p_e}\,. \tag{71}$$

Therefore, Eq. (70) is equivalent to the biquadratic equation

$$(c_1^2 + c_2^2)z^4 - (c_1^2 + A)z^2 + A(c_1^2 + c_3^2) = 0\,. \tag{72}$$

Thus, we arrive at a compact exact analytical form of the conversion surface $n = n(\mathbf{c})$:

$$n = n_{\pm}(\mathbf{c}) = \sqrt{\frac{\varepsilon_o}{2(1-c_3^2)}\left(c_1^2 + A \pm \sqrt{(c_1^2 - A)^2 - 4Ac_2^2 c_3^2}\right)}\,, \tag{73}$$

on which, however, additional condition I (37) was superimposed. As we see from this solution, the surface under study is very symmetric. It is invariant with respect to the change of sign of any component of vector **c**, i.e., it has three mutually orthogonal symmetry planes coincident with the coordinate planes. However, the main property of this surface is that each point of it describes two different conversion reflections, $o \to e$ and $e \to o$. In other words, the following assertion is valid:

If a two-partial wave reflection with the change of the refraction sheet (e.g., $o \to e$) exists in the geometry under consideration, then the other conversion reflection ($e \to o$) should also be at the same orientations of the boundary (θ) and the reflection plane (φ) and the same refractive index (n).

Analysis shows that the shape of the mode conversion surface is significantly different for optically positive ($\gamma > 1$) and negative ($\gamma < 1$) crystals. This is clearly seen from Fig. 4, where surface (73) was constructed numerically at $\gamma = 1.5$ (a) and 0.8 (b). Because of symmetry, we show only the upper halves ($c_2 > 0$) of the corresponding surfaces from which the quarters ($0 < \varphi < \pi/2$) were cut out for clarity. For an optically positive crystal, the conversion surface has the shape of an axially asymmetric torus in which the hole shrinks to the point coincident with the coordinate origin. This surface is a single-sheet one in the sense that the ray along each vector **c** crosses it once.

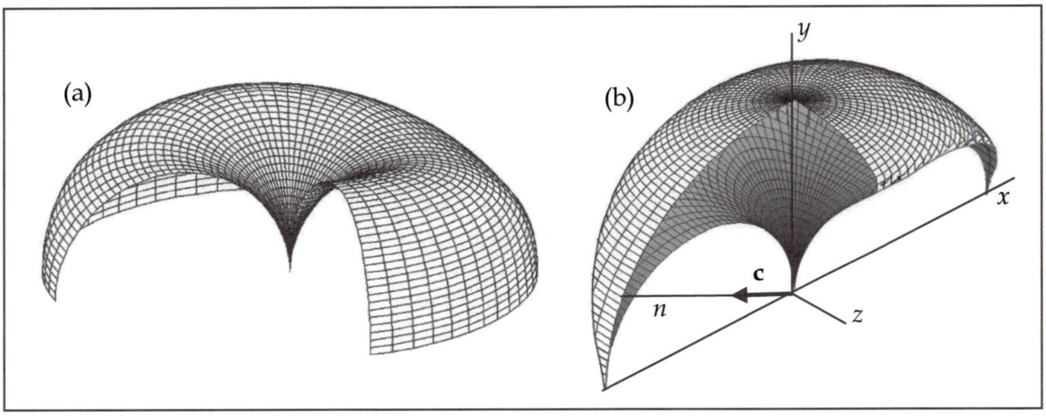

Fig. 4. Conversion surfaces for an optically positive crystal with $\gamma = 1.5$ (a) and an optically negative crystal with $\gamma = 0.8$ (b)

The conversion surface for a negative crystal is radically different: it is a two-sheet one (the ray along **c** crosses it twice); the upper half of the surface resembles a mushroom with a cap that descends to the x axis on both sides and a stipe that sharpens downward.

To analyze in more detail the geometry of the conversion surfaces of the above two types, it is convenient to consider their sections by the $c_1 = 0$, $c_2 = 0$, and $c_3 = 0$ coordinate planes.

4.3.2 The section of the conversion surface by the $c_1 = 0$ coordinate plane

In this section, the azimuth φ is fixed ($\varphi = \pi/2$ and $3\pi/2$) and solution (73) takes the form

$$n = n_{\pm}(\theta) = \sqrt{\frac{\varepsilon_o}{2\sin^2\theta}\left(A \pm \sqrt{A(A - \sin^2 2\theta)}\right)}, \qquad A = 1 + (\gamma - 1)\sin^2\theta . \tag{74}$$

According to condition I (37), the closed $n = n_{\pm}(\theta)$ curves should not go beyond the circle:

$$n \le \min\{\sqrt{\varepsilon_o}, \sqrt{\varepsilon_e}\} . \tag{75}$$

It is easy to verify that n_+ lies outside region (75) for optically positive crystals ($\gamma > 1$) at any angle θ, i.e., $n_+(\theta) > \sqrt{\varepsilon_o}$, while n_- belongs to this region at all θ: $n_-(\theta) < \sqrt{\varepsilon_o}$. Thus, the root n_+ should be discarded at $\gamma > 1$; this explains why the surface $n(\mathbf{c})$ is a single-sheet one at $\gamma > 1$. Thus, the sought-for section of the conversion surface by the $c_1 = 0$ plane is

$$n = n_-(\theta) = \sqrt{\frac{\varepsilon_o A}{2\sin^2\theta}\left(1 - \sqrt{1 - \frac{\sin^2 2\theta}{A}}\right)}\,. \tag{76}$$

Curiously enough, as $\gamma = \varepsilon_e/\varepsilon_o$ (the crystal anisotropy) increases, solution (76) tends to a limiting function that does not depend on γ,

$$n \approx n_-^{\gamma>>1}(\theta) = \sqrt{\varepsilon_o}\,|\cos\theta|\,, \tag{77}$$

and that is represented by two circumferences with radius $\sqrt{\varepsilon_o}/2$ touching one another at the coordinate origin with a common tangent along the y axis (see Fig. 5a). However, most crystals have a small rather than large anisotropy, when $\gamma - 1 << 1$, and solution (76) is close to another limiting function:

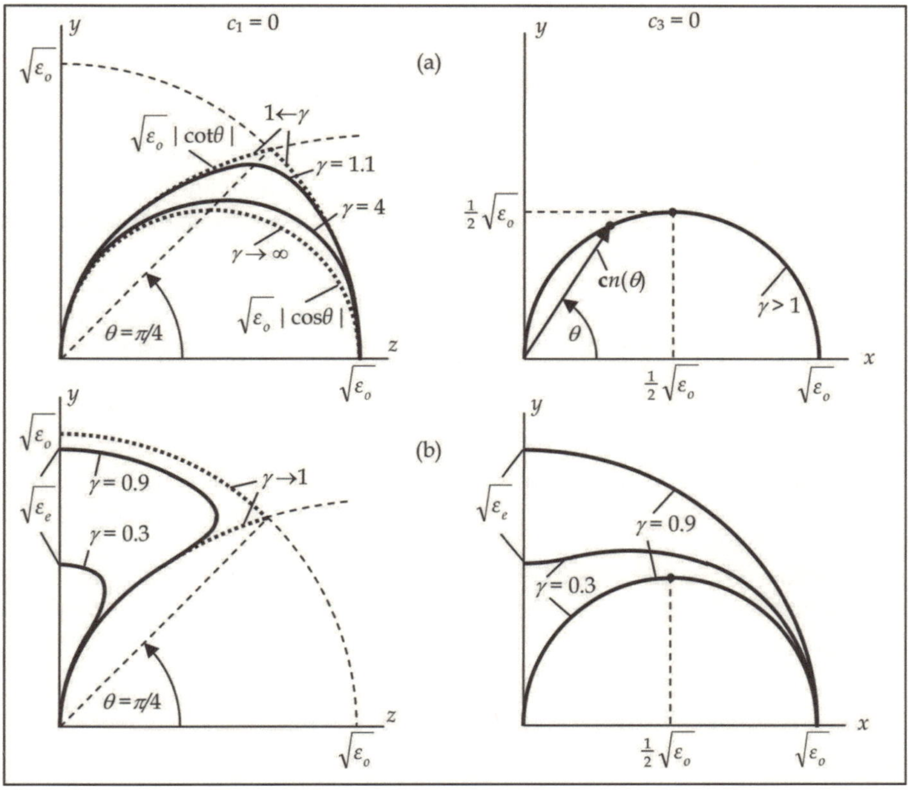

Fig. 5. Sections of the conversion surfaces $n(\mathbf{c})$ by the $c_1 = 0$ and $c_3 = 0$ planes for optically positive (a) and optically negative (b) crystals (given the symmetry of the surfaces $n(\mathbf{c})$ relative to the coordinate planes, the region is bounded by the interval $0 \le \theta \le \pi/4$)

$$n \approx n_-^{\gamma \approx 1}(\theta) = \begin{cases} \sqrt{\varepsilon_o}, & |\theta| \le \pi/4; \\ \sqrt{\varepsilon_o}\,|\cot\theta|, & |\theta| \ge \pi/4, \end{cases} \tag{78}$$

which coincides with the large circumference $\hat{n}_o(\theta) = \sqrt{\varepsilon_o}$ in the region $|\theta| \le \pi/4$ and approaches to the small circumferences (77) near the coordinate origin (see Fig. 5a). As we see from the same figure, the $n(\theta)$ curves corresponding to $\gamma = 4$ and 1.1 actually approach the limiting curves (77) and (78).

For an optically negative crystal ($\gamma < 1$), by no means any orientation of vector **c** satisfies the requirement that (74) be real and condition (75), which is now reduced to the inequality $n_\pm(\theta) \le \hat{n}_e = \sqrt{\varepsilon_e}$. On the other hand, both branches of solution (74), $n_-(\theta)$ and $n_+(\theta)$, are valid in this case in the region of allowed angles θ, so the surface becomes a two-sheet one. Naturally, the configuration of the section of the conversion surface differs significantly from that in the case of $\gamma > 1$ considered above (see the $c_1 = 0$ sections in Fig. 5b).

Let us now consider Eq. (74) in the special case of a small anisotropy, $0 < 1 - \gamma << 1$, which is practically realized in most crystals. Formally, this solution is very similar to (78)

$$n = n_\pm(\theta) \approx \begin{cases} \sqrt{\varepsilon_o}, & |\theta| \ge \pi/4, \\ \sqrt{\varepsilon_o}\,|\cot\theta|, & |\theta| \ge \pi/4. \end{cases} \tag{79}$$

This time, however, the limiting solution exists only in the region $|\theta| \ge \pi/4$, where the inner branch $n_-(\theta)$ closely follows Eq. (78). At $|\theta| = \pi/4$, the transition to the large circumference again occurs (Fig. 5b), but upward, to the outer branch $n_+(\theta)$, rather than downward. Here, it should be noted that we do not distinguish ε_o from ε_e in (78) with our accuracy of the zero-order approximation in parameter $1 - \gamma$. However, we can show that including the next expansion terms ensures that the inequality $n_\pm \le \hat{n}_e$ is satisfied.

Note also yet another circumstance important for the subsequent analysis. A check indicates that the curvature of the $c_1 = 0$ section under consideration remains positive for a negative crystal at the upper and lower points ($\theta = \pm\pi/2$) in the entire region $0 < \gamma < 1$.

4.3.3 The sections of the conversion surface by the planes of simple reflection

Curiously enough, the conversion surface under study also intersects with the other two coordinate planes, $c_3 = 0$ and $c_2 = 0$, which, as we know, are the planes of simple reflection. The physical interpretation of the paradoxical existence of such "antagonistic" reflection modes will be given below.

The $c_3 = 0$ section ($\varphi = 0$ and $\varphi = \pi$). In the $c_3 = 0$ coordinate plane, Eqs. (66) and (67) are simplified via factorization, although they appear different. However, given the identity

$$c_1 g \pm c_2 p_e = \frac{p_e(c_1^2 p_o^2 - c_2^2)}{c_1 p_e \mp c_2 \gamma / A}, \tag{80}$$

which is valid at $c_3 = 0$, both Eqs. (66) and (67) in this plane can be reduced to

$$p_e(c_1^2 p_o^2 - c_2^2) = 0. \tag{81}$$

The first root of this equation that corresponds to the requirement $p_e = 0$ and that is equal to

$$n = \hat{n}_e(\theta) = \sqrt{\varepsilon_o \cos^2\theta + \varepsilon_e \sin^2\theta}\ , \tag{82}$$

belongs to region I (37) (being its boundary) only at $\gamma < 1$. We are talking about the one-partial eigensolutions (54), which can turn into the two-partial superposition corresponding either to a pure reflection at grazing incidence or to a conversion reflection at certain small perturbations of the geometry of the problem. In the latter case, an ordinary bulk (incident or reflected) wave with small amplitude is admixed to a weakly perturbed original extraordinary component.

For optically positive crystals ($\gamma > 1$), solution (82) lies outside region I (37) bounded by the circumference with radius $n = \hat{n}_o$ and is extraneous, because p_o is a purely imaginary parameter at $n > \hat{n}_o$ and the exceptional bulk polariton under consideration cannot turn into a conversion reflection at any small perturbation (i.e., (82) is not part of the conversion surface). The fact that the other two factors in Eq. (82) become zero, $c_1 p_o - c_2 = 0$ and $c_1 p_o + c_2 = 0$, leads to the combined solution

$$n = \hat{n}_o \,|\cos\theta\,|, \qquad \varphi = 0,\ \pi. \tag{83}$$

It coincides with the asymptotic solution (77) in the $c_1 = 0$ section and has a simple graphical representation in polar coordinates (n, θ) in the form of two circumferences with radius $\hat{n}_o/2$ symmetric relative to the x and y coordinate axes and touching one another at the coordinate origin (see the $c_3 = 0$ section in Fig. 5a). Thus, at large γ the conversion surface has identical circular sections in the $c_1 = 0$ and $c_3 = 0$ sections. A check shows that at $\gamma >> 1$ any other sections of the conversion surface by the planes passing through the y axis are described by the same Eq. (83). In other words, the corresponding limiting surface $n(\mathbf{c})$ should have the shape of an ideal circular torus.

It is easy to verify that the pair of circumferences (83) entirely belongs to region I (37) at any γ. For optically positive crystals ($\gamma > 1$), this trivially follows from the comparison of Eq. (83) with the equality $\hat{n}_o = \sqrt{\varepsilon_o}$ (see (35)). However, it is also satisfied at any θ for optically negative crystals ($\gamma < 1$), for which condition I (37) is reduced to the more binding inequality $n < \hat{n}_e(\theta)$. This becomes obvious if we write this condition as

$$\sqrt{\varepsilon_o}\,|\cos\theta\,| < \sqrt{\varepsilon_o \cos^2\theta + \varepsilon_e \sin^2\theta}\ . \tag{84}$$

Certainly, at $\gamma < 1$ solution (83) should be complemented by root (72) obtained above, which is in agreement with the conversion surface being a two-sheet one. As we see from Fig. 5b ($c_3 = 0$), the outer part of the section described by function (82) changes the sign of the curvature at the upper and lower points ($\theta = \pm\pi/2$) as the parameter γ passes through a certain critical value γ_0. A simple calculation gives $\gamma = 1/2$ for the zero-curvature parameter. As we saw, the curvature remains universally positive at any γ for optically negative crystals in the $c_1 = 0$ section in the vicinity of the same direction. This means that the conversion surface in the direction $\mathbf{c} \,||\, y$ remains convex only at $1/2 < y < 1$ and has a saddle point at $0 < \gamma < 1/2$.

The physical meanings of solutions (82) and (83) differ significantly. In contrast to the boundary one-partial solution corresponding to (82), Eq. (83) specifies a two-partial reflection, which however, has its own peculiarities. The condition $c_3 = 0$ implies that the optical axis belongs to the plane of incidence and that the reflections described by Eq. (83) correspond to the two specific situations where the direction of either the incident wave or the reflected one coincides with the optical axis (see Fig. 6). The equations $c_1 p_o - c_2 = 0$ and $c_1 p_o + c_2 = 0$ are satisfied in the former and latter cases, respectively. In a sense, the two-partial reflections of this kind with one of the components belonging to both refraction sheets are simultaneously simple and conversion ones. Therefore, the intersection of the conversion surface with the $c_3 = 0$ plane along lines (83) contains no paradox.

As we see from Fig. 6a, at the fixed refractive index n corresponding to a wave incident along the optical axis, two different two-partial combinations can be realized. These differ by the choice of one reflected wave from the two possible waves that belong to different refraction sheets and that have different propagation directions and polarizations. This requires only properly choosing the polarization of the wave incident along the optical axis.

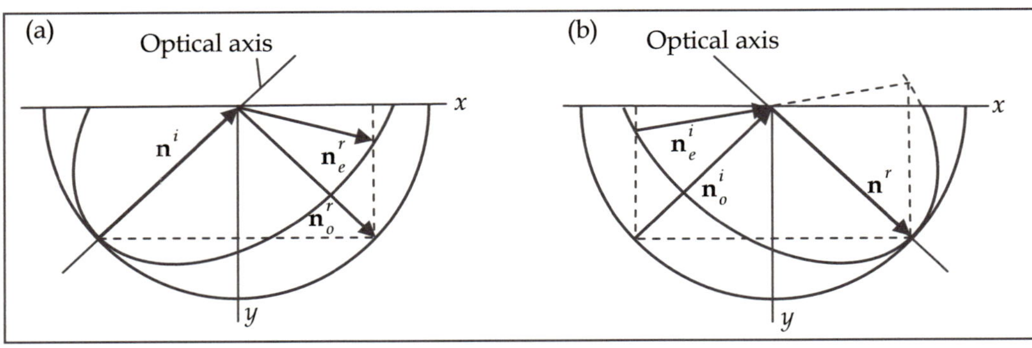

Fig. 6. Two-partial reflections in the $c_3 = 0$ plane when the incident (a) and reflected (b) waves coincide with the optical axis

In such degenerate directions, a crystal is known to allow the propagation of waves with a polarization arbitrarily oriented in the plane orthogonal to the optical axis. In particular, a wave with a polarization different from the two possible values corresponding to the two-partial reflections mentioned above can also be thrown on the surface in this direction. In this case, an ordinary three-partial reflection is realized in the crystal.

Any of the two possible incident waves corresponding to the same factor n (83) can also accompany the wave reflected along the optical axis (Fig. 6b). In this case, the crystal itself will choose such a polarization of the reflected degenerate wave that the total superposition of the two-partial reflection satisfies the boundary condition (19).

Clearly, four different two-partial reflections can be obtained at certain small perturbations of the reflection geometry shown in Fig. 6: two simple, $o \to o$ and $e \to e$, and two conversion, $o \to e$ and $e \to o$, ones. This ultimately resolves the paradox of the surprising coexistence of simple and conversion reflections in the $c_3 = 0$ plane.

The $c_2 = 0$ section ($\theta = 0$). In the $c_2 = 0$ coordinate plane, Eqs. (66) and (67) again coincide:

$$p_o(p_o^2 + c_3^2) = 0 . \tag{85}$$

Equation (85) has only one solution, $p_o = 0$, i.e.,

$$n = \hat{n}_o = \sqrt{\varepsilon_o} , \tag{86}$$

which corresponds to the one-partial ordinary eigenmode (54). At $\gamma > 1$, circumference (86) is the boundary of region I (37). We previously obtained the exceptional bulk polariton (54) by passing to the limit of grazing incidence, $n \to \hat{n}_o$, on the family of pure reflections. At the reverse perturbation of the refractive index, $n \leftarrow \hat{n}_o$, with the optical axis retained in the $c_2 = 0$ plane, polariton (54) naturally again splits into the two-partial solution corresponding to a pure reflection. However, at certain matched changes of the optical-axis orientation and the refractive index n corresponding to the motion along the conversion surface, the same ordinary polariton, being slightly distorted, will attach an extraordinary (incident or reflected) component with a small amplitude, which relates to a mode conversion.

In an optically negative crystal ($\gamma < 1$), circumference (86) lies outside region I (37), except for its two points at the intersection with the x axis (where the "mushroom cap" in Fig. 4b touches the "ground"). In the latter case, we are talking about the bulk polariton propagating along the optical axis parallel to x and characterized by parameters (59). The same points are also seen on the $c_3 = 0$ section (see Fig. 5b).

5. Resonance reflection from the metal coating with nonzero impedance

In this section we turn to a more practical problem considering wave reflections in a crystal from the metal coating which is not perfect anymore. The impedance of the metal will be supposed finite ($\zeta \neq 0$) but small. We shall consider the physical consequences of this new feature of the boundary problem on the example of rather nontrivial resonance of reflection which may be interpreted as an excitation of strong polariton by means of a weak incident pump wave. The idea of the resonance was already discussed in the Introduction. Now we are prepared to start a new stage of studies, following to (Depine & Gigli, 1995; Alshits & Lyubimov, 2010).

In this section, we consider in detail the specific features of resonance excitation of an extraordinary polariton in an optically negative crystal and of the accompanying plasmon in a metal by an incident ordinary pump wave. Similar results for an optically positive crystal will be shortly presented separately.

5.1 The structure of wave fields in a crystal

As we have seen, in crystals covered by a perfect metal there are two special geometries admitting simple reflections. They occur in the both sheets of the refraction surface when the optical axis is parallel to either the interface ($c_2 = 0$) or the sagittal plane ($c_3 = 0$). At grazing incidence simple reflections may transform into exceptional bulk waves: the ordinary wave - in the $c_2 = 0$ plane and the extraordinary wave - in the $c_3 = 0$ plane. In optically negative crystals ($\gamma < 1$) the extraordinary exceptional wave (56)-(58) belongs to an internal sheet of the refraction surface and is accompanied by the independent simple reflection of the ordinary waves at the same refractive index $n = \hat{n}_e^0$ (Fig. 7a). The wave characteristics of this simple reflection in accordance with Eqs. (49)-(51) are given by

$$\mathbf{n}_o^{0i,r} = (1,\ \mp\hat{p}_o^0,\ 0)\hat{n}_e^0, \qquad \hat{p}_o^0 = |c_2|(\hat{n}_o/\hat{n}_e^0)\sqrt{1-\gamma}, \tag{87}$$

$$\begin{pmatrix} \mathbf{e}_o^{0i,r} \\ \mathbf{h}_o^{0i,r} \end{pmatrix} = \begin{pmatrix} (0,\ 0,\ 1) \\ (\mp\hat{p}_o^0,\ -1,\ 0)\hat{n}_e^0 \end{pmatrix}, \qquad C_o^r = C_o^i. \tag{88}$$

Here we stress that in the $c_3 = 0$ plane these independent solutions also coexist even at $\zeta \neq 0$. That follows from Eqs. (33), (34) providing simple reflections in both wave branches ($D_{eo} = D_{oe} = 0$) in any symmetry plane ($c_3 = 0$) containing the optical axis. One of them at grazing incidence creates an extraordinary bulk polariton. Below we consider the mechanism of its excitation by means of a resonance reflection in the vicinity of this exceptional orientation.
Assuming that $\zeta \neq 0$, let us deviate the plane of incidence from the direction of the optical axis **c** by a small angle θ. Naturally this leads to a change in the orientation of the surface and in the coordinate system connected to the geometry, in which now

$$\mathbf{c} = (c_1,\ c_2,\ \widehat{c}_3)^T\,, \tag{89}$$

where $\widehat{c}_3 = \sin\theta$ (an arc over the parameter indicates that this parameter is small: $|\widehat{c}_3| << 1$). When $\widehat{c}_3 \neq 0$, the section of the refraction surface by the new plane of incidence xy (Fig. 7b) does not show any tangency of sheets, in contrast to the unperturbed situation of Fig. 7a.

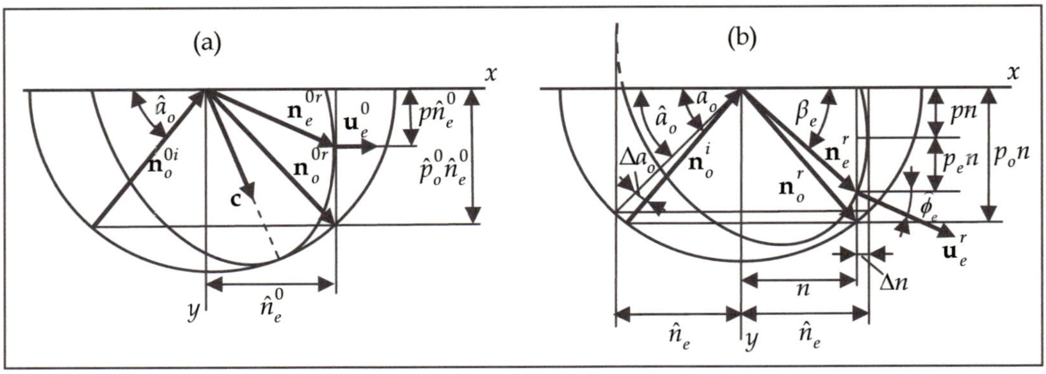

Fig. 7. The wave characteristics of reflections in an optically negative crystal; (a) in the absence of perturbation, $c_3 = 0$, and (b) with perturbation $c_3 \neq 0$. The refraction vectors are shown together with the sections of the refraction sheets by the plane of incidence xy

A perturbed expression for the limiting parameter $\hat{n}_e$ is given by the exact formula (35). This parameter corresponds to the perturbed limiting refraction vector

$$\hat{\mathbf{n}}_o^i = (1,\ -\hat{p}_o,\ 0)\hat{n}_e\,. \tag{90}$$

By varying the propagation direction of the incident ordinary wave with the refraction vector $\mathbf{n}_o^i$ near the limiting vector (90), we can conveniently represent the vector $\mathbf{n}_o^i$ as

$$\mathbf{n}_o^i = (1,\ -p_o,\ 0)n\,, \quad p_o = \hat{p}_o + \Delta p_o\,, \quad n = \hat{n}_e + \Delta n\,. \tag{91}$$

In other words, the angle of incidence $a_o = \arctan p_o$ (see Fig. 7b) of this wave becomes a free parameter close to the limiting angle $\hat{a}_o = \arctan\hat{p}_o$, where $\hat{p}_o = \sqrt{B/A - 1}$. Taking into account (11), one can easily find for ordinary waves the following relation

$$\Delta p_o \approx -[\varepsilon_o / (\hat{n}_e^0)^3]\Delta n\,, \tag{92}$$

which gives a relationship between Δn and $\Delta a_o = a_o - \hat{a}_o$:

$$\Delta n \approx -\hat{n}_e^0 \hat{p}_o^0 \Delta a_o\,, \tag{93}$$

where $\hat{n}_e^0$ and $\hat{p}_o^0$ are unperturbed limiting parameters defined by (58) and (87).
The wave field structure of the reflection is determined by the superposition of three partial waves given by Eqs. (4)-(7) with $C_e^i = 0$. A concretization of the parameter p_o in these formulae yields the expression

$$p_o \approx \hat{p}_o + \frac{\varepsilon_o}{(\hat{n}_e^0)^2}\Delta a_o\,, \tag{94}$$

p is defined by the exact Eq. (11) and the parameter p_e is determined by Eq. (36):

$$p_e = p_e(\Delta a_o) = \sqrt{\frac{\varepsilon_o \varepsilon_e}{(\hat{n}_e^0)^2}\left(\frac{1}{n^2} - \frac{1}{\hat{n}_e^2}\right)} \approx \frac{\sqrt{2\hat{p}_o^0 \varepsilon_o \varepsilon_e}}{(\hat{n}_e^0)^2}\begin{cases} \sqrt{\Delta a_o}\,, & \Delta a_o \ge 0, \\ i\sqrt{-\Delta a_o}\,, & \Delta a_o < 0. \end{cases} \tag{95}$$

As expected, the variation of the angle of incidence near the limiting position $\hat{a}_o$ leads to the transformation of the extraordinary partial wave from bulk reflected for $\Delta a_o \ge 0$ to the accompanying localized wave for $\Delta a_o < 0$. However, for sufficiently small $|\Delta a_o|$ in (31), when $|p_e| << 1$, in the first case we have nearly grazing refection, whereas, in the second case, the localized mode should be a deeply penetrating (quasibulk) wave. Thus, in either case the extraordinary wave remains a weakly perturbed initial bulk polariton (56).

5.2 Reflection coefficients and the excitation factor of a polariton-plasmon

An analysis of the considered resonance may be done basing on general solution (30)-(34) of the reflection problem specifying it to a small region of geometrical parameters $|\Delta a_o| << 1$, $\hat{c}_3^2/c_2^2 << 1$ and limiting ourselves to the linear approximation in the impedance ζ. In this case one can replace in (32)-(34) $R \to 1$ retaining only terms $\sim \zeta$. The results are

$$r_{eo}(\delta, \Delta a_o) \equiv \frac{C_e^r}{C_o^i} = -\frac{\delta \kappa_o \gamma \hat{n}_o}{\delta^2 \kappa_o \hat{n}_e^0 / 2 + \sqrt{\kappa_o \gamma \Delta a_o} + \zeta}\,, \tag{96}$$

$$r_{oo}(\delta^2, \Delta a_o) \equiv \frac{C_o^r}{C_o^i} = \frac{\delta^2 \kappa_o \hat{n}_e^0 / 2 - \sqrt{\kappa_o \gamma \Delta a_o} - \zeta}{\delta^2 \kappa_o \hat{n}_e^0 / 2 + \sqrt{\kappa_o \gamma \Delta a_o} + \zeta}\,, \tag{97}$$

where the notation is introduced

$$\kappa_o = 2\hat{p}_o^0 (\hat{n}_e^0 / \varepsilon_e)^2\,, \qquad \delta = \hat{c}_3 / c_2\,. \tag{98}$$

Eqs. (96), (97) are valid for either sign of Δa_o; however one should remember that at the change of this sign all square roots in these equations become imaginary instead of real, or vise versa, as in (95). Below we will see that the reflection coefficients exhibit completely different behavior in the domains with real and imaginary parameter p_e.
Both the numerators and denominators in the functions (96) and (97) are combinations of small quantities, and even a small variation of these quantities may lead to large variations

of the functions themselves. It is this fact that is responsible for the resonance properties of the reflection: a sharp increase or, conversely, decrease in the relative intensity of the partial components in the related wave superposition for appropriate combinations of the small parameters Δa_o, δ, and ζ.

As the measure of efficiency of a resonance, introduce the excitation factor of an extraordinary polariton as the ratio of the moduli of Poynting vectors:

$$K_{eo} = \left|\mathbf{P}_e^r\right| / \left|\mathbf{P}_o^i\right|\Big|_{y=0} . \tag{99}$$

The energy fluxes $\mathbf{P}_o^i$ and $\mathbf{P}_o^r$ entering (99) are related to the energy densities w_o^i and w_e^r of the corresponding waves by the formulas

$$|\mathbf{P}_o^i| = w_o^i\,|\mathbf{u}_o^i| = w_o^i c / \hat{n}_o, \quad |\mathbf{P}_e^r| = w_e^r\,|\mathbf{u}_e^r| = w_e^r c / \hat{n}_e^0 . \tag{100}$$

Here $\mathbf{u}_o^i$ and $\mathbf{u}_e^r$ are the group velocities (3), which can be calculated near the peak in zero approximation; moreover, by definition we have

$$w_o^i = \frac{1}{8\pi}|\mathbf{H}_o^i|^2 = \frac{1}{8\pi}|\mathbf{h}_o^i C_o^i|^2 = \frac{1}{8\pi}|C_o^i|^2, \quad w_e^r = \frac{1}{8\pi}|C_e^r|^2 . \tag{101}$$

Taking into account these relations and Eq. (96), we obtain

$$K_{eo}(\delta^2, \Delta a_o) = \frac{\hat{n}_o}{\hat{n}_e^0}\left|r_{eo}(\delta, \Delta a_o)\right|^2 . \tag{102}$$

Figure 8 shows a three-dimensional picture of the maximum of the excitation factor $K_{eo}(\delta^2, \Delta a_o)$ of an extraordinary polariton and the minimum of $|r_{oo}(\delta^2, \Delta a_o)|^2$ for the same combination of the control parameters δ^2 and Δa_o for a sodium nitrate crystal with aluminum coating at a wavelength in vacuum of $\lambda_0 = 0.85$ μm. Below, these characteristics of the resonance will be studied in more detail.

Let us show that the factor $K_{eo}(\delta^2, \Delta a_o)$ simultaneously characterizes the excitation of the accompanying surface plasmon in the metal ($y < 0$), whose field is given by Eqs. (8), (9). In zero approximation (when $c_3 = 0$ and $\zeta = 0$), the polarization of the excited extraordinary polariton (56), (57) is close to the *TM* type, while the polarizations of the incident and reflected ordinary waves $(50)_1$, to the *TE* type. In this case, the ratio of the amplitudes of these components at resonance, which is characterized by the excitation factor K_{eo}, shows that the dominant polarization of the whole wave superposition is the *TM* polarization, when the field **H** is parallel to the z axis. Since the tangential components $\mathbf{H}_t$ must be continuous on the interface (Landau & Lifshitz, 1993), the plasmon amplitudes can be estimated as $C_{TM} \approx C_e$ and $|C_{TE}| << |C_{TM}|$. Taking into account these relations and the normalization condition $|\mathbf{h}_{TM}| = |\mathbf{h}_e| = 1$, we obtain the following expression in zero approximation:

$$|C_{TM} / C_o^i|^2 \approx |C_e / C_o^i|^2 = |r_{eo}(\delta, \Delta a_o)|^2 . \tag{103}$$

Thus, the factor K_{eo} (102) describes the resonance excitation of both a polariton in the crystal and a localized plasmon in the metal coating.

5.3 Resonance excitation of a surface polariton

When the pumping ordinary wave is incident on the crystal boundary at angle $a_o < \hat{a}_o$, an extraordinary surface polariton is excited. In this case formulas (96) and (102) yield

$$K_{eo}(\delta^2, \Delta a_o) = \frac{(\delta\kappa_o\varepsilon_e)^2 / n_o\hat{n}_e^0}{\left(\delta^2\kappa_o\hat{n}_e^0 / 2 + \zeta'\right)^2 + \left(\sqrt{-\kappa_o\gamma\Delta a_o} - |\zeta''|\right)^2}. \tag{104}$$

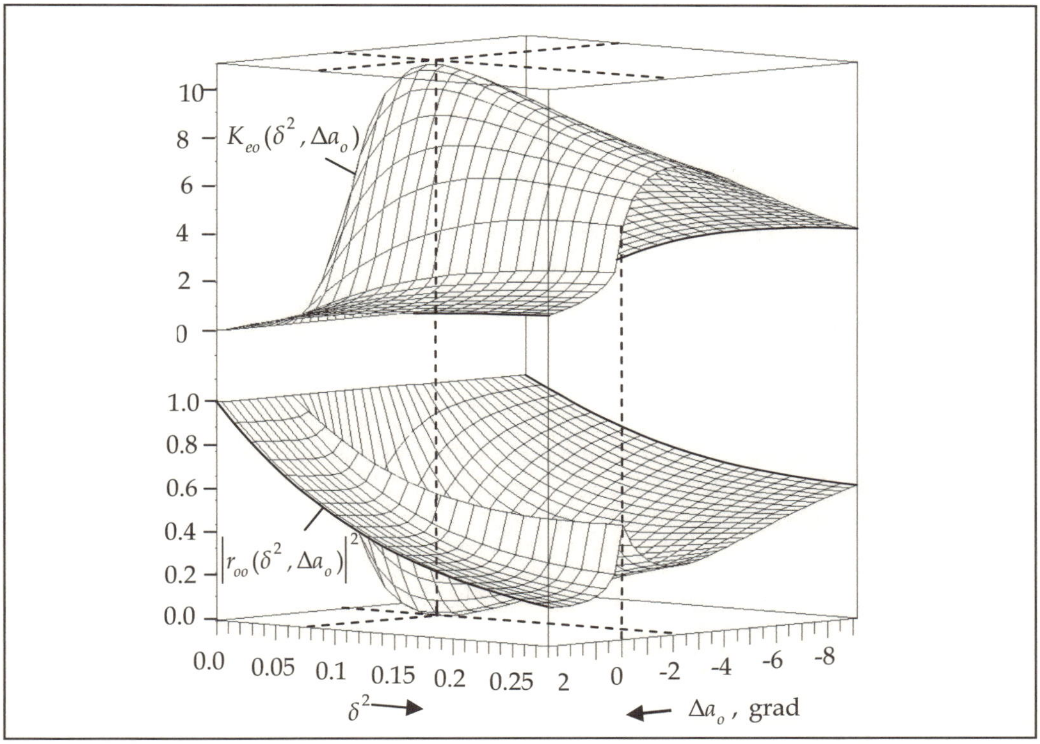

Fig. 8. Surfaces (a) $K_{eo}(\delta^2, \Delta a_o)$ and (b) $|\, r_{oo}(\delta^2, \Delta a_o)\,|^2$ for a sodium nitrate crystal with aluminum coating for $\lambda_0 = 0.85$ μm.

As a function of the deviation of the incidence angle Δa_o, this expression has an obvious maximum for

$$\Delta a_o^{max} = -|\zeta''|^2 / \kappa_o\gamma . \tag{105}$$

For a fixed value of $\Delta a_o = \Delta a_o^{max}$, the excitation factor remains a function of the value δ^2:

$$K_{eo}\left(\delta^2, \Delta a_o^{max}\right) = \frac{(\delta\kappa_o\varepsilon_e)^2 / n_o\hat{n}_e^0}{\left(\delta^2\kappa_o\hat{n}_e^0 / 2 + \zeta'\right)^2}, \tag{106}$$

which also has a maximum for an appropriate choice of $\delta \equiv \hat{c}_3 / c_2$:

$$\delta_{max}^2 = 2\zeta' / \kappa_o \hat{n}_e^0 . \tag{107}$$

Substituting (107) into (106), we find the absolute maximum of the excitation factor

$$K_{eo}^{max} = K_{eo}\left(\delta_{max}^2, \Delta a_o^{max}\right) = \frac{\kappa_o \gamma \varepsilon_e \hat{n}_o}{2(\hat{n}_e^0)^2 \zeta'} = \frac{\hat{p}_o^0}{n_o \zeta'} \tag{108}$$

which is inversely proportional to the small parameter ζ'; this guarantees the efficiency of the resonance, especially in the infrared region. According to (87) and (58), the numerator in (108) is expressed as

$$\hat{p}_o^0 = \frac{|c_2|\sqrt{1-\gamma}}{\sqrt{1-(1-\gamma)c_2^2}} . \tag{109}$$

This shows that the coefficient K_{eo}^{max} can be additionally increased by choosing a crystal with high anisotropy factor $(1-\gamma)$ and the orientation of the optical axis in the yz plane ($c_1 = 0$) corresponding to the maximum possible component $|c_2| = 1$. As a result, we obtain $\hat{p}_o^0 = \sqrt{1/\gamma - 1}$ and, instead of (108), we have the optimized value

$$K_{eo}^{max} = \frac{1}{\zeta'}\sqrt{\frac{\varepsilon_o - \varepsilon_e}{\varepsilon_o \varepsilon_e}} . \tag{110}$$

Below we will assume that $c_1 = 0$ in all numerical estimates and figures.

In terms of the ratios K_{eo} / K_{eo}^{max}, $\delta^2 / \delta_{max}^2$, and $\Delta a_o / \Delta a_o^{max}$, the sections of the peak (104) for a fixed value of the parameter $\Delta a_o = \Delta a_o^{max}$ (105) or $\delta^2 = \delta_{max}^2$ (107) are given by

$$K_{eo}\left(\delta^2, \Delta a_o^{max}\right) = \frac{4\delta^2 / \delta_{max}^2}{(\delta^2 / \delta_{max}^2 + 1)^2} K_{eo}^{max}, \quad K_{eo}(\delta_{max}^2, \Delta a_o) = \frac{K_{eo}^{max}}{\left(\dfrac{\sqrt{\Delta a_o / \Delta a_o^{max}} - 1}{2\zeta' / \zeta''}\right)^2 + 1} . \tag{111}$$

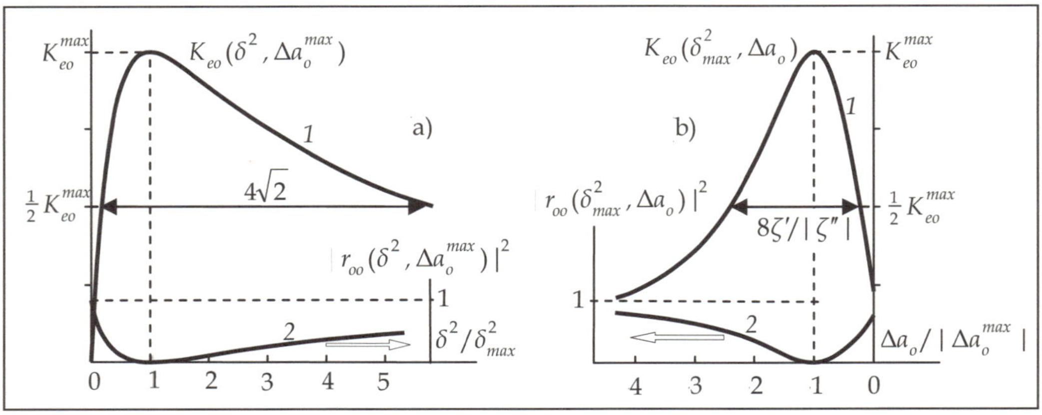

Fig. 9. Two sections of the surfaces $K_{eo}(\delta^2, \Delta a_o)$ and $|r_{oo}(\delta^2, \Delta a_o)|^2$ shown in Fig. 8 when (a) $\Delta a_o \equiv \Delta a_o^{max} \approx$ --2.1 or (b) $\delta^2 \equiv \delta_{max}^2 \approx 0.078$; $\lambda_0 = 0.85$ μm and $K_{eo}^{max} \approx 10.8$

Figure 9a (curve *1*) shows that the section of the peak for $\Delta a_o = \Delta a_o^{max}$ rapidly reaches a maximum and then slowly decreases as the parameter δ increases. Of course, this is advantageous for applications but restricts (at least, in the visible range) the applicability of the approximation based on the inequality $\delta^2 << 1$. The half-width of this peak is

$$(\Delta\delta^2)_{1/2} = 4\sqrt{2}\delta_{max}^2 = 8\sqrt{2}\zeta' / \kappa_o \hat{n}_e^0 . \tag{112}$$

Away from the section $\Delta a_o = \Delta a_o^{max}$, the coordinate of the maximum and the half-width of the peak with respect to δ noticeably increase, which is clearly shown in the three-dimensional picture of the peak in Fig. 8.

Another section of the same peak (for $\delta^2 = \delta_{max}^2$) is shown in Fig. 9b (curve *1*). According to $(111)_2$, its half-width is

$$(\Delta\alpha_o)_{1/2} = \frac{8\zeta'}{|\zeta''|} |\Delta\alpha_o^{\max}| = \frac{8\zeta' |\zeta''|}{\kappa_o \gamma} . \tag{113}$$

Compared with (112), this quantity contains an additional small parameter $|\zeta''|$, which accounts for the relatively small width in this section of the peak in the region $|\Delta a_o| << 1$.

The penetration depth d_e of a polariton into a crystal is limited by the parameter p_e and, according to (95), depends on the angle Δa_o. At the maximum point $\Delta a_o = \Delta a_o^{max}$ (105), the penetration depth is

$$d_e = \frac{1}{k \operatorname{Im} p_e} \approx \frac{1}{k_0 \hat{n}_e^0 \operatorname{Im} p_e} = \frac{\lambda_0 (\hat{n}_e^0)^2}{2\pi \varepsilon_o \varepsilon_e |\zeta''|} . \tag{114}$$

The plasmon penetration depth into the metal is found quite similarly

$$d_m = \frac{1}{k \operatorname{Im} p_m} \approx \frac{\lambda_0 |\zeta''|}{2\pi} , \tag{115}$$

where we have made use of Eq. $(11)_4$ by expressing $\operatorname{Im} p_m \approx 1/|\zeta''| \hat{n}_e^0$. Comparing Eqs. (114) and (115), we can see that the plasmon in metal is localized much stronger than the polariton in the crystal: $d_m/d_e \sim |\zeta''|^2$.

In Fig. 9, the material characteristics of the crystal ε_o and ε_e, as well as the geometric parameters c_1 and c_2 are "hidden" in the normalizing factors δ_{max}^2, Δa_o^{max}, and K_{eo}^{max}. The first section (Fig. 9a) is independent of other parameters and represents a universal characteristic in a wide range of wavelengths, whereas the second section (Fig. 9b) depends on the ratio $\zeta'/|\zeta''|$ obtained from Table 1 for aluminum at a vacuum wavelength of $\lambda_0 = 0.85$ μm.

λ_0, μm	0.4	0.5	0.6	0.85	1.2	2.5	5.0
ζ'	0.0229	0.0234	0.0253	0.0373	0.0092	0.0060	0.0046
$-\zeta''$	0.267	0.215	0.180	0.135	0.108	0.050	0.026

Table 1. Components of the surface impedance $\zeta = \zeta' + i\zeta''$ for aluminum in the visible and infrared ranges at room temperature, obtained from the data of (Motulevich, 1969)

The absolute values of the main parameters of the peak are shown in Table 2 for a sodium nitrate crystal $NaNO_3$ for various wavelengths. In our calculations (including those related to Fig. 8), we neglected a not too essential dispersion of permittivities and used fixed values of $\varepsilon_o = 2.515$, $\varepsilon_e = 1.785$, and $\gamma = 0.711$ (Sirotin & Shaskolskaya, 1979, 1982) at $\lambda_0 = 0.589$ μm.
First of all, it is worth noting that, in the visible range of wavelengths of $\lambda_0 = 0.4$--0.6 μm, the maximal excitation factor (110) relatively slowly decreases as λ_0 increases, although remains rather large ($K_{eo}^{max} \approx 16$--18). With a further increase in the wavelength to the infrared region of the spectrum, the factor first continues to decrease down to a point of $\lambda_0 = 0.85$ μm and then rather rapidly increases and reaches a value of about 90 at $\lambda_0 = 5$ μm. The half-width of the peak $(\Delta a_c)_{1/2}$ (113), starting from the value of $(\Delta a_o)_{1/2} \approx 5°$, rapidly decreases as the wavelength increases and becomes as small as about 0.1° at $\lambda_0 = 5$ μm, which, however, is greater than the usual angular widths of laser beams. The half- width $(\Delta\delta^2)_{1/2}$ (112) differs from δ^2_{max} (107) only by a numerical factor of $4\sqrt{2}$ and therefore is not presented in the table. The penetration depth d_e (114) of a polariton into the crystal at the point of absolute maximum of the resonance peak is comparable with the wavelength of the polariton and remains small even in the infrared region, although being much greater than the localization depth d_m of the plasmon (115). However, as $\Delta a_o \to 0$, $p_e \to 0$ (95), the penetration depth d_e rapidly increases, and the polariton becomes a quasibulk wave. The optimized perturbation δ_{max} corresponding to the angle $\theta_{max} = \arctan\delta_{max}$ remains small over the entire range of wavelengths and varies from 0.05 to 0.01, which certainly guarantees the correctness of the approximate formulas obtained.

	λ_0, μm	0.4	0.6	0.85	1.2	2.5	5.0
Surface polariton (a pumped mode) $\Delta a_o^{max} < 0$	K_{eo}^{max}	17.6	15.9	10.8	43.8	67.1	87.5
	$(\Delta a_o)_{1/2}$	5.5°	4.1°	4.5°	0.9°	0.3°	0.11°
	d_e, μm	0.090	0.225	0.399	0.719	3.18	12.4
	$-\Delta a_o^{max}$	8.0°	3.6°	2.1°	1.3°	0.3°	0.08°
	δ^2_{max}	0.048	0.053	0.078	0.019	0.013	0.010
	θ_{max}	12°	13°	16°	7.8°	6.5°	5.7°
	ψ_o	15°	15°	18°	9.3°	7.7°	6.8°
Bulk polariton $\Delta a_o^{max} = 0$	$\bar{K}_{eo}^{max}$	2.7	3.9	4.6	6.9	14.4	26.3
	$\bar{\delta}^2_{max}$	0.56	0.38	0.28	0.23	0.11	0.05
	$\bar{\theta}_{max}$	37°	32°	28°	26°	18°	13°
Plasmon	d_m, μm	0.017	0.017	0.018	0.021	0.020	0.021

Table 2. Parameters of polaritons excited in an optically negative sodium nitrate crystal with aluminum coating for various wavelengths ($c_1 = 0$, $\hat{a}_o = 32.5°$)

5.4. Conversion reflection and a pumped surface mode

Now we consider the reflection coefficient (97) in more detail for $\Delta a_o < 0$:

$$r_{oo}(\delta^2, \Delta a_o) \equiv \frac{C_o^r}{C_o^i} = \frac{\delta^2 \kappa_o \hat{n}_e^0 / 2 - \zeta' - i\left(\sqrt{-\kappa_o \gamma \Delta a_o} - |\zeta''|\right)}{\delta^2 \kappa_o \hat{n}_e^0 / 2 + \zeta' + i\left(\sqrt{-\kappa_o \gamma \Delta a_o} - |\zeta''|\right)} . \tag{116}$$

In contrast to the excitation factor K_{eo} (104) of the extraordinary polariton, the reflection coefficient (116) does not "promise" any amplification peaks. Conversely, it follows from (116) that the amplitude of the ordinary reflected wave never exceeds in absolute value the amplitude of the incident wave. Moreover, the substitution of the coordinates Δa_o^{max} (105) and δ_{max} (107) of the absolute maximum of the excitation factor (104) into (116) gives the absolute minimum (see Fig. 8 and curves 2 in Fig. 9):

$$|r_{oo}(\delta_{max}^2, \Delta a_o^{max})|^2 = 0 \tag{117}$$

Thus, the resonance reflection in the optimized geometry is a conversion reflection (i.e., a two-partial reflection with a change of branch) and a quite nontrivial one at that. Indeed, in this case the incident ordinary partial wave in the crystal is accompanied by a unique wave, which, being an extraordinary wave belongs to the other refraction sheet and is not a bulk reflected wave. This wave is localized at the interface between the crystal and metal and transfers energy along the interface (Fig. 10).

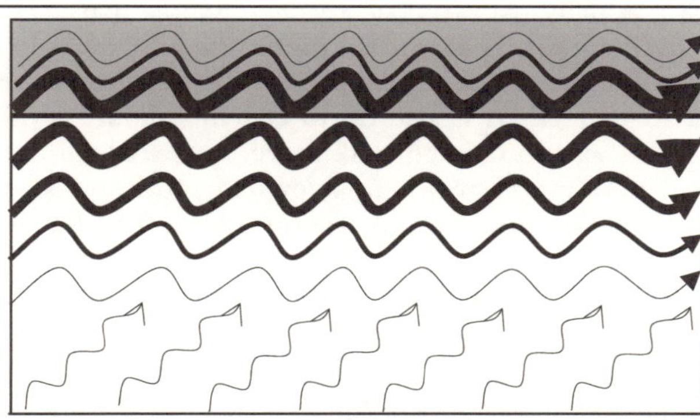

Fig. 10. Schematic picture of the pumped polariton-plasmon near the interface between a crystal and metal coating

Naturally, in this case the absence of the reflected wave does not imply the violation of the energy conservation law; just the propagation geometry corresponding to the minimum (117) is chosen so that the normal component of the Poynting vector of the incident wave is completely absorbed in the metal. This component is estimated by means of $(100)_1$:

$$|\mathbf{P}_{o\perp}^i| = |\mathbf{P}_o^i| \sin a_o = w_o^i (c / \hat{n}_o) \hat{p}_o^0 (\hat{n}_e^0 / \hat{n}_o) = c w_o^i \hat{p}_o^0 \hat{n}_e^0 / \varepsilon_o . \tag{118}$$

Following (Landau & Lifshitz, 1993) and relations (100), (101), and (103), we can easily verify that this component is equal the normal energy flux absorbed by the metal coating at resonance:

$$|\mathbf{P}_{m\perp}| = \frac{c\zeta'}{8\pi}|\mathbf{H}_t|_S^2 = c\zeta'|r_{eo}|_{max}^2 w_o^i = cw_o^i \frac{\hat{p}_o^0 \hat{n}_e^0}{\varepsilon_o}. \tag{119}$$

It is not incidental that the final expression for $|\mathbf{P}_{m\perp}|$ does not contain the components of the impedance. Indeed, according to the energy conservation law, in this case dissipation should completely compensate the normal energy flux in the incident wave, which "knows" nothing about the metallization of the crystal surface. It is essential that the dissipation (119), remaining comparable with the energy flux density in the incident wave, is very small compared with the intensity of the polariton--plasmon localized at the interface:

$$|\mathbf{P}_e|_{max} = w_e|_S \frac{c}{\hat{n}_e^0} = \frac{1}{8\pi}|\mathbf{H}_e|_S^2 \frac{c}{\hat{n}_e^0} = |r_{eo}|_{max}^2 \frac{cw_o^i}{\hat{n}_e^0} >> |\mathbf{P}_{m\perp}|. \tag{120}$$

The fact that the energy flux of the polariton--plasmon at the interface is considerably greater than the intensity of the pumping wave in no way contradicts either the energy conservation law or the common sense. We consider a steady-state problem on the propagation of infinitely long plane waves. In this statement, the superposition of waves jointly transfers energy along the surface from --∞ to +∞. These waves exist only together, and the question of the redistribution of energy between the partial waves can be solved only within a non-stationary approach. Indeed, suppose that, starting from a certain instant, a plane wave coinciding with our ordinary wave is incident on the surface of a crystal. Upon reaching the boundary, this wave generates an extraordinary wave whose amplitude increases in time and gradually reaches a steady-state regime that we describe. Naturally, the time of reaching this regime is the larger, the higher the peak of the excitation factor.

In fact, the conversion reflection considered represents an eigenwave mode that arises due to the anisotropy of the crystal. It is natural to call this mode, consisting of a surface polariton--plasmon and a weak pumping bulk wave, a *pumped* surface wave by analogy with the known *leaky* surface waves, which are known in optics and acoustics (Alshits et al., 1999, 2001). The latter waves also consist of a surface wave and the accompanying weak bulk wave, which, in contrast to our case, removes energy from the surface to infinity, rather than brings it to the surface; i.e., it is a leak, rather than a pump, partial wave.

Numerical analysis of the exact expression for the reflection coefficient r_{oo}, Eqs. $(30)_1$, (32), has shown (Lyubimov et al., 2010) that the conversion phenomenon (117) retains independently of the magnitude of the impedance ζ. However it turns out that for not too small ζ, positions of the maximum of the excitation factor K_{eo} and the minimum (117) of the reflection coefficient r_{oo} do not exactly coincide anymore, as they do in our approximation.

5.5. Resonance excitation of a bulk polariton

When a pump wave of the ordinary branch is incident at angle $a_o \geq \hat{a}_o$ on the boundary of the crystal, a bulk extraordinary polariton is generated. The expression, following from (96) and (102), for the excitation factor K_{eo} of such a polariton is significantly different from expression (104), which is valid for $\Delta a_o < 0$:

$$K_{eo}(\delta^2, \Delta a_o) = \frac{(\delta\kappa_o\varepsilon_e)^2 / \hat{n}_o\hat{n}_e^0}{\left(\delta^2\kappa_o\hat{n}_e^0 / 2 + \sqrt{\kappa_o\gamma\Delta a_o} + \zeta'\right)^2 + \zeta''^2}. \tag{121}$$

As the angle Δa_o increases, the function (121) monotonically decreases, so that the excitation factor attains its maximum for $\Delta a_o = 0$, i.e., for $a_o = \hat{a}_o$:

$$K_{eo}(\delta^2, 0) = \frac{(\delta\kappa_o\varepsilon_e)^2 / \hat{n}_o\hat{n}_e^0}{\left(\delta^2\kappa_o\hat{n}_e^0 / 2 + \zeta'\right)^2 + \zeta''^2}. \tag{122}$$

In turn, formula (122), as a function of the parameter δ^2 , forms a peak with the coordinate of the maximum

$$\bar{\delta}_{max}^2 = \frac{2|\zeta|}{\kappa_o\hat{n}_e^0} \approx \frac{2|\zeta''|}{\kappa_o\hat{n}_e^0}. \tag{123}$$

Note that the optimized parameters $\bar{\delta}_{max}^2$ (123) and δ_{max}^2 (107) for the excitation of bulk and localized polaritons are substantially different (see Table 1):

$$\bar{\delta}_{max}^2 / \delta_{max}^2 \approx |\zeta''| / \zeta'. \tag{124}$$

With regard to (123), the absolute maximum of the excitation factor (121) is expressed as

$$\bar{K}_{eo}^{max} = K_{eo}(\bar{\delta}_{max}^2, 0) = \frac{2\hat{p}_o^0}{\hat{n}_o(|\zeta| + \zeta')} \approx \frac{2\hat{p}_o^0}{\hat{n}_o(|\zeta''| + \zeta')}. \tag{125}$$

Next, by analogy with (110) and with regard to (109), for $c_1 = 0$ we obtain the following optimized value:

$$\bar{K}_{eo}^{max} \approx \frac{2}{|\zeta''| + \zeta'}\sqrt{\frac{\varepsilon_o - \varepsilon_e}{\varepsilon_o\varepsilon_e}}. \tag{126}$$

The approximate equality in formulas (123)--(126) implies that the terms of order $\sim(\zeta'/\zeta'')^2 \ll 1$ are omitted.
The three-dimensional picture of the excitation peak (121) is shown in Fig. 8 as a slope of a ridge in the region $\Delta a_o \geq 0$. The figure shows that, in the domain $\delta \sim \bar{\delta}_{max}$, $\Delta a_o \approx 0$, the factor $K_{eo}(\delta^2, \Delta a_o)$ rather weakly depends on δ and can be estimated at $\delta = \bar{\delta}_{max}$ as

$$K_{eo}(\bar{\delta}_{max}^2, \Delta a_o) \approx \frac{\bar{K}_{eo}^{max}}{1 + \sqrt{\Delta a_o / |\Delta a_o^{max}|}}. \tag{127}$$

The half-width of this one-sided peak is obviously given by $(\Delta a_o)_{1/2} = |\Delta a_o^{max}|$. In Fig. 8, the section (127) is shown as the edge of the surface $K_{eo}(\delta^2, \Delta a_o)$ that reaches the plane $\delta^2 = \bar{\delta}_{max}^2 \approx 0.28$ (see Table 2).
Note that, in the domain $\Delta a_o \geq 0$, conversion is impossible ($r_{oo} \neq 0$) for $\zeta \neq 0$; thus, along with the extraordinary reflected wave, an ordinary reflected wave always exists, such that

$$|r_{oo}(\bar{\delta}_{max}^2, \Delta a_o)|^2 \approx \frac{1 - \sqrt{\Delta a_o / |\Delta a_o^{max}|} - \zeta'/|\zeta''|}{1 + \sqrt{\Delta a_o / |\Delta a_o^{max}|} + \zeta'/|\zeta''|}. \tag{128}$$

where, just as in (127), the terms quadratic in ζ' and linear in Δa_o are omitted. Formula (128) shows that, for $\Delta a_o << |\Delta a_o^{max}|$, $\zeta' << |\zeta''|$, the absolute values of the amplitudes of the incident and ordinary reflected waves are rather close to each other; hence, if we neglect the dissipation in the metal, nearly all the energy of the incident wave is passed to the ordinary reflected wave. In this situation, the presence of additional quite intense extraordinary reflected wave looks paradoxical.

This result can be more clearly interpreted in terms of wave beams rather than plane waves (Fig. 11). Let us take into consideration that plane waves are an idealization of rather wide (compared to the wavelength) beams of small divergence. Of course, it is senseless to choose the angle Δa_o smaller than the angle of natural divergence of a beam. However, this angle can be very small (10^{-4}--10^{-3} rad) for laser beams. If the width of an incident beam of an ordinary wave is l, then the reflected beam of the same branch of polarization has the same width. However, the beam of an extraordinary wave is reflected at a small angle $\widehat{\varphi}_e$ to the surface, and its width $\widehat{l}$ should also be small: $\widehat{l} = \widehat{\varphi}_e l / \sin a_o$ (Fig. 11). It can easily be shown that this width decreases so that even a small amount of energy in a narrow beam ensures a high intensity of this wave. The consideration would be quite similar to our analysis of the energy balance in the previous sub-section.

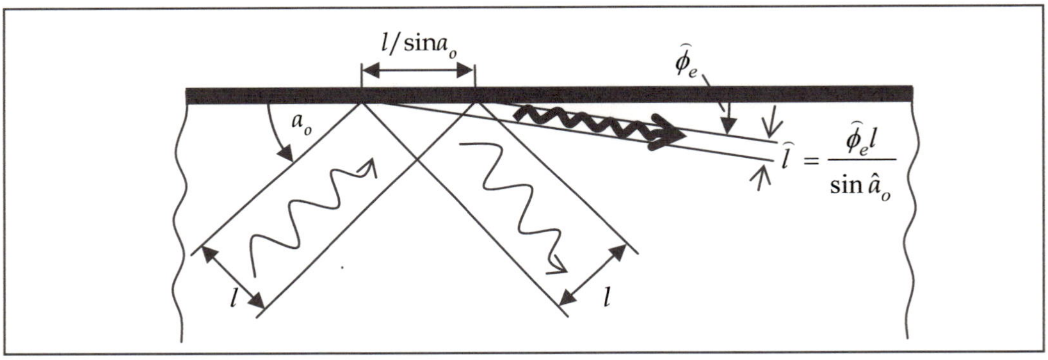

Fig. 11. The scheme of the resonance excitation of a bulk polariton by a finite-width beam

Fortunately, even a small deviation of Δa_o from zero easily provides a compromise that allows one, at the expense of the maximum possible intensity in the extraordinary reflected wave, to keep this intensity high enough and, moreover, to direct a significant part of the energy of the incident wave to this reflected wave. Indeed, formulas (127) and (128) show that, say, at $\Delta a_o \approx 0.1\,|\Delta a_o^{max}|$, the energy is roughly halved between the reflected waves, and $K_{eo} \approx 0.76\,\overline{K}_{eo}^{max}$. For $\Delta a_o \approx 0.2\,|\Delta a_o^{max}|$, we obtain $|r_{oo}|^2 \approx 0.3$ and $K_{eo} \approx 0.7\,\overline{K}_{eo}^{max}$.

The ratio of the absolute maxima (110) and (126) taken for different optimizing parameters δ_{max}^2 and $\overline{\delta}_{max}^2$, respectively, is usually much greater than unity:

$$\frac{K_{eo}(\delta_{max}^2, \Delta a_o^{max})}{K_{eo}(\overline{\delta}_{max}^2, 0)} = \frac{K_{eo}^{max}}{\overline{K}_{eo}^{max}} = \frac{1}{2}\left(\frac{|\zeta''|}{\zeta'} + 1\right). \tag{129}$$

In other words, the excitation efficiency of bulk polaritons is less than that of surface polaritons (see Table 2). Nevertheless, the attainable values of the excitation factor $\overline{K}_{eo}^{max}$ of a bulk polariton are in no way small. According to Table 2, when $\Delta a_o = 0$, the intensity of the reflected extraordinary wave is three or four times greater than that of the incident ordinary wave even in the visible range of wavelengths of 0.4--0.6 μm (however, since the parameter $\overline{\delta}^2_{max}$ in this part of the table is not small enough, the accuracy of these estimates is low). Toward the infrared region, the surface impedance ζ of the aluminum coating decreases (see Table 1), while the excitation constant sharply increases, reaching values of tens.

5.6 Anormalous reflection of an extraordinary wave

Now we touch upon the specific features of the resonance excitation of an ordinary polariton by an incident extraordinary pumping wave. As mentioned above, such an excitation is possible only in optically positive crystals ($\gamma > 1$). The resonance arises under the perturbation of the geometry in which a bulk polariton of the ordinary branch (54) and simple reflection (44)-(46) in the extraordinary branch exist independently of each other.

Let us slightly "perturb" the orientation of the crystal surface by rotating it through a small angle $\theta = \arcsin \hat{c}_2$ with respect to the optical axis: $\mathbf{c} = (c_1, \ \hat{c}_2, \ c_3)$. The structure of the corresponding perturbed wave field is determined by formula (5) at $C_o^i = 0$ in which the appropriate vector amplitudes (6), (7) are substituted. The perturbed polarization vectors are found from formulas (14), (15), and the geometrical meaning of the parameters p, p_e, and p_o is illustrated in Fig. 2a. The refraction vectors, which determine the propagation direction of the incident and reflected waves, are present in (10). In the considered case the horizontal component n of the refraction vector is close to the limiting parameter $\hat{n}_o = \sqrt{\varepsilon_o}$ (Fig. 3), and the parameter p_e is close to the limiting value of $\hat{p}_e$: $n = \hat{n}_o + \Delta n$, $p_e = \hat{p}_e + \Delta p_e$. Here the parameter $\hat{p}_e$ is given by the exact expression $\hat{p}_e^2 = (\gamma - 1)(A - c_1^2)/A^2$ and p is defined by Eq. (11) as before. The angle of incidence a_e of the extraordinary wave (Fig. 2a) is now close to the angle $\hat{a}_e = \arctan(\hat{p}_e - p)$: $a_e = \hat{a}_e + \Delta a_e$. The relation between the increments Δn, Δp_e, and Δa_e has the form

$$\Delta n = -\hat{p}_e^0 n_o (c_1^2 / \gamma + c_3^2)\Delta a_e , \quad \Delta p_e = (c_1^2 + \gamma c_3^2)\Delta a_e , \tag{130}$$

where $\hat{p}_e^0$ relates to the unperturbed $c_2 = 0$: $\hat{p}_e^0 = |c_3|\sqrt{\gamma - 1}$. Another important characteristic of the resonance is the angle of reflection β_o,

$$\beta_o = \arctan p_o, \qquad p_o \approx \sqrt{\varepsilon_o \kappa_e} \begin{cases} \sqrt{\Delta a_e}, & \Delta a_e \ge 0, \\ i\sqrt{-\Delta a_e}, & \Delta a_e < 0, \end{cases} \tag{131}$$

$$\kappa_e = 2\hat{p}_e^0 (c_1^2 / \varepsilon_e + c_3^2 / \varepsilon_o) . \tag{132}$$

Introduce a small parameter $\delta = \hat{c}_2 / c_3$, which is the inverse of $(98)_2$. Now, instead of (96) and (97), we have the following expressions for the reflection coefficients:

$$r_{oe}(\delta, \Delta a_e) \equiv \frac{C_o^r}{C_e^i} = \frac{2\hat{p}_e^0 \delta / \sqrt{\varepsilon_e}}{\delta^2 \hat{p}_e^0 / \hat{n}_o + \sqrt{\kappa_e \Delta a_e} + \zeta} , \tag{133}$$

$$r_{ee}(\delta^2, \Delta a_e) \equiv \frac{C_e^r}{C_e^i} = \frac{\delta^2 \hat{p}_e^0 / \hat{n}_o - \sqrt{\kappa_e \Delta a_e} - \zeta}{\delta^2 \hat{p}_e^0 / \hat{n}_o + \sqrt{\kappa_e \Delta a_e} + \zeta} . \tag{134}$$

These expressions exhibit the same structure of dependence on the small parameters δ and Δa_e as formulas (96) and (97) for optically negative crystals. Naturally, the main features of the reflection resonance considered above nearly completely persist under new conditions. By analogy with (99), let us introduce the excitation factor of an ordinary polariton,

$$K_{oe}(\delta^2, \Delta a_e) = \left|\mathbf{P}_o^r\right| / \left|\mathbf{P}_e^i\right|\Big|_{y=0} = (|\mathbf{u}_o^r| / |\mathbf{u}_o^i|)\left|r_{oe}(\delta, \Delta a_e)\right|^2, \tag{135}$$

where $\mathbf{u}_o^r$ and $\mathbf{u}_e^i$ are the group velocities (3) of the excited and incident waves (in zero approximation): $|\mathbf{u}_o^r| = c / \hat{n}_o$, $|\mathbf{u}_e^i| = c\sqrt{B} / \hat{n}_o$.
The analysis of expressions (133)--(135) shows that, when

$$\delta_{max}^2 = \hat{n}_o \zeta' / \hat{p}_e^0, \qquad \Delta a_e^{max} = -|\zeta''|^2 / \kappa_e, \tag{136}$$

a conversion occurs ($r_{ee} = 0$); i.e., the amplitude of the extraordinary reflected wave strictly vanishes. As a result, again a pumped polariton--plasmon arises in which the primary mode is the localized mode (an ordinary polariton in the crystal and a plasmon in the metal) whose intensity on the interface is much greater than the intensity of the incident pumping wave, which is clear from the expression for the absolute maximum of the excitation factor:

$$K_{oe}\left(\delta_{max}^2, \Delta a_e^{max}\right) \equiv K_{oe}^{max} = \hat{p}_e^0 \hat{n}_o / \varepsilon_e \zeta' \sqrt{B}. \tag{137}$$

Substituting here $\hat{p}_e^0 = |c_3| \sqrt{\gamma - 1}$, we can easily see that again the factor K_{oe}^{max} is optimized for $c_1 = 0$ when $c_3 \approx 1$. In this case,

$$K_{eo}^{max} = \frac{1}{\zeta'} \sqrt{\frac{\varepsilon_e - \varepsilon_o}{\varepsilon_o \varepsilon_e}}. \tag{138}$$

Formulas (138) and (126) turn into each other under the interchange $e \leftrightarrow o$.
The penetration depth of the polariton into the crystal in the pumped configuration is

$$d_o = \lambda_0 / 2\pi\varepsilon_o |\zeta''|. \tag{139}$$

In the neighborhood of coordinates (136) of the absolute maximum (137), a peak of the excitation factor $K_{oe}(\delta^2, \Delta a_e)$ is formed whose configuration is qualitatively correctly illustrated in Figs. 8 and 9. The half-widths of the curves that arise in two sections of this peak $\Delta a_e \equiv \Delta a_e^{max}$ and $\delta^2 \equiv \delta_{max}^2$ are, respectively, given by

$$(\Delta\delta^2)_{1/2} = 4\sqrt{2}\hat{n}_o \zeta' / \hat{p}_e^0, \quad (\Delta a_e)_{1/2} = 8\zeta' |\zeta''| / \kappa_e. \tag{140}$$

The excitation resonance of a bulk polariton in the crystal for $\Delta a_e \geq 0$ is also completely analogous to the resonance described above. Again the excitation factor is the larger, the smaller is the deviation angle Δa_e, and again a peak arises with respect to δ^2:

$$K_{oe}(\delta^2, 0) = \frac{4\delta^2 (\hat{p}_e^0)^2 / \varepsilon_e \sqrt{B}}{(\delta^2 \hat{p}_e^0 / \hat{n}_o + \zeta')^2 + \zeta''^2}, \tag{141}$$

the coordinate of whose maximum is given by

$$\overline{\delta}^2_{max} = \hat{n}_o |\zeta| / \hat{p}^0_e \approx \hat{n}_o |\zeta''| / \hat{p}^0_e \,, \tag{142}$$

and the peak height (the absolute maximum) is given by an analog of (125):

$$\overline{K}^{max}_{oe} = K_{oe}(\overline{\delta}^2_{max}, 0) \approx 2\hat{n}_o \hat{p}^0_e / \varepsilon_e \sqrt{B}(|\zeta''| + \zeta') \,. \tag{143}$$

As above, the choice of the geometry $c_1 = 0$ optimizes the factor $\overline{K}^{max}_{eo}$ and reduces (143) to the following analog of (126):

$$\overline{K}^{max}_{eo} = \frac{2}{|\zeta''| + \zeta'} \sqrt{\frac{\varepsilon_e - \varepsilon_o}{\varepsilon_o \varepsilon_e}} \,. \tag{144}$$

The maximum intensity (143), (144) of the bulk wave attained for $\Delta a_e = 0$ is again accompanied by zero integral energy in this wave, because the main part of the incident extraordinary wave (except for the absorption in metal) is transferred to a reflected extraordinary wave. However, as is shown in Subsection 5.5, even a small increase in the angle of incidence from the value $\Delta a_e = 0$ substantially improves the energy distribution between reflected waves with a small loss in the amplitude of the excitation factor. This fact can easily be verified quantitatively by analyzing formulas (127) and (128) upon the interchange of the indices $o \leftrightarrow e$.

6. Recommendations for setting up an experiment

The resonance discussed is completely attributed to the anisotropy of the crystal and the shielding of the wave field in the crystal by metallization of the surface. Therefore, one should choose a crystal with large anisotropy factor $|\gamma - 1|$ and a metal with low surface impedance ζ. This will guarantee the maximum intensity of the wave excited during reflection (see formulas (112), (140) and (128), (145)).

The orientation of the working surfaces of a sample is determined by the optical sign and the permittivities of the crystal and by the impedance of the metal coating at a given wavelength. As shown above, the optical axis should be chosen to be orthogonal to the propagation direction x: $c_1 = 0$ (Fig. 1). In optically positive and negative crystals, this axis should make angles of θ_{max} and 90° -- θ_{max}, respectively, with the metallized surface. When a *surface* polariton--plasmon is excited in an optically *positive* crystal, we have

$$\theta_{max} = \arctan \delta_{max} \,, \qquad \delta^2_{max} = \frac{\zeta' \varepsilon_o}{\sqrt{\varepsilon_e - \varepsilon_o}} \,. \tag{145}$$

If the goal of the experiment is to obtain an intense *bulk* reflected wave, then one should change δ^2_{max} to $\overline{\delta}^2_{max} = \delta^2_{max} \zeta'' | / \zeta'$ (i.e., $\zeta' \to |\zeta''|$) and θ_{max} to $\overline{\theta}_{max}$ in (145). For optically *negative* crystals, appropriate angles θ_{max} and $\overline{\theta}_{max}$ are defined by the same formulas (145) in which the indices o and e should be interchanged. For sodium nitrate crystals, the angles θ_{max} and $\overline{\theta}_{max}$ are given in Table 2.

In an optically *positive* crystal in which a *surface* polariton--plasmon is excited, the input surface for a normally incident initial wave should be cut at the angle

$$a_e = \hat{a}_e + \Delta a_e^{max}, \qquad \hat{a}_e = \arctan\sqrt{\frac{(\gamma - 1)(1+\delta_{max}^2)}{1+\gamma\delta_{max}^2}}, \qquad \Delta a_e^{max} = -\frac{|\zeta''|^2 \varepsilon_o}{2\sqrt{\gamma - 1}}. \tag{146}$$

In the case of excitation of a *bulk* polariton, one should make the following changes in (146): $\hat{a}_e \to \bar{\hat{a}}_e$ and $\delta_{max}^2 \to \bar{\delta}_{max}^2$. The expressions for $\hat{a}_e$ and $\bar{\hat{a}}_e$ following from (146) are exact. We did not decompose them with respect to the parameters δ_{max}^2 and $\bar{\delta}_{max}^2$, because they are not small enough at some wavelengths. To successfully observe a resonance, one should determine the angles of incidence as precisely as possible, especially when the angular width of the resonance is small.
In an optically *negative* crystal, instead of (146) we have

$$a_o = \hat{a}_o + \Delta a_o^{max}, \qquad \hat{a}_o = \arctan\sqrt{1/\gamma - 1}, \qquad \Delta a_o^{max} = -\frac{|\zeta''|^2 \varepsilon_o}{2\sqrt{1/\gamma - 1}}. \tag{147}$$

Here the limiting angle $\hat{a}_o$ is insensitive to the perturbation of c_3, being the same for the excitation of localized and bulk polaritons (see Table 2).
The output surface for the excited *bulk* wave should be orthogonal to its refraction vector, determined in an optically positive or negative crystal by the angle β_o or β_e (Figs. 2a and 7b):

$$\beta_o = \arctan\left(|\zeta''|\sqrt{\varepsilon_o}\right), \qquad \beta_e = \arctan\left(|\zeta''|\sqrt{\varepsilon_e}/\gamma\right). \tag{148}$$

For optically negative crystals, the angle β_e is naturally different from the slope angle ϕ_e of its ray velocity $\mathbf{u}_e$ in the reflected beam (see Figs. 7b and 11).
A correct choice of the polarization of the incident laser beam allows one to avoid the occurrence of a parasitic beam as a result of birefringence at the input of the crystal, i.e., additional loss of the energy of the incident beam. According to (45) and (50) for $c_1 = 0$, the polarization of the wave at the input should be of *TE* type in zero approximation $\delta = 0$): the field $\mathbf{e}^i$ is parallel to the z axis for crystals of both optical signs. In a more precise analysis ($\delta = \delta_{max}$), the polarization vector $\mathbf{e}^i$ should be turned (about the vector $\mathbf{n}^i$) through an angle ψ. When exciting a *surface* polariton--plasmon, in the first approximation this angle is given by

$$\psi \approx \arctan(\delta_{max}/\sqrt{\gamma}); \tag{149}$$

in optically negative crystals, this rotation is clockwise, whereas, in optically positive crystals, counterclockwise. Table 2 shows that the angle ψ is small.
The situation is changed when one deals with the excitation of a *bulk* wave. Now the optimized polarization of the incident wave is defined by the same Eq. (149) in which δ_{max} is replaced by $\bar{\delta}_{max}$. In this case, the rotation angle ψ sharply increases, while the accuracy of approximation substantially degrades (at least for the visible range). It seems that in this case it is better to choose an optimal polarization of the initial wave experimentally.
As we have seen, the resonance width with respect to the angle of incidence sharply decreases when passing to the infrared region to values of $(\Delta a_{o,e})_{1/2} \approx 0.1$. This imposes a constraint on the divergence of the initial laser beam: the higher the divergence of a beam, the larger part of this beam goes out of resonance. One should also take into account that, by narrowing down the beam at the input, we increase its natural diffraction divergence.

7. Acknowledgements

This work was supported by the Polish Foundation MNiSW, project no. NN501252334. One of the authors (V.I.A.) acknowledges the support of the Polish--Japanese Institute of Information Technology, Warsaw, and the Kielce University of Technology, Poland.

8. References

Agranovich, V.M. (1975). Crystal optics of surface polaritons and the properties of surfaces. *Usp. Fiz. Nauk*, Vol. 115, No. 2 (Feb., 1975) 199-237, ISSN 0042-1294 [*Sov. Phys. Usp.*, Vol. 18, No. 2, (1975) 99-117, ISSN 1063-7859]

Agranovich, V.M. & Mills, D.L. (Eds.) (1982). *Surface Polaritons: Electromagnetic Waves at Surfaces and Interfaces*, North-Holland, ISBN 0444861653, Amsterdam

Alshits, V.I.; Gorkunova, A.S.; Lyubimov, V.N.; Gierulski, W.; Radowicz, A. & Kotowski, R.K. (1999). Methods of resonant excitation of surface waves in crystals, In: *Trends in Continuum Physics (TRECOP '88)*, B.T. Maruszewski, W. Muschik & A. Radowicz, (Eds.), pp. 28-34, World Scientific, ISBN 981023760X, Singapore.

Alshits, V.I. & Lyubimov, V.N. (2002a). Dispersionless surface polaritons in the vicinity of different sections of optically uniaxial crystals. *Fiz. Tverd. Tela (St. Petersburg)*, Vol. 44, No. 2 (Feb., 2002) 371-374, ISSN 0367-3294 [*Phys. Solid State*, Vol. 44, No. 2 (2002) 386-390, ISSN 1063-7834]

Alshits, V.I. & Lyubimov, V.N. (2002b). Dispersionless polaritons on symmetrically oriented surfaces of biaxial crystals. *Fiz. Tverd. Tela (St. Petersburg)*, Vol. 44, No. 10 (Oct., 2002) 1895-1899, ISSN 0367-3294 [*Phys. Solid State*, Vol. 44, No. 10 (2002) 1988-1992, ISSN 1063-7834]

Alshits, V.I. & Lyubimov, V.N. (2005). Dispersion polaritons on metallized surfaces of optically uniaxial crystals. *Zh. Eksp. Teor. Fiz.*, Vol. 128, No. 5 (May, 2005) 904-912, ISSN 0044-4510 [*JETP*, Vol. 101, No. 5 (2005) 779-787, ISSN 1063-7761]

Alshits, V.I. & Lyubimov, V.N. (2009a). Generalization of the Leontovich approximation for electromagnetic fields on a dielectric - metal interface. *Usp. Fiz. Nauk*, Vol. 179, No. 8, (Aug., 2009) 865-871, ISSN 0042-1294 [Physics - Uspekhi, Vol. 52, No.8 (2009) 815-820, ISSN 1063-7859]

Alshits, V.I. & Lyubimov, V.N. (2009b). Bulk polaritons in a biaxial crystal at the interface with a perfect metal. *Kristallografiya*, Vol. 54, No. 6 (Nov., 2009) 989-993, ISSN 0023-4761 [*Crystallography Reports*, Vol. 54, No. 6 (2009) 941-945, ISSN 1063-7745]

Alshits, V.I.; Lyubimov, V.N. (2010). Resonance excitation of polaritons and plasmons at the interface between a uniaxial crystal and a metal. *Zh. Eksp. Teor. Fiz.*, Vol. 138, No. 4 (*Oct.*, 2010) 669-686, ISSN 0044-4510 [*JETP*, Vol. 111, No. 4 (2010) 590-606, ISSN 1063-7761]

Alshits, V.I.; Lyubimov, V.N. & Radowicz, A. (2007). Electromagnetic waves in uniaxial crystals with metallized boundaries: mode conversion, simple reflections, and bulk polaritons. *Zh. Eksp. Teor. Fiz.*, Vol. 131, No. 1 (Jan., 2007) 14-29, ISSN 0044-4510 [*JETP*, Vol. 104, No. 1 (2007) 9-23, ISSN 1063-7761]

Alshits, V.I. ; Lyubimov, V.N. & Shuvalov, L.A. (2001). Pseudosurface dispersion polaritons and their resonance excitation. *Fiz. Tverd. Tela (St. Petersburg)*, Vol. 43, No. 7 (Jul., 2001) 1322-1326, ISSN 0367-3294 [*Phys. Solid State*, Vol. 43, No. 7 (2001) 1377-1381, ISSN 1063-7834]

Born, M. & Wolf, E. (1986). *Principles of Optics*, Pergamon press, ISBN 0.08-026482.4, Oxford

Depine, R.A. & Gigli, M.L. (1995). Excitation of surface plasmons and total absorption of light at the flat boundary between a metal and a uniaxial crystal. *Optics Letters*, Vol. 20, No. 21 (Nov., 1995) 2243-2245, ISSN 0146-9592

D'yakonov, M.I. (1988). New type of electromagnetic wave propagating at the interface. *Zh. Eksp. Teor. Fiz.*, Vol. 94, No. 4 (Apr., 1988) 119-123, ISSN 0044-4510 [*Sov. Phys. JETP*, Vol. 67, No. 4 (1988) 714-716, ISSN 1063-7761]

Fedorov, F.I. (2004). *Optics of Anisotropic Media* (in Russian), Editorial URSS, ISBN 5-354-00432-2, Moscow

Fedorov, F.I. & Filippov, V.V. (1976). *Reflection and Refraction of Light by Transparent Crystals* (in Russian), Nauka I Tekhnika, Minsk

Furs, A.N. & Barkovsky, L.M. (1999). General existence conditions for polaritons in anisotropic, superconductive and isotropic systems *J. Opt. A: Pure Appl. Opt.*, Vol. 1 (Jan., 1999) 109-115, ISSN 1464-4258

Landau, L.D. & Lifshitz, E.M. (1993) *Electrodynamics of Continuous Media*, Butterworth-Heinemann, ISBN, Oxford

Lyubimov, V.N.; Alshits, V.I.; Golovina, T.G.; Konstantinova, A.F. & Evdischenko, E.A. (2010). Resonance and conversion reflections from the interface between a crystal and a metal. *Kristallografiya*, Vol. 55, No. 6 (Nov., 2010) 968-974, ISSN 0023-4761 [*Crystallography Reports*, Vol. 55, No. 6 (2010) 910-916 ISSN 1063-7745]

Marchevskii, F.N.; Strizhevskii, V.L. & Strizhevskii, S.V. (1984). Singular electromagnetic waves in bounded anisotropic media *Fiz. Tverd. Tela (St. Petersburg)*, Vol. 26, No. 5 (*May*, 1984) 1501-1503, ISSN 0367-3294 [*Sov. Phys. Solid State*, Vol. 26, No. 5 (1984) 911-913, ISSN 1063-7834]

Motulevich G.P. (1969). Optical properties of polyvalent non-transition metals. *Usp. Fiz. Nauk*, Vol. 97, No. 2 (Jan., 1969) 211-256, ISSN 0042-1294 [*Sov. Phys. Usp.*, Vol. 12, North-Holland, No. 1, 80-104, ISSN 1063-7859]

Sirotin, Yu.I. & Shaskol'skaya, M.P. (1979). *Fundamentals of Crystal Physics* (in Russian), Nauka, Moscow [(1982) translation into English, Mir, ISBN , Moscow]

4

Electromagnetic Waves Propagation Characteristics in Superconducting Photonic Crystals

Arafa H Aly
Physics department, Faculty of Sciences, Beni-suef University
Egypt

1. Introduction

Photonic crystals (PCs) are structures with periodically modulated dielectric constants whose distribution follows a periodicity of the order of a fraction of the optical wavelength. Since the first pioneering work in this field, many new interesting ideas have been developed dealing with one-dimensional (1D), two-dimensional (2D), and three-dimensional (3D) PCs. Researchers have proposed many new and unique applications of photonic devices which may revolutionize the field of photonics in much the same way as semiconductors revolutionized electronics. They can generate spectral regions named photonic band gaps (PBGs) where light cannot propagate in a manner analogous to the formation of electronic band gaps in semiconductors [1,2]. There are several studies of metallic [3-7] and superconducting photonic crystals [7,8] which are mostly concentrated at microwave, millimeterwave, and far-infrared frequencies. In those frequencies, metals act like nearly perfect reflectors with no significant absorption problems.

Yablonovitch [1] main motivation was to engineer the photonic density of states in order to control the spontaneous emission of materials embedded with photonic crystal while John's idea was to use photonic crystals to affect the localization and control of light. However due to the difficulty of actually fabricating the structures at optical scales early studies were either theoretical or in the microwave regime where photonic crystals can be built on the far more reading accessible centimeter scale. This fact is due to the property of the electromagnetic fields known as scale invariance in essence, the electromagnetic fields as the solutions to Maxwell's equations has no natural length scale and so solutions for centimeter scale structure at microwave frequencies as the same for nanometer scale structures at optical frequencies.

The optical analogue of light is the photonic crystals in which atoms or molecules are replaced by macroscopic media with different dielectric constants and the periodic potential is replaced by a periodic dielectric function. if the dielectric constants of the materials is sufficiently different and also if the absorption of light by the material is minimal then the refractions and reflections of light from all various interfaces can produce many of the same phenomena for photons like that the atomic potential produced for electrons[9].

The previous details can guide us to the meaning of photonic crystals that can control the propagation of light since it can simply defined as a dielectric media with a periodic

modulation of refractive index in which the dielectric constant varies periodically in a specific directions. Also it can be constructed at least from two component materials with different refractive index due to the dielectric contrast between the component materials of the crystal .it's characterized by the existence of photonic band gap (PBG) in which the electromagnetic radiation is forbidden from the propagation through it.
Optical properties of low dimensional metallic structures have also been examined recently. For example, the optical transmission through a nanoslit collection structure shaped on a metal layer with thin film thickness was analyzed in Refs. [10,11]. The photonic band structures of a square lattice array of metal or semiconductor cylinders, and of an array of metal or semiconductor spheres, were enumerated numerically in Ref. [12]. In addition, superconducting (SC) photonic crystals also attract much attention recently [13,14]. In new experiments superconducting metals (in exact, Nb) have been used as components in optical transmission nanomaterials. Dielectric losses are substantially reduced in the SC metals relative to analogous structures made of normal metals. The dielectric losses of such a SC nanomaterial are reduced by a factor of 6 upon penetrating into the SC state [15]. Indeed, studies of the optical properties of superconductor metal/dielectric multilayers are not numerous, may be the results have been used in the design of high reflection mirrors, beam splitters, and bandpass filters [16]. The superiority of a photonic crystal with superconducting particles is that the scattering of the incident electromagnetic wave due to the imaginary part of the dielectric function is much less than for normal metallic particles at frequencies smaller than the superconducting gap. The loss caused by a superconducting photonic crystal is thus expected to be much less than that by a metallic photonic crystal. For a one-dimensional superconductor–dielectric photonic crystal (SuperDPC), it is seen like in an MDPC that there exists a low-frequency photonic band gap (PBG). This low frequency gap is not seen in a usual DDPC. This low frequency PBG is found to be about one third of the threshold frequency of a bulk superconducting material [12]. In this paper, based on the transfer matrix method, two fluid models, we have investigated the effect of the different parameters on transmittance and PBG in a one-dimensional superconductor-dielectric photonic crystals.

2. Numerical methods

We will explain in brief a mathematical treatment with a simple one dimensional photonic crystal structure (1DPC) (see fig.1) which is composed of two materials with thicknesses (d_2 and d_3) and refractive indices (n_2 and n_3) respectively. The analysis of the incident electromagnetic radiation on this structure will be performed using the transfer matrix method (TMM).
A one-dimensional nonmagnetic conventional and high tempeature superconductor-dielectric photonic crystal will be modelled as a periodic superconductor-dielectric multilayer structure with a large number of periods N » 1, Such an N-period superlattice is shown in Fig. 1, where $d = d_2 + d_3$ is the spatial periodicity, where d_2 is the thickness of the superconducting layer and d_3 denotes the thickness of the dielectric layer. We consider that the electromagnetic wave is incident from the top medium which is taken to be free space with a refractive index, $n_1 = 1$. The index of refraction of the lossless dielectric is given by $n_3 = \sqrt{\varepsilon_{r3}}$, n_2 the index of refraction of the superconductor material, which can be described

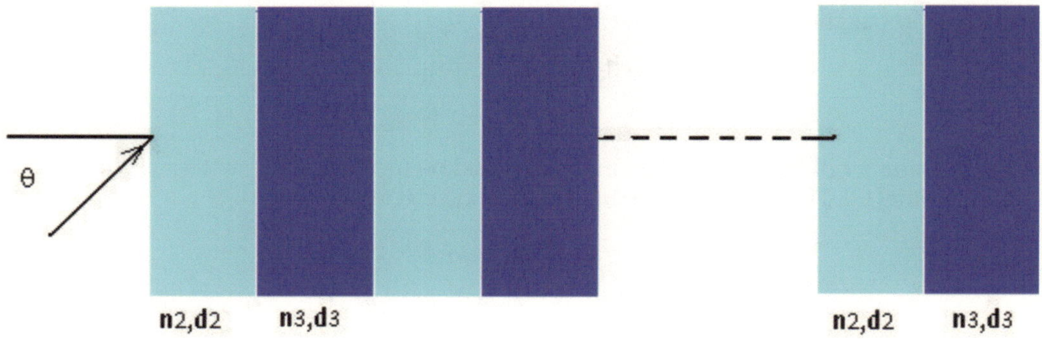

Fig. 1. A superconductor--dielectric structure. The thicknesses of superconducting and dielectric are denoted by d_2 and d_3, respectively, and the corresponding refractive indices are separately indicated by n_1, n_2 n_3, where n_1=1 and n_4 is the indexof substrate layer.

on the basis of the conventional two- fluid model [18].Accordingly to the two fluid model the electromagnetic response of a superconductor can be described in terms of the complex conductivity, $\sigma = \sigma_1 - i\sigma_2$, where the real part indicating the loss contributed by normal electrons, and the imaginary part is due to superelectrons, the imaginary part is expressed as [19,20] $\sigma_2 = \ 1/\omega\mu_0\lambda_l^2$, where the temperature-dependent penetration depth is given by $\lambda_l = \lambda_l(T) = \lambda_0 / \sqrt{1-f(T)}$, where Gorter-Casimir expression for f (T) is given for low and conventional superconductor by f (T)= $(T/T_c)^4$, and for high temperature superconductor f (T)= $(T/T_c)^2$ [13,18].

We shall consider the lossless case, meaning that the real part of the complex conductivity of the superconductor can be neglected and consequently it becomes $\sigma = -i\sigma_2 = -i(1/\omega\mu_0\lambda_l^2)$. The relative permittivity as well as its associated index of refraction can be obtained by,

$$\varepsilon_{r2} = 1 - \frac{c^2}{\omega^2\lambda_l^2} \text{ and } n_2 = \sqrt{\varepsilon_{r2}} = \sqrt{1 - \frac{c^2}{\omega^2\lambda_l^2}} \tag{1}$$

We will go to mention the mathematical form of the dynamical matrices and for the propagation matrix to obtain an expressions for the reflection and transmission, the dynamical matrices take the form [17]:-

$$D_\alpha = \begin{pmatrix} 1 & 1 \\ n_\alpha\, Cos\theta_\alpha & -n_\alpha\, Cos\theta_\alpha \end{pmatrix} \text{ for S - wave} \tag{2}$$

$$D_\alpha = \begin{pmatrix} Cos\theta_\alpha & Cos\theta_\alpha \\ n_\alpha & -n_\alpha \end{pmatrix} \text{ for P - wave} \tag{3}$$

with

$$\beta = n_\alpha \frac{\omega}{c} Sin\theta_\alpha \text{ , and } k_{\alpha x} = n_\alpha \frac{\omega}{c} Cos\theta_\alpha$$

while the propagation matrix take the form:-

$$P_{\eta} = \begin{pmatrix} \exp(i\phi_{\alpha}) & 0 \\ 0 & \exp(-i\phi_{\alpha}) \end{pmatrix} \tag{4}$$

Since the number of the propagation matrix depend on the number of materials which build our structure [17]. Finally the transfer matrix method can take the form:-

$$M = \begin{pmatrix} M_{11} & M_{12} \\ M_{21} & M_{22} \end{pmatrix} \tag{5}$$

$$P_2 = \begin{pmatrix} Cos\phi_1 + i\,Sin\phi_1 & 0 \\ 0 & Cos\phi_1 - i\,Sin\phi_1 \end{pmatrix}$$

$$P_3 = \begin{pmatrix} Cos\phi_2 + i\,Sin\phi_2 & 0 \\ 0 & Cos\phi_2 - i\,Sin\phi_2 \end{pmatrix}$$

Since;

$$\phi_1 = \frac{2\pi d_2}{\lambda} n_2\, Cos\theta_2 \text{, and } \phi_2 = \frac{2\pi d_3}{\lambda} n_3\, Cos\theta_3 \;.$$

The components of the transfer matrix method can be written in a detailed form for an S - wave as:-

$$M_{11} = \frac{1}{2}\left[\left(1 + \frac{n_4\, Cos\ \theta_4}{n_1\, Cos\ \theta_1}\right) Cos\ \phi_2 + \left(\frac{n_3\, Cos\ \theta_3}{n_1\, Cos\ \theta_1} + \frac{n_4\, Cos\ \theta_4}{n_3\, Cos\ \theta_3}\right)(i\, Sin\ \phi_2)\right] Cos\ \phi_1 +$$
$$\frac{1}{2}\left[\left(\frac{n_2\, Cos\ \theta_2}{n_1\, Cos\ \theta_1} + \frac{n_4\, Cos\ \theta_4}{n_2\, Cos\ \theta_2}\right) Cos\ \phi_2 + \left(\frac{n_3\, Cos\ \theta_3}{n_2\, Cos\ \theta_2} + \frac{n_2 n_4 Cos\ \theta_2\, Cos\ \theta_4}{n_1 n_3\, Cos\ \theta_1\, Cos\ \theta_3}\right)(i\, Sin\ \phi_2)\right](i\, Sin\ \phi_1) \tag{6}$$

$$M_{12} = \frac{1}{2}\left[\left(1 - \frac{n_4\, Cos\ \theta_4}{n_1\, Cos\ \theta_1}\right) Cos\ \phi_2 + \left(\frac{n_3\, Cos\ \theta_3}{n_1\, Cos\ \theta_1} - \frac{n_4\, Cos\ \theta_4}{n_3\, Cos\ \theta_3}\right)(i\, Sin\ \phi_2)\right] Cos\ \phi_1 +$$
$$\frac{1}{2}\left[\left(\frac{n_2\, Cos\ \theta_2}{n_1\, Cos\ \theta_1} - \frac{n_4\, Cos\ \theta_4}{n_2\, Cos\ \theta_2}\right) Cos\ \phi_2 + \left(\frac{n_3\, Cos\ \theta_3}{n_2\, Cos\ \theta_2} - \frac{n_2 n_4 Cos\ \theta_2\, Cos\ \theta_4}{n_1 n_3\, Cos\ \theta_1\, Cos\ \theta_3}\right)(i\, Sin\ \phi_2)\right](i\, Sin\ \phi_1) \tag{7}$$

$$M_{21} = \frac{1}{2}\left[\left(1 - \frac{n_4\, Cos\ \theta_4}{n_1\, Cos\ \theta_1}\right) Cos\ \phi_2 - \left(\frac{n_3\, Cos\ \theta_3}{n_1\, Cos\ \theta_1} - \frac{n_4\, Cos\ \theta_4}{n_3\, Cos\ \theta_3}\right)(i\, Sin\ \phi_2)\right] Cos\ \phi_1 -$$
$$\frac{1}{2}\left[\left(\frac{n_2\, Cos\ \theta_2}{n_1\, Cos\ \theta_1} - \frac{n_4\, Cos\ \theta_4}{n_2\, Cos\ \theta_2}\right) Cos\ \phi_2 - \left(\frac{n_3\, Cos\ \theta_3}{n_2\, Cos\ \theta_2} - \frac{n_2 n_4 Cos\ \theta_2\, Cos\ \theta_4}{n_1 n_3\, Cos\ \theta_1\, Cos\ \theta_3}\right)(i\, Sin\ \phi_2)\right](i\, Sin\ \phi_1) \tag{8}$$

$$M_{22}=\frac{1}{2}\left[\left(1+\frac{n_4 Cos\ \theta_4}{n_1 Cos\ \theta_1}\right)Cos\ \phi_2-\left(\frac{n_3 Cos\ \theta_3}{n_1 Cos\ \theta_1}+\frac{n_4 Cos\ \theta_4}{n_3 Cos\ \theta_3}\right)(i Sin\ \phi_2)\right]Cos\ \phi_1-$$
$$\frac{1}{2}\left[\begin{matrix}\left(\frac{n_2 Cos\ \theta_2}{n_1 Cos\ \theta_1}+\frac{n_4 Cos\ \theta_4}{n_2 Cos\ \theta_2}\right)Cos\ \phi_2-\left(\frac{n_3 Cos\ \theta_3}{n_2 Cos\ \theta_2}+\frac{n_2 n_4 Cos\ \theta_2 Cos\ \theta_4}{n_1 n_3 Cos\ \theta_1 Cos\ \theta_3}\right)\\ (i Sin\ \phi_2)\end{matrix}\right](i Sin\ \phi_1) \tag{9}$$

Where the reflectance and transmittance can be written as:-

$$R=\left|r^2\right|=\left|\frac{M_{21}}{M_{11}}\right|^2 \tag{10}$$

$$T=\frac{n_4 Cos\ \theta_4}{n_1 Cos\ \theta_1}|t|^2=\frac{n_4 Cos\ \theta_4}{n_1 Cos\ \theta_1}\left|\frac{1}{M_{11}}\right|^2 \tag{11}$$

Where r and t is the reflection and transmission and we can also obtain by the same method the components of the transfer matrix method (TMM) for P- wave.

3. Results and discussions

The periodicity of the permittivity plays the same role for the photons that propagate inside the structure than the atomic potential for the electrons. Leading further this analogy, the thicknesses and the index contrast of the photonic crystal determinate many of its optical properties as it does for conduction properties of semiconductors. Playing on these two parameters, we can obtain frequency ranges for which light propagation is forbidden in the material and others ranges for which light can propagate. These frequency ranges are also scale dependent. Reducing the size of the elementary cell of the periodic lattice shifts the whole frequency range to higher values. The consequence of this property is the possibility to transpose a photonic crystal design from the microwave domain to infrared or visible wavelengths. In our results we have studied one dimensional superconducting (Super)/dielectric (Na3AIF6) photonic crystals (SuperDPC's). In all our figures we have used the thickness of Na3AIF6 layer is 320nm and the thickness layer of superconducting material is 80nm. Also we have used different periods equal to 7 and the incidence of angle is 48° for the all our results (Fig's 2 and 3).

In fig. 2a we have examined the transmittance in the case of s-polarized depend on the wavelengths in the range of ultraviolet (UV), visible (VI) and near infrared (NIR). We have obtained the magnitude of transmittance 100% from 100nm to 350nm UV range) and we have obtained the PBG from 600nm to 1050nm. At the 730 nm we have got unique peak explaining as a defect localized mode which can be used as Fabry-Port micro cavity, this is a good application. In the case of p-polarized (fig.2b), we can show different results and there are about seven PBG's in the range from 100nm to 900nm. The width of each PBG is widest at the long wavelength as from 700nm to 900nm and is narrowest at the short wavelengths as at 100 nm. Also we have examined the angle dependence on wavelengths to Super/Nas3AIF6 structure (fig.3).

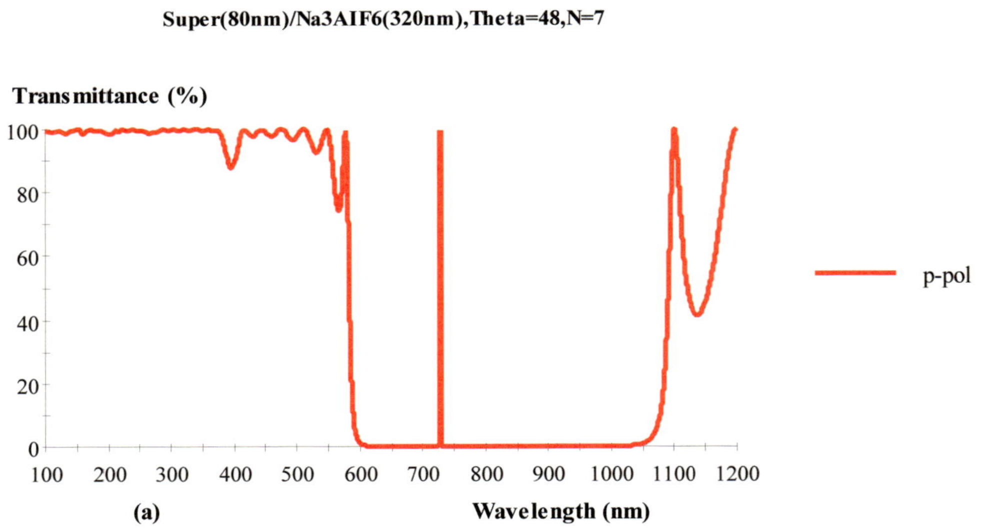

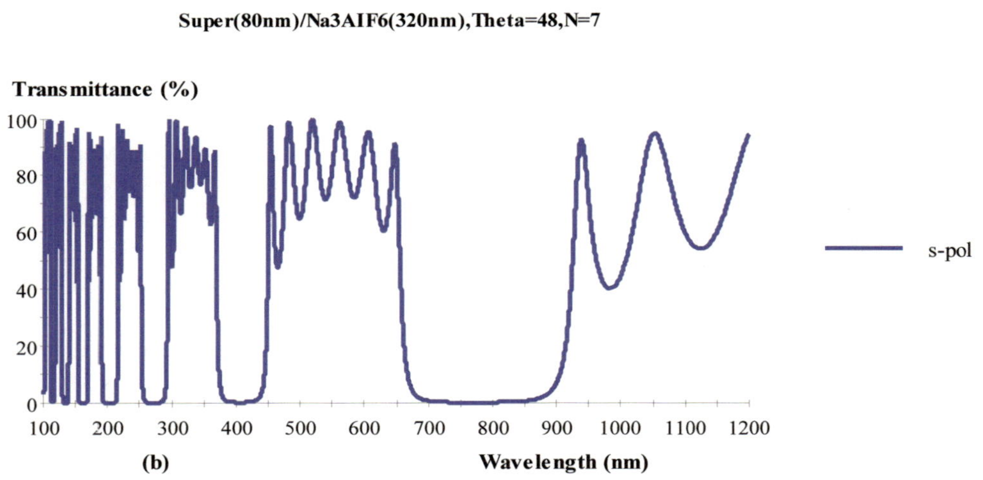

Fig. 2. The transmittance spectra in Super (**80**nm)/Na3AIF6 (329nm) structure**,** **N=7**,a) p-pol and b)s-pol

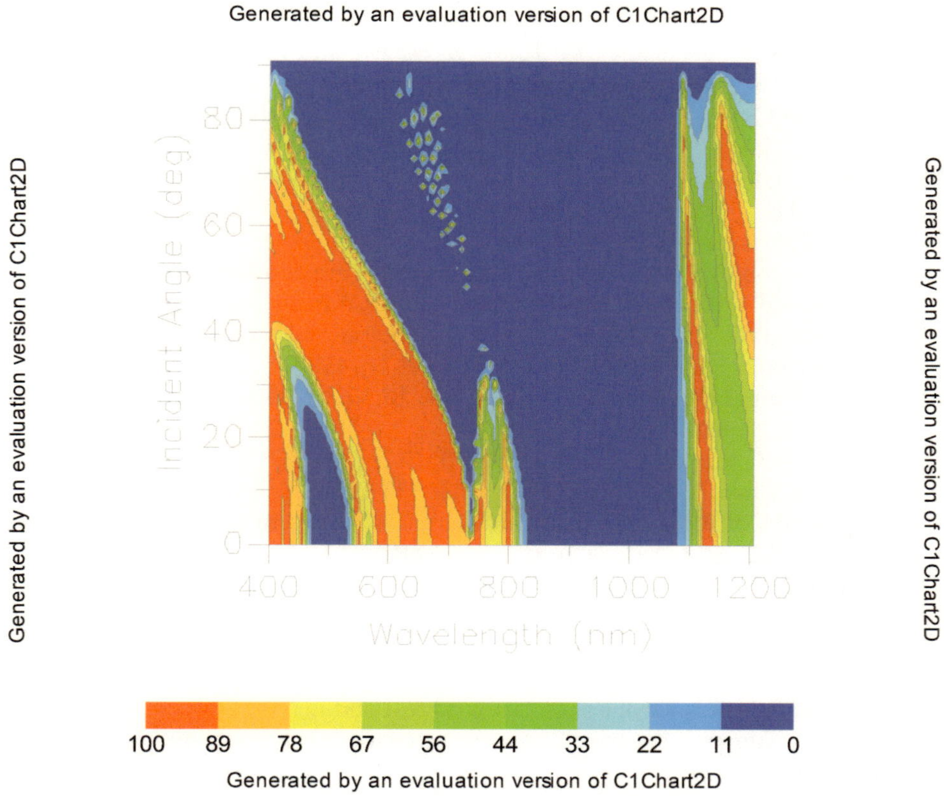

Fig. 3. The incident angle dependence on wavelength to Super/ Na3AIF6 structure

4. Conclusion

We performed numerical analyses to investigate the wave propagation characteristics of a simple-one dimensional superconducting(Super)-dielectric Na3AIF6 structure.The advantage of a photonic crystal with superconducting particles is that the dissipation of the incident electromagnetic wave due to the imaginary part of the dielectric function is much greater for normal metallic than for superconducting particles, because the imaginary part of the dielectric function for superconducting particles is negligible in comparison with the imaginary part of the dielectric function for normal metal particles at frequencies smaller than the superconducting gap. We have obtained good applications at the 730 nm and we have got unique peak explaining as a defect localized mode which can be used as Fabry-Port micro cavity.

5. References

[1] E. Yablonovitch, Phys. Rev. Lett. 58, 2059, (1987).
[2] S. John, Phys. Rev. Lett. 58,2487, (1987).
[3] A.R. McGurn, A.A. Maradudin, Phys. Rev. B 48 (1993) 17576

[4] M. M. Sigalas, C. T. Chan, K. M. Ho, and C. M. Soukoulis, Phys. Rev. B 52, 11 744 , (1995).
[5] J. S. McCalmont, M. M. Sigalas, G. Tuttle, K. M. Ho, and C. M. Soukoulis, Appl. Phys. Lett. 68, 2759, (1995).
[6] Arafa H Aly and Sang-Wan Ryu, J. of Computational and Theoretical Nannoscience, Vol 5, 1-15, (2008).
[7] Arafa H Aly, Materials Chemistry and Physics, 115, 391, (2009).
[8] Arafa H. Aly, Heng-Tung Hsu, Tzong-Jer Yang, Chien-Jang Wu, and C-K Hwangbo, J. of Applied Physics, 105, 083917,(2009).
[9] John D. Joannopoulos,Steven G. Johnson,Joshua N. Winn,and,Robert D. Meade, Photonic crystals moleding the flow of light. 2nd Edition. Princeton University Press, (2008).
[10] J.D. Joannopoulos, R.D. Meade, J.N. Winn, Photonic Crystals: Molding the Flow of Light, Princeton University Press, Princeton, NJ, 1995.
[11] Z. Sun, Y.S. Jung, H.K. Kim, Appl. Phys. Lett. 83,3021 (2003); Z. Sun, H.K. Kim, Appl. Phys. Lett. 85, 642 (2004).
[12] V. Kuzmiak, A.A. Maradudin, Phys. Rev. B 55,7427 (1997).
[13] C.-J.Wu, M.-S. Chen, T.-J. Yang, Physica C 432,133 (2005).
[14] C.H. Raymond Ooi, T.C. Au Yeung, C.H. Kam, T.K. Lam, Phys. Rev. B 61 5920 (2000).
[15] M. Ricci, N. Orloff, S.M. Anlage, Appl. Phys. Lett. 87,034102 (2005).
[16] H.A. Macleod, Thin-Film Optical Filters, 3rd ed., Institute of Publishing, Bristol, 2001, (Chapter 7).
[17] P. Yeh., "Optical Waves in Layered Media", J. Wiley a& Sons, Inc., Hoboken, New Jersey, (2005).
[18] M Bom ,Wolf E, Principles of optics, Cambridge, London,(1999).
[19] Van Duver.T.,Tumer C.W, Princibles of Superconductor Devices and Circuits, Edward Arnold,London,(1981)
[20] Hung-Ming Lee, and Jong-Ching Wu, J. Appl. Phys. 107, 09E149 (2010) .

5

Electromagnetic Wave Propagation in Two-Dimensional Photonic Crystals

Oleg L. Berman[1], Vladimir S. Boyko[1],
Roman Ya. Kezerashvili[1,2] and Yurii E. Lozovik[3]
[1]*Physics Department, New York City College of Technology,
The City University of New York, Brooklyn, NY 11201*
[2]*The Graduate School and University Center,
The City University of New York, New York, NY 10016*
[3]*Institute of Spectroscopy, Russian Academy of Sciences, 142190 Troitsk*
[1,2]*USA*
[3]*Russia*

1. Introduction

Photonic crystals are media with a spatially periodical dielectric function (Yablonovitch, 1987; John, 1987; Joannopoulos et al., 1995; 2008). This periodicity can be achieved by embedding a periodic array of constituent elements with dielectric constant ε_1 in a background medium characterized by dielectric constant ε_2. Photonic crystals were first discussed by Yablonovitch (Yablonovitch, 1987) and John (John, 1987). Different materials have been used for the corresponding constituent elements including dielectrics (Joannopoulos et al., 1995; 2008), semiconductors, metals (McGurn & Maradudin, 1993; Kuzmiak & Maradudin, 1997), and superconductors (Takeda & Yoshino, 2003; Takeda et al., 2004; Berman et al., 2006; Lozovik et al., 2007; Berman et al., 2008; 2009). Photonic crystals attract the growing interest due to various modern applications (Chigrin & Sotomayor Torres, 2003). For example, they can be used as the frequency filters and waveguides (Joannopoulos et al., 2008).
The photonic band gap (PBG) in photonic crystals was derived from studies of electromagnetic waves in periodic media. The idea of band gap originates from solid-state physics. There are analogies between conventional crystals and photonic crystals. Normal crystals have a periodic structure at the atomic level, which creates periodic potentials for electrons with the same modulation. In photonic crystals, the dielectrics are periodically arranged and the propagation of photons is largely affected by the structure. The properties of the photons in the photonic crystals have the common properties with the electrons in the conventional crystals, since the wave equations in the medium with the periodic dielectric constant have the band spectrum and the Bloch wave solution similarly to the electrons described by the Schrödinger equation with the periodic potential (see (Berman et al., 2006) and references therein). Photonic crystals can be either one-, two- or three-dimensional as shown in Fig. 1.
In normal crystals there are valence and conduction bands due to the periodic field. Electrons cannot move inside the completely filled valence band due to the Pauli exclusion principle for electrons as fermions. Electrons can move inside the crystal if they are excited to the

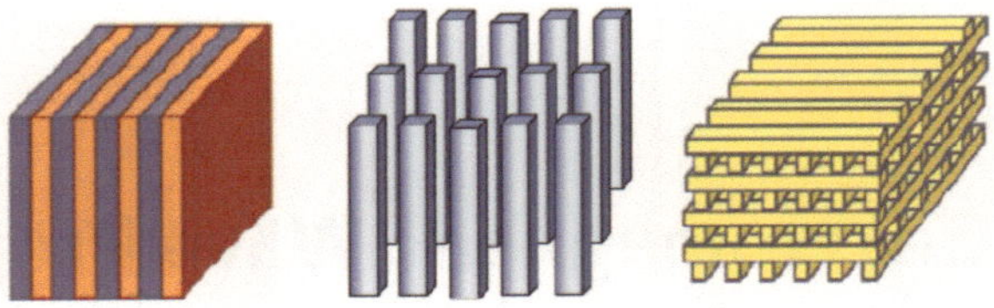

(a) (b) (c)

Fig. 1. Example of 1D, 2D and 3D photonic crystals. All of the photonic crystals shown above have two different dielectric media. (a) 1D multilayer; (b) 2D array of dielectric rods; (c) 3D woodpile.

conduction band. Because the photons are bosons, all bands in the photonic crystals' band structure are conduction bands. If the frequency corresponds to the allowed band, the photon can travel through the media. If the photonic gap exists only in the part of Brillouin zone, then this gap corresponds to the stop band. By other words, photons cannot propagate with frequencies inside the gap at the wavevectors, where this gap exists. Of particular interest is a photonic crystal whose band structure possesses a complete photonic band gap. A PBG defines a range of frequencies for which light is forbidden to exist inside the crystal.

The photonic crystals with the dielectric, metallic, semiconductor, and superconducting constituent elements have different photonic band and transmittance spectra. The dissipation of the electromagnetic wave in all these photonic crystals is different. The photonic crystals with the metallic and superconducting constituent elements can be used as the frequency filters and waveguides for the far infrared region of the spectrum, while the dielectric photonic crystals can be applied for the devices only for the optical region of the spectrum.

In this Chapter we discuss the photonic band structure of two-dimensional (2D) photonic crystals formed by dielectric, metallic, and superconducting constituent elements and graphene layers. The Chapter is organized in the following way. In Sec. 2 we present the description of 2D dielectric photonic crystals. In Sec. 3 we review the 2D photonic crystals with metallic and semiconductor constituent elements. In Sec. 4 we consider the photonic band structure of the photonic crystals with the superconducting constituents. A novel type of the graphene-based photonic crystal formed by embedding a periodic array of constituent stacks of alternating graphene and dielectric discs into a background dielectric medium is studied in Sec. 5. Finally, the discussion of the results presented in this Chapter follows in Sec. 6.

2. Dielectric photonic crystals

The 2D photonic crystals with the dielectric constituent elements were discussed in Ref. (Joannopoulos et al., 2008). Maxwell's equations, in the absence of external currents and sources, result in a form which is reminiscent of the Schrödinger equation for magnetic field $\mathbf{H}(\mathbf{r})$ (Joannopoulos et al., 2008):

$$\nabla \times \left(\frac{1}{\varepsilon(\mathbf{r})} \nabla \times \mathbf{H}(\mathbf{r}) \right) = \left(\frac{\omega}{c} \right)^2 \mathbf{H}(\mathbf{r}) , \qquad (1)$$

where ω is the frequency of the electromagnetic wave, c is the speed of light, $\varepsilon(\mathbf{r})$ is the dielectric constant, which is the periodic function of the radius vector in the photonic crystal. Eq. (1) represents a linear Hermitian eigenvalue problem whose solutions are determined entirely by the properties of the macroscopic dielectric function $\varepsilon(\mathbf{r})$. Therefore, for a crystal

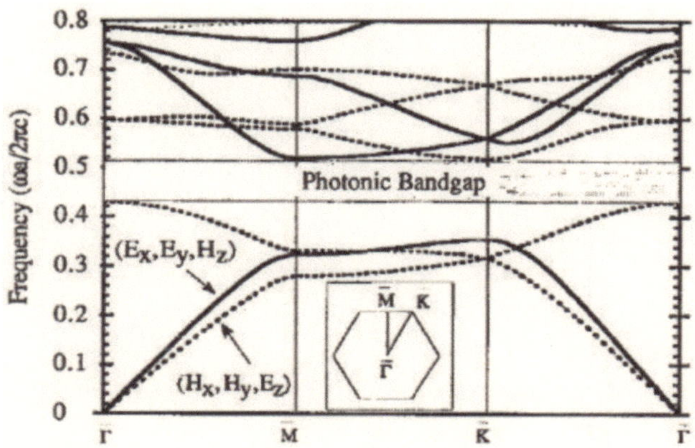

Fig. 2. Frequencies of the lowest photonic bands for a triangular lattice of air columns ($\varepsilon_{air} = 1$) drilled in dielectric ($\varepsilon = 13$). The band structure is plotted along special directions of the in-plane Brillouin zone ($k_z = 0$), as shown in the lower inset. The radius of the air columns is $r = 0.48a$, where a is the in-plane lattice constant. The solid (dashed) lines show the frequencies of bands which have the electric field parallel (perpendicular) to the plane. Notice the PBG between the third and fourth bands.

consisting of a periodic array of macroscopic uniform dielectric constituent elements, the photons in this photonic crystal could be described in terms of a band structure, as in the case of electrons. Of particular interest is a photonic crystal whose band structure possesses a complete photonic band gap.

All various kinds of 2D dielectric photonic crystals were analyzed including square, triangular, and honeycomb 2D lattices (Joannopoulos et al., 2008; Meade et al., 1992). Dielectric rods in air, as well as air columns drilled in dielectric were considered. At the dielectric contrast of GaAs ($\varepsilon = 13$), the only combination which was found to have a PBG in both polarizations was the triangular lattice of air columns in dielectric. Fig. 2 (Meade et al., 1992) represents the eigenvalues of the master equation (1) for a triangular lattice of air columns ($\varepsilon_{air} = 1$) drilled in dielectric ($\varepsilon = 13$).

The photonic band structure in a 2D dielectric array was investigated using the coherent microwave transient spectroscopy (COMITS) technique (Robertson et al., 1992). The array studied in (Robertson et al., 1992) consists of alumina-ceramic rods was arranged in a regular square lattice. The dispersion relation for electromagnetic waves in this photonic crystal was determined directly using the phase sensitivity of COMITS. The dielectric photonic crystals can be applied as the frequency filters for the optical region of spectrum, since the propagation of light is forbidden in the photonic crystal at the frequencies, corresponding to the PBG, which corresponds to the optical frequencies.

3. Photonic crystals with metallic and semiconductor components

The photonic band structures of a square lattice array of metal or semiconductor cylinders, and of a face centered cubic lattices array of metal or semiconductor spheres, were studies in Refs. (McGurn & Maradudin, 1993; Kuzmiak & Maradudin, 1997). The frequency-dependent dielectric function of the metal or semiconductor is assumed to have the free-electron Drude form $\varepsilon(\omega) = 1 - \omega_p^2/\omega^2$, where ω_p is the plasma frequency of the charge carriers. A

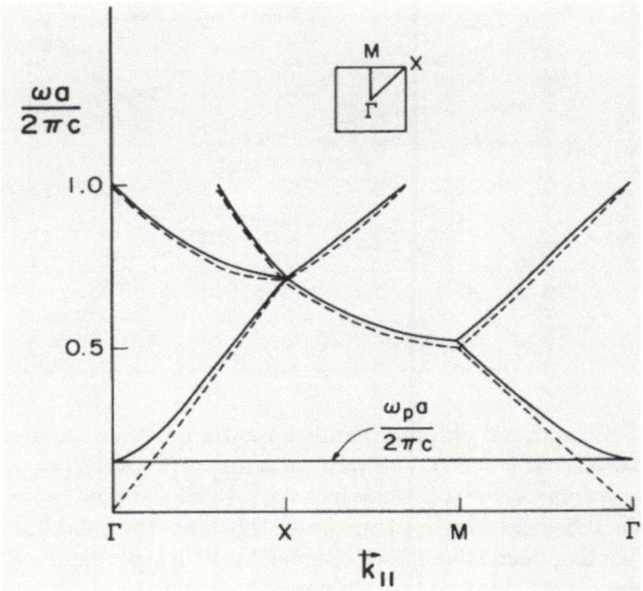

Fig. 3. Band structure for a square lattice of metal cylinders with a filling factor $f = 70\%$. Only results for $\omega \geq \omega_p$ are shown. Results for the dispersion curve in vacuum are shown as dashed lines.

plane-wave expansion is used to transform the vector electromagnetic wave equation into a matrix equation. The frequencies of the electromagnetic modes are found as the zeros of the determinant of the matrix.

The results of the numerical calculations of the photonic band structure for 2D photonic crystal formed by a square lattice of metal cylinders with a filling factor $f = 70\%$ are shown in Fig. 3 (McGurn & Maradudin, 1993). Here the filling factor f is defined as $f \equiv S_{cyl}/S = \pi r_0^2/a^2$, where S_{cyl} is the cross-sectional area of the cylinder in the plane perpendicular to the cylinder axis, S is the total area occupied by the real space unit cell, and r_0 is the cylinder radius.

The photonic crystals with the metallic and semiconductor constituent elements can be used as the frequency filters and waveguides for the far infrared range of spectrum, since the PBG in these photonic crystals corresponds to the frequencies in the far infrared range (McGurn & Maradudin, 1993; Kuzmiak & Maradudin, 1997).

Photonic gaps are formed at frequencies ω at which the dielectric contrast $\omega^2(\varepsilon_1(\omega) - \varepsilon_2(\omega))$ is sufficiently large. Since the quantity $\omega^2\varepsilon(\omega)$ enters in the electromagnetic wave equation (Joannopoulos et al., 1995; 2008), only metal-containing photonic crystals can maintain the necessary dielectric contrast at small frequencies due to their Drude-like behavior $\varepsilon_{Met}(\omega) \sim -1/\omega^2$ (McGurn & Maradudin, 1993; Kuzmiak & Maradudin, 1997). However, the damping of electromagnetic waves in metals due to the skin effect (Abrikosov, 1988) can suppress many potentially useful properties of metallic photonic crystals.

4. Superconducting photonic crystals

4.1 Photonic band structure of superconducting photonic crystals

Photonic crystals consisting of superconducting elements embedded in a dielectric medium was studied in Ref. (Berman et al., 2006). The equation for the electric field in the ideal lattice of parallel cylinders embedded in medium has the form (Berman et al., 2009)

$$-\nabla^2 E_z(x,y) = \frac{\omega^2}{c^2}\left[\Lambda \pm (\varepsilon(\omega) - \epsilon) \sum_{\{\mathbf{n}^{(l)}\}} \eta(\mathbf{r} \in S)\right] E_z(x,y), \quad (2)$$

where ϵ is the dielectric constant of dielectric, $\varepsilon(\omega)$ is a dielectric function of the superconductor component. This equation describes the electric field in the ideal lattice of parallel superconducting cylinders (SCCs) in dielectric medium (DM) when within brackets on the right side is taken $\Lambda = \epsilon$ and sign "+" and the electric field in the ideal lattice of parallel DCs cylinders in a superconducting medium when $\Lambda = \varepsilon(\omega)$ and sign "-". In Eq. (2) $\eta(\mathbf{r} \in S)$ is the Heaviside step function. $\eta(\mathbf{r} \in S) = 1$ if $\mathbf{r}$ is inside of the cylinders S, and otherwise $\eta(\mathbf{r} \in S) = 0$, $\mathbf{n}^{(l)}$ is a vector of integers that gives the location of scatterer l at $\mathbf{a}(\mathbf{n}^{(l)}) \equiv \sum_{i=1}^{d} n_i^{(l)} \mathbf{a}_i$ ($\mathbf{a}_i$ are real space lattice vectors and d is the dimension of the lattice). The summation in Eq. (2) goes over all lattice nodes characterizing positions of cylinders. Eq. (2) describes the lattice of parallel cylinders as the two-component 2D photonic crystal. The first term within the bracket is associated to the medium, while the second one is related to the cylinders. Here and below the system described by Eq. (2) will be defined as an ideal photonic crystal. The ideal photonic crystal based on the 2D square lattice of the parallel superconducting cylinders was studied in Refs. (Berman et al., 2006; Lozovik et al., 2007).

Let us describe the dielectric constant for the system superconductor-dielectric. We describe the dielectric function of the superconductor within the Kazimir–Gorther model (Lozovik et al., 2007). In the framework of this model, it is assumed that far from the critical temperature point of the superconducting transition there are two independent carrier liquids inside a superconductor: superconducting with density $n_s(T,B)$ and normal one with density $n_n(T,B)$. The total density of electrons is given by $n_{tot} = n_n(T,B) + n_s(T,B)$. The density of the superfluid component $n_s(T,B)$ drops and the density of the normal component $n_n(T,B)$ grows when the temperature T or magnetic field B increases. The dielectric function in the Kazimir-Gorther model of superconductor is defined as

$$\varepsilon(\omega) = 1 - \frac{\omega_{ps}^2}{\omega^2} - \frac{\omega_{pn}^2}{\omega(\omega + i\gamma)}, \quad (3)$$

where ω is the frequency and γ represents the damping parameter in the normal conducting states. In Eq. (1) ω_{ps} and ω_{pn} are the plasma frequencies of superconducting and normal conducting electrons, respectively and defined as

$$\omega_{pn} = \left(\frac{4\pi n_n e^2}{m}\right)^{1/2}, \quad \omega_{ps} = \left(\frac{4\pi n_s e^2}{m}\right)^{1/2}, \quad (4)$$

where m and e are the mass and charge of an electron, respectively. The plasma frequency ω_{p0} is given by

$$\omega_{p0} = \left(\frac{4\pi n_{tot} e^2}{m}\right)^{1/2}. \quad (5)$$

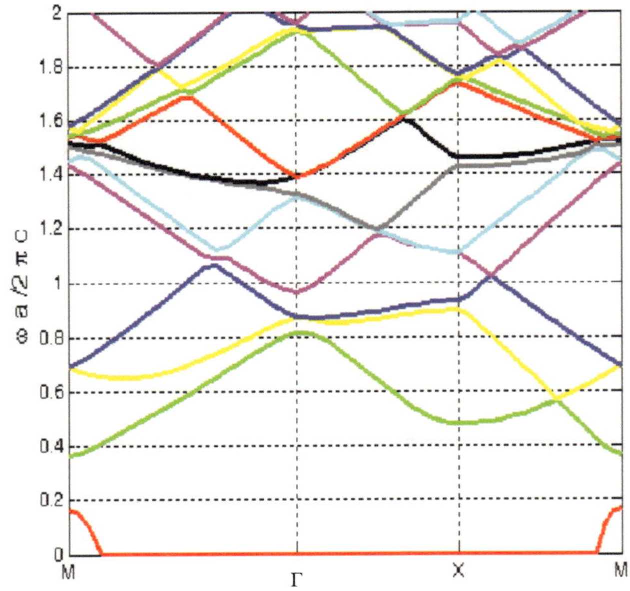

Fig. 4. Dispersion relation for a 2D photonic crystal consisting of a square lattice of parallel infinite superconducting cylinders with the filling factor $f = 0.3$. The ordinate plots frequencies in lattice units $2\pi c/a$. A band gap is clearly apparent in the frequency range $0.18 < \omega < 0.38$ in units of $2\pi c/a$.

From Eqs. (4) and (5) it is obvious that

$$\omega_{p0}^2 = \omega_{ps}^2 + \omega_{pn}^2 \,. \tag{6}$$

For the such superconductor, when the condition $\gamma \ll \omega_{p0}$ is valid, the damping parameter in the dielectric function can be neglected. Therefore, taking into account Eq. (6), Eq. (3) can be reduced to the following expression

$$\varepsilon(\omega) = 1 - \frac{\omega_{p0}^2}{\omega^2} \,. \tag{7}$$

Thus, Eq. (7) defines the dependence of the dielectric function of the superconductor on the frequency.

The photonic band structure for a 2D photonic crystal consisting of a square lattice of parallel infinite superconducting cylinders in vacuum was calculated using the eigenfrequencies of Eq. (2) obtained by the plane wave expansion method (Berman et al., 2006). The dielectric function of the superconductor $\varepsilon(\omega)$ in Eq. (2) was substituted from Eq. (7). The photonic band structure of the photonic crystal build up from the $YBa_2Cu_3O_{7-\delta}$ (YBCO) superconducting cylinders embedded in vacuum is presented in Fig. 4. Since for the YBCO the plasma frequency $\omega_{p0} = 1.67 \times 10^{15}$rad/s and the damping parameter $\gamma = 1.84 \times 10^{13}$rad/s, then the condition $\gamma \ll \omega_{p0}$ is valid for the YBCO superconductor, and Eq. (7) can used for the dielectric function of the YBCO superconductor.

The advantage of a photonic crystal with superconducting constituents is that the dissipation of the incident electromagnetic wave due to the imaginary part of the dielectric function is much less than for normal metallic constituents at frequencies smaller than the superconducting gap. Thus, in this frequency regime, for a photonic crystal consisting of several layers of scatterers the dissipation of the incident electromagnetic wave by an array of superconducting constituents is expected to be less than that associated with an analogous array composed of normal metallic constituents.

4.2 Monochromatic infrared wave propagation in 2D superconductor-dielectric photonic crystal

The dielectric function in the ideal photonic crystal is a spatially periodic. This periodicity can be achieved by the symmetry of a periodic array of constituent elements with one kind of the dielectric constant embedded in a background medium characterized by the other kind of the dielectric constant. The localized photonic mode can be achieved in the photonic crystals whose symmetry is broken by a defect (Yablonovitch et al., 1991; Meade et al., 1991; McCall et al., 1991; Meade et al., 1993). There are at least two ways to break up this symmetry: (i) to remove one constituent element from the node of the photonic crystal ("vacancy"); (ii) to insert one extra constituent element placed out of the node of photonic crystal ("interstitial"). We consider two types of 2D photonic crystals: the periodical system of parallel SCC in dielectric medium and the periodical system of parallel dielectric cylinders (DC) in superconducting medium. The symmetry of the SCCs in DM can be broken by two ways: (i) to remove one SCC out of the node, and (ii) to insert one extra SCC in DM out of the node. We will show below that only the first way of breaking symmetry results in the localized photonic state with the frequency inside the band gap for the SCCs in the DM. The second way does not result in the localized photonic state inside the band gap. The symmetry of DCs in SCM can be broken also by two ways: (i) to remove one DC out of the node, and (ii) to insert one extra DC in SCM out of the node. We will show below that only the second way of breaking symmetry results in the localized photonic state inside the band gap for DCs in SCM. The first way does not result in the localized photonic state.
Let us consider now the real photonic crystal when one SCC of the radius ξ removed from the node of the square lattice located at the position $\mathbf{r}_0$ or one extra DC of the same radius is placed out of the node of the square lattice located at the position $\mathbf{r}_0$ presented in Fig. 5. The free space in the lattice corresponding to the removed SCC in DM or to placed the extra DC in SCM contributes to the dielectric contrast by the adding the term $-(\varepsilon(\omega)-\epsilon)\eta(\xi-|\mathbf{r}-\mathbf{r}_0|)E_z(x,y)$ to the right-hand side in Eq. (2):

$$-\nabla^2 E_z(x,y)=\frac{\omega^2}{c^2}\left[\Lambda\pm(\varepsilon(\omega)-\epsilon)\sum_{\{\mathbf{n}^{(l)}\}}\eta(\mathbf{r}\in S)-(\varepsilon(\omega)-\epsilon)\eta(\xi-|\mathbf{r}-\mathbf{r}_0|)\right]E_z(x,y)\,, \quad (8)$$

Eq. (8) describes the photonic crystal implying a removed SCC from the node of the lattice ($\Lambda=\epsilon$ and sign "+") shown in Fig. 5a or implying the extra DC inserted and placed out of the node of the lattice ($\Lambda=\varepsilon(\omega)$ and sign "-") shown in Fig. 5b, which are defined as a real photonic crystal.
Substituting Eq. (7) into Eq. (8), we obtain for the wave equation for the real photonic crystal:

$$-\nabla^2 E_z(x,y)-\frac{\omega^2}{c^2}\left[\pm(1-\epsilon)\mp(\frac{\omega_{p0}^2}{\omega^2})\left(\sum_{\{\mathbf{n}^{(l)}\}}\eta(\mathbf{r}\in S)-\eta(\xi-|\mathbf{r}-\mathbf{r}_0|)\right)\right]E_z(x,y)=\Lambda\frac{\omega^2}{c^2}E_z(x,y)\,. \quad (9)$$

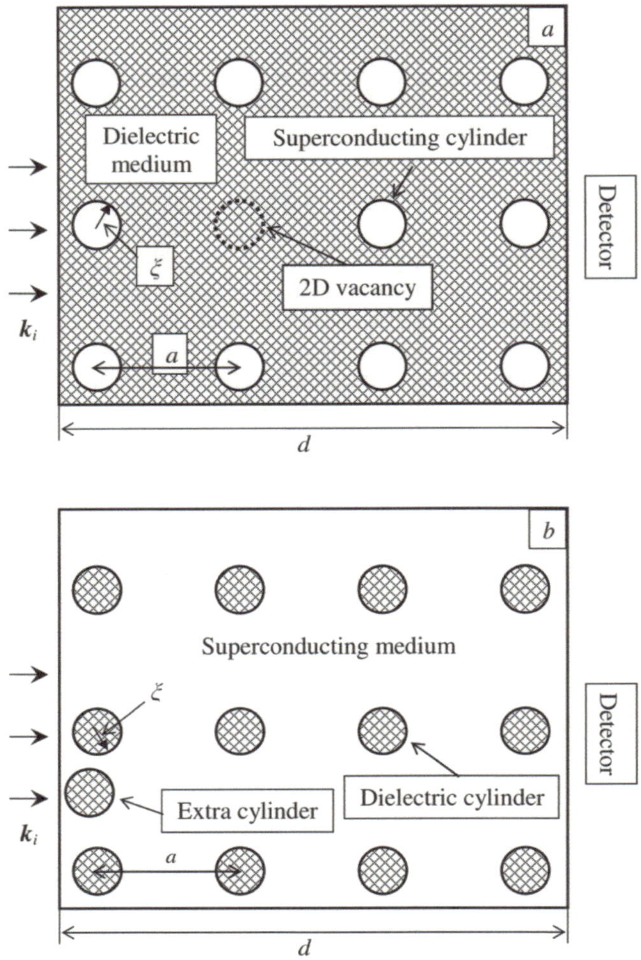

Fig. 5. Anomalous far infrared monochromatic transmission through a a lattice of (a) parallel SCCs embedded in DM and (b) parallel DCs embedded in SCM. a is the equilateral lattice spacing. ξ is the the radius of the cylinder. d denotes the length of the film. The dashed cylinder is removed out of the node of the lattice.

Follow Ref. (Berman et al., 2008) Eq. (9) can be mapped onto the Schrödinger equation for an "electron" with the effective electron mass in the periodic potential in the presence of the potential of the "impurity"Therefore, the eigenvalue problem formulated by Eq. (9) can be solved in two steps: i. we recall the procedure of the solution the eigenvalue problem for the calculation of the photonic-band spectrum of the ideal superconducting photonic crystal (for SCCs in DM see Ref. (Berman et al., 2006; Lozovik et al., 2007)); ii. we apply the Kohn-Luttinger two-band model (Luttinger & Kohn, 1955; Kohn, 1957; Keldysh, 1964; Takhtamirov & Volkov, 1999) to calculate the eigenfrequency spectrum of the real photonic crystal with the symmetry broken by defect.

Eq. (9) when the dielectric is vacuum ($\epsilon = 1$) has the following form

$$- \nabla^2 E_z(x,y) + \left[\pm \frac{\omega_{p0}^2}{c^2} \sum_{\{\mathbf{n}^{(l)}\}} \eta(\mathbf{r} \in S) - \frac{\omega_{p0}^2}{c^2} \eta(\xi - |\mathbf{r} - \mathbf{r}_0|) \right] E_z(x,y) = \Lambda \frac{\omega^2}{c^2} E_z(x,y) \, . \quad (10)$$

In the framework of the Kohn-Luttinger two-band model the eigenvalue and localized eigenfunction of Eq. (10) are presented in Ref. (Berman et al., 2008). Since Eq. (10) is analogous to the wave equation for the electric field in Ref. (Berman et al., 2008), we can follow and apply the procedure of the solution of the eigenvalue problem formulated in Ref. (Berman et al., 2008) to the eigenvalue problem presented by Eq. (10). This procedure is reduced to the 2D Schrödinger-type equation for the particle in the negative potential given by step-function representing a 2D well. This equation has the localized eigenfunction representing the localized electric field as discussed in Ref. (Berman et al., 2008). Let us emphasize that this localized solution for the SCCs in DM can be obtained only if one cylinder is removed. If we insert the extra cylinder in 2D lattice of the SCCs, the corresponding wave equation will be represented by adding the term $(\varepsilon(\omega) - \epsilon)\eta(\xi - |\mathbf{r} - \mathbf{r}_0|)E_z(x,y)$ to the right-hand side of Eq. (2). This term results in the positive potential in the effective 2D Schrödinger-type equation. This positive potential does not lead to the localized eigenfunction. Therefore, inserting extra cylinder in the 2D ideal photonic crystal of the SCCs in DM does not result in the localized photonic mode causing the anomalous transmission.

In terms of the initial quantities of the superconducting photonic crystal the eigenfrequency ω of the localized photonic state is given by (Berman et al., 2008)

$$\omega = \left(\Delta^4 - A \right)^{1/4} , \quad (11)$$

where Δ is photonic band gap of the ideal superconducting photonic crystal calculated in Refs. (Berman et al., 2006; Lozovik et al., 2007) and A is given by

$$A = \frac{16 c^4 k_0^2}{3 \xi^2} \exp \left[- \frac{8 k_0^2 c^4}{3 \Delta^2 \xi^2 \omega_{p0}^2} \right] . \quad (12)$$

In Eq. (12) $k_0 = 2\pi/a$, where a is the period of the 2D square lattice of the SCCs. In the square lattice the radius of the cylinder ξ and the period of lattice a are related as $\xi = \sqrt{f/\pi}a$, where f is the filling factor of the superconductor defined as the ratio of the cross-sectional area of all superconducting cylinders in the plane perpendicular to the cylinder axis $S_{supercond}$ and the total area occupied by the real space unit cell S: $f \equiv S_{supercond}/S = \pi\xi^2/a^2$.

According to Refs. (Berman et al., 2006; Lozovik et al., 2007), the photonic band gap Δ for the ideal superconducting square photonic crystal at different temperatures and filling factors ($f = 0.3$ and $T = 0$ K, $f = 0.05$ and $T = 85$ K, $f = 0.05$ and $T = 10$ K) is given by $0.6c/a$ in Hz. The period of lattice in these calculations is given by $a = 150$ μm.

According to Eqs. (11) and (12), the eigenfrequency ν corresponding to the localized photonic mode for the parameters listed above for the YBCO is calculated as $\nu = \omega/(2\pi) = 117.3$ THz. The corresponding wavelength is given by $\lambda = 2.56 \times 10^{-6}$ m. Let us emphasize that the frequency corresponding to the localized mode does not depend on temperature and magnetic field at $\gamma \ll \omega_{p0}$, since in this limit we neglect damping parameter, and, according to Eq. (7), the dielectric constant depends only on ω_{p0}, which is determined by the total electron density n_{tot}, and n_{tot} does not depend on the temperature and magnetic field. Therefore, for given parameters of the system at $f = 0.05$ the anomalous transmission appears at frequency $\nu = 117.3$ THz inside the forbidden photonic gap $0 \leq \Delta/(2\pi) \leq 0.6c/(2\pi a) = 119.9$ THz.

Let us emphasize that this localized solution for the DCs embedded in the SCM can be obtained only if one extra cylinder is inserted. If we remove the extra cylinder from 2D lattice of the DCs in the SCM, the corresponding wave equation will be represented by adding the term $-(\epsilon - \varepsilon(\omega))\eta(\xi - |\mathbf{r} - \mathbf{r}_0|)E_z(x,y)$ to the right-hand side of Eq. (2). This term results in the positive potential in the effective 2D Schrödinger-type equation. This positive potential does not lead to the localized eigenfunction. Therefore, removed DC in the 2D ideal photonic crystal in the lattice of DCs embedded in the SCM does not result in the localized photonic mode causing the anomalous transmission.

Let us mention that the localization of the photonic mode at the frequency given by Eqs. (11) causes the anomalous infrared transmission inside the forbidden photonic band of the ideal photonic crystal. This allows us to use the SCCs in the DM and DC in the SCM with the symmetry broken by a defect as the infrared monochromatic filter.

Based on the results of our calculations we can conclude that it is possible to obtain a different type of a infrared monochromatic filter constructed as real photonic crystal formed by the SCCs embedded in DM and DCs embedded in SCM. The symmetry in these two systems can be broken by the defect of these photonic crystal, constructed by removing one cylinder out of the node of the ideal photonic crystal lattice and by inserting the extra cylinder, respectively, in other words, by making the "2D vacancy" and "2D interstitial" in the ideal photonic crystal lattice, respectively. Finally, we can conclude that the symmetry breaking resulting in the breakup of spacial periodicity of the dielectric function by removal of the SCC from periodic structure of SCCs embedded in DM, or inserting extra DC in SCM results in transmitted infrared frequency in the forbidden photonic band.

4.3 Far infrared monochromatic transmission through a film of type-II superconductor in magnetic field

Let us consider a system of Abrikosov vortices in a type-II superconductor that are arranged in a triangular lattice. We treat Abrikosov vortices in a superconductor as the parallel cylinders of the normal metal phase in the superconducting medium. The axes of the vortices, which are directed along the $\hat{z}$ axis, are perpendicular to the surface of the superconductor. We assume the $\hat{x}$ and $\hat{y}$ axes to be parallel to the two real-space lattice vectors that characterize the 2D triangular lattice of Abrikosov vortices in the film and the angle between $\hat{x}$ and $\hat{y}$ is equal $\pi/3$. The nodes of the 2D triangular lattice of Abrikosov vortices are assumed to be situated on the $\hat{x}$ and $\hat{y}$ axes.

For simplicity, we consider the superconductor in the London approximation (Abrikosov, 1988) i.e. assuming that the London penetration depth λ of the bulk superconductor is much greater than the coherence length ξ: $\lambda \gg \xi$. Here the London penetration depth is $\lambda = [m_e c^2/(4\pi n_e e^2)]^{1/2}$, where n_e is electron density. The coherence length is defined as $\xi = c/(\omega_{p0}\sqrt{\epsilon})$, where $\omega_{p0} = 2\pi c\omega_0$ is the plasma frequency. A schematic diagram of Abrikosov lattices in type-II superconductors is shown in Fig. 6 As it is seen from Fig. 6 the Abrikosov vortices of radius ξ arrange themselves into a 2D triangular lattice with lattice spacing $a(B,T) = 2\xi(T)\left[\pi B_{c2}/(\sqrt{3}B)\right]^{1/2}$ (Takeda et al., 2004) at the fixed magnetic field B and temperature T. Here B_{c2} is the critical magnetic field for the superconductor. We assume the wavevector of the incident electromagnetic wave vector $\mathbf{k}_i$ to be perpendicular to the direction of the Abrikosov vortices and the transmitted wave can be detected by using the detector D.

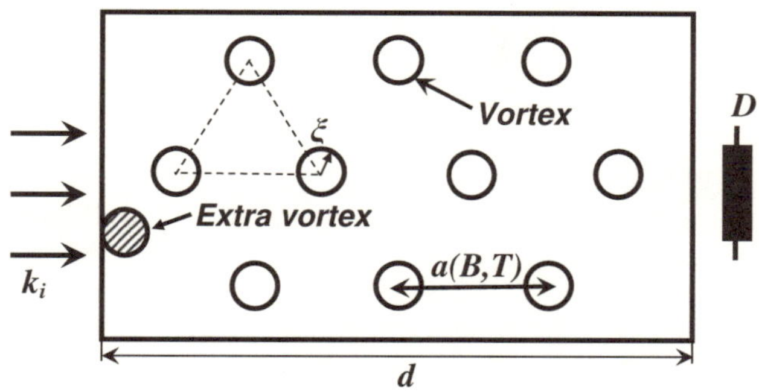

Fig. 6. Anomalous far infrared monochromatic transmission through a film of type-II superconductor in the magnetic field parallel to the vortices. $a(B,T)$ is the equilateral triangular Abrikosov lattice spacing. ξ is the coherence length and the radius of the vortex. d denotes the length of the film. The shaded extra vortex placed near the boundary of the film and situated outside of the node of the lattice denotes the defect of the Abrikosov lattice.

Now let us follow the procedure used in Ref. (Berman et al., 2006; 2008) to obtain the wave equation for Abrikosov lattice treated as a two-component photonic crystal. In Ref. (Berman et al., 2006), a system consisting of superconducting cylinders in vacuum is studied. By contrast, the system under study in the present manuscript consists of the cylindrical vortices in a superconductor, which is a complementary case (inverse structure) to what was treated in Ref. (Berman et al., 2006). For this system of the cylindrical vortices in the superconductor, we write the wave equation for the electric field $\mathbf{E}(x,y,t)$ parallel to the vortices in the form of 2D partial differential equation. The corresponding wave equation for the electric field is

$$-\nabla^2\mathbf{E} = -\frac{1}{c^2}\epsilon \sum_{\{\mathbf{n}^{(l)}\}} \eta(\mathbf{r})\frac{\partial^2\mathbf{E}}{\partial t^2} - \frac{4\pi}{c^2}\frac{\partial\mathbf{J}(\mathbf{r})}{\partial t}, \tag{13}$$

where ϵ is a dielectric constant of the normal metal component inside the vortices, $\eta(\mathbf{r})$ is the Heaviside step function which is $\eta(\mathbf{r}) = 1$ inside of the vortices and otherwise $\eta(\mathbf{r}) = 0$. In Eq. (13) $\mathbf{n}^{(l)}$ is a vector of integers that gives the location of a scatterer l at $\mathbf{a}(\mathbf{n}^{(l)}) \equiv \sum_{i=1}^{d} n_i^{(l)}\mathbf{a}_i$ ($\mathbf{a}_i$ are real space lattice vectors situated in the nodes of the 2D triangular lattice and d is the dimension of Abrikosov lattice).

At $(T_c - T)/T_c \ll 1$ and $\hbar\omega \ll \Delta \ll T_c$, where T_c is the critical temperature and Δ is the superconducting gap, a simple relation for the current density holds (Abrikosov, 1988):

$$\mathbf{J}(\mathbf{r}) = \left[-\frac{c}{4\pi\delta_L^2} + \frac{i\omega\sigma}{c}\right]\mathbf{A}(\mathbf{r}). \tag{14}$$

In Eq. (14) σ is the conductivity of the normal metal component.

The important property determining the band structure of the photonic crystal is the dielectric constant. The dielectric constant, which depends on the frequency, inside and outside of the vortex is considered in the framework of the two-fluid model. For a normal metal phase inside of the vortex it is $\epsilon_{in}(\omega)$ and for a superconducting phase outside of the vortex it is $\epsilon_{out}(\omega)$ and can be described via a simple Drude model. Following Ref. (Takeda et al., 2004) the dielectric constant can be written in the form:

$$\epsilon_{in}(\omega) = \epsilon, \qquad \epsilon_{out}(\omega) = \epsilon\left(1 - \frac{\omega_{p0}^2}{\omega^2}\right). \tag{15}$$

Eqs. (15) are obtained in Ref. (Takeda et al., 2004) from a phenomenological two-component fluid model by applying the following condition: $\omega_{pn} \ll \omega \ll \gamma$. Here ω_{pn} is the plasma frequency of normal conducting electrons, and γ is the damping term in the normal conducting states.
Let's neglect a damping in the superconductor. After that substituting Eq. (14) into Eq. (13), considering Eqs. (15) for the dielectric constant, and seeking a solution in the form with harmonic time variation of the electric field, i.e., $\mathbf{E}(\mathbf{r},t) = \mathbf{E}_0(\mathbf{r})e^{i\omega t}$, $\mathbf{E} = i\omega\mathbf{A}/c$, we finally obtain the following equation

$$-\nabla^2 E_z(x,y) = \frac{\omega^2\epsilon}{c^2}\left[1 - \frac{\omega_{p0}^2}{\omega^2} + \frac{\omega_{p0}^2}{\omega^2}\sum_{\{\mathbf{n}^{(l)}\}}\eta(\mathbf{r})\right]E_z(x,y)\,. \tag{16}$$

The summation in Eq. (16) goes over all lattice nodes characterizing positions of the Abrikosov vortices. Eq. (16) describes Abrikosov lattice as the two-component 2D photonic crystal. The first two terms within the bracket are associated to the superconducting medium, while the last term is related to vortices (normal metal phase). Here and below the system described by Eq. (16) will be defined as an ideal photonic crystal. The ideal photonic crystal based on the Abrikosov lattice in type-II superconductor was studied in Refs. (Takeda et al., 2004). The wave equation (16) describing the Abrikosov lattice has been solved in Ref. (Takeda et al., 2004) where the photonic band frequency spectrum $\omega = \omega(\mathbf{k})$ of the ideal photonic crystal of the vortices has been calculated.
Let us consider an extra Abrikosov vortex pinned by some defect in the type-II superconducting material, as shown in Fig. 6. This extra vortex contributes to the dielectric contrast by the adding the term $\epsilon\omega_{p0}^2/c^2\eta(\xi - |\mathbf{r}-\mathbf{r}_0|)E_z(x,y)$, where $\mathbf{r}_0$ points out the position of the extra vortex, to the r.h.s. in Eq. (16):

$$-\nabla^2 E_z(x,y) = \frac{\omega^2\epsilon}{c^2}\left[1 - \frac{\omega_{p0}^2}{\omega^2} + \frac{\omega_{p0}^2}{\omega^2}\sum_{\{\mathbf{n}^{(l)}\}}\eta(\mathbf{r}) + \frac{\omega_{p0}^2}{\omega^2}\eta(\xi - |\mathbf{r}-\mathbf{r}_0|)\right]E_z(x,y), \tag{17}$$

Eq. (17) describes the type-II superconducting medium with the extra Abrikosov vortex pinned by a defect in the superconductor. We define the photonic crystal implying an extra Abrikosov vortex pinned by a defect as a real photonic crystal and it is described by Eq. (17). The eigenfrequency ω of the localized photonic state due to the extra Abrikosov vortex pinned by a defect is obtained from Eq. (17) as (Berman et al., 2008)

$$\omega(x) = \left(\omega_{up}^4(x) - A(x)\right)^{1/4}, \tag{18}$$

where $x = B/B_{c2}$ and function $A(x)$ is given by

$$A(x) = \frac{16c^4k_0^2(x)}{3\epsilon^2\xi^2}\exp\left[-\frac{8k_0^2(x)c^4}{3\epsilon^2\omega_{up}^2(x)\xi^2\omega_{p0}^2}\right] \tag{19}$$

and $k_0(x)$ is defined below through the electric field of the lower and higher photonic bands of the ideal Abrikosov lattice.
The electric field $E_z(x,y)$ corresponding to this localized photonic mode can be obtained from Eq. (17) as (Berman et al., 2008)

$$E_z^{(00)}(\mathbf{r}) = \begin{cases} \tilde{C}_1, & |\mathbf{r}-\mathbf{r}_0| < \xi\,, \\ \tilde{C}_2 B(x), & |\mathbf{r}-\mathbf{r}_0| > \xi\,, \end{cases} \tag{20}$$

where the constants $\tilde{C}_1$ and $\tilde{C}_2$ can be obtained from the condition of the continuity of the function $E_z^{(00)}(\mathbf{r})$ and its derivative at the point $|\mathbf{r}-\mathbf{r}_0| = \xi$ and

$$B(x) = \log\left[\xi\left(|\mathbf{r}-\mathbf{r}_0|\times \exp\left[-4k_0^2(x)c^4/\left(3\epsilon^2\left(\tilde{\Delta}^2(x)-\omega_{p0}^2\right)\xi^2\omega_{p0}^2\right)\right]\right)^{-1}\right]. \tag{21}$$

The function $k_0(x)$ is given as

$$k_0\delta_{\alpha\beta} = -i\left(\left[\int E_{zc0}^*(\mathbf{r})\nabla E_{zv0}(\mathbf{r})d^2r\right]_{cv\alpha}\right)_\beta, \tag{22}$$

where $E_{zc0}(\mathbf{r})$ and $E_{zv0}(\mathbf{r})$ are defined by the electric field of the up and down photonic bands of the ideal Abrikosov lattice. The exact value of k_0 can be calculated by substituting the electric field $E_{zc0}(\mathbf{r})$ and $E_{zv0}(\mathbf{r})$ from Ref. (Takeda et al., 2004). Applying the weak coupling model (Abrikosov, 1988) corresponding to the weak dielectric contrast between the vortices and the superconductive media $\omega_{p0}^2/\omega^2\left|\sum_{\{\mathbf{n}^{(l)}\}}\eta(\mathbf{r})-1\right| \ll 1$ we use the approximate estimation of k_0 in our calculations as $k_0(x) \approx 2\pi/a(x) = \pi\xi^{-1}\sqrt{\sqrt{3}x/\pi}$.
We consider the Abrikosov lattice formed in the YBCO and study the dependence of the photonic band structure on the magnetic field. For the YBCO the characteristic critical magnetic field $B_{c2} = 5$ T at temperature $T = 85$ K is determined experimentally in Ref. (Safar et al., 1994). So we obtained the frequency corresponding to the localized wave for the YBCO in the magnetic field range from $B = 0.72B_{c2} = 3.6$ T up to $B = 0.85B_{c2} = 4.25$ T at $T = 85$ K. Following Ref. (Takeda et al., 2004), in our calculations we use the estimation $\epsilon = 10$ inside the vortices and for the YBCO $\omega_0/c = 77$ cm^{-1}. The dielectric contrast between the normal phase in the core of the Abrikosov vortex and the superconducting phase given by Eq. (15) is valid only for the frequencies below ω_{c1}: $\omega < \omega_{c1}$, where $\omega_{c1} = 2\Delta_S/(2\pi\hbar)$, $\Delta_S = 1.76k_BT_c$ is the superconducting gap, k_B is the Boltzmann constant, and T_c is the critical temperature. For the YBCO we have $T_c = 90$ K, and $\omega < \omega_{c1} = 6.601$ THz. It can be seen from Eqs. (18) and (20), that there is a photonic state localized on the extra Abrikosov vortex, since the discrete eigenfrequency corresponds to the electric field decreasing as logarithm of the distance from an extra vortex. This logarithmical behavior of the electric field follows from the fact that it comes from the solution of 2D Dirac equation. The calculations of the eigenfrequency ω dependence on the ratio B/B_{c2}, where B_{c2} is the critical magnetic field, is presented in Fig. 7. According to Fig. 7, our expectation that the the eigenfrequency level ω corresponding to the extra vortex is situated inside the photonic band gap is true. We calculated the frequency corresponding to the localized mode, which satisfies to the condition of the validity of the dielectric contrast given by Eq. (15). According to Eqs. (20) and (21), the localized field is decreasing proportionally to $\log\xi|\mathbf{r}-\mathbf{r}_0|^{-1}$ as the distance from an extra vortex increases. Therefore, in order to detect this localized mode, the length of the film d

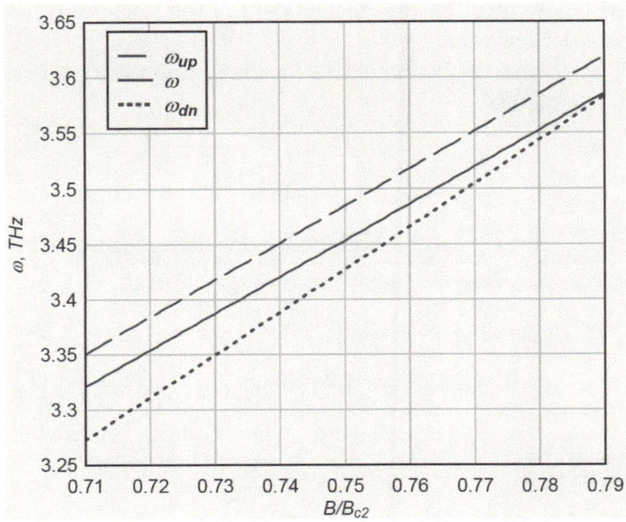

Fig. 7. The dependence of the photonic band structure of the real Abrikosov lattice on B/B_{c2}. Solid line represents the eigenfrequency ω corresponding to the localized mode near the extra vortex in the real Abrikosov lattice given by Eq. (18). The dashed and dotted lines represent, respectively, the top ω_{up} and bottom ω_{dn} boundaries of the photonic band gap of the ideal Abrikosov lattice according to Ref. (Takeda et al., 2004).

should not exceed approximately $10a(B/B_{c2})$, which corresponds to $d \lesssim 200$ μm for the range of magnetic fields for the YBCO presented by Fig. 7. For these magnetic fields $a \approx 20$ μm. Since the frequency corresponding to the localized mode is situated inside the photonic band gap, the extra vortex should be placed near the surface of the film as shown in Fig. 6. Otherwise, the electromagnetic wave cannot reach this extra vortex. Besides, we assume that this localized photonic state is situated outside the one-dimensional band of the surface states of two-dimensional photonic crystal. It should be mentioned, that in a case of several extra vortices separated at the distance greater than the size of one vortex (it is the coherence length estimated for the YBCO by $\xi \approx 6.5$ μm) the localized mode frequency for the both vortices is also going to be determined by Eq. (18). The intensity of the localized mode in the latter case is going to be enhanced due to the superposition of the modes localized by the different vortices. In the case of far separated extra vortices we have neglected by the vortex-vortex interaction. Thus, the existence of other pinned by crystal defects vortices increases the intensity of the transmitted mode and improves the possibility of this signal detection. Note that at the frequencies ω inside the photonic band gap $\omega_{dn} < \omega < \omega_{up}$ the transmittance and reflectance of electromagnetic waves would be close to zero and one, correspondingly, everywhere except the resonant frequency ω related to an extra vortex. The calculation of the transmittance and reflectance of electromagnetic waves at this resonant frequency ω is a very interesting problem, which will be analyzed elsewhere.

We considered a type-II superconducting medium with an extra Abricosov vortex pinned by a defect in a superconductor. The discrete photonic eigenfrequency corresponding to the localized photonic mode, is calculated as a function of the ratio B/B_{c2}, which parametrically depends on temperature. This photonic frequency increases as the ratio B/B_{c2} and temperature T increase. Moreover, since the localized field and the corresponding photonic eigenfrequency depend on the distance between the nearest Abrikosov vortices $a(B,T)$, the resonant properties of the system can be tuned by control of the external magnetic field B and temperature T. Based on the results of our calculations we can conclude that it is possible to obtain a new type of a tunable far infrared monochromatic filter consisting of extra vortices placed out of the nodes of the ideal Abrikosov lattice, which can be considered as real photonic crystals. These extra vortices are pinned by a crystal defects in a type-II superconductor in strong magnetic field. As a result of change of an external magnetic field B and temperature T the resonant transmitted frequencies can be controlled.

5. Graphene-based photonic crystal

A novel type of 2D electron system was experimentally observed in graphene, which is a 2D honeycomb lattice of the carbon atoms that form the basic planar structure in graphite (Novoselov et al., 2004; Luk'yanchuk & Kopelevich, 2004; Zhang et al., 2005). Due to unusual properties of the band structure, electronic properties of graphene became the object of many recent experimental and theoretical studies (Novoselov et al., 2004; Luk'yanchuk & Kopelevich, 2004; Zhang et al., 2005; Novoselov et al., 2005; Zhang et al., 2005; Kechezhdi et al., 2008; Katsnelson, 2008; Castro Neto et al., 2009). Graphene is a gapless semiconductor with massless electrons and holes which have been described as Dirac-fermions (Novoselov et al., 2004; Luk'yanchuk & Kopelevich, 2004; Das Sarma et al., 2007). The unique electronic properties of graphene in a magnetic field have been studied recently (Nomura & MacDonald, 2006; Tőke et al., 2006; Gusynin & Sharapov, 2005;?). It was shown that in infrared and at larger wavelengths transparency of graphene is defined by the fine structure constant (Nair et al., 2008). Thus, graphene has unique optical properties. The space-time dispersion of graphene conductivity was analyzed in Ref. (Falkovsky & Varlamov, 2007) and the optical properties of graphene were studied in Refs. (Falkovsky & Pershoguba, 2007; Falkovsky, 2008).

In this Section, we consider a 2D photonic crystal formed by stacks of periodically placed graphene discs embedded into the dielectric film proposed in Ref. (Berman et al., 2010). The stack is formed by graphene discs placed one on top of another separated by the dielectric placed between them as shown in Fig. 8. We calculate the photonic band structure and transmittance of this graphene-based photonic crystal. We will show that the graphene-based photonic crystals can be applied for the devices for the far infrared region of spectrum.

Let us consider polarized electromagnetic waves with the electric field $\mathbf{E}$ parallel to the graphene discs. The wave equation for the electric field in a dielectric media has the form (Landau & Lifshitz, 1984)

$$-\triangle \mathbf{E}(\mathbf{r},t) + \nabla(\nabla \cdot \mathbf{E}(\mathbf{r},t)) - \frac{\varepsilon(\mathbf{r})}{c^2}\frac{\partial^2 \mathbf{E}(\mathbf{r},t)}{\partial t^2} = 0\,, \tag{23}$$

where $\varepsilon(\mathbf{r},t)$ is the dielectric constant of the media.

In photonic crystals, dielectric susceptibility is a periodical function and it can be expanded in the Fourier series:

$$\varepsilon(\mathbf{r}) = \sum_{\mathbf{G}} \varepsilon(\mathbf{G}) e^{i\mathbf{G}\mathbf{r}}\,, \tag{24}$$

where $\mathbf{G}$ is the reciprocal photonic lattice vector.

Expanding the electric field on the Bloch waves inside a photonic crystal, and seeking solutions with harmonic time variation of the electric field, i.e., $\mathbf{E}(\mathbf{r},t) = \mathbf{E}(\mathbf{r})e^{i\omega t}$, one obtains

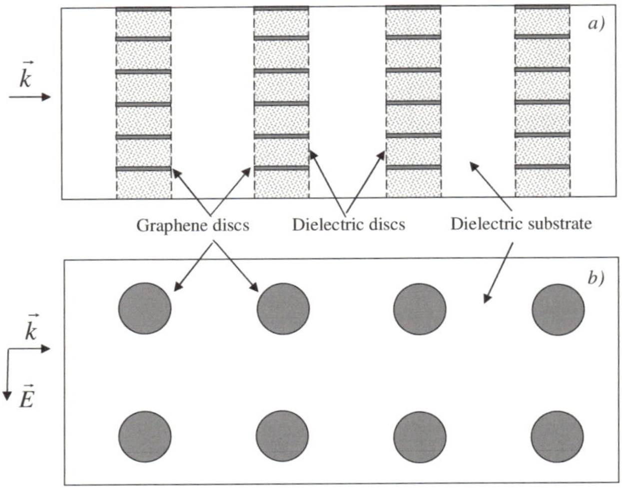

Fig. 8. Graphene-based photonic crystal: a) the side view. The material of the dielectric between graphene discs can be the same as the material of the dielectric substrate; b) the top view.

from Eq. (23) using Eq. (24) the system of equations for Fourier components of the electric field (Joannopoulos et al., 2008; McGurn & Maradudin, 1993):

$$(\mathbf{k}+\mathbf{G})^2 E_{\mathbf{k}}(\mathbf{G}) = \frac{\omega^2(\mathbf{k})}{c^2} \sum_{\mathbf{G}'} \varepsilon(\mathbf{G}-\mathbf{G}') E_{\mathbf{k}}(\mathbf{G}') \,, \tag{25}$$

which presents the eigenvalue problem for finding photon dispersion curves $\omega(\mathbf{k})$. In Eq. (25) the coefficients of the Fourier expansion for the dielectric constant are given by

$$\varepsilon(\mathbf{G}-\mathbf{G}') = \varepsilon_0 \delta_{\mathbf{G}\mathbf{G}'} + (\varepsilon_1 - \varepsilon_0) M_{\mathbf{G}\mathbf{G}'} \,. \tag{26}$$

In Eq. (26) ε_0 is the dielectric constant of the dielectric, ε_1 is the dielectric constant of graphene multilayers separated by the dielectric material, and $M_{\mathbf{G}\mathbf{G}'}$ for the geometry considered above is

$$\begin{aligned} M_{\mathbf{G}\mathbf{G}'} &= 2f \frac{J_1(|\mathbf{G}-\mathbf{G}'|r)}{(|\mathbf{G}-\mathbf{G}'|r)} \,, \quad \mathbf{G} \neq \mathbf{G}' \,, \\ M_{\mathbf{G}\mathbf{G}'} &= f \,, \quad \mathbf{G} = \mathbf{G}' \,, \end{aligned} \tag{27}$$

where J_1 is the Bessel function of the first order, and $f = S_g/S$ is the filling factor of 2D photonic crystal.

In our consideration the size of the graphene discs was assumed to be much larger than the period of the graphene lattice, and we applied the expressions for the dielectric constant of the infinite graphene layer for the graphene discs, neglecting the effects related to their finite size.

The dielectric constant $\varepsilon_1(\omega)$ of graphene multilayers system separated by the dielectric layers with the dielectric constant ε_0 and the thickness d is given by (Falkovsky & Pershoguba, 2007;

Falkovsky, 2008)

$$\varepsilon_1(\omega) = \varepsilon_0 + \frac{4\pi i \sigma_g(\omega)}{\omega d}, \tag{28}$$

where $\sigma_g(\omega)$ is the dynamical conductivity of the doped graphene for the high frequencies ($\omega \gg kv_F$, $\omega \gg \tau^{-1}$) at temperature T given by (Falkovsky & Pershoguba, 2007; Falkovsky, 2008)

$$\begin{aligned}\sigma_g(\omega) = &\frac{e^2}{4\hbar}\left[\eta(\hbar\omega - 2\mu)\right.\\ &+\frac{i}{2\pi}\left(\frac{16k_BT}{\hbar\omega}\log\left[2\cosh\left(\frac{\mu}{2k_BT}\right)\right]\right.\\ &\left.\left.-\log\frac{(\hbar\omega+2\mu)^2}{(\hbar\omega-2\mu)^2+(2k_BT)^2}\right)\right].\end{aligned} \tag{29}$$

Here τ^{-1} is the electron collision rate, k is the wavevector, $v_F = 10^8$ cm/s is the Fermi velocity of electrons in graphene (Falkovsky, 2008), and μ is the the chemical potential determined by the electron concentration $n_0 = (\mu/(\hbar v_F))^2/\pi$, which is controlled by the doping. The chemical potential can be calculated as $\mu = (\pi n_0)^{1/2}\hbar v_F$. In the calculations below we assume $n_0 = 10^{11}$ cm^{-2}. For simplicity, we assume that the dielectric material is the same for the dielectric discs between the graphene disks and between the stacks. As the dielectric material we consider SiO_2 with the dielectric constant $\varepsilon_0 = 4.5$.

To illustrate the effect let us, for example, consider the 2D square lattice formed by the graphene based metamaterial embedded in the dielectric. The photonic band structure for the graphene based 2D photonic crystal with the array of cylinders arranged in a square lattice with the filling factor $f = 0.3927$ is presented in Fig. 9. The cylinders consist of the metamaterial stacks of alternating graphene and dielectric discs. The period of photonic crystal is $a = 25$ μm, the diameter of discs is $D = 12.5$ μm, the width of the dielectric layers $d = 10^{-3}$ μm . Thus the lattice frequency is $\omega_a = 2\pi c/a = 7.54 \times 10^{13}$ rad/s. The results of the plane wave calculation for the graphene based photonic crystal are shown in Fig. 9, and the transmittance spectrum obtained using the Finite-Difference Time-Domain (FDTD) method (Taflove, 1995) is presented in Fig. 10. Let us mention that plane wave computation has been made for extended photonic crystal, and FDTD calculation of the transmittance have been performed for five graphene layers. A band gap is clearly apparent in the frequency range $0 < \omega < 0.6$ and $0.75 < \omega < 0.95$ in units of $2\pi c/a$. The first gap is originated from the electronic structure of the doped graphene, which prevents absorbtion at $\hbar\omega < 2\mu$ (see also Eq. (29)). The photonic crystal structure manifests itself in the dependence of the lower photonic band on the wave vector k. In contrast, the second gap $0.75 < \omega < 0.95$ is caused by the photonic crystal structure and dielectric contrast.

According to Fig. 10, the transmittance T is almost zero for the frequency lower than $0.6\omega_a$, which corresponds to the first band gap shown in Fig. 9. The second gap in Fig. 9 (at the point G) corresponds to $\omega = 0.89\omega_a$, and it also corresponds to the transmittance spectrum minimum on Fig. 10.

Let us mention that at $\hbar\omega < 2\mu$ the dissipation of the electromagnetic wave in graphene is suppressed. In the long wavelength (low frequency) limit, the skin penetration depth is given by $\delta_0(\omega) = c/Re\left[2\pi\omega\sigma_g(\omega)\right]^{1/2}$ (Landau & Lifshitz, 1984). According to Eq. (29), $Re[\sigma_g(\omega < 2\mu)] = 0$, therefore, $\delta_0(\omega) \to +\infty$, and the electromagnetic wave penetrates along the graphene

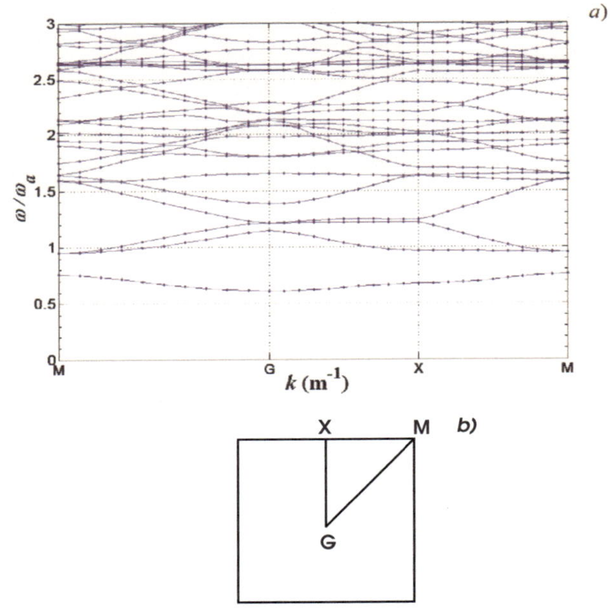

Fig. 9. a) Band structure of graphene based 2D square photonic crystal of cylinder array arranged in a square lattice. The cylinders consist of stack of graphene monolayer discs separated by the dielectric discs. The filling factor $f = 0.3927$. M, G, X, M are points of symmetry in the first (square) Brillouin zone. b) The first Brillouin zone of the 2D photonic crystal.

layer without damping. For the carrier densities $n_0 = 10^{11}$ cm^{-2} the chemical potential is $\mu = 0.022$ eV (Falkovsky & Pershoguba, 2007), and for the frequencies $\nu < \nu_0 = 10.42$ THz we have $Re[\sigma_g(\omega)] = 0$ at $\omega \gg 1/\tau$ the electromagnetic wave penetrates along the graphene

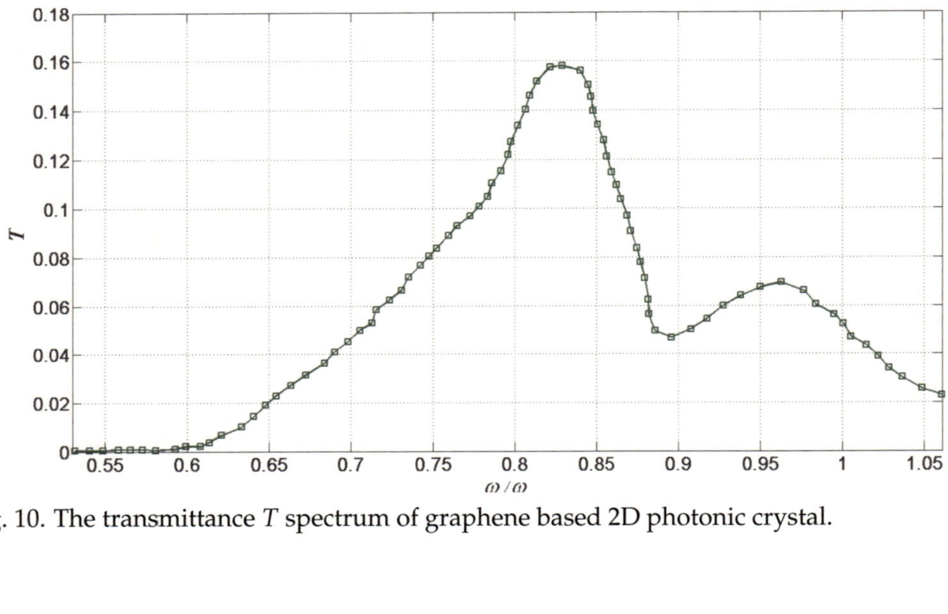

Fig. 10. The transmittance T spectrum of graphene based 2D photonic crystal.

layer almost without damping, which makes the graphene multilayer based photonic crystal to be distinguished from the metallic photonic crystal, where the electromagnetic wave is essentially damped. As a result, the graphene-based photonic crystals can have the sizes much larger than the metallic photonic crystals. The scattering of the electrons on the impurities can result in non-zero $Re[\sigma_g(\omega)]$, which can cause the dissipation of the electromagnetic wave. Since the electron mobility in graphene can be much higher than in typical semiconductors, one can expect that the scattering of the electrons on the impurities does not change the results significantly.

The physical properties of graphene-based photonic crystals are different from the physical properties of other types of photonic crystals, since the dielectric constant of graphene has the unique frequency dependence (Falkovsky & Pershoguba, 2007; Falkovsky, 2008). According to the results presented above, the graphene-based photonic crystal has completely different photonic band structure in comparison to the photonic crystals based on the other materials. The photonic band structure of the photonic crystal with graphene multilayer can be tuned by changing the distance d between graphene discs in the r.h.s. of Eq. (28). The photonic band structure of the graphene-based photonic crystals can also be controlled by the doping, which determines the chemical potential μ entering the expressions for the conductivity and dielectric constant of graphene multilayer (29).

6. Discussion and conclusions

Comparing the photonic band structure for graphene-based photonic crystal presented in Fig. 9 with the dielectric (Joannopoulos et al., 2008), metallic (McGurn & Maradudin, 1993; Kuzmiak & Maradudin, 1997), semiconductor (McGurn & Maradudin, 1993) and superconductor-based (Berman et al., 2006; Lozovik et al., 2007) photonic crystals, we conclude that only graphene- and superconductor-based photonic crystals have essential photonic band gap at low frequencies starting $\omega = 0$, and the manifestation of the gap in the transmittance spectra is almost not suppressed by the damping effects. Therefore, only graphene-based and superconducting photonic crystals can be used effectively as the frequency filters and waveguides in low-frequency for the far infrared region of spectrum, while the devices based on the dielectric photonic crystals can be used only in the optical region of electromagnetic waves spectrum. The graphene based-photonic crystal can be used at room temperatures, while the superconductor-based photonic crystal can be used only at low temperatures below the critical temperature T_c, which is about 90 K for the YBCO superconductors.

In summary, photonic crystals are artificial media with a spatially periodical dielectric function. Photonic crystals can be used, for example, as the optical filters and waveguides. The dielectric- and metal-based photonic crystals have different photonic bands and transmittance spectrum. It was shown that the photonic band structure of superconducting photonic crystal leads to their applications as optical filters for far infrared frequencies . It is known that the dielectric- and metal-based photonnic crystals with defects can be used as the waveguides for the frequencies corresponding to the media forming the photonic crystals. Far infrared monochromatic transmission across a lattice of Abrikosov vortices with defects in a type-II superconducting film is predicted. The transmitted frequency corresponds to the photonic mode localized by the defects of the Abrokosov lattice. These defects are formed by extra vortices placed out of the nodes of the ideal Abrokosov lattice. The extra vortices can be pinned by crystal lattice defects of a superconductor. The corresponding frequency is studied as a function of magnetic field and temperature. The control of the transmitted frequency by varying magnetic field and/or temperature is analyzed. It is suggested that found transmitted localized mode can be utilized in the far infrared monochromatic filters. Besides, infrared monochromatic transmission through a superconducting multiple conductor system consisting of parallel superconducting cylinders is found. The transmitted frequency corresponds to the localized photonic mode in the forbidden photonic band, when one superconducting cylinder is removed from the node of the ideal two-dimensional lattice of superconducting cylinders. A novel type of photonic crystal formed by embedding a periodic array of constituent stacks of alternating graphene and dielectric discs into a background dielectric medium is proposed. The frequency band structure of a 2D photonic crystal with the square lattice of the metamaterial stacks of the alternating graphene and dielectric discs is obtained. The electromagnetic wave transmittance of such photonic crystal is calculated. The graphene-based photonic crystals have the following advantages that distinguish them from the other types of photonic crystals. They can be used as the frequency filters for the far-infrared region of spectrum at the wide range of the temperatures including the room temperatures. The photonic band structure of the graphene-based photonic crystals can be controlled by changing the thickness of the dielectric layers between the graphene discs and by the doping. The sizes of the graphene-based photonic crystals can be much larger than the sizes of metallic photonic crystals due to the small dissipation of the electromagnetic wave. The graphene-based photonic crystals can be used effectively as the frequency filters and waveguides for the far infrared region of electromagnetic spectrum. Let us also mention that above for simplicity we assume that the dielectric material is the same between the graphene disks and between the stacks. This assumption has some technological advantage for the most easier possible experimental realization of the graphene-based photonic crystal.

7. References

Abrikosov, A. A. (1988). *Fundamentals of the Theory of Metals* (North Holland, Amsterdam).

Berman, O. L., Boyko, V. S., Kezerashvili, R. Ya., and Lozovik, Yu .E. (2008). Anomalous far-infrared monochromatic transmission through a film of type-II superconductor in magnetic field. Phys. Rev. B, 78, 094506.

Berman, O. L., Boyko, V. S., Kezerashvili, R. Ya., and Lozovik, Yu .E. (2009). Monochromatic Infrared Wave Propagation in 2D Superconductor–Dielectric Photonic Crystal. Laser Physics, 19, No. 10, pp. 2035–2040.

Berman, O. L., Boyko, V. S., Kezerashvili, R. Ya., Kolesnikov, A. A., and Lozovik, Yu .E. (2010). Graphene-based photonic crystal. Physics Letters A, 374, pp. 4784–4786.

Berman, O. L., Lozovik, Yu. E., Eiderman, S. L., and Coalson, R. D. (2006). Superconducting photonic crystals: Numerical calculations of the band structure. Phys. Rev. B, 74, 092505.

Castro Neto, A. H., Guinea, F.,, Peres, N. M. R., Novoselov, K. S. and Geim, A. K. (2009). The electronic properties of graphene, Reviews of Modern Physics, 81, pp. 109–162.

Chigrin, D. N. and Sotomayor Torres, C. M.. (2003). Self-guiding in two-dimensional photonic crystals. Optics Express, 11, No 10, pp. 1203–1211.

Das Sarma, S., Hwang, E. H., and Tse, W. K. (2007). Many-body interaction effects in doped and undoped graphene: Fermi liquid versus non-Fermi liquid. Phys. Rev. B, 75, 121406(R).

Falkovsky, L. A. and Pershoguba, S. S. (2007). Optical far-infrared properties of a graphene monolayer and multilayer. Phys. Rev. B, 76, 153410.

Falkovsky, L. A. and Varlamov, A A. (2007). Space-time dispersion of graphene conductivity.

Eur. Phys. J. B 56, pp. 281–284.

Falkovsky, L. A. (2008). Optical properties of graphene. *J.* Phys.: Conf. Ser., 129, 012004.

Gusynin, V. P. and Sharapov, S. G. (2005). Magnetic oscillations in planar systems with the Dirac-like spectrum of quasiparticle excitations. II. Transport properties, Phys. Rev. B, 71, 125124.

Gusynin, V. P. and Sharapov, S. G. (2005). Unconventional Integer Quantum Hall Effect in Graphene, Phys. Rev. Lett., 95, 146801.

Joannopoulos, J. D., Meade, R. D., and Winn, J. N. (1995). *Photonic Crystals: The Road from Theory to Practice* (Princeton University Press, Princeton, NJ).

Joannopoulos, J. D., Johnson, S. G., Winn, J. N., and Meade, R. D. (2008). *Photonic Crystals: Molding the Flow of Light* (Second Edition, Princeton University Press, Princeton, NJ).

John, S. (1987). Strong localization of photons in certain disordered dielectric superlattices. Phys. Rev. Lett., 58, pp. 2486–2489.

Katsnelson, M. I. (2008). Optical properties of graphene: The Fermi liquid approach. Europhys. Lett., 84, 37001.

Kechedzhi, K., Kashuba O., and Fal'ko, V. I. (2008). Quantum kinetic equation and universal conductance fluctuations in graphene. Phys. Rev. B, 77, 193403.

Keldysh, L. V. (1964). Deep levels in semiconductors. Sov. Phys. JETP 18, 253.

Kohn, W. (1957). In *Solid State Physics*, edited by F. Seitz and D. Turnbull, vol. 5, pp. 257–320 (Academic, New York).

Kuzmiak, V. and Maradudin, A. A. (1997). Photonic band structures of one- and two-dimensional periodic systems with metallic components in the presence of dissipation. Phys. Rev. B 55, pp. 7427–7444.

Landau. L. D. and Lifshitz, E. M. (1984). *Electrodynamics of continuous media.* (Second Edition, Pergamon Press, Oxford).

Lozovik, Yu. E., Eiderman, S. I., and Willander, M. (2007). The two-dimensional superconducting photonic crystal. Laser physics, 9, No 17, pp. 1183–1186.

Luk'yanchuk, I. A. and Kopelevich, Y. (2004). Phase Analysis of Quantum Oscillations in Graphite. Phys. Rev. Lett., 93, 166402.

Luttinger, J. M. and Kohn, W. (1955). Motion of Electrons and Holes in Perturbed Periodic Fields. Phys. Rev., 97, pp. 869–883.

McGurn, A. R. and Maradudin, A. A. (1993). Photonic band structures of two- and three-dimensional periodic metal or semiconductor arrays. Phys. Rev. B, 48, pp. 17576–17579.

Meade, R. D., Brommer, K. D., Rappe, A. M., and Joannopoulos, J. D. (1991). Photonic bound states in periodic dielectric materials. Phys. Rev. B, 44, pp. 13772–13774.

Meade, R. D., Brommer, K. D., Rappe, A. M. and Joannopoulos, J. D. (1992). Existence of a photonic band gap in two dimensions. Appl. Phys. Lett., 61, pp. 495–497.

Meade, R. D., Rappe, A. M., Brommer K. D., Joannopoulos, J. D., and Alerhand, O. L. (1993). Accurate theoretical analysis of photonic band-gap materials. Phys. Rev. B 48, pp. 8434–8437.

McCall, S. L., Platzmann, P. M., Dalichaouch R., Smith, D. and Schultz, S. (1991). Microwave propagation in two-dimensional dielectric lattices. Phys. Rev. Lett. 67, pp. 2017–2020.

Nair, R. R., Blake, P., Grigorenko, A. N., Novoselov, K. S., Booth, T. J., Stauber, T., Peres, N. M. R., and Geim, A. K. (2008). Fine Structure Constant Defines Visual Transparency of Graphene. Science, 320, no. 5881, 1308.

Nomura, K. and MacDonald, A. H. (2006). Quantum Hall Ferromagnetism in Graphene.

Phys. Rev. Lett., 96, 256602.
Novoselov, K. S., Geim, A. K., Morozov, S. V., Jiang, D., Zhang, Y., Dubonos, S. V., Grigorieva, I. V., and Firsov, A. A. (2004). Electric Field Effect in Atomically Thin Carbon Films. Science, 306, no. 5696, pp. 666–669.
Novoselov, K. S., Geim, A. K., Morozov, S. V., Jiang, D., Katsnelson, M. I., Grigorieva, I. V. and Dubonos, S. V., (2005). Two-Dimensional Gas of Massless Dirac Fermions in Graphene. Nature (London), 438, pp. 197–200.
Robertson, W. M., Arjavalingam, G., Meade, R. D., Brommer, K. D., Rappe, A. M., and Joannopoulos, J. D. (1992). Measurement of photonic band structure in a two-dimensional periodic dielectric array, Phys. Rev. Lett., 68, pp. 2023–2026.
Safar, H., Gammel, P. L., Huse, D. A., Majumdar, S. N., Schneemeyer, L. F., Bishop, D. J., López, D., Nieva, G., and de la Cruz, F. (1994). Observation of a nonlocal conductivity in the mixed state of $YBa_2Cu_3O_{7-\delta}$: Experimental evidence for a vortex line liquid. Phys. Rev. Lett., 72, pp. 1272–1275.
Taflove, A. (1995). *Computational Electrodynamics: The Finite-Difference Time-Domain Method* (MA: Artech House).
Takeda, H. and Yoshino, K. (2003). Tunable light propagation in Y-shaped waveguides in two-dimensional photonic crystals utilizing liquid crystals as linear defects. Phys. Rev. B, 67, 073106.
Takeda, H. and Yoshino, K. (2003). Tunable photonic band schemes in two-dimensional photonic crystals composed of copper oxide high-temperature superconductors. Phys. Rev. B, 67, 245109.
Takeda, H., Yoshino, K., and Zakhidov, A. A. (2004). Properties of Abrikosov lattices as photonic crystals. Phys. Rev. B, 70, 085109.
Takhtamirov, E. E. and Volkov, V. A. (1999). Generalization of the effective mass method for semiconductor structures with atomically sharp heterojunctions. JETP, 89, No 5, pp. 1000–1014.
Tőke, C., Lammert, P. E., Crespi, V. H., and Jain, J. K. (2006). Fractional quantum Hall effect in graphene. Phys. Rev. B, 74, 235417.
Yablonovitch, E. (1987). Inhibited Spontaneous Emission in Solid-State Physics and Electronics, Phys. Rev. Lett., 58, pp. 2059–2062.
Yablonovitch, E., Gmitter, T. J., Meade, R. D., Brommer, K. D., Rappe, A. M., and Joannopoulos, J. D. (1991). Donor and acceptor modes in photonic band structure. Phys. Rev. Lett., 67, pp. 3380–3383.
Zhang, Y., Small, J. P., Amori, M. E. S., and Kim P. (2005). Electric Field Modulation of Galvanomagnetic Properties of Mesoscopic Graphite. Phys. Rev. Lett., 94, 176803.
Zhang, Y., Tan, Y.-W., Stormer, H. L., and Kim, P. (2005). Experimental observation of the quantum Hall effect and Berry's phase in graphene. Nature, 438, pp. 201–204.

6

Terahertz Electromagnetic Waves from Semiconductor Epitaxial Layer Structures: Small Energy Phenomena with a Large Amount of Information

Hideo Takeuchi
Department of Electronic Systems Engineering, School of Engineering,
The University of Shiga Prefecture
Japan

1. Introduction

Terahertz electromagnetic waves have a frequency range between infrared light and microwaves: the frequency of 1 THz corresponds to the photon energy of 4.1 meV (33 cm^{-1}) and to the wavelength of 300 μm. It is well known that the terahertz waves have a high sensitivity to the water concentration in materials. For example, Hu and Nuss compared the terahertz-wave transmittance image of the freshly cut leaf with that of the same leaf after 48 hours (Hu & Nuss, 1995). They demonstrated the freshness between the two leaves can be clearly evaluated from the terahertz-wave transmittance images. In addition, the terahertz waves are sensitive to explosive chemical materials (Yamamoto et al., 2004). Accordingly, the terahertz waves are applicable to a security system in airports because conventional x-ray inspection systems are insensitive to chemical materials. The above-mentioned characteristics of the terahertz wave lead to the reason why terahertz-wave spectroscopy is attractive. We note that the terahertz waves are useful to investigate the vibration of biological molecules, dielectric constant of materials, and so on (Nishizawa et al., 2005).
In the present chapter, we focus our attention on the time-domain terahertz-wave measurements based on the femtosecond-pulse-laser technology. Most of the terahertz-wave measurement systems employ photoconductive antenna devices (Auston, 1975; Nuss & Orenstein, 1999) as an emitter of terahertz waves. As mentioned later, the antenna-based terahertz emitters, which are categorized into a lateral/planer structure type emitter, have various disadvantages. For the progress in terahertz-wave spectroscopy, it is still required to develop convenient terahertz-wave emitters. Compound semiconductors with a surface electric field, by being irradiated by femtosecond-laser pulses, emit the terahertz wave originating from the surge current of the photogenerated carriers flowing from the surface to the internal side in the surface depletion layer. This phenomenon provides us a convenient terahertz emitter free from a device fabrication for an external applied bias. In the above terahertz emission mechanism, the doping concentration is a major factor determining the depletion-layer width and surface electric field, which are in the relation of trade-off. In order to obtain intense terahertz wave emission, earlier works focused on

searching a suitable compound semiconductor and subsequently adjusted the doping concentration (Gu & Tani, 2005). Moreover, external magnetic fields, which are of the order of 1 T, were used for enhancing the terahertz emission (Sarukura et al., 1998; Ohtake et al., 2005); however, the terahertz spectroscopic system with use of the magnetic field generator lacks the advantage of being convenient. In the above-mentioned earlier studies, bulk crystals were employed as emitters. From the viewpoint of utilizing the advantage of compound semiconductors, it should be emphasized that compound semiconductors are rich in a degree of freedom in designing their structures with use of the technology for epitaxial layer growth. Accordingly, we have focused our attention on the epitaxial layer structure, and have explored the feasibility of controlling characteristics of the terahertz waves by appropriately designing an epitaxial layer structure.

The purpose of the present chapter is to demonstrate the fact that the appropriate design of epitaxial layer structures is effective to control the characteristics of the terahertz wave. The above-mentioned epitaxial layer structure design is based on the fundamental semiconductor physics, so that the results demonstrated here contain a large amount of information on ultrafast carrier dynamics. We organize the present chapter based on our recent works (Takeuchi et al., 2008; 2009; 2010). In Section 2, we review the current status of terahertz-wave spectroscopy, and discuss the hidden problems. In Section 3, we approach the enhancement of the terahertz-wave emission using the way different from those of the earlier works; namely, we explore the feasibility of enhancing the terahertz-emission intensity by appropriately designing an epitaxial layer structure. We demonstrate that an undoped GaAs/*n*-type GaAs (*i*-GaAs/*n*-GaAs) epitaxial layer structure is effective to enhance the terahertz emission and that the emission intensity from the *i*-GaAs/*n*-GaAs sample can exceed the emission intensity from *i*-InAs that is known as one of the most intense terahertz emitters. In Section 4, we demonstrate frequency tunable terahertz emitters based on *i*-GaAs/*n*-GaAs epitaxial layer structures with various *i*-GaAs-layer thicknesses d, using the sub-picosecond-range carrier-transport processes. The observed time-domain terahertz waveform consists of the following two components: the intense monocycle oscillation, the so-called first burst, around the time delay of 0 ps originating from the surge current and oscillation patterns from the coherent GaAs longitudinal optical (LO) phonon. From the Fourier power spectrum of the terahertz waveform, it is elucidated that the enhancement of the built-in electric field in the *i*-GaAs layer, which is controlled by changing d, causes a high frequency shift of the first burst band. Based on the above-mentioned phenomenon, we discuss the photogenerated carrier transport in the sub-picosecond range. We also find that the intensity of the coherent LO phonon band increases with a decrease in d.

In the research field of terahertz waves, most of the efforts have been focused on how to enhance the terahertz intensity of emitters or on how to improve the quality of the terahertz-wave images, which indicates that the terahertz wave has been attracting attention only as a tool for probing a given sample under test. On the other hand, we emphasize that the terahertz waves contain a lot of information on physics of the terahertz-wave sources, which is also pointed out in Section 4. In Section 5, we especially focus our attention on this viewpoint, and investigate the direction of the surface band bending using terahertz-wave measurements. We utilize the polarity of the terahertz wave. We investigate the polarity of the terahertz wave from GaAs-based dilute nitride ($GaAs_{1-x}N_x$ and $In_yGa_{1-y}As_{1-x}N_x$) epitaxial layers in order to clarify the effects of nitrogen incorporation on the direction of the surface band bending. The *i*-GaAs/*n*-GaAs sample has an upward band bending at the surface

region, which indicates that photogenerated electrons flow into the inside. In the $GaAs_{1-x}N_x$ samples, the terahertz-wave polarity is reversed; namely, the $GaAs_{1-x}N_x$ sample has a downward band bending. The reversal of the terahertz-wave polarity is attributed to the phenomenon that the conduction band bottom is considerably lowered by the band anticrossing peculiar to $GaAs_{1-x}N_x$, which results in approaching the conduction band bottom to the surface Fermi level. This modifies the direction of the surface band bending connected with the polarity of the terahertz wave. We also investigate the terahertz wave from an $In_yGa_{1-y}As_{1-x}N_x$ epitaxial layer. It is found that the terahertz-wave polarity is also reversed in the $In_yGa_{1-y}As_{1-x}N_x$ sample. Thus, we conclude that the direction reversal of the surface potential bending induced by the band anticrossing is universal in GaAs-based dilute nitrides.
Finally, we summarize this chapter and point out what should be elucidated next for the progress in terahertz-wave spectroscopy.

2. Problems hidden in current terahertz-wave spectroscopic measurements with use of photoconductive antenna devices

Recently, terahertz-wave spectroscopy shifts from the research stage to the commercial stage: terahertz-wave spectroscopic systems, which employ the femtosecond-pulse-laser technology, are commercially available. Most of the terahertz-wave measurement systems equip with a photoconductive antenna (Auston, 1975; Nuss & Orenstein, 1999) as a terahertz-wave emitter. Figures 1(a) and 1(b) show a typical bow-tie antenna and dipole antenna, respectively. The metal electrodes of the antennas are formed on the low-temperature-grown GaAs epitaxial layer (Othonos 1998). The application of the low-temperature-grown GaAs epitaxial layer to the antenna was proposed by Gupta et al., 1991. The gap of the antenna is about 5 μm. The generation mechanism of the terahertz wave is schematically shown in Fig. 1(c). The illumination of the laser pulses (the pump beam) induces the surge current of the photogenerated carriers j. According to Faraday's law of induction, the electric field of the terahertz wave $E_{THz}(t)$ is expressed by the following equation (Bolivar, 1999):

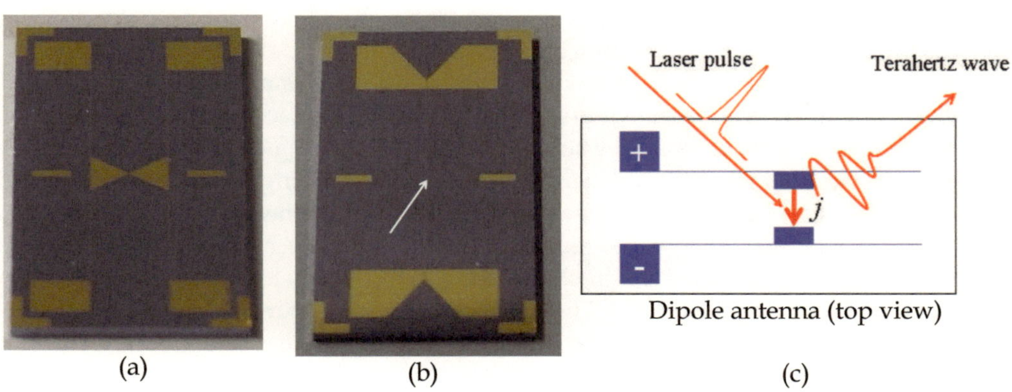

Fig. 1. (a) Optical photograph of a bow-tie antenna. The gap between the triangles is the area emitting the terahertz wave. (b) Optical photograph of a dipole antenna. The position of the arrow corresponds to the gap area emitting terahertz wave. (c) Generation mechanism of the terahertz wave in the dipole antenna under the external bias voltage.

$$E_{\mathrm{THz}}(t) \propto \frac{\partial}{\partial t} j(t) \quad (1)$$

Equation (1) is a basic relation between the surge current and terahertz wave. In addition, it should be mentioned that the emission of the terahertz wave occurs in the reflection and transmission directions of the pump beam. The emission of the terahertz wave along the reflection direction is expressed with use of the generalized Fresnel equation (Bolivar, 1999):

$$n_{\mathrm{pump}}^{\mathrm{air}} \sin(\theta_{\mathrm{pump}}^{\mathrm{air}}) = n_{\mathrm{terahertz}}^{\mathrm{air}} \sin(\theta_{\mathrm{terahertz}}^{\mathrm{air}}) . \quad (2)$$

Here, $n(\omega_{\mathrm{pump}}^{\mathrm{air}})$ and $n(\omega_{\mathrm{terahertz}}^{\mathrm{air}})$ are the refractive indices of the pump beam and of the terahertz wave, respectively. The quantities of $\theta_{\mathrm{pump}}^{\mathrm{air}}$ and $\theta_{\mathrm{terahertz}}^{\mathrm{air}}$ denote the angle of incidence of the pump beam and the emission angle of the terahertz wave, respectively. In general, the refractive index of the pump beam can be approximated to that of the terahertz wave. Accordingly, the emission direction of the terahertz wave is almost the same as the reflection direction of the pump beam.

As mentioned above, the photoconductive antennas are usually used for the terahertz-wave measurement; however, the antennas are fragile and less controllable. The one factor originates from the narrow gap of the antenna placed on the surface, which is weak against the static electricity. Another factor arises from the fact that the low-temperature-grown GaAs epitaxial layer is unstable in principle. Furthermore, the transport process of the photogenerated carriers is remarkably sensitive to the growth condition of the epitaxial layer (Abe et al., 1996; Othonos, 1998). We also point out the disadvantage that the lateral carrier transport along the surface is influenced by the surface degradation arising from humidity and oxidisation. These problems of the planer photoconductive antennas lead to the reason why the terahertz emitters with use of the vertical carrier transport are attractive.

3. Intense terahertz emission from an *i*-GaAs/*n*-GaAs structure

3.1 Terahertz waves from the bulk crystals of compound semiconductors

In advance to describing the terahertz emission from the *i*-GaAs/*n*-GaAs structure, we briefly describe the terahertz wave from bulk crystals.

Figures 2(a) and 2(b) schematically show the generation mechanism of the terahertz wave in the bulk crystal. In general, as shown in Fig. 2(a), compound semiconductors have a surface band bending, which results from the surface Fermi level pinning owing to the presence of the surface states (Aspnes, 1983; Wieder, 1983). The band bending forms a built-in electric field and surface depletion layer. In the case where the photon energy of the pump beam is larger than the band-gap energy, the real excitation of carriers occurs. The photogenerated carriers are accelerated by the built-in electric field, which leads to the drift motion and to the generation of the surge current j. The surge current is the source of the terahertz wave, as expressed by Eq. (1). The terahertz wave is, according to Eq. (2), emitted in the same direction as that of the reflected pump beam, as shown in Fig. 2(b). We also note that the other generation mechanism of the surge current exists: the photo-Dember effect induced by the carrier diffusion dominating in narrow gap semiconductors such as InAs. In GaAs, the drift motion is dominant because, even in the presence of the negligibly low electric field (several kV/cm), the photo-Dember effect is quite small owing to the slight excess energy

(Heyman et al., 2003). The photo-Dember effect is out of the scope of the present chapter. The details of the photo-Dember effect are described by Gu and Tani, 2005.

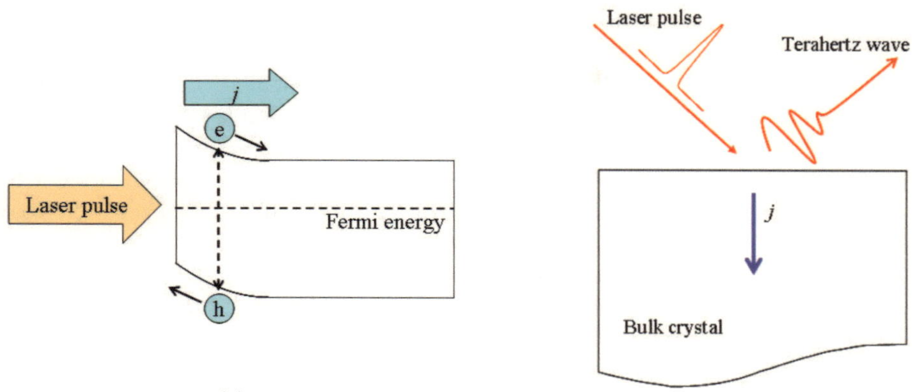

Fig. 2. (a) Band diagram of a semiconductor crystal. The surface Fermi level pinning causes the band bending and forms the surface depletion layer. (b) Generation mechanism of the terahertz wave. The flow of the photogenerated electrons (*e*) and holes (*h*) which are shown in Fig. 2(a), leads to the surge current *j* resulting in the emission of the terahertz wave.

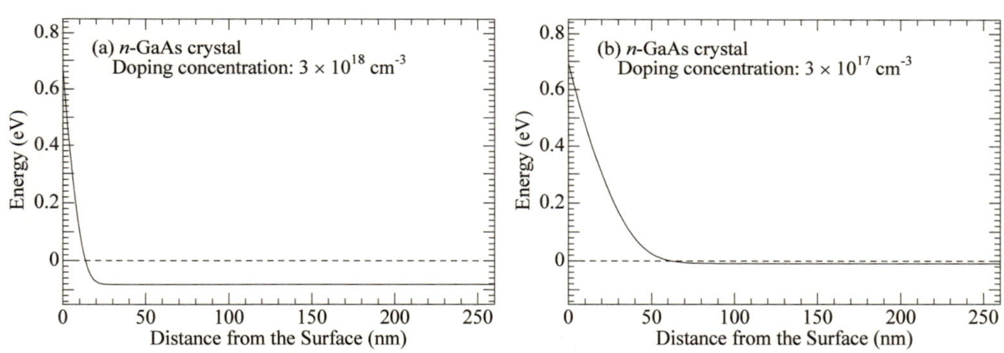

Fig. 3. Potential energies of the *n*-GaAs bulk crystals as a function of distance from the surface calculated on the basis of the Boltzmann-Poisson model. The solid lines indicate the conduction-band energy. The origin of the energy axis corresponds to the Fermi level, which is denoted as the dashed lines. (a) *n*-GaAs bulk crystal with a doping concentration of 3×10^{18} cm^{-3}. (b) *n*-GaAs bulk crystal with a doping concentration of 3×10^{17} cm^{-3}.

As mentioned above, the built-in electric field is a major factor dominating the surge current. It should be mentioned that the built-in electric field is in the trade-off with the thickness of the surface depletion layer. Figures 3(a) shows the potential energy of the *n*-GaAs crystal with a doping concentration of 3×10^{18} cm^{-3} as a function of distance from the surface calculated on the basis of the Boltzmann-Poisson model (Basore, 1990; Clugston & Basore, 1998). In the calculation, the parameters used are the same as those employed in our earlier work (Takeuchi et al., 2005). In addition, the effect of the band-gap shrinkage (Huang et al., 1990) is also taken into account. The solid and dashed lines are the conduction band energy and Fermi energy, respectively. The pinning position of the surface Fermi level locates at the center of the band gap in GaAs (Shen et al., 1990). As shown in Fig. 3(a), the conduction-band energy remarkably bends around the surface owing to the surface Fermi-level pinning. The electric field at the surface is estimated to be 775 kV/cm, which is relatively large. The thickness of the surface depletion layer is, however, estimated to be about 10 nm at most; namely, the region for the surge current flow is quite limited. In contrast, as shown in Fig. 3(b), the thickness of the surface depletion layer is estimated to be about 60 nm in the *n*-GaAs crystal with a doping concentration of 3×10^{17} cm^{-3}; however, the electric field at the surface is estimated to be 250 kV/cm. This value is much smaller than that in the *n*-GaAs crystal with a doping concentration of 3×10^{18} cm^{-3}. From the above-mentioned discussion, it is apparent that the built-in electric field is actually in the trade-off with the thickness of the surface depletion layer. We also note that the depletion-layer thickness is much smaller than the penetration length of the typical femtosecond laser pulse with a center wave length of 800 nm: the value of the above-mentioned penetration depth is estimated to be about 740 nm (Madelung, 2004). This fact means that the photogenerated carrier density in the depletion layer is negligibly small in comparison with the photogenerated carrier density in the internal side (bulk crystal region). It is, therefore, conclude that the adjustment of the doping concentration of the bulk crystal is an inappropriate strategy for controlling the characteristics of the terahertz wave.

3.2 Design of the epitaxial layer structure: potential structure of the *i*-GaAs/*n*-GaAs structure

From the present subsection, we describe the main theme of the present chapter: terahertz electromagnetic waves from semiconductor epitaxial layer structures. Initially, we describe the strategy for designing the epitaxial layer structure on the basis of the potential structure. The calculated potential structure of the *i*-GaAs (200 nm)/*n*-GaAs (3 μm, 3×10^{18} cm^{-3}) epitaxial layer structure is shown in Fig. 4, where the values in the parentheses denote the layer thickness and doping concentration. The conduction-band energy of the *i*-GaAs/*n*-GaAs structure, which is indicated by the solid line, has a linear potential slope in the whole

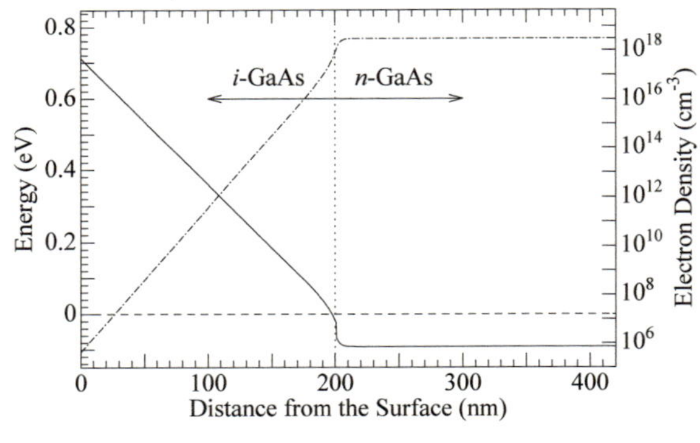

Fig. 4. Potential energy of the *i*-GaAs(200 nm)/*n*-GaAs (3 μm 3×10^{18} cm^{-3}) structure as a function of distance from the surface calculated on the basis of the Boltzmann-Poisson model. The solid and dashed lines indicate the conduction-band energy and Fermi level, respectively. The dashed-and-dotted line denotes the electron density in the *i*-GaAs/*n*-GaAs structure calculated as a function of distance from the surface.

i-GaAs layer with a thickness of 200 nm, which results from the fact that the *i*-GaAs top layer is free from dopants. The *i*-GaAs layer has a uniform built-in electric field of 35 kV/cm owing to the linear potential slope. Note that the *i*-GaAs layer thickness is much larger than the surface-depletion-layer thicknesses of the *n*-GaAs crystals shown in Figs. 3(a) and 3(b). Accordingly, the photogeneration of the carriers are more effective than those in *n*-GaAs crystals. We discuss in Subsection 3.4 and Section 4 the details whether the built-in electric field in the *i*-GaAs/*n*-GaAs sample is sufficient or not.

It should be pointed out that the *i*-GaAs/*n*-GaAs structure has another merit. The dashed-and-dotted line denotes the calculated electron density. The electron density in the *i*-GaAs layer is much smaller than that in the *n*-GaAs layer, which is advantageous for emitting intense terahertz wave because the terahertz wave is strongly absorbed by the free carriers (Nishizawa et al., 2005). From the above discussion, it is expected that the intense terahertz emission can be obtained with use of the *i*-GaAs/*n*-GaAs structure.

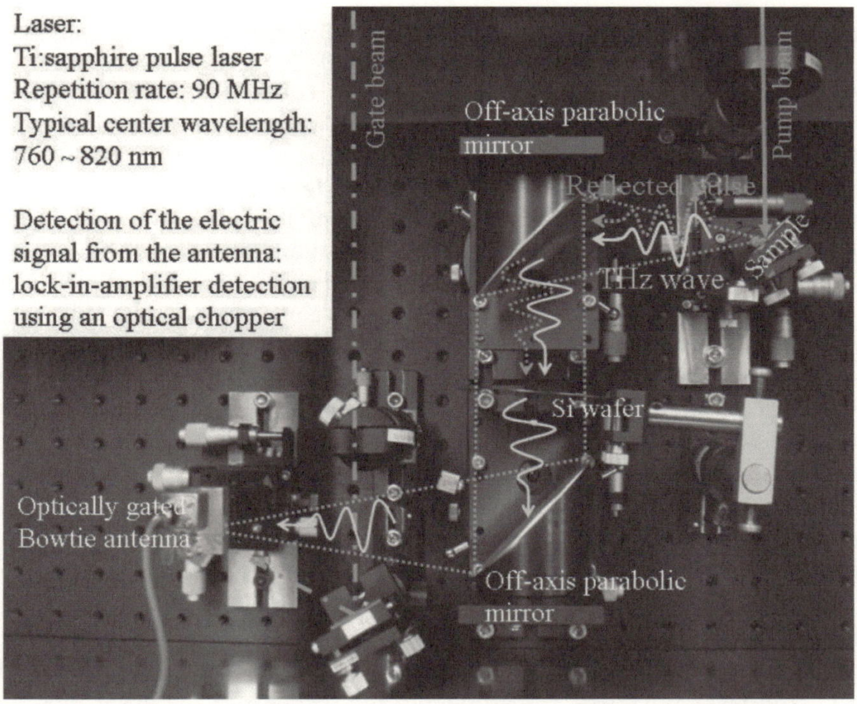

Fig. 5. Photograph of the layout of the optical components for the terahertz-wave measurement.

3.3 Samples and experimental procedure

The present sample was an *i*-GaAs (200 nm)/*n*-GaAs (3 μm, 3 × 10^{18} cm^{-3}) structure grown on a 2°-off semi-insulating (001)-oriented GaAs substrate by metal organic vapour phase epitaxy, where the values in the parentheses denote the individual layer thickness and doping concentration.

The time-domain terahertz-wave signals from the samples were measured at room temperature with use of laser pulses with a duration time of about 70 fs. The measurement system for the terahertz wave is shown in Fig. 5 as a photograph. The sample, by being illuminated by the pump beam, emits a terahertz wave along the reflection direction of the pump beam as described in Section 2. The emitted terahertz wave was collected with use of two off-axis parabolic mirrors. The high resistivity silicon wafer was placed as a filter for the pump beam. The collected terahertz wave was focused on the bow-tie antenna with a gap of 5.0 μm formed on a low-temperature-grown GaAs. The bow-tie antenna was optically gated with use of the laser-pulse beam (gate beam), which was controlled by the mechanical delay line, the so-called stepper. Consequently, the terahertz wave was detected only in the case where the bow-tie antenna was illuminated by the gate beam. The above-mentioned method for the detection of the terahertz wave is the so-called optically gating technique (Nuss & Orenstein, 1999; Bolivar, 1999). In the present experiment, the power of the gate beam was fixed to 4.0 mW. For the reference samples, a (001) *n*-GaAs (about 2×10^{18} cm^{-3}) crystal and a (001) *i*-InAs crystal were examined.

3.4 Intense terahertz emission caused by the surge current in the *i*-GaAs/*n*-GaAs structure

Figure 6(a) shows the terahertz waveforms of the *i*-GaAs/*n*-GaAs (solid line), *n*-GaAs (dotted line), and *i*-InAs (dashed line) samples at the pump-beam energies of 1.531, 1.589, and 1.621 eV. All the samples show a monocycle oscillation around the time delay of 0 ps, the so-called first burst. It is obvious that the amplitude of the first-burst of the *i*-GaAs/*n*-GaAs sample is larger by a factor of 10 than that of the *n*-GaAs crystal. It should be emphasized that the *i*-GaAs/*n*-GaAs sample emits the more intense terahertz wave, in spite of the fact that the built-in electric field is much weaker than the surface electric fields of the *n*-GaAs crystals shown in Figs. 3(a) and 3(b). The above-mentioned results indicate that the presence of the relatively thick *i*-GaAs layer, which is depleted, actually leads to the enhancement of the emission intensity. Thus, it is concluded that the appropriate epitaxial layer structure plays an important role for enhancing the terahertz-emission intensity.

Next, we discuss the pump-beam energy dependence of the terahertz emission, comparing the first-burst amplitude of the *i*-GaAs/*n*-GaAs sample with that of the *i*-InAs crystal. The increase in the pump-beam energy corresponds to an increase in the absorption coefficient. The absorption coefficients of GaAs (InAs) at 1.531, 1.589, and 1.621 eV are 1.41×10^{-3} (6.95×10^{-3}), 1.77×10^{-3} (7.69×10^{-3}), and 1.96×10^{-3} (8.09×10^{-3}) nm^{-1}, respectively (Madelung 2004); namely, the increase in the pump-beam energy from 1.531 to 1.621 eV magnifies the absorption coefficient of GaAs (InAs) by 1.39 (1.16). In the present *i*-GaAs/*n*-GaAs sample, the penetration depth, which is the reciprocal of the absorption coefficient, is much longer than the *i*-GaAs layer thickness. Consequently, the increase in the absorption coefficient leads to the enhancement of the terahertz emission efficiency because the total carrier number accelerated in the *i*-GaAs layer increases. The absorption coefficients of InAs are relatively insensitive to the change in the photon energy because the fundamental transition energy of InAs (0.354 eV) is much smaller than that of GaAs (1.424 eV) (Madelung, 2004).

The effect of an increase in the absorption coefficient on the emission intensity clearly appears in Fig. 6(a). At 1.531 eV, the first-burst amplitude of the *i*-GaAs/*n*-GaAs sample is slightly smaller than that of the *i*-InAs crystal, while, at 1.589 and 1.621 eV, the first-burst amplitudes of the *i*-GaAs/*n*-GaAs sample are remarkably larger than those of the *i*-InAs crystal; namely, the first-burst amplitude of the *i*-GaAs/*n*-GaAs sample is enhanced by the increase in the photogenerated carriers. Thus, it is experimentally confirmed that the

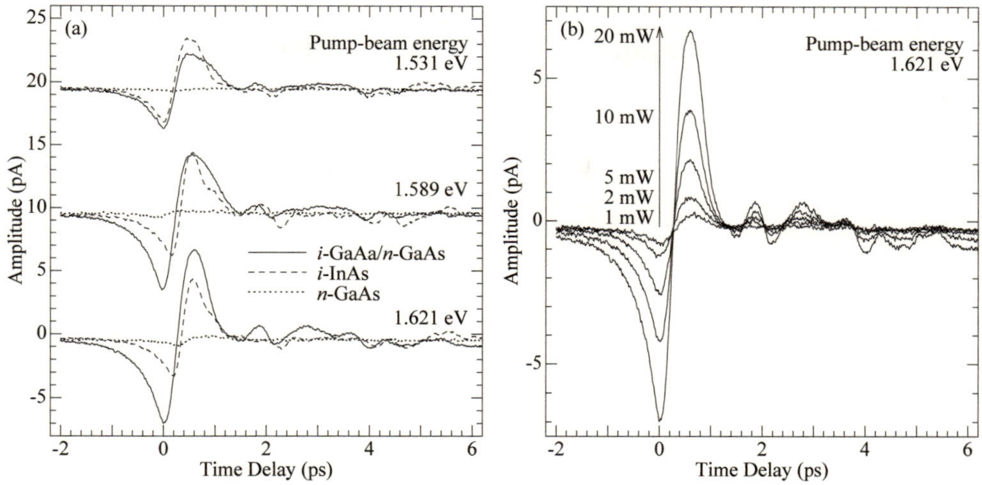

Fig. 6. (a) Amplitude of the terahertz waveform as a function of time delay at room temperature. The solid, dotted, and dashed lines indicate the time domain signals of the *i*-GaAs/*n*-GaAs, *n*-GaAs, and *i*-InAs samples, respectively. The pump-beam power was constant: 20 mW, while the pump-beam energies were varied: 1.531, 1.589, and 1.621 eV. For clarity, each waveform is vertically shifted. (b) Terahertz waveforms as a function of time delay in the *i*-GaAs/*n*-GaAs sample at the various pump-beam powers. The pump-beam energy was 1.621 eV.

dominant generation mechanism of the terahertz emission is attributed to the surge current of the photogenerated carriers flowing through the *i*-GaAs layer. It was reported that the terahertz emission intensity from GaAs is weaker by a factor of 10 than that of InAs (Ohtake et al., 2005). Taking this report into account, we conclude that the *i*-GaAs/*n*-GaAs structure is a solution for enhancing the terahertz emission intensity.

We also investigated the pump-beam-power dependence of the terahertz wave from the *i*-GaAs/*n*-GaAs structure. Figure 6(b) shows the terahertz waveforms of the *i*-GaAs/*n*-GaAs sample as a function of time delay at various pump-beam powers. The pump-beam energy was 1.621 eV. Except for the amplitude, all the waveforms have the same pattern. Taking account of the fact that the pattern of the waveform is a response from the surge current flowing in the *i*-GaAs layer, we conclude that the flow of the surge current does not depend on the pump-beam power in the *i*-GaAs/*n*-GaAs sample.

4. Frequency control of the terahertz waves using *i*-GaAs(*d* nm)/*n*-GaAs structures

4.1 Relation among the electric field, carrier-transport process, and terahertz wave

In section 3, we focused our attention on the terahertz wave from the *i*-GaAs (200 nm)/*n*-GaAs structure from the viewpoint of how to enhance the emission intensity. It is also worthy to investigate the characteristics of the terahertz waves from the *i*-GaAs(*d* nm)/*n*-GaAs structures with various *i*-GaAs layer thicknesses *d* because the *i*-GaAs(*d* nm)/*n*-GaAs structure has the ability to control the built-in electric field of the *i*-GaAs layer. The potential energies of the *i*-GaAs(200 nm)/*n*-GaAs structure and those of the *i*-GaAs(500 nm)/*n*-GaAs

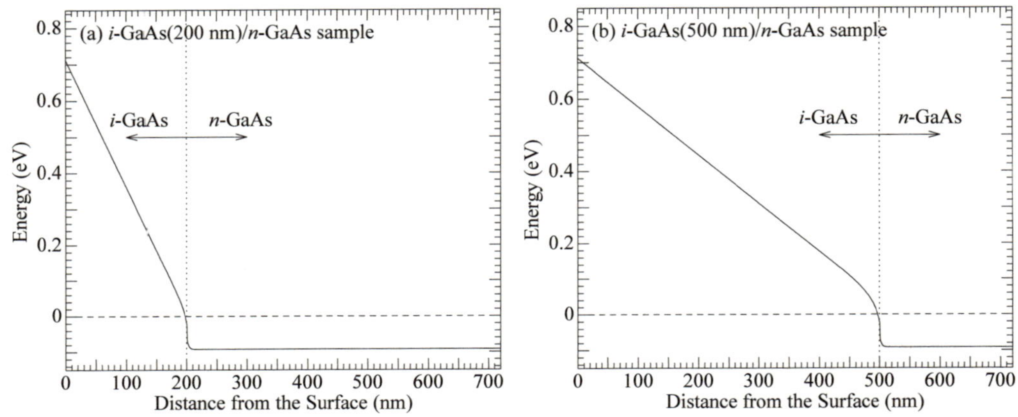

Fig. 7. Potential energy of the *i*-GaAs(d nm)/n-GaAs structure as a function of distance from the surface calculated on the basis of the Boltzmann-Poisson model. The doping concentration and layer thickness of the n-GaAs layer are 3×10^{18} cm^{-3} and 3 μm, respectively. The solid and dashed lines indicate the conduction-band energy and Fermi level, respectively. (a) $d = 200$ nm. (b) $d = 500$ nm.

structure are depicted in Figs. 7(a) and 7(b), respectively, as a function of distance from the surface. Comparing Fig. 7(a) with Fig. 7(b), it is evident that the potential slope increases with a decrease in d, which means the built-in electric field in the *i*-GaAs layer can be controlled by d. The values of the built-in electric field are calculated to be 35 and 13 kV/cm for the *i*-GaAs(d nm)/n-GaAs samples with $d = 200$ and 500 nm, respectively.

From the viewpoint of semiconductor physics, the *i*-GaAs(d nm)/n-GaAs structures are suitable for the investigation of the carrier-transport process in the presence of an electric field. In GaAs under the steady state condition, the electron drift velocity increases with an increase in an electric field and reaches the maximum velocity at the electric field of 4 kV/cm (Blakemore, 1982). Above 4 kV/cm, the electron velocity decreases in spite of an increase in an electric field and almost saturates at 10 kV/cm. The electron-velocity saturation is attributed to the effects of the intervalley scattering (Blakemore, 1982). The progress in the femtosecond-pulse-laser technology enables the transient photogeneration of carriers within the sub-picosecond range. This progress gives a chance to clarify whether the carrier-transport processes in the steady state are also valid in the sub-picosecond range. In order to investigate the carrier-transport process in sub-picosecond range, the terahertz-wave measurements are suitable because, as shown in Eq. (1), the electric field of the terahertz wave is proportional to the time derivative of the surge current of the photogenerated carriers; namely, the electric field of the terahertz wave is connected with the acceleration of the photogenerated carriers. Consequently, in the Fourier power spectrum of the terahertz waveform, the band originating from the surge current of the photogenerated carriers shifts to a high frequency side in the case where the photogenerated carriers are monotonously accelerated without being affected by the intervalley scattering. This theme is not only scientifically interesting but also technologically important because it leads to the realization of frequency tunable terahertz-wave emitters that enable the spectrally resolved time-domain terahertz measurement.

In Section 4, we explore the sub-picosecond-range carrier-transport processes in the *i*-GaAs (*d* nm)/*n*-GaAs structures with various *i*-GaAs-layer thicknesses *d* ranging from 200 to 2000 nm and present the realization of the frequency tunable terahertz-wave emitters. In addition, we discuss the intense terahertz emission from the coherent GaAs LO phonons, which leads to the monochromatic terahertz-wave source.

4.2 Confirmation of the controllability on the built-in electric field

The present samples were the *i*-GaAs (*d* nm)/*n*-GaAs structures grown on semi-insulating (001) GaAs substrates by metal organic vapour phase epitaxy. The layer thickness and doping concentration of the *n*-GaAs layer were 3 μm and 3×10^{18} cm^{-3}, respectively. The values of *d* were 200, 500, 800, 1200, and 2000 nm. The sheet resistances of all the samples are the same value of 3.1 Ω per square, which indicates that the doping process was well controlled.

In the present experiment, it is essential to experimentally confirm the change in the built-in electric field. In order to estimate the built-in electric field, we applied the photoreflectance measurement, which is a convenient and non-destructive method to estimate the built-in electric field. The details of the photoreflectance measurements are described in the review paper by Pollak and Shen, 1993.

Figure 8(a) shows the photoreflectance spectra of the *i*-GaAs(200 nm)/*n*-GaAs and *i*-GaAs(500 nm)/*n*-GaAs samples. In the photoreflectance measurement, the pump beam was the laser light with a photon energy of 1.96 eV chopped at the frequency of 630 Hz. The pump-beam power was 2.0 mW. The probe beam was obtained from a tungsten-halogen lamp dispersed by a monochromator with a resolution of 0.5 nm. The probe-beam power was about 4 μW. As shown in Fig. 8(a), the oscillation patterns, the so-called Franz-Keldysh oscillations (FKOs), are observed. Since the FKOs are caused by an electric field, the appearance of the FKOs indicates the presence of the built-in electric field in the *i*-GaAs layer. In order to estimate the built-in electric field, as shown in Fig. 8(b), the extrema of the FKOs from the *i*-GaAs layer are plotted as a function of quasi-index $\xi \equiv [(3\pi/4) \cdot (j-1/2)]^{2/3}$, where j denotes the index of each extremum numbered from the fundamental transition energy position (Aspnes, 1974). The slope of the solid line is the electro-optic energy $\hbar\Theta$ given by $(e^2\hbar^2F^2/2\mu)^{1/3}$, where F and μ are the built-in electric field and interband reduced effective mass, respectively (Aspnes, 1974). From the slope of the solid line, it is evident that the built-in electric field of the *i*-GaAs(200 nm)/*n*-GaAs sample is higher than that of the *i*-GaAs(500 nm)/*n*-GaAs sample. It is, therefore, confirmed that the built-in electric field of the *i*-GaAs(*d* nm)/*n*-GaAs sample is enhanced by decreasing *d*. The built-in electric fields of the *i*-GaAs(*d* nm)/*n*-GaAs samples are estimated and listed in Table 1, using the relation of $\hbar\Theta \equiv (e^2\hbar^2F^2/2\mu)^{1/3}$. In this estimation of the built-in electric field, the used value of μ, which is the reduced effective mass of a GaAs bulk crystal, is 0.0556 in units of the electron rest mass m_0 in vacuum (Nelson et al., 1987). In Table 1, the results of the numerical calculation are also indicated. The estimated built-in electric fields are almost in good agreement with the calculated value, which means that the present samples are appropriately designed. Note that the built-in electric fields of the *i*-GaAs(*d* nm)/*n*-GaAs samples with d = 200 and 500 nm are in the range where the electron velocity saturates under the steady state condition.

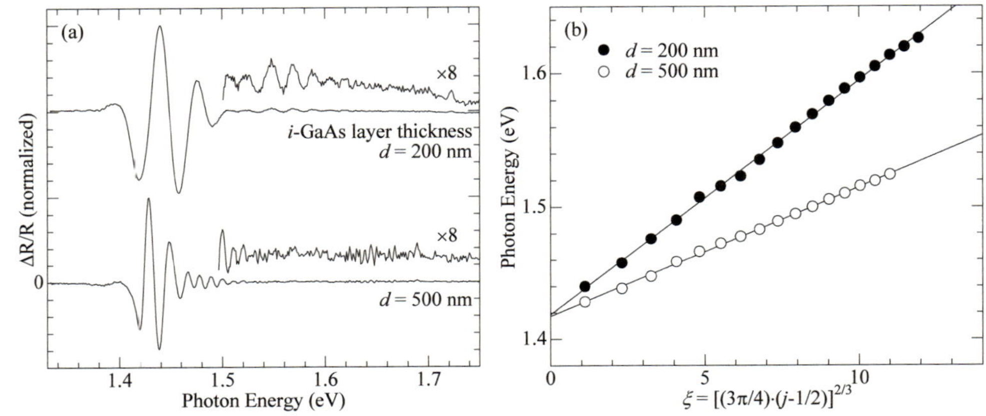

Fig. 8. (a) Photoreflectance spectra of the *i*-GaAs(200 nm)/*n*-GaAs and *i*-GaAs(500 nm) /*n*-GaAs samples at room temperature. (b) Plots of the extrema of the FKOs from the *i*-GaAs(200 nm)/*n*-GaAs sample (closed circles) and *i*-GaAs(500 nm)/*n*-GaAs sample (open circles) as a function of quasi-index ξ.

Structure	F (kV/cm)*1	F (kV/cm)*2
i-GaAs(200 nm)/*n*-GaAs	28	35
i-GaAs(500 nm)/*n*-GaAs	12	13
i-GaAs(800 nm)/*n*-GaAs	8.2	8.1
i-GaAs(1200 nm)/*n*-GaAs	6.1	5.2
i-GaAs(2000 nm)/*n*-GaAs	4.7	3.1

Table 1. Built-in Electric field F in the *i*-GaAs layer. *1: Estimated value from the electro-optic energy $\hbar\Theta$. *2: Calculated value on the basis of the Boltzmann-Poisson model.

4.3 Frequency tunability of the terahertz wave originating from the surge current

In the present terahertz-wave measurement, the experimental apparatus was almost the same as that described in Section 3, though there were some improvements. In the present experiment, we used a dipole antenna with a gap of 6.0 μm formed on a low-temperature-grown GaAs because the range of the frequency-dependent sensitivity of the dipole antenna is wider than that of the bow-tie antenna. In addition, to remove the effects of the water vapour absorption on the terahertz wave, the humidity was suppressed to be 10% during the measurement under a nitrogen-gas-purge condition. The powers of the pump and gate beams were fixed to 40 and 10 mW, respectively. The photon energies of the pump and gate beams were the same: 1.57 eV. The scan range of the time delay was from -2 to 8 ps. All the measurements were performed at room temperature.

The time-domain terahertz waveforms of the samples are shown in Fig. 9(a). All the samples exhibit an intense monocycle oscillation around the time delay of 0 ps, the so-called first burst resulting from the surge current of the photogenerated carriers. The amplitude of the first burst is relatively pronounced, which results from the fact that the *i*-GaAs layer of the *i*-GaAs(*d* nm)/*n*-GaAs samples is depleted by its built-in electric field. Accordingly, the terahertz wave from the *i*-GaAs(*d* nm)/*n*-GaAs structures provides the more precise information on the first burst related to the surge current.

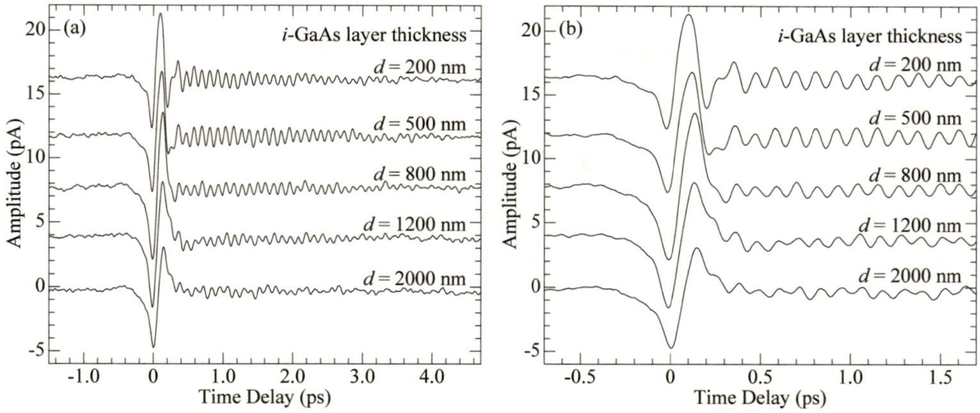

Fig. 9. (a) Amplitudes of the terahertz waveforms of the *i*-GaAs(*d* nm)/*n*-GaAs samples as a function of time delay at room temperature. The pump-beam power was fixed to be 40 mW. (b) Terahertz waveforms within the time delay from -0.5 to 1.5 ps.

The first burst shown in Fig. 9(a) is followed by an oscillatory profile with a period of 113 fs. The period corresponds to the frequency of the GaAs LO phonon (8.8 THz); namely, the terahertz emission from the coherent LO phonon is also detected. The details of terahertz emission from the coherent LO phonon are discussed in Subsection 4.5.

For the clarity of the shape of the first burst signal, Fig. 9(b) shows the first burst signal within the time delay from -0.5 to 1.5 ps. The amplitudes of the first burst signals are almost same in all samples, while the width of the first burst signal exhibits a gradual narrowing with a decrease in the *i*-GaAs layer thickness. Taking account of the fact that the decrease in the *i*-GaAs layer thickness results in the enhancement of the built-in electric field accelerating photogenerated carriers, the increase in the electron velocity, which corresponds to the enhancement of the surge current, has a tendency not to enhance of the amplitude of the first burst signal but to cause the change in the frequency components forming the first burst signal.

In order to analyze the frequency components, we transformed the terahertz waveforms to the Fourier power spectra, which are shown in Fig. 10(a). The Fourier power spectrum of each sample exhibits the two bands. Judging from the oscillation period in the time-domain signal shown in Fig. 9(a), the low frequency band is assigned to the band originating from the first burst. The band of the first burst gradually shifts to a high frequency side with a decrease in *d*. For example, the peak frequency of the first burst band locates at 1.5 THz in the *i*-GaAs(2000 nm)/*n*-GaAs sample, while the peak frequency of the first burst band locates at 4.0 THz in the *i*-GaAs(200 nm)/*n*-GaAs sample. This is the significant finding in the present work; namely, the frequency of the first burst band is tunable by changing the *i*-GaAs layer thickness *d*.

Next, we discuss the mechanism of the frequency shift of the first burst band. Since a decrease in the *i*-GaAs layer thickness leads to the enhancement of the built-in electric field as shown in Table 1, the high frequency shift of the first burst band indicates that the photo-generated carriers are monotonously accelerated by the built-in electric field. Thus, it is concluded that the intervalley scattering, which dominates the carrier-transport process under the steady state condition in a high electric field range, hardly influences in the sub-picosecond range. It should be noted that the frequency shift of the first burst band is not related to plasmons because the pump power was fixed to 40 mW in the present experiment.

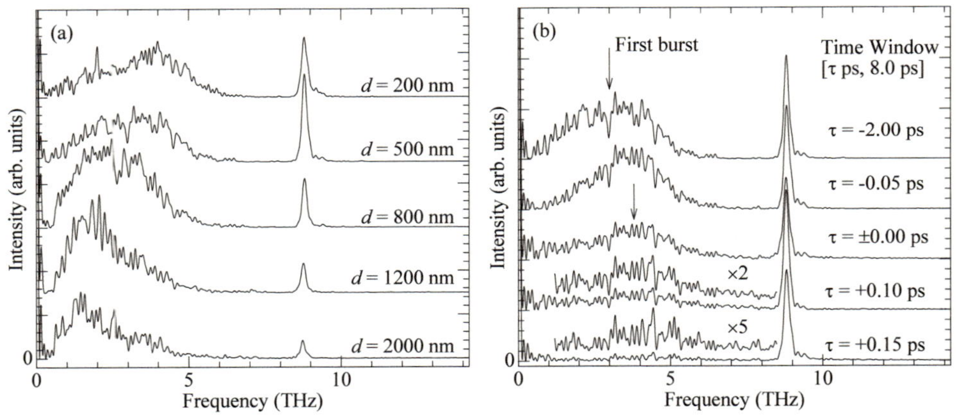

Fig. 10. (a) Fourier power spectra of the terahertz waveforms of the *i*-GaAs(*d* nm)/*n*-GaAs samples shown in Fig. 10(a). (b) Time-partitioning Fourier power spectra of the terahertz waveforms of the *i*-GaAs(500 nm)/*n*-GaAs sample. The values of τ are -2.00, ±0.00, +0.10 and +0.15 ps.

We also confirmed that the photogenerated carriers are monotonously accelerated by the built-in electric field, using a time-partitioning Fourier transform method, which is a powerful way to investigate the time evolution of the signal. The time-partitioning Fourier power spectrum $I(\omega)$, where ω is a frequency, is given by

$$I(\omega) \propto \left| \int_{\tau}^{8\,\mathrm{ps}} A(t)\exp(-i\omega t)dt \right|^2 . \tag{3}$$

Here, $A(t)$ is the time-domain terahertz waveform and τ is the time delay (-2 ps $\leq \tau <$ 8 ps) that determines the time widow of the Fourier transform.

Figure 10(b) shows the time-partitioning Fourier power spectra of the *i*-GaAs(500 nm)/*n*-GaAs sample at various time windows. The peak frequency of the first burst is shifted to the high frequency side with an increase in τ. In general, the frequency of the electromagnetic wave reflects with an increase in the electron velocity, which is the well-known fundamental concept on the high-frequency devices for microwave generation. Consequently, it is confirmed that the monotonous acceleration of the photogenerated carriers is responsible for the high frequency shift of the first burst band shown in Fig. 10(a).

It is interesting to compare the present results with those of the Monte Carlo simulation. According to the Monte Carlo simulation, the transient electron velocity in a GaAs crystal is accelerated by the electric field and reaches the maximum value of 5.5×10^7 cm/s at 0.5 ps in the condition of the electric field of 10 kV/cm. In the electric field of 20 kV/cm, the maximum transient electron velocity reaches 7×10^7 cm/s at 0.3 ps (Tomizawa, 1993). Taking account of the above-mentioned simulation, our experimental results are reasonable.

4.4 Intense terahertz emission from the coherent LO phonon

In advance to discuss the present results of the intense terahertz emission from the coherent LO phonon, we briefly describe the reason why it is desired to generate terahertz emission from coherent LO phonons with use of simpler methods. The terahertz emission from coherent optical phonons has been attracting much attention since the development of monochromatic terahertz emitters is an important issue in terahertz-wave spectroscopy. In general, however, the intensity of the terahertz emission from coherent optical phonons is weak in bulk semiconductors (Gu & Tani, 2005). As a solution of this problem, the application of the multiple quantum wells was proposed (Mizoguchi et al., 2005; Nakayama et al., 2008; Nakayama & Mizoguchi, 2008). In the multiple quantum wells, the terahertz emission from the coherent LO phonon is enhanced in the case where the fundamental heavy-hole and light-hole exciton energy spacing is equal to the LO phonon frequency. In addition, the photon energy of the pump beam should be tuned to the center energy between the heavy-hole and light-hole exciton energy spacing: The quantum interference between the heavy-hole and light-hole excitons is a driving force for the coherent LO phonon. The above method requires the strict sample growth and limits the photon energy of the pump beam. In contrast to the former strategy for enhancing terahertz emission, the present strategy is quite simple. As mentioned in Subsection 4.3, the value of the photon energy of the pump beam is 1.57 eV, which is much higher than the fundamental transition energy of GaAs (1. 424 eV) at room temperature, so that the excitation process by the pump beam is under the off-resonance condition. In addition, the sample structure consists of just two layers.

As shown in Fig. 10(a), the intensity of the LO phonon band increases with a decrease in d that enhances the built-in electric field of the *i*-GaAs layer. In d = 500 nm, the peak intensity of the LO phonon band exceeds that of the first burst band peaking at about 3.0 THz. It should be noted that, in the present experiment, the frequency-dependent sensitivity of the dipole antenna (the detector) was not calibrated. In general, the sensitivity of the dipole antenna remarkably lowers in a high frequency range. Actually, the sensitivity at 1 THz remarkably drops above 5.0 THz by the factor of 10^{-3} at least (Bolivar, 1999). The intensity of the coherent LO phonon band, therefore, has a possibility of drastically exceeding that of the first burst band. The present observation of the intense terahertz wave from the coherent LO phonons results from the following two factors. The one factor is the sweeping-out effects on carriers owing to the presence of the built-in electric field in the *i*-GaAs layer, which reduces the free-carrier absorption of the terahertz wave. The second factor is an increase in initial displacements of the constituent atoms. From the viewpoint of the polarization dynamics, we explain the generation mechanism of the terahertz wave from coherent LO phonon in detail together with its relation with initial displacements of the constituent atoms. As shown in Table 1, the increase in the built-in electric field enlarges the initial displacements of the constituent Ga and As atoms; namely, the static polarization due to the initial displacements is enhanced. The initial displacements are released by the instantaneous change in the built-in electric field by the surge current, which launches the coherent oscillation of the constituent atoms, i.e., the coherent LO phonon (Cho et al., 1990; Dekorsy et al., 2000). This phenomenon leads to the oscillation of the LO-phonon polarization producing the terahertz wave. It is noted that the enlargement of the initial displacement results in the enhancement of the amplitude of the coherent LO phonon. Consequently, taking account of the generation mechanism of the coherent LO phonon mentioned above, it is apparent that the terahertz-wave intensity from the coherent LO phonon is increased with a decrease in d.

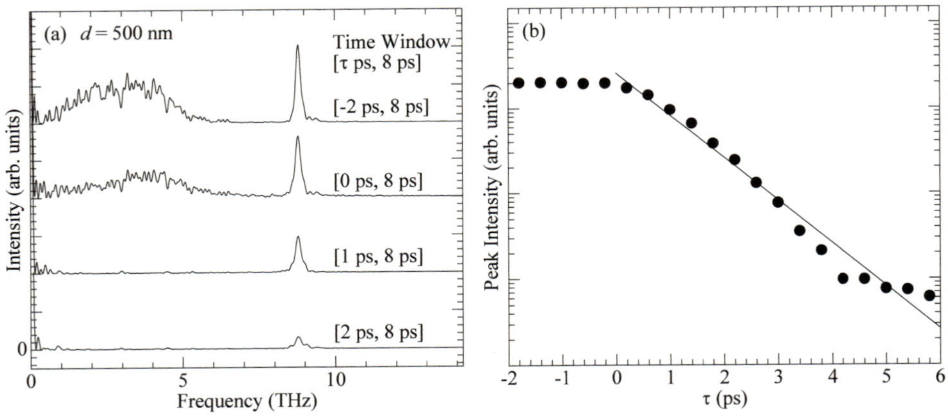

Fig. 11. (a) Time-partitioning Fourier power spectra of the terahertz waveforms of the *i*-GaAs(500 nm)/*n*-GaAs sample. The values of τ are -2, 0, 1 and 2 ps. (b) Peak intensities of the LO phonon bands of the *i*-GaAs(500 nm)/*n*-GaAs sample plotted as a function of τ. The solid line indicates the fitting results of a single exponential function.

In the *i*-GaAs(200 nm)/*n*-GaAs sample, the intensity of the coherent LO phonon band is slightly reduced though the built-in electric field is the highest in all the samples. The reduction of the intensity of the coherent LO phonon in the *i*-GaAs(200 nm)/*n*-GaAs sample in comparison with that in the *i*-GaAs(500 nm)/*n*-GaAs sample mainly results from the decrease of the *i*-GaAs layer thickness, i.e. the volume effect.

We also analyzed the time evolution of the terahertz wave from the coherent LO phonon with use of the time-partitioning Fourier transform method. Figure 11(a) shows the time-partitioning Fourier transform spectra of the *i*-GaAs(500 nm)/*n*-GaAs sample. The band of the first burst between 0 to 5 THz rapidly decays and disappears at $\tau = 1$ ps, which coincides with the fact that the first burst appears around the time delay of 0 ps in the terahertz waveform shown in Fig. 10(a). In contrast, the LO phonon band at 8.8 THz still remains at $\tau = 2$ ps, which is consistent with the fact that the oscillatory profile of the coherent LO phonon signals is observed up to 5 ps in the THz waveforms. The decay rate of the coherent LO phonon is estimated from Fig. 11(b) to be 1.1 ps^{-1} using a single exponential function fitting. As indicated in Eq. (3), the decay rate is estimated from the Fourier power spectrum that corresponds to the square of the amplitude. Consequently, the decay rate of the amplitude of the terahertz wave from the coherent LO phonon, which is shown in Fig. 9(a), is a half value of the decay rate estimated from the time-partitioning Fourier transform method. The decay rate is about 0.5 ps^{-1}, so that the decay time is 2.0 ps. Note that the decay time of terahertz wave from the coherent LO phonon is longer than that of the decay time of terahertz wave from the first burst. This is advantageous to control the mechanical delay line, the stepper, which is explained in Subsection 3.3.

The above-mentioned result opens the way to the novel terahertz-wave imaging system. In general, the spatial resolution of the terahertz-wave image is 1 mm at most (Herman et al., 2005). This fact originates from the diffraction limit of the terahertz wave emitted from the conventional dipole antenna. The dipole antenna emits a terahertz wave with a frequency range from 0 to 5 THz. The position of the peak intensity of the band locates at about 1.0 THz. This frequency corresponds to the wavelength λ of 300 μm. In addition, the frequency

range of the terahertz wave from the dipole antenna is relatively wide (Bolivar, 1999; Sakai & Tani, 2005), which leads to the possibility of causing the chromatic aberration.
The schematic view of the imaging system that we propose is shown in Fig. 12. Figure 12 is similar to the conventional terahertz-wave-transmittance imaging system (Herman et al., 2005). In the conventional system, a mirror is equipped at the position of the *i*-GaAs/*n*-GaAs terahertz emitter in Fig. 12. The expanded pump beam is reflected by the mirror and the dipole antenna in front of the sample is illuminated by the reflected pump beam. The terahertz wave emitted from the dipole antenna is transmitted and detected with use of electro-optic crystals, for example, ZnTe. At each point of the crystal, the electric field of the terahertz wave modulates the refractive index through the electro-optic effects. This phenomenon modifies the polarization of the probe beam, and only the component of the probe beam with the modified polarization passes the analyzer. The charge-coupled-device (CCD) camera makes a terahertz-wave transmittance image by detecting these components. This is the so-called electro-optic sampling method, which is a well-established technique. The imaging system that we propose is based on the above-mentioned electro-optic sampling method. The difference is the position of the terahertz emitter, which is equipped instead of the mirror providing the expanded pump beam for the dipole antenna. In addition, the scan range, which is controlled by the delay line, is limited within the range that the terahertz wave from coherent LO phonon appears. The frequency of the GaAs LO phonon (8.8 THz) corresponds to the wavelength of 34 μm, which is much shorter than that of the terahertz wave from the dipole antenna (300 μm). Thus, the resolution of the image is improved by a factor of 10: the spatial resolution is estimated to be 0.1 mm. In the field of medical science, it was reported that the terahertz wave is sensitive to skin cancers (Wallace et al., 2004). Accordingly, the spatial resolution improvement contributes to the detection of the cancer at the earlier stage.

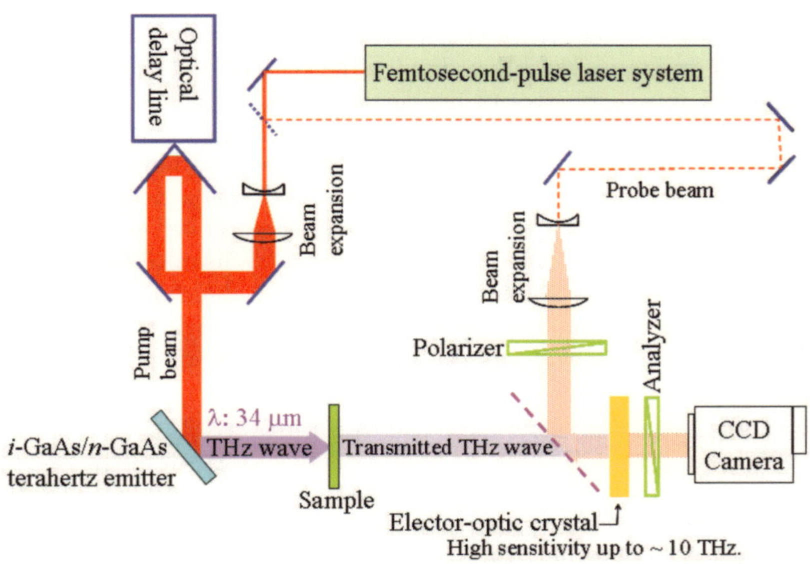

Fig. 12. Schematic view of the terahertz-wave-transmittance imaging system with use of the terahertz emission from the coherent LO phonon. The abbreviation "THz wave" corresponds to "terahertz wave".

5. Analysis of the epitaxial layer structures emitting the terahertz wave: direction reversal of the surface band bending in GaAs-based dilute nitride epitaxial layers

5.1 Relation between the polarity of the terahertz wave and surface band bending

In Section 4, we describe the relation between the photogenerated carrier transport process and terahertz-wave frequency. The results of Section 4 indicate that the emitted terahertz wave itself contains a large amount of information on an epitaxial layer structure for its source. In Section 5, we focus our attention on the relation between the polarity of the terahertz wave and the surface band bending.

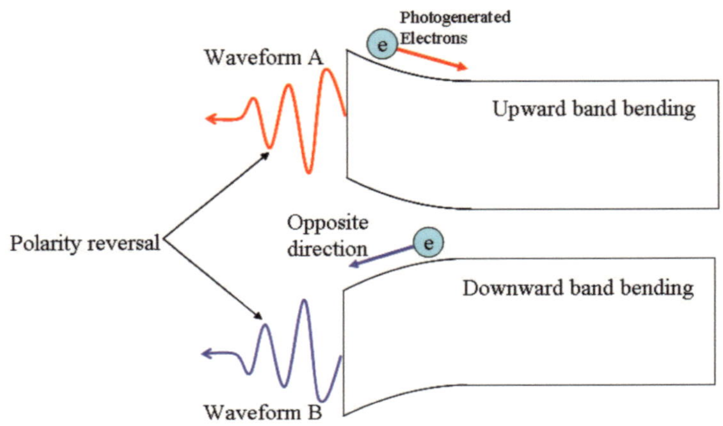

Fig. 13. Relation between the surface band bending and terahertz-wave polarity.

Figure 13 schematically shows the relation between the direction of the surge current and the surface band bending. According to Eq. (1), the electric field of the terahertz wave is proportional to the time derivative of the surge current. Equation (1) indicates that the polarity of the terahertz wave reflects the direction of the surge current. In the sample with the upward surface band bending, the photogenerated electrons flow into the inside of the crystal, and the terahertz wave is emitted with the waveform labeled by A. In contrast, in the sample with the downward band bending, the photogenerated electrons flows toward the surface, and the terahertz wave is emitted with the waveform labeled by B. Comparing the waveform A with the waveform B, it is apparent that the change in an electron-flow direction causes the reversal of the polarity of the terahertz wave; therefore, the terahertz-wave polarity is sensitive to the direction of the surface band bending.

5.2 Energy band structure of GaAs-based dilute nitrides

In this subsection, we briefly describe the reason why we focus our attention on the surface band bending of GaAs-based dilute nitrides. GaAs-based dilute nitrides, e.g. $GaAs_{1-x}N_x$ and $In_yGa_{1-y}As_{1-x}N_x$, have an interesting property that they exhibit giant negative bowing of the band-gap energy as a function of nitrogen content. For example, in $GaAs_{1-x}N_x$, the reduction of the band-gap energy is estimated to be 180 meV per nitrogen mole fraction of 1% (Walukiewicz et al., 2008). The mechanism of the band-gap energy reduction has been intensively investigated. According to the earlier work in GaAs-based dilute nitrides (Shan et al., 1999), a strong interaction between the conduction band of the host material and the nitrogen energy level causes a band anticrossing, which produces the lower and upper subbands. The lower subband corresponds to the band edge, and the band-gap energy exhibits a large negative bowing. Thus, the origin of the band-gap energy bowing has been identified.

It is considered that, like other compound semiconductors, the GaAs-based dilute nitride epitaxial layers also have a surface band bending, which is dominated by numerous deep levels on the surface. The above consideration gives rise to an interesting issue whether the incorporation of nitrogen modifies the surface band bending, which is the present motivation.

5.3 Samples and experimental procedures

The present samples were the undoped $GaAs_{1-x}N_x$ epitaxial layers with $x = 0.43\%$ and with $x = 1.53\%$, and an $In_yGa_{1-y}As_{1-x}N_x$ epitaxial layer with $x = 5.0\%$ and $y = 14\%$ grown by metal organic vapour phase epitaxy. The thicknesses of the $GaAs_{1-x}N_x$ and $In_yGa_{1-y}As_{1-x}N_x$ epitaxial layers were 500 nm. The *i*-GaAs(200 nm)/*n*-GaAs(3 μm, 3×10^{18} cm^{-3}) structure was also used as a reference sample. This is because, as described in Sections 3 and 4, the potential structure of the *i*-GaAs(200 nm)/*n*-GaAs sample has a linear potential slope categorized into an upward surface band bending. We also used a CrO-doped semi-insulating GaAs bulk crystal as a sample to evaluate of the effects of the incorporation of nitrogen.

The time-domain terahertz waves from the samples were measured at room temperature with use of laser pulses with a duration time of about 70 fs. The measurement system for the terahertz wave was the same shown in Fig. 5. The emitted terahertz beam was received by an optically gated bow-tie antenna with a gap of 5.0 μm formed on a low-temperature-grown GaAs. The power of the gate beam was fixed to 4.0 mW. A typical pump-beam power was 40 mW, and the wavelength was 800 nm. The phase of the lock-in amplifier was tuned with use of the signal of the *i*-GaAs(200 nm)/*n*-GaAs sample and was fixed in all measurements.

5.4 Polarity reversal and the origin of the modification of the surface band bending in GaAs-based diluten

Figure 14 shows the terahertz waveforms of the *i*-GaAs(200 nm)/*n*-GaAs, semi-insulating GaAs, $GaAs_{1-x}N_x$, and $In_yGa_{1-y}As_{1-x}N_x$ samples. All the samples show the first burst around the time delay of 0 ps. As the time delay increases, the polarity of the terahertz waveform changes from the negative to the positive in the *i*-GaAs(200 nm)/*n*-GaAs sample. The terahertz-waveform polarity of the semi-insulating GaAs sample is the same as that of the *i*-GaAs(200 nm)/*n*-GaAs sample, which indicates that the direction of the photogenerated current producing the terahertz wave is the same between the *i*-GaAs(200 nm)/*n*-GaAs and semi-insulating GaAs samples. Accordingly, it is considered that the present semi-insulating GaAs sample has an upward band bending at the surface region. As shown in Fig. 14, the amplitude of the first-burst of the semi-insulating GaAs sample is the smallest, which suggests that the surface band bending is relatively small.

In contrast, the terahertz-waveform polarities of the $GaAs_{1-x}N_x$ samples reverse in comparison with that of the *i*-GaAs(200 nm)/*n*-GaAs sample, which means that the surge current direction in the $GaAs_{1-x}N_x$ samples is opposite to that in the *i*-GaAs(200 nm)/*n*-GaAs sample. The $GaAs_{1-x}N_x$ samples, therefore, have a downward band bending at the surface

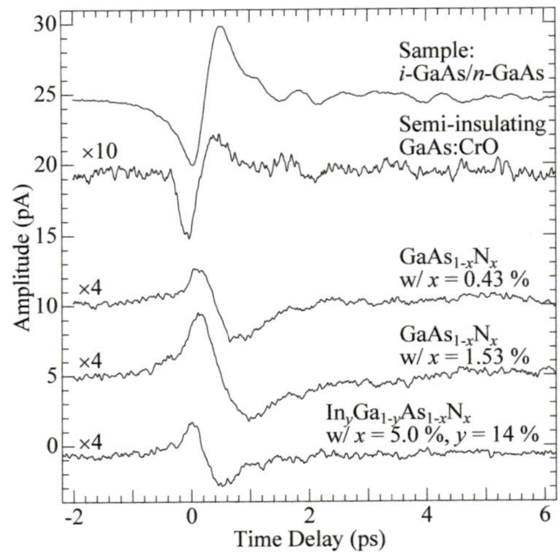

Fig. 14. Amplitudes of the terahertz waveforms of the *i*-GaAs(200 nm)/*n*-GaAs, semi-insulating GaAs, and $GaAs_{1-x}N_x$ (x = 0.43% and 1.53%), and $In_yGa_{1-y}As_{1-x}N_x$ samples as a function of time delay at room temperature. The pump-beam power was 40 mW.

region. It is noteworthy that, even in the $GaAs_{1-x}N_x$ sample with x = 0.43%, the polarity is inverted in comparison with that of the *i*-GaAs(200 nm)/*n*-GaAs sample. Comparing the polarity of the semi-insulating GaAs sample with the polarity of the $GaAs_{1-x}N_x$ sample with x = 0.43%, it is evident that the incorporation of the small amount of nitrogen changes the direction of the surface band bending. In addition, the amplitude of the $GaAs_{1-x}N_x$ sample with x = 1.53% is larger than the amplitude of the $GaAs_{1-x}N_x$ sample with x = 0.43%, which indicates that, in the $GaAs_{1-x}N_x$ samples, the magnitude of the downward band bending is enhanced with an increase in the incorporation of nitrogen. We also note that, in the $In_yGa_{1-y}As_{1-x}N_x$ sample, the polarity reversal of the terahertz waveform is observed. It is, therefore, concluded that the direction reversal of the surface band bending induced by the incorporation of nitrogen is universal in GaAs-based dilute nitrides.

Next, we discuss the mechanism causing the reversal of the direction of the surface band bending in the $GaAs_{1-x}N_x$ samples. In general, the surface Fermi level pinning originates from a large amount of deep levels at the surface locating within the forbidden band. The electronic wave functions of the deep levels are strongly localized in the atomic-order region. The average distance between the nitrogen atoms is estimated to be several ten nanometers in a $GaAs_{1-x}N_x$ epitaxial layer with x = 1%, taking account of the fact that the atomic monolayer thickness is 0.283 nm for the (001) direction in GaAs (Madelung, 2004). In addition, the energy of the nitrogen level locates above the conduction-band bottom. These facts suggest that the nitrogen incorporation can not disturb the electronic wave functions of the deep levels; therefore, the energies of the deep levels relative to the vacuum level are not influenced by the nitrogen incorporation.

The nitrogen incorporation, however, influences the conduction band according to the band anticrossing model. In the framework of this model, the conduction band of GaAs strongly interacts with the energy level of the incorporated nitrogen. As a result, the conduction band of GaAs splits into the upper (E_+) and lower (E_-) subbands, whose energies at Γ-point are expressed by the following equation (Walukiewicz et al., 2008):

$$E_{\pm} = \frac{1}{2}\left[\left(E_{g,GaAs} + E_N\right) \pm \sqrt{\left(E_{g,GaAs} - E_N\right)^2 + 4xC_{GaAs,N}^2}\right]. \quad (4)$$

The quantities of $E_{g,GaAs}$, E_N, and $C_{GaAs,N}$ are the fundamental transition energy of GaAs, the energy position of nitrogen-related level, and the hybridization matrix element, respectively (Walukiewicz et al., 2008). The energy level of the incorporated nitrogen locates above the conduction-band bottom by 226 meV, where the value of 226 meV corresponds to the energy difference between E_N and the conduction-band bottom of GaAs (Walukiewicz et al., 2008). The above interaction generates the E_+ and E_- subbands through the band anticrossing. We, for example, calculated the energies of the E_+ and E_- subbands in the $GaAs_{1-x}N_x$ sample with x = 1.53% using Eq. (4). In the calculation, the values of $E_{g,GaAs}$, E_N, and $C_{GaAs,N}$ are 1.424 (Madelung, 2004), 1.65, and 2.7 eV (Walukiewicz et al., 2008), respectively. The E_- and E_+ subband energies at the Γ point are estimated to be 1.890 and 1.184 eV, respectively. This result indicates that the incorporation of the nitrogen lowers the energy of the conduction band bottom by 240 meV in the $GaAs_{1-x}N_x$ sample with x = 1.53%. The bottom energy of the E_- subband relative to the vacuum level, therefore, becomes larger than the conduction bottom energy of GaAs relative to the vacuum level; namely, the E_--subband bottom approaches the deep levels responsible for the surface Fermi level pinning.

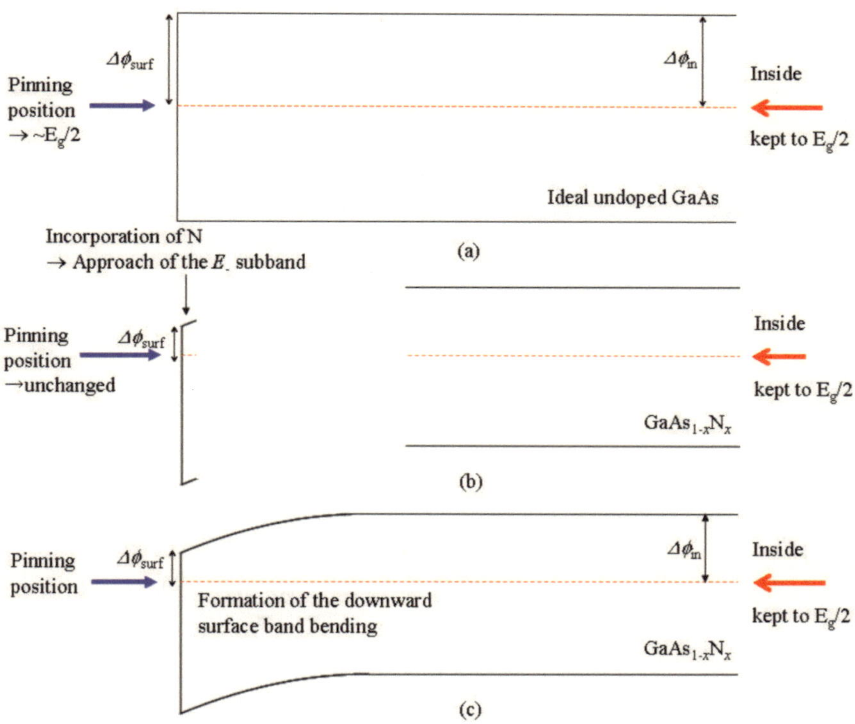

Fig. 15. Formation process of the downward surface band bending in $GaAs_{1-x}N_x$. (a) Before the incorporation of nitrogen. (b) Difference in the effects of the nitrogen incorporation between the inside and surface of the epitaxial layer. (c) Formation of the downward surface band bending.

In contrast, in the inside far away from the surface, the Fermi level in $GaAs_{1-x}N_x$ locates at almost the center of the band gap according to thermal statistical mechanics.
Here, using Fig. 15, we discuss the process the formation of the downward surface band bending in $GaAs_{1-x}N_x$. We denote the difference in energy between the conduction-band bottom and the surface Fermi level as $\Delta\phi_{surf}$. Figure 15(a) shows the band structure of ideal undoped GaAs. In the case where the nitrogen incorporation causes the downward shift of the E_- subband, the quantity of $\Delta\phi_{surf}$ becomes smaller, as shown in Fig. 15(b). In the same manner, we denote the energy difference between the conduction-band bottom and the Fermi level in the inside of the crystal as $\Delta\phi_{in}$. From the above discussion on the Fermi level in the inside of the crystal, it is apparent that $\Delta\phi_{in}$ is constant and that the value of $\Delta\phi_{in}$ is almost equal to the half of the band-gap energy. Thermal statistical mechanics tells us that, at equilibrium, the Fermi level is uniform from the surface to the inside in the crystal. It should be reminded that the surface Fermi level locates at the center of the band gap in GaAs, which means that the surface band bending is negligibly small in the ideal GaAs crystal (Fig. 15(a)): $\Delta\phi_{surf} - \Delta\phi_{in} \approx 0$. In $GaAs_{1-x}N_x$, even the small nitrogen incorporation makes the value of $\Delta\phi_{surf} - \Delta\phi_{in}$ negative owing to the band anticrossing producing E_- subband in the conduction band, which bends the conduction band downward as shown in Fig 15(c). Thus, it is concluded that the downward surface band bending in the $GaAs_{1-x}N_x$ epitaxial layer results from the shift of the E_- subband toward the surface Fermi level at the surface region. This is the origin of the polarity reversal of the terahertz waveform.

5.5 Precise evaluation of the $GaAs_{1-x}N_x$ epitaxial layer with use of the combination of the photoreflectance and terahertz-wave measurements

In Subsection 5.4, we demonstrate that the terahertz-wave measurement is applicable to the investigation of the surface band bending. As described in Section 4, the photoreflectance spectroscopic measurement is useful to estimate the built-in electric field, though this measurement provides only the electric field *strength*. Accordingly, the combination of the photoreflectance ant terahertz-wave measurements has the ability to precisely evaluate the surface band bending.

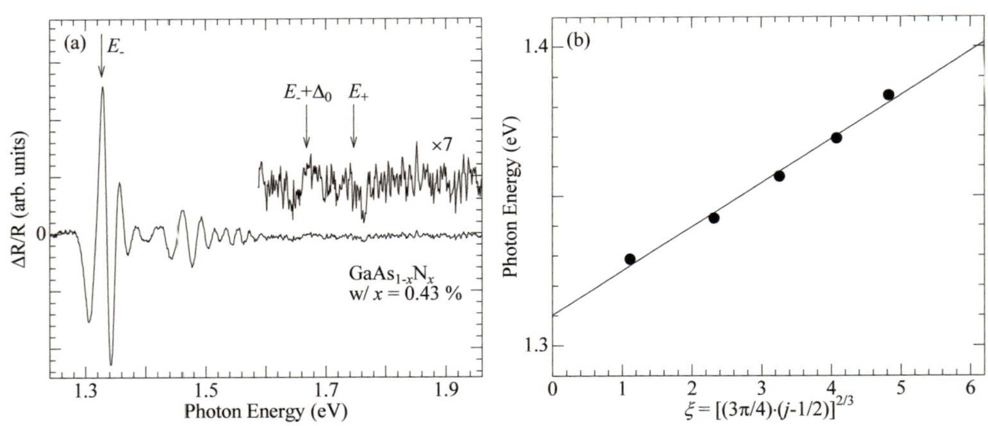

Fig. 16. (a) Photoreflectance spectrum of the $GaAs_{1-x}N_x$ sample with $x = 0.43\%$ at room temperature. (b) Linear plot of the extrema of the FKOs from the $GaAs_{1-x}N_x$ sample with $x = 0.43\%$ as a function of the quasi-index ξ.

Figure 16(a) shows the photoreflectance spectrum of the $GaAs_{1-x}N_x$ sample with $x = 0.43\%$. The position of the arrow labeled E_- at 1.327 eV (E_+ at 1.747 eV) corresponds to the energy of the transition between the lower (upper) conduction subband and heavy hole and/or light hole bands at the Γ-point, and the position of the arrow labeled $E_-+\Delta_0$ at 1.668 eV corresponds to the E_--subband--split-off-hole-band transition at the Γ-point, where the energies of E_-, E_+, and $E_-+\Delta_0$ are calculated on the basis of the band anticrossing model (Walukiewicz et al., 2008). Consequently, the oscillation pattern starting from about 1.3 eV is assigned to the FKOs from the $GaAs_{1-x}N_x$ layer, while the oscillation pattern starting from about 1.4 eV is assigned to the FKOs from the GaAs buffer layer. In order to estimate the surface electric field, as shown in Fig. 16(b), the extrema of the FKOs from the $GaAs_{1-x}N_x$ sample with $x = 0.43\%$ are plotted as a function of the quasi-index ξ. In the same manner applied in Section 4, the surface electric field is estimated to be 24 kV/cm using an interband reduced mass of $0.0689m_0$ (Walukiewicz et al., 2008). In general, undoped GaAs crystals have a surface electric field of several kV/cm at most. The present results indicate that the nitrogen incorporation enhances the surface electric field.

6. Conclusion

In the present chapter, we have described the terahertz electromagnetic wave from semiconductor epitaxial layer structures. In Section 2, we have pointed out the problems in photoconductive-antenna-based terahertz emitters. In Sections 3 and 4, we have focused our attention on the *i*-GaAs(*d* nm)/*n*-GaAs(3 μm, 3×10^{18} cm^{-3}) epitaxial layer structures. It has been clarified that the appropriate design of the epitaxial layer, which is based on semiconductor physics, is effective to control the characteristics of the terahertz wave: the emission intensity and frequency. In addition, we have elucidated the photogenerated carrier transport process in the sub-picosecond range. Furthermore, we have demonstrated the relatively intense terahertz emission from the coherent LO phonon and have proposed an example of its application. In Section 5, we have indicated that the terahertz wave contains the information of its source materials. From the polarity reversal of the terahertz waveform, it has been clarified that the GaAs-based dilute nitride ($GaAs_{1-x}N_x$ and $In_yGa_{1-y}As_{1-x}N_x$) epitaxial layers have a downward band bending owing to the band anticrossing induced by the nitrogen incorporation. This finding suggests that the terahertz wave has the ability not only to the probe the samples under test but also to reveal the physical properties of the materials emitting the terahertz wave.

The present approach to the terahertz wave is quite different from the approach used in the research field of microwaves because antenna structures are out of the scope. This is because the terahertz waves have a frequency range between infrared light and microwaves; namely, the terahertz waves have the properties both of light and of microwaves. However, on the basis of Faraday's law of induction, the generation mechanism of the terahertz wave is the same as that of the microwave. In fact, the terahertz wave from the coherent LO phonon is categorized into the typical dipole radiation. Accordingly, we conclude that the progress in the terahertz-wave technology depends on the fusion of the conventional concepts in the research fields of optics and of microwaves.

In the present chapter, we have focused our attention on the terahertz emitters. It is also important to precisely investigate the detection mechanism/method of the terahertz wave. To finalize the present chapter, we point out the above-mentioned issue.

7. Acknowledgements

The author acknowledges Professor Masaaki Nakayama, Mr. Syuichi Tsuruta (Osaka City University), and Dr. Takayuki Hasegawa (Hyogo Prefecture University) for their aid and useful discussion. The author also appreciates Dr. Hisashi Yamada and Dr. Masahiko Hata (Tsukuba Research Laboratory, Sumitomo Chemical Co., Ltd.) for the growth of the *i*-GaAs/*n*-GaAs structures. The $GaAs_{1-x}N_x$ and $In_yGa_{1-y}As_{1-x}N_x$ epitaxial layers were provided from Sumitomo Electric Industries. The present work is partially supported by Grant-in-Aid for Young Scientists (B) No. 22760010 from Japan Society for the Promotion of Science.

8. References

8.1 Journal papers

Abe, H.; Harima, H.; Nakashima, S.; Tani, M.; Sakai, K.; Tokuda, Y.; Kanamoto, K. & Abe, Y. (1996). Characterization of Crystallinity in Low-Temperature-Grown GaAs Layers by Raman Scattering and Time-Resolved Photoreflectance Measurements, *Japanese Journal of Applied Physics,* Vol. 35, Part 1, No. 12A, pp. 5955-5963, ISSN: 00214922.

Aspnes, D. E. (1974). Band nonparabolicities, broadening, and internal field distributions: The spectroscopy of Franz-Keldysh oscillations, *Physical Review B,* Vol. 10, Issue 10, pp. 4228-4238, ISSN: 10980102.

Aspnes, D. E. (1983). Recombination at semiconductor surfaces and interfaces, *Surface Science,* Vol. 132, Issue 1-3, pp. 406-421, ISSN: 00396028.

Auston, D. H. (1975). Picosecond optoelectronic switching and gating in silicon, *Applied Physics Letters,* Vol. 26, Issue 3, pp. 101-103, ISSN: 00036951.

Basore, P. A. (1990). Numerical modeling of textured silicon solar cells using PC-1D, *IEEE Transaction on Electron Devices,* Vol. 37, No. 2, pp. 337-343, ISSN: 00189383.

Blakemore, J. S. (1982). Semiconducting and other major properties of gallium arsenide, *Journal of Applied Physics,* Vol. 53, Issue 10, pp. R123-R181, ISSN: 00218979.

Cho, G. C.; Kütt, W. & Kurz, H. (1990). Subpicosecond Time-Resolved Coherent-Phonon Oscillations in GaAs, *Physical Review Letters,* Vol. 65, Issue 6, pp. 764-766, ISSN: 00319007.

Gupta, S.; Frankel, M. Y.; Valdmanis, J. A.; Whitaker, J. F.; Mourou, G. A.; Smith, F. W. & Calawa, A. R. (1991). Subpicosecond carrier lifetime in GaAs grown by molecular beam epitaxy at low temperatures, *Applied Physics Letters,* Vol. 59, Issue 25, pp. 3276-3278, ISSN: 0036951.

Heyman, J. N.; Coates, N.; Reinhardt, A. & Strasser, G. (2003). Diffusion and drift in terahertz emission at GaAs surfaces, *Applied Physics Letters,* Vol. 83, Issue 26, pp. 5476-5478, ISSN: 0036951.

Hu, B. B. & Nuss, M. C. (1995). Imaging with terahertz waves, *Optics Letters,* Vol. 20, Issue 16, pp. 1716-1718, ISSN: 01469592.

Huang, H. C.; Yee, S. & Soma, M. (1990). The carrier effects on the change of the refractive index for *n*-type GaAs at λ = 1.06, 1.3, and 1.55 μm, *Journal of Applied Physics,* Vol. 67, Issue 3, pp. 1497-1503, ISSN: 00218979.

Mizoguchi, K.; Furuichi, T.; Kojima, O.; Nakayama, M.; Saito, S.; Syouji, A. & Sakai, K. (2005). Intense terahertz radiation from longitudinal optical phonons in GaAs/AlAs multiple quantum wells, *Applied Physics Letters,* Vol. 87, Issue 9, pp. 093102 1-3, ISSN: 0036951.

Nakayama, M.; Ito, S.; Mizoguchi, K.; Saito, S. and Sakai, K. (2008). Generation of Intense and Monochromatic Terahertz Radiation from Coherent Longitudinal Optical Phonons in GaAs/AlAs Multiple Quantum Wells at Room Temperature, *Applied Physics Express*, Vol. 1, No. 1, pp. 012004 1-3, ISSN: 18829778.

Nakayama, M. & Mizoguchi, K. (2008). Interactions between coherent optical phonons and excitonic quantum beats in GaAs/AlAs multiple quantum wells: strategy for enhancement of terahertz radiation from coherent optical phonons, *Physica Status Solidi C*, Vol. 5, Issue 9, pp. 2911-2916, ISSN: 16101634.

Nelson, F. R.; Miller, C.; Tu, C. W. & Sputz, S. K. (1987). Exciton binding energies from an envelope-function analysis of data on narrow quantum wells of integral monolayer widths in $Al_{0.4}Ga_{0.6}As$/GaAs, *Physical Review B*, Vol. 36, Issue 15, pp. 8063–8070, ISSN: 10980102.

Othonos, A. (1998). Probing ultrafast carriers and phonon dynamics in semiconductors, *Journal of Applied Physics*, Vol. 83, Issue 4, pp. 1789-1830, ISSN: 00218979.

Pollak, F. H. & Shen, H. (1993). Modulation spectroscopy of semiconductors: bulk/thin film, microstructures, surfaces/interfaces and devices, *Materials Science and Engineering R*, Vol. 10, Issue 7-8, pp. 275-374, ISSN: 0927796X.

Sarukura, N.; Ohtake, H.; Izumida, S. & Liu, Z. (1998). High average-power THz radiation from femtosecond laser-irradiated InAs in a magnetic field and its elliptical polarization characteristics, *Journal of Applied Physics*, Vol. 84, Issue 1, pp. 654-656, ISSN: 00218979.

Shan, W.; Walukiewics, W.; Ager, J. W., III; Haller, E. E.; Geisz, J. F.; Friedman, D. J.; Olson, J. M. & Kurz, S. R. (1999). Band Anticrossing in GaInNAs Alloys, *Physical Review Letters*, Vol. 82, Issue 6, pp. 1221-1224, ISSN: 00319007.

Shen, H.; Dutta, M.; Fotiadis L.; Newman, P. G.; Moerkirk, R. P.; Chang, W. H. & Sacks, R. N. (1990). Photoreflectance study of surface Fermi Level in GaAs and GaAlAs, *Applied Physics Letters*, Vol. 57, Issue 20, pp. 2118-2120, ISSN: 0036951.

Takeuchi, H.; Kamo, Y.; Yamamoto, Y.; Oku, T.; Totsuka, M. & Nakayama, M. (2005). Photovoltaic effects on Franz-Keldysh oscillations in photoreflectance spectra: Application to determination of surface Fermi level and surface recombination velocity in undoped GaAs/*n*-type GaAs epitaxial layer structures, *Journal of Applied Physics*, Vol. 97, Issue 6, pp. 093539 1-16, ISSN: 00218979.

Takeuchi, H.; Yanagisawa, J.; Hasegawa, T. & Nakayama, M. (2008). Enhancement of terahertz electromagnetic wave emission from an undoped GaAs/*n*-type GaAs epitaxial layer structure, *Applied Physics Letters*, Vol. 93, Issue 8, pp. 081916 1-3, ISSN: 0036951.

Takeuchi, H.; Yanagisawa, J.; Hashimoto, J. & Nakayama, M. (2009). Effects of nitrogen incorporation on a direction of a surface band bending investigated by polarity of terahertz electromagnetic waves in $GaAs_{1-x}N_x$ epitaxial layers, *Journal of Applied Physics*, Vol. 105, Issue 9, pp. 093539 1-4, ISSN: 00218979.

Takeuchi, H.; Yanagisawa, J.; Tsuruta, S.; Yamada, H.; Hata, M. & Nakayama, M. (2010). Frequency Shift of Terahertz Electromagnetic Waves Originating from Sub-Picosecond-Range Carrier Transport in Undoped GaAs/*n*-type GaAs Epitaxial Layer Structures, *Japanese Journal of Applied Physics*, Vol. 49, No. 8, pp. 082001 1-5, ISSN: 00214922.

Wallace, V. P.; Fitzgerald, A. J.; Shankar, S.; Flanagan, N.; Pye, R.; Cluff, J. & Arnone D. D. (2004). Terahertz pulsed imaging of basal cell carcinoma *ex vivo* and *in vivo*, *British Journal of Dermatology*, Vol. 151, Issue 2, pp. 424-432, ISSN: 00070963.

Wieder, H. H. (1983). Surface Fermi level of III–V compound semiconductor-dielectric interfaces, *Surface Science*, Vol. 132, Issue 1-3, pp. 390-405, ISSN: 00396028.

Yamamoto, K.; Yamaguchi, M.; Miyamaru, F.; Tani, M.; Hangyo, M.; Ikeda, T.; Matsushita, A.; Koide, K.; Tatsuno, M. & Minami, Y. (2004). Noninvasive Inspection of C-4 Explosive in Mails by Terahertz Time-Domain Spectroscopy, *Japanese Journal of Applied Physics*, Vol. 43, No. 3B, pp. L414 -L417, ISSN: 00214922.

8.2 Conference papers

Clugston, D. A. & Basore P. A. (1998). PC1D VERSION 5: 32-BIT SOLAR CELL MODELING ON PERSONAL COMPUTERS, *Conference Record of the 26th IEEE Photovoltaic Specialists Conference*, pp. 207-210, ISBN: 0780337670, Anaheim, CA, U.S.A., September 1997, IEEE, Piscataway, NJ.

8.3 Books and book chapters

Bolivar, P. H. (1999). Coherent terahertz spectroscopy, In: *SEMICONDUCTOR QUANTUM OPOELECTRONICS*, Millar, A.; Ebrahimzadeh, M & Finlayson, D. M. (Eds.), pp. 151-192, Institute of Physics, ISBN: 0750306297, Bristol.

Dekorsy, T.; Cho, G. C. & Kurtz, H. (2000). Coherent Phonons in Condensed Media, In: *Light Scattering in Solids VIII*, Cardona, M. & Güntherodt, G (Eds.), pp. 169-209, Springer, ISBN: 3540660852, Berlin.

Gu, P. & Tani, M. (2005). Terahertz Radiation from Semiconductor Surface, In: *Terahertz Optoelectronics*, Sakai, K. (Ed.), pp. 63-97, Springer, ISBN: 3540200130, Berlin.

Herman, M.; Fukasawa, R. & Morikawa, O. (2005). Terahertz Imaging, In: *Terahertz Optoelectronics*, Sakai, K. (Ed.), pp. 203-270, Springer, ISBN: 3540200130, Berlin.

Madelung, O. (2004). *Semiconductors -Basic Data (2nd ed.)*, Springer, ISBN: 3540608834, Berlin.

Nishizawa, S.; Sakai, K.; Hangyo, M.; Nagashima, T.; Wada, M.; Tominaga, K.; Oka, A; Tanaka, K, & Morikawa, O. (2005). Terahertz Time-Domain Spectroscopy, In: *Terahertz Optoelectronics*, Sakai, K. (Ed.), pp. 203-270, Springer, ISBN: 3540200130, Berlin.

Nuss, M. C. & Orenstein, J. (1999). Terahertz Time-Domain Spectroscopy, In: *Millimeter and Submillimeter Wave Spectroscopy of Solids*, Grüner, G. (Ed.), pp. 7-50, Springer, ISBN: 3540628606, Berlin.

Ohtake, H.; Ono, S. & Sarukura, N. (2005). Enhanced Generation of Terahertz Radiation from Semiconductor Surface with External Magnetic Field, In: *Terahertz Optoelectronics*, Sakai, K. (Ed.), pp. 99-116, Springer, ISBN: 3540200130, Berlin.

Sakai, K. & Tani, M. (2005). Introduction to Terahertz Pulse, In: *Terahertz Optoelectronics*, Sakai, K. (Ed.), pp. 1-30, Springer, ISBN: 3540200130, Berlin.

Tomizawa, K. (1993). *Numerical Simulation of Submicron Semiconductor Devices*, Artech House, ISBN: 0890066205, Boston.

Walukiewicz, W.; Alberi, K.; Wu, J.; Shan, W.; Yu, K. M. & Ager, J. W., III. (2008). Electronic Band Structure of Highly Mismatched Semiconductor Alloys, In: *Dilute III-V Nitride Semiconductors and Material Systems*, Erol, A. (Ed.), pp. 65-89, Springer, ISBN: 9783540745280, Berlin.

7

Wave Propagation in Dielectric Medium Thin Film Medium

E. I. Ugwu, Senior Lecturer
Department of Industrial Physics, Ebonyi State University, Abakaliki, Nigeria

1. Introduction

Various tools have been employed in studying and computing beam or field propagation in a medium with variation of small refractive index Feit and (Fleck,et al,1976)(Fleck,1978)(Ugwu et al,2007) some researchers had employed beam propagation method based on diagonalization of the Hermitian operator that generates the solution of the Helmholtz equation in media with real refractive indices (Thylen and Lee,1992), while some had used 2x2 propagation matrix formalism for finding the obliquely propagated electromagnetic fields in layered inhomogeneous un-axial structure (Ong,1993)

Recently, we have looked at the propagation of electromagnetic field through a conducting surface (Ugwu, 2005) and we observed the behaviour of such a material. The effect of variation of refractive index of F_eS_2 had also been carried out (Ugwu, 2005)

The parameters of the film that were paramount in this work are dielectric constants and the thickness of the thin film.

The dielectric constants were obtained from a computation using pseudo-dielectric function in conjunction with experimentally measured extinction co-efficient and the refractive indices of the film and the thickness of the film which was assumed to range from 0.1μm to 0.7μm [100nm to 700nm] based on the experimentally measured value, at the wavelength, 450μm (Cox, 1978)(Lee and Brook,1978)

This work is based on a method that involves propagating an input field over a small distance through the thin film medium and then iterating the computation over and over through the propagation distance using Lippmann-Schwinger equation and its counterpart, Dyson's equation (Economou, 1979) here, we first derived Lippmann-Schwinger equation using a specific Hamiltonian from where the field function ψ_k (z) was obtained. From this, it was observed that to ease out the solution of the Lippmann-Schwinger equation, it was discretized. After this, Born approximation was applied in order to obtain the solution. The formalism is logically built up step-by-step, which allowed point-to-point observation of the behaviour of the field propagating through the film inorder to analyze the influence of the field behavior as it propagates through small thin film thickness with consideration given to solid state properties such as dielectric function/refractive index, determine the absorption characteristics within the wavelength regions of electromagnetic wave spectrum and then to have a view of the influence of the dielectric function on the amplitude of the propagated field from the theoretical solution of the scalar wave equation considered within the ultra violet, optical and near infrared. The advantage of this approach is clearly glaring as it provides a good picture of the field in a medium with variation of dielectric constant, refractive index and above all, the method requires no resolution of a system of equations as it can accommodate multiple layers easily.

2. Theoretical procedure

This our method is to find a solution $\psi(r)$ of the scalar wave equation

$$\nabla^2\psi(r)+\omega^2\varepsilon_o\mu_o\varepsilon(r)\psi(r) = 0 \tag{1}$$

for arbitrary complex dielectric medium permittivity $\varepsilon_p(r)$ of homogeneous permeability μ_o starting with Halmitonian. In equation (1) we assume the usual time dependence, exp(-iωt), for the electromagnetic field $\psi(r)$. Such a scalar field describes, for instance, the transverse electric modes propagating in thin media deposited on glass slide using solution growth technique (Ugwu, 2005).

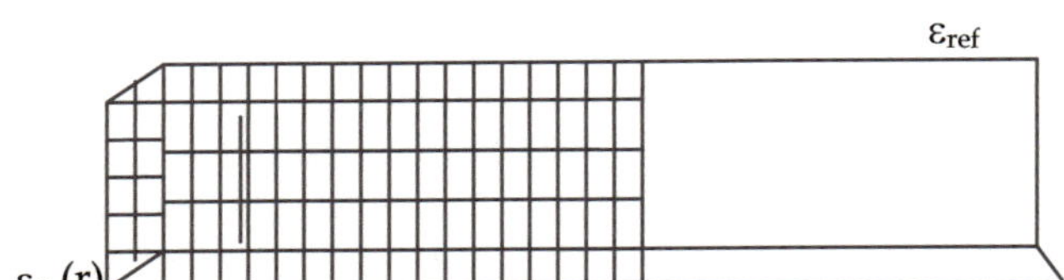

ε_{ref} is reference medium
$\varepsilon_p(r)$ is perturbed medium.

Fig. 1. Geometry used in the model. The dielectric medium for which we see a solution of the wave equation can be split into two parts; reference homogeneous medium, ε_{ref}, and a perturbed medium where the film is deposited $\varepsilon_p(r)$

The assumption made here regarding the dielectric medium is that it is split into two parts; a homogenous reference medium of dielectric constant ε_{ref} and a perturbation $\varepsilon_p(r)$ confined to the reference medium. Hence, the dielectric function of the system can be written as

$$\varepsilon_p(r)=\varepsilon_{ref}+\Delta\varepsilon_p(r) \tag{2}$$

Where $\Delta\varepsilon_p(r) = \varepsilon_p(r) - \varepsilon_{ref}$. The assumption here can be fulfilled easily where both reference medium and the perturbation depend on the problem we are investigating. For example, if one is studying an optical fiber in vacuum, the reference medium is the vacuum and the perturbation describes the fiber. For a ridge wedge, wave guide the reference medium is the substrate and the perturbation is the ridge, In our own case in this work, the reference medium is air and the perturbing medium is thin film deposited on glass slide. Lippman-Schwinger equation is associated with the Hamiltonian H which goes with $H_0 + V$
Where H_0 is the Hamiltonian before the field penetrates the thin field and V is the interaction.

$$H_0\Phi_k\rangle = E_k\Phi_k\rangle \tag{3}$$

The eigenstate of $H_0 + V$ is the solution of

$$(E_k - H_0)\Psi_k(z)\rangle = V \mid \Psi_k(z)\rangle \tag{4a}$$

Here, z is the propagation distance as defined in the problem

$$\mid \Psi_k(z)\rangle = \left| \Phi_k + \frac{1}{E_k - H_0} \right|_n V \mid \Psi_k(z)\rangle \tag{4b}$$

Where η is the boundary condition placed on the Green's function $(E_k\text{-}H_0)^{-1}$. Since energy is conserved, the propagation field component of the solutions will have energy E_n with the boundary conditions that only handle the singularity when the eigenvalue of H_0 is equal to E_k. Thus we write;

$$\mid \Psi_k(z)\rangle = \left| \Phi_n + \frac{1}{E_k - H_0 + i\eta} \right|_n V \mid \Psi_k^f(z)\rangle \tag{5}$$

as the Lippman-Schwinger equation without singularity; where η is a positively infinitesimal, $\left|\Psi_k^f(z)\right\rangle$ is the propagating field in the film while $\left|\Psi_k^f(z)\right.$ is the reflected. With the above equation (4) and (5) one can calculate the matrix elements with (z) and insert a complete set of z and Φ_k states as shown in equation (6).

$$\langle z \mid \psi_k(z) = \langle z \mid \Phi_k\rangle + \int d^3z^! \int \frac{d^3k^!}{(2\pi)^3} \langle z \mid \frac{1}{E_k - H + in} \mid \Phi_{k^!}\rangle\langle\Phi_{k^!} \mid J^! \mid z^!\rangle\langle z^! \mid \psi_z \tag{6}$$

$$\Psi_k(z)\rangle = e^{fk} + \int d^3z' \int \frac{d^3k'}{(2\pi)^3} \frac{e^{-fk(z-z')}}{E_k - E_k + i\eta} V(z')\Psi_k(z) \tag{7}$$

$$G(z) = \int \frac{d^3k^!}{(2\pi)^3} \frac{e^{-fk(z-z!)}}{E_k - E_k + i\eta} \tag{8}$$

is the Green's Function for the problem, which is simplified as:

$$G(z) = \int \frac{m}{2\pi^2 h^2 z} \int_{-\infty}^{\infty} dk^! \frac{\sin k^! z}{K^{!2} - (k + i\eta)^2} \tag{9}$$

When η ≈0 is substituted in equation (9) we have

$$\Psi_k(z) = e^{ikz} - \frac{m}{2\pi^2 h^2} \int_{-\infty}^{\infty} d^3z^! \frac{e^{ik(z-z^!)}}{\left|z - z^!\right|} V(z^!)\Psi_k(z^!) \tag{10}$$

The perturbated term of the propagated field due to the inhomogeneous nature of the film occasioned by the solid-state properties of the film is:

$$\Psi_k(z) = -\frac{m}{2\pi^2 h^2} \int_{-\infty}^{+\infty} d^3z' \frac{e^{fk(z-z')}}{\left|z - z'\right|} V(z')\Psi_k(z') \tag{11a}$$

$$\Psi_k(z) = -\frac{1}{4\pi}\frac{2}{h^2}\Delta_{k'k} \tag{11b}$$

Where $\Delta_{kk'}$ is determined by variation of thickness of the thin film medium and the variation of the refractive index (Ugwu,et al 2007) at various boundary of propagation distance. As the field passes through the layers of the propagation distance, reflection and absorption of the field occurs thereby leading to the attenuation of the propagating field on the film medium. Blatt, 1968

3. Iterative application

Lippman-Schwinger equation can be written as

$$\Psi_k(z) = \Psi_k^o + \int dz' G^o \Delta\varepsilon_p(z')\Psi(z') \tag{12}$$

Where $G^o(z,z')$ is associated with the homogeneous reference system, (Yaghjian1980) (Hanson,1996)(Gao et al,2006)(Gao and Kong1983)
The function

$$V(z) = -k_o^2 \Delta\varepsilon_p(z) \tag{13}$$

define the perturbation
Where

$$k_o^2 = \frac{c^2}{\lambda^2}\varepsilon_o \mu_o \tag{14}$$

The integration domain of equation (12) is limited to the perturbation. Thus we observe that equation (12) is implicit in nature for all points located inside the perturbation. Once the field inside the perturbation is computed, it can be generated explicitly for any point of the reference medium. This can be done by defining a grid over the propagation distance of the film that is the thickness. We assume that the discretized system contains $\Delta_{kk'}$ defined by T/N.
Where T is thickness and N is integer
(N= 1, 2, 3, N - 1). The discretized form of equation (12) leads to large system of linear equation for the field;

$$\Psi_i = \Psi_i^o + \sum_{k=1}^{\Delta} G_{i,k}^o V_k \Delta_k \Psi_k \tag{15}$$

$$\Psi_i = \Psi_i^o + \sum_{k=1}^{\Delta} G_{i,k}^o V_k \Delta_k \Psi_i \tag{16}$$

However, the direct numerical resolution of equation (15) is time consuming and difficult due to singular behaviour of $G_{i,k}^o$. As a result, we use iterative scheme of Dyson's equation, which is the counter part of Lippman-Schwinger equation to obtain $G_{i,k}^o$. Equation 10 is easily solved by using Born approximation, which consists of taking the incident field in place of the total field as the driving field at each point of the propagation distance. With this, the propagated field through the film thickness was computed and analyzed.

$$\Psi(z)=\Psi_o(z)e^t(\lambda.z-\omega t) \tag{17}$$

From equation (1),

Let $\lambda^2 = \mu_O\varepsilon_O\omega^2\varepsilon_{ref} + i\mu_O\varepsilon_O\omega^2\varepsilon_{ref}\Delta\varepsilon_p(z)$

$$|\lambda^2 = \left[\varepsilon_{ref} + i\Delta\varepsilon_p(z)\right]^{1/2} \omega[\mu_O\varepsilon_O]^{1/2} \tag{18}$$

$$= k[\varepsilon_{ref} + \Delta\varepsilon_p(z)]^{1/2} \mu_O\varepsilon_O\left[\frac{1}{o} = (\mu_o\varepsilon_O)^{1/2}\right]$$

$$= k\left[\varepsilon_{ref} + \Delta\varepsilon_p(z)\right]^{1/2} \tag{19}$$

Expanding the expression up to 2 terms, we have

$$= k\left[\varepsilon_{ref} + \frac{1}{2}i\Delta\varepsilon_p\right] \tag{20}$$

Where $\Delta\varepsilon_p$ gives rise to exponential damping for all frequencies of field radiation of which its damping effect will be analyzed for various radiation wavelength ranging from optical to near infra-red

The relative amplitude

$$\frac{\psi(z)}{\psi_o(z)} = \exp\left(-\frac{K}{2}\Delta\varepsilon_p(z)\right)z\ \exp[ik\varepsilon_{ref} - \omega t] \tag{21}$$

Decomposing equation (3.18) into real and complex parts, we have the following

$$\frac{\psi(z)}{\psi^o(z)} = \left(\exp -\frac{K}{2}\Delta\varepsilon_p(z)\right)z\ \cos k\varepsilon_{ref} - \omega t \tag{22}$$

$$\frac{\psi(z)}{\psi^o(z)} = \left(\exp -\frac{K}{2}\Delta\varepsilon_p(z)\right)z\ \sin k\varepsilon_{ref} - \omega t \tag{23}$$

Considering a generalised solution of the wave equation with a damping factor

$$\psi(r) = \psi_o(r)\ \exp\text{-}\ \beta\ \exp\text{-}i(\alpha z - \omega t) \tag{24}$$

[Smith et al, 1982].

in which k_{ref} is the wave number β is the barrier whose values describe our model. With this, we obtain the expression for a plane wave propagating normally on the surface of the material in the direction of z inside the dielectric film material. Where $-\beta$ describes the barrier as considered in our model.

When a plane wave $\psi_o(z) = \exp(i\ k_{ref}\ z)$ with a wave number corresponding to the reference medium impinges upon the barrier, one part is transmitted, the other is reflected or in some cases absorbed (Martin et al 1994). This is easily obtained with our method as can be illustrated in Fig 2 present the relative amplitude of the computed field accordingly.

Three different cases are investigated:

a. When the thin film medium is non absorbing, in this case $\Delta\varepsilon_p(z)$ is considered to be relatively very small.
b. When the film medium has a limited absorption, $\Delta\varepsilon_p(z)$ is assumed to have a value slightly greater than that of (a)
c. When the absorption is very strong, $\Delta\varepsilon_p(z)$ has high value. In this first consideration, λ_{eff} = 0.4μm, 0.70μm, 0.80μm and 0.90μm while z =0.5μm as a propagation distance in each case. In each case, the attenuation of the wave as a function of the absorption in the barrier is clearly visible in graphs as would be shown in the result.

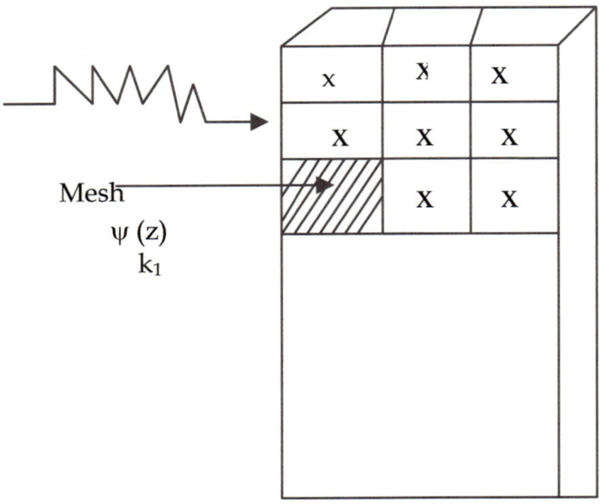

Fig. 2. Geometrical configuration of the model on which a wave propagates. The description is the same as in fig 1

4. Beam propagated on a mesh of the thin film

We consider the propagation of a high frequency beam through an inhomogeneous medium; the beam propagation method will be derived for a scalar field. This restricts the theory to cases in which the variation in refractive index is small or in which a scalar wave equation can be derived for the transverse electric, (T.E) or transverse magnetic, (T.M) modes separately. We start with the wave equation (Martin et al, 1994; Ugwu et al, 2007):

$$\nabla^2\psi + K^2 n^2 (z)\, \psi = 0 \tag{26a}$$

where ψ represents the scalar field, n(z) the refractive index and K the wave number in vacuum. In equation 26a, the refractive index n^2 is split into an unperturbed part n_o^2 and a perturbed part Δn^2; this expression is given as

$$n^2 z(=)\, n_o^2 + \Delta n^2(z) \tag{26b}$$

Thus

$$\nabla^2\psi + k^2 n_o^2 (z) = \rho(z) \tag{27}$$

where ρ (z) is considered the source function. The refractive index is $n^2_o + \Delta n^2(z)$ and the refractive index $n_o^2(z)$ is chosen in such a way that the wave equation

$$\nabla^2\psi + k^2 n_o^2(z)\,\psi = 0 \tag{29}$$

together with the radiation at infinity, can be solved. If a solution ψ for $z = z_o$ the field ψ and its derivative $\frac{\partial \Psi}{\partial z}$ can be calculated for all values of z by means of an operator â;

$$\frac{\partial \Psi}{\partial z} = \hat{a}\,\psi(x, y, z_o) \tag{30}$$

where the operator â acts with respect to the transverse co-ordinate (x, y) only (VanRoey et al 1981)

We considered a beam propagating toward increasing z and assuming no paraxiality, for a given co-coordinate z, we split the field ψ into a part ψ_1 generated by the sources in the region where $z_1 < z$ and a part of ψ_2 that is due to sources with $z_1 > z$;

$$\Psi = \psi_1 + \psi_2 \tag{31}$$

An explicit expression for ψ_1 and ψ_2 can be obtained by using the function (Van Roey et al, 1981).

$$e_1(z / z_1) = \begin{cases} {}^{o}\, for z < z_1 \\ 1/2\, for z = z_1 \\ 1\, for z > z_1 \end{cases} \tag{32}$$

If G is the Green's function of equation 26a ψ_1 can be formally expressed as follows:

$$\psi_1(z) = \iiint_{\infty}^{+\infty} G(z, z_1)\; e_1(z, z_1)\,\rho(z_1)\,d\,z_1 \tag{33}$$

that leads to

$$\frac{\partial \Psi_1}{\partial z} = \iiint_{\infty}^{+\infty} \frac{\partial G}{\partial z}(z, z_1)\,\rho(z_1)\,d^3 z_1$$

$$+ \iiint_{-\infty}^{+\infty} G(z, z_1)\delta(z - z_1)\,\rho(z_1)\,d^3 z_1 \tag{34}$$

The first integral represents the propagation in the unperturbed medium, which can be written in terms of the operator â defined in equation 34 as

$$\frac{\partial \psi(z)}{\partial z} = \hat{a}\,\psi_1 \tag{35}$$

and ψ_1 being generated by sources situated to the left of z.

The second part of the integral is written with assignment of an operator **ƀ** acting on ψ with respect to the transverse coordinate (x, y) only. Such that we have

$$\frac{\partial \Psi}{\partial z} = \hat{a}\Psi_1 + \hat{b}\Psi_2 \tag{36}$$

(Now neglecting the influence of the reflected field ψ_2 on ψ_1) we use ψ_1 instead of ψ_2 and equation 36 becomes

$$\frac{\partial^{\psi_1}}{\partial z} = \hat{a}\psi_1 + \boldsymbol{b}\ \boldsymbol{\Psi} \tag{37}$$

Equation 37 is an important approximation and restricts the use of the B.P.M to the analysis of structures for which the influence of the reflected fields on the forward propagation field can be neglected in equation (36) ψ_1 describes the propagation in an unperturbed medium and a correction term representing the influence of Δn. Since equation (30) is a first order differential equation, it is important as it allows us to compute the field ψ_1 for $z > z_o$ starting from the input beam on a plane of our model $z = z_o$. Simplifying the correction term (Van Roey et al, 1981), we have $\boldsymbol{b}$

$$\boldsymbol{b}\boldsymbol{\Psi}_1 = -\frac{ik}{2n_o}\Delta n^2 \boldsymbol{\Psi}_1 \tag{38}$$

Equation 3.18 becomes

$$\frac{\partial \psi_1}{\partial z} = a\psi_1 = \frac{ik}{2n_o}\Delta n^2 \psi_1 \tag{39}$$

The solution of this equation gives the propagation formalism that allows one propagate light beam in small steps through an inhomogeneous medium both in one and two dimensional cases which usually may extend to three dimension.

5. Analytical solution of the propagating wave with step-index

Equation 37 is an important approximation, though it restricts the use of the beam propagation method in analyzing the structures of matters for which only the forward propagating wave is considered. However, this excludes the use of the method in cases where the refractive index changes abruptly as a function of z or in which reflections add to equation (26). The propagation of the field ψ_1 is given by the term describing the propagation in an unperturbed medium and the correction term-representing the influence of Δn^2 (z) (Ugwu et al, 2007).

As the beam is propagated through a thin film showing a large step in refractive index of an imperfectly homogeneous thin film, this condition presents the enabling provisions for the use of a constant refractive index n_o of the thin film. One then chooses arbitrarily two different refractive indices n_1 and n_2 at the two sides of the step so that

$$\left.\begin{aligned} n_o(z) = n_1 \quad z < 0 \\ n_o(z) = n_2 \quad z > 0 \end{aligned}\right\} \tag{40}$$

with $\dfrac{n(z) - n_o(z)}{n_o(z)} >> 1 \ \textit{for all z}$

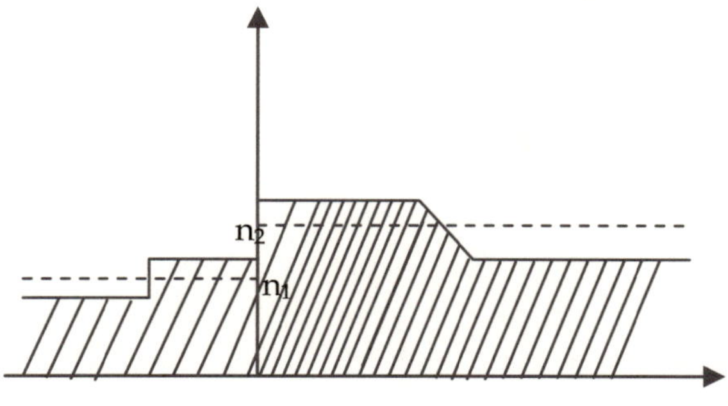

Fig. 3. Refractive index profile showing a step

The refractive index distribution of the thin film was assumed to obey the Fermi distribution that is an extensively good technique for calculating the mode index using the well known WKB approximation (Miyazowa et al,1975). The calculation is adjusted for the best fit to the value according to

$$n(z) = n_o + \Delta n\left[\exp[\frac{z-d}{a}] - 1\right]^{-1} \quad (41)$$

Small change in the refractive index over the film thickness can be obtained

Equation (41) represents the Fermi distribution. where n(z) is the refractive index at a depth z below the thin surface, n_o is the refractive index of the surface ,Δn is the step change in the film thickness, z is average film thickness and "a" is the measure of the sharpness of the transition region (Ugwu et al, 2005).

With smoothly changing refractive index at both sides of the step, we assume that the sensitivity to polarization is due mainly to interface and hence in propagating a field ψ through such a medium, one has to decompose the field into (T.E) and (TM) polarized fields in which we neglect the coupling between the E and H fields due to small variation ($n-n_o$). When the interface condition ψ_m and $\partial\Psi_m/\partial z$ are continuous at z = 0 are satisfied, the T.E field could be propagated by virtue of these decomposition. Similarly, TM fields were also propagated by considering that ψ_m and $\partial\Psi_m/\partial z$ were continuous at z = 0.

When we use a set or discrete modes, different sets of ψ_m can be obtained by the application of the periodic extension of the field. To obtain a square wave function for $n_o(z)$ as in fig 3.4, n_o has to be considered periodic. We were primarily interested in the field guided at the interface $z = z_1$. The field radiated away from the interface was assumed not to influence the field in the adjacent region because of the presence of suitable absorber at $z = z - z_1$. The correction operator ∂ contains the perturbation term Δn^2 and as we considered it to be periodic function without any constant part as in equation 3.18. The phase variation of the correction term is the same such that the correction term provided a coupling between the two waves.

The Green's function as obtained in the equation (29) satisfies (1)

$$\left[\frac{\partial^2}{\partial x^2} + \frac{\partial^2}{\partial y^2} + \Delta n(y)\right] G(x,y) = \delta(x-x_1)\delta(y-y_1) \quad (42)$$

at the source point and satisfies (Ugwu et al, 2007) the impedance boundary condition.

$$G + B_o \frac{\partial G}{\partial n} = 0 \tag{43}$$

where $B_o = -\frac{iR_s}{\kappa R_o}$ s

and $R_o = \left[\frac{\mu}{\varepsilon_o}\right]^{\frac{1}{2}}$

is the free space characteristic impedance, and $\partial/\partial n$ is the normal derivative. The impedance R_s offered to the propagating wave by the thin film is given by

$$R_s = \frac{R_o}{n}\left[1 - \frac{\kappa^2}{(\kappa_o n)^2}\right]^{\frac{1}{2}} \tag{44}$$

$$R = \frac{R_0}{n}\left[1 - \frac{1}{n^2}\right] \tag{45}$$

where n is the average refractive index of the film (Wait, 1998; Bass et al, 1979) κ is the wave-number of the wave in the thin film where κ_o is the wave number of the wave in the free space. For every given wave with a wavelength say λ propagating through the film with the appropriate refractive index n, the impedance R of the film can be computed using equation (44) When κ is equal to κ_o we have equation (45) and when n is relatively large $|n|$

6. Integral method

The integral form of Lippmann-Schwinger as given in equation 12 may be solved analytically as Fredholm problem if the kernel $k(z,z') = G(z,z')V(z')$ is separable, but due to the implicit nature of the equation as $\psi(z')$ is unknown, we now use Born approximation method to make the equation explicit. This simply implies using $\psi_o(z)$ in place of $\psi(z')$ to start the numerical integration that would enable us to obtain the field propagating through the film.

In the solution of the problem, we considered $\Psi_0(z)$ as the field corresponding to homogeneous medium without perturbation and then work to obtain the field $\Psi_0(z)$ corresponding to the perturbed system (6) is facilitated by introducing the dyadic Green's function associated with the reference system is as written in equation (12). However, to do this we first all introduce Dyson's equation, the counterpart of Lippmann-Schwinger equation to enable us compute the value of the Green's function over the perturbation for own ward use in the computation of the propagation field.

Now, we note that both Lippmann-Schwinger and Dyson's equation are implicit in nature for all points located inside the perturbation and as a result, the solution is handle by applying Born Approximation method as already mentioned before now

7. Numerical consideration

In this method, we defined a grid over the system as in fig 3 this description, we can now write Lippmann-Schwinger and Dyson's equations as:

$$\psi_1 = \psi_1^0 + \sum_{k=1}^{N_p} G^o_{1,k}\ V_K \Delta_K \psi_K \qquad (45)$$

$$G_{i,j} = G^0_{i,j} + \sum_{k=1}^{Np} G^0_{i,k}\ v_k \Delta_k G_k \qquad (46)$$

The descretization procedure is identical in one, two, or three dimensions; for clarity, we use only one segment to designate the position of a mesh and we assume it to be k and we assume also that the discretized system contains N meshes from which N_p describes the meshes, $N_p \leq N$, we denote the discretized field. The formulation of the matrix in equations (46) leads to the solution Gs that would be used in building up the matrix in (46) which eventually makes it possible to obtain the propagating field. Also the number of the matrices obtained will depend on the number meshes considered
For instant if 3 meshes are considered, 9 algebraic equations as

$$G_{1,1} = G^o_{1,1} + G^o_{1,1} V_1 \Delta_1 G_{1,1} + G^o_{2,2} V_2 G_{2,1} + G^o_{1,3} V_3 \Delta_{3,1} \qquad a$$

$$G_{2,1} = G^o_{2,1} + G^o_{2,1} V_1 \Delta_1 G_{1,1} + G^o_{2,2} v_2 \Delta_2 G_{2,I} + G^o_{2,3} V_{3,} \Delta G_{3,1} \qquad b$$

$$G_{3,1} = G^o_{2,2} + G^o_{31} V_1 \Delta_1 G_{1,1} + G^o_{3,2} V_2 \Delta_{2,1} + G^o_{3,3} V_3 \Delta_3 G_{3,1} \qquad c$$

$$G_{1,2} = G^o_{1,2} + G^o_{1,1} V_1 \Delta_1 G_{1,2} + G^o_{1,2} V_2 \Delta_2 G_{2,2} + G^o_{1,2} V_3 \Delta_3 G_{3,2}\ s \qquad d$$

$$G_{2,2} = G^o_{2,2} + G^o_{2,1} V_1 \Delta G_{1,2} + G^o_{2,2} V_2 \Delta_2 G_{2,2} + G_{2,3} V_3 \Delta_3 G_{3,2} \qquad e$$

$$G_{3,2} = G^o_{3,2} + G_{3,1} V_1 \Delta_1 G_{1,2} + G^o_{3,2} V_2 \Delta_2 G_{2,2} + G^o_{3,3} V_3 \Delta_3 G_{3,2} \qquad f$$

$$G_{1,3} + G^o_{1,3} + G^o_{1,1} V_1 \Delta_1 G_{1,3} + G^o_{1,2}\ G^o_{1,2} V_2 \Delta_2 G_{2,3} + G_{1,3} V_3 \Delta_3 G_{3,3} \qquad g$$

$$G_{2,3} + G^o_{2,3} + G_{2,1} V_1 \Delta_1 G_{1,3} + G^o_{1,2} V_2 \Delta_2 G_{2,3} + G^o_{2,3} V_3 \Delta_3 G_{3,3} \qquad h$$

$$G_{3,3} = G^o_{3,3} + G^o_{3,1} + {}_1V_1 \Delta_1 G_{1,2} + G^o_{3,2} V_3 \Delta_3 G_{2,3} + G^0_{3,3} V_3 \Delta_3 G_{3,3} \qquad i\ (47)$$

Are generated and from that 9x9 matrix is formulated from where one can obtain nine values of G_{ij}
$G_{1,1}, G_{1,2}, G_{1,3}, G_{2,1}, G_{2,2}, G_{2,3}, G_{1,3}, G_{2,3}$ and $G_{3,3}$. Using these Gs, one generates three algebraic equations in terms of field as shown below.

$$\begin{aligned}
\Psi_1 &= \Psi_1^0 + G^0_{1,1} V_1 \Delta_1 \Psi_1 + G^0_{1,2} V_2 \Delta_2 \Psi_2 + G^0_{1,3} V_3 \Delta_3 \Psi_3 \\
\Psi_2 &= \Psi_2^0 + G^0_{2,1} V_1 \Delta_1 \Psi_1 + G^0_{2,2} V_2 \Delta_2 \Psi_2 + G^0_{3,3} V_3 \Delta_3 \Psi_3 \\
\Psi_3 &= \Psi_3^0 + G^0_{3,1} V_1 \Delta_1 \Psi_1 + G^0_{3,2} V_2 \Delta_2 \Psi_2 + G^0_{3,3} V_3 \Delta_3 \Psi_3
\end{aligned}$$

The compact matrix form of the equation can be written thus,

$$\begin{bmatrix} 1 - G^0_{1,1} V_1 \Delta_1 - G^0_{1,2} V_2 \Delta_2 - G^0_{1,3} V_3 \Delta_3 \\ -G^0_{2,1} V_1 \Delta_1 1 - G^0_{2,2} V_2 \Delta_2 - G^0_{2,3} V_3 \Delta_3 \\ -G^0_{3,1} V_1 \Delta_1 - G^0_{3,2} V_2 \Delta_2 1 - G^0_{3,3} V_3 \Delta_3 \end{bmatrix} \begin{bmatrix} \Psi_1 \\ \Psi_2 \\ \Psi_3 \end{bmatrix} = \begin{bmatrix} \Psi_1^0 \\ \Psi_2^0 \\ \Psi_3^0 \end{bmatrix} \qquad (48)$$

8. Result/discussion

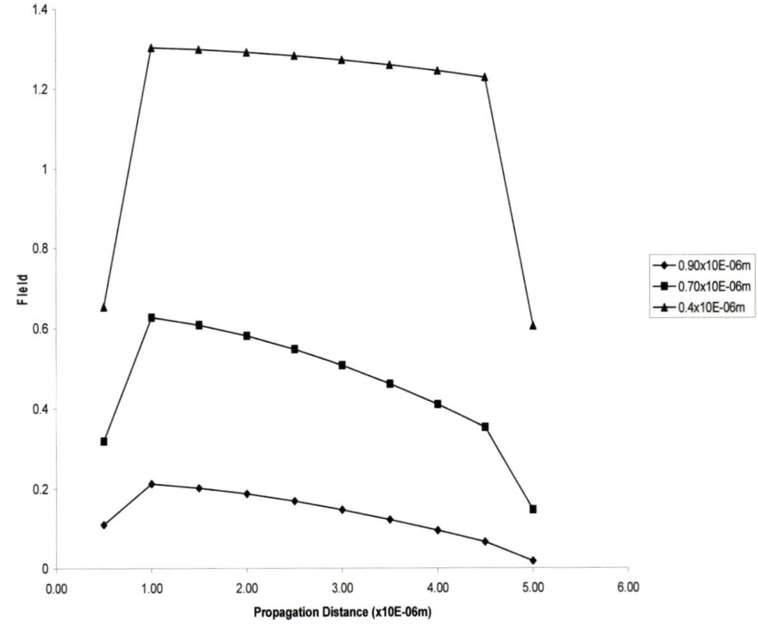

Fig. 1. The field behaviour as it propagates through the film thickness Zµm for mesh size = 10 when λ =0.4µm 0.7µm and 0.9µm.

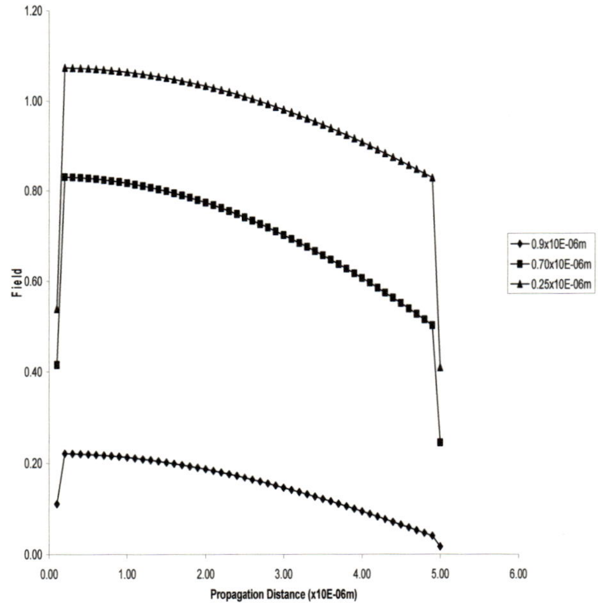

Fig. 2. The field behaviour as it propagates through the film thickness Zµm for mesh size = 50 when λ = 0.25µm, 0.7µm and 0.9µm.

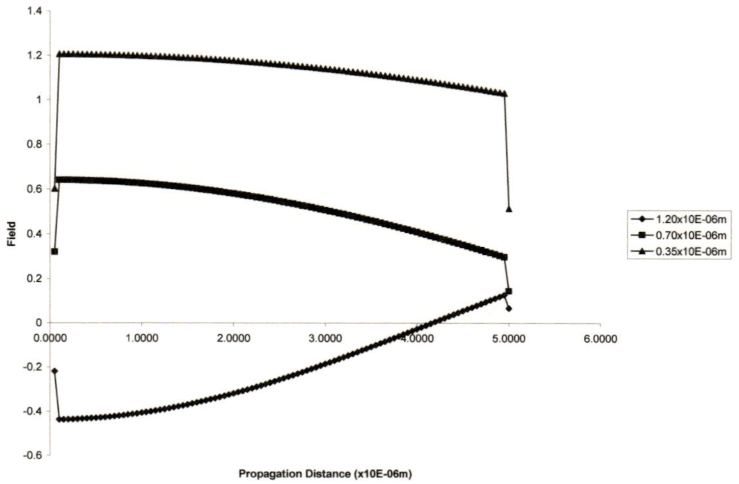

Fig. 3. The field behavour as it propagates through the film thickness Zµm for mesh size = 50 when λ = 0.25µm, 0.7µm and 0.9µm.

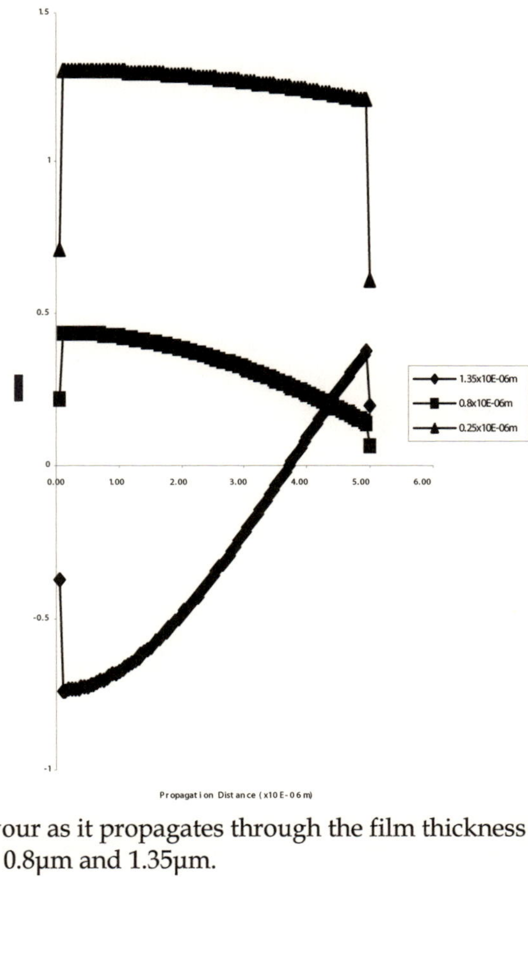

Fig. 4. The field behavour as it propagates through the film thickness Zµm for mesh size = 100 when λ = 0.25µm, 0.8µm and 1.35µm.

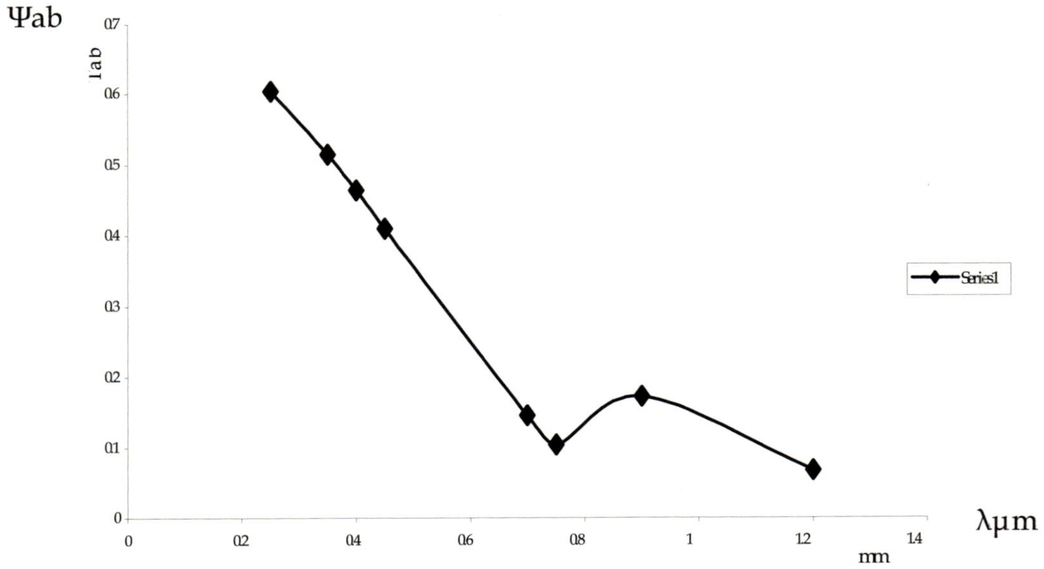

Fig. 5. The filed absorbance as a function wavelength.

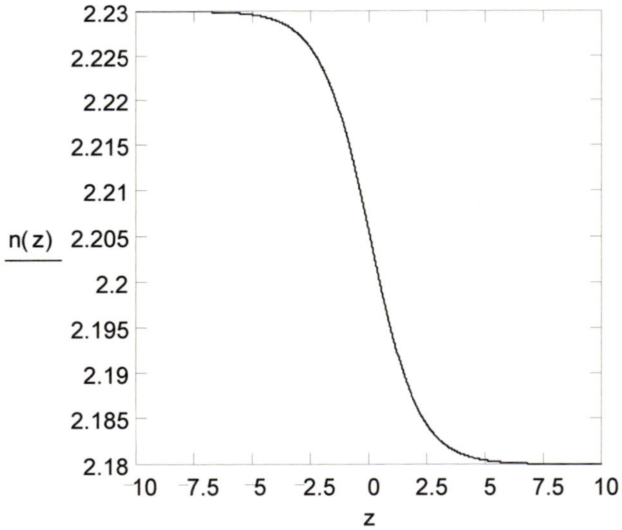

Fig. 6. Refractive index profile using Fermi distribution

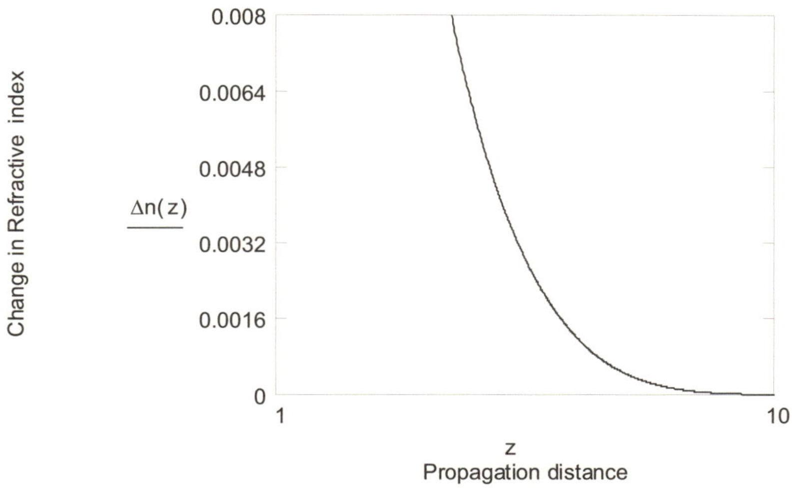

Fig. 7. Graph of change in Refractive Index as a function of a propagation distance

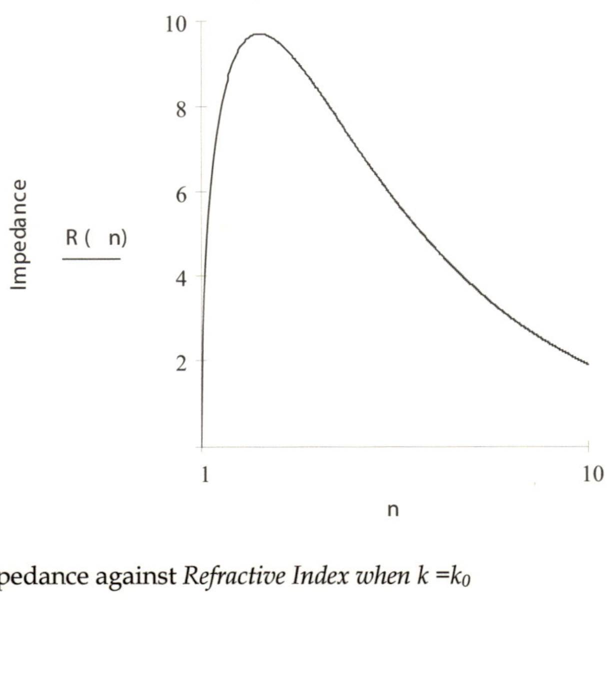

Fig. 8. Graph of Impedance against *Refractive Index when k =k_0*

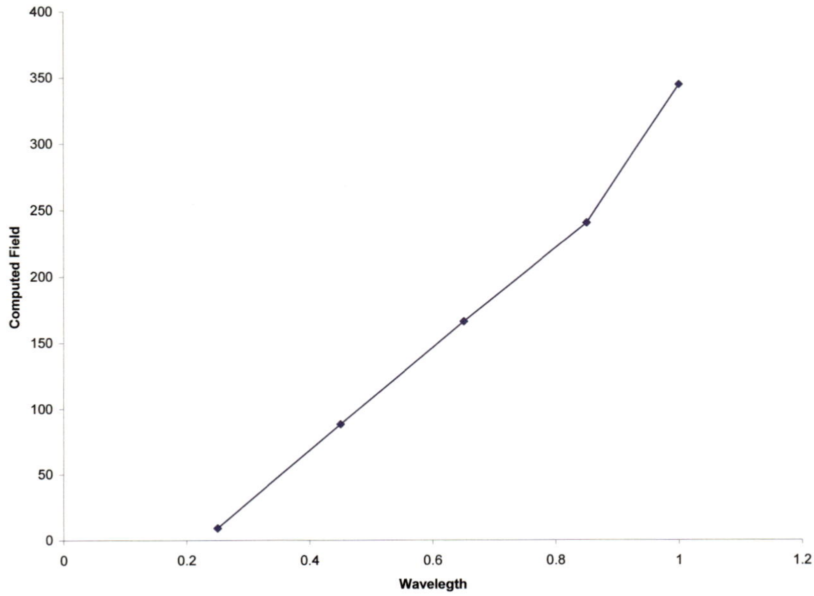

Fig. 9. Computed field against wavelength when the mesh size is constant

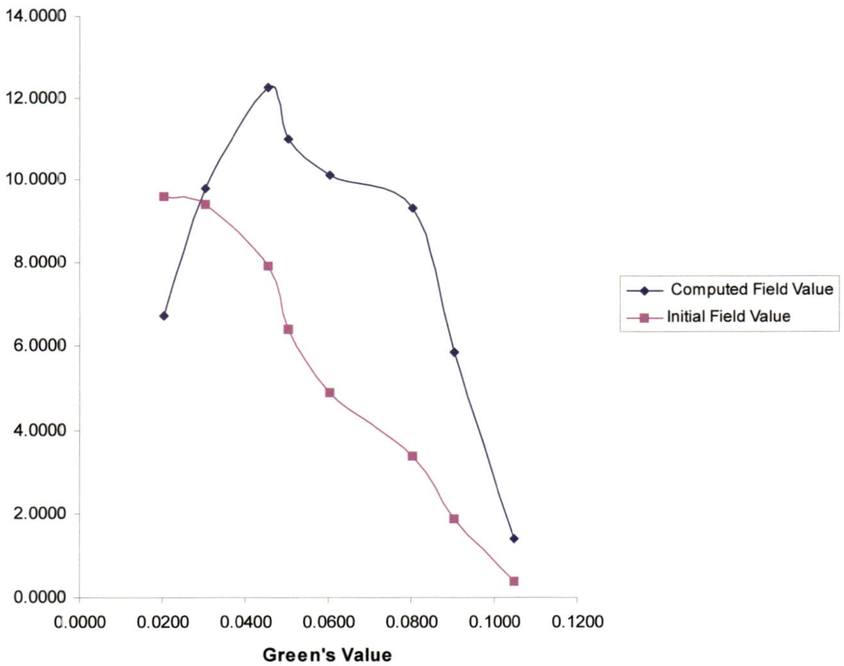

Fig. 10. Computed and Initial field values in relation to the Green's value within the uv region

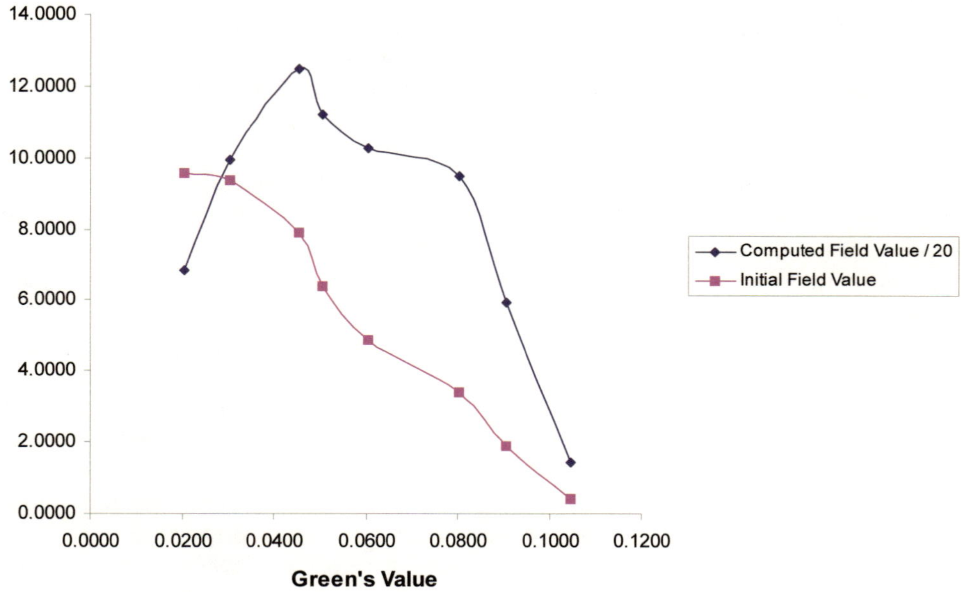

Fig. 11. Computed and Initial field values in relation to the Green's value within the near infrared region

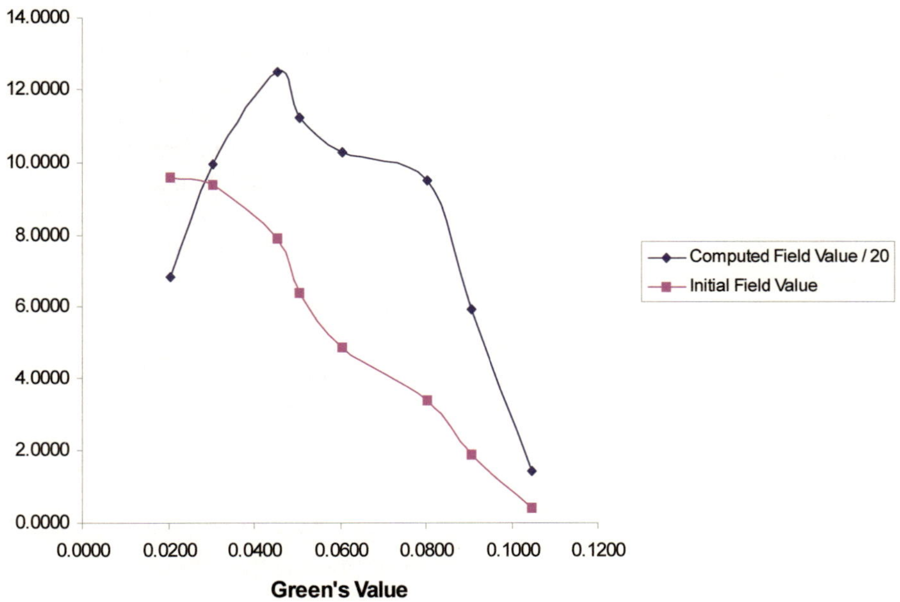

Fig. 12. Computed and Initial field values in relation to the Green's value within the visible region

From the result obtained using this formalism, the field behaviour over a finite distance was contained and analyzed by applying born approximation method in Lippman-Schwinger equation involving step by step process. The result yielded reasonable values in relation to the experimental result of the absorption behaviour of the thin film (Ugwu, 2001).

The splitting of the thickness into more finite size had not much affected on the behaviour of the field as regarded the absorption trends.

The trend of the graph obtained from the result indicated that the field behaviour have the same pattern for all mesh size used in the computation. Though, there is slight fall in absorption within the optical region, the trend of the graph look alike when the thickness is 1.0 μ m with minimum absorption occurring when the thickness is 0.5 μ m. within the near infrared range and ultraviolet range, (0.25 μ m) the absorption rose sharply, reaching a maximum of1.48 and 1.42 respectively when thickness is 1.0 μ m having value greater than unity.

From the behaviour of the propagated field for the specified region, UV, Visible and Near infrared, (Ugwu, 2001) the propagation characteristic within the optical and near infrared regions was lower when compared to UV region counterpart irrespective of the mesh size and the number of points the thickness is divided. The field behaviour was unique within the thin film as observed in fig. 3 and fig 4: for wavelength 1.2μm and 1.35μm while that of fig,1 and fig.2 were different as the wave patterns were shown within the positive portion of the graph. The field unique behavior within the film medium as observed in the graphs in fig.1 to fig.4 for all the wave length and N_{max} suggests the influence of scattering and reflection of the propagated field produced by the particles of the thin field medium. The peak as seen in the graphs is as a result of the first encounter of the individual molecules of the thin film with the incident radiation. The radiation experiences scattering by the individual molecules at first conforming to Born and Huang, 1954 where it was explained that when a molecule initially in a normal state is excited, it generates spontaneous radiation of a given frequency that goes on to enhance the incident radiation This is because small part of the scattered incident radiation combines with the primary incident wave resulting in phase change that is tantamount to alternation of the wave velocity in the thin film medium. One expects this peat to be maintained, but it stabilized as the propagation continued due to fact that non-forward scattered radiation is lost from the transmitted wave(Sanders,19980) since the thin film medium is considered to be optically homogeneous, non-forward scattered wave is lost on the account of destructive interference. In contrast, the radiation scattered into the forward direction from any point in the medium interferes constructively (Fabelinskii, 1968)

We also observed in each case that the initial value of the propagation distance zμm, initial valve of the propagating field is low, but increase sharply as the propagation distance increases within the medium suggesting the influence of scattering and reflection of propagating field produced by the particles of the thin film as it propagates.Again, as high absorption is observed within the ultraviolet (UV) range as depicted in fig.5, the thin film could be used as UV filter on any system the film is coated with as it showed high absorption. On other hand, it was seen that the absorption within the optical (VIS) and near infrared (NIR) regions of solar radiation was low. Fig.6 depicts the refractive index profile according to equation (41) while that of the change in refractive index with propagation distant is shown on figs.7. The impedance appears to have a peak at lower refractive index as shown in fig. 8. Fig. 9 shows the field profile for a constant mesh size while that of Fig.10

to fig.12 are profile for the three considered regions of electromagnetic radiation as obtained from the numerical consideration.

9. Conclusion

A theoretical approach to the computation and analysis of the optical properties of thin film were presented using beam propagation method where Green's function, Lippmann-Schwinger and Dyson's equations were used to solve scalar wave equation that was considered to be incident to the thin film medium with three considerations of the thin film behaviour These includes within the three regions of the electromagnetic radiation namely: ultra violet, visible and infrared regions of the electromagnetic radiation with a consideration of the impedance offered to the propagation of the field by the thin film medium.

Also, a situation where the thin film had a small variation of refractive index profile that was to have effect on behaviour of the propagated field was analyzed with the small variation in the refractive index. The refractive index was presented as a small perturbation. This problem was solved using series solution on Green's function by considering some boundary conditions (Ugwu et al 2007). Fermi distribution function was used to illustrate the refractive index profile variation from where one drew a close relation that facilitated an expression that led to the analysis of the impedance of the thin film

The computational technique facilitated the solution of field values associated first with the reference medium using the appropriate boundary conditions on Lippmann-Schwinger equation on which dyadic Green's operator was introduced and born approximation method was applied both Lippmann-Schwinger and Dyson's equations. These led to the analysis of the propagated field profile through the thin film medium step by step.

10. Reference

[1] A.B Cody, G. Brook and Abele 1982 "Optical Absorption above the Optical Gap of Amorphous Silicon Hydride". Solar Energy material, 231-240.

[2] A.D Yaghjian 1980 "Electric dynamic green's functions in the source region's Proc IEEE 68,248-263.

[3] Abeles F. 1950 "Investigations on Propagation of Sinusoidal Electromagnetic Waves in Stratified Media Application to Thin Films", Ann Phy (Paris) 5 596- 640.

[4] Born M and Huang K 1954, Dynamical theory of crystal lattice Oxford Clarendon

[5] Born, M and Wolf E, 1980, "Principle of optics" 6th Ed, Pergamon N Y.

[6] Brykhovestskii, A.S, Tigrov,M and I.M Fuks 1985 "Effective Impendence Tension Of Computing Exactly the Total Field Propagating in Dielectric Structure of arbitrary shape". J. opt soc Am A vol 11, No3 1073-1080.

[7] E.I. Ugwu 2005 "Effects of the electrical conductivity of thin film on electromagnetic wave propagation. JICCOTECH Maiden Edition. 121-127.

[8] E.I. Ugwu, C.E Okeke and S.I Okeke 2001."Study of the UV/optical properties of FeS_2 thin film Deposited by solution Growth techniques JEAS Vol1 No. 13-20.

[9] E.I. Ugwu, P.C Uduh and G.A Agbo 2007 "The effect of change in refractive index on wave propagation through (feS_2) thin film". Journal of Applied Sc.7 (4). 570-574.

[10] E.N Economou 1979 "Green's functions in Quantum physics", 1st. Ed. Springer. Verlag, Berlin.

[11] F.J Blatt 1968 "Physics of Electronic conduction in solid". Mc Graw - Hill Book Co Ltd New York, 335-350.

[12] Fablinskii I. L, 1968 Molecular scattering of light New York Plenum Press.

[13] Fitzpatrick, .R, (2002), "Electromagnetic wave propagation in dielectrics". http: // farside. Ph. UTexas. Edu/teaching/jkl/lectures/node 79 htmil. Pp 130 - 138.

[14] G. Gao, C Tores - Verdin and T.M Hat 2005 "Analytical Techniques toe valuate the integration of 3D and 2D spatial Dyadic Green's function" *progress in Electromagnetic Research PIER 52,* 47-80.

[15] G.W. Hanson 1996 "A Numerical formation of Dyadic Green's functions for planar Bianisotropic Media with Application to printed Transmission line" IEEE Transaction on Microwave theory and techniques, 44(1).

[16] H.L Ong 1993 "2x2 propagation matrix for electromagnetic waves propagating obliquely in layered inhomogeneous unaxial media" *J.Optical Science* A/10(2). 283-393.

[17] Hanson, G W, (1996), "A numerical formulation of Dyadic Green's functions for Planar Bianisotropic Media with Application to Printed Transmission lines". S 0018 - 9480 (96) 00469-3 lEEE pp144 - 151.

[18] J.A Fleck, J.R Morris and M.D. Feit 1976 "Time - dependent propagation of high energy laser beans through the atmosphere" *Applied phys 10,*129-160.

[19] L. Thylen and C.M Lee 1992 "Beam propagation method based on matrix digitalization" *J. optical science* A/9 (1). 142-146.

[20] Lee, J.K and Kong J.A 1983 Dyadic Green's Functions for layered an isotropic medium. Electromagn. Vol 3 pp 111-130.

[21] M.D Feit and J.A Fleck 1978 "Light propagation in graded - index optical fibers" *Applied optical17,* 3990-3998.

[22] Martin J F Oliver, Alain Dereux and Christian Girard 1994 "Alternative Scheme of

[23] P.A. Cox 1978 "The electroni c structure and Chemistry of solids "Oxford University Press Ch. 1-3. Plenum Press ; New York Press.

[24] Sanders P.G.H,1980 Fundamental Interaction and Structure of matters: 1st edition

[25] Smith E.G. and Thomos J.H., 1982. "Optics ELBS and John Wiley and Sons Ltd London. Statically Rough Ideally Conductive Surface. Radioplys. Quantum Electro 703 -708

8

The Electrodynamic Properties of Structures with Thin Superconducting Film in Mixed State

Mariya Golovkina
Povolzhskiy State University of Telecommunacations and Informatics
Russia

1. Introduction

Thin superconducting films are used in many areas of microwave technics. The discovery of high temperature superconductors in 1986 (Bednorz & Muller, 1986) was a powerful incentive to application of superconductors in science and engineering. High-temperature superconductors have a lot of necessary microwave properties, for example: low insertion loss, wide frequency band, low noise, high sensibility, low power loss and high reliability. High-temperature superconductors have significant potential for applications in various devices in microelectronics because of the ability to carry large amount of current by high temperature (Zhao et al., 2002). The widely applicable high-temperature superconductor $YBa_2Cu_3O_7$ with critical temperature T_c=90 K keeps the superconductivity above the boiling point of nitrogen. One of the applications of high-temperature superconductors is the passive microwave device because of it extremely small resistance and low insertion loss. In recent years new techniques have been developed for production of superconducting layered systems. The superconducting films are indispensable for manufacture of resonators and filters with the technical parameters that significantly surpass the traditional materials. The progress in microsystem technologies and nanotechnologies enables the fabrication of thin superconducting films with thickness about several atomic layers (Koster et al., 2008), (Chiang, 2004). Due to new technologies scientists produce a thin film which exhibits a nanometer-thick region of superconductivity (Gozar et al., 2008).

The thin superconducting films are more attractive for scientists and engineers than the bulk superconducting ceramics. The thin films allow to solve a problem of heat think. The application of thin films increases with the growth of critical current density J_c. Nowadays it is known a large number of superconducting materials with critical temperature above 77 K. But despite of the bundle of different high-temperature superconducting compounds, only three of group have been widely used in thin film form: $YBa_2Cu_3O_7$, $Bi_vSr_wCa_xCu_yO_z$, $Tl_vBa_wCa_xCu_yO_z$ (Phillips, 1995). $YBa_2Cu_3O_7$ has critical temperature T_c=90 K (Wu et al., 1987) and critical current density $J_c=5\cdot10^{10}$ A/m^2 at 77 K (Yang et al., 1991), (Schauer et al., 1990). The critical temperature of $Bi_vSr_wCa_xCu_yO_z$ films T_c is 110 K (Gunji et al., 2005), that makes these films more attractive than $YBa_2Cu_3O_7$. But single-phase films with necessary phase with T_c=110 K have not been grown successfully (Phillips, 1995). Also the $Bi_vSr_wCa_xCu_yO_z$ films have lower critical current density than $YBa_2Cu_3O_7$. The $Tl_vBa_wCa_xCu_yO_z$ films with T_c=125 K and critical current density above 10^{10} A/m^2 and $HgBa_2Ca_2Cu_3O_{8.5+x}$ films with T_c=135 K are attractive for application in microwave devices (Itozaki et al., 1989), (Schilling et al., 1993).

Now thin high-temperature superconducting films can find application in active and passive microelectronic devices (Hohenwarter et al., 1999), (Hein, 1999), (Kwak et al., 2005). The superconductors based on complex oxide ceramic are the type-II superconductors. It means that the magnetic field can penetrate in the thickness of superconducting film in the form of Abrikosov vortex lattice (Abrikosov, 2004). If we transmit the electrical current along the superconducting film, the Abricosov vortex lattice will come in the movement under the influence of Lorentz force. The presence of moving vortex lattice in the film leads to additional dissipation of energy and increase of losses (Artemov et al., 1997). But we can observe the amplification of electromagnetic waves by the interaction with moving Abrikosov vortex lattice. The mechanism of this amplification is the same as in a traveling-wave tube and backward-wave-tube (Gilmour, 1994). The amplification will be possible if the velocity of electromagnetic wave becomes comparable to the velocity of moving vortex lattice. Due to the energy of moving Abricosov vortex lattice the electromagnetic waves amplification can be observed in thin high-temperature superconducting film on the ferromagnetic substrate (Popkov, 1989), in structures superconductor - dielectric and superconductor - semiconductor (Glushchenko & Golovkina, 1998 a), (Golovkina, 2009 a). The moving vortex structure can generate and amplify the ultrasonic waves (Gutliansky, 2005). Thus, the thin superconducting films can be successfully used in both passive and active structures.

2. Thin superconducting film in planar structure

2.1 The method of surface current

The calculation of electromagnetic waves characteristics after the interaction with thin films is possible by various methods. These methods match the fields outside and inside of thin film. The method of two-sided boundary conditions belongs to this methods (Kurushin et al., 1975), (Kurushin & Nefedov, 1983). The calculation of electromagnetic waves characteristics with help of this method is rigorous. The thin film is considered as a layer of final thickness with complex dielectric permeability. The method of two-sided boundary conditions can be used with any parameters of the film. However, this method is rather difficult. From the point of view of optimization of calculations the approximate methods are more preferable. The method of surface current can be applied for research of electrodynamic parameters of thin superconducting films when the thin film is considered as a current carrying surface. In the framework of this method the influence of thin resistive film can be considered by introduction of special boundary conditions for tangential components of electric and magnetic field (Veselov & Rajevsky, 1988).

The HTSC are the type-II superconductors. If we place the type-II superconductor in the magnetic film $B_{c1}<B<B_{c2}$, where B_{c1} and B_{c2} are first and second critical fields for superconductor respectively, the superconductor will pass in the mixed state (Schmidt, 2002). In the mixed state the superconductor has small resistance which value is on some orders less than resistance of pure metals. Let us consider the thin superconducting film in resistive state. The tangential components of electric field will be continuous, if the following conditions are satisfied (Veselov & Rajevsky, 1988)

$$\Delta\sqrt{\frac{\mu\omega(\sigma^2+\varepsilon^2\omega^2+\varepsilon\omega\sqrt{\varepsilon^2\omega^2+\sigma^2})}{2(\varepsilon\omega+\sqrt{\varepsilon^2\omega^2+\sigma^2})}} << 1 , \tag{1}$$

where Δ is the thickness, σ - conductivity of film, ε is permittivity and μ is permeability of superconductor, ω is the angular frequency of applied electromagnetic wave. If the inequality σ>>ε ω is carried out, the condition (1) can be written in the form

$$\Delta\sqrt{\mu\omega\sigma/2} = \Delta/d << 1 \quad ,(2)$$

where d is skin depth of superconducting material. In the following consideration the condition (2) is carried out in all cases.
And now let's consider the magnetic field. If the condition (1) and (2) are satisfied, the boundary conditions for tangential components will be given by

$$H_x^I - H_x^{II} = j_z \ , \quad (3)$$

$$H_z^I - H_z^{II} = -j_x \ , \quad (4)$$

where j is current density.
Thus if condition σ>>ε ω is satisfied, the tangential components of electric field will be continuous and the boundary conditions for tangential components of magnetic field will be written in the form (3-4). This condition is satisfied for superconducting films for microwave and in some cases for infrared and optical range.

2.2 The boundary conditions for thin type-II superconducting film in mixed state

Let us consider the thin type-II superconducting film with thickness t<<λ, where λ is a microwave penetration depth.

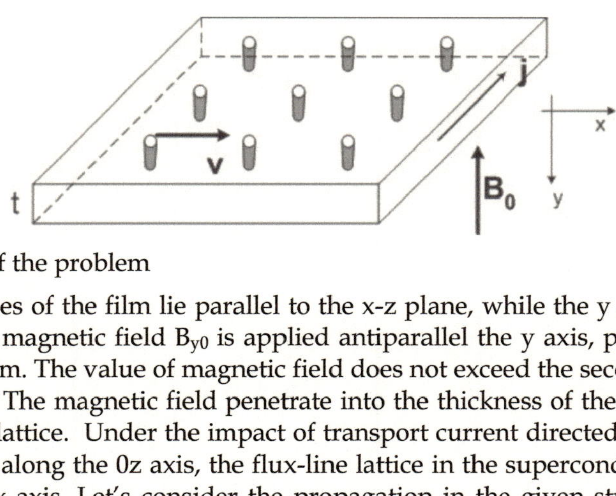

Fig. 1. Geometry of the problem

We let the interfaces of the film lie parallel to the x-z plane, while the y axis points into the structure. A static magnetic field B_{y0} is applied antiparallel the y axis, perpendicular to the interfaces of the film. The value of magnetic field does not exceed the second critical field for a superconductor. The magnetic field penetrate into the thickness of the film in the form of Abrikosov vortex lattice. Under the impact of transport current directed perpendicularly to magnetic field B_{y0} along the 0z axis, the flux-line lattice in the superconductor film starts to move along the 0x axis. Let's consider the propagation in the given structure p-polarized wave being incident with angle θ in the x0y plane. It can be assumed that $\partial/\partial z=0$.
The presence of a thin superconductor layer with the thickness of t<<l is reasonable to be accounted by introduction of a special boundary condition because of a small amount of thickness. Let's consider the superconductor layer at the boundary y=0. At the inertia-free approximation and without account of elasticity of fluxon lattice (the presence of elastic forces in the fluxon lattice at its deformation results in non-linear relation of the wave to the

lattice, that is insignificant at the given linear approximation) the boundary condition is written in the following way (Popkov, 1989):

$$\frac{\partial B_y}{\partial t}(y=t)+\frac{j_{z0}\Phi_0}{\eta}\frac{\partial B_y}{\partial t}(y=t)=\frac{B_{y0}\Phi_0}{\eta\, t}\frac{\partial}{\partial x}[H_x(y=t)-H_x(y=0)], \tag{5}$$

where j_{z0} is the current density in the superconducting film and η is the vortex viscosity. The method of account of thin superconducting film in the form of boundary condition enables to reduce the complexity of computations and makes it possible to understand the mechanism of interaction of electromagnetic wave and thin superconducting film.

3. The periodic structures with thin superconducting film

3.1 Dispersion relation for one-dimensional periodic structure superconductor – dielectric

Let's consider the infinite one-dimensional periodic structure shown in Fig. 2 (Glushchenko & Golovkina, 1998 b). The structure consists of alternating dielectric layers with thickness d_1 and type-II superconductor layers with thickness $t \ll \lambda$. An external magnetic field B_{y0} is applied antiparallel the y axis, perpendicular to the interfaces of the layers. The flux-line lattice in the superconductor layers moves along the 0x axis with the velocity v. Let's consider the propagation in the given structure p-polarized wave being incident with angle θ in the x0y plane.

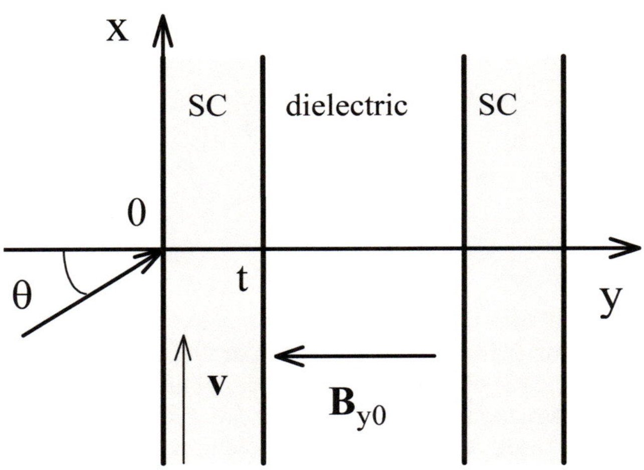

Fig. 2. Periodic structure superconductor (SC) - dielectric

Let's write the boundary condition (5) in the form of matrix M_s, binding fields at the boundaries y=0 and y=t:

$$\begin{pmatrix} E_z(t) \\ H_x(t) \end{pmatrix} = M_s \begin{pmatrix} E_z(0) \\ H_x(0) \end{pmatrix}, \tag{6}$$

$$M_s = \begin{pmatrix} 1 & 0 \\ \dfrac{t}{B_{y0}}\left(\dfrac{\eta}{\Phi_0} - \dfrac{j_{z0}\,k_x}{\omega}\right) & 1 \end{pmatrix}, \quad (7)$$

where k_x is the projection of the passing wave vector onto the 0x axis and ω is the angular frequency of the passing wave.

Using matrix method we found dispersion relation for H-wave:

$$\cos Kd = \cos k_y d_1 + \frac{i\,\omega\,\mu_0\,t}{2\,k_y B_{y0}}\left(\frac{\eta}{\Phi_0} - \frac{j_{z0}\,k_x}{\omega}\right)\sin k_y d_1, \quad (8)$$

where $K=K'-iK''$ is the Bloch wave number and k_y is the projection of passing wave vector onto the 0y axis. The imaginary pert of Bloch wave number K'' acts as coefficient of attenuation.

The interaction of electromagnetic wave with thin superconducting film leads to emergence of the imaginary unit in the dispersion equation. The presence of imaginary part of the Bloch wave number indicates that electromagnetic wave will damp exponentially while passing into the periodic system even if the dielectric layers are lossless (Golovkina, 2009 b). However, when one of the conditions

$$\sin k_y\, d_1 = 0, \quad (9)$$

$$\frac{\eta}{\Phi_0} - \frac{j_{z0}\,k_x}{\omega} = 0 \quad (10)$$

is executed, the Bloch wave vector becomes purely real and electromagnetic wave may penetrate into the periodic structure (Golovkina, 2009 a).

The implementation of condition (9) depends on the relation between the parameters of layers and the frequency of electromagnetic wave, while the implementation of condition (10) depends on parameters of superconducting film only, namely on current density j_{z0}. Still, we are able to manage the attenuation and propagation of electromagnetic waves by changing the value of transport current density j_{z0}. Moreover, the electromagnetic wave can implement the amplification in such structure (Golovkina, 2009 b).

When the medium is lossless and the imaginary part in dispersion relation is absent, the dispersion relation allows to find the stop bands for electromagnetic wave. If the condition $|\cos Kd|<1$ fulfils, than the Bloch wave number K will be real and electromagnetic wave will propagate into the periodic structure. This is the pass band. If the condition $|\cos Kd|>1$ fulfils, than the Bloch wave number will be complex and the electromagnetic wave will attenuate at the propagating through the layers. This is the stop band. The dispersion characteristics for the pass band calculated on the base of the condition $|\cos Kd|<1$ are presented in Fig. 3. These characteristics are plotted for the first Brillouin zone. We can see that the attenuation coefficient K'' decreases by the growth of magnetic field. But this method of definition of pass band is unacceptable when there is the active medium in considered structure. Even if there are the losses in the periodic structure and the imaginary unit is presents in the dispersion relation we should draw the graph in the whole Brillouin zone, including the parts on which the condition $|\cos Kd|>1$ is executed. Then the stop band will correspond to the big values of attenuation coefficient K''.

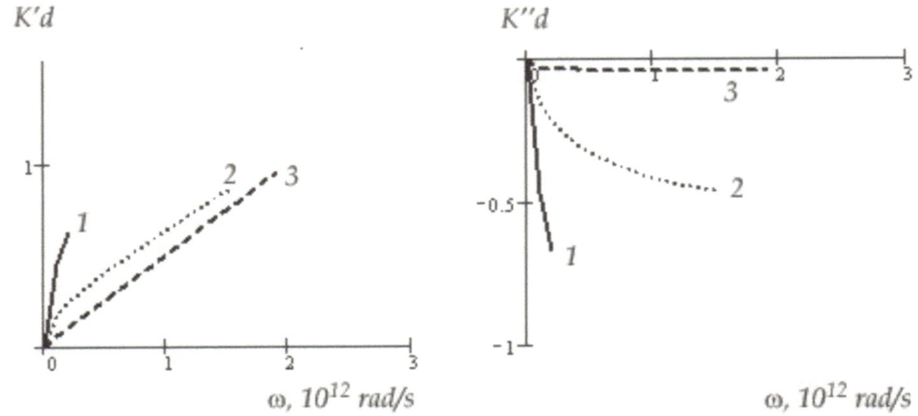

Fig. 3. The real and imaginary part of dispersion characteristic for one-dimensional periodic structure superconductor - dielectric for different values of external magnetic field B. Curve 1: B=0.05 T, curve 2 :B=0.4 T, curve 3: B=5 T. Parameters: d_1=6 μm, t=70 nm, $j_{z0}=10^9$ A/m^2, $\eta=10^{-8}$ N·s/m^2, θ=0.1

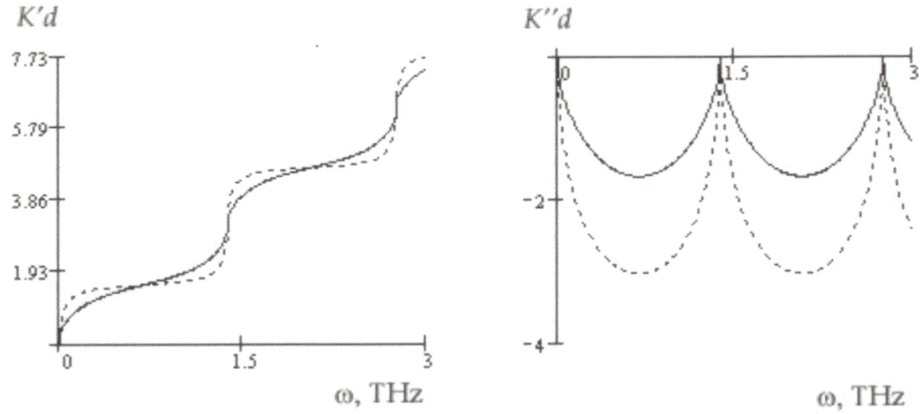

Fig. 4. The real and imaginary part of dispersion characteristic for one-dimensional periodic structure superconductor – dielectric. Curve 1: t=20 nm, curve 2: t=80 nm. Parameters: d_1=100 μm, $j_{z0}=10^8$ A/m^2, $\eta=10^{-8}$ N·s/m^2, B_{y0}=1 T, θ=1.1

The dispersion characteristics for whole Brillouin zone are presented in the Fig. 4. We can't see the stop band in the explicit form in these figures. The band edge can be found from the condition of the big values of attenuation coefficient. This definition of band edge contains an element of indeterminacy. The required attenuation coefficient can accept various values depending on application. For the purposes of our study we must investigate the dynamics of change of the attenuation coefficient K''. If the structure contains an active element (the thin superconducting layer with moving vortex structure for example) the attenuation coefficient K'' can change its sign. And the positive values of K'' indicate that the electromagnetic wave amplifies at the expense of energy reserved in the active element.

3.2 Larkin-Ovchinnikov state

We have considered the superconductor for the case of linear dependence of its characteristics. The differential resistance of superconductor is given by following expression (Schmidt, 2002)

$$\rho_f = \frac{\Phi_0 B}{\eta}, \tag{11}$$

where η is the vortex viscosity depending of temperature T and magnetic field B. This case corresponds to the linear part of voltage-current characteristic of superconductor. Such linear part of voltage-current characteristic exists only in the narrow area of currents exceeding a critical current. With further increase of the transport current in the thin superconducting film the nonlinear area containing jumps of voltage of voltage-current characteristic appears. The theory of Larkin-Ovchinnikov gives the explanation of these phenomena (Larkin & Ovchinnikov, 1975).

Let's suppose, that there is the good heat sink in thin superconducting film, the lattice is in the thermal equilibrum with thermostat and the relaxation time, determined by interelectronic collisions one order greater than the time of electron-phonon interaction. That means that the time of a power relaxation is big. Theory of Larkin-Ovchinnikov gives the following basic expressions (Dmitrenko, 1996):

$$\eta(v) = \eta(0)\,\frac{1}{1+(v/v^*)^2}, \tag{12}$$

$$v^{*2} = \frac{D\sqrt{14\zeta(3)}\,\sqrt{1-T/T_c}}{\pi\,\tau_\varepsilon}, \tag{13}$$

$$D = \frac{1}{3} v_F\, l, \tag{14}$$

$$\eta(0) = 0{,}45\,\frac{\sigma_n T_c}{D}\,\sqrt{1-T/T_c}\,. \tag{15}$$

Here v^* is the critical velocity corresponding to the maximum of viscous friction, D is the diffusion coefficient, v_F is the Fermi velocity, l is the free electrons length, τ_ε is the electron relaxation time, σ_n is the conductivity of superconductor in normal state, $\zeta(3)$ is Riemann zeta-function for 3. This expressions are valid near the critical temperature T_c for small magnetic field $B/B_{c2}<0.4$.

The boundary condition (5) for the superconductor in Larkin-Ovchinnikov state can be written in the following form (Glushchenko & Golovkina, 2007)

$$\begin{aligned} &\frac{2\Phi_0}{\eta(0)\,v^{*2}}\frac{\partial B_y}{\partial t} + \left(\frac{1}{j_z} \pm \sqrt{\frac{1}{j_z^2} - \frac{4\Phi_0^2}{\eta(0)^2 v^{*2}}}\right)\frac{\partial B_y}{\partial x} = \\ &= \frac{B_{y0}}{t}\left(\frac{1}{j_z} \pm \sqrt{\frac{1}{j_z^2} - \frac{4\Phi_0^2}{\eta(0)^2 v^{*2}}}\right)\frac{\partial}{\partial x}\left[H_x(y=0) - H_x(y=t)\right] \end{aligned} \tag{16}$$

The dispersion relation for H-wave is given by

$$\cos Kd = \cos k_y d_1 + C \sin k_y d_1 \ , \tag{17}$$

$$C = \frac{i\omega\mu_0\, t\, j_z^{\,2}}{2\, k_y B_{y0}} \sqrt{1 - \frac{4\Phi_0^{\,2} j_z^{\,2}}{\eta(0)^2 v^{*2}}} \left[\frac{2\Phi_0}{\eta(0) v^{*2}} \left(\sqrt{1 - \frac{4\Phi_0^{\,2} j_z^{\,2}}{\eta(0)^2 v^{*2}}} \pm 1 \right)^{-1} \mp \frac{k_x}{\omega\, j_z} \right] , \tag{18}$$

The top sign corresponds to a wave propagating in a positive direction of the y axis, bottom corresponds to a wave propagating in the opposite direction along the motion of vortex structure.

The dependence of vortex viscosity from magnetic field, temperature and vortex velocity leads to the origination of new control methods, which could operate on parameters of electromagnetic waves. Let's compare the structure dielectric - superconductor in linear case with the structure dielectric - superconductor in Larkin-Ovchinnikov state. If the structure with superconductor in Larkin-Ovchinnikov state has the same parameters as the structure with superconductor in linear case, then the imaginary part of Bloch wave number will be less for superconductor in Larkin-Ovchinnikov state (see Fig. 5). Therefore structure with superconductor in Larkin-Ovchinnikov state demonstrates small attenuation in addition to new control methods.

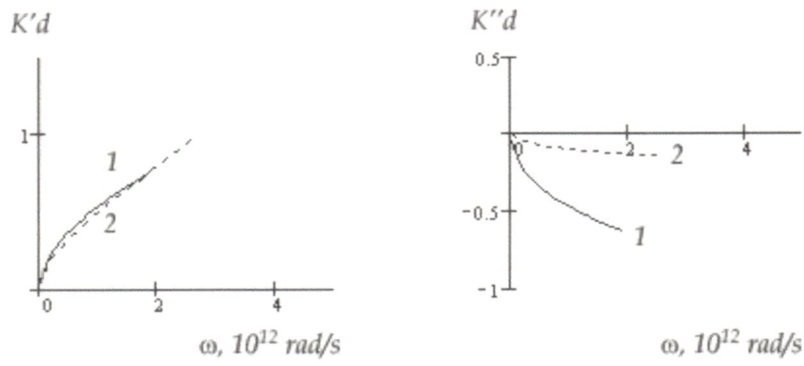

Fig. 5. The dispersion characteristics of periodic structure superconductor – dielectric. Curve 1 : superconductor in linear case, curve 2: superconductor in Larkin-Ovchinnikov state. Parameters: d_1=6 μm, t=70 nm, η=10^{-8} N·s/m², B_{y0}=5 T, θ=0.5, v^*=1750 m/s

Let us consider the expression (17). The imaginary part of Bloch wave number K equals zero if C=0. That corresponds to two values of transport current density:

$$j_{z1} = \frac{\eta(0)\, v^*}{2\Phi_0} \ \ at\ \ v = v^* \ , \tag{19}$$

$$j_{z2} = \frac{\omega \eta(0)}{k_x \Phi_0 [1 + \omega^2 / (v^{*2} k_x^{\,2})]} \ . \tag{20}$$

The calculated under the formula (20) transport current density j_{z2} is frequency-independent for structure superconductor - dielectric.

The value of the j_{z1} depends only on parameters of superconductor; the value of the j_{z2} depends on parameters of superconductor and dielectric. For epitaxial films $YBa_2Cu_3O_7$ on substrate MgO the velocity reaches the value $v^*=2000$ m/s in magnetic field $B_y=1$ T at the temperature T=79.65 K (Dmitrenko, 1996). For this parameters the transport current density j_{z1} reaches the value $j_{z1}=4.8\cdot10^9$ A/m² for viscosity coefficient $\eta=10^{-8}$ N·s/m². At these parameters of superconducting film the transport current density j_{z2} varies with the angle θ from $2\cdot10^5$ A/m² for big θ to $2\cdot10^5$ A/m² for $\theta=0.01$ (Glushchenko & Golovkina, 2007). Thus if the superconductor is found in the Larkin-Ovchinnikov state the amplification electromagnetic wave could be observed at the lower values of transport current density.

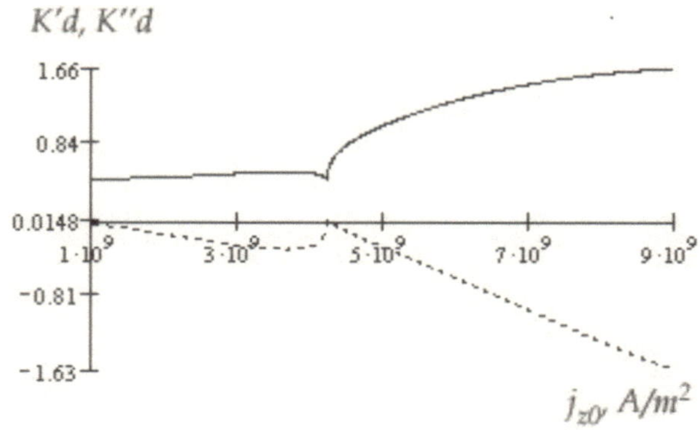

Fig. 6. The normalized Bloch wave number *K'd* (solid line) and attenuation coefficient *K''d* (dotted line) versus transport current density. The case of opposite direction of electromagnetic wave and vortex structure, $\omega=10^9$ rad/s

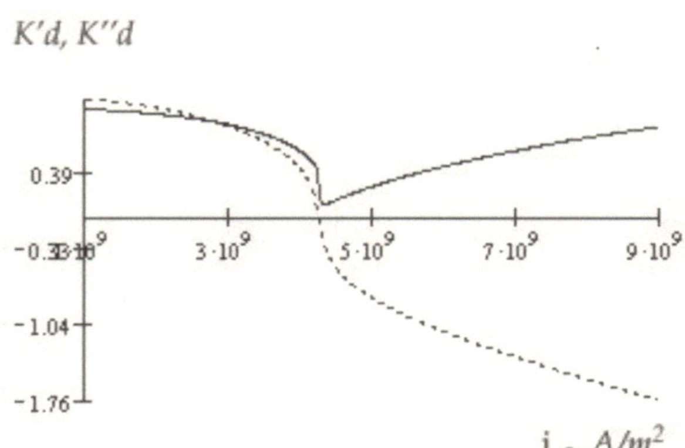

Fig. 7. The normalized Bloch wave number *K'd* (solid line) and attenuation coefficient *K''d* (dotted line) versus transport current density. The electromagnetic waves and vortex structure propagate in the same direction, $\omega=10^9$ rad/s

The dependence of Bloch wave number and attenuation coefficient from the transport current density is presented on Fig. 6 and 7. The parameters of superconducting film and dielectric layers are following: thickness of the dielectric layers d_1=6 μm, thickness of the superconducting layers t=50 nm, $\eta=10^{-8}$ N·s/m^2, B_{y0}=1 T, θ=0.5, v^*=1750 m/s. The Fig. 6 corresponds to the choice of the top sign in formula (18). The Fig. 7 corresponds to the bottom sign in (18), when the electromagnetic wave propagates along the moving vortex lattice. We can see that the attenuation coefficient K'' changes its sign at transport current density $j_z = j_{z1}$ (see Fig. 7). The amplification of electromagnetic waves could be observed at positive values of attenuation coefficient. Thus we can manage the process of amplification or attenuation by changing the transport current density.

Let us examine the behavior of attenuation coefficient K'' for the case when the electromagnetic waves and vortex structure propagate in the same direction (see Fig. 7, Fig. 8, Fig.9). The parameters of the structure in these figures are the same as in the Fig. 6.

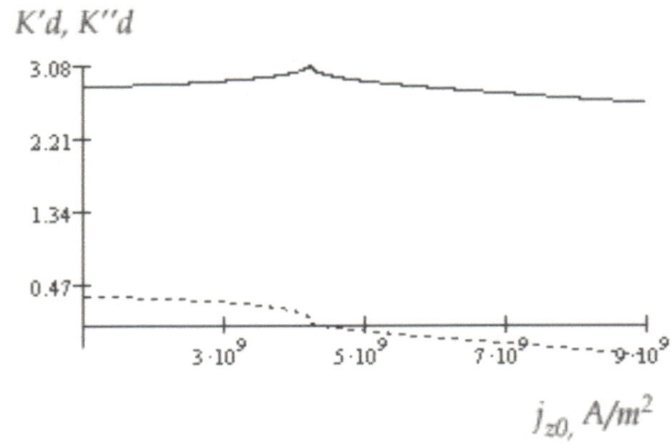

Fig. 8. The normalized Bloch wave number *K'd* (solid line) and attenuation coefficient *K''d* (dotted line) versus transport current density, ω=7·10^9 rad/s

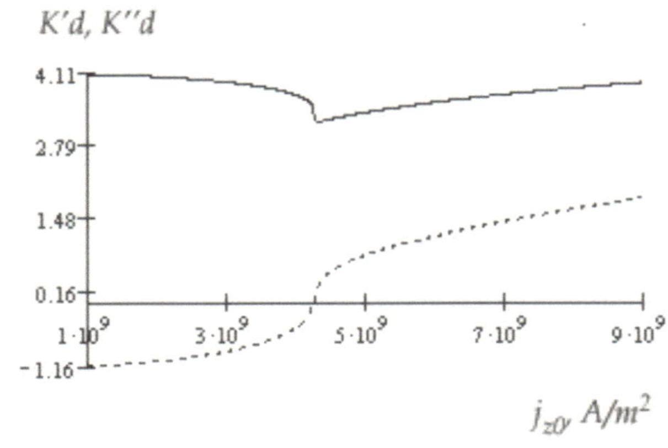

Fig. 9. The normalized Bloch wave number *K'd* (solid line) and attenuation coefficient *K''d* (dotted line) versus transport current density, ω=8·10^9 rad/s

The value of attenuation coefficient depends on the angular frequency ω. At the frequency change from $\omega=10^9$ rad/s (Fig. 7) up to $7\cdot10^9$ rad/s (Fig. 8) the absolute value of K'' decreases. And at the further growth of angular frequency up to $\omega=8\cdot10^9$ rad/s the areas of attenuation and amplification change their places (Fig. 9).
Thus the periodic structure with thin superconducting layers in Larkin-Ovchinnikov state demonstrates the new features in comparison with the superconducting structure in mixed state. Firstly, the amplification becomes possible at lower values of transport current density. Secondly, the coefficient of attenuation K'' can change its sign depending on j_z. Therefore we can manage amplification by means of transport current. Thirdly, the value of attenuation coefficient K'' depends on the angular frequency ω. All these features allow us to design new broadband amplifiers and filters. We can change the parameters of this devices not only by external magnetic field but also by transport current.

3.3 One-dimensional periodic structure superconductor - semiconductor

Let us consider the properties of one-dimensional periodic structure superconductor-semiconductor. There are essential distinctions between the structure superconductor-dielectric and the structure superconductor-semiconductor. The presence of a frequency dispersion of permeability in semiconductor layers leads to the appearance of new types of waves, propagating with various phase velocities. Also under the action of external electric field the free charged particle drift appears in semiconductor. As the result the medium gains active properties with new types of instabilities of electromagnetic waves. It is necessary to consider, that various dissipative processes exert significant influence on the electromagnetic wave propagation. That leads to the increase of attenuation and change of dispersion characteristics.
Let's consider the semiconductor plasma as a set of mobile electrons and holes which exist in a crystal. Let's use the hydrodynamic model in which electronic plasma is described as the charged liquid. The effective permittivity of superconductor can be written in following form:

$$\varepsilon_{eff} = \frac{2\cos^2\theta + \sin^2\theta\left(1+\dfrac{\varepsilon_\perp}{\varepsilon_{||}}\right)}{2\left(\dfrac{\cos^2\theta}{\varepsilon}+\dfrac{\sin^2\theta}{\varepsilon_{||}}\right)} \pm \frac{\sqrt{\left[2\cos^2\theta + \sin^2\theta\left(1+\dfrac{\varepsilon_\perp}{\varepsilon_{||}}\right)\right]^2 - 4\varepsilon_\perp\left(\dfrac{\cos^2\theta}{\varepsilon}+\dfrac{\sin^2\theta}{\varepsilon_{||}}\right)}}{2\left(\dfrac{\cos^2\theta}{\varepsilon}+\dfrac{\sin^2\theta}{\varepsilon_{||}}\right)}, \quad (21)$$

$$\varepsilon = 1 - \frac{\omega_p^2(\omega - i\nu_e)}{\omega\,[(\omega - i\nu_e)^2 - \omega_c^2]}, \quad (22)$$

$$\varepsilon_{||} = 1 - \frac{\omega_p^2}{\omega\,(\omega - i\nu_e)}, \quad (23)$$

$$\varepsilon_\perp = \frac{\omega_p^2\,\omega_c}{\omega\,[(\omega - i\nu_e)^2 - \omega_c^2]}, \quad (24)$$

where ν_e is the effective collision frequency, ω_p is the plasma frequency, ω_c is the cyclotron frequency of charge carriers. The dispersion equation for structure superconductor-semiconductor coincides with the dispersion equation for structure superconductor-dielectric (8). The difference is that we can not separate the independent E-and H-waves. In periodic structure with semiconductor layers two elliptically polarized waves can propagate. Each polarization corresponds to one of signs in the equation (21). The frequency dependence of permittivity leads to the appearance of new stop bands and new amplification bands. The presence of an imaginary part at Bloch wave number K indicates that the electromagnetic wave will attenuate exponentially when they pass through the periodic structure. However the Bloch wave number K becomes real when the condition *Im(K)=0* is fulfilled and the electromagnetic wave can penetrate deep into the structure. The amplification is observed if *Im(K)>0*. The equality of *Im(K)* to zero is possible if two condition are fulfilled:

$$\frac{\eta\,\omega}{\Phi_0} - j_{z0}k_x = 0 \tag{25}$$

or

$$\sin k_y d_1 = 0\,. \tag{26}$$

Taking into account the formulas (21) - (24) we can write the expressions (25) and (26) in the following form

$$\varepsilon_{eff} = \frac{\eta^2 \cdot c^2}{\Phi_0^{\,2}\, j_{z0}^{\,2} \sin^2\theta} \tag{27}$$

or

$$\varepsilon_{eff} = \frac{\pi^2 c^2 n^2}{d_1^{\,2} \cos^2\theta\, \omega^2},\; n = 0,1,2... \tag{28}$$

The solution of equations (27) and (28) is difficult. To simplify the solution, we consider an extreme case of collisionless plasma (when the effective collision frequency ν_e=0). This yields to the following expression for effective permittivity of semiconductor (Vural& Steele, 1973)

$$\varepsilon_{eff} = 1 - \frac{2\left(y^2-1\right)}{2\left(y^2-1\right)y^2 - \frac{\omega_c^{\,2}}{\omega_p^{\,2}} y^2 \sin^2\theta \pm \sqrt{y^4 \frac{\omega_c^{\,4}}{\omega_p^{\,4}} \sin^4\theta + 4\frac{\omega_c^{\,2}}{\omega_p^{\,2}} y^2\left(y^2-1\right)\cos^2\theta}}, \tag{29}$$

where $y=\omega/\omega_p$. In the expression (29) the top sign "+" in a denominator corresponds to an ordinary wave, and the bottom sign "-" to an extraordinary wave. In the further for the designation of effective permittivity of the extraordinary wave we shall use index 1, and for the ordinary wave - index 2. The effective permittivity of the ordinary wave vanishes when

$$y_{20} = 1\,. \tag{30}$$

The effective permittivity of the extraordinary wave vanishes when

$$y_{10} = \sqrt{1 \pm \frac{\omega_{ce}^{2}}{\omega_{pe}^{2}}}. \tag{31}$$

As it has been shown in (Golovkina, 2009 a), the solution of equation (27) corresponds to the resonance frequencies of ε_{eff} at value of vortex viscosity $\eta=10^{-8}$ N·s/m^2 and transport current density $j_{z0}=10^{10}$ A/m^2. The solutions of (27) on the frequencies which are not equal to the resonance frequencies of ε_{eff} appear when the vortex viscosity decreases and the transport current density increases (Bespyatykh et al., 1993), (Ye et al., 1995). The appropriate dispersion characteristic is shown in the Fig. 10.

We can see from the Fig. 10 that the imaginary part of Bloch wave number K is equal to zero at the frequencies $\omega_1=0.025\,\omega_p$, $\omega_2=0.15\,\omega_p$ and $\omega_3=0.19\,\omega_p$. These frequencies are the solutions of equation (27). At the frequencies $\omega<\omega_1$ and $\omega_2<\omega<\omega_3$ the electromagnetic wave attenuates, and at the frequencies $\omega_1<\omega<\omega_2$ the electromagnetic wave amplifies.

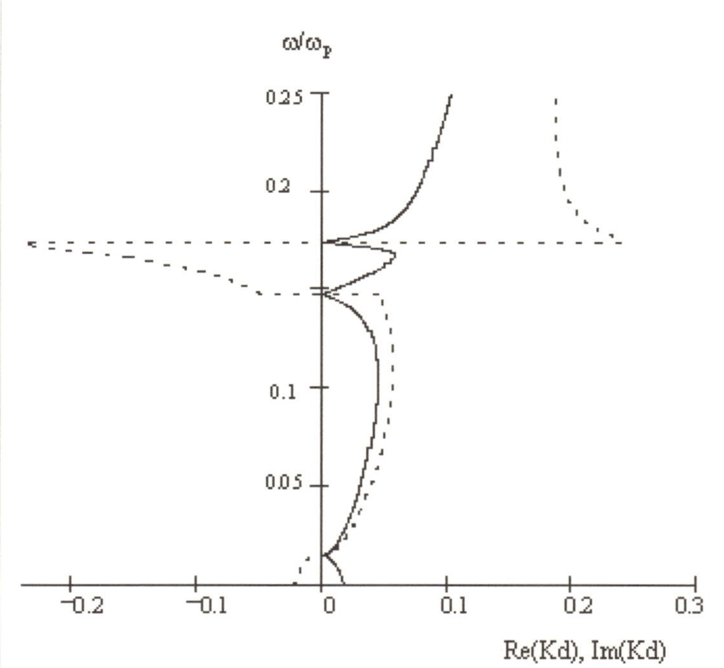

Fig. 10. The dispersion characteristics of periodic structure superconductor – semiconductor (the ordinary wave). The solid line: Re(Kd) , the dotted line: Im(Kd). Parameters: $\omega_p=1.2\cdot10^{12}$ s^{-1}, $\omega_c=10^{12}$ s^{-1}, $\nu_e=10^{10}$ s^{-1}, $d_1=3$ μm, t=60 nm, $\eta=10^{-8}$ N·s/m^2, $j_{z0}=10^{10}$ A/m^2

Thus the amplification of electromagnetic waves can be observed in the periodic structure superconductor - semiconductor as well as in the structure superconductor - dielectric. The amplification realizes at the expense of energy of moving Abikosov vortex lattice. The presence of frequency dispersion in semiconductor layers leads to the appearance of additional stop bands and amplification bands.

4. The structures with thin superconducting film and negative-index material

4.1 Periodic structure with combination of dielectric layer and layer with negative refractive index

In this section we consider the dispersion relations for electromagnetic wave propagation in an infinite periodic structure containing thin superconducting film and combination of two layers - dielectric and negative index material. The negative index materials or metamaterials are artificially structured materials featuring properties that can not be acquired in nature (Engheta & Ziolkowski, 2006). The new materials with negative index of refraction were theoretically predicted in 1968 by Veselago (Veselago, 1967). In these materials both the permittivity and the permeability take on simultaneously negative values at certain frequencies. In metamaterials with the negative refractive index the direction of the Pointing vector is antiparallel to the one of the phase velocity, as contrasted to the case of plane wave propagation in conventional media. The metamaterials with negative index of refraction are demonstrated experimentally first in the beginning of 20 century (Smith et al., 2000), (Shelby et al., 2001). In negative-index materials we can observe many interesting phenomena that do not appear in natural media. To unusual effects in negative-index materials concern the modification of the Snell's law, the reversal Cherenkov effect, the reversal Doppler shift (Jakšić, 2006). The most important effect is that wavevector and Pointing vector in negative-index material are antiparallel. Therefore the phase and group velocities are directed opposite each other. The unusual properties of negative-index materials are demonstrated especially strongly in its combination with usual medium.

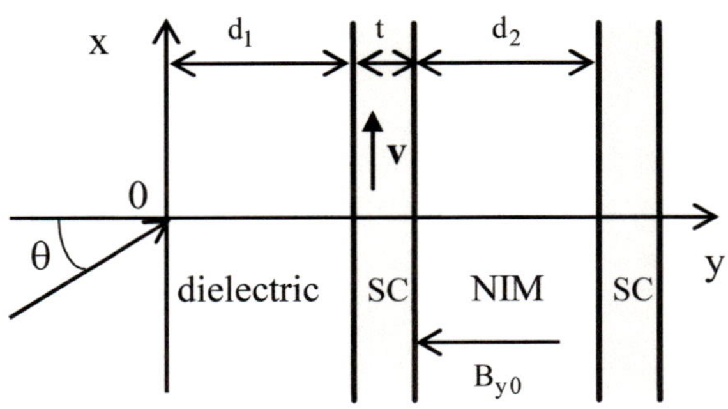

Fig. 11. Geometry of the problem. One-dimensional structure dielectric – superconductor – negative-index material

Let's consider the periodic structure containing the layer of usual dielectric with thickness d_1, the layer of negative-index material with thickness d_2 and the thin superconducting film with thickness t (see Fig. 11). By usage of matrix method we expressed dispersion relation for H-wave for considered structure in the following way (Golovkina, 2009 b):

$$\cos Kd = \cos k_{y1}d_1 \cos k_{y2}d_2 - \frac{1}{2}\left(\frac{k_{y1}\mu_2}{k_{y2}\mu_1} + \frac{k_{y2}\mu_1}{k_{y1}\mu_2}\right)\sin k_{y1}d_1 \sin k_{y2}d_2 - \\ -\frac{1}{2}\frac{i\omega\mu_0 t}{B_{y0}}\left(\frac{j_{z0}k_x}{\omega} - \frac{\eta}{\Phi_0}\right)\left(\frac{\mu_1}{k_{y1}}\sin k_{y1}d_1 \cos k_{y2}d_2 + \frac{\mu_2}{k_{y2}}\cos k_{y1}d_1 \sin k_{y2}d_2\right), \quad (32)$$

where ε_1 and μ_1 are the permittivity and permeability of usual dielectric ($\varepsilon_1>0$, $\mu_1>0$), ε_2 and μ_2 are the permittivity and permeability of negative-index material ($\varepsilon_2<0$, $\mu_2<0$).
The study of dispersion characteristics of electromagnetic wave in considered periodic structure with thin superconducting film and combination of two layers - dielectric and negative-index material has shown that these characteristics don't differ qualitative from the dispersion characteristics of periodic structure without negative index material layer.
The explanation of this fact consists in following. When electromagnetic wave propagates through the infinite one - dimensional periodic structure the phase velocities in dielectric and negative-index material are directed opposite each other (see Fig. 12). But the group velocities are co-directional in projection to axis z. Therefore the presence of negative index material layer in infinite structure does not affect on the resulting group velocity of electromagnetic wave.

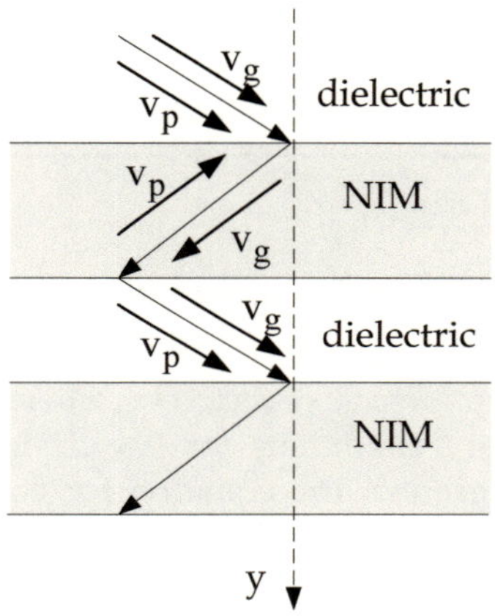

Fig. 12. The directions of phase and group velocities in periodic structure dielectric - negative-index material (NIM)

The important distinctive feature of negative index material (opposite direction of phase and group velocities) can be revealed only in the limited structures. The waveguide structures containing combination of dielectric and negative index material can excite the nondispersive modes and super - slow waves (Nefedov & Tretyakov, 2003), (Golovkina, 2007). Such slow waves can interact efficiently with moving Abrikosov vortex lattice.

4.2 Nonlinear pulses in waveguide with negative index material and thin superconducting film

The electromagnetic wave can amplify at the interaction with moving vortex structure when the velocities of electromagnetic wave and vortex lattice are approximately equal. For implementation of amplify condition it is necessary to slow down the electromagnetic wave. The slow waves can exist in two layered waveguide with negative index material slab (Nefedov & Tretyakov, 2003), in two layered waveguide with negative index material and with resistive film (Golovkina, 2007). The combination of two layers: dielectric and negative index material acts the role of slow-wave structure. The presence in waveguide of dielectric with negative index material can lead to amplification of evanescent electromagnetic waves (Baena et al., 2005). The amplification can be observed also in waveguide with negative-index material and thin superconducting film (Golovkina, 2009 c). If we add thin

superconducting film in waveguide with nonlinear dielectric we can manage the process of electromagnetic waves propagation (Golovkina, 2008).
Let us consider the wave propagation in two-layered waveguide. On layer of thickness a is a negative-index material ($\varepsilon_1<0$, $\mu_1<0$) and the other one of thickness b is an usual dielectric ($\varepsilon_2>0$, $\mu_2>0$) (see Fig. 15). The thin film of type-II superconductor with thickness t and thin film of Kerr nonlinear dielectric with thickness δ are placed in the plane y0z. The thickness of superconductor is $t<<\lambda$, where λ is magnetic field penetration depth, $\delta<<\Lambda$, where Λ is wavelength. The transport current in superconductor is directed along the 0y axis.

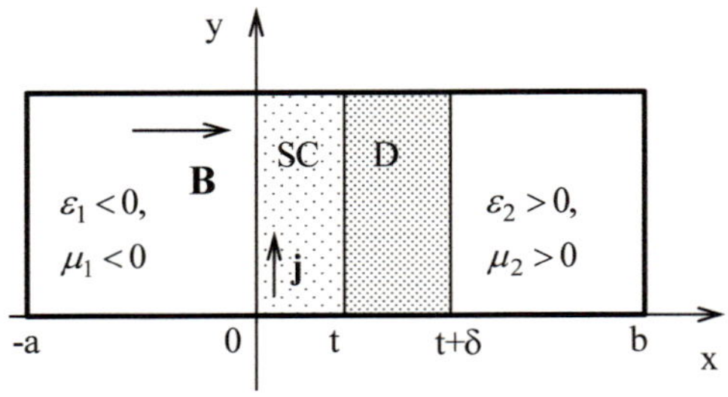

Fig. 13. The geometry of the structure. SC - superconductor, D - dielectric with Kerr-type nonlinearity

Let us consider the H-wave which effectively interacts with flux-line lattice in superconductor. Using boundary conditions for two-layered thin film superconductor - nonlinear dielectric we have received the equation for component $E_y(z,t)$ of the electric intensity and its Fourier transform $E_y(\omega, \beta)$ in the form (Glushchenko & Golovkina, 2006)

$$\int_{-\infty}^{+\infty}\int_{-\infty}^{+\infty}(2\pi)^{-2}d\omega d\beta R(\omega,\beta)E_y(\omega,\beta)\exp[i(\omega t-\beta z)]=\frac{\partial}{\partial t}(P_N(E_y)), \tag{33}$$

where $\mathbf{P}=\alpha_1\mathbf{E}+\alpha_3|\mathbf{E}|^2\mathbf{E}+\ldots$ is the vector of polarization of Kerr nonlinear dielectric media, $R((\omega, \beta)$ is function of parameters of the film, magnetic field and input impedances of negative index material and usual dielectric. After carrying out the $E_y(\omega, \beta)$ in form of high-frequency pulse with slowly varying complex envelope e(z, t) we obtain from (33) the nonlinear Schrödinger equation (Korn & Korn, 2000). The solution of this equation is represented by lattice of nonlinear pulses or by lattice of dark pulses.
The group velocity of pulses depends on parameters of thin films, on amplitude of pulse and on external magnetic field:

$$v=[\frac{\partial R(\omega=\omega_0,\beta=\beta_0)}{\partial \beta}\left(-\frac{\partial R(\omega=\omega_0,\beta=\beta_0)}{\partial \omega}\right)+i2\pi\delta k^{-2}\alpha_3 E_s^{\,2}(2k^2-1)]^{-1}. \tag{34}$$

Here E_s is the amplitude of pulse, k is module, α_3 is the coefficient in the nonlinear series expansion of vector of polarization P .
The numerical calculation demonstrates that the group velocity can change the magnitude and the sign by the variation of magnetic field (see Fig. 14). The calculated results offer the opportunity to considered waveguide to operate as an effective control structure at optical, IR and microwave frequencies. It should be noted that each layer of the structure is of the

great importance: thin Kerr nonlinear dielectric produces the soliton-like pulses, thin superconducting film gives the possibility of control, negative index material slab reduces the pulse velocity providing large interaction of pulses with a flux-line lattice in superconductor.

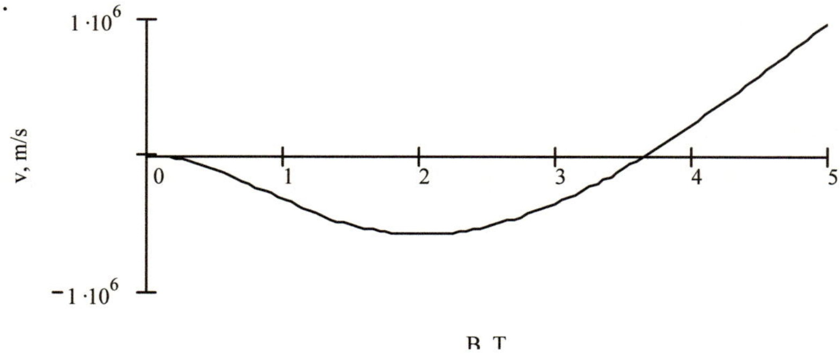

Fig. 14. Magnitude of the pulse velocity as function of external magnetic field. The parameters of the structure are: t=40 nm, $\eta=10^{-8}$ N·s/m^2, $j_{y0}=10^9$ A/m^2, $\omega=20\cdot10^{11}$ rad/s, a=0.025 m, b=0/02 m, δ=100 μm, $E_s=10^3$ V/m, $\alpha_3=10^{-15}$ C·m/V^2

7. Conclusion

In this chapter the propagation of electromagnetic waves in structures with thin high temperature superconducting film is investigated. The interaction of electromagnetic wave with thin superconductor in the mixed state is studied. It is shown, that the presence of the moving magnetic vortex structure in superconductor can lead not only to attenuation, but also to amplification of electromagnetic waves. The condition of amplification consists in equality of velocity of electromagnetic wave and the moving vortex structure. It is shown, that the electromagnetic wave amplification takes place at the expense of energy of Abrikosov vortex lattice. The representation of thin superconducting film in the form of boundary condition has enabled us to understand the mechanism of electromagnetic wave interaction with moving vortex structure. This method allowed us to simplify the numerical calculation.

In this study also the propagation of electromagnetic waves in periodic structures superconductor - dielectric is examined. The peculiarities of periodic structures with thin superconducting film in Larkin-Ovchinnikov state are revealed. The features of periodic structures superconductor - semiconductor are studied. The new pass bands and amplification bands are found. The possibility of the control of processes of attenuation and amplification is shown. The control can be realized by means of change of external magnetic field and transport current density. The dependence of coefficients of attenuation and amplification on the thickness of superconducting film and frequency enables us to make active devices which parameters can vary widely.

The structures with thin superconducting film in mixed state and combination of dielectric layer and negative index material layer are considered. It is shown, that the combination of dielectric and negative index material acts as the slow-wave structure in limited structures. The combination of dielectric and negative index material with thin superconducting film can be used in different devices such as waveguides and resonators as the control section. As the example of such application the waveguide with nonlinear thin film is considered. It is shown, that the nonlinear film with Kerr like nonlinearity excites the nonlinear soliton-like pulses. And the presence in such waveguide of the control section on the base of thin superconducting film permits to change not only attenuation coefficient, but also the direction of pulse propagation.

The properties of structures with thin high temperature superconducting films in the mixed state open the promissory perspective for their application in modern devices with control of parameters.

8. References

Abrikosov, A. A. (2004). Type-II superconductors and the vortex lattice. *Sov. Phys. Uspekhi,* Vol. 174, No. 11, pp. 1234–1239.

Artemov, Y. V.; Genkin, V. M.; Leviev, G. I. & Ovchinnikova, V. (1997). Nonlinear microwave losses in thin superconducting YBCO films. *Superconductor Science and Technology,* Vol. 10, No. 8, pp. 590-593.

Baena, J. D.; Jelinek, L.; Marqués, R. & Medina, F. (2005). Near-perfect tunneling and amplifcation of evanescent electromagnetic waves in a waveguide filled by a metamaterial: Theory and experiment. Physical Review B, Vol. 72, pp. 075116–1–8.

Bednorz, J.G.; Muller, K.A. (1986). Possible high T_c superconductivity in the Ba-La-Cu-O system. *Zeitschrift für Physik B,* Vol. 64, No. 2, pp. 189-193.

Bespyatykh, Y. I.; Wasilewski, W.; Gajdek, M.; Simonov, A. D. & Kharitonov, V. D. (1993). Dispersion and damping of surface magnetostatic waves in a ferromagnet-type II superconductor structure, *Physics of the Solid State,* Vol. 35, No. 11, pp.1466-1470.

Bulgakov, A. A. & Shramkova, O. V. Dispersion and instability of electromagnetic waves in layered periodic semiconductor structures, *Technical Physics,* Vol. 48, No. 3, pp. 361-369.

Chiang, T.-C. (2004). Superconductivity in thin films. *Science,* Vol. 306, No. 5703, pp. 1900 - 1901.

Dmitrenko, I. M. (1996). Resistive state of broad superconducting films and phase-slip lines, *Low Temperature Physics,* Vol. 22, pp. 648-665.

Engheta, N. & Ziolkowski, R.W. (2006). Introduction, history, and selected topics in fundamental theories of metamaterials, In: *Metamaterials,* Ed. by Engheta, N. & Ziolkowski, R.W., John Wiley & Sons, ISBN 13978-0-47176102-0, New York.

Gilmour, A. S. (1994). *Principles of traveling wave tubes,* Artech House, ISBN 0890067201, Boston.

Glushchenko, A.G. & Golovkina, M.V. (1998 a). Electromagnetic wave propagation in superconductor-dielectric multilayers, *Proceedings of International symposium on electromagnetic compatibility EMC'98,* pp. 430-432, Italy, Rome. [25]

Glushchenko, A.G. & Golovkina, M.V. (1998 b). Reflection of an electromagnetic wave by a layered superconductor-dielectric structure. *Technical Physics Letters,* Vol. 24, No. 4, pp.9-12.

Glushchenko, A.G. & Golovkina, M.V. (2006). A nonlinear pulse propagation in a waveguide thin-layer superconductor-insulator structure with Kerr nonlinearity. *Physics of Wave Processes and Radio Systems (in Russian),* Vol. 9, No. 2, pp. 12-17.

Glushchenko, A.G. & Golovkina, M.V. (2007). Propagation of electromagnetic waves in periodic structures with superconducting layers having electrodynamic parameters in the nonlinearity range of the dynamic mixed state, *Technical Physics*, Vol. 52, No. 10, pp. 1366-1368.

Golovkina, M.V. (2007). Two-layered waveguide containing a negative index material slab with resistive film., *Proceedings of Metamaterials'2007, First International Congress on Advanced Electromagnetic Materials in Microwaves in Microwaves and Optics*, pp. 377-379, Italy , October 2007, Rome.

Golovkina, M.V. (2008). Properties of pulses in the layered structure with negative index material slab and two-layered thin film superconductor-nonlinear dielectric, *Proceedings of Second International Congress of Advanced Electromagnetic Materials in Microwaves and Optics Metamaterials*, Spain , September 2008, Pamplona.

Golovkina, M.V. (2009 a). Characteristics of electromagnetic waves propagation in multilayered structure semiconductor - superconductor, *Vestnic Pomorskogo Universiteta*, Vol. 3, pp.70-75, ISSN 1728-7340.

Golovkina, M.V. (2009 b). Electromagnetic wave propagation in multilayered structures with negative index material, In: *Wave propagation in materials for modern applications*, Petrin, A. (Ed.), pp. 149-162, Intech, ISBN 978-953-7619-65-7.

Golovkina, M.V. (2009 c). Electromagnetic wave propagation in waveguide with thin superconducting film and metamaterial slab, *Progress in Electromagnetics Research Symposium*, Russia, August 2009, Moscow.

Gozar, A.; Logvenov, G.; Kourkoutis, L.F.; Bollinger, A. T.; Giannuzzi, L. A.; Muller, D. A. & Bozovic, I. (2008). High-temperature interface superconductivity between metallic and insulating copper oxides. *Nature*, Vol. 455, pp. 782-785.

Gunji, T.; Unno, M.; Arimitsu, K.; Abe, Y.; Long, N. & Bubendorfer, A. (2005). Preparation of YBCO and BSCCO superconducting thin films by a new chemical precursor method. *Bulletin of the Chemical Society of Japan*, Vol. 78, , No. 1, pp.187-191.

Gutliansky, E. D. (2005). Amplification of longitudinal ultrasonic waves by a moving vortex structure in type II superconductors. *JETP Letters*, Vol. 82, No. 2, pp. 72-76.

Hein, M. (1999). *High-Temperature-Superconductor Thin Films at Microwave Frequencies*, Springer; ISBN 3540656464.

Hohenwarter, G. K. G.; Martens, J. S.; Beyer, J. B.; Nordman, J. E. & McGinnis, D. P.(1989). Single superconducting thin film devices for applications in high Tc materials circuits. *IEEE Transactions on Magnetics*, Vol. 25, pp. 954-956.

Jakšić, Z.; Dalarsson, N. & Maksimović, M. (2006) Negative refractive index metamaterials: principles and applications. *Microwave Review*, Vol. 12, №o. 1, pp. 36–49.

Itozaki, H.; Higaki, K.; Harada, K.; Tanaka, S.; Yazu, S.& Tada, K. (1989). Properties of high Jc BiSrCaCuO and TlBaCaCuO thin film. *Physica C: Superconductivity*, Vol. 162-164, Part 1, pp. 367-368.

Korn, G. A. & Korn, T. M. (2000). *Mathematical Handbook for Scientists and Engineers*, Dover Publications, ISBN 0486411478.

Koster, G.; Brinkman, A.; Hilgenkamp H.; Rijnders A.J.H.M. & Blank D.H.A. (2008). High-Tc superconducting thin films with composition control on a sub-unit cell level; the effect of the polar nature of the cuprates. *Journal of Physics: Condensed Matter*, Vol. 20, No. 26, pp. 264007-1-264007-6.

Kurushin, E. P. & Nefedov, E. I. (1983). *Electrodynamics of anisotropic waveguiding structures*, Nauka, Moscow.

Kurushin, E. P.; Nefedov, E. I. & Fialkovsky, A. T. (1975). *Diffraction of electromagnetic waves in anisotropic structures*, Nauka, Moscow.

Kwak, M. H.; Tae Kim, Y. T.; Moon, S E.; Ryu, H.-C; Lee, S.-J. & Kang, K. Y. (2005). Microwave properties of tunable phase shifter using superconductor/ ferroelectric thin films. *Integrated Ferroelectrics*, Vol. 77, No. 1, pp. 79 - 85.

Larkin, A.I. & Ovchinnikov, Y.N. (1975). Nonlinear conductivity of superconductors in the mixed state, *Sov. Phys. JETP,* Vol. 41, No. 5, pp. 960-965.

Nefedov, I. S. & Tretyakov, S. A. (2003). Waveguide containing a backward-wave slab. *Radio Science*, Vol. 38, No. 6, pp. 9–1.

Phillips, J. M. (1995). High temperature superconducting thin films, In: *High-temperature superconducting materials science and engineering: new concepts and technology*, Shi, D. (Ed.), pp. 305-325, ISBN 0080421512, Pergamon, Great Britain.

Popkov, A.F. (1989). The magnetostatic wave amplification by magnetic vortex flux in structure ferrite - superconductor. *Technical Physics Letters*, Vol. 15, pp. 9-14.

Schauer, W.; Xia, X.X.; Windte, V.; Meyera, O.; Linkera, G.; Lia, Q. & Geerk, J. (1990). Growth quality and critical current density of sputtered YBaCuO thin films. *Cryogenics*, Vol. 30,No. 7, pp. 586-592.

Schilling, A.; Cantoni, M., Guo, J. C. & Ott, H. R. (1993)., Superconductivity above 130 K in the HgBaCaCuO system. *Nature*, Vol. 363, pp. 56-58.

Schmidt, V.V. (2002). *The physics of superconductors: introduction to fundamentals and applications,* Springer, ISBN 3540612432.

Shelby, A.; Smith, D. R. & Schultz, S. (2001). Experimental verification of a negative index of refraction. *Science*, Vol. 292, 77-79.

Smith, D. R.; Padilla, W. J.; Vier, D. C.; Nemat-Nasser, S.C. & Schultz, S. (2000). Composite medium with simultaneously negative permeability and permittivity. *Phys. Rev. Lett.*, Vol. 84, 4184-4187.

Veselago, V. G. (1968) The electrodynamics of substances with simultaneously negative values of epsilon and mu. *Sov. Phys. Uspekhi*, Vol. 10, 509-514.

Veselov, G. I. & Rajevsky, S. B. (1988). Layered metal-dielectric waveguides, Radio and Svjaz, Moscow.

Vural, B. & Steele, M. (1973). *Wave Interactions in Solid State Plasmas*, McGraw- Hill, New York.

Wu, M. K.; Ashburn, J. R.; Torng, C. J.; Hor, P. H.; Meng, R. L.; Gao, L.; Huang, Z. J.; Wang, Y. Q. & Chu, C. W. (1987). Superconductivity at 93 K in a new mixed-phase Y-Ba-Cu-O compound system at ambient pressure. *Physics Review Letters*, Vol. 58, pp. 908–910.

Yang , B. C.; Wang, X. P.; Wang, C. Q.; Wang, R. K.; Cui, C. G. & Li, S. L. (1991). Single-crystal YBaCuO thin films with high critical current density on Zr(Y)O2, SrTiO3 and LaAlO3 deposited by DC-magnetron sputtering. *Superconductor Science and Technology* , Vol. 4, No. 4, pp. 143-148.

Ye, M.; Mehbood, M. & Deltour, R. (1995). High critical current density in epitaxial $YBa_2Cu_3O_7$ thin films. *Physica B*, Vol. 204, No. 1-4, pp. 200206–1–8.

Zhao, X.; Li, L.; Lei, C. & Tian, Y. (2002). High temperature superconducting thin films for microwave filters. *Science in China (Series A)*, Vol. 45, No. 9, pp. 1183-1191.

Part 2

Light Wave Propagation and Nanofocusing

Detection and Characterization of Nano-Defects Located on Micro-Structured Substrates by Means of Light Scattering

Pablo Albella,[1] Francisco González,[1] Fernando Moreno,[1]
José María Saiz[1] and Gorden Videen[2]
[1]*University of Cantabria*
[2]*Army Research Laboratory*
[1]*Spain*
[2]*USA*

1. Introduction

Detection and characterization of microstructures is important in many research fields such as metrology, biology, astronomy, atmospheric contamination, etc. These structures include micro/nano particles deposited on surfaces or embedded in different media and their presence is typical, for instance, as a defect in the semiconductor industry or on optical surfaces. They also contribute to SERS and may contribute to solar cell performance [Sonnichsen *et al.*, 2005; Stuart *et al.*, 2005; Lee *et al.*, 2007]. The central problem related to the study of morphological properties of microstructures (size, shape, composition, density, volume, etc.) is often lumped into the category of "Particle Sizing" and has been a primary research topic [Peña *et al.*, 1999; Moreno and Gonzalez, 2000; Stuart *et al.*, 2005; Lee *et al.*, 2007].

There are a great variety of techniques available for the study of micro- and nano-structures, including profilometry and microscopy of any type: optical, electron, atomic force microscopy (AFM), etc. Those based on the analysis of the scattered light have become widely recognized as a powerful tool for the inspection of optical and non-optical surfaces, components, and systems. Light-scattering methods are fast, flexible and robust. Even more important, they are generally less expensive and non-invasive; that is, they do not require altering or destroying the sample under study [Germer *et al.*, 2005; Johnson *et al.*, 2002; Mulholland *et al.*, 2003].

In this chapter we will focus on contaminated surfaces composed of scattering objects on or above smooth, flat substrates. When a scattering system gets altered either by the presence of a defect or by any kind of irregularity on its surface, the scattering pattern changes in a way that depends on the shape, size and material of the defect. Here, the interest lies not only in the characterization of the defect (shape, size, composition, etc.), but also on the mere detection of its presence. We will show in detail how the analysis of the backscattering patterns produced by such systems can be used in their characterization. This may be useful in practical situations, like the fabrication of a chip in the semiconductor industry in the case of serial-made microstructures, the performance of solar cells, for detection and characterization of contaminants in optical surfaces like telescope mirrors or other sophisticated optics, and for assessing surface roughness, etc. [Liswith, 1996; Chen, 2003]. Before considering the first practical situation, we find it convenient to describe the backscattering detection concept.

Backscattering detection

In a typical scattering experiment, a beam of radiation is sent onto a target and the properties of the scattered radiation are detected. Information about the target is then extracted from the scattered radiation. All situations considered in this work exploit this detection scenario in the backscattering direction. Although backscattered light may be the only possible measurement that can be made in some situations, especially when samples are crowded with other apparati, it also does have some advantages that make it a useful approach in other situations. Backscattering detection can be very sensitive to small variations in the geometry and/or optical properties of scattering systems with structures comparable to the incident wavelength. It will be shown how an integration of these, over either the positive or negative quadrant, corresponding to the defect side or the opposite one, respectively, yields a parameter that allows one not only to deduce the existence of a defect, but also to provide some information about its size and location on the surface, constituting a non-invasive method for detecting irregularities in different scattering systems.

2. System description

Figure 1 shows an example of a typical practical situation of a microstructure that may or may not contain defects. In this case, the microstructure is an infinitely long cylinder, or fiber. Together with the real sample, we show the 2D modelling we use to simulate this situation and provide a 3D interpretation.

This basic design consists on an infinitely long metallic cylinder of diameter D, placed on a flat substrate. We define two configurations: the Non-Perturbed Cylinder (NPC) configuration, where the cylinder has no defect and the Perturbed Cylinder (PC) configuration, which is a replica of the NPC except for a defect that can be either metallic or dielectric and can be located either on the cylindrical microstructure itself or at its side, lying on the flat substrate underneath. We consider the spatial profile of this defect to be cylindrical, but other defect shapes can be considered without difficulty. The cylinder axis is parallel to the Y direction and the X-Z plane corresponds to both the incidence and scattering planes. This restricts the geometry to the two-dimensional case, which is adequate for the purpose of our study [Valle *et al.*, 1994; Moreno *et al.*, 2006; Albella *et al.*,2006; Albella *et al.*, 2007]. The scattering system is illuminated by a monochromatic Gaussian beam of wavelength λ (*633nm*) and width $2\omega_0$, linearly polarized perpendicular to the plane of incidence (S-polarized).

In order to account for the modifications introduced by the presence of a defect in the scattering patterns of the whole system, we use the Extinction theorem, which is one of the bases of modern theories developed for solving Maxwell's Equations. The primary reason for this choice is that it has been proven a reliable and effective method for solving 2D light-scattering problems of rounded particles in close proximity to many kinds of substrates [Nieto-Vesperinas *et al.*, 1992; Sanchez-Gil *et al.*, 1992; Ripoll *et al.*, 1997; Saiz *et al.*, 1996].

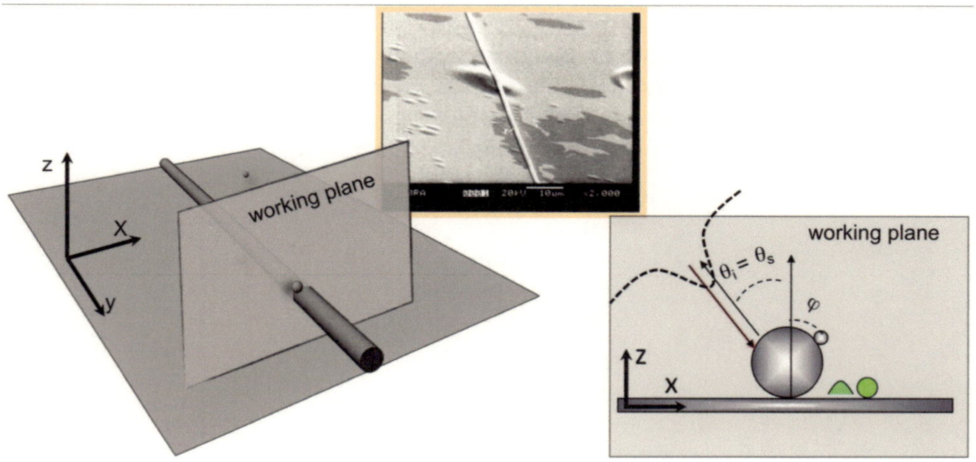

Fig. 1. Example of a contaminated microstructure (top figure) and its corresponding 2D and 3D models.

The Extinction theorem is a numerical algorithm. To perform the calculations, it is necessary to discretize the entire surface contour profile (substrate, cylinder and defect) into an array of segments whose length is much smaller than any other length scale of the system, including the wavelength of light and the defect. Bear in mind that it is important to have a partition fine enough to assure a good resolution in the high curvature regions of the surface containing the lower portion of the cylinder and defect. Furthermore, and due to obvious computing limitations, the surface has to be finite and the incident Gaussian beam has to be wide enough to guarantee homogeneity in the incident beam but not so wide as to produce undesirable edge effects at the end of the flat surface. Consequently, in our calculations, the length of the substrate has been fixed to 80λ and the width ($2\omega_0$) of the Gaussian beam to 8λ.

3. Metallic substrates

In this section we initially discuss the case of metallic cylinders, or fibers, deposited on metallic substrates and with the defect either on the cylinder itself or on the substrate but near the cylinder.

Defect on the Cylinder

As a first practical situation, Figure 2 shows the backscattered intensity pattern, as a function of the incident angle θ_i, for a metallic cylinder of diameter $D = 2\lambda$. We consider two different types of defect materials of either silver or glass and having diameter $d = 0.15\lambda$.
It can be seen how the backscattering patterns measured on the unperturbed side of the cylinder (corresponding to $\theta_s < 0$) remain almost unchanged from the reference pattern. In this case, we could say that the defect was hidden or shadowed by the incident beam. If the scattering angle is such that the light illuminates the defect directly, a noticeable change in the positions and intensity values of the maxima and minima results. The number of maxima and minima observed may even change if the defect is larger than 0.4λ. This means that there is no change in the effective size of the cylinder due to the presence of the defect. This result can be explained using a phase-difference model [Nahm & Wolf, 1987; Albella *et al.*, 2007], where the substrate is replaced by an image cylinder located opposite the

substrate from the real one. Then, we can consider the two cylinders as two coherent scatterers. The resultant backscattered field is the linear superposition of the scattered fields from each cylinder, which only differ by a phase corresponding to the difference in their optical paths and reflectance shifts. This phase difference is directly related to the diameter of the cylinder. See the Appendix for more detail on this model.

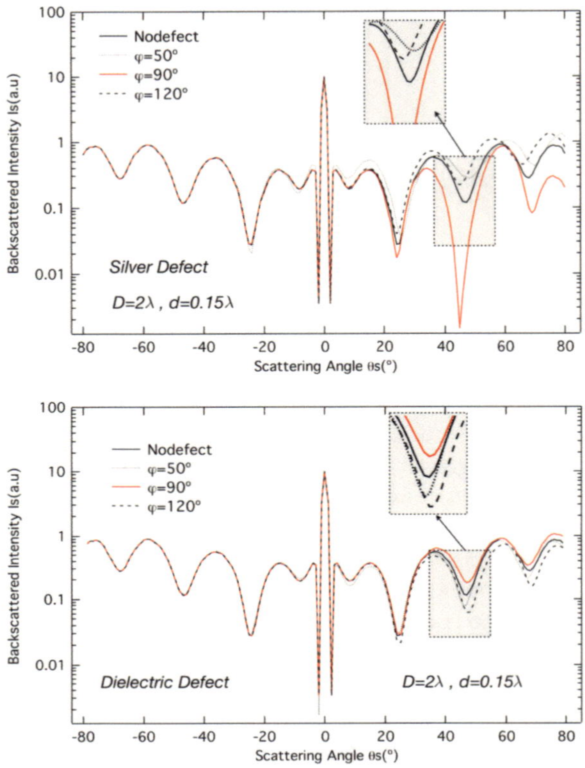

Fig. 2. Backscattered intensity pattern I_{back} as a function of the incident angle θ_i for a metallic cylinder of diameter $D = 2\lambda$. Two different types of defect material (either silver or glass) have been considered, with a diameter $d = 0.15\lambda$ [Albella *et al.*, 2007].

One other interesting point is that the backscattering pattern has nearly the same shape in terms of intensity values and minima positions, regardless of the nature of the defect. Perhaps, the small differences observed manifest themselves better when the defect is metallic. However for $\theta_s > 0$, a difference in the backscattered intensity can be noticed when comparing the results for silver and glass defects. If we observe a minimum in close detail (magnified regions), we see how I_{back} increases with respect to the cylinder without the defect when it is made of silver. In the case of glass located at the same position, I_{back} decreases. These differences can be analyzed by considering an incremental integrated backscattering parameter σ_{br}, which is the topic of the next section.

Parameter σ_{br}

One of the objectives outlined in the introduction of this chapter was to show how the backscattering pattern changes when the size and the position of a defect are changed, and whether it is possible to find a relationship between those changes and the defect properties of size and position. A systematic analysis of the pattern evolution is necessary. The possibility of using the shift in the minima to obtain the required information [Peña *et al.*, 1999] is not suitable in this case because there is no consistency in the behaviour of the angular positions of the minima with defect change. Based on the loss of symmetry in the backscattering patterns introduced by the cylinder defect, a more suitable parameter can be introduced to account for these variations. We have defined it as

$$\sigma_{br}^{\pm} = \frac{\sigma_b^{\pm} - \sigma_{b0}^{\pm}}{\sigma_{b0}^{\pm}} = \frac{\sigma_b^{\pm}}{\sigma_{b0}^{\pm}} - 1$$

where

$$\sigma_b^{\pm} = \int_0^{\pm 90^\circ} I_{back}(\theta_s) \cdot d\theta_s$$

is the backscattering intensity (I_{back}) integrated over either the positive (0° *to* 90°) or the negative (0° *to* -90°) quadrant. Subscript *0* stands for the NPC configuration. One of the reasons for using $\sigma_{br}^{\pm}$ is that integrating over θ_s allows us to account for changes produced by the defect in the backscattering efficiencies associated with an entire backscattering quadrant, not just in a fixed direction.
Figure 3 shows a comparison of σ_{br} calculated from the scattering patterns shown in Figure 2, as a function of the angular position of the defect on the main cylinder. It can be seen that the maximum value of $\sigma_{br}^{\pm}$ has an approximate linear dependence on the defect size *d*. As an example, for the case of a metallic defect near a $D = 2\lambda$ cylinder, $[\sigma_{br}]_{max} = 2.51d - 0.14$ with a regression coefficient of *0.99* and *d* expressed in units of λ. For $d \in [0.05\lambda, 0.2\lambda]$ and cylinder sizes comparable to λ, it is found that the positions $[\sigma_{br}]_{max}$ and $[\sigma_{br}]_{min}$ are independent of the cylinder size *D*. Another characteristic of the evolution of σ_{br}^{+} is the presence of a minimum or a maximum around $\varphi = 90^o$ for metallic and dielectric defects, respectively. Examining the behaviour of this minimum allows us to conclude that $[\sigma_{br}]_{min}$ also changes linearly with defect size; however, the slope is no longer independent of the cylinder size.

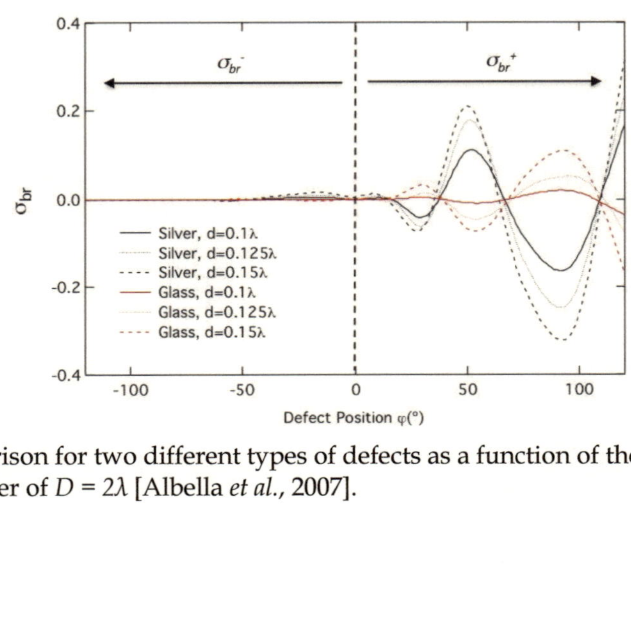

Fig. 3. $\sigma_{br}^{\pm}$ comparison for two different types of defects as a function of the defect position for a silver cylinder of $D = 2\lambda$ [Albella *et al.*, 2007].

The most interesting feature shown in Figure 3 is that in all cases considered, σ_{br} for a glass defect has the opposite behaviour of that observed for a silver defect. That is, when the behaviour of the dielectric is maximal, the behaviour of the conductor is minimal, and *vice versa*. This behaviour suggests a way to discriminate metallic from dielectric defects. We shall focus now on the evolution of parameter $\sigma^{\pm}$ with the optical properties of the defect and in particular for a dielectric defect around the regions where the oscillating behaviour of σ_{br} reaches the maximum amplitude, that is, $\varphi = 50°$ and $\varphi = 90°$.

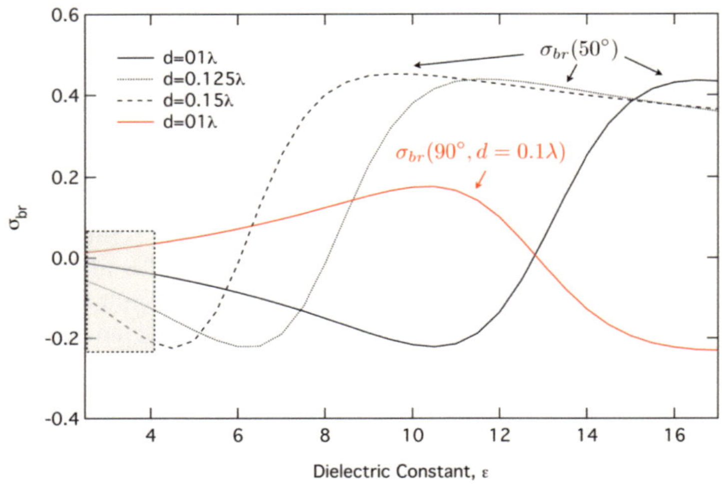

Fig. 4. Evolution of $\sigma_{br}^{\pm}$ with ε for a fixed defect position (50°). The behaviour for the glass defect (red) is opposite that of the silver defect (black) [Albella *et al.*, 2007].

Figure 4, shows three curves of σ_{br}^{+} (50°) as a function of the dielectric constant ε (ranging from *2.5* to *17*), for three different defect sizes. For each defect size, σ_{br}^{+}(50°) begins negative and with negative slope; it reaches a minimum (-0.22) and then undergoes a transition to a positive slope to a maximum (approximately 0.45). This means that for each size, there is a value of ε large enough to produce values of σ_{br}^{+}(50°) similar to those obtained for silver defects. The zero value would correspond to a situation where σ_{br}^{+}(50°) cannot be used to discriminate the original defect. When the former analysis is repeated for $\varphi = 90°$ similar behaviour is found, although σ_{br}^{+}(90°) has the opposite sign, as expected. As an example, σ_{br}^{+}(90°) for a $d = 0.1\lambda$ defect is shown in Figure 4. Analogue calculations have been carried out for different values of D ranging from λ to 2λ, leading to similar results, i.e., the same $\sigma_{br}(\varepsilon)$ with zero values is obtained for different values of ε. The region shadowed in Figure 4, typically a glass defect, can be fit linearly and could produce a direct estimation of the dielectric constant of the defect. As an example, $\sigma_{br}(50°) = -0.03\varepsilon + 0.02$ for the case of $d = 0.1\lambda$.

4. Defect on the substrate

We now consider the defect located on the substrate close to the main cylinder, within 1 or 2 wavelengths. In Figure 5 we see that there remains a clear difference in the backscattering patterns obtained in each of the two hemispheres, thus making it possible to predict which side of the cylinder the defect is located. Results shown in Figure 4 correspond to a cylinder of $D = 2\lambda$ and defect positions: $x = 1.2\lambda, 2\lambda$, while the defect size is fixed at $d = 0.2\lambda$.

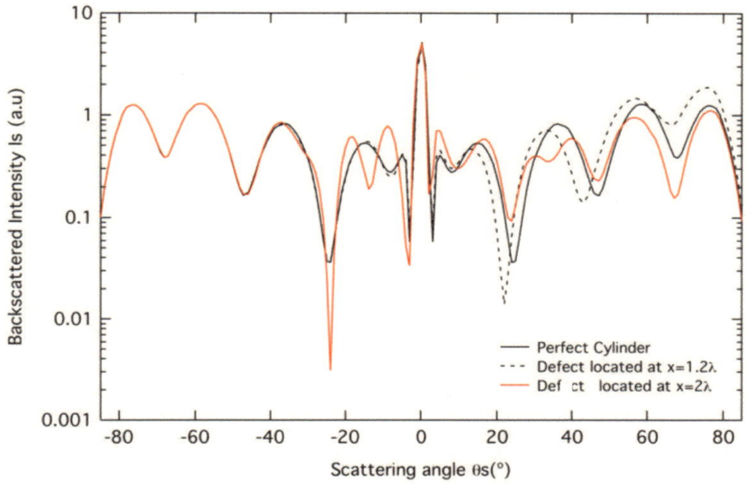

Fig. 5. Backscattering patterns obtained for when the defect is located on the substrate, together with the perfect, isolated cylinder case.

We observe again that the backscattering pattern changes more on the side where the defect is located. This behaviour is similar to that observed in the case of a defect on the cylinder. However, if we look at the side opposite the defect, the backscattering pattern does change if the defect is located outside the shadow cast by the cylinder. When the defect is near the cylinder (x = 1.2λ), that is, within the shadow region, the change in the scattering pattern is negligible at any incident angle. Nevertheless, when the defect is located further from the cylinder (x = 2λ), the change in the left-hand side can be noticed for incidences as large as $\theta_i = 40^o$.

Figure 6 shows the evolution of σ_{br} for two different cylinders of $D = 1\lambda$ and $D = 2\lambda$, and for three different sized defects, $d = 0.1\lambda$, 0.15λ and 0.2λ. The shadowed area represents defect positions beneath the cylinder, not considered in the calculations. The smaller shadow produced in the $D = \lambda$ case causes oscillations in σ_{br}^- for smaller values of x.

It is worth noting that for a given cylinder size, the presence and location of a defect can be monitored. For a $D = \lambda$ cylinder with x as great as 2λ, σ_{br}^- can increase as much as 10% for a defect of $d = 0.2\lambda$. When the defect is closer than $x = 1.5\lambda$, σ_{br}^+ becomes negative while σ_{br}^- is not significant. Finally, when $x < \lambda$, σ_{br}^+ is very sensitive and strongly tends to zero. In the case of a $D = 2\lambda$ cylinder, the most interesting feature is the combination of high absolute values and the strong oscillation of σ_{br}^+ for x within the interval $[\lambda, 2\lambda]$. Here the absolute value of $|\sigma_{br}^+|$ indicates the proximity of the defect, and the sign designates the location within the interval.

Although the size of the defect does not change the general behaviour, it is interesting to notice that when comparing both cases, σ_{br}^+ is sensitive to the defect size and also dependent on the size of the cylinder, something that did not occur in the former configuration when the defect was located on the cylinder. Both situations can be considered as intrinsically different scattering problems: with the defect on the substrate, there are two distinct scattering particles, but with the defect on the cylinder, the defect is only modifying slightly the shape of the cylinder and consequently the overall scattering pattern.

To illustrate this difference, Figure 7 shows some examples of the near-field and far-field patterns produced by both situations for different defect positions. Figure 7(a) shows the

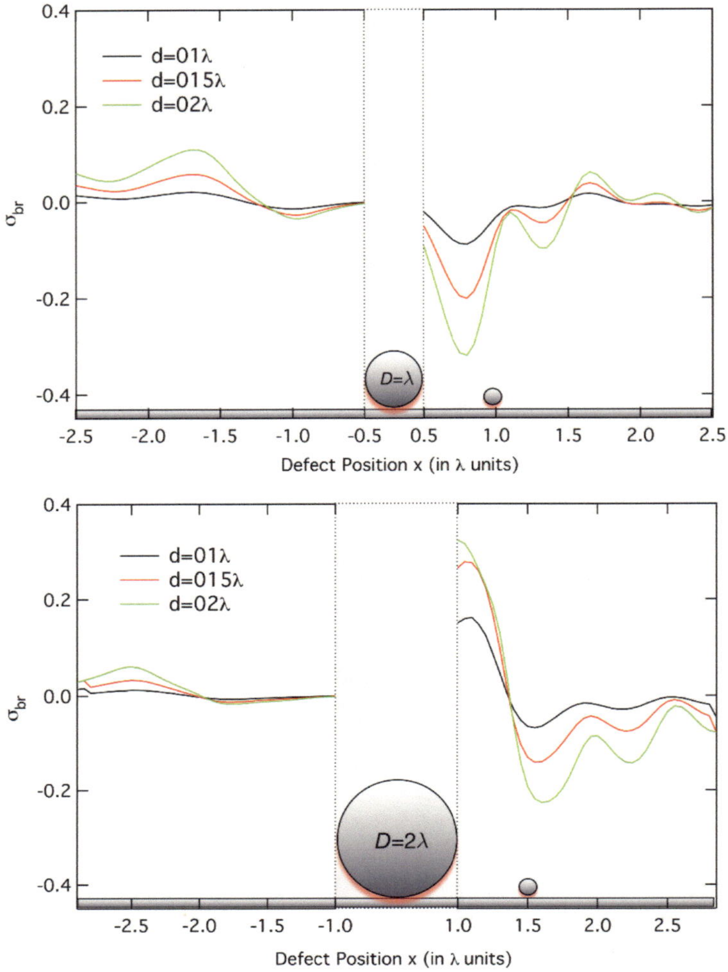

Fig. 6. σ_{br} for two different cylinders sizes, $D = \lambda$, 2λ and three defect sizes $d = 0.1\lambda$, 0.15λ and 0.2λ. Defect position x ranges from 0.5λ to 3λ from the center of the Cylinder [Albella *et al.*, 2007].

near-field plot of the perfect cylinder and will be used as a reference. Figure 7(b) and 7(c) correspond to the cases of a metallic defect on the cylinder. The outline of the defect is visible on these panels. It can be seen that the defect does not change significantly the shape of the near field when compared to the perfect cylinder case. Figure 7(d) and 7(e) correspond to the cases of a metallic defect on the substrate. In Figure 7(d), the defect is farthest from the cylinder, outside the shadow region, and we observe a significantly different field distribution around the micron-sized particle located in what initially was a maximum of the local field produced by the main cylinder. The same feature can be found in the far-field plot. This case corresponds to the maximum change with respect to the non-defect case. Finally, Figure 7(e) corresponds to the case of a metallic defect on the substrate and close to the cylinder, very close to the position shown in Figure 7(c). As expected, both cases are almost indistinguishable.

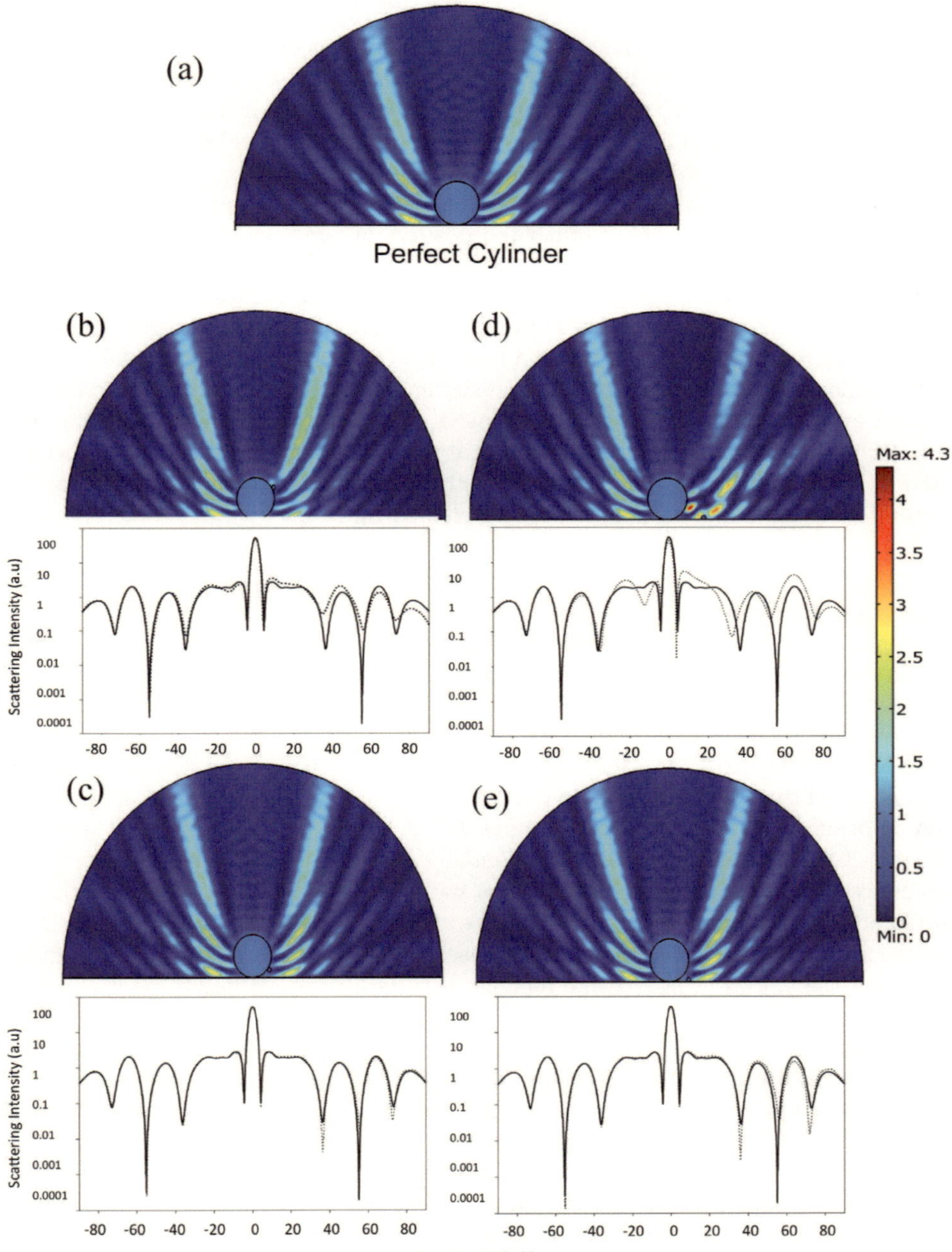

Fig. 7. (a) Near field plots for a silver cylinder sized $D = 2\lambda$ located on a silver substrate and illuminated at normal incidence. (b) and (c) show patterns for the defect located on the cylinder and (d) and (e) show patterns for the defect located on the substrate. The main cylinder and defect are outlined in black. Each plot has its corresponding far-field at the bottom compared to the NPC case pattern.

5. Influence of the optical properties of the substrate

In this section, we discuss the sensitivity of the defect detection technique to the optical properties of the substrate in the two possible situations described before: (A) with the defect on the cylinder and (B) with the defect on the substrate near the cylinder. We will see how the substrate can affect the detection capabilities in each particular situation.
In the previous sections, we described how a small defect located on a micron-sized silver cylinder on a substrate changes the backscattered intensity. We showed that an integration of the backscattered intensity over either the positive or negative quadrant, corresponding to the defect side or the opposite side, yields a parameter σ_{br} sensitive not only to the existence of the defect but also to its size and location on the microstructure. These results were initially obtained for perfectly conducting systems and later on, for more realistic systems: dielectric or metallic defects on a metallic cylinder located on a metallic substrate. From a practical point of view, detection and sizing of very small defects on microstructures located on any kind of substrate by non-invasive methods could be very useful in quality-control technology and in nano-scale monitoring processes. This section is focused on examining the sensitivity of this technique to the optical properties of the substrate in the aforementioned situations. In this section we also consider the cylinder to be composed of gold.

5.1 Defect on the cylinder

Figure 8 shows a comparison between the backscattered intensity pattern for a perturbed gold cylinder located on a metallic gold substrate and the backscattering pattern obtained for the same system located on other substrates having different optical properties. The diameter of the main cylinder and of the defect are $D = \lambda$ and $d = 0.1\lambda$, respectively, thus keeping constant the ratio $d/D = 0.1$. The defect position has been fixed on the cylinder at $\varphi = 50^o$ as it is one of the most representative cases.
As can be observed for backscattering angles $\theta_s < 0$, that is when the cylinder shadows the defect, the shape of the pattern remains essentially the same. We also do see slightly smaller values as we increase the dielectric constant of the substrate. When we illuminate the system on the same side as the defect, $\theta_s > 0$, we tend to see the opposite behaviour: in the locations where the values of I_{back} increased as we increase the dielectric constant of the substrate, and approaching the values of I_{back} for the metallic substrate case. Although the changes in the backscattering induced by the defect may seem negligible, we will see that these differences can be monitored with appropriate integrating parameters. In particular, we use the integrated backscattering parameter σ_{br} as defined in the previous sections.
Figure 9(a) shows the behaviour of σ_{br} for different dielectric substrates as a function of the angular position of the defect. An interesting result is the increase of $|\sigma_{br}^+|$ as we increase ε, reaching a maximum for the case of a metal. The opposite behaviour is observed for $|\sigma_{br}^-|$.
The maximum absolute value of σ_{br}^+ and σ_{br}^- for pure dielectric substrates is plotted in Figure 9(b) as a function of the substrate dielectric constant ε. We notice that the quantity σ_{br} is more sensitive to ε within the interval $\varepsilon \in [1.2, 4]$ and it saturates for high values of ε, tending to the metal substrate case. Absorption has not been considered for the case of real dielectrics as it is very small in the visible range. Another interesting result is that for a defect on the upper part of the cylinder $\varphi < 50^o$, σ_{br}^- is very sensitive to the metal/dielectric nature of the substrate. On the other hand, these remarkable values of σ_{br}^- make it more difficult to locate the position of the defect.

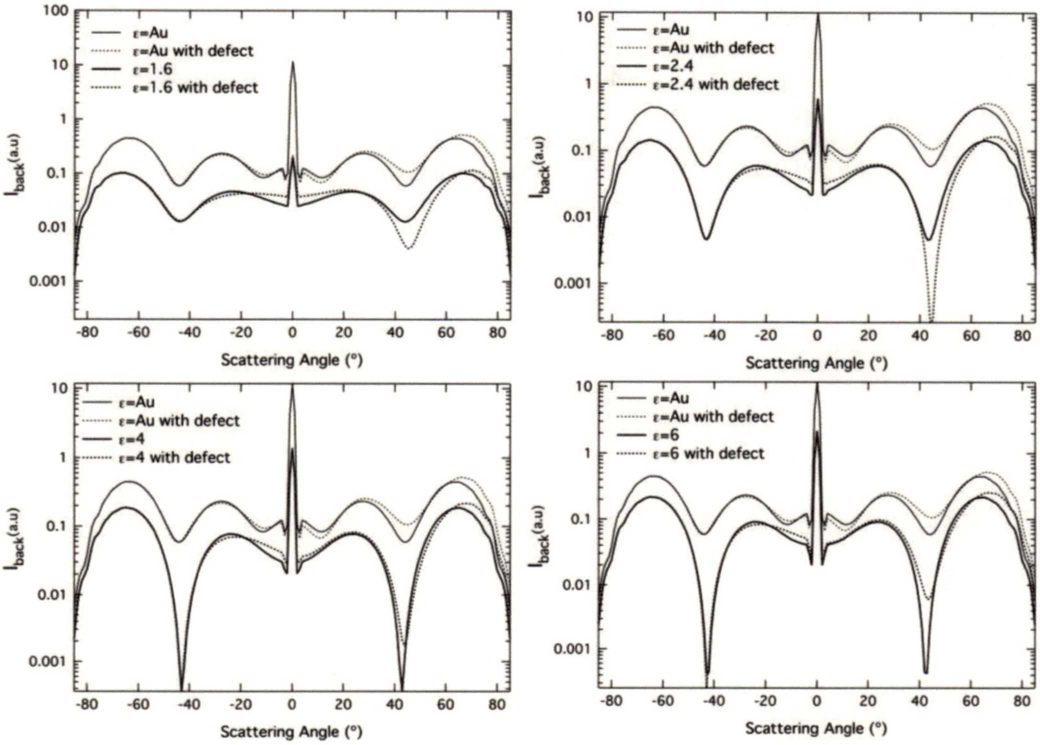

Fig. 8. Backscattering patterns for a defect on a cylinder placed on a substrate for different substrate optical properties (ε). The defect and cylinder are made of gold and of size $d = 0.1\lambda$ and $D=\lambda$ respectively.

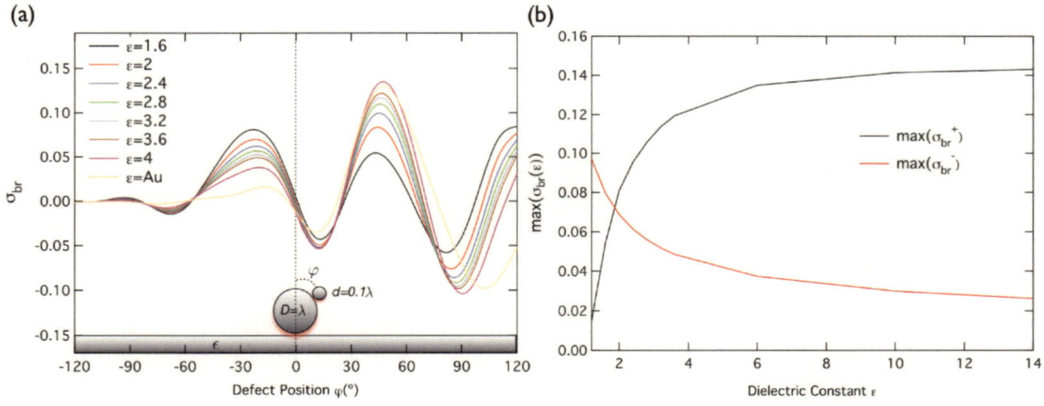

Fig. 9. (a) Evolution of σ_{br} for a defect on a cylinder as a function of the optical properties of the substrate. (b) Evolution of $max(\sigma^{+})$ and $max(\sigma^{-})$ for pure dielectric substrates as a function of the substrate dielectric constant, ε [Albella *et al.*, 2008].

5.2 Defect on the substrate

Figure 10 shows a series of graphs comparing the patterns obtained for different dielectric substrates with those obtained for a metallic substrate for a fixed defect position $x = 3\lambda/4$. For $\theta_s < 0$, i.e. the region opposite to the defect side, the change induced by the defect in the backscattering pattern is not significant, independent of the kind of substrate material. However, when $\theta_s > 0$, the backscattering is strongly affected by the defect, especially for the dielectric substrate. For increasing values of the dielectric constant, the change induced by the defect becomes smaller. This is the opposite of what was found when the defect was on the cylinder.

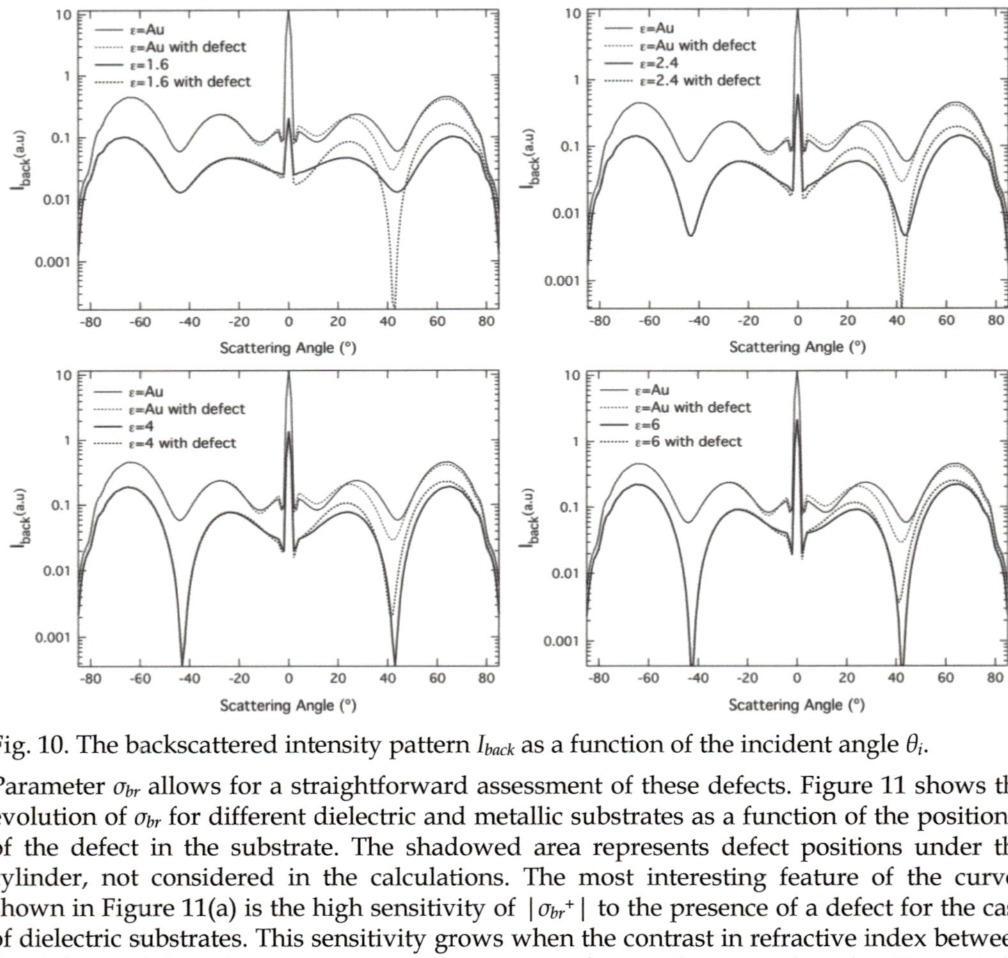

Fig. 10. The backscattered intensity pattern I_{back} as a function of the incident angle θ_i.

Parameter σ_{br} allows for a straightforward assessment of these defects. Figure 11 shows the evolution of σ_{br} for different dielectric and metallic substrates as a function of the position x of the defect in the substrate. The shadowed area represents defect positions under the cylinder, not considered in the calculations. The most interesting feature of the curves shown in Figure 11(a) is the high sensitivity of $|\sigma_{br}^+|$ to the presence of a defect for the case of dielectric substrates. This sensitivity grows when the contrast in refractive index between the defect and the substrate increases. The maxima of σ_{br}^+ and σ_{br}^- are plotted in Figure 11(b) as a function of ε for the dielectric substrate case. Both decrease and saturate for large values of ε. It is also worth remarking that when the defect is on the substrate, $\sigma_{br} > 0$, except for some positions corresponding to the metallic substrate case. This means that, on average, the backscattering is enhanced by a particle on the dielectric substrate, but it can be reduced when the defect lies on a metallic substrate. However, when the defect is on the cylinder, negative values are found for either kind of substrate.

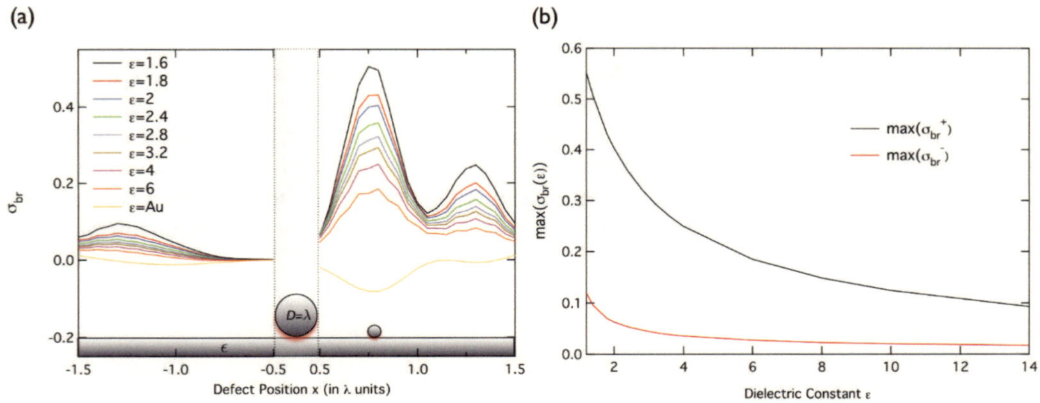

Fig. 11. (a) Evolution of σ_{br} as a function of the substrate optical properties. A defect lies on the substrate. (b) Evolution of *max(σ⁺)* and *max(σ⁻)* for pure dielectric substrates as a function of the substrate dielectric constant ε [Albella *et al.*, 2008].

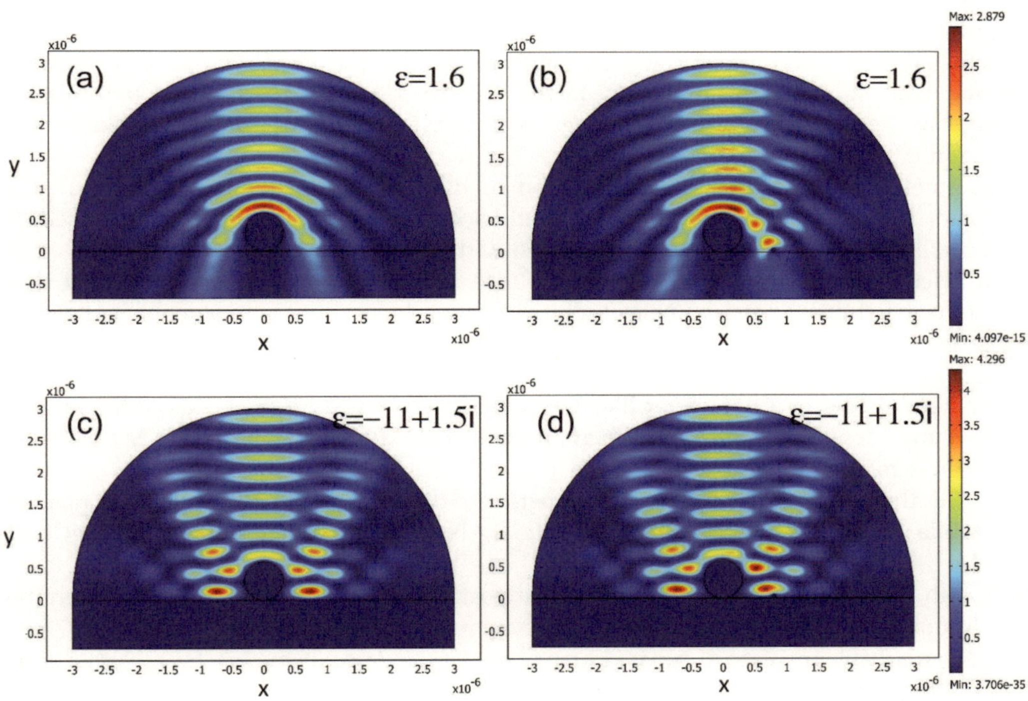

Fig. 12. Near-field plots corresponding to two different substrates illuminated at normal incidence. The figures on the left correspond to the reference case having no defect and on the right to a defect on the substrate. On the top are results for a dielectric substrate ε = 1.6 and on the bottom for a gold substrate ε = 11 + 1.5i.

Finally, to illustrate this enhancement, Figure 12 shows some examples of the near-field pattern obtained for two different substrates illuminated at normal incidence. The figures on the left correspond to the reference case having no defect and on the right to a defect on the substrate. On the top are results for a dielectric substrate $\varepsilon = 1.6$ and on the bottom for a gold substrate. We see that for low values of ε, we observe a new spatial distribution around the micron-sized particle resulting in a new local maximum of intensity, which produces the maximum change with respect to the non-defect case. This strong change in the near-field distribution is not surprising since this defect position corresponds to a maximum in σ_{br}^{+} as observed in Figure 11(b). Near-field plots (Figure 12) show a close correlation among the maxima located near the microstructure and the angular (φ) and linear (x) positions of the maxima obtained for σ_{br} plots.

6. Conclusion

The results in this chapter suggest that the measurement of σ_{br} can be a useful means of monitoring, sizing and characterizing small defects adhered to microstructures or substrates. From a practical point of view, detection and sizing of very small defects on microstructures by some reliable and non-invasive method is useful in quality control technology. In this context, the objective was to study the sensitivity of this technique to the optical properties of the substrate in two situations: (A) when the defect was located on a cylinder and (B) when the defect was located on the substrate near a cylinder. We also have anticipated how choosing the appropriate material for the substrate in each particular situation can influence this detection. In this chapter a system structured at two different levels has been analyzed by studying its scattering properties in the backscattering direction.

A parameter $\sigma_b^{\pm}$ defined as the integration of the backscattered intensity over a given quadrant is a quantitative measure of overall backscattering variations. It can be used, for instance, by measuring the relative variation with respect to an initial system $\sigma_{br}^{\pm}$ as has been discussed in this chapter, as a simple measure of asymmetry through $|\sigma_b^{+} - \sigma_b^{-}|$, or through a backscattering asymmetry index $2|\sigma_b^{+} - \sigma_b^{-}|/(\sigma_b^{+} - \sigma_b^{-})$, presumably suitable for experimental situations. Parameter σ_b itself can be experimentally obtained through different configurations [Peña *et al.*, 2000].

The potential uses of a parameter like $\sigma_{br}^{\pm}$ depend very much on the needs of the research and on any previous knowledge. When applied to the double-cylinder case, the following has been demonstrated:

i. With the defect on the cylinder, parameter $\sigma_{br}^{\pm}$ depends on the defect size and position, while when the defect is on the substrate, $\sigma_{br}^{\pm}$ is also dependent on the main particle size D;
ii. With the defect on the cylinder, its composition (metal/dielectric) may be identified from the sign of $\sigma_{br}^{\pm}$ for a given defect position;
iii. A remarkable increase of σ_{br}^{-} is characteristic of the defect being on the substrate and is not found when the defect is on the principle particle;
iv. Strong oscillations in σ_{br}^{+} observed when the defect is on the substrate for small values of x can identify very precisely the position of the defect;
v. The use of metallic or dielectric substrates does not affect significantly the behaviour of σ_{br} and therefore does not present a limitation for this parameter.

This is not the first time that changes in the backscattering pattern obtained from a simple configuration allows for the characterization of some geometrical or material changes produced in the scattering object, but we think the procedure shown here is applicable to a wide range of 2D structures whose defects often require rapid identification. Finally, these results have a natural extension to 3D geometries that would enlarge the range of practical situations and the scope of this work. This would require more powerful computing techniques and tools applied over an extensive caustic. Everything shown in this chapter can be seen as the basis for analyzing more complex geometrical systems.

7. Appendix

A particle sizing method is proposed using a double-interaction model (DIM) for the light scattered by particles on substrates. This model, based on that proposed by Nahm and Wolfe [Nahm & Wolfe, 1987], accounts for the reflection of both the incident and the scattered beams and reproduces the scattering patterns produced by particles on substrates, provided that the angle of incidence, the polarization and the isolated particle scattering pattern are known.

In this context, we show the results obtained using a simple model to assess the effect of the presence of a nano-defect on a microstructure located on a substrate. This three-object system can be modeled with an extension of the double interaction model, which was shown to be useful for obtaining the electric field scattered from a single particle resting on a substrate. In this chapter, we extend that model in a 2D frame, to a system where there are two metallic cylinders, one being a nano-defect lying on a micron-sized structure that rests on a flat substrate. We also show how this simple model reproduces the scattering pattern variations with respect to the defect-free system when compared to that given by an exact method. It is important to remark that this model has two interesting features: (1) Transparency, in that it is easy to understand the mechanisms involved in the scattering; and (2) Easy numerical implementation that can lead to fast computation.

Model description

The system consists of an infinitely long metallic cylinder of radius *R* placed on a flat substrate (see Figure 13) while another, much smaller, cylinder rests on top of the first. Both cylinders are assumed metallic, and in our calculations we give them the properties of silver. The cylinder axis is parallel to the Y direction, and the X-Z plane is the scattering plane. This reduces the geometry to 2D. The model we propose in this work can be described as an application of the DIM [Nahm & Wolfe, 1987] for normal incidence in two steps: (1) to the large cylindrical microstructure located on the flat substrate and (2) to the small cylindrical nano-defect, assuming that the underlying cylinder approximates a flat substrate for the small cylinder. Of course, the accuracy of this second approximation improves as the ratio r/R approaches zero. We shall limit our solution to the range $0 < r/R < 0.1$.

The DIM is based on the standard *T*-matrix solution for the isolated scatterer at normal incidence. Two contributions to the total scattered field are generated in any direction: the light directly scattered from the particle, and that scattered and reflected off the substrate. The latter is affected by a complex Fresnel coefficient, thus having its phase shifted because of its additional path.

If we now consider the two-particle system, the total scattered far field at a fixed scattering direction given by the scattering angle θ_s is the coherent superposition of four contributions, two of them due to the large cylinder and two due to the small one. Looking at Figure 13, arrows labelled (1) and (3) represent the components directly scattered by the particles to

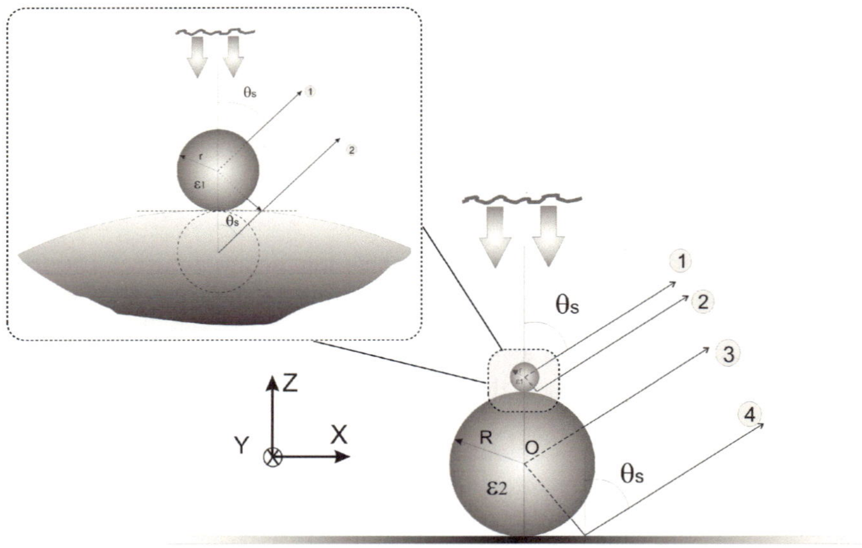

Fig. 13. Multiple interaction model for two structures illuminated at normal incidence. Inset shows the flat substrate approximation and image theory applied to the defect.

the detector; and arrows labelled (2) and (4) correspond to the components scattered downwards by the particles and then reflected towards the observation angle.
Simple calculations allow us to account for the phase shifts produced by the extra-path of each contribution from a plane normal to the incidence direction to a plane normal to the observation. Taking the direct contribution (3) as the reference beam having zero phase ($\delta_3 = 0$):

$$\delta_1(\theta_s) = -\frac{2\pi(R+r).(1+\cos\theta_s)}{\lambda} \tag{A.1}$$

$$\delta_2(\theta_s) = \delta_1 + \frac{4\pi.\, r\cos\theta_s}{\lambda} \tag{A.2}$$

$$\delta_4(\theta_s) = \frac{4\pi.\, R\cos\theta_s}{\lambda} \tag{A.3}$$

where θ_s is the scattering angle, R and r are the radii of the large and small cylinders respectively, and λ is the incident wavelength. Reflection from the substrate is simplified using the plane wave approximation and considering S-polarization, and is given by the Fresnel reflection coefficient in its complex form,

$$\hat{r}_s(\theta) = \frac{\cos\theta - \sqrt{\hat{\varepsilon}_{sub} - \sin^2(\theta)}}{\cos\theta + \sqrt{\hat{\varepsilon}_{sub} - \sin^2(\theta)}} \tag{A.4}$$

The total scattered electric field under the assumptions of this Combined Double Interaction Model (CDIM) is given by the sum of the four contributions,

$$E_1(\theta_s) = A_0 \cdot F\left(180^o - \theta_s\right) \cdot e^{i\cdot(\phi_1 + \delta_1)} \tag{A.5}$$

$$E_2(\theta_s) = A_0 \cdot F(\theta_s) \cdot \left|r(\theta_s)\right| \cdot e^{i\cdot(\phi_2 + \delta_2 + \alpha_s)} \tag{A.6}$$

$$E_3(\theta_s) = A_0 \cdot F\left(180^o - \theta_s\right) \cdot e^{i\cdot\phi_3} \tag{A.7}$$

$$E_4(\theta_s) = A_0 \cdot F(\theta_s) \cdot \left|r(\theta_s)\right| \cdot e^{i\cdot(\phi_4 + \delta_4 + \alpha_s)} \tag{A.8}$$

where A_0 is the amplitude of the incident beam, and the complex terms have been expressed in their polar form, φ_i is the Mie phase corresponding to the ith contribution, $F(\theta_s)$ is the amplitude of the scattered electric far field in the θ_s direction for an isolated particle, which are given by the *T*-matrix amplitudes applied to the case of a cylinder, and a_s is the phase introduced by the Fresnel reflection.

Testing the model

Results obtained by using the CDIM are compared with others obtained from the ET method, which is an exact rendering of the Maxwell equations in their integral form. Figure 14 shows the scattering intensity patterns obtained from the CDIM approximation (top) and from the ET (bottom) in semi-logarithmic scale. CDIM results have been shifted upwards for an easy visualization. The continuous line plots correspond to the defect-free situation (single cylinder) in all cases, while the dashed line corresponds to the case of a defect located on the top of the cylinder.

In Figure 15 we consider the range of validity of this model. CDIM and ET are compared for $R = \lambda$ and for $r = 0.05\lambda$, 0.1λ and 0.15λ. We observe that both patterns show the same outer minima positions for a defect/cylinder aspect ratio (r/R) up to 0.1, while for the case $r/R= 0.15$, distortions in the pattern become important. The CDIM reproduces the changes produced in the lobed structure of the scattering patterns and particularly the minima positions. If we look at Figure 14(a) as an example, we can see that in the case of the exact solution (shown at the bottom part of the graph), there is a shift in the minima produced by the presence of the defect, and this shift is accurately reproduced by the model.

One limit imposed on the ratio r/R is due to assuming the underlying main cylinder is flat. For values of $r/R < 0.1$ the changes introduced by the real curvature in the Fresnel coefficients and in the *T*-matrix scattering amplitude are very small and the changes in the phase of the contribution due to the optical path difference are negligible as long as R is of the order of the wavelength.

Although this problem may be overcome by numerically calculating the exact angular reflections and paths, it is worthwhile to consider this model for its simplicity and transparency. Because of its simplicity, the model can be extended to other situations, for instance, a 3D implementation by introducing the Mie coefficients for spheres, different contaminating particles, different defect locations, etc.

8. Acknowledgements

The authors wish to acknowledge the funds provided by the Ministry of Education of Spain under project #FIS2007-60158. We also thank the computer resources provided by the Spanish Supercomputing Network (RES) node at Universidad de Cantabria.

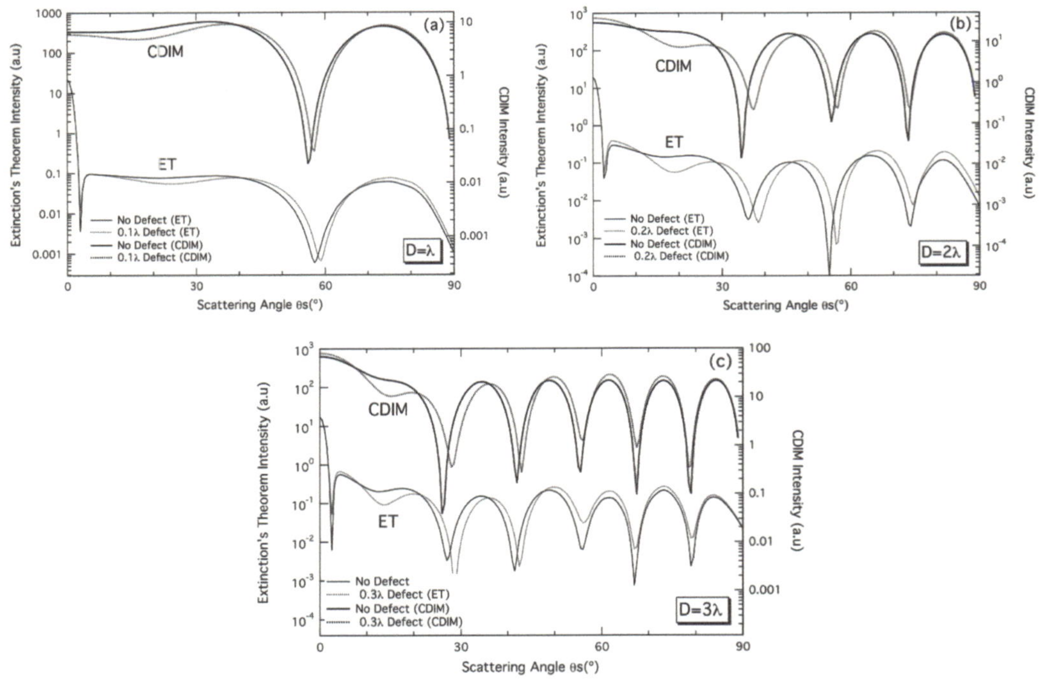

Fig. 14. Scattering patterns calculated using CDIM and ET for three different sizes of the underlying cylinder, keeping a constant defect/cylinder aspect ratio of 0.1: (a). $R = \lambda/2$ and $r = \lambda/20$; (b). $R = \lambda$ and $r = \lambda/10$; (c). $R = 1.5\lambda$ and $r = 0.15\lambda$. This corresponds with defect sizes of 60, 120 and 180 nm for an incident wavelength of 0.6 µm [Albella *et al.*, 2007].

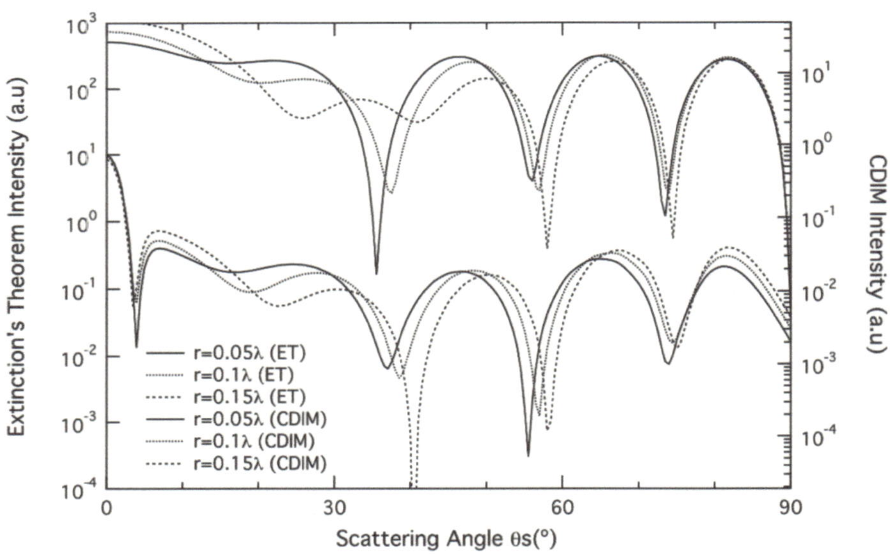

Fig. 15. Scattering pattern of an $R = \lambda$ cylinder with a defect of $r = 0.05\lambda$, 0.1λ and 0.15λ calculated using CDIM (top) and ET (bottom) [Albella *et al.*, 2007].

9. References

Albella, P., F. Moreno, J. M. Saiz, and F. González, "Monitoring small defects on surface microstructures through backscattering measurements." *Opt. Lett.* 31, 1744-1746, (2006).

Albella, P., F. Moreno, J. M. Saiz, and F. González, "Backscattering of metallic microstructures will small defects located on flat substrates." *Opt. Exp.* 15, (2007) 6857–6867.

Albella, P., F. Moreno, J. M. Saiz, and F. González, "2D double interaction method for modeling small particles contaminating microstructures located on substrates." *J. Quant. Spectrosc. Radiative Trans.* 106, 4–10 (2007).

Albella, P., F. Moreno, J. M. Saiz, and F. González, "Influence of the Substrate Optical Properties on the backscattering of contaminated microstructures." *J. Quant. Spectrosc. Radiative Trans.* 109, (2008) 1339–1346.

Chen, H. T., R. Kersting and G. C. Cho, "Terahertz imaging with nanometer resolution." *Appl. Phys. Lett.* 83, (2003) 3.

Germer, T. A., and G. W. Mulholland. "Size metrology comparison between aerosol electrical mobility and laser surface light scattering," Characterization and Metrology for ULSI Technology, (2005), 579–583.

Germer, T. A., "Light scattering by slightly non-spherical particles on surfaces," *Opt. Lett.* 27, 1159-1161 (2002).

Johnson, B. R, "Light scattering from a spherical particle on a conducting plane, in normal incidence," *J. Opt. Soc. Am. A* 19, 11 (2002).

Lee, K. G., H. W. Kihm, J. E. Kihm, W. J. Choi, H. Kim, C. Ropers, D. J. Park, Y. C. Yoon, S. B. Choi, D. H. Woo, J. Kim, B. Lee, Q. H. Park, C. Lienau, and D. S. Kim, "Vector field microscopic imaging of light," *Nature Photonics* 1, (2007) 53–56.

Liswith, M. L., E. J. Bawolek, and E. D. Hirleman, "Modeling of light scattering by submicrometer spherical particles on silicon and oxidized silicon surfaces," *Opt. Eng.* 35, 858-869 (1996).

Mittal, K. L. (editor). Particles on surfaces: Detection, Adhesion and Removal. VSP, Utrech (1999).

Moreno, F and F. González Eds. Light scattering from microstructures. *Springer Verlag* (2000).

Moreno, F., F. González, and J. M. Saiz, "Plasmon spectroscopy of metallic nanoparticles above flat dielectric substrates," *Opt. Lett.*, 31, (2006), 1902-1904.

Mulholland, G. W, T. A. Germer, and J. C. Stover, "Modeling, measurement and standards for wafer inspection," *Proceedings of the Government Microcircuits Applications and Critical Technologies*. (2003), 1–4.

Nahm, K. B., and W. L. Wolfe, "Light-scattering models for spheres on a conducting plane," *Appl. Opt.* 26, (1987), 2995–2999.

Nieto-Vesperinas, M., and J. A. Sánchez-Gil, "Light scattering from a random rough interface with total internal reflection," *J. Opt. Soc. Àm. A* 9, (1992) 424–436.

Peña, J.L., J. M. Saiz, P. Valle, F. González, and F. Moreno, "Tracking scattering minima to size metallic particles on flat substrates," *Particle & Particle Systems Characterization* 16, (1999) 113–118.

Ripoll, J., A. Madrazo, and M. Nieto-Vesperinas, "Scattering of electromagnetic waves from a body over a random rough surface," *Opt. Comm.* 142, (1997) 173–178.

Saiz, J.M., P. J. Valle, F. González, E. M. Ortiz, and F. Moreno, "Scattering by a metallic cylinder on a substrate: burying effects," *Opt. Lett.*, 21, (1996) 1330–1332.

Sánchez-Gil, J. A., and M. Nieto-Vesperinas, "Resonance effects in multiple light scattering from statistically rough metallic surfaces," *Phys. Rev. B* 45 8623-8633 (1992).

Sonnichsen, C., B. M. Reinhard, J. Liphardt, and A. P. Alivisatos, "A molecular ruler based on plasmon coupling of single gold and silver nanoparticles," *Nature Biotechnology* 23, (2005), 741–745.

Stuart, D., A. J. Haes, C. R. Yonzon, E. Hicks, and R. V. Duyne, "Biological applications of localised surface plasmonic phenomenae," *IEE Proc., Nanobiotechnology* 152, (2005) 13–32.

Valle, P., F. González, and F. Moreno, "Electromagnetic wave scattering from conducting cylindrical structures on flat substrates: study by means of the extinction theorem," *Appl. Opt.* 33, (1994), 512–523.

10

Nanofocusing of Surface Plasmons at the Apex of Metallic Tips and at the Sharp Metallic Wedges. Importance of Electric Field Singularity

Andrey Petrin
Joint Institute for High Temperatures of Russian Academy of Science
Russia

1. Introduction

Nanofocusing of light is localization of electromagnetic energy in regions with dimensions that are significantly smaller than the wavelength of visible light (of the order of one nanometer). This is one of the central problems of modern near-field optical microscopy that takes the resolution of optical imaging beyond the Raleigh's diffraction limit for common optical instruments [Zayats (2003), Pohl (1984), Novotny (1994), Bouhelier (2003), Keilmann (1999), Frey (2002), Stockman (2004), Kawata (2001), Naber (2002), Babadjanyan (2000), Nerkararyan (2006), Novotny (1995), Mehtani (2006), Anderson (2006)]. It is also important for the development of new optical sensors and delivery of strongly localized photons to tested molecules and atoms (for local spectroscopic measurements [Mehtani (2006), Anderson (2006), Kneipp (1997), Pettinger (2004), Ichimura (2004), Nie (1997), Hillenbrand (2002)]). Nanofocusing is also one of the major tools for efficient delivery of light energy into subwavelength waveguides, interconnectors, and nanooptical devices [Gramotnev (2005)].

There are two phenomena of exceptional importance which make it possible nanofocusing. The first is the phenomenon of propagation with small attenuation of electromagnetic energy of light along metal-vacuum or metal-dielectric boundaries. This propagation exists in the form of strictly localized electromagnetic wave which rapidly decreases in the directions perpendicular to the boundary. Remembering the quantum character of the surface wave they say about surface plasmons and surface plasmon polaritons (SPPs) as quasi-particles associated with the wave. The dispersion of the surface wave has the following important feature [Economou (1969), Barnes (2006)]: the wavelength tends to zero when the frequency of the SPPs tends to some critical (cut off) frequency above which the SPPs cannot propagate. For SPPs propagating along metal-vacuum plane boundary this critical frequency is equal to $\omega_p/\sqrt{2}$ (we use Drude model without absorption in metal). For spherical boundary this critical frequency [Bohren, Huffman (1983)] is equal to $\omega_p/\sqrt{3}$. So, the SPP critical frequency depends on the form of the boundary. By changing the frequency of SPPs it is possible to decrease the wavelength of the SPPs to the values substantially smaller than the wavelength of visible light in vacuum and use the SPPs for trivial focusing by creation a converging wave [Bezus (2010)]. In this case there is no breaking the diffraction Raleigh's limitation and the energy of the wave is focused into the region with dimensions of the order of wavelength of the SPPs. These dimensions may be substantially smaller than the wavelength of light in vacuum corresponding to the same frequency. As a result we have nanofocusing of light energy. The second phenomenon is electrostatic electric field strengthening at the apex of conducting tip (at the apex of geometrically ideal tip there is electrostatic field singularity, i.e. the electrostatic field tends to infinity at the apex). This phenomenon exists not only in electrostatics. For alternating electric field in the region with the apex of the tip at the center (with dimensions smaller than wavelength) the quasi-static approximation is applicable and there is a singularity of the time varying electric field (if the frequency is low enough as we will see below). Surely, at the apex of a real tip there is no singularity of electrostatic field since the apex is rounded. But near the apex the electric field increases in accordance with power (negative) law of the singularity and the electric field saturation at the apex is defined by the radius of the apex. This radius may be very small, of the order of atomic size.

Nanofocusing of SPPs at the apex of metal tip is considered in [Stockman (2004), De Angelis (2010)]. SPPs are created symmetrically at the basement of the tip and this surface wave converges along the surface of the metal to the tip's apex where surface wave energy is focused. But conditions for existence of electric field singularity are considered in [Stockman (2004), De Angelis (2010)] only for very sharp conical metal tips with small angle at the apex. In [Petrin (2010)] it is shown that due to frequency dependence of metal permittivity in optic frequency range the singularity of electric field at the tip's of not very sharp apex may exist in different forms.

The goal of the present chapter is investigation of the factors defined the type of singular concentration of electromagnetic energy at the geometrically singular metallic elements (such as apexes and edges) as one of the important condition for optimal nanofocusing.

In the next sections of this chapter we discuss the following:

electric field singularities in the vicinity of metallic tip's apex immersed into a uniform dielectric medium;

electric field singularities in the vicinity of metallic tip's apex touched a dielectric plate;

electric field singularities in the vicinity of edge of metallic wedge.

2. Nanofocusing of surface plasmons at the apex of metallic probe microtip. Conditions for electric field singularity at the apex of microtip immersed into a uniform dielectric medium.

In this section of the chapter we focus our attention on finding the condition for electric field singularity of focused SPP electric field at the apex of a metal tip which is used as a probe in a uniform dielectric medium. As we have discussed above this singularity is an important feature of optimal SPP nanofocusing.

2.1 Condition for electric field singularity at the apex

Consider the cone surface of metal tip (see Fig. 1).

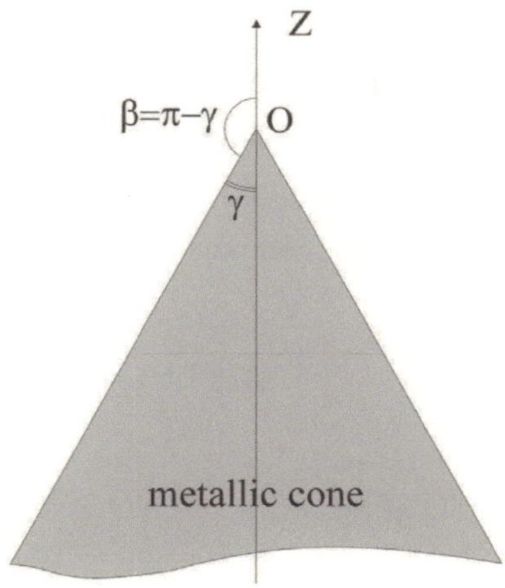

Fig. 1. Geometry of the problem.

Let calculate the electric field distribution near the tip's apex. In spherical coordinates with origin O at the apex and polar angle θ (see Fig. 1), an axially symmetric potential Ψ obeys Laplace's equation (we are looking for singular solutions, so in the vicinity of the apex the quasistatic approximation for electric field is applicable)

$$\frac{\partial}{\partial r}\left(r^2\frac{\partial\Psi}{\partial r}\right)+\frac{1}{\sin\theta}\frac{\partial}{\partial\theta}\left(\sin\theta\frac{\partial\Psi}{\partial\theta}\right)=0 .$$

Representing the solution as $\Psi = r^{\alpha} f(\theta)$, where α is a constant parameter, we have

$$\frac{\partial}{\partial\theta}\left(\sin\theta\frac{\partial f}{\partial\theta}\right)=-\alpha(\alpha+1)f\sin\theta .$$

Changing to the function g defined by the relation $f(\theta)=g(\cos\theta)$, which entails

$$\frac{df}{d\theta}=-\sin\theta\frac{dg}{d\cos\theta},$$

we obtain Legendre's differential equation

$$\left(1-\cos^2\theta\right)\frac{d^2g}{d\cos^2\theta}-2\cos\theta\frac{dg}{d\cos\theta}+\alpha(\alpha+1)g=0 .$$

It's solution is the Legendre polynomial $P_{\alpha}(\cos\theta)$ of degree α, which can be conveniently represented as

$$P_{\alpha}(\cos\theta)=F\left(-\alpha,\alpha+1,1,\frac{1-\cos\theta}{2}\right),$$

where F is the hypergeometric function. This representation is equally valid whether α is integer or not.

Let denote the permittivity of metal cone as ε_m and the permittivity of external uniform dielectric medium as ε_d. The electric potentials outside and inside the cone can be represented, respectively, as $\Psi_d = Ar^{\alpha} P_{\alpha}(\cos\theta)$ and $\Psi_m = Br^{\alpha} P_{\alpha}(-\cos\theta)$, where A and B are constant parameters. Since the functions $P_{\alpha}(\cos\theta)$ and $P_{\alpha}(-\cos\theta)$ are linearly independent [Angot (1962)], the field components outside and inside the cone are, respectively,

$$E_{d,\tau} = -\frac{\partial \Psi_d}{\partial r} = -A\alpha r^{\alpha-1} P_{\alpha}(\cos\theta),$$

$$E_{d,n} = -\frac{1}{r}\frac{\partial \Psi_d}{\partial \theta} = Ar^{\alpha-1} \sin\theta P'_{\alpha}(\cos\theta);$$

and

$$E_{m,\tau} = -\frac{\partial \Psi_m}{\partial r} = -B\alpha r^{\alpha-1} P_{\alpha}(-\cos\theta),$$

$$E_{m,n} = -\frac{1}{r}\frac{\partial \Psi_m}{\partial \theta} = -Br^{\alpha-1} \sin\theta P'_{\alpha}(-\cos\theta).$$

At $\theta = \pi - \gamma$, the tangential and normal electric field components satisfy the boundary conditions

$$E_{d,\tau} = E_{m,\tau}, \quad \varepsilon_d E_{d,n} = \varepsilon_m E_{m,n},$$

which yield the system of equations

$$A\alpha r^{\alpha-1} P_{\alpha}(\cos(\pi-\gamma)) - B\alpha r^{\alpha-1} P_{\alpha}(-\cos(\pi-\gamma)) = 0,$$

$$\varepsilon_d A r^{\alpha-1} \sin(\pi-\gamma) P'_{\alpha}(\cos(\pi-\gamma)) + \varepsilon_m B r^{\alpha-1} \sin(\pi-\gamma) P'_{\alpha}(-\cos(\pi-\gamma)) = 0,$$

where the primes denote derivatives with respect to the arguments.
A nontrivial solution of the system exists when the determinant is equal to zero,

$$(\varepsilon_m/\varepsilon_d)^{-1} P_{\alpha}(\cos\gamma)\frac{d}{d\gamma} P_{\alpha}(-\cos\gamma) - P_{\alpha}(\cos(\pi-\gamma))\frac{d}{d\gamma} P_{\alpha}(\cos\gamma) = 0. \quad (1)$$

Numerical calculation of the functions in Eq.(1) may be carried out with the aid of the hypergeometric function. Taking into account the following identity [Olver (1974)]

$$\frac{\partial}{\partial z} F(a,b,c,z) = abF(a+1,b+1,c+1,z),$$

we find

$$\frac{\partial}{\partial \theta} P_{\alpha}(\cos\theta) = -\alpha(\alpha+1)\frac{\sin\theta}{2} F\left(-\alpha+1, \alpha+2, 2, \frac{1-\cos\theta}{2}\right),$$

$$\frac{\partial}{\partial\theta}P_\alpha(-\cos\theta)=\alpha(\alpha+1)\frac{\sin\theta}{2}F\left(-\alpha+1,\alpha+2,2,\frac{1+\cos\theta}{2}\right).$$

Therefore Eq.(1) may be rewritten as

$$\varepsilon_m F\left(-\alpha,\alpha+1,1,\frac{1+\cos\gamma}{2}\right)F\left(-\alpha+1,\alpha+2,2,\frac{1-\cos\gamma}{2}\right)+$$
$$+\varepsilon_d F\left(-\alpha,\alpha+1,1,\frac{1-\cos\gamma}{2}\right)F\left(-\alpha+1,\alpha+2,2,\frac{1+\cos\gamma}{2}\right)=0.$$

For a given γ Eq.(1) implicitly defines the function $\alpha=\alpha(\varepsilon_m/\varepsilon_d)$. Let use Drude model without losses for permittivity of metal $\varepsilon_m=1-\omega_p^2/\omega^2$ and consider that the metal tip surrounded with vacuum with permittivity $\varepsilon_d=1$, then $\varepsilon_m/\varepsilon_d=1-\omega_p^2/\omega^2$. In this case, $\alpha=\alpha(\omega/\omega_p)$ and remembering that $\Psi_d\sim r^\alpha$ we have $E_d\sim r^{\alpha-1}$. So, the function $\alpha=\alpha(\omega/\omega_p)$ defines the character of electric field singularity at the apex.
Note, that for given γ and ω/ω_p Eq.(1) has many roots but not all of them give solution which have physical sense or solution with electric field singularity at the apex of the tip. Obviously that only roots with $\mathrm{Re}(\alpha)<1$ ($E_d\sim r^{\alpha-1}$) will give singular solutions. So, we have interest only in these solutions. To find the lower boundary of the roots of interest it is necessary to remember that in the vicinity of a singular point the density of electric field power must be integrable value. It means that in the limit $r\to 0$ the electric field must increase slower than $r^{-3/2}$ and, therefore, the density of electric field power must increase slower than r^{-3}. It gives us the lower boundary for the roots. I.e. the appropriate roots must satisfy the inequity $\mathrm{Re}(\alpha)>-1/2$.
Eq.(1) was solved numerically. As an example of such calculations Fig. 2 shows real and image values of all roots of Eq.(1) in the interval $-1/2\le\mathrm{Re}(\alpha)<1$ as functions of ω/ω_p for $\gamma=15°$, $\varepsilon_d=1$ (solid curves). The dielectric constant of the metal was calculated by Drude formula without losses $\varepsilon_m=1-\omega_p^2/\omega^2$. From Fig. 2 we may see that there is a critical frequency of SPP ω_{cr} which separates the pure real solution (for $\omega<\omega_{cr}$) from essentially complex solution (for $\omega>\omega_{cr}$). It follows that there are two essentially different types of the SPP nanofocusing. First type takes place when $\omega<\omega_{cr}$ and is characterized by the electric field singularity without oscillations along radius in the vicinity of the apex. Second type takes place when $\omega_{cr}<\omega<\omega_p/\sqrt{2}$ and is characterized by the electric field singularity with oscillations along radius and the wave number of these oscillations tends to infinity when $r\to 0$. Note, that in the second type of nanofocusing $\mathrm{Re}(\alpha)\equiv-1/2$ if there is no losses in metal. The integral of the total electric power in the vicinity of the apex diverges. But if we take into account the metal losses considering the Drude's formula with losses

$$\varepsilon_m=1-\frac{\omega_p^2}{\omega^2+i\Gamma\omega},\ \text{where } \omega_p \text{ and } \Gamma \text{ are constants,}$$

then in the second type of nanofocusing the integral of the total electric power in the vicinity of the apex converges since $\mathrm{Re}(\alpha)>-1/2$. As an example of the metal with losses was made silver ($\omega_p=1.36\times10^{16}$ s^{-1} и $\Gamma=2\times10^{14}$ s^{-1} [Gay (2007)]). In Fig. 2 the curves for silver are shown by dotted lines.

Note, that here there is an essential difference of the quasistatic approach to the problem from the pure static one. In the static problem physical sense have electric potentials with $\mathrm{Re}(\alpha) > 0$ since only under this condition the potential on the surface of conducting cone may be made constant. In the quasistatic problem the equipotentiality of the metallic cone is not a necessary condition since the metal of the cone is considered as a dielectric with frequency dependent permittivity. The dielectric surface is not obligatory equipotential.

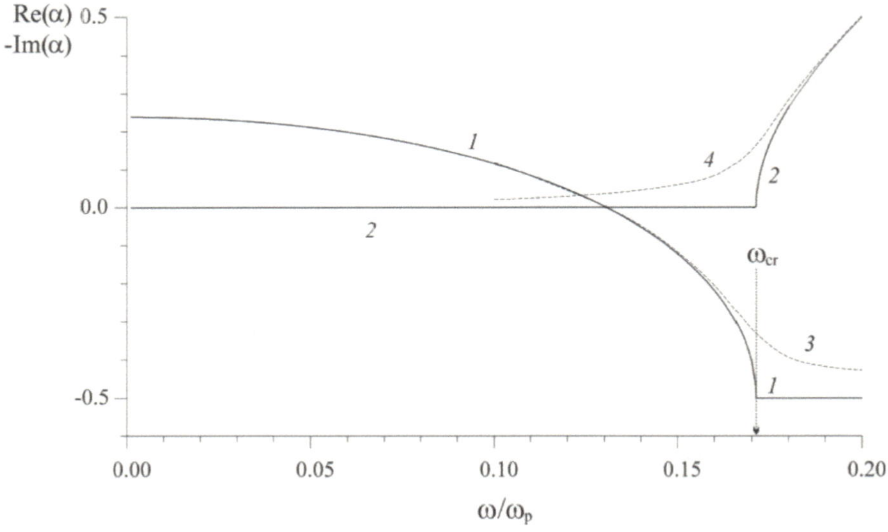

Fig. 2. Real (curve 1) and image (curve 2) parts of index α as a function of normalized frequency ω/ω_p for of tip's angle $\gamma = 15°$ (no losses in metal). Dotted lines (curve 3 and 4) show analogues dependences for silver (metal with losses).

From the functions of Fig. 2, the normalized critical frequency ω_{cr} as a function of angle γ was found numerically (see Fig.3). This function may be found analytically. If we take in Eq.(1) $\alpha = -1/2$ и $\varepsilon_m/\varepsilon_d = 1 - \omega_p^2/\omega_{cr}^2$ (the characteristics of critical points) then

$$\omega_{cr}/\omega_p = \left[1 + \frac{F\left(1/2,1/2,1,\frac{1-\cos\gamma}{2}\right)F\left(3/2,3/2,2,\frac{1+\cos\gamma}{2}\right)}{F\left(1/2,1/2,1,\frac{1+\cos\gamma}{2}\right)F\left(3/2,3/2,2,\frac{1-\cos\gamma}{2}\right)}\right]^{-\frac{1}{2}}.$$

In Fig. 3 the frequency $\omega_p/\sqrt{2}$ is shown. SPPs can not exist above this frequency on the plane boundary metal-vacuum. We may see that when $\gamma \to 90°$ the critical frequency tends to $\omega_p/\sqrt{2}$. It is absolutely unexpectable that the utmost frequency of SPP existence arises in quasistatic formulation of the problem on electric field singularity finding.

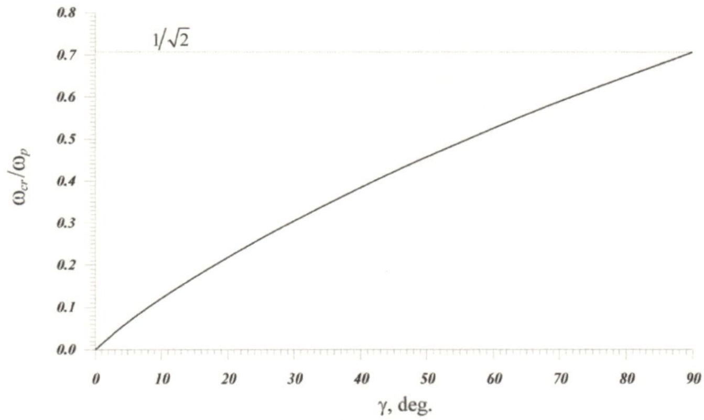

Fig. 3. Normalized critical frequency as a function of γ .

2.2 Application of the theory to a silver tip.

Consider a silver tip with plasma frequency [Fox (2003)] $\omega_p = 1.36 \times 10^{16}$ s-1. Based on the function of Fig.3 we may find the function $\gamma_{cr} = \gamma_{cr}(\lambda_0)$ for silver by simple recalculating (see Fig. 4). The wavelength λ_0 is defined through the critical frequency ω_{cr} as $\lambda_0 = 2\pi c/\omega_{cr}$, where c is the speed of light in vacuum The curve plotted in Fig. 4 separates the region with the first type of singularity (above the curve) and the region with the second type of singularity (below the curve).

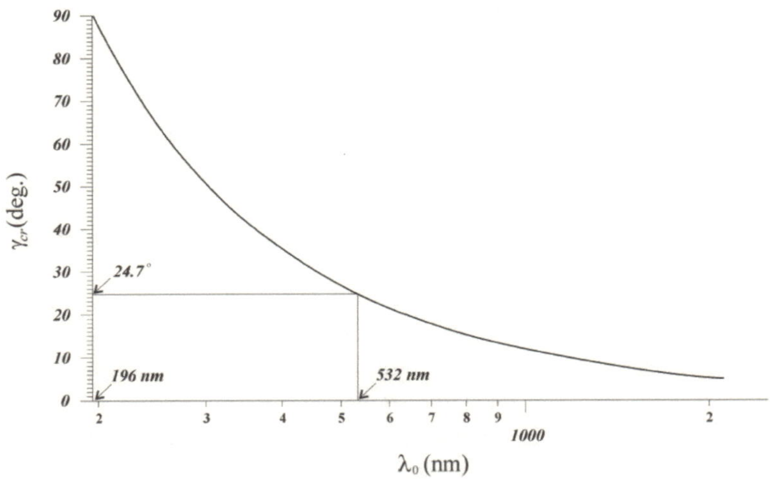

Fig. 4. Critical angle γ_{cr} as a function of wavelength in vacuum of exciting laser. The curve starts from the lowest wavelength (in vacuum) of SPPs spectrum for silver: $\left(\lambda_{sp}\right)_{cr} = 2\sqrt{2}\pi c/\omega_p \approx 196\,nm$.

As an example of application of the obtained results we consider the experiment on local Raman's microscopy [De Angelis (2010)]. A silver tip with angle $\gamma \approx 12°$ in the vicinity of the apex was used. The wavelength of laser used in the experiment was equal to $\lambda_0 = 532nm$. The excited SPPs propagated along the surface of the tip to the focus at the apex. But from the above results we may conclude that the angle is not optimal for the best focusing. Indeed, from Fig. 4 we may see that the critical angle for $\lambda_0 = 532nm$ is $\gamma_{cr} \approx 24.7°$. So, there is electric field singularity of the second type near the apex of the tip with $\gamma \approx 12°$. So, it was obtained ***the fundamental result: the field singularity of alternative electric field at the apex of geometrically ideal metal cone tip exists in two form. First – with pure real index of singularity α, and second – with essentially complex index of singularity with constant real part*** $\mathrm{Re}(\alpha) = -1/2$.

But it is remained the following unresolved problem yet. If the metal microtip is used as a SPP probe of the surface properties it is obvious that the value of the microtip's angle depends on the dielectric constant of the surface which the microtip's apex is touched and on the dielectric constant of the surrounding medium. It is clear that it is necessary to consider the influence of the probed surface on the electric field singularity at the microtip's apex and therefore on nanofocusing. The next section of this chapter gives the answer to the following question: what happens with the considered phenomenon when the apex of metal cone touches a dielectric plane plate?

3. Nanofocusing of surface plasmons at the apex of metallic microtip probe touching a dielectric plane. Conditions for electric field singularity at the apex of a microtip immersed into uniform dielectric medium and touched a probed dielectric plane.

In experiments the apex of the microtip may touch the surface of a probed dielectric plane plate (see Fig. 5). In this connection the following question arises: how the dielectric constant of the plate affects the electric field singularity index at the apex? By another words, if the microtip used as a concentrator of SPPs for Raman's spectroscopy of a dielectric surface [De Angelis (2010)] how the investigated material affects the electric field singularity at the apex and therefore the efficiency of SPPs focusing?

3.1 Method of electric field singularity finding at the apex of metal microtip touching a dielectric plane

Consider the cone metal microtip touching a dielectric plane plate (see Fig. 5). The space between the metal tip and dielectric plane is filled by a uniform dielectric. The dielectric constants of the metal, the dielectric plane and the filling uniform dielectric are equal to ε_m, ε_p and ε_d respectively. The frequency of SPP wave which is focused at the apex is equal to ω. As in the previous part, in the quasistatic formulation of the problem the electric field potential must obeys Laplace's equation and normal and tangential components of electric field must obey the following boundary conditions: at the boundary of metal cone and free space $E_{m,\tau} = E_{d,\tau}$, $\varepsilon_m E_{m,n} = \varepsilon_d E_{d,n}$ and at the boundary of free space and dielectric plate $E_{d,\tau} = E_{p,\tau}$, $\varepsilon_d E_{d,n} = \varepsilon_p E_{p,n}$.

Based on the general solutions of Laplace's equation obtained in the previous part of this chapter, the electric potential in the considering three regions may be written as:

in the metal tip $\Psi_m = Ar^{\alpha} P_{\alpha}(-\cos\theta)$;

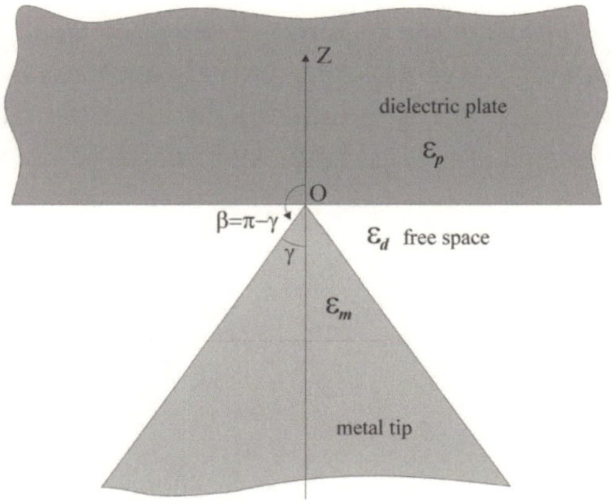

Fig. 5. Geometry of the problem.

in dielectric plate $\Psi_p = Br^{\alpha} P_{\alpha}(\cos\theta)$;

in free space $\Psi_d = Cr^{\alpha} P_{\alpha}(\cos\theta) + Dr^{\alpha} P_{\alpha}(-\cos\theta)$.

In these expressions we use the facts that for noninteger α the function $P_{\alpha}(\cos\theta)$ tends to infinity when $\theta \to \pi$ and $P_{\alpha}(-\cos\theta)$ tends to infinity when $\theta \to 0$.
Therefore, the field components in the considering three regions may be written by the following expressions:
tangential components:

$$E_{m,\tau} = -\frac{\partial \Psi_m}{\partial r} = -A\alpha r^{\alpha-1} P_{\alpha}(-\cos\theta),$$

$$E_{p,\tau} = -\frac{\partial \Psi_p}{\partial r} = -B\alpha r^{\alpha-1} P_{\alpha}(\cos\theta),$$

$$E_{d,\tau} = -\frac{\partial \Psi_d}{\partial r} = -C\alpha r^{\alpha-1} P_{\alpha}(\cos\theta) - D\alpha r^{\alpha-1} P_{\alpha}(-\cos\theta);$$

normal components:

$$E_{m,n} = -\frac{1}{r}\frac{\partial \Psi_m}{\partial \theta} = -Ar^{\alpha-1} \sin\theta P'_{\alpha}(-\cos\theta),$$

$$E_{p,n} = -\frac{1}{r}\frac{\partial \Psi_p}{\partial \theta} = Br^{\alpha-1} \sin\theta P'_{\alpha}(\cos\theta),$$

$$E_{d,n} = -\frac{1}{r}\frac{\partial \Psi_d}{\partial \theta} = Cr^{\alpha-1} \sin\theta P'_{\alpha}(\cos\theta) - Dr^{\alpha-1} \sin\theta P'_{\alpha}(-\cos\theta),$$

where we use the natural notations:

$$P'_{\alpha}(\cos\theta) = dP_{\alpha}(\mu)/d\mu\big|_{\mu=\cos\theta} \text{ and } P'_{\alpha}(-\cos\theta) = dP_{\alpha}(\mu)/d\mu\big|_{\mu=-\cos\theta}.$$

At the boundaries of the considering regions (at $\theta = \pi - \gamma$ and $\theta = \pi/2$) the mentioned above boundary conditions for tangential and normal components of electric field must be satisfied. Substituting into the boundary conditions the expressions for the field components and taking into account that $-\cos(\pi - \gamma) = \cos\gamma$ we have the following four equations:

$$P_\alpha(\cos\gamma)A - P_\alpha(\cos(\pi-\gamma))C - P_\alpha(\cos\gamma)D = 0\,,$$

$$\varepsilon_m P'_\alpha(\cos\gamma)A + \varepsilon_d P'_\alpha(\cos(\pi-\gamma))C - \varepsilon_d P'_\alpha(\cos\gamma)D = 0\,,$$

$$B - C - D = 0\,,$$

$$-\varepsilon_p B + \varepsilon_d C - \varepsilon_d D = 0\,,$$

where $P'_\alpha(\cos(\pi-\gamma)) = dP_\alpha(\mu)/d\mu\big|_{\mu=\cos(\pi-\gamma)}$.

A nontrivial solution of the system exists when the determinant is equal to zero,

$$\begin{vmatrix} P_\alpha(\cos\gamma) & 0 & -P_\alpha(\cos(\pi-\gamma)) & -P_\alpha(\cos\gamma) \\ (\varepsilon_m/\varepsilon_d)P'_\alpha(\cos\gamma) & 0 & P'_\alpha(\cos(\pi-\gamma)) & -P'_\alpha(\cos\gamma) \\ 0 & 1 & -1 & -1 \\ 0 & -(\varepsilon_p/\varepsilon_d) & 1 & -1 \end{vmatrix} = 0\,.$$

By expansion the determinant we have

$$\begin{aligned} &(\varepsilon_m/\varepsilon_d)P'_\alpha(\cos\gamma)\left\{P_\alpha(\cos(\pi-\gamma)) + P_\alpha(\cos\gamma) + (\varepsilon_p/\varepsilon_d)\left[P_\alpha(\cos(\pi-\gamma)) - P_\alpha(\cos\gamma)\right]\right\} + \\ &\quad + P_\alpha(\cos\gamma)\left\{P'_\alpha(\cos(\pi-\gamma)) - P'_\alpha(\cos\gamma) + (\varepsilon_p/\varepsilon_d)\left[P'_\alpha(\cos(\pi-\gamma)) + P'_\alpha(\cos\gamma)\right]\right\} = 0. \end{aligned} \tag{2}$$

Numerical calculations of the functions in (2) were carried out with the aid of the hypergeometric function. Taking into account the identity [Olver (1974)]

$$\frac{\partial}{\partial z}F(a,b,c,z) = abF(a+1,b+1,c+1,z)\,,$$

we find

$$\frac{\partial}{\partial\theta}P_\alpha(\cos\theta) = -\sin\theta\, dP_\alpha(\mu)/d\mu\big|_{\mu=\cos\theta} = -\alpha(\alpha+1)\frac{\sin\theta}{2}F\left(-\alpha+1,\alpha+2,2,\frac{1-\cos\theta}{2}\right),$$

$$\frac{\partial}{\partial\theta}P_\alpha(-\cos\theta) = \sin\theta\, dP_\alpha(\mu)/d\mu\big|_{\mu=-\cos\theta} = \alpha(\alpha+1)\frac{\sin\theta}{2}F\left(-\alpha+1,\alpha+2,2,\frac{1+\cos\theta}{2}\right),$$

and therefore

$$dP_\alpha(\mu)/d\mu\big|_{\mu=\cos\theta} = \frac{\alpha(\alpha+1)}{2}F\left(-\alpha+1,\alpha+2,2,\frac{1-\cos\theta}{2}\right),$$

$$dP_\alpha(\mu)/d\mu\big|_{\mu=-\cos\theta} = \frac{\alpha(\alpha+1)}{2} F\left(-\alpha+1,\alpha+2,2,\frac{1+\cos\theta}{2}\right).$$

By substituting these expressions into (2) we have

$$\frac{\varepsilon_m}{\varepsilon_d} F\left(-\alpha+1,\alpha+2,2,\frac{1-\cos\gamma}{2}\right)\times$$
$$\left\{\left(1+\frac{\varepsilon_p}{\varepsilon_d}\right)F\left(-\alpha,\alpha+1,1,\frac{1-\cos(\pi-\gamma)}{2}\right)+\left(1-\frac{\varepsilon_p}{\varepsilon_d}\right)F\left(-\alpha,\alpha+1,1,\frac{1-\cos\gamma}{2}\right)\right\}+$$
$$F\left(-\alpha,\alpha+1,1,\frac{1-\cos\gamma}{2}\right)\times$$
$$\left\{\left(1+\frac{\varepsilon_p}{\varepsilon_d}\right)F\left(-\alpha+1,\alpha+2,2,\frac{1-\cos(\pi-\gamma)}{2}\right)-\left(1-\frac{\varepsilon_p}{\varepsilon_d}\right)F\left(-\alpha+1,\alpha+2,2,\frac{1-\cos\gamma}{2}\right)\right\}=0.$$

If the particular case when $\varepsilon_p/\varepsilon_d = 1$ is considered, this equation is transformed into

$$\varepsilon_d P_\alpha(\cos\gamma)P'_\alpha(-\cos\gamma)+\varepsilon_m P_\alpha(-\cos\gamma)P'_\alpha(\cos\gamma)=0\,.$$

Since $P'_\alpha(\cos\gamma) = -\sin^{-1}\gamma \frac{d}{d\gamma}P_\alpha(\cos\gamma)$ and $P'_\alpha(-\cos\gamma) = \sin^{-1}\gamma \frac{d}{d\gamma}P_\alpha(-\cos\gamma)$ we may write

$$\varepsilon_d P_\alpha(\cos\gamma)\frac{d}{d\gamma}P_\alpha(-\cos\gamma) - \varepsilon_m P_\alpha(\cos(\pi-\gamma))\frac{d}{d\gamma}P_\alpha(\cos\gamma) = 0\,. \tag{3}$$

Eq.(3) is identical to the corresponding Eq.(1) for the geometry without dielectric plate. In this case the cone tip with dielectric constant ε_m is immersed into the uniform dielectric with constant ε_d. This problem has been solved for example in [Petrin (2007)].

The minimal root α of Eq.(2) (which corresponds to the physically correct solution) defines the character of electric field singularity in the vicinity of the cone apex. From Eq.(2) it follows that α is a function of three independent variables: the angle γ and the ratios of dielectric constants $\varepsilon_m/\varepsilon_d$ and $\varepsilon_p/\varepsilon_d$, i.e. $\alpha = \alpha(\gamma,\varepsilon_m/\varepsilon_d,\varepsilon_p/\varepsilon_d)$. As it was shown below $\alpha = \alpha(\gamma,\varepsilon_m/\varepsilon_d,\varepsilon_p/\varepsilon_d)$ is a complex function even for real arguments (it is important).

Let use again Drude's model for permittivity of metal without absorption $\varepsilon_m = 1-\omega_p^2/\omega^2$, where ω_p is the plasma frequency of the metal. Therefore, for fixed values of γ, ε_d and ε_p, we may find the dependence $\alpha(\omega/\omega_p)$. Taking into account that $\Psi_d \sim r^\alpha$, we have $E_{ex} \sim r^{\alpha-1}$.

Note, that for fixed values of γ, ε_d, ε_p and ω/ω_p Eq.(1) has many roots α_i but not all of the roots have physical sense or represent the singular electric field at the cone apex. Obviously, that only roots smaller than unit ($E_{ex} \sim r^{\alpha-1}$) give the singular electric field. So, we will be interested by the solutions of Eq.(2) in the interval $\mathrm{Re}(\alpha)<1$. To define the lower boundary of the solution's interval it is necessary to remind that in the vicinity of the apex the electric field density must be integrable. It means that the electric field and density must increase slower than $r^{-3/2}$ and r^{-3} respectively when $r \to 0$. So, the lower boundary of the roots interval is equal to $1/2$ and the total roots interval of interest is $-1/2 < \mathrm{Re}(\alpha) < 1$.

Eq.(2) was solved numerically. For $\gamma = 15°$, $\varepsilon_d = 1$ and $\varepsilon_p = 1$ the results of calculations are the same as in Fig. 2 obtained from Eq.(1). The plots for other values of dielectric constant of the plate ε_p are analogous to the plot of Fig. 2.
Using the same approach as in the case of Fig. 3 it were calculated numerically (see Fig. 6) the dependences of the critical frequency ω_{cr} (normalized on the plasma frequency of the metal) on the cone angle γ for $\varepsilon_d = 1$ and several values of ε_p.
These dependences may be found analytically. The critical frequency ω_{cr} corresponds to the root value $\alpha = -1/2$. Substituting $\alpha = -1/2$ and $\varepsilon_m = 1 - \omega_p^2 / \omega_{cr}^2$ into Eq.(2) we find the following expression which is valid for any values of ε_d and ε_p:

$$\frac{\omega_{cr}}{\omega_p} = \left\{ 1 + \varepsilon_d \frac{F\left(1/2,1/2,1,\frac{1-\cos\gamma}{2}\right)}{F\left(3/2,3/2,2,\frac{1-\cos\gamma}{2}\right)} \times \right. \\ \left. \times \frac{\left(1+\frac{\varepsilon_p}{\varepsilon_d}\right) F\left(3/2,3/2,2,\frac{1+\cos\gamma}{2}\right) - \left(1-\frac{\varepsilon_p}{\varepsilon_d}\right) F\left(3/2,3/2,2,\frac{1-\cos\gamma}{2}\right)}{\left(1+\frac{\varepsilon_p}{\varepsilon_d}\right) F\left(1/2,1/2,1,\frac{1+\cos\gamma}{2}\right) + \left(1-\frac{\varepsilon_p}{\varepsilon_d}\right) F\left(1/2,1/2,1,\frac{1-\cos\gamma}{2}\right)} \right\}^{-\frac{1}{2}}$$

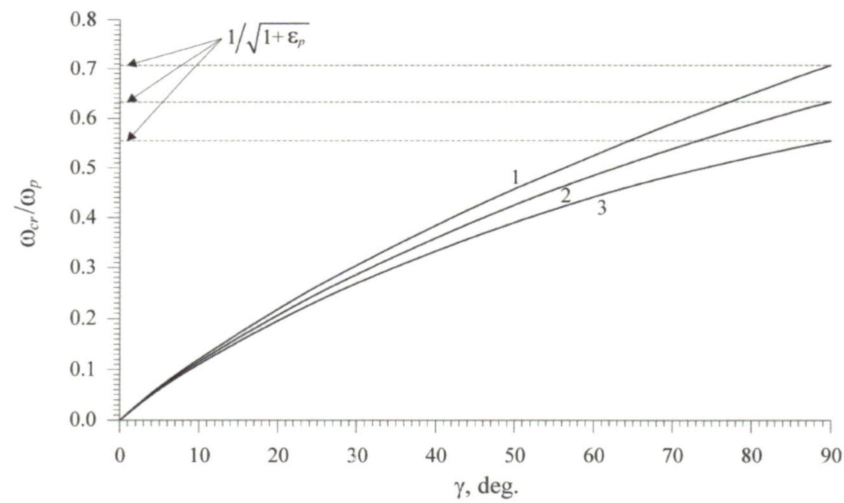

Fig. 6. Normalized critical frequency ω_{cr}/ω_p as a function of γ for $\varepsilon_d = 1$ and several values of ε_p. Curve 1 for $\varepsilon_p = 1$, 2 - $\varepsilon_p = 1.5$, 3 - $\varepsilon_p = 2.25$. It is shown the asymptotic of the curve when $\gamma \to 90°$.

It was found that when $\gamma \to 90°$ (the metal cone turns into metal plane and the free space between the cone and the dielectric plate disappears) the curves of the critical frequencies tends to the value $\omega_p / \sqrt{1+\varepsilon_p}$ - the utmost frequencies of SPP's existence on the boundary metal-dielectric plate [Stern (1960)]. As in the previous part of the chapter we see that it is absolutely unexpected that the utmost maximal frequency of SPP's existence arises in the quastatic statement of the singularity existence problem.

3.2 Application of the theory to a silver tip

So, as in the section 2.2, we see that if the working frequency is fixed, then there are two different types of singularity. In this case there is a critical angle γ_{cr} which separates the regime with the first type of singularity from the regime with the second type of singularity. For $\varepsilon_d = 1$ and several values of ε_p the dependences $\gamma_{cr} = \gamma_{cr}(\lambda_0)$ of the critical angle on the wavelength of light in vacuum of the focused SPPs with frequency ω may be found from Fig. 6 by a recalculation as it was made (section 2.2) for microtip immersed into uniform medium. The plots $\gamma_{cr} = \gamma_{cr}(\lambda_0)$ for silver, $\varepsilon_d = 1$ and for three values $\varepsilon_p = 1$, $\varepsilon_p = 1.5$ and $\varepsilon_p = 2.25$ are shown in Fig. 7. Calculating the plots $\gamma_{cr} = \gamma_{cr}(\lambda_0)$ we neglect by losses in silver. If the angle of the cone γ is more than γ_{cr}, then the singularity at the apex is of the first type. If the angle γ is smaller than γ_{cr}, then the singularity is of the second type.

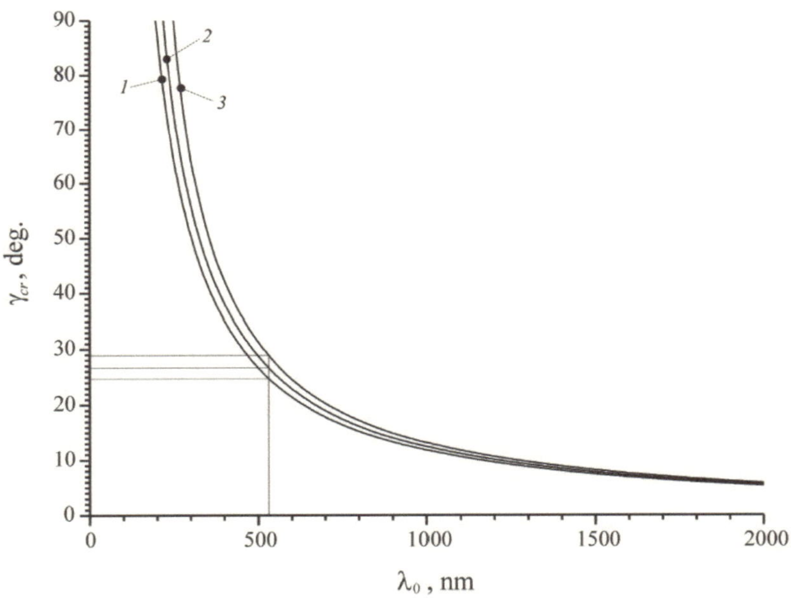

Fig. 7. Critical angle γ_{cr} of silver cone as a function of wavelength in vacuum λ_0 of exciting laser for $\varepsilon_d = 1$ and several values of ε_p. Curve 1 for $\varepsilon_p = 1$, 2 - $\varepsilon_p = 1.5$, 3 - $\varepsilon_p = 2.25$. The left boundaries of the plots $(\lambda_{sp})_{cr} = 2\pi c\sqrt{1+\varepsilon_p}/\omega_p$ are the limits of the spectrums of SPP on plane surface.

If (as in section 2.2) the wavelength of the laser λ_0, the dielectric constant of the working medium ε_d and the dielectric constant of the dielectric plate ε_p are given, then the cone angle γ at the apex of focusing SPPs microtip defines the type of singularity. From Fig. 7 it may be seen that the more ε_p the more γ_{cr} under the other things being equal.

As in the previous section of the chapter, consider the setup of the work [De Angelis (2010)] on local Raman's microscopy. The waveength of the laser excited the focused SPPs is equal to $\lambda_0 = 532$ nm. The SPPs travel along the surface of the microtip cone and focus on its apex. From Fig. 7 it may be seen that the critical angle for $\varepsilon_d = 1$ and $\varepsilon_p = 1$ is equal to $\gamma_{cr} \approx 24.7°$ (as in the previous section). If the dielectric constant ε_p is equal to 1.5 or 2.25 then the critical angles are $\gamma_{cr} \approx 26.7°$ and $\gamma_{cr} \approx 28.9°$, respectively.

4. Nanofocusing of surface plasmons at the edge of metallic wedge. Conditions for electric field singularity existence at the edge immersed into a uniform dielectric medium.

In this part of the chapter we focus our attention on finding the condition for electric field singularity of focused SPP electric field at the edge of a metal wedge immersed into a uniform dielectric medium.

SPP nanofocusing at the apex of microtip (considered in the previous sections) corresponds (based on the analogy with conventional optics) to the focusing by spherical lens. Thus, SPP nanofocusing at the edge of microwedge corresponds to the focusing by cylindrical lens at the edge [Gramotnev (2007)]. The main advantage of the wedge SPP waveguide in nanoscale is the localization of plasmon wave energy in substantially smaller volume [Moreno (2008)] due to the electric field singularity at the edge of the microwedge. This advantage is fundamentally important for miniaturization of optical computing devices which have principally greater data processing rates in comparison with today state of the art electronic components [Ogawa (2008), Bozhevolnyi (2006)].

As it will be shown below the electric field singularity at the edge of the microwedge may be of two types due to frequency dependence of dielectric constant of metal in optical frequency range. This phenomenon is analogues to the same phenomenon for microtips which was considered in the previous parts of this chapter. The investigation of these types of electric field singularities at the edge of metal microwedge is the goal of this chapter section.

4.1 Condition for electric field singularity at the edge of metallic wedge

Let consider the metal microwedge (see Fig. 8) with dielectric constant of the metal ε_m. The frequency of the SPP is ω. The wedge is immersed into a medium with dielectric costant ε_d.

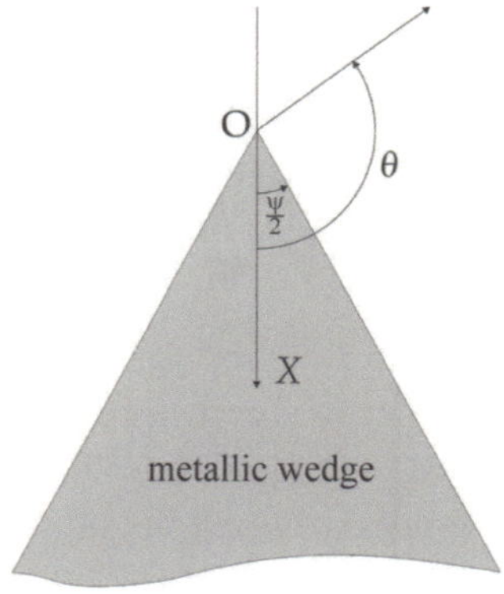

Fig. 8. Geometry of the wedge.

Let calculate the electric field distribution near the edge. In cylindrical system of coordinates with origin O and angle θ (see Fig. 8), a symmetric quasistatic potential φ obeys Laplace's equation. The two independent solutions of the Laplace's equation are the functions $r^{\alpha}\sin(\alpha\theta)$ and $r^{\alpha}\cos(\alpha\theta)$ [Landau, Lifshitz (1982)] where α is a constant; θ is the angle from the axis OX; r is the radial coordinate from the origin O. Taking into account that the electric potential in the metal and dielectric depends on r as the same power we may write the following expressions for potential in metal and dielectric respectively

$$\varphi = S_m r^{\alpha}\cos(\alpha\theta)\text{, where } -\psi/2 \le \theta \le \psi/2\,,$$

$$\varphi = S_d r^{\alpha}\cos(\alpha(\theta-\pi))\text{, where } \psi/2 \le \theta \le (2\pi-\psi/2)\,,$$

where S_m and S_d are constants, ψ is the total angle of the metallic wedge.

The boundary conditions for tangential and normal components of electric field may be written as

$$E_{m,\tau} = E_{d,\tau} \text{ and } \varepsilon_m E_{m,n} = \varepsilon_d E_{d,n}\,.$$

Using the above expressions for electric potential in the two media the boundary conditions may be rewritten in the following form

$$E_{m,\tau} = -\frac{\partial\varphi}{\partial r} = -\alpha r^{\alpha-1} S_m \cos(\alpha\theta)\,,$$

$$E_{d,\tau} = -\frac{\partial\varphi}{\partial r} = -\alpha r^{\alpha-1} S_d \cos(\alpha(\theta-\pi))\,,$$

$$E_{m,n} = -\frac{1}{r}\frac{\partial\varphi}{\partial\theta} = \alpha r^{\alpha-1} S_m \sin(\alpha\theta)\,,$$

$$E_{d,n} = -\frac{1}{r}\frac{\partial\varphi}{\partial\theta} = \alpha r^{\alpha-1} S_d \sin(\alpha(\theta-\pi))\,.$$

At the first boundary of the wedge (where $\theta = \psi/2$) the boundary conditions give two equations

$$-\alpha r^{\alpha-1} S_m \cos(\alpha\theta) = -\alpha r^{\alpha-1} S_d \cos(\alpha(\theta-\pi))\Big|_{\theta=\psi/2}\,,$$

$$\varepsilon_m \alpha r^{\alpha-1} S_m \sin(\alpha\theta) = \varepsilon_d \alpha r^{\alpha-1} S_d \sin(\alpha(\theta-\pi))\Big|_{\theta=\psi/2}\,,$$

or

$$S_m \cos(\alpha\psi/2) = S_d \cos(\alpha(\psi/2-\pi))\,,$$

$$\varepsilon_m S_m \sin(\alpha\psi/2) = \varepsilon_d S_d \sin(\alpha(\psi/2-\pi))\,.$$

Note, that at the second boundary of the wedge (where $\theta = -\psi/2$) the boundary conditions give absolutely identical equations due to symmetry of the problem.
A nontrivial solution of the system exists when the determinant is equal to zero,

$$\varepsilon_m \cos\left(\alpha(\psi/2-\pi)\right)\sin\left(\alpha\psi/2\right) - \varepsilon_d \sin\left(\alpha(\psi/2-\pi)\right)\cos\left(\alpha\psi/2\right) = 0 \ . \tag{4}$$

From this equation it follows that the index of singularity α is a function of two variables: angle ψ and the ratio $\varepsilon_m/\varepsilon_d$, i.e. $\alpha = \alpha_{\min}\left(\psi, \varepsilon_m/\varepsilon_d\right)$.
Note, that in electrostatic field $\varepsilon_m \to \infty$ and, therefore, in the limit we have

$$\cos\left(\alpha(\psi/2-\pi)\right)\sin\left(\alpha\psi/2\right) = 0 .$$

The minimal root of this equation will be when $\cos\left(\alpha(\psi/2-\pi)\right) = 0$ (we are interested in the interval $\psi < \pi$). Therefore, the minimal value of α is defined by equation $\alpha(\psi/2-\pi) = -\pi/2$ or $\alpha = \pi/(2\pi-\psi)$ (it is well-known result [Landau, Lifshitz (1982)]). When $\psi \to 0$, we have $\alpha \to 1/2$.
Using Drude's model for permittivity of metal without absorption $\varepsilon_m = 1-\omega_p^2/\omega^2$ and considering that the wedge is immersed into vacuum ($\varepsilon_d = 1$) we have $\varepsilon_m/\varepsilon_d = 1-\omega_p^2/\omega^2$. Therefore, for fixed value of we may find the dependence $\alpha = \alpha_{\min}\left(\omega/\omega_p\right)$. Taking into account that $\varphi \sim r^{\alpha}$, we have $E_{ex} \sim r^{\alpha-1}$. Note, that from the physical sense of the electric potential φ it follows that allways $\mathrm{Re}(\alpha) \geq 0$. Therefore, the interval α of singularity existence is $0 \leq \mathrm{Re}(\alpha) < 1$.
Fig. 9 shows the dependences $\alpha = \alpha_{\min}\left(\omega/\omega_p\right)$ obtained from Eq.(4) for the wedge angle $\psi = 30°$. We can see that as in the case of cone tip in the case of wedge there are two types of electric field singularity at the edge of metallic wedges. The first type of electric field singularity takes place when $\omega < \omega_{cr}$. Here, the index α has a pure real value. The second type of electric field singularity takes place when $\omega > \omega_{cr}$ and the index α has a pure image value.

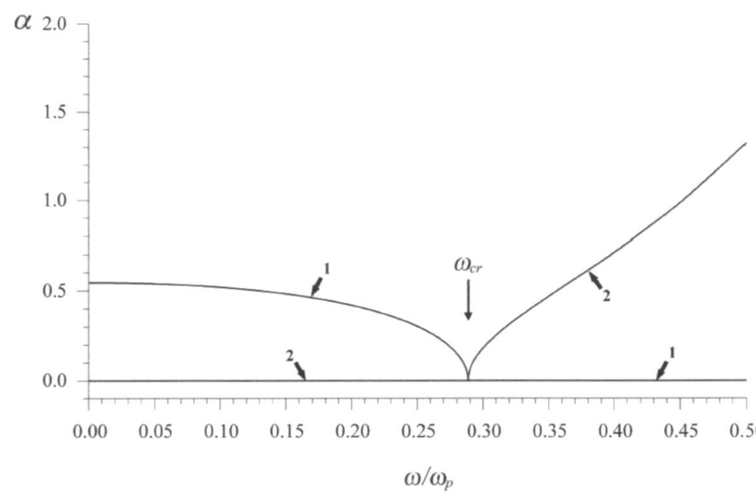

Fig. 9. Real (curve 1) and image (curve 2) part of the index α as a function of normalized frequency ω/ω_p for the wedge angle $\psi = 30°$.

Fig. 10 shows the plots of $\mathrm{Re}(\alpha)$ and $\mathrm{Im}(\alpha)$ as functions of ω/ω_p (the same functions depicted in Fig.9) in the vicinity of the critical frequency ω_{cr} without losses. For comparison in Fig. 10 the plots of $\mathrm{Re}(\alpha)$ and $\mathrm{Im}(\alpha)$ as functions of ω/ω_p for silver (the metal with losses) are shown.

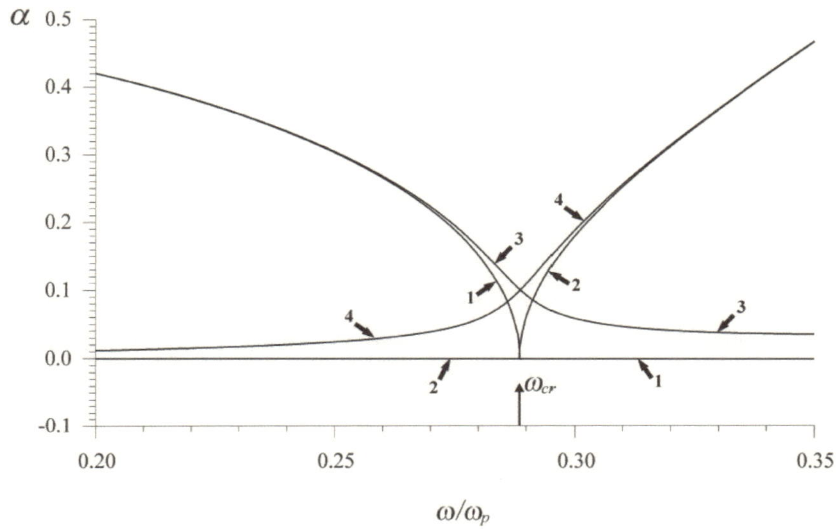

Fig. 10. The vicinity of the critical frequency ω_{cr}. Real (curve 1) and image (curve 2) parts of the index α as a function of normalized frequency ω/ω_p for the wedge angle $\psi = 30°$ and no losses in the metal. For comparison, the analogous curves 3 and 4 for silver wedge (metal with losses).

From the dependences like of Fig. 9 it was numerically found ω_{cr} (normalized on the plasma frequency ω_p) as a function of the wedge angle ψ (see Fig. 11). The obtained function is excellently approximated by the elementary function $\omega_{cr}/\omega_p \approx 0.05255\sqrt{\psi[\text{deg.}]}$. This is not a coincidence. Indeed, the condition $\omega = \omega_{cr}$ implies that $\alpha = 0$ and therefore from Eq.(4) it follows

$$(\varepsilon_m)_{cr} = \lim_{\alpha\to 0}\frac{\sin(\alpha(\psi/2-\pi))\cos(\alpha\psi/2)}{\cos(\alpha(\psi/2-\pi))\sin(\alpha\psi/2)} = 1 - \frac{2\pi}{\psi}.$$

Thus, $\omega_{cr} = \omega_p\sqrt{\psi/2\pi}$ or $\omega_{cr}/\omega_p = \sqrt{\psi[\text{deg.}]/360} \approx 0.0527046\sqrt{\psi[\text{deg.}]}$ exactly (analytical solution)! We can see that the accuracy of the numerical results is substantially high.

Fig. 11 also shows the frequency $\omega_p/\sqrt{2}$, above which SPP can not exist. Note, that when $\psi \to 180°$ (the wedge turns into a plane) the critical frequency tends to $\omega_p/\sqrt{2}$ - the utmost frequency of SPP existence on plane surface. Again it is absolutely unexpectable that the utmost frequency of SPP existence arises in the quasistatic problem of singularity existence.

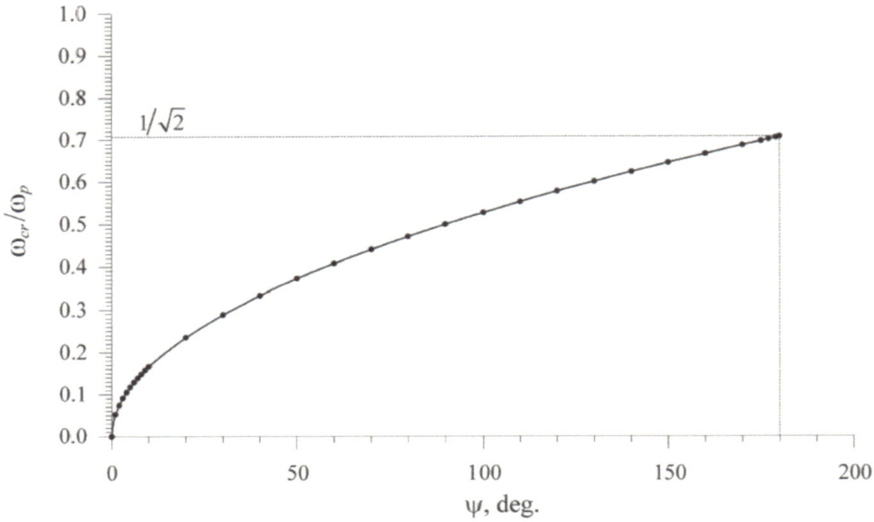

Fig. 11. Normalized critical frequency of singularity existence ω_{cr}/ω_p as a function of the wedge angle ψ. Solid line is the approximating function $\omega_{cr}/\omega_p \approx 0.05255\sqrt{\psi[\text{deg.}]}$.

4.2 Results of calculation for a silver microwegde

From the results of the previous section 4.1 we see that the problem of finding of the SPP nanofocusing properties of microwedge is the following. On the one hand the wavelength of SPP must be possibly smaller. So, the SPP frequency must be close (but smaller) to the critical frequency of SPP existence $\omega_p/\sqrt{2}$. On the other hand it is necessary to use the effect of additional increasing of the SPP electric field at the edge of the microwedge due to electric field singularity at the edge. As we have seen the electric field singularity at the edge exists for any frequency of SPP, but there two different types of electric field singularity. The choice of the singularity depends on the particular technical problem (in this work this problems do not discuss).

Consider the following problem. Let there is a microwedge on the edge of which SPPs with frequency ω are focused (the wavelength in vacuum of a laser exciting the SPP is equal to λ_0). What is the value of wedge angle ψ_{cr} which separates regimes of nanofocusing with different types of singularities?

Consider a microwedge made from silver (plasma frequency of silver is equal to $\omega_p = 1.36\times10^{16}\ s^{-1}$ [Fox (2003)] and no losses). Based on the dependences of Fig. 11 it is elementary to find the function $\psi_{cr} = \psi_{cr}(\lambda_0)$ for silver (see Fig. 12) which is the solution of the considering problem. Indeed,

$$\omega_{cr} = \omega_p\sqrt{\psi/2\pi} \Rightarrow 2\pi c/\lambda_0 = \omega_p\sqrt{\psi_{cr}/2\pi} \Rightarrow \psi_{cr} = \left(8\pi^3c^2/\omega_p^2\right)/\lambda_0^2$$

or for the angle in degrees

$$\psi_{cr}[\text{deg.}] = \left(1440\pi^2c^2/\omega_p^2\right)/\lambda_0^2 .$$

Thus, we have obtained for this problem the exact analytical solution

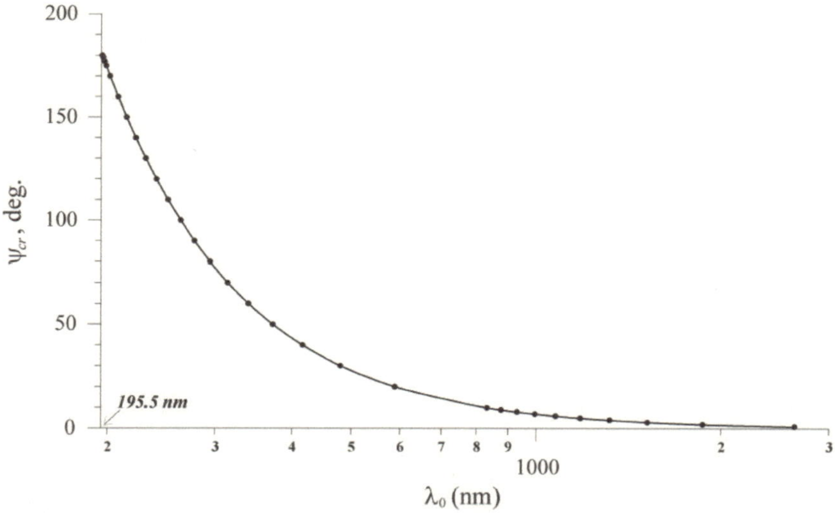

Fig. 12. Critical wedge angle ψ_{cr} as a function of the SPP wavelength in vacuum. The left limit of the graphic $\left(\lambda_{sp}\right)_{cr} = 2\sqrt{2}\pi c/\omega_p \approx 195.5nm$ - the utmost point of SPP spectrum for silver.

Thus, we may finish this section by the following fundamental statement: ***the field singularity of alternative electric field at the edge of geometrically ideal metal wedge exists in two forms. First - with pure real index of singularity*** α***, and second - with essentially image index of singularity with constant real part*** $\mathrm{Re}(\alpha)=0$***. This phenomenon is due to the strong frequency dependence of metal permittivity at the range of light frequencies.***

6. Conclusion

It was obtained the fundamental result: the electric field singularity of alternative electric field of SPP wave at the apex of geometrically ideal metal conical tip or at the edge of geometrically ideal metal wedge may exist in two forms depending on the type of the index of singularity. For a given SPP frequency and surrounding media the type the singularity defines by the angle of the tip (or wedge). This remains true for the case when the apex of metal cone touches a dielectric plane plate.

7. References

Anderson N., Bouhelier A., Novotny L. (2006) *J. Opt. A* Vol. 8, p. S227.

Angot A. (1962), *Complements de mathematiques a l'usage des ingenieurs de l'electrotechnique et des telecommunications*, La Revue d'Optique, Paris, 2nd ed.

Babadjanyan A. J., Margaryan N. L., Nerkararyan K. V. (2000) *J. Appl. Phys.* Vol. 87, p. 8.

Barnes W. L. (2006) *J. Opt. A: Pure Appl. Opt.* Vol. 8, p. S87.

Bezus E. A., Doskolovich L. L., Kazanskiy N. L. et al. (2010) *J. Opt.* Vol. 12, p. 015001.

Bohren C. F., Huffman D. R. (1983), *Absorption and Scattering of Light by Small Particles*, John Wiley & Sons, New York.

Bouhelier A., Renger J., Beversluis M. R., Novotny L. (2003) *J. Microsc.* Vol. 210, p. 220.

Bozhevolnyi S. I., Volkov V. S., Devaux E., et al. (2006) *Nature*, Vol. 440, p. 508.
De Angelis F., Das G., Candeloro P., et al. (2010) *Nature Nanotech.* Vol. 5, p. 67.
Economou E. N. (1969) *Phys. Rev.* Vol. 182, p.539.
Fox M. (2003), *Optical Properties of Solids*, Oxford University Press Inc., New York.
Frey H. G., Keilmann F., Kriele A., Guckenberger R. (2002) *Appl. Phys. Lett.* Vol. 81, p. 5030.
Gay G., Alloschery O., Weiner J., et al. (2007) *Phys. Rev. E*, Vol. 75, p. 016612.
Gramotnev D. K. (2005) *J. Appl. Phys.* Vol. 98, p. 104302.
Gramotnev D. K., Vernon K. C. (2007) *Appl. Phys. B.* Vol. 86, p. 7.
Hillenbrand R., Taubner T., Kellmann F. (2002) *Nature* Vol. 418, p. 159.
Ichimura T., Hayazawa N., Hashimoto M., et al. (2004) *Phys. Rev. Lett.* Vol. 92, p. 220801.
Kawata S. (2001), *Near-Field Optics and Surface Plasmon-Polaritons*, Springer, Berlin.
Keilmann F. (1999) *J. Microsc.* Vol. 194, p.567.
Kneipp K., Wang Y., Kneipp H., et al. (1997) *Phys. Rev. Lett.* Vol. 78, p. 1667.
Mehtani D., Lee N., Hartschuh R. D., et al. (2006) *J. Opt. A* Vol. 8, p. S183.
Moreno E., Rodrigo S. G., Bozhevolnyi S. I., et al. (2008) *Phys. Rev. Lett.* Vol. 100, p. 023901.
Landau L. D., Lifshitz E. M. (1982), Electrodynamics of continuos media, Nauka, Moscow.
Naber A., Molenda D., Fischer U. C., et al. (2002) *Phys. Rev. Lett.* Vol. 89, p. 210801.
Nerkararyan K. V., Abrahamyan T., Janunts E., et al. (2006) *Phys. Lett. A* Vol. 350, p. 147.
Nie S. M., Emory S. R. (1997) *Science* Vol. 275, p. 1102.
Novotny L., Hafner C. (1994) *Phys. Rev. E* Vol. 50, p. 4094.
Novotny L., Pohl D. W., Hecht B. (1995) *Ultramicroscopy* Vol. 61, p. 1.
Petrin A. B. (2007), *Journal of Experimental and Theoretical Physics*, Vol. 105, №6, p. 1236.
Petrin A. B. (2010), *J. Nanoelectron. Optoelectron.*, Vol. 5, № 1, p. 55.
Petrin A. B. (2010), *High Temperature* (in press).
Pettinger B., Ren B., Picardi G., et al. (2004) *Phys. Rev. Lett.* Vol. 92, p. 096101.
Pohl D. W., Denk W., Lanz M. (1984) *Appl. Phys. Lett.* Vol. 44, p.651.
Ogawa T., Pile D. F. P., Okamoto T., et al. (2008) *J. Appl. Phys.* Vol. 104, p. 033102.
Olver F. W. J. (1974), Asymptotics and Special Functions, L.: Academic Press, New York.
Stern E. A., Ferrell R. A. (1960) *Phys. Rev.* Vol. 120, p. 130.
Stockman M. I. (2004) *Phys. Rev. Lett.* Vol. 93, p. 137404.
Zayats A. V., Smolyaninov I. (2003) *J. Opt. A.* Vol. 5, p. S16.

11

Radiative Transfer Theory for Layered Random Media

Saba Mudaliar
Sensors Directorate, Air Force Research Laboratory, Hanscom AFB
USA

1. Introduction

When a random medium is sparse and the extent or size of the random medium is small, then a single scattering theory is sufficient; multiple scattering effects are negligible (Tatarskii, 1971; Ishimaru, 1997; Tsang et al., 1985). However, when the medium is not sparse or when the extent of the scattering medium is large, then multiple scattering becomes important. In principle, one can use the wave equations or Maxwell's equations to carry out a multiple scattering analysis (Foldy, 1945; Lax, 1951; Twersky, 1980). This procedure, also known as the statistical wave approach, is quite rigorous and takes into consideration all multiple scattering processes involved in the problem. However, the methods of analysis and solution techniques are rather complicated. One is forced to impose various approximations in order to perform numerical computations and arrive at useful results. On the other hand, the radiative transfer theory (RTT), another approach to this problem, is conceptually simple and at the same time very efficient in modelling multiple scattering processes. Furthermore, there are well-established techniques for numerical analysis of the radiative transfer equations (Clough et al., 2005; Stamnes et al., 1988; Berk et al., 1998; Lenoble, 1985).
However,the RTT is heuristic and lacks the rigour of the statistical wave theory. The fundamental quantity in the RTT is the specific intensity, which is a measure of energy flux density per unit area, per unit steradian. Although the concept of specific intensity has many desirable properties, the fact that RTT deals entirely with intensities means that it does not possess phase information and it cannot adequately describe wave phenomena such as diffraction and interference. The basic equation of the RTT is the radiative transfer equation, given as (Chandrasekar, 1960; Sobolev, 1963; Ishimaru, 1997)

$$\hat{s} \cdot \nabla I(\mathbf{r},\hat{s}) + \gamma I(\mathbf{r},\hat{s}) = \int P(\hat{s},\hat{s}')I(\mathbf{r},\hat{s}')d\Omega', \qquad (1)$$

where I is the radiant intensity, which is a phase-space quantity at position $\mathbf{r}$ and direction $\hat{s}$; γ is the extinction coefficient, which is a measure of loss of energy in direction $\hat{s}$ due to scattering in other directions. P is the phase function, representing the increase in energy density in direction $\hat{s}$ due to scattering from neighbouring elements. $d\Omega'$ is the solid angle element subtended by the radiant intensity in direction $\hat{s}$. Equation (1) is the radiative transfer equation, which may be regarded as a statement of conservation of radiant intensity. This scalar transport equation is inappropriate when the scattering medium has anisotropic fluctuations or if it involves boundaries. Even for models with spherical scatterers the scalar approach is inaccurate (Kattawar & Adams, 1990; Stammes, 1994; Hasekamp et al.,

2002; Stam & Hovenier, 2005; Levy et al., 2004; Mishchenko et al., 2006). It is important in such situations to use the following vector version of the transport equation

$$\hat{s} \cdot \nabla \mathbf{I}(\mathbf{r},\hat{s}) + \overline{\overline{\gamma}}\, \mathbf{I}(\mathbf{r},\hat{s}) = \int \bar{\mathbf{P}}(\hat{s},\hat{s}')\mathbf{I}(\mathbf{r},\hat{s}')d\Omega'. \tag{2}$$

where $\mathbf{I}$ is the Stokes vector, $\overline{\overline{\gamma}}$ is the extinction matrix, and $\bar{\mathbf{P}}$ is the phase matrix. Quite often in applications these quantities are modelled using empirical data. One may also calculate these quantities (Tsang et al., 1985; Ulaby et al., 1986) using wave scattering theory if one knows the statistical characteristics of the medium.

Most problems encountered in applications involve boundaries. Hence the radiative transfer (RT) equations must be supplemented by boundary conditions. Among the very early applications of the RTT, the plane parallel geometry has been thoroughly studied (Chandrasekar, 1960). However, in those applications (e.g., atmosphere) the boundaries are nonscattering and hence do not significantly impact the scattering process. There are, indeed, several other applications such as subsurface sensing (Moghaddam et al., 2007), optical mirrors (Amra, 1994; Elson, 1995), and seismology (Sato & Fehler, 1998) where the boundaries do scatter, thereby influencing the multiple scattering process.

Consider the problem of two scattering media separated by a boundary. The geometry of this problem is shown in Figure 1. Scattering media 1 and 2 are separated by a boundary Σ. The permittivities of the media have a deterministic part ϵ_j and a randomly fluctuating part $\tilde{\epsilon}_j$. Note that there is an index mismatch between the background permittivities of medium 1 and medium 2. Thus the boundary is, indeed, a scattering boundary. Let I_1 and I_2 be the radiant intensities in medium 1 and medium 2, respectively. We use the superscript "in" to denote that part of the radiant intensity that goes towards the boundary and the superscript "out" to denote the part of radiant intensity that goes away from the boundary. The boundary conditions used for this kind of problem are

$$I_1^{out} = \mathcal{R}_{21} I_1^{in} + \mathcal{T}_{12} I_2^{in} \tag{3a}$$

$$I_2^{out} = \mathcal{R}_{12} I_2^{in} + \mathcal{T}_{21} I_1^{in} \tag{3b}$$

where $\mathcal{R}$ and $\mathcal{T}$ symbolically represent the reflection and transmission processes that take place at the boundary. The first subscript indicates the region where the scattered beam travels. The second subscript indicates the region where the incident beam originates. For instance, $\mathcal{R}_{12}$ represents the reflection at the boundary for a beam incident from below. Note that these boundary conditions are based on energy conservation at the boundary. For bounded geometries the system of equations that needs to be solved comprises the RT equation (1) along with the equation associated with boundary conditions (3).

One should point out that the RTT as applied to a particular problem is a model constructed on certain hypotheses and assumptions. In most papers on applications using the RTT the conditions and assumptions involved are rarely stated or discussed. Since energy balance considerations are employed in constructing the RT equation people often take it as a fundamental axiom that requires no further explanation or justification. Even in a few works where the underlying assumptions are mentioned the particular approximations involved are described in terms of special technical terminologies specific to the discipline where it is used. One good way to understand in more general terms the RT approach and its underlying assumptions is to connect it with the more rigorous statistical wave approach. For the case of an unbounded random medium this kind of study was carried out in the 1970s

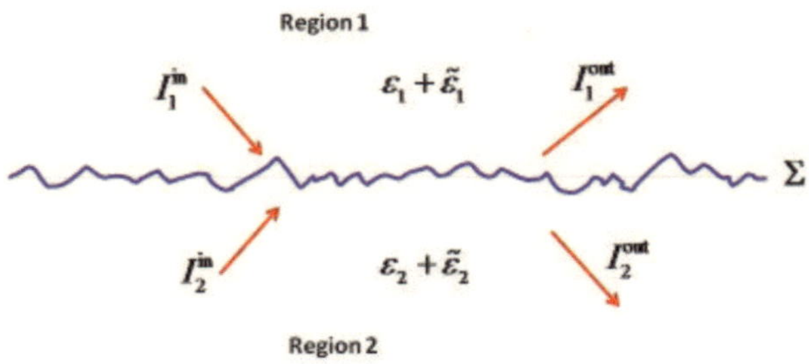

Fig. 1. Boundary separating two scattering media.

(Barabanenkov et al., 1972). From that study we learn that the radiative transfer theory can be applied under the following conditions:
1. Quasi-stationary field approximation.
2. Weak fluctuations (first-order approximation to Mass and Intensity operators)
3. Statistical homogeneity of the medium fluctuations.

However, our problem has bounded structures which may be planar or randomly rough. Therefore it remains to be seen whether the conditions arrived at in the case of unbounded random media will be sufficient for our problem.

In this work we employ a statistical wave approach using surface scattering operators (Voronovich, 1999; Mudaliar, 2005) to derive the coherence functions, and hence make a transition (using Wigner transforms) to transport equations for our multilayer problem. In this process we find that there are more conditions implied when we choose to apply the RT approach to our problem than it is widely believed to be necessary. One such condition is the weak surface correlation approximation. This means that the RT approach places certain restrictions on the type of rough interfaces that it can model accurately.

This chapter is organized as follows. In Section 2 we consider layered random media with planar boundaries. In Section 3 we consider the corresponding problem with rough interfaces. This chapter concludes in Section 4 with a summary and a discussion of our main findings.

2. Layered random medium with planar interfaces

Multiple scattering in layered scattering media with planar boundaries has been studied for nearly 100 years (Chandrasekar, 1960; Ambartsumian, 1943; de Hulst, 1980). This has been the model used for radiation processes in atmosphere. However, the boundaries involved in such problems are nonscattering in nature. Hence the fact that the scattering medium is confined to boundaries does not significantly affect the scattering processes. In several other situations where the boundaries are of scattering type, as in remote sensing of the earth (Elachi & van Zyl, 2006; Kuo & Moghaddam, 2007), seismology (Sato & Fehler, 1998), ground penetrating radar (Daniels, 2004; Urbini et al., 2001), optical devices (Amra, 1994; Elson, 1995), and medical tomography (Arridge & Hebden, 1997) the multiple scattering processes do get influenced by the boundaries. We study these processes in the context of a multilayer geometry in the following sections

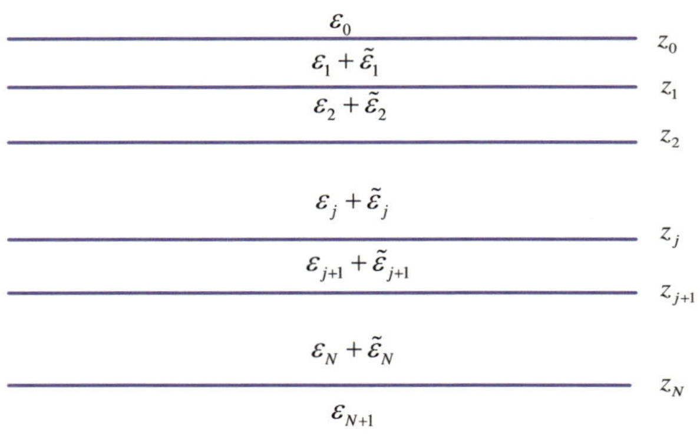

Fig. 2. Geometry of the problem with planar interfaces.

2.1 Description of the problem

The geometry of the problem is shown in Figure 2. We have an N-layer random media stack whose interfaces are parallel planes defined by $z = z_0, z_1, z_2, \ldots, z_N$. The permittivity of the j-th layer is $\epsilon_j + \tilde{\epsilon}_j(\mathbf{r})$, where ϵ_j is the deterministic part and $\tilde{\epsilon}_j$ is the randomly fluctuating part. The permeability of each of the layers is that of free space. It is assumed that $\tilde{\epsilon}_j$s are zero-mean isotropic stationary random processes independent of each other. Let $z_0 = 0$, and let d_j be the thickness of the j-th layer. The media above and below the stack are homogeneous with parameters ϵ_0, k_0, and ϵ_{N+1}, k_{N+1}, respectively. This system is excited by a monochromatic electromagnetic plane wave and we are interested in the resulting multiply scattered fields.

2.2 Radiative transfer approach

Since our layer problem has translational invariance in azimuth the RT equation for the m-th layer takes the following form (Chandrasekar, 1960; Lenoble, 1993),

$$\cos\theta_m \frac{d}{dz}\mathbf{I}_m(z,\hat{s}) + \overline{\overline{\gamma}}_m \, \mathbf{I}_m(z,\hat{s}) = \int_{\Omega_m} \bar{\mathbf{P}}_m(\hat{s},\hat{s}')\mathbf{I}_m(z,\hat{s}')d\Omega', \tag{4}$$

where $\Omega_m = \{\mathbf{r}'_\perp; z_m < z' < z_{m-1}\}$. The subscript m denotes that the quantity corresponds to the m-th layer and θ_m is the elevation angle of $\hat{s}$ in the m-th layer. This set of RT equations is complemented by a set of boundary conditions (Karam & Fung, 1982; Caron et al., 2004) which are in turn based on energy conservation considerations. In other words, we impose the condition that the energy flux density at each interface is conserved. This leads to the following boundary condition on the m-th interface

$$\mathbf{I}^u_m(z_m,\hat{s}) = \mathcal{R}_{m+1,m}(\hat{s})\mathbf{I}^d_m(z_m,\hat{s}) + \mathcal{T}_{m,m+1}(\hat{s})\mathbf{I}^u_{m+1}(z_m,\hat{s}). \tag{5}$$

The boundary condition on the $(m-1)$-th interface is given as

$$\mathbf{I}^d_m(z_{m-1},\hat{s}) = \mathcal{R}_{m-1,m}(\hat{s})\mathbf{I}^u_m(z_{m-1},\hat{s})\mathcal{T}_{m,m-1}(\hat{s})\mathbf{I}^d_{m-1}(z_{m-1},\hat{s}), \tag{6}$$

where $\mathcal{R}_{mn}$ and $\mathcal{T}_{mn}$ are the local reflection and transmission Müller matrices. To be more specific, $\mathcal{R}_{mn}$ represents the reflection Müller matrix of waves incident from medium n on the interface that separates medium m and medium n. The superscripts u and d indicate whether the intensity corresponds to a wave travelling upwards or downwards. Suppose we have a plane wave incident on this stack from above. Then the downward travelling intensity in Region 0 is

$$\mathbf{I}_0^d(z,\hat{s}) = \mathbf{B}_0\delta(\cos\theta_0 - \cos\theta_i)\,\delta(\phi_0 - \phi_i), \tag{7}$$

where $\mathbf{B}_0$ is the intensity of the incident plane wave and $\{\theta_i, \phi_i\}$ describes its direction. Since there is no source or scatterer in Region $N+1$,

$$\mathbf{I}_{N+1}^u(z,\hat{s}) = 0.$$

Notice again that these boundary conditions represent conservation of intensity at the interfaces. In order to better understand this procedure we now relate this with the statistical wave approach to this problem.

2.3 Statistical wave approach

Following are the equations that govern the waves in the layer structure:

$$\nabla \times \nabla \times \mathbf{E}_j - k_j^2\mathbf{E}_j = v_j\mathbf{E}_j \qquad j = 1,\cdots,N, \tag{8}$$

where $v_j \equiv \omega^2\mu\tilde{\epsilon}_j(\mathbf{r})$ represents the volumetric fluctuation in Region j. For the homogeneous regions above and below we have

$$\nabla \times \nabla \times \mathbf{E}_0 - k_0^2\mathbf{E}_0 = 0, \tag{9a}$$

$$\nabla \times \nabla \times \mathbf{E}_{N+1} - k_{N+1}^2\mathbf{E}_{N+1} = 0. \tag{9b}$$

The boundary conditions at the j-th interface are

$$\hat{\mathbf{z}} \times \mathbf{E}_j(\mathbf{r}_\perp, z_j) = \hat{\mathbf{z}} \times \mathbf{E}_{j+1}(\mathbf{r}_\perp, z_j), \tag{10a}$$

$$\hat{\mathbf{z}} \times \nabla \times \mathbf{E}_j(\mathbf{r}_\perp, z_j) = \hat{\mathbf{z}} \times \nabla \times \mathbf{E}_{j+1}(\mathbf{r}_\perp, z_j). \tag{10b}$$

This system is complemented by the radiation conditions well away from the stack. We assume that we know the solution to the unperturbed problem, and denote it as $\mathbf{E}^o$ (Chew, 1995). The corresponding Green's functions, denoted as $\bar{\mathbf{G}}_{ij}^o$, are governed by the following set of equations:

$$\nabla \times \nabla \times \bar{\mathbf{G}}_{jk}^o(\mathbf{r},\mathbf{r}') - k_j^2\bar{\mathbf{G}}_{jk}^o(\mathbf{r},\mathbf{r}') = \bar{\mathbf{I}}\delta_{jk}\delta(\mathbf{r}-\mathbf{r}'), \tag{11a}$$

$$\hat{\mathbf{z}} \times \bar{\mathbf{G}}_{jk}^o(\mathbf{r}_\perp, z_j; \mathbf{r}') = \hat{\mathbf{z}} \times \bar{\mathbf{G}}_{(j+1)k}^o(\mathbf{r}_\perp, z_j; \mathbf{r}'), \tag{11b}$$

$$\hat{\mathbf{z}} \times \nabla \times \bar{\mathbf{G}}_{jk}^o(\mathbf{r}_\perp, z_j; \mathbf{r}') = \hat{\mathbf{z}} \times \nabla \times \bar{\mathbf{G}}_{(j+1)k}^o(\mathbf{r}_\perp, z_j; \mathbf{r}'), \tag{11c}$$

$$\hat{\mathbf{z}} \times \bar{\mathbf{G}}_{jk}^o(\mathbf{r}_\perp, z_{j-1}; \mathbf{r}') = \hat{\mathbf{z}} \times \bar{\mathbf{G}}_{(j-1)k}^o(\mathbf{r}_\perp, z_{j-1}; \mathbf{r}'), \tag{11d}$$

$$\hat{\mathbf{z}} \times \nabla \times \bar{\mathbf{G}}_{jk}^o(\mathbf{r}_\perp, z_{j-1}; \mathbf{r}') = \hat{\mathbf{z}} \times \nabla \times \bar{\mathbf{G}}_{(j-1)k}^o(\mathbf{r}_\perp, z_{j-1}; \mathbf{r}'). \tag{11e}$$

$\bar{\mathbf{I}}$ represents the unit dyad. Using these Green's functions and the radiation conditions the electric field in Region j is represented as

$$\mathbf{E}_j(\mathbf{r}) = \mathbf{E}_j^o(\mathbf{r}) + \sum_{k=1}^{N}\int_{\Omega_k}\bar{\mathbf{G}}_{jk}^o(\mathbf{r},\mathbf{r}')v_k(\mathbf{r}')\mathbf{E}_k(\mathbf{r}')d\mathbf{r}' \qquad j = 0,1,\cdots,N+1 \tag{12}$$

Note that $v_0 = v_{N+1} = 0$. We first average (10) w.r.t. fluctuation in permittivities to get

$$\langle\mathbf{E}_j(\mathbf{r})\rangle_v = \mathbf{E}_j^o(\mathbf{r}) + \sum_{k=1}^{N}\sum_{l=1}^{N}\int_{\Omega_k}d\mathbf{r}'\int_{\Omega_l}d\mathbf{r}''\bar{\mathbf{G}}_{jk}^o(\mathbf{r},\mathbf{r}')\langle\bar{\mathbf{G}}_{kl}(\mathbf{r}',\mathbf{r}'')\rangle_v\langle v_k(\mathbf{r}')v_l(\mathbf{r}'')\rangle\langle\mathbf{E}_l(\mathbf{r}'')\rangle_v. \tag{13}$$

where $\bar{\mathbf{G}}_{kl}$ is governed by the following system of equations

$$\nabla \times \nabla \times \bar{\mathbf{G}}_{kl}(\mathbf{r},\mathbf{r}') - k_k^2 \bar{\mathbf{G}}_{kl}(\mathbf{r},\mathbf{r}') = \bar{\bar{\mathbf{I}}}\delta_{kl}\delta(\mathbf{r}-\mathbf{r}') + v_k \bar{\mathbf{G}}_{kl}(\mathbf{r},\mathbf{r}'),$$

$$\hat{\mathbf{z}} \times \bar{\mathbf{G}}_{kl}(\mathbf{r}_\perp, z_k; \mathbf{r}') = \hat{\mathbf{z}} \times \bar{\mathbf{G}}_{(k+1)l}(\mathbf{r}_\perp, z_k; \mathbf{r}'),$$

$$\hat{\mathbf{z}} \times \nabla \times \bar{\mathbf{G}}_{kl}(\mathbf{r}_\perp, z_k; \mathbf{r}') = \hat{\mathbf{z}} \times \nabla \times \bar{\mathbf{G}}_{(k+1)l}(\mathbf{r}_\perp, z_k; \mathbf{r}').$$

We also have a similar set of boundary conditions on the $(k-1)$-th interface. Here we have used a first-order approximation to the mass operator based on weak fluctuations. The fluctuations in permittivity in different regions are assumed to be uncorrelated, which means that

$$\langle v_k(\mathbf{r}') v_l(\mathbf{r}'') \rangle = \delta_{kl} C_k(\mathbf{r}' - \mathbf{r}''), \tag{14}$$

where C_k is the correlation function of the volumetric fluctuations in Region k. We have assumed that the fluctuations of the parameters of our problem are Gaussian and statistically homogeneous. Inserting (14) in (13) and employing $\nabla \times \nabla \times \bar{\bar{\mathbf{I}}} - k_j^2$ on (13) we get

$$\nabla \times \nabla \times \langle \mathbf{E}_j(\mathbf{r}) \rangle - k_j^2 \langle \mathbf{E}_j(\mathbf{r}) \rangle_v = \int_{\Omega_j} d\mathbf{r}' \langle \bar{\mathbf{G}}_{jj}(\mathbf{r},\mathbf{r}') \rangle C_j(\mathbf{r}-\mathbf{r}') \langle \mathbf{E}_j(\mathbf{r}') \rangle. \tag{15}$$

First note from (14) that $(\nabla \times \nabla \times \bar{\bar{\mathbf{I}}} - k_j^2)\langle \mathbf{E}_j(\mathbf{r}) \rangle = 0$ for $j = 0, N+1$. This means that the coherent propagation constants in regions above and below the layer stack are unaffected by the fluctuations of the problem. However, they indeed get modified within the stack region. On writing (15) as $(\nabla \times \nabla \times \bar{\bar{\mathbf{I}}} - k_j^2 - \mathcal{L})\langle \psi_j \rangle = 0$, where $\mathcal{L}$ denotes the integral operator $\int_{\bar{\Omega}_j} d\mathbf{r}' \left\langle \bar{\mathbf{G}}_{jj}(\mathbf{r},\mathbf{r}') \right\rangle C_j(\mathbf{r}-\mathbf{r}')$, we infer that $\chi_j \equiv \sqrt{k_j^2 + \mathcal{L}}$ represents the mean propagation constant in $\bar{\Omega}_j$. Observe that χ_j depends explicitly on the volumetric fluctuations in Region j and implicitly on the fluctuations of the stack. This is in contrast to the RT approach where $\bar{\bar{\gamma}}_j$ depends exclusively on the volumetric fluctuations in Region j. Moreover, χ_j depends on the polarization if the fluctuations of the problem are anisotropic. Further, even if the volumetric fluctuations are isotropic, χ_j will be polarization-dependent because of surface reflections. This is in contrast to the RT approach where $\bar{\bar{\gamma}}_j$ is polarization-dependent only when the volumetric fluctuations are anisotropic. Therefore the question is this: when do the effects of boundaries on the mean propagation constants become negligible? A first-order solution to the above dispersion relation shows that in situations where the thickness of the layer is larger than the corresponding mean free path, the influence of the boundaries on the mean propagation constants becomes negligible, as in the case of the RT system.

Since the problem is invariant under translations in azimuth, the mean wave functions for our problem have the following form:

$$\langle \mathbf{E}_j^p(\mathbf{r}) \rangle = \exp(i\mathbf{k}_{\perp i} \cdot \mathbf{r}) \left\{ A_j^p(\mathbf{k}_{\perp i}) \mathbf{p}_j^+ \exp[iq_j^p z] + B_j^p(\mathbf{k}_{\perp i}) \mathbf{p}_j^- \exp[-iq_j^p z] \right\} \quad j = 1, 2, \cdots, N, \tag{16}$$

$$\langle \mathbf{E}_0^p(\mathbf{r}) \rangle = \exp(i\mathbf{k}_{\perp i} \cdot \mathbf{r}) \left\{ \mathbf{p}_0^- \exp[-ik_{0zi} z] + R^p(\mathbf{k}_{\perp i}) \mathbf{p}_0^+ \exp[ik_{0zi} z] \right\}, \tag{17}$$

and

$$\langle \mathbf{E}_{N+1}^p(\mathbf{r}) \rangle = \exp(i\mathbf{k}_{\perp i} \cdot \mathbf{r}) T^p(\mathbf{k}_{\perp i}) \mathbf{p}_{N+1}^- \exp[-ik_{(N+1)zi} z], \tag{18}$$

where the superscript p stands for the polarization, either horizontal or vertical. $\mathbf{p}$ is the unit vector representing polarization. q_j is the z-component of $\boldsymbol{\chi}_j$. The subscript i is used to indicate that the wave vector is in the incident direction. R and T denote, respectively, the mean reflection and transmission coefficients of the stack. A_j and B_j denote, respectively, the mean coefficients of up-going and down-going waves in the j-th layer. The boundary conditions associated with the above equations at the j-th interface are

$$\hat{\mathbf{z}} \times \langle \mathbf{E}_j(\mathbf{r}_\perp, z_j) \rangle = \hat{\mathbf{z}} \times \langle \mathbf{E}_{j+1}(\mathbf{r}_\perp, z_j) \rangle \qquad j = 1, 2, \cdots, N \tag{19a}$$

and

$$\hat{\mathbf{z}} \times \nabla \times \langle \mathbf{E}_j(\mathbf{r}_\perp, z_j) \rangle = \hat{\mathbf{z}} \times \nabla \times \langle \mathbf{E}_{j+1}(\mathbf{r}_\perp, z_j) \rangle \qquad j = 1, 2, \cdots, N. \tag{19b}$$

The above system may be solved to evaluate the mean coefficients that appear in (16)-(18). We proceed now to the analysis of the second moments, by starting with (10). For convenience we write it in symbolic form as

$$\mathbf{E}_j = \mathbf{E}_j^o + \sum_{k=1}^{N} \bar{\mathbf{G}}_{jk}^o v_k \mathbf{E}_k. \tag{20}$$

We take the tensor product of this equation with its complex conjugate and average w.r.t. fluctuations in permittivity and obtain

$$\left\langle \mathbf{E}_j \otimes \mathbf{E}_j^* \right\rangle = \langle \mathbf{E}_j \rangle \otimes \langle \mathbf{E}_j^* \rangle + \sum_{k=1}^{N} \sum_{k'=1}^{N} \sum_{l=1}^{N} \sum_{l'=1}^{N} \langle \bar{\mathbf{G}}_{jk} \rangle \otimes \langle \bar{\mathbf{G}}_{jk'}^* \rangle \hat{\mathbf{K}}_{kk'll'} \langle \mathbf{E}_l \otimes \mathbf{E}_{l'}^* \rangle, \tag{21}$$

where $\hat{\mathbf{K}}$ is the intensity operator of the permittivity fluctuations. Employing the weak fluctuation approximation we approximate $\hat{\mathbf{K}}$ by its leading term

$$\hat{\mathbf{K}}_{kk'll'} \simeq \langle v_k \otimes v_k^* \rangle \delta_{kk'll'} \bar{\mathbf{I}}. \tag{22}$$

The above is an equation for the second moment of the wave function $\mathbf{E}$, which can be decomposed into a coherent part $\bar{\mathbf{E}}$ and a diffuse part $\tilde{\mathbf{E}}$. Therefore,

$$\langle \mathbf{E} \otimes \mathbf{E}^* \rangle = \langle \mathbf{E} \rangle \otimes \langle \mathbf{E}^* \rangle + \langle \tilde{\mathbf{E}} \otimes \tilde{\mathbf{E}}^* \rangle. \tag{23}$$

The coherent part is not of much interest, we know that it is specular for our problem. The diffuse or the incoherent part is of more interest. Therefore we write (21) in terms of the diffuse fields:

$$\left\langle \tilde{\mathbf{E}}_j \otimes \tilde{\mathbf{E}}_j^* \right\rangle = \sum_{k=1}^{N} \left\langle \bar{\mathbf{G}}_{jk} \right\rangle_v \otimes \left\langle \bar{\mathbf{G}}_{jk}^* \right\rangle_v \langle v_k \otimes v_k^* \rangle \langle \mathbf{E}_k \otimes \mathbf{E}_k^* \rangle, \tag{24}$$

Let us now write (24) in more detail as:

$$\left\langle \tilde{\mathbf{E}}_j(\mathbf{r}) \otimes \tilde{\mathbf{E}}_j^*(\mathbf{r}') \right\rangle = \sum_{k=1}^{N} \int_{\bar{\Omega}_k} d\mathbf{r}_1 \int_{\bar{\Omega}_k} d\mathbf{r}_1' \left\langle \bar{\mathbf{G}}_{jk}(\mathbf{r}, \mathbf{r}_1) \right\rangle_v \otimes \left\langle \bar{\mathbf{G}}_{jk}^*(\mathbf{r}', \mathbf{r}_1') \right\rangle_v |k_k|^4 C_k(\mathbf{r}_1 - \mathbf{r}_1')\rangle \langle \mathbf{E}_k(\mathbf{r}_1) \otimes \mathbf{E}_k^*(\mathbf{r}_1') \rangle. \tag{25}$$

As it stands, this equation is not convenient for seeking a solution, either analytically or numerically. Besides, one important goal for us is to investigate the conditions needed for employing the radiative transfer approach. With this in mind, we introduce Wigner transforms. Note that (25) is an equation for the coherence function. On the other hand, the RT

equation, as we saw earlier, is an equation for the specific intensity, which is a 'phase-space' quantity. Wigner transforms serve as a bridge to link these two quantities (Yoshimori, 1998; Friberg, 1986; Marchand & Wolf, 1974; Pederson & Stamnes, 2000).
We introduce the Wigner transforms of waves and Green's functions as

$$\mathcal{E}_m\left(\frac{\mathbf{r}+\mathbf{r}'}{2},\mathbf{k}\right) = \int \langle \mathbf{E}_m(\mathbf{r}) \otimes \mathbf{E}_m^*(\mathbf{r}')\rangle e^{-i\mathbf{k}\cdot(\mathbf{r}-\mathbf{r}')} d(\mathbf{r}-\mathbf{r}'), \tag{26}$$

$$\mathcal{G}_{mn}\left(\frac{\mathbf{r}+\mathbf{r}'}{2},\mathbf{k}\left|\frac{\mathbf{r}_1+\mathbf{r}_1'}{2},\mathbf{l}\right.\right) = \int d(\mathbf{r}-\mathbf{r}') \int d(\mathbf{r}_1-\mathbf{r}_1')$$
$$e^{-i\mathbf{k}\cdot(\mathbf{r}-\mathbf{r}')} e^{i\mathbf{l}\cdot(\mathbf{r}_1-\mathbf{r}_1')} \langle \bar{\mathbf{G}}_{mn}(\mathbf{r},\mathbf{r}_1)\rangle \otimes \langle \bar{\mathbf{G}}_{mn}^*(\mathbf{r}',\mathbf{r}_1')\rangle. \tag{27}$$

In terms of these transforms, (25) becomes

$$\tilde{\mathcal{E}}_m(\mathbf{r},\mathbf{k}) = \frac{1}{(2\pi)^6}\sum_{n=1}^{N} |k_n|^4 \int_{\Omega_n} d\mathbf{r}' \int d\alpha \int d\beta\, \mathcal{G}_{mn}(\mathbf{r},\mathbf{k}|\mathbf{r}',\boldsymbol{\alpha})\Phi_n(\boldsymbol{\alpha}-\boldsymbol{\beta})\mathcal{E}_n(\mathbf{r}',\beta), \tag{28}$$

where Φ_n is the spectral density of the permittivity fluctuations in the n-th layer.
The fact that our problem has translational invariance in azimuth implies the following:

$$\mathcal{E}_m(\mathbf{r},\mathbf{k}) = \mathcal{E}_m(z,\mathbf{k}), \tag{29a}$$

$$\mathcal{G}_{mn}(\mathbf{r},\mathbf{k}|\mathbf{r}',\mathbf{l}) = \mathcal{G}_{mn}(z,\mathbf{k}|z',\mathbf{l};\mathbf{r}_\perp - \mathbf{r}_\perp'). \tag{29b}$$

Using these relations in (28) we have

$$\tilde{\mathcal{E}}_m(z,\mathbf{k}) = \frac{1}{(2\pi)^6}\sum_{n=1}^{N} |k_n|^4 \int_{z_n}^{z_{n-1}} dz' \int d\boldsymbol{\alpha} \int d\boldsymbol{\beta} \mathcal{G}_{mn}(z,\mathbf{k}|z',\boldsymbol{\alpha};0)\Phi_n(\boldsymbol{\alpha}-\boldsymbol{\beta})\mathcal{E}_n(z',\boldsymbol{\beta}), \tag{30}$$

where $\mathcal{G}_{mn}(z,\mathbf{k}|z',\boldsymbol{\alpha};0)$ is the Fourier transform of $\mathcal{G}_{mn}(z,\mathbf{k}|z',\boldsymbol{\alpha};\mathbf{r}_\perp - \mathbf{r}_\perp')$ w.r.t. $\mathbf{r}_\perp - \mathbf{r}_\perp'$ evaluated at the origin of the spectral space. To proceed further we need to evaluate $\mathcal{G}_{mn}$. Furthermore, we need to relate this system to that of the RT, which involves the boundary conditions at the interfaces. In view of this we need to identify the coherence functions corresponding to up- and down-going wave functions. To facilitate this, we decompose $\langle\bar{\mathbf{G}}_{mn}\rangle$ into its components,

$$\langle\bar{\mathbf{G}}_{mn}\rangle = \delta_{mn}\bar{\mathbf{G}}_m^o + \bar{\mathbf{G}}_{mn}^{uu} + \bar{\mathbf{G}}_{mn}^{ud} + \bar{\mathbf{G}}_{mn}^{du} + \bar{\mathbf{G}}_{mn}^{dd}, \tag{31}$$

where the first term is the singular part of the Green's function. The superscripts u and d indicate up- and down-going elements of the waves. The other components are due to reflections from boundaries. These are formally constructed using the concept of surface scattering operators as (Voronovich, 1999),

$$\left\langle G_{mn}^{ab}(\mathbf{r},\mathbf{r}')\right\rangle^{\mu\nu} = \frac{1}{(2\pi)^4}\int \{S_{mn}^{ab}(\mathbf{k}_\perp)\}^{\mu\nu} e^{i\mathbf{k}_\perp\cdot\mathbf{r}+iaq_m^\mu(\mathbf{k}_\perp)z} e^{-i\mathbf{k}_\perp\cdot\mathbf{r}'-ibq_n^\nu(\mathbf{k}_\perp)z'} d\mathbf{k}_\perp, \tag{32}$$

where $\bar{\mathbf{S}}_{mn}^{ab}$ is the surface scattering operator. The superscripts a and b on S are used to indicate whether the waves are up-going or down-going. In the exponents, $a,b=1$ if the waves are up-going. We let $a,b=-1$ if the waves are down-going. The z-component of the mean propagation constant in the n-th layer is denoted as q_n.

We recall that $\mathcal{G}_{mn}$ is the Wigner transform of $\langle \bar{\mathbf{G}}_{mn} \rangle \otimes \langle \bar{\mathbf{G}}^*_{mn} \rangle$. The superscripts μ, ν stand for polarization, either h or v. It is important to note that only the in the quasi-uniform limit does the Wigner transform of the coherence function lead to the specific intensity of the RT equation. For our layer geometry, the Green's function is nonuniform. However, each of its components given in (31) is quasi-uniform. When we use (31) to perform the Wigner transform we ignore all cross terms. In other words, we make the approximation

$$\mathcal{G}_{mn} \simeq \delta_{mn}\mathcal{G}^o_m + \mathcal{G}^{uu}_{mn} + \mathcal{G}^{ud}_{mn} + \mathcal{G}^{du}_{mn} + \mathcal{G}^{dd}_{mn},$$

where $\mathcal{G}^{ab}_{mn}$ is the Wigner transform of $\left\langle \bar{\mathbf{G}}^{ab}_{mn} \right\rangle \otimes \left\langle \bar{\mathbf{G}}^{ab*}_{mn} \right\rangle$. Most of the cross terms are nonuniform and may be neglected under the quasi-uniform field assumption. Two of the cross terms are quasi-uniform and their inclusion leads to phase matrices that are different from those of the RT system. It turns out that such additional coherence terms become negligible when the layer thickness is of the same order or greater than the mean free path of the corresponding layer. It is under these conditions that the approximate expression for $\mathcal{G}_{mn}$ given above is good.

With the introduction of this representation for $\mathcal{G}_{mn}$ in (28), we can trace the upward travelling and downward travelling waves to obtain the following equations for the coherence function:

$$\tilde{\mathcal{E}}^u_m(z,\mathbf{k}) = \frac{1}{(2\pi)^6}|k_m|^4 \int_{z_m}^{z} dz' \int d\boldsymbol{\alpha} \int d\boldsymbol{\beta}\ \mathcal{G}^>_m(z,\mathbf{k}|z',\boldsymbol{\alpha};0)\Phi_m(\boldsymbol{\alpha}-\boldsymbol{\beta})\mathcal{E}_m(z',\boldsymbol{\beta})$$

$$+\frac{1}{(2\pi)^6}\sum_{n=1}^{N}|k_n|^4 \int_{z_n}^{z_{n-1}} dz' \int d\boldsymbol{\alpha} \int d\boldsymbol{\beta}\ \mathcal{G}^{ua}_{mn}(z,\mathbf{k}|z',\boldsymbol{\alpha};0)\Phi_n(\boldsymbol{\alpha}-\boldsymbol{\beta})\mathcal{E}^a_n(z',\boldsymbol{\beta}) \quad (33a)$$

$$\tilde{\mathcal{E}}^d_m(z,\mathbf{k}) = \frac{1}{(2\pi)^6}|k_m|^4 \int_{z}^{z_{m-1}} dz' \int d\boldsymbol{\alpha} \int d\boldsymbol{\beta}\ \mathcal{G}^<_m(z,\mathbf{k}|z',\boldsymbol{\alpha};0)\Phi_m(\boldsymbol{\alpha}-\boldsymbol{\beta})\mathcal{E}_m(z',\boldsymbol{\beta})$$

$$+\frac{1}{(2\pi)^6}\sum_{n=1}^{N}|k_n|^4 \int_{z_n}^{z_{n-1}} dz' \int d\boldsymbol{\alpha} \int d\boldsymbol{\beta}\ \mathcal{G}^{da}_{mn}(z,\mathbf{k}|z',\boldsymbol{\alpha};0)\Phi_n(\boldsymbol{\alpha}-\boldsymbol{\beta})\mathcal{E}^a_n(z',\boldsymbol{\beta}) \quad (33b)$$

where $\mathcal{G}^<_m$ and $\mathcal{G}^>_m$ are the Wigner transforms of the tensor product of $\bar{\mathbf{G}}^o_m$ when $z < z'$ and $z > z'$, respectively. Note that summation over $a = \{u, d\}$ is implied in the above equations. When we substitute the expressions for $\mathcal{G}_{mn}$ in (33) we find that

$$\{\tilde{\mathcal{E}}^a_m(z,\mathbf{k})\}_{\mu\nu} = 2\pi\delta\left\{k_z - \frac{1}{2}a[q^\mu_m(\mathbf{k}_\perp) + q^{\nu*}_m(\mathbf{k}_\perp)]\right\} e^{ia\left[q^\mu_m - q^{\nu*}_m\right]z}\{\tilde{\mathcal{E}}^a_m(z,\mathbf{k}_\perp)\}_{\mu\nu}. \quad (34)$$

On substituting this in (33) and differentiating w.r.t. z we obtain the following transport equations:

$$\left\{\frac{d}{dz} - i\left[q_\mu(\mathbf{k}_\perp) - q^*_\nu(\mathbf{k}_\perp)\right]\right\}\tilde{\mathcal{E}}^u_{\mu\nu}(z,\mathbf{k}_\perp) = \bar{\mathcal{E}}^u_{\mu\nu}(z,\mathbf{k}_\perp) + \frac{|k_m|^4}{(2\pi)^2}\int d\alpha_\perp S^>_\mu S^{>*}_\nu$$

$$\Phi_m\left\{\mathbf{k}_\perp - \alpha_\perp; \frac{1}{2}\left[q_\mu(\mathbf{k}_\perp) + q^*_\nu(\mathbf{k}_\perp)\right] - \frac{1}{2}a\left[q_{\mu'}(\alpha_\perp) + q^*_{\nu'}(\alpha_\perp)\right]\right\}(\boldsymbol{\mu}\cdot\boldsymbol{\mu}')(\boldsymbol{\nu}\cdot\boldsymbol{\nu}')\tilde{\mathcal{E}}^a_{\mu'\nu'}(z,\alpha_\perp), \quad (35a)$$

$$\left\{-\frac{d}{dz} - i\left[q_\mu(\mathbf{k}_\perp) - q^*_\nu(\mathbf{k}_\perp)\right]\right\}\tilde{\mathcal{E}}^d_{\mu\nu}(z,\mathbf{k}_\perp) = \bar{\mathcal{E}}^d_{\mu\nu}(z,\mathbf{k}_\perp) + \frac{|k_m|^4}{(2\pi)^2}\int d\alpha_\perp S^<_\mu S^{<*}_\nu$$

$$\Phi_m\left\{\mathbf{k}_\perp - \alpha_\perp; -\frac{1}{2}\left[q_\mu(\mathbf{k}_\perp) + q^*_\nu(\mathbf{k}_\perp)\right] - \frac{1}{2}a\left[q_{\mu'}(\alpha_\perp) + q^*_{\nu'}(\alpha_\perp)\right]\right\}(\boldsymbol{\mu}\cdot\boldsymbol{\mu}')(\boldsymbol{\nu}\cdot\boldsymbol{\nu}')\tilde{\mathcal{E}}^a_{\mu'\nu'}(z,\alpha_\perp), \quad (35b)$$

where $\bar{\mathcal{E}}^a_{\mu\nu}$ represents scattering due to the coherent part of $\mathcal{E}$, whereas the integral term in (35) represents scattering due to the diffuse part of $\mathcal{E}$. We may also regard $\bar{\mathcal{E}}^a_{\mu\nu}$ as the source to our transport equations and calculate it to obtain

$$\bar{\mathcal{E}}^a_{\mu\nu} = |k_m|^4 \Phi_m \left\{ \mathbf{k}_\perp - \mathbf{k}_{\perp i}; \frac{1}{2} a \left[q_\mu(\mathbf{k}_\perp) + q^*_\nu(\mathbf{k}_\perp) \right] - \frac{1}{2} b \left[q_\mu(\mathbf{k}_{\perp i}) + q^*_\nu(\mathbf{k}_{\perp i}) \right] \right\} \times$$

$$\times \left(S^a_\mu S^{a*}_\mu \right) (\boldsymbol{\mu} \cdot \boldsymbol{\mu}_i)(\boldsymbol{\nu} \cdot \boldsymbol{\nu}_i) \left((\mathcal{S}^b_{m0})_{\mu i} E_{\mu i} \right) \left((\mathcal{S}^b_{m0})_{\nu i} E_{\nu i} \right)^*, \qquad (36)$$

where summation over b is implied. Note that $\tilde{\mathcal{E}}^a$ in (35) includes both $\tilde{\mathcal{E}}^u$ and $\tilde{\mathcal{E}}^d$ (corresponding to up- and down-going waves). When the superscripts $\{a,b\}$ correspond to u, the value of $\{a,b\}$ in the argument of Φ_m takes the value $+1$; on the other hand, when the superscripts $\{a,b\}$ correspond to d the value of a in the argument of Φ_m takes the value -1. Since all quantities in (35) and (36) correspond to the same layer m, we have dropped the subscript m to avoid cumbersome notations. Summation over μ' and ν' is implicit in (35). Similarly, summation over μ_i and ν_i is implicit in (36). $\mathcal{S}^b_{m0}$ is the scattering amplitude of waves with direction b in m-th layer due to the wave incident in Region 0. To obtain appropriate boundary conditions we have to go back to the integral equation representations for $\tilde{\mathcal{E}}^u_{\mu\nu}$ and $\tilde{\mathcal{E}}^d_{\mu\nu}$, observe their behaviour at the interfaces, and try to find a relation between them.
After some manipulations we arrived at the following boundary conditions. At the $(m-1)$-th interface we have

$$\tilde{\mathcal{E}}^d_m(z_{m-1}, \mathbf{k}_\perp) = \dddot{\mathcal{R}}_{m-1,m}(\mathbf{k}_\perp) \tilde{\mathcal{E}}^u_m(z_{m-1}, \mathbf{k}_\perp), \qquad (37a)$$

with $\dddot{\mathcal{R}} = \dddot{\mathbf{R}} \otimes \dddot{\mathbf{R}}^*$ where $\dddot{\mathbf{R}}_{m-1,m}$ is the stack reflection matrix (not the local reflection matrix) for a wave incident from below on the $(m-1)$-th interface. Similarly,

$$\tilde{\mathcal{E}}^u_m(z_{m-1}, \mathbf{k}_\perp) = \dddot{\mathcal{R}}_{m+1,m}(\mathbf{k}_\perp) \tilde{\mathcal{E}}^d_m(z_{m-1}, \mathbf{k}_\perp), \qquad (37b)$$

where $\dddot{\mathcal{R}}_{m+1,m}$ is the tensor product of the stack reflection matrix for a wave incident from above on the m-th interface.
We were able to obtain the boundary conditions only after imposing certain approximations as given below. Consider the following identity:

$$\bar{\mathbf{S}}^{du}_{mm} = \bar{\mathbf{F}}_m \dddot{\mathbf{R}}_{m-1,m} \left\{ \bar{\mathbf{S}}^{>}_m + \bar{\mathbf{S}}^{uu}_{mm} \right\} \bar{\mathbf{F}}_m \qquad (38)$$

where $\bar{\mathbf{F}}_m = \mathrm{diag}\left\{ e^{iq_h d_m}, e^{iq_v d_m} \right\}$. Taking the tensor product of (38) with its complex conjugate we have

$$\bar{\mathbf{S}}^{du}_{mm} \otimes \bar{\mathbf{S}}^{du*}_{mm} = (\bar{\mathbf{F}}_m \otimes \bar{\mathbf{F}}^*_m) \left(\dddot{\mathbf{R}}_{m-1,m} \otimes \dddot{\mathbf{R}}^*_{m-1,m} \right) \left(\left\{ \bar{\mathbf{S}}^{>}_m + \bar{\mathbf{S}}^{uu}_{mm} \right\} \otimes \left\{ \bar{\mathbf{S}}^{>}_m + \bar{\mathbf{S}}^{uu}_{mm} \right\}^* \right) (\bar{\mathbf{F}}_m \otimes \bar{\mathbf{F}}^*_m). \qquad (39)$$

A further approximation that we impose is given as follows:

$$\left\{ \bar{\mathbf{S}}^{>}_m + \bar{\mathbf{S}}^{uu}_{mm} \right\} \otimes \left\{ \bar{\mathbf{S}}^{>}_m + \bar{\mathbf{S}}^{uu}_{mm} \right\}^* \simeq \bar{\mathbf{S}}^{>}_m \otimes \bar{\mathbf{S}}^{>*}_m + \bar{\mathbf{S}}^{uu}_{mm} \otimes \bar{\mathbf{S}}^{uu*}_{mm}. \qquad (40)$$

This is similar to the approximation we used while computing the Wigner transforms of the Green's functions. We again need to use this approximation to arrive at our boundary conditions.

2.4 Transition to radiative transfer

Now we have to transition from this transport equation (40) to the phenomenological radiative transfer equation discussed earlier. To accomplish this we have to link the key quantities of waves and radiative transfer, viz., coherence function and specific intensity. The relation between them is obtained by computing the energy density using the two concepts. Thus we have

$$\frac{1}{2}\epsilon\left\{\langle|E_v(\mathbf{r})|^2\rangle+\langle|E_h(\mathbf{r})|^2\rangle\right\}=\frac{1}{c}\int d\Omega_s I(\mathbf{r},\hat{s}). \tag{41}$$

The Wigner transform provides us with the following relation:

$$\langle E_\mu(\mathbf{r})E_\nu^*(\mathbf{r})\rangle=\frac{1}{(2\pi)^2}\int d\mathbf{k}_\perp \mathcal{E}_{\mu\nu}(z,\mathbf{k}_\perp). \tag{42}$$

Defining $\mathcal{I}_{\mu\nu}$ as

$$\mathcal{I}_{\mu\nu}(z,\hat{s})=\frac{1}{2\eta}\frac{k'^2}{(2\pi)^2}\cos\theta\,\mathcal{E}_{\mu\nu}(z,\mathbf{k}_\perp), \tag{43}$$

where η is the intrinsic impedance of the medium, we have from (41) and (42) $I=\mathcal{I}_{vv}+\mathcal{I}_{hh}$. To facilitate comparison with the results of Ulaby et al.(Ulaby et al., 1986), and Lam and Ishimaru (Lam & Ishimaru, 1993) we use a modified version of the Stokes vector (Ishimaru, 1997). Instead of the standard form $\{I,Q,U,V\}$ we use $\{(I+Q)/2,(I-Q)/2,U,V\}$. Thus, in terms of $\mathcal{I}_{\mu\nu}$ defined in (43), our modified Stokes vector is $\left\{\mathcal{I}_{vv},\mathcal{I}_{hh},\frac{1}{2}(\mathcal{I}_{vh}+\mathcal{I}_{hv}),-\frac{i}{2}(\mathcal{I}_{vh}-\mathcal{I}_{hv})\right\}$. There is still one difference that needs to be ironed out before we transition to the RT equations. Notice that in our wave approach we obtained transport equations for $\tilde{\mathcal{E}}$, which is the fluctuating part of the coherence function. On the other hand, the phenomenological RT equations are traditionally written for total intensities. Therefore, we have to express our transport equations in terms of $\mathcal{E}$. Notice that $\mathcal{E}=\bar{\mathcal{E}}+\tilde{\mathcal{E}}$, where $\bar{\mathcal{E}}$, the average part of $\mathcal{E}$, satisfies:

$$\left\{\frac{d}{dz}-ia(q_\mu-q_\nu^*)\right\}\bar{\mathcal{E}}^a_m(z,\mathbf{k}_\perp)=0. \tag{44}$$

Using (44) in (35) we obtain

$$\left\{\frac{d}{dz}-i\left[q_\mu(\mathbf{k}_\perp)-q_\nu^*(\mathbf{k}_\perp)\right]\right\}\mathcal{E}^u_{\mu\nu}(z,\mathbf{k}_\perp)=\frac{|k_m|^4}{(2\pi)^2}\int d\alpha_\perp S^>_\mu S^{>*}_\nu(\boldsymbol{\mu}\cdot\boldsymbol{\mu}')(\boldsymbol{\nu}\cdot\boldsymbol{\nu}')\times$$
$$\times\Phi_m\left\{\mathbf{k}_\perp-\alpha_\perp;\frac{1}{2}\left[q_\mu(\mathbf{k}_\perp)-q_\nu^*(\mathbf{k}_\perp)\right]-\frac{1}{2}a\left[q_{\mu'}(\alpha_\perp)-q_{\nu'}^*(\alpha_\perp)\right]\right\}\mathcal{E}^a_{\mu'\nu'}(z,\alpha_\perp), \tag{45a}$$

$$\left\{-\frac{d}{dz}-i\left[q_\mu(\mathbf{k}_\perp)-q_\nu^*(\mathbf{k}_\perp)\right]\right\}\mathcal{E}^d_{\mu\nu}(z,\mathbf{k}_\perp)=\frac{|k_m|^4}{(2\pi)^2}\int d\alpha_\perp S^<_\mu S^{<*}_\nu(\boldsymbol{\mu}\cdot\boldsymbol{\mu}')(\boldsymbol{\nu}\cdot\boldsymbol{\nu}')\times$$
$$\times\Phi_m\left\{\mathbf{k}_\perp-\alpha_\perp;-\frac{1}{2}\left[q_\mu(\mathbf{k}_\perp)-q_\nu^*(\mathbf{k}_\perp)\right]-\frac{1}{2}a\left[q_{\mu'}(\alpha_\perp)-q_{\nu'}^*(\alpha_\perp)\right]\right\}\mathcal{E}^a_{\mu'\nu'}(z,\alpha_\perp). \tag{45b}$$

Notice that this equation is expressed entirely in total intensity. Now we can transition to the phenomenological RT equations. Using the relation between $\mathcal{E}$ and I, we change the integration variable to solid angle and arrive at

$$\left\{\cos\theta\frac{d}{dz}+\gamma_{ij}\right\}I^u_j(z,\hat{s})=\int P^{ua}_{ij}(\Omega,\Omega')I^a_j(z,\hat{s}')d\Omega', \tag{46a}$$

$$\left\{-\cos\theta\frac{d}{dz}+\gamma_{ij}\right\}I_j^d(z,\hat{s})=\int P_{ij}^{da}(\Omega,\Omega')I_j^a(z,\hat{s}')d\Omega', \tag{46b}$$

where $\overline{\overline{\gamma}}$ is the extinction matrix and $\overline{\overline{\mathbf{P}}}$ is the phase matrix. Implicit summation over superscripts a and subscript j is assumed in (46). Although the structure of this equation is identical to that of the RT (Equation (2)), the elements of the phase matrix and the extinction matrices are not the same primarily because of coherence induced by the boundaries. As mentioned earlier, we assume that the layer thickness is greater than the mean free path of the corresponding medium. If, in addition, we assume the quasi-homogeneous field approximation we obtain the following expressions for the extinction and phase matrices:

$$\overline{\overline{\gamma}}=\cos\theta\ \mathrm{diag}\left\{2q_v'',2q_h'',q_v''+q_h'',q_v''+q_h''\right\} \tag{47a}$$

$$P_{\mu\nu}^{ab}=\frac{1}{(2\pi)^2}\frac{1}{4}|k_m|^4\Phi_m\left\{\mathbf{k}_\perp-\mathbf{k}_\perp';k_m'[a\cos\theta-b\cos\theta']\right\}\mathcal{P}_{\mu\nu}^{ab}. \tag{47b}$$

The double primes are used in (47) to denote imaginary parts. For $\mu=\{v,h\}$,

$$\mathcal{P}_{\mu v}^{ab}=\left(\boldsymbol{\mu}^a\cdot\mathbf{v}'^b\right)^2 \qquad \mathcal{P}_{\mu h}^{ab}=\left(\boldsymbol{\mu}^a\cdot\mathbf{h}'^b\right)^2$$

$$\mathcal{P}_{\mu U}^{ab}=\left(\boldsymbol{\mu}^a\cdot\mathbf{v}'^b\right)\left(\boldsymbol{\mu}^a\cdot\mathbf{h}'^b\right) \qquad \mathcal{P}_{\mu V}^{ab}=0 \tag{48}$$

Similarly,

$$\mathcal{P}_{Uv}^{ab}=2\left(\mathbf{v}^a\cdot\mathbf{v}'^b\right)\left(\mathbf{h}^a\cdot\mathbf{v}'^b\right) \qquad \mathcal{P}_{Uh}^{ab}=2\left(\mathbf{v}^a\cdot\mathbf{h}'^b\right)\left(\mathbf{h}^a\cdot\mathbf{h}'^b\right)$$

$$\mathcal{P}_{UU}^{ab}=\left(\mathbf{v}^a\cdot\mathbf{v}'^b\right)\left(\mathbf{h}^a\cdot\mathbf{h}'^b\right)+\left(\mathbf{v}^a\cdot\mathbf{h}'^b\right)\left(\mathbf{h}^a\cdot\mathbf{v}'^b\right) \qquad \mathcal{P}_{UV}^{ab}=0 \tag{49}$$

$$\mathcal{P}_{Vv}^{ab}=\mathcal{P}_{Vh}^{ab}=\mathcal{P}_{VU}^{ab}=0$$

$$\mathcal{P}_{VV}^{ab}=\left(\mathbf{v}^a\cdot\mathbf{v}'^b\right)\left(\mathbf{h}^a\cdot\mathbf{h}'^b\right)-\left(\mathbf{v}^a\cdot\mathbf{h}'^b\right)\left(\mathbf{h}^a\cdot\mathbf{v}'^b\right) \tag{50}$$

Noting the implied summation over a in (46) we see that they are identical to the RT equations given in Section 2. Now we have explicit expressions for the extinction matrix and phase matrix in terms of the statistical parameters of the problem, thanks to our wave approach.

We next turn our attention to the boundary conditions (BC). In our statistical wave approach we obtained BCs in terms of the 'stack' reflection matrix $\dddot{\bar{\mathbf{R}}}$, whereas in the RT approach the BCs are given in terms of the local interface reflection matrices. We can readily reconcile this apparent difference. Note that the BC in the wave approach forms a closed system whereas in the RT approach it is 'open' (linked to adjacent layer intensities). Let us take a look at the BC at the $(m-1)$-th interface. $\dddot{\bar{\mathbf{R}}}_{m-2,m}$ can be expressed in terms of $\bar{\mathbf{R}}_{m-2,m-1}$ as follows,

$$\dddot{\bar{\mathbf{R}}}_{m-1,m}=\bar{\mathbf{R}}_{m-1,m}+\bar{\mathbf{T}}_{m,m-1}\left\{I-\dddot{\bar{\mathbf{R}}}_{m-2,m-1}\bar{\mathbf{F}}_{m-1}\bar{\mathbf{R}}_{m,m-1}\right\}^{-1}\dddot{\bar{\mathbf{R}}}_{m-2,m-1}\bar{\mathbf{F}}_{m-1}\bar{\mathbf{T}}_{m-1,m}. \tag{51}$$

This is the relation between the stack reflection coefficients of adjacent interfaces. The $\bar{\mathbf{R}}$ and $\bar{\mathbf{T}}$ are local (single interface) reflection and transmission matrices at the $(m-1)$-th interface. On operating $\mathbf{E}_m^u$ with (51) we get

$$\mathbf{E}_m^d=\bar{\mathbf{R}}_{m-1,m}\mathbf{E}_m^u+\bar{\mathbf{T}}_{m,m-1}\mathbf{E}_{m-1}^d. \tag{52}$$

Notice that this boundary condition now involves only local interface Fresnel coefficients. Similarly we write $\dddot{\mathbf{R}}_{m+1,m}$ in terms of $\dddot{\mathbf{R}}_{m+2,m+1}$ and hence obtain the BC at the m-th interface as

$$\mathbf{E}_m^u = \bar{\mathbf{R}}_{m+1,m}\mathbf{E}_m^d + \bar{\mathbf{T}}_{m,m+1}\mathbf{E}_{m+1}^u. \tag{53}$$

Next we take the tensor product of (52) with its complex conjugate. Employing the Wigner transform operator to that product, we obtain the boundary condition at the $(m-1)$-th interface, which is similar to that of the RT system. However, the reflection and transmission matrices used in the RT system correspond to the unperturbed medium, as opposed to the average medium as in the case of the statistical wave approach. Similarly we can obtain the transport-theoretic boundary conditions at the m-th interface using (53).

2.5 Remarks

Now that we have made the transition from statistical wave theory to radiative transfer theory it is instructive to itemize the assumptions implicitly involved in the RT approach.
1. Quasi-stationary field approximation.
2. Weak fluctuations.
3. Statistical homogeneity of fluctuations.

These are the three well-known conditions necessary for the unbounded random media problem. However, if the medium is bounded we need to impose additional conditions. We found that the extinction coefficients calculated in the wave approach and the RT approach are different and only after applying further approximations can they be made to agree with each other. The following two additional conditions are required for our bounded random media problem:
4. Layer thickness must be of the same order or greater than the corresponding mean free path.
5. All fluctuations of the problem are statistically independent.

In the next section, we turn our attention to the problem where the interfaces are randomly rough.

3. Layered random media with rough interfaces

The model of layered random media with rough interfaces is often encountered in many applications in various disciplines. A simple approach is to incoherently add the contributions of volumetric and surface fluctuations(Zuniga et al., 1979; Lee & Kong, 1985). However, this is valid only when we are in the single-scattering regime(Elson, 1997; Mudaliar, 1994). There are some other hybrid approaches (Papa & Tamasanis, 1991; Chauhan et al., 1991) which take into consideration some multiple scattering effects. Brown (Brown, 1988) outlines an iterative procedure which properly includes all multiple scattering interactions. However, it does not appear feasible to carry out the calculation beyond one or two iterations. Among the other methods currently used, perhaps the most widely used approach is the radiative transfer (RT) approach (Ulaby et al., 1986; Lam & Ishimaru, 1993; Karam & Fung, 1982; Shin & Kong, 1989; Caron et al., 2004; Ulaby et al., 1990; Liang et al., 2005; Fung & Chen, 1981). Here, one formulates the scattering and propagation in each layer by using the radiative transfer equation, which involves only the parameters of the medium of that layer. The boundary conditions are derived separately and independently using some asymptotic procedure developed in rough surface scattering theory (Beckmann & Spizzichino, 1987; Bass & Fuks, 1979; Voronovich, 1999). The RT equations, along with the boundary conditions, comprise the

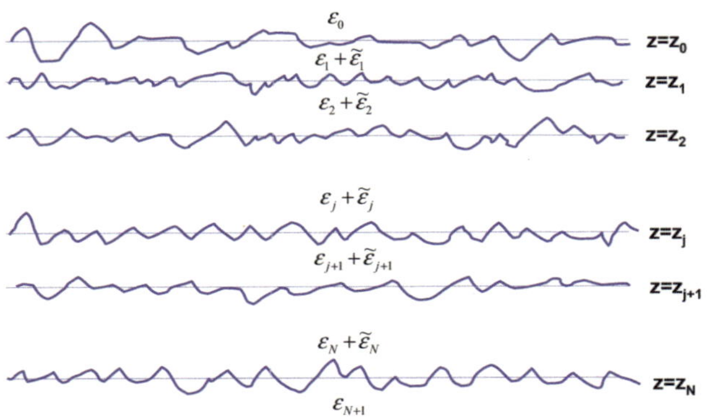

Fig. 3. Geometry of the problem with rough interfaces.

system that describes the problem. In order to better understand the conditions under which this procedure is good we study the following multi-layer problem.

3.1 Description of the problem

The geometry of the problem is shown in Figure 3. All the parameters of this problem are the same as in Section 2 except the interfaces are randomly rough now. Thus we have an N-layer random media stack with rough interfaces. The randomly rough interfaces are defined as $z = \zeta_j \equiv z_j + \tilde{\zeta}_j(\mathbf{r}_\perp)$. It is assumed that $\tilde{\epsilon}_j$ and $\tilde{\zeta}_j$ are zero-mean isotropic stationary random processes independent of each other. Thus, on the average the interfaces are parallel planes defined as $z = z_0, z_1, z_2, \cdots, z_N$. Let $z_0 = 0$, and let d_j be the thickness of the j-th layer. As before, this system is excited by a monochromatic electromagnetic plane wave and we are interested in formulating the resulting multiple scattering process.

3.2 Radiative transfer approach

The radiative transfer equations for this problem are the same as in the planar interface case (Equation (4)). However, the boundary conditions are different. On the m-th interface it is

$$\mathbf{I}^u_m(z_m,\hat{s}) = \int \langle \mathcal{R}_{m+1,m}(\hat{s},\hat{s}')\rangle \, \mathbf{I}^d_m(z_m,\hat{s}')d\Omega' + \int \langle \mathcal{T}_{m,m+1}(\hat{s},\hat{s}')\rangle \, \mathbf{I}^u_{m+1}(z_m,\hat{s}')d\Omega'. \tag{54}$$

The boundary condition on the $(m-1)$-th interface is given as

$$\mathbf{I}^d_m(z_{m-1},\hat{s}) = \int \langle \mathcal{R}_{m-1,m}(\hat{s},\hat{s}')\rangle \, \mathbf{I}^u_m(z_{m-1},\hat{s}')d\Omega' + \int \langle \mathcal{T}_{m,m-1}(\hat{s},\hat{s}')\rangle \, \mathbf{I}^d_{m-1}(z_{m-1},\hat{s}')d\Omega', \tag{55}$$

where $\mathcal{R}_{mn}$ and $\mathcal{T}_{mn}$ are the local reflection and transmission Müller matrices. To be more specific, $\mathcal{R}_{mn}$ represents the reflection Müller matrix of waves incident from medium n on the interface that separates medium m and medium n. The superscripts u and d indicate whether the intensity corresponds to a wave travelling upwards or downwards. The integrations in these expressions are over a solid angle (hemisphere) corresponding to $\hat{s}'$. For the time-harmonic plane wave incident on this stack from above the downward going intensity in Region 0 and the upward going intensity in Region $N+1$ are given as

$$\mathbf{I}^d_0(z,\hat{s}) = \mathbf{B}_0 \delta\left(\cos\theta_0 - \cos\theta_i\right)\delta(\phi_0 - \phi_i), \tag{56}$$

$$\mathbf{I}^u_{N+1}(z,\hat{s}) = 0.$$

This is the system that is widely used for this kind of problem Ulaby et al. (1986). The validity conditions for this approach have never been clearly stated. Let us now see what we can learn from adopting a statistical wave approach to this problem.

3.3 Statistical wave approach

The governing equations for the electric fields are the same as in Section 2. The primary difference is in the boundary conditions, which on the j-th interface are

$$\hat{\mathbf{n}}_j \times \mathbf{E}_j(\mathbf{r}_\perp,\zeta_j) = \hat{\mathbf{n}}_j \times \mathbf{E}_{j+1}(\mathbf{r}_\perp,\zeta_j), \tag{57a}$$

$$\hat{\mathbf{n}}_j \times \nabla \times \mathbf{E}_j(\mathbf{r}_\perp,\zeta_j) = \hat{\mathbf{n}}_j \times \nabla \times \mathbf{E}_{j+1}(\mathbf{r}_\perp,\zeta_j). \tag{57b}$$

where $\hat{\mathbf{n}}_j$ is the unit vector normal to the j-th interface with normal pointing into the medium j. This system is complemented by the radiation conditions well away from the stack. We assume that we know the solution to the problem without volumetric fluctuations, and denote it as $\check{\mathbf{E}}$. The corresponding Green's functions, denoted as $\check{\mathbf{G}}_{ij}$, are governed by the following set of equations:

$$\nabla \times \nabla \times \check{\mathbf{G}}_{jk}(\mathbf{r},\mathbf{r}') - k_j^2 \check{\mathbf{G}}_{jk}(\mathbf{r},\mathbf{r}') = \bar{\mathbf{I}}\delta_{jk}\delta(\mathbf{r}-\mathbf{r}'), \tag{58a}$$

$$\hat{\mathbf{n}}_j \times \check{\mathbf{G}}_{jk}(\mathbf{r}_\perp,\zeta_j;\mathbf{r}') = \hat{\mathbf{n}}_j \times \check{\mathbf{G}}_{(j+1)k}(\mathbf{r}_\perp,\zeta_j;\mathbf{r}'), \tag{58b}$$

$$\hat{\mathbf{n}}_j \times \nabla \times \check{\mathbf{G}}_{jk}(\mathbf{r}_\perp,\zeta_j;\mathbf{r}') = \hat{\mathbf{n}}_j \times \nabla \times \check{\mathbf{G}}_{(j+1)k}(\mathbf{r}_\perp,\zeta_j;\mathbf{r}'). \tag{58c}$$

Another pair of equations similar to (58b) and (58c) corresponding to the $(j-1)$-th interface must be added to this list. Using these Green's functions and the radiation conditions the wave functions can be represented as

$$\mathbf{E}_j(\mathbf{r}) = \check{\mathbf{E}}_j(\mathbf{r}) + \sum_{k=1}^{N} \int_{\Omega_k} d\mathbf{r}' \check{\mathbf{G}}_{jk}(\mathbf{r},\mathbf{r}') v_k(\mathbf{r}') \mathbf{E}_k(\mathbf{r}') \qquad j = 0,1,\cdots,N+1 \tag{59}$$

where $\Omega_k = \{\mathbf{r}'_\perp; \zeta_k < z' < \zeta_{k-1}\}$. We first average (59) w.r.t. volumetric fluctuations to get

$$\langle \mathbf{E}_j(\mathbf{r})\rangle_v = \check{\mathbf{E}}_j(\mathbf{r}) + \sum_{k=1}^{N}\sum_{l=1}^{N} \int_{\Omega_k} d\mathbf{r}' \int_{\Omega_l} d\mathbf{r}'' \check{\mathbf{G}}_{jk}(\mathbf{r},\mathbf{r}') \langle \bar{\mathbf{G}}_{kl}(\mathbf{r}',\mathbf{r}'')\rangle_v \langle v_k(\mathbf{r}') v_l(\mathbf{r}'')\rangle \langle \mathbf{E}_l(\mathbf{r}'')\rangle_v, \tag{60}$$

where $\bar{\mathbf{G}}_{kl}$ is governed by the following system of equations:

$$\nabla \times \nabla \times \bar{\mathbf{G}}_{kl}(\mathbf{r},\mathbf{r}') - k_k^2 \bar{\mathbf{G}}_{kl}(\mathbf{r},\mathbf{r}') = \bar{\mathbf{I}}\delta_{kl}\delta(\mathbf{r}-\mathbf{r}') + v_k \bar{\mathbf{G}}_{kl}(\mathbf{r},\mathbf{r}'),$$

$$\hat{\mathbf{n}}_k \times \bar{\mathbf{G}}_{kl}(\mathbf{r}_\perp,\zeta_k;\mathbf{r}') = \hat{\mathbf{n}}_k \times \bar{\mathbf{G}}_{(k+1)l}(\mathbf{r}_\perp,\zeta_k;\mathbf{r}'),$$

$$\hat{\mathbf{n}}_k \times \nabla \times \bar{\mathbf{G}}_{kl}(\mathbf{r}_\perp,\zeta_k;\mathbf{r}') = \hat{\mathbf{n}}_k \times \nabla \times \bar{\mathbf{G}}_{(k+1)l}(\mathbf{r}_\perp,\zeta_k;\mathbf{r}').$$

Here, $\hat{\mathbf{n}}_k$ is the unit vector normal to the k-th interface. We also have a similar set of boundary conditions on the $(k-1)$-th interface. The subscript v is used to denote averaging with respect to volumetric fluctuations. Here, we have used a first-order approximation to the mass operator based on weak permittivity fluctuations. We have assumed that the fluctuations of the parameters of our problem are Gaussian and statistically homogeneous. Imposing the condition that the volumetric fluctuations in different regions are uncorrelated, by using (14) in (60) and employing $\nabla \times \nabla \times \bar{\bar{\mathbf{I}}} - k_j^2$ on (60) we get

$$\nabla \times \nabla \times \langle \mathbf{E}_j(\mathbf{r}) \rangle_v - k_j^2 \langle \mathbf{E}_j(\mathbf{r}) \rangle_v = \int_{\Omega_j} \langle \bar{\mathbf{G}}_{jj}(\mathbf{r}, \mathbf{r}') \rangle_v C_j(\mathbf{r} - \mathbf{r}') \langle \mathbf{E}_j(\mathbf{r}') \rangle_v d\mathbf{r}'. \tag{61}$$

Next we average (61) over the surface fluctuations,

$$\nabla \times \nabla \times \langle \mathbf{E}_j(\mathbf{r}) \rangle_{vs} - k_j^2 \langle \mathbf{E}_j(\mathbf{r}) \rangle_{vs} = \int_{\bar{\Omega}_j} \left\langle \langle \bar{\mathbf{G}}_{jj}(\mathbf{r}, \mathbf{r}') \rangle_v C_j(\mathbf{r} - \mathbf{r}') \langle \mathbf{E}_j(\mathbf{r}') \rangle_v \right\rangle_s d\mathbf{r}',$$

where the subscript s denotes averaging over surface fluctuations and $\bar{\Omega}_j = \{\mathbf{r}'_\perp; \hat{\mathbf{z}}_j < z' < \hat{\mathbf{z}}_{j-1}\}$. We approximate $\left\langle \langle \bar{\mathbf{G}}_{jj}(\mathbf{r}, \mathbf{r}') \rangle_v C_j(\mathbf{r} - \mathbf{r}') \langle \mathbf{E}_j(\mathbf{r}') \rangle_v \right\rangle_s$ as $\left\langle \bar{\mathbf{G}}_{jj}(\mathbf{r}, \mathbf{r}') \right\rangle_{vs} C_j(\mathbf{r} - \mathbf{r}') \left\langle \mathbf{E}_j(\mathbf{r}') \right\rangle_{vs}$ and obtain

$$\nabla \times \nabla \times \langle \mathbf{E}_j(\mathbf{r}) \rangle_{vs} - k_j^2 \langle \mathbf{E}_j(\mathbf{r}) \rangle_{vs} = \int_{\bar{\Omega}_j} \langle \bar{\mathbf{G}}_{jj}(\mathbf{r}, \mathbf{r}') \rangle_{vs} C_j(\mathbf{r} - \mathbf{r}') \langle \mathbf{E}_j(\mathbf{r}') \rangle_{vs} d\mathbf{r}'. \tag{62}$$

We call this the weak surface correlation approximation. We will later see that this is one additional approximation necessary to arrive at the RT system. As before, the result that $(\nabla \times \nabla \times \bar{\bar{\mathbf{I}}} - k_j^2) \langle \mathbf{E}_j(\mathbf{r}) \rangle_{vs} = 0$ for $j = 0, N+1$ implies that the coherent propagation constants in regions above and below the layer stack are unaffected by the fluctuations of the problem. However, they indeed get modified within the stack region. On writing (62) as $(\nabla \times \nabla \times \bar{\bar{\mathbf{I}}} - k_j^2 - \mathcal{L}) \langle \psi_j \rangle = 0$, where $\mathcal{L}$ denotes the integral operator $\int_{\bar{\Omega}_j} d\mathbf{r}' \left\langle \bar{\mathbf{G}}_{jj}(\mathbf{r}, \mathbf{r}') \right\rangle_{vs} C_j(\mathbf{r} - \mathbf{r}')$, we infer that $\chi_j \equiv \sqrt{k_j^2 + \mathcal{L}}$ represents the mean propagation constant in $\bar{\Omega}_j$. Observe that χ_j depends explicitly on the volumetric fluctuations in Region j and implicitly on the fluctuations of the stack, both volumetric and surface. This is in contrast to the RT approach where $\bar{\bar{\gamma}}_j$ depends exclusively on the volumetric fluctuations in Region j. Moreover, χ_j depends on the polarization if the fluctuations of the problem are anisotropic. Further, even if the volumetric fluctuations are isotropic χ_j will be polarization-dependent because of surface reflections. This is in contrast to the RT approach where $\bar{\bar{\gamma}}_j$ is polarization-dependent only when the volumetric fluctuations are anisotropic. A first-order solution to the above dispersion relation shows that in situations where the thickness of the layer is larger than the corresponding mean free path the influence of the boundaries on the mean propagation constants become negligible.

Since the problem is invariant under translations in azimuth the mean wave functions for our problem have the following form:

$$\langle \mathbf{E}_j^p(\mathbf{r}) \rangle_{vs} = \exp(i\mathbf{k}_{\perp i} \cdot \mathbf{r}) \left\{ A_j^p(\mathbf{k}_{\perp i}) \mathbf{p}_j^+ \exp[iq_j^p z] + B_j^p(\mathbf{k}_{\perp i}) \mathbf{p}_j^- \exp[-iq_j^p z] \right\} \qquad j = 1, 2, \cdots, N, \tag{63}$$

$$\langle \mathbf{E}_0^p(\mathbf{r}) \rangle_{vs} = \exp(i\mathbf{k}_{\perp i} \cdot \mathbf{r}) \left\{ \mathbf{p}_0^- \exp[-ik_{0zi} z] + R^p(\mathbf{k}_{\perp i}) \mathbf{p}_0^+ \exp[ik_{0zi} z] \right\}, \tag{64}$$

and

$$\langle \mathbf{E}_{N+1}^p(\mathbf{r}) \rangle_{vs} = \exp(i\mathbf{k}_{\perp i} \cdot \mathbf{r}) T^p(\mathbf{k}_{\perp i}) \mathbf{p}_{N+1}^- \exp[-ik_{(N+1)zi} z], \tag{65}$$

where the superscript p stands for the polarization, either horizontal or vertical. $\mathbf{p}$ is the unit vector representing polarization. R and T denote, respectively, the mean reflection and transmission coefficients of the stack. A_j and B_j denote, respectively, the mean coefficients of up-going and down-going waves in the j-th layer. Based on this we can formulate the waves averaged w.r.t. volumetric fluctuations as

$$\langle \mathbf{E}_j(\mathbf{r})\rangle_v^p = \frac{1}{4\pi^2}\int d\mathbf{k}_\perp \exp(i\mathbf{k}_\perp\cdot\mathbf{r})\Big\{A_j^{pq}(\mathbf{k}_\perp,\mathbf{k}_{\perp i})\mathbf{q}_j^+\exp[iq_jz] + B_j^{pq}(\mathbf{k}_\perp,\mathbf{k}_{\perp i})\mathbf{q}_j^-\exp[-iq_jz]\Big\} \quad j=1,2,\cdots,N, \tag{66}$$

$$\langle \mathbf{E}_0(\mathbf{r})\rangle_v^p = \exp(i\mathbf{k}_{\perp i}\cdot\mathbf{r})\exp[-ik_{0zi}z]\mathbf{p}_0^- + \frac{1}{4\pi^2}\int d\mathbf{k}_\perp \exp(i\mathbf{k}_\perp\cdot\mathbf{r})R^{pq}(\mathbf{k}_\perp,\mathbf{k}_{\perp i})\mathbf{q}_0^+\exp[ik_{0z}z], \tag{67}$$

and

$$\langle \mathbf{E}_{N+1}(\mathbf{r})\rangle_v^p = \frac{1}{4\pi^2}\int d\mathbf{k}_\perp \exp(i\mathbf{k}_\perp\cdot\mathbf{r})T^{pq}(\mathbf{k}_\perp,\mathbf{k}_{\perp i})\mathbf{q}_{N+1}^-\exp[-ik_{(N+1)z}z], \tag{68}$$

where A_j, B_j, R, and T are now integral operators representing scattering from rough interfaces. The boundary conditions associated with the above equations at the j-th interface are

$$\hat{\mathbf{n}}_j \times \langle \mathbf{E}_j(\mathbf{r}_\perp,\zeta_j)\rangle_v = \hat{\mathbf{n}}_j \times \langle \mathbf{E}_{j+1}(\mathbf{r}_\perp,\zeta_j)\rangle_v \qquad j=1,2,\cdots,N \tag{69a}$$

and

$$\hat{\mathbf{n}}_j \times \nabla \times \langle \mathbf{E}_j(\mathbf{r}_\perp,\zeta_j)\rangle_v = \hat{\mathbf{n}}_j \times \nabla \times \langle \mathbf{E}_{j+1}(\mathbf{r}_\perp,\zeta_j)\rangle_v \qquad j=1,2,\cdots,N. \tag{69b}$$

The above system may be solved either numerically or by any of the asymptotic methods[1] available in rough surface scattering theory (Beckmann & Spizzichino, 1987; Bass & Fuks, 1979; Voronovich, 1999) to evaluate the mean coefficients that appear in (63)-(65).
We proceed now to the analysis of the second moments, by starting with (59) represented in symbolic form as

$$\mathbf{E}_j = \check{\mathbf{E}}_j + \sum_{k=1}^{N} \check{\mathbf{G}}_{jk} v_k \mathbf{E}_k. \tag{70}$$

We take the tensor product of this equation with its complex conjugate and average w.r.t. volumetric fluctuations and obtain

$$\left\langle \mathbf{E}_j \otimes \mathbf{E}_j^* \right\rangle_v = \langle \mathbf{E}_j\rangle_v \otimes \langle \mathbf{E}_j^*\rangle_v + \sum_{k=1}^{N}\sum_{k'=1}^{N}\sum_{l=1}^{N}\sum_{l'=1}^{N} \langle \bar{\mathbf{G}}_{jk}\rangle_v \otimes \langle \bar{\mathbf{G}}_{jk'}^*\rangle_v \hat{\mathbf{K}}_{kk'll'} \langle \mathbf{E}_l \otimes \mathbf{E}_{l'}^*\rangle_v, \tag{71}$$

where $\hat{\mathbf{K}}$ is the intensity operator of the volumetric fluctuations. Employing the weak fluctuation approximation we approximate $\hat{\mathbf{K}}$ by its leading term

$$\hat{\mathbf{K}}_{kk'll'} \simeq \langle v_k \otimes v_k^*\rangle \delta_{kk'll'} \bar{\mathbf{I}}. \tag{72}$$

Next, we average (71) w.r.t. the surface fluctuations and employ the weak surface correlation approximation, as before, to get

$$\left\langle \mathbf{E}_j \otimes \mathbf{E}_j^* \right\rangle_{vs} = \left\langle \langle \mathbf{E}_j\rangle_v \otimes \langle \mathbf{E}_j^*\rangle_v \right\rangle_s + \sum_{k=1}^{N} \left\langle \left\langle \bar{\mathbf{G}}_{jk}\right\rangle_v \otimes \left\langle \bar{\mathbf{G}}_{jk}^*\right\rangle_v \right\rangle_s \langle v_k \otimes v_k^*\rangle \langle \mathbf{E}_k \otimes \mathbf{E}_k^*\rangle_{vs}. \tag{73}$$

[1]It is necessary, however, to meet the weak surface correlation approximation employed earlier

The above is an equation for the second moment of the wave function $\mathbf{E}$, which can be decomposed into a coherent part $\bar{\mathbf{E}}$ and a diffuse part $\tilde{\mathbf{E}}$. Therefore,

$$\langle \mathbf{E} \otimes \mathbf{E}^* \rangle = \langle \mathbf{E} \rangle \otimes \langle \mathbf{E}^* \rangle + \langle \tilde{\mathbf{E}} \otimes \tilde{\mathbf{E}}^* \rangle. \tag{74}$$

The coherent part is indeed known for our problem. Our primary interest is in the diffuse part. Therefore, we write (73) in terms of diffuse fields:

$$\left\langle \tilde{\mathbf{E}}_j \otimes \tilde{\mathbf{E}}_j^* \right\rangle = \left\langle \widetilde{\langle \mathbf{E}_j \rangle}_v \otimes \widetilde{\langle \mathbf{E}_j^* \rangle}_v \right\rangle_s + \sum_{k=1}^{N} \left\langle \left\langle \bar{\mathbf{G}}_{jk} \right\rangle_v \otimes \left\langle \bar{\mathbf{G}}_{jk}^* \right\rangle_v \right\rangle_s \langle v_k \otimes v_k^* \rangle \left\langle \mathbf{E}_k \otimes \mathbf{E}_k^* \right\rangle_{vs}, \tag{75}$$

where $\widetilde{\langle \mathbf{E}_j \rangle}_v$ is the fluctuating part of $\langle \mathbf{E}_j \rangle_v$. Let us now write (75) in more detail as:

$$\left\langle \tilde{\mathbf{E}}_j(\mathbf{r}) \otimes \tilde{\mathbf{E}}_j^*(\mathbf{r}') \right\rangle = \left\langle \widetilde{\langle \mathbf{E}_j(\mathbf{r}) \rangle}_v \otimes \widetilde{\langle \mathbf{E}_j^*(\mathbf{r}') \rangle}_v \right\rangle_s + \sum_{k=1}^{N} \int_{\bar{\Omega}_k} d\mathbf{r}_1 \int_{\bar{\Omega}_k} d\mathbf{r}_1'$$

$$\left\langle \left\langle \bar{\mathbf{G}}_{jk}(\mathbf{r},\mathbf{r}_1) \right\rangle_v \otimes \left\langle \bar{\mathbf{G}}_{jk}^*(\mathbf{r}',\mathbf{r}_1') \right\rangle_v \right\rangle_s |k_k|^4 C_k(\mathbf{r}_1 - \mathbf{r}_1') \left\langle \mathbf{E}_k(\mathbf{r}_1) \otimes \mathbf{E}_k^*(\mathbf{r}_1') \right\rangle_{vs}. \tag{76}$$

To obtain the RT equations, we introduce the Wigner transforms of waves and Green's functions as

$$\mathcal{E}_m\left(\frac{\mathbf{r}+\mathbf{r}'}{2}, \mathbf{k}\right) = \int \langle \mathbf{E}_m(\mathbf{r}) \otimes \mathbf{E}_m^*(\mathbf{r}') \rangle e^{-i\mathbf{k}\cdot(\mathbf{r}-\mathbf{r}')} d(\mathbf{r}-\mathbf{r}'), \tag{77}$$

$$\mathcal{G}_{mn}\left(\frac{\mathbf{r}+\mathbf{r}'}{2}, \mathbf{k} \middle| \frac{\mathbf{r}_1+\mathbf{r}_1'}{2}, \mathbf{l}\right) = \int d(\mathbf{r}-\mathbf{r}') \int d(\mathbf{r}_1 - \mathbf{r}_1')$$

$$e^{-i\mathbf{k}\cdot(\mathbf{r}-\mathbf{r}')} e^{i\mathbf{l}\cdot(\mathbf{r}_1-\mathbf{r}_1')} \left\langle \langle \bar{\mathbf{G}}_{mn}(\mathbf{r},\mathbf{r}_1) \rangle_v \otimes \langle \bar{\mathbf{G}}_{mn}^*(\mathbf{r}',\mathbf{r}_1') \rangle_v \right\rangle_s. \tag{78}$$

In terms of these transforms (76) becomes

$$\tilde{\mathcal{E}}_m(\mathbf{r},\mathbf{k}) = \tilde{\mathcal{E}}_m^s(\mathbf{r},\mathbf{k}) + \frac{1}{(2\pi)^6} \sum_{n=1}^{N} |k_n|^4 \int_{\bar{\Omega}_n} d\mathbf{r}' \int \int \mathcal{G}_{mn}(\mathbf{r},\mathbf{k}|\mathbf{r}',\boldsymbol{\alpha}) \Phi_n(\boldsymbol{\alpha}-\boldsymbol{\beta}) \mathcal{E}_n(\mathbf{r}',\beta) d\beta d\alpha, \tag{79}$$

where Φ_n is the spectral density of the volumetric fluctuations in the n-th layer. We have used the superscript s in the first term to indicate that this is due to surface scattering as defined by the first term in (76).

As before translational invariance in azimuth implies the following:

$$\mathcal{E}_m(\mathbf{r},\mathbf{k}) = \mathcal{E}_m(z,\mathbf{k}), \tag{80a}$$

$$\mathcal{G}_{mn}(\mathbf{r},\mathbf{k}|\mathbf{r}',\mathbf{l}) = \mathcal{G}_{mn}(z,\mathbf{k}|z',\mathbf{l};\mathbf{r}_\perp - \mathbf{r}_\perp'). \tag{80b}$$

Using these relations in (79) we have

$$\tilde{\mathcal{E}}_m(z,\mathbf{k}) = \tilde{\mathcal{E}}_m^s(z,\mathbf{k}) + \frac{1}{(2\pi)^6} \sum_{n=1}^{N} |k_n|^4 \int_{z_n}^{z_{n-1}} dz' \int d\boldsymbol{\alpha} \int d\boldsymbol{\beta}\, \mathcal{G}_{mn}(z,\mathbf{k}|z',\boldsymbol{\alpha};0) \Phi_n(\boldsymbol{\alpha}-\boldsymbol{\beta}) \mathcal{E}_n(z',\boldsymbol{\beta}), \tag{81}$$

where $\mathcal{G}_{mn}(z,\mathbf{k}|z',\boldsymbol{\alpha};0)$ is the Fourier transform of $\mathcal{G}_{mn}(z,\mathbf{k}|z',\boldsymbol{\alpha};\mathbf{r}_\perp - \mathbf{r}_\perp')$ w.r.t. $\mathbf{r}_\perp - \mathbf{r}_\perp'$ evaluated at the origin of the spectral space. To proceed further, we decompose $\langle G_{mn} \rangle_v$ into its components,

$$\langle \bar{\mathbf{G}}_{mn} \rangle_v = \delta_{mn} \bar{\mathbf{G}}_m^o + \bar{\mathbf{G}}_{mn}^{uu} + \bar{\mathbf{G}}_{mn}^{ud} + \bar{\mathbf{G}}_{mn}^{du} + \bar{\mathbf{G}}_{mn}^{dd}, \tag{82}$$

where the first term is the singular part of the Green's function. The superscripts u and d indicate up- and down-going elements of the waves. The other components are due to reflections from boundaries. These are formally constructed using the concept of surface scattering operators as follows (Voronovich, 1999),

$$\left\langle G_{mn}^{ab}(\mathbf{r},\mathbf{r}')\right\rangle_v^{\mu\nu} = \frac{1}{(2\pi)^4}\int\int \{S_{mn}^{ab}(\mathbf{k}_\perp,\mathbf{k}'_\perp)\}^{\mu\nu} e^{i\mathbf{k}_\perp\cdot\mathbf{r}+iaq_m^\mu(\mathbf{k}_\perp)z} e^{-i\mathbf{k}'_\perp\cdot\mathbf{r}'-ibq_n^\nu(\mathbf{k}'_\perp)z'} d\mathbf{k}_\perp d\mathbf{k}'_\perp, \tag{83}$$

where $\bar{\mathbf{S}}_{mn}^{ab}$ is the surface scattering operator. The superscript and superscript notations have the same meaning as in Section 2.

We recall that $\mathcal{G}_{mn}$ is the Wigner transform of $\left\langle \langle\bar{\mathbf{G}}_{mn}\rangle_v \otimes \langle\bar{\mathbf{G}}_{mn}^*\rangle_v \right\rangle_s$. As mentioned before, it is only in the quasi-uniform limit does the Wigner transform of the coherence function lead to the specific intensity of the RT equation. For our layer geometry, the Green's function is nonuniform. However, each of its components given in (82) is quasi-uniform. When we use (82) to perform the Wigner transform we ignore all cross terms. In other words, we make the following approximation,

$$\mathcal{G}_{mn} \simeq \delta_{mn}\mathcal{G}_m^o + \mathcal{G}_{mn}^{uu} + \mathcal{G}_{mn}^{ud} + \mathcal{G}_{mn}^{du} + \mathcal{G}_{mn}^{dd},$$

where $\mathcal{G}_{mn}^{ab}$ is the Wigner transform of $\left\langle \left\langle\bar{\mathbf{G}}_{mn}^{ab}\right\rangle_v \otimes \left\langle\bar{\mathbf{G}}_{mn}^{ab*}\right\rangle_v \right\rangle_s$. Most of the cross terms are nonuniform and may be neglected under the quasi-uniform field assumption. A few cross terms turn out to be quasi-uniform and their inclusion lead to phase matrices that are different from those of the RT system. It turns out that such additional coherence terms become negligible when the layer thickness is of the same order or greater than the mean free path of the corresponding layer. It is under these conditions, the approximate expression for $\mathcal{G}_{mn}$ given above is good.

With the introduction of this representation for $\mathcal{G}_{mn}$ in (81), we can trace up- and down-going waves to obtain the following equations for the coherence function:

$$\tilde{\mathcal{E}}_m^u(z,\mathbf{k}) = \tilde{\mathcal{E}}_m^{su}(z,\mathbf{k}) + \frac{1}{(2\pi)^6}|k_m|^4\int_{z_m}^{z} dz'\int d\boldsymbol{\alpha}\int d\boldsymbol{\beta}\; \mathcal{G}_m^{>}(z,\mathbf{k}|z',\boldsymbol{\alpha};0)\Phi_m(\boldsymbol{\alpha}-\boldsymbol{\beta})\mathcal{E}_m(z',\boldsymbol{\beta})$$
$$+\frac{1}{(2\pi)^6}\sum_{n=1}^{N}|k_n|^4\int_{z_n}^{z_{n-1}} dz'\int d\boldsymbol{\alpha}\int d\boldsymbol{\beta}\; \mathcal{G}_{mn}^{ua}(z,\mathbf{k}|z',\boldsymbol{\alpha};0)\Phi_n(\boldsymbol{\alpha}-\boldsymbol{\beta})\mathcal{E}_n^a(z',\boldsymbol{\beta}) \tag{84a}$$

$$\tilde{\mathcal{E}}_m^d(z,\mathbf{k}) = \tilde{\mathcal{E}}_m^{sd}(z,\mathbf{k}) + \frac{1}{(2\pi)^6}|k_m|^4\int_{z}^{z_{m-1}} dz'\int d\boldsymbol{\alpha}\int d\boldsymbol{\beta}\; \mathcal{G}_m^{<}(z,\mathbf{k}|z',\boldsymbol{\alpha};0)\Phi_m(\boldsymbol{\alpha}-\boldsymbol{\beta})\mathcal{E}_m(z',\boldsymbol{\beta})$$
$$+\frac{1}{(2\pi)^6}\sum_{n=1}^{N}|k_n|^4\int_{z_n}^{z_{n-1}} dz'\int d\boldsymbol{\alpha}\int d\boldsymbol{\beta}\; \mathcal{G}_{mn}^{da}(z,\mathbf{k}|z',\boldsymbol{\alpha};0)\Phi_n(\boldsymbol{\alpha}-\boldsymbol{\beta})\mathcal{E}_n^a(z',\boldsymbol{\beta}) \tag{84b}$$

Note that summation over $a = \{u,d\}$ is implied in the above equations. The first term in these equations, $\tilde{\mathcal{E}}^{sa}$, represents the contribution due exclusively to surface scattering, and has the following form:

$$\{\tilde{\mathcal{E}}_m^{sa}(z,\mathbf{k})\}^{\mu\nu} = 2\pi\delta\left\{k_z - \frac{1}{2}a\left[q_m^\mu(\mathbf{k}_\perp) + q_m^{\nu*}(\mathbf{k}_\perp)\right]\right\}\times$$
$$\times e^{ia\left[q_m^\mu - q_m^{\nu*}\right]z}\left\langle \{\tilde{\Sigma}_m^a\}^{\mu\mu'}\{\tilde{\Sigma}_m^{a*}\}^{\nu\nu'}(\mathbf{k}_\perp,\mathbf{k}_{\perp i})\right\rangle_s E_{\mu' i}E_{\nu' i}^*, \tag{85}$$

where Σ_m^a is the amplitude of the up-going wave in the m-th layer after volumetric averaging is performed. This means that it is a random function of surface fluctuations. When we substitute (85) and the expressions for $\mathcal{G}_{mn}$ in (84) we find that

$$\left\{\tilde{\mathcal{E}}_m^a(z,\mathbf{k})\right\}_{\mu\nu} = 2\pi\delta\left\{k_z - \frac{1}{2}a[q_m^\mu(\mathbf{k}_\perp) + q_m^{\nu*}(\mathbf{k}_\perp)]\right\} e^{ia\left[q_m^\mu - q_m^{\nu*}\right]z}\left\{\tilde{\mathcal{E}}_m^a(z,\mathbf{k}_\perp)\right\}_{\mu\nu}. \tag{86}$$

On substituting this in (84) and differentiating w.r.t. z we obtain the following transport equations:

$$\left\{\frac{d}{dz} - i\left[q_\mu(\mathbf{k}_\perp) - q_\nu^*(\mathbf{k}_\perp)\right]\right\}\tilde{\mathcal{E}}_{\mu\nu}^u(z,\mathbf{k}_\perp) = \tilde{\tilde{\mathcal{E}}}_{\mu\nu}^u(z,\mathbf{k}_\perp) + \frac{|k_m|^4}{(2\pi)^2}\int d\alpha_\perp S_\mu^> S_\nu^{>*}$$
$$\Phi_m\left\{\mathbf{k}_\perp - \alpha_\perp; \frac{1}{2}\left[q_\mu(\mathbf{k}_\perp) + q_\nu^*(\mathbf{k}_\perp)\right] - \frac{1}{2}a\left[q_{\mu'}(\alpha_\perp) + q_{\nu'}^*(\alpha_\perp)\right]\right\}(\boldsymbol{\mu}\cdot\boldsymbol{\mu}')(\boldsymbol{\nu}\cdot\boldsymbol{\nu}')\tilde{\mathcal{E}}_{\mu'\nu'}^a(z,\alpha_\perp), \tag{87a}$$

$$\left\{-\frac{d}{dz} - i\left[q_\mu(\mathbf{k}_\perp) - q_\nu^*(\mathbf{k}_\perp)\right]\right\}\tilde{\mathcal{E}}_{\mu\nu}^d(z,\mathbf{k}_\perp) = \tilde{\tilde{\mathcal{E}}}_{\mu\nu}^d(z,\mathbf{k}_\perp) + \frac{|k_m|^4}{(2\pi)^2}\int d\alpha_\perp S_\mu^< S_\nu^{<*}$$
$$\Phi_m\left\{\mathbf{k}_\perp - \alpha_\perp; -\frac{1}{2}\left[q_\mu(\mathbf{k}_\perp) + q_\nu^*(\mathbf{k}_\perp)\right] - \frac{1}{2}a\left[q_{\mu'}(\alpha_\perp) + q_{\nu'}^*(\alpha_\perp)\right]\right\}(\boldsymbol{\mu}\cdot\boldsymbol{\mu}')(\boldsymbol{\nu}\cdot\boldsymbol{\nu}')\tilde{\mathcal{E}}_{\mu'\nu'}^a(z,\alpha_\perp), \tag{87b}$$

where $\tilde{\tilde{\mathcal{E}}}_{\mu\nu}^a$ represents scattering due to the coherent part of $\mathcal{E}$, whereas the integral terms in (87) represent scattering due to the diffuse part of $\mathcal{E}$. We may also regard $\tilde{\tilde{\mathcal{E}}}_{\mu\nu}^a$ as the source to our transport equations; it is given as

$$\tilde{\tilde{\mathcal{E}}}_{\mu\nu}^a = |k_m|^4\Phi_m\left\{\mathbf{k}_\perp - \mathbf{k}_{\perp i}; \frac{1}{2}a\left[q_\mu(\mathbf{k}_\perp) + q_\nu^*(\mathbf{k}_\perp)\right] - \frac{1}{2}b\left[q_\mu(\mathbf{k}_{\perp i}) + q_\nu^*(\mathbf{k}_{\perp i})\right]\right\}\times$$
$$\times\left(S_\mu^a S_\mu^{a*}\right)(\boldsymbol{\mu}\cdot\boldsymbol{\mu}_i)(\boldsymbol{\nu}\cdot\boldsymbol{\nu}_i)\left(\langle S_{m0}^b\rangle_{\mu i}E_{\mu i}\right)\left(\langle S_{m0}^b\rangle_{\nu i}E_{\nu i}\right)^*, \tag{88}$$

where summation over b is implied. Note that $\tilde{\mathcal{E}}^a$ in (87) includes both $\tilde{\mathcal{E}}^u$ and $\tilde{\mathcal{E}}^d$ (corresponding to up- and down-going waves). When the superscripts $\{a,b\}$ correspond to u, the value of $\{a,b\}$ in the argument of Φ_m takes the value $+1$; on the other hand, when the superscripts $\{a,b\}$ correspond to d the value of a in the argument of Φ_m takes the value -1. Since all quantities in (87) and (88) correspond to the same layer m we have dropped the subscript m to avoid cumbersome notations. Summation over μ' and ν' is implicit in (87). Similarly, summation over μ_i and ν_i is implicit in (88). $\mathcal{S}_{m0}^b$ is the scattering amplitude of waves with direction b in m-th layer due to wave incident in Region 0. To obtain appropriate boundary conditions we have to go back to the integral equation representations for $\tilde{\mathcal{E}}_{\mu\nu}^u$ and $\tilde{\mathcal{E}}_{\mu\nu}^d$ and observe their behaviour at the interfaces and try to find a relation between them.
After some manipulations we arrived at the following boundary conditions. At the $(m-1)$-th interface we have

$$\tilde{\mathcal{E}}_m^d(z_{m-1},\mathbf{k}_\perp) = \int d\mathbf{k}'_\perp\left\langle\dddot{\mathcal{R}}_{m-1,m}(\mathbf{k}_\perp,\mathbf{k}'_\perp)\right\rangle\tilde{\mathcal{E}}_m^u(z_{m-1},\mathbf{k}'_\perp), \tag{89a}$$

with $\dddot{\mathcal{R}} = \dddot{\mathbf{R}}\otimes\dddot{\mathbf{R}}^*$ where $\dddot{\mathbf{R}}_{m-1,m}$ is the stack reflection matrix (not the local reflection matrix) for a wave incident from below on the $(m-1)$-th interface. Similarly

$$\tilde{\mathcal{E}}_m^u(z_{m-1},\mathbf{k}_\perp) = \int d\mathbf{k}'_\perp\left\langle\dddot{\mathcal{R}}_{m+1,m}(\mathbf{k}_\perp,\mathbf{k}'_\perp)\right\rangle\tilde{\mathcal{E}}_m^d(z_{m-1},\mathbf{k}'_\perp), \tag{89b}$$

where $\dddot{\mathcal{R}}_{m+1,m}$ is the tensor product of the stack reflection matrix for a wave incident from above on the m-th interface.
We were able to obtain the boundary conditions only after imposing certain approximations as given below. Consider the following identity:

$$\bar{\mathbf{S}}_{mm}^{du} = \bar{\mathbf{F}}_m \,\dddot{\bar{\mathbf{R}}}_{m-1,m} \left\{\bar{\mathbf{S}}_m^{>} + \bar{\mathbf{S}}_{mm}^{uu}\right\} \bar{\mathbf{F}}_m \tag{90}$$

where $\bar{\mathbf{F}}_m = \mathrm{diag}\left\{e^{iq_h d_m}, e^{iq_v d_m}\right\}$. Notice that this is an operator relation where all elements are operators. Taking the tensor product of (90) with its complex conjugate we have

$$\bar{\mathbf{S}}_{mm}^{du} \otimes \bar{\mathbf{S}}_{mm}^{du*} = (\bar{\mathbf{F}}_m \otimes \bar{\mathbf{F}}_m^*) \left(\dddot{\bar{\mathbf{R}}}_{m-1,m} \otimes \dddot{\bar{\mathbf{R}}}_{m-1,m}^{*}\right) \left(\left\{\bar{\mathbf{S}}_m^{>} + \bar{\mathbf{S}}_{mm}^{uu}\right\} \otimes \left\{\bar{\mathbf{S}}_m^{>} + \bar{\mathbf{S}}_{mm}^{uu}\right\}^*\right) (\bar{\mathbf{F}}_m \otimes \bar{\mathbf{F}}_m^*). \tag{91}$$

Next we average (91) w.r.t. surface fluctuations and get

$$\left\langle \bar{\mathbf{S}}_{mm}^{du} \otimes \bar{\mathbf{S}}_{mm}^{du*} \right\rangle \simeq (\bar{\mathbf{F}}_m \otimes \bar{\mathbf{F}}_m^*) \left\langle \dddot{\bar{\mathbf{R}}}_{m-1,m} \otimes \dddot{\bar{\mathbf{R}}}_{m-1,m}^{*} \right\rangle \left\langle \left\{\bar{\mathbf{S}}_m^{>} + \bar{\mathbf{S}}_{mm}^{uu}\right\} \otimes \left\{\bar{\mathbf{S}}_m^{>} + \bar{\mathbf{S}}_{mm}^{uu}\right\}^* \right\rangle (\bar{\mathbf{F}}_m \otimes \bar{\mathbf{F}}_m^*) \tag{92a}$$

where the two tensor products in the middle are assumed to be weakly correlated. A further approximation that we impose is given as follows

$$\left\langle \left\{\bar{\mathbf{S}}_m^{>} + \bar{\mathbf{S}}_{mm}^{uu}\right\} \otimes \left\{\bar{\mathbf{S}}_m^{>} + \bar{\mathbf{S}}_{mm}^{uu}\right\}^* \right\rangle \simeq \bar{\mathbf{S}}_m^{>} \otimes \bar{\mathbf{S}}_m^{>*} + \left\langle \bar{\mathbf{S}}_{mm}^{uu} \otimes \bar{\mathbf{S}}_{mm}^{uu*} \right\rangle. \tag{92b}$$

This is similar to the approximation we used while computing the Wigner transforms of the Green's functions. These are the kinds of approximations required to arrive at our boundary conditions.

3.4 Transition to radiative transfer

The procedure for transition to radiative transfer is identical to the planar interface problem and hence we do not repeat it here. We find that the conditions necessary to connect the RTT with statistical wave theory are:
(a) layer thickness is greater than the mean free path of the corresponding medium,
(b) quasi-homogeneous field approximation.
By following the same procedure as in Section 2 we obtain the transport theoretic boundary conditions from (89).

3.5 Remarks

Now that we have made the transition from statistical wave theory to radiative transfer theory, we itemize the assumptionsimplicitly involved in the RT approach.
1. Quasi-stationary field approximation.
2. Weak fluctuations.
3. Statistical homogeneity of fluctuations.
These are the three well-known conditions necessary for the unbounded random media problem. However, if the medium is bounded with rough interfaces we need to impose additional conditions. We found that the extinction coefficients calculated in the statistical wave approach and the RT approach are different and only after applying further approximations can they be made to agree with each other. The following additional

conditions are required for our bounded random medium problem:
4. Layer thickness must be of the same order or greater than the corresponding mean free path.
5. Weak surface correlation approximation.
6. All fluctuations of the problem are statistically independent.

4. Conclusion

To summarize, we have enquired into the assumptions involved in adopting the radiative transfer approach to scattering from layered random media with rough interfaces. To facilitate this enquiry we adopted a statistical wave approach to this problem and derived the governing equations for the first and second moments of the wave fields. We employed Wigner transforms and transitioned to the system corresponding to that of radiative transfer approach. In this process we found that there are more conditions implicitly involved in the RT approach to this problem than it is widely believed to be sufficient. With the recent development of fast and efficient algorithms for scattering computations and the enormous increase in computer resources it is now feasible to take an entirely numerical approach to this problem without imposing any approximations. In spite of such developments, to keep the size of the problem manageable only special cases have been studied thus far (Giovannini et al., 1998; Peloci & Coccioli, 1997; Pak et al., 1993; Sarabandi et al., 1996). Hence it is very much of relevance, interest, and convenience to apply the RT approach to these problems. However, one should keep in mind the assumptions involved in such an approach. Otherwise interpretations of results based on RT theory can be misleading.

In this work we have modelled the random media as random continua. Another approach to this problem is the discrete random medium model (Foldy, 1945; Lax, 1951; Twersky, 1964; Ishimaru, 1997; Tsang et al., 1985; Mishchenko et al., 2006). Recently Mishchenko (Mishchenko, 2002) (hereafter referred to as MTL for brevity) derived the vector radiative transfer equation (VRTE) for a bounded discrete random medium using a rigorous microphysical approach. This enabled them to identify the following assumptions embedded in the VRTE.

1. Scattering medium is illuminated by a plane wave.
2. Each particle is located in the far-field zone of all other particles and the observation point is also located in the far-field zones of all the particles forming the scattering medium.
3. Neglect all scattering paths going through a particle two or more times (Twersky approximation).
4. Assume that the scattering system is ergodic and averaging over time can be replaced by averaging over particle positions and states.
5. Assume that (i) the position and state of each particle are statistically independent of each other and of those of all other particles and (ii) the spatial distribution of the particles throughout the medium is random and statistically uniform.
6. Assume that the scattering medium is convex.
7. Assume that the number of particles N forming the scattering medium is very large.
8. Ignore all the diagrams with crossing connections in the diagrammatic expansion of the coherency dyadic.

It is apparent that there are distinct differences in the analyses for scattering from discrete and continuous random media. Hence it is not possible to make a one-to-one correspondence between the conditions of MTL and those in this work. Below is an attempt to make a connection between the two by considering each condition derived by MTL and relating it to ours. We will denote the condition numbers derived by MTL as MTL # and those obtained in this work as SM #.

MTL 1:- We also have a plane electromagnetic wave illuminating our system, although as pointed out by MTL it can be a quasi-plane wave.

MTL 2:- We also have implicitly employed the far-field approximation. It is embedded in SM 1.

MTL 3:- This is embedded in SM 2. Although not explicitly stated, the scattering processes as mentioned in MTL 3 are avoided.

MTL 4:- In this work we have restricted our attention to the time-independent problem and hence did not encounter the issue of ergodicity.

MTL 5:- This condition is embedded in SM 3.

MTL 6:- In our problem we have distinct scattering boundaries and the character of the waves exiting or entering them are explicitly contained in the boundary conditions. Hence convexity of the scattering medium is not a necessary condition for us.

MTL 7:- This condition is embedded in SM 4.

MTL 8:- This condition is embedded in SM 2. Under weak fluctuation approximation we only take into consideration the leading term of the intensity operator.

Since the problem that we considered in this work involved scattering boundaries we have some additional conditions beyond those of MTL. Although the main conclusions obtained are the same for the problem with random continuum and discrete random medium there are some peculiarities with the discrete random medium case and hence there are some differences in the assumptions implied in the RT approach.

There are a few more remarks that we would like to make before closing.

(a) In RT theory the medium is assumed to be sparse and hence the "refraction effects" of the fluctuations are ignored. Thus in the boundary conditions we should use the background medium parameters rather than the effective medium parameters as derived in our statistical wave theory.

(b) To arrive at (46) from (45) we have ignored the contribution of evanescent modes.

(c) The condition about statistical homogeneity of fluctuations may be relaxed by assuming it to be statistically quasi-homogeneous and we still can arrive at our results without much difficulty.

(d) The assumption regarding the underlying statistics to be Gaussian is not only a convenience but also a reasonably good approximation in many applications. However, there are indeed certain situations where the statistics are not Gaussian. Similar analysis for such more general statistics are more complex and involved.

5. Acknowledgment

The author thanks the US Air Force of Scientific Research for support.

6. References

Ambartsumian, V. A. (1943). Diffuse scattering of light by a turbid medium, *Dokl. Akad. Nauk SSSR* 38: 85–91.

Amra, C. (1994). Light scattering from multilayer optics I. Tools of investigation, *J. Opt. Soc. Am.* A 11: 197–210.

Arridge, S. R. & Hebden, J. C. (1997). Optical imaging in medicine: II. Modelling and reconstruction, *Phys. Med. Biol.* 42: 841–853.

Barabanenkov, Y. N., Vinogradov, A. G., Kravtsov, Y. A. & Tatarskii, V. I. (1972). Application of the theory of multiple scattering of waves to the derivation of the radiation transfer equation for a statistically inhomogeneous medium, *Izv. VUZ Radiofiz.* 15: 1852–1860.

Bass, F. G. & Fuks, I. M. (1979). *Wave Scattering from Statistically Rough Surfaces*, Pergamon, Oxford.

Beckmann, P. & Spizzichino, A. (1987). *The Scattering of Electromagnetic Waves from Rough Surfaces*, Artech House, Norwood, Massachusetts.

Berk, A., Bernstein, L. S., G. P. Anderson, P. K. A., Robertson, D. C., Chetwynd, J. H. & Adler-Golden, S. H. (1998). MODTRAN cloud and multiple scattering upgrades with application to AVIRIS, *Remote Sensing of Environment* 65: 367–375.

Brown, G. S. (1988). A theoretical study of the effects of vegetation on terrain scattering, *Technical Report RADC-TR-89-64*, Rome Air Development Center, Rome, NY.

Caron, J., Andraud, C. & Lafait, J. (2004). Radiative transfer calculations in multilayer system with smooth or rough interfaces, *J. Mod. Opt.* 51: 575–595.

Chandrasekar, S. (1960). *Radiative Transfer*, Dover, New York.

Chauhan, N. S., Lang, R. H. & Ranson, K. J. (1991). Radar modelling of boreal forest, *IEEE Trans. Geosci. Rem. Sens.* 29: 627–638.

Chew, W. (1995). *Waves and Fields in Inhomogeneous Media*, IEEE Press, Piscataway, New Jersey.

Clough, S. A., Shephard, M. W., Milawer, E. J., Delamere, J. S., Iacono, M. J., Cady-Pereira, K., Boukabara, S. & Brown, P. D. (2005). Atmospheric radiative transfer modeling: A summary of the AER codes, *J. Quant. Spectrosc. Radiat. Transfer* 91: 233–244.

Daniels, D. J. (2004). *Ground Penetrating Radar*, 2nd edn, IEE, Herts, United Kingdom.

de Hulst, H. C. V. (1980). *Multiple Light Scattering: Tables, Formulas, and Applications*, Academic Press, London, United Kingdom.

Elachi, C. & van Zyl, J. J. (2006). *Introduction to the Physics and Techniques of Remote Sensing*, 2nd edn, John Wiley and Sons, Hobokens, New Jersey.

Elson, J. M. (1995). Multilayer-coated optics: Guided-wave coupling and scattering by means of interface roughness, *J. Opt. Soc. Am.* A 12: 729–742.

Elson, J. M. (1997). Characteristics of far field scattering by means of surface roughness and variations in subsurface permittivity, *Waves Random Media* 7: 303–317.

Foldy, L. L. (1945). The multiple scattering of waves, *Phys. Rev.* 67: 107–117.

Friberg, A. T. (1986). Energy transport in optical systems with partially coherent light, *Applied Optics* 25: 4547–4556.

Fung, A. K. & Chen, M. (1981). Scattering from a Rayleigh layer with an irregular interface, *Radio Sci.* 16: 1337–1347.

Giovannini, H., Saillard, M. & Sentenac, A. (1998). Numerical study of scattering from rough inhomogeneous films, *J. Opt. Soc. Am. A* 15: 1182–1191.

Hasekamp, O. P., Landgraf, J. & van Oss, R. (2002). The need of polarization modelling for ozone profile retrieval from backscattered sunlight, *J. Geophys. Res.* 107: 4692.

Ishimaru, A. (1997). *Wave Propagation and Scattering in Random Media*, IEEE Press, New York.

Karam, M. A. & Fung, A. K. (1982). Propagation and scattering in multi-layer random media with rough interfaces, *Electromagnetics* 2: 239–256.

Kattawar, G. W. & Adams, C. N. (1990). Errors in radiance calculations induced by using scalar rather than Stokes vector theory in realistic atmosphere-ocean system, *in* R. W. Spinrad (ed.), *Ocean Optics X*, Vol. 1302 of *Proc. of SPIE*, pp. 2–12.

Kuo, C. H. & Moghaddam, M. (2007). Electromagnetic scattering from multilayer rough surfaces with arbitrary dielectric profiles for remote sensing of subsurface soil moisture, *IEEE Trans. Geosci. Rem. Sens.* 45: 349–366.

Lam, C. M. & Ishimaru, A. (1993). Muller matrix representation for a slab of random medium with discrete particles and random rough surfaces, *Waves Random Media* 3: 111–125.

Lax, M. (1951). Multiple scattering of waves, *Rev. Mod. Phys.* 23: 287–310.

Lee, J. K. & Kong, J. A. (1985). Active microwave remote sensing of an anisotropic layered random medium, *IEEE Trans. Geosci. Rem. Sens.* 23: 910–923.

Lenoble, J. (1993). *Atmospheric Radiative Transfer*, A. Deepak Publishing, Hampton, Virginia.

Lenoble, J. (ed.) (1985). *Radiative Transfer in Scattering and Absorbing Atmospheres: Standard Computational Procedures*, A. Deepak Publishing, Hampton, Virginia.

Levy, R. C., Remer, L. A. & Kaufman, Y. J. (2004). Effects of neglecting polarization on the MODIS aerosol retrieval over land, *IEEE Trans. Geosci. Rem. Sens.* 42: 2576–2583.

Liang, P., Moghaddam, M., Pierce, L. E. & Lucas, R. M. (2005). Radar backscattering model for multilayer mixed species, *IEEE Trans. Geosci. Rem. Sens.* 43: 2612–2626.

Marchand, E. W. & Wolf, E. (1974). Radiometry with sources in any state of coherence, *J. Opt. Soc. Am.* 64: 1219–1226.

Mishchenko, M. I. (2002). Vector radiative transfer equation for arbitrarily shaped and arbitrarily oriented particles: A microphysical derivation from statistical electrodynamics, *Appl. Opt.* 41: 7114–7134.

Mishchenko, M. I., Travis, L. D. & Lacis, A. A. (2006). *Multiple Scattering of Light by Particles*, Cambridge University Press, New York.

Moghaddam, M., Rahmat-Samii, Y., E.Rodriguez, Entekhabi, D., Hoffman, J., Moller, D., Pierce, L. E., Saatchi, S. & Thomson, M. (2007). Microwave Observatory of Subcanopy and Subsurface (MOSS): A mission concept of global deep soil moisture observations, *IEEE Trans. Geosci. Remote Sens.* 45: 2630–2643.

Mudaliar, S. (1994). Electromagnetic wave scattering from a random medium layer with a rough interface, *Waves Random Media* 4: 167–176.

Mudaliar, S. (2005). Statistical wave theory for a random medium layer over a two-scale rough surface, Proc. IASTED Intl. Conf. Antennas, Radars and Wave Propagation, pp. 246–253.

Pak, K., Tsang, L., Li, L. & Chan, C. H. (1993). Combined random rough surface and volume scattering based on Monte-Carlo simulation of Maxwell equations, *Radio Sci.* 23: 331–338.

Papa, R. J. & Tamasanis, D. (1991). A model for scattering of electromagnetic waves from foliage-covered terrain, Vol. 46 of *AGARD Conf. Proc.*, pp. 4.1–4.12.

Pederson, H. M. & Stamnes, J. J. (2000). Radiometric theory of spatial coherence in free space propagation, *J. Opt. Soc. Am. A* 17: 1413–1420.

Peloci, G. & Coccioli, K. (1997). A finite element approach for scattering from inhomogeneous media with a rough interface, *Waves Random Media* 7: 119–127.

Sarabandi, K., Yisok, O. & Ulaby, F. (1996). A numerical simulation of scattering from one-dimensional inhomogeneous dielectric rough surfaces, *IEEE Trans. Geosci. Remote Sens.* 34: 425–432.

Sato, H. & Fehler, M. C. (1998). *Seismic Wave Propagation and Scattering in the Heterogeneous Earth*, Springer-Verlag, New York.

Shin, R. T. & Kong, J. A. (1989). Radiative transfer theory for active remote sensing of two layer random medium, *PIER* 3: 359–419.

Sobolev, V. V. (1963). *A Treatise on Radiative Transfer*, Van Nostrand, Princeton, New Jersey.

Stam, D. M. & Hovenier, J. W. (2005). Errors in calculated planetary phase functions and albedos due to neglecting polarization, *Astron. Astrophys.* 444: 275–286.

Stammes, P. (1994). Errors in UV reflectivity and albedo calculations due to neglecting polarization, *in* R. P. Santer (ed.), *Atmospheric Gases*, Vol. 2311 of *Proc. of SPIE*, pp. 227–235.

Stamnes, K., Tsay, S. C., Wiscombe, W. & Jayweera, K. (1988). Numerically stable algorithm for discrte-ordinate-method radiative transfer in multiple scattering and emitting layered media, *Appl. Opt.* 27: 2502–2509.

Tatarskii, V. I. (1971). *The Effects of Turbulent Atmosphere on Wave Propagation*, Israel Program of Scientific Translations, Jerusalem, Israel.

Tsang, L., Kong, J. A. & Shin, R. T. (1985). *Theory of Microwave Remote Sensing*, John Wiley, New York.

Twersky, V. (1964). On propagation in random media of discrete scatterers, Vol. 16 of *Proc. of the Symposium of Applied Mathematics*, pp. 84–116.

Twersky, V. (1980). Multiple scattering of waves by correlated distributions, *in* J. A. DeSanto, A. W. Saenz & W. W. Zachary (eds), *Mathematical Methods and Apllications in Scattering Theory*, Vol. 130 of *Lecture Notes in Physics*, Springer Verlag, New York, USA, pp. 1–9.

Ulaby, F. T., Moore, R. K. & Fung, A. K. (1986). *Microwave Remote Sensing: Active and Passive*, Vol. 3, Artech House, Norwood, Massachusetts.

Ulaby, F. T., Sarabandi, K., McDonald, K., Whitt, K. M. & Dobson, M. C. (1990). Michigan microwave canopy model, *Int. J. Rem. Sens.* 11: 1223–1253.

Urbini, S., Vitterari, L. & Gandolfi, S. (2001). GPR and GPS data integration: example of application in Antarctica, *Annali di Geofisica* 44: 687–701.

Voronovich, A. G. (1999). *Wave Scattering from Rough Surfaces*, 2nd edn, Springer-Verlag, Berlin.

Yoshimori, K. (1998). Radiometry and coherence in a nonstationary optical field, *J. Opt. Soc. Am. A* 15: 2730–2734.

Zuniga, M. A., Habashy, T. M. & Kong, J. A. (1979). Active remote sensing of layered random media, *IEEE Trans. Geosci. Rem. Sens.* 17: 296–302.

Part 3

Antennas and Waveguides

12

Metamaterial Waveguides and Antennas

Alexey A. Basharin, Nikolay P. Balabukha,
Vladimir N. Semenenko and Nikolay L. Menshikh
Institute for Theoretical and Applied Electromagnetics RAS
Russia

1. Introduction

In 1967, Veselago (1967) predicted the realizability of materials with negative refractive index. Thirty years later, metamaterials were created by Smith et al. (2000), Lagarkov et al. (2003) and a new line in the development of the electromagnetics of continuous media started. Recently, a large number of studies related to the investigation of electrophysical properties of metamaterials and wave refraction in metamaterials as well as and development of devices on the basis of metamaterials appeared Pendry (2000), Lagarkov and Kissel (2004). Nefedov and Tretyakov (2003) analyze features of electromagnetic waves propagating in a waveguide consisting of two layers with positive and negative constitutive parameters, respectively. In review by Caloz and Itoh (2006), the problems of radiation from structures with metamaterials are analyzed. In particular, the authors of this study have demonstrated the realizability of a scanning antenna consisting of a metamaterial placed on a metal substrate and radiating in two different directions. If the refractive index of the metamaterial is negative, the antenna radiates in an angular sector ranging from -90 ° to 0°; if the refractive index is positive, the antenna radiates in an angular sector ranging from 0° to 90° . Grbic and Elefttheriades (2002) for the first time have shown the backward radiation of CPW- based NRI metamaterials. A. Alu et al.(2007), leaky modes of a tubular waveguide made of a metamaterial whose relative permittivity is close to zero are analyzed. Thus, an interest in the problems of radiation and propagation of structures with metamaterials is evident. The purposes of this study is to analyze propagation of electromagnetic waves in waveguides manufactured from metamaterials and demonstrate unusual radiation properties of antennas based on such waveguides.

2. Planar metamaterial waveguide. Eigenmodes of a planar metamaterial waveguide

For the analysis of the eigenmodes of metamaterial waveguides, we consider a planar magnetodielectric waveguide. The study of such a waveguide is of interest because, if the values of waveguide parameters are close to limiting, solutions for the waveguides with more complicated cross sections are very close to the solution obtained for a planar waveguide, Markuvitz (1951). Moreover, calculation of the dispersion characteristics of a rectangular magnetodielectric waveguide with an arbitrary cross section can be approximately reduced to calculation of the characteristics of a planar waveguide.

Let us consider a perfect (lossless) planar magnetodielectric waveguide (Fig. 1), Wu et al. (2003). A magnetodielectric (metamaterial) layer with a thickness of $2a_1$ is infinite along the y and z axes. The field is independent of coordinate y.

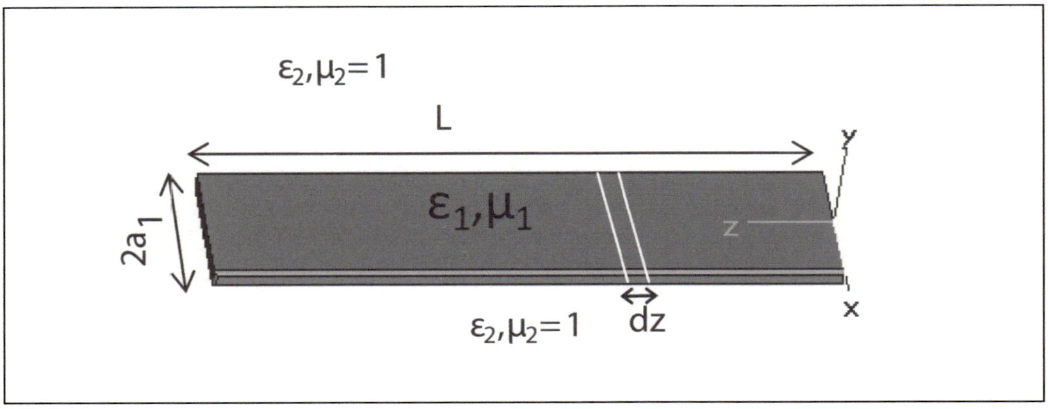

Fig. 1. A planar metamaterial waveguide.

The relative permittivity of this layer is ε_1 and the relative permeability is μ_1. The relative permittivity and relative permeability of the ambient space are ε_2 and μ_2, respectively. Let us represent the field in the waveguide in terms of the longitudinal components of the electric Hertz vector. For even TM modes, we write

$$\Pi_z^e = A_1 \sin(k_1 x)\exp(ihz) \text{ for } |x| < a_1 \tag{1}$$

$$\Pi_z^e = B\exp(-k_2|x|)\exp(ihz) \text{ for } |x| > a_1, \tag{2}$$

For odd TM modes, we have

$$\Pi_z^e = A_2 \cos(k_1 x)\exp(ihz) \text{ for } |x| < a_1 \tag{3}$$

$$\Pi_z^e = B\exp(-k_2|x|)\exp(ihz) \text{ for } |x| > a_1, \tag{4}$$

Where $k_0 = 2\pi/\lambda$ is the wave number in vacuum, λ is the wavelength, h is the longitudinal wave number,

$$k_1 = \sqrt{k_0^2 \varepsilon_1 \mu_1 - h^2}, \tag{5}$$

and

$$k_2 = \sqrt{h^2 - k_0^2 \varepsilon_2 \mu_2}. \tag{6}$$

Hereinafter, time factor $\exp(-i\omega t)$ is omitted.

The field components are expressed through the Hertz vectors as

$$E_x = \frac{\partial}{\partial x}\left(\frac{\partial \Pi_z^e}{\partial z}\right)$$

$$E_z = k_0^2 \varepsilon \mu \Pi_z^e + \frac{\partial}{\partial z}\left(\frac{\partial \Pi_z^e}{\partial z}\right). \qquad (7)$$

$$H_y = ik_0 \varepsilon \frac{\partial \Pi_z^e}{\partial x}$$

Using the continuity boundary condition for tangential components *Ez* and *Hy* calculated from formulas (1), (2), and (7) at the material-free-space interface (at $x=a_1$), we obtain the characteristic equation for even TM modes

$$k_2 a_1 = k_1 a_1 \frac{\varepsilon_2}{\varepsilon_1} \mathrm{tg}(k_1 a_1) \qquad (8)$$

It follows from formulas (3), (4), and (7) that, for odd TM modes,

$$k_2 a_1 = -k_1 a_1 \frac{\varepsilon_2}{\varepsilon_1} \frac{1}{\mathrm{tg}(k_1 a_1)} \qquad (9)$$

Using also the continuity condition for longitudinal wave number *h*, we find from (5) and (6) that

$$(k_1 a_1)^2 + (k_2 a_1)^2 = (k_0 a_1)^2 (\varepsilon_1 \mu_1 - \varepsilon_2 \mu_2) > 0 \qquad (10)$$

Equations (8), (10) and (9), (10) can be used to determine transverse wave numbers k_1 and k_2 for even and odd TM modes, respectively. Solutions to systems (8), (10) and (9), (10) will be obtained using a graphical procedure. Figure 2 presents the values of $k_2 a_1$ as a function of $k_1 a_1$. Solid curves *1,2,3*, etc., correspond to Eqs. (8) and dashed curves *A,B,C*, etc., correspond to Eq. (10), for even TM modes.

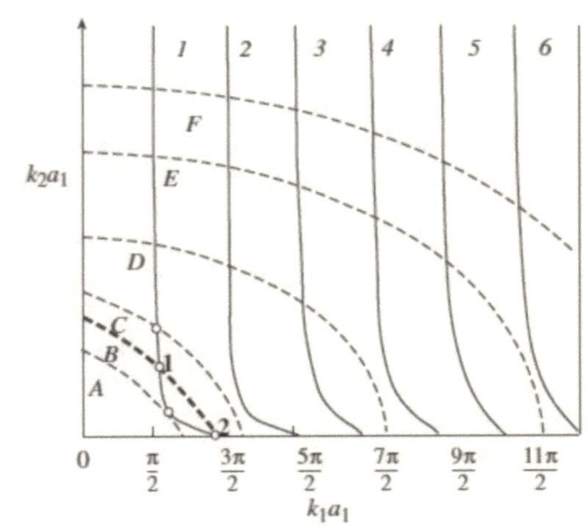

Fig. 2. Solution of the characteristic equations (8),(10).

Some curves (for example, curves *1* and *A* and curves *1* and *C*) may intersect at one point. These curves will be referred to as solutions of types I and II. Some curves (for example, curves *1* and *3*) may intersect at two points. These curves will be referred to as solutions of type III.

Let us consider the behavior of the power flux in such a waveguide for solutions I, II, and III. For the TM modes, the Poynting vector in the direction of the z axis is

$$S_z = \mathrm{Re}(E_x H_y^*) \tag{11}$$

For the layer of metamaterial,

$$S_{z1}(x) = A^2 \cos^2(k_1 x) h k_1^2 k_0 \varepsilon_1 \tag{12}$$

For the ambient space,

$$S_{z2}(x) = B^2 \exp(-2k_2 x) h k_2^2 k_0 \varepsilon_2 \tag{13}$$

The power flux is an integral over the waveguide cross section. The total power flux is

$$S_\Sigma = A^2 k_1^2 h k_0 \varepsilon_1 [\frac{a_1}{2} + \frac{\sin(2k_1 a_1)}{4k_1}] + B^2 k_2^2 h k_0 \varepsilon_2 \frac{\exp(-2k_2 a_1)}{2k_2}, \tag{14}$$

where the first term corresponds to the power flux in the metamaterial and the second term corresponds to the power flux in the ambient space.

Using the boundary condition, we can express coefficient A in terms of coefficient B. As a result, expression (14) transforms to

$$S_\Sigma = B^2 h k_0 \exp(-2k_2 a_1)[\frac{k_2^2 \varepsilon_1}{k_1^2 \sin^2(k_1 a)}(\frac{a_1}{2} + \frac{\sin(2k_1 a_1)}{4k_1}) + \frac{\varepsilon_2}{2k_2}] \tag{15}$$

For $\varepsilon_1 < 0$ and a chosen value of parameter h, the total flux can be either positive or negative. In the case of solution II, the total flux is negative, i.e., a backward wave propagates in the negative direction of the z axis, Shevchenko V.V (2005). Unlike dielectric waveguides, there are frequencies at which metamaterial waveguides can support two simultaneously propagating modes. Point **1** of solution III (see Fig. 2) corresponds to a negative power flux (a backward wave) and point **2** corresponds to a positive power flux (a forward wave). Solution I corresponds to zero flux, i.e., formation of a standing wave.

If the total flux takes a negative value, the negative value of wave number h should be chosen, Shevchenko V.V (2005).

Thus, depending on frequency, a planar metamaterial waveguide can support forward, backward, or standing waves. If constitutive parameters are negative, flux (12) and (13) are opposite to each other along the z axis. Accordingly, it can be expected that an antenna based on a planar metamaterial waveguide will radiate in the forward direction when flux $S_{z2}(x)$ (13) takes positive values; otherwise, it will radiate in the backward direction when $S_{z2}(x)$ takes negative values. Here, the total power flux is assumed to be positive.

3. Radiation of antenna based on planar metamaterial waveguide

Antennas manufactured on the basis of planar waveguides belong to the class of traveling-wave antennas, Balanis (1997). In calculation of the radiation patterns of such antennas, we can approximately assume that the field structure in the antenna is the same as the field structure in an infinitely long planar waveguide. A planar waveguide supports TM and TE propagating modes. These modes are reflected at the end of the antenna rod, Balanis (1997), Aizenberg (1977). Under given assumptions, the radiation pattern of an antenna based on a metamaterial waveguide (the dependence of the field intensity measured in dB on azimuth angle θ measured in degrees) is calculated from the following formula:

$$f(\theta)=\frac{2}{k_0 L}\left(\frac{\sin(k_0\frac{L}{2}(\frac{h}{k_0}-\cos\theta))}{\frac{h}{k_0}-\cos\theta}+p\exp(iLh)\frac{\sin(k_0\frac{L}{2}(\frac{h}{k_0}+\cos\theta))}{\frac{h}{k_0}+\cos\theta}\right)f_1(\theta)\,, \tag{16}$$

where L is the antenna length, $f_1(\theta)$ is the radiation pattern of a antenna element dz (Fig. 1), obtained by a Hyugens` principle, and p is the reflection coefficient for reflection from the antenna end. p is defined under Fresnel formula. The second term (16) considers radiation of the reflected wave from a end of a waveguide, traveling in a negative direction z. Usually, the sizes and an antenna configuration chooses in such a manner that intensity of the reflected wave is small, Volakis J. A. (2007). And in practice the formula 17 is used:

$$f(\theta)=\frac{2}{k_0 L}\left(\frac{\sin(k_0\frac{L}{2}(\frac{h}{k_0}-\cos\theta))}{\frac{h}{k_0}-\cos\theta}\right)f_1(\theta)\,. \tag{17}$$

Formulas 16 and 17 are lawful for a case of small difference of a field in a vicinity of the end of a waveguide from a field in a waveguide. These formulas yield exact enough results only at small values of $a_1\sqrt{\varepsilon_1\mu_1}$, Angulo (1957).

Figures 3–7 show calculated H-plane radiation patterns of the antenna (based on planar metamaterial waveguide) for the TM modes. The patterns are normalized by their maximum values. The patterns obtained for the TE modes are qualitatively identical to the patterns corresponding to the TM modes and are not presented. The waveguide dimensions are a_1 = 50 mm and L = 400 mm, the relative permittivity of the metamaterial is $\varepsilon_1 = -2$, and the relative permeability of the metamaterial is $\mu_1 = -1$. Calculation was performed at the following frequencies: 2.5, 2.8, 3.0, and 3.5 GHz.

Let us consider evolution of the radiation pattern of this antenna with frequency. A frequency of 2.5 GHz corresponds to solution I and zero total power flux. A standing wave is formed in the waveguide. This wave results from the interference of the forward and backward waves with wave numbers h having equal absolute values and opposite signs. The radiation pattern corresponding to the forward wave ($h > 0$) is shown in Fig. 3a.

The radiation pattern corresponding to the backward wave ($h < 0$) is the radiation pattern of the forward wave rotated through 180° (Fig. 3b). The radiation pattern corresponding to the interference of the forward and backward waves is shown in Fig. 4.

As frequency increases, solution I splits into two solutions. The upper part (point **1**) of solution III corresponds to a backward wave. In this case, the negative value of parameter h should be chosen. The total power flux (15) is negative. Figure 5a presents the radiation pattern at a frequency of 2.8 GHz. The back lobe of this pattern exceeds the main lobe; the

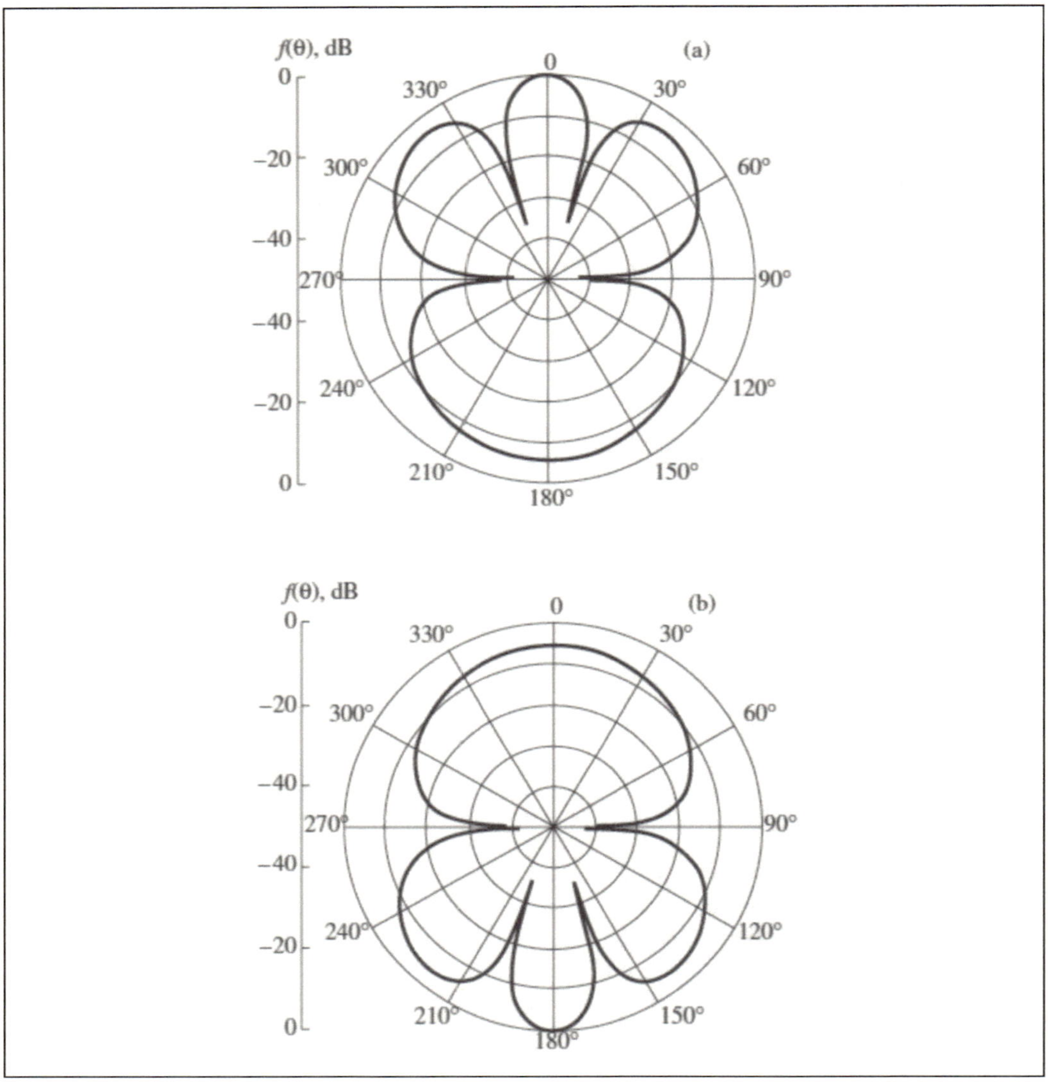

Fig. 3. Radiation patterns $f(\theta)$ on an antenna based on a planar metamaterial waveguide at a frequency of 2.5 GHz for solution I in the cases of (a) forward and (b) backward waves.

antenna radiation direction is 180°. The lower part (point **2**) of solution III corresponds to a forward wave. The positive value of longitudinal wave number h is chosen. The field is localized out of a waveguide and the total power flux (15) is positive. The maximum of the radiation pattern is located at 0° (Fig. 5b).

A frequency of 3.0 GHz corresponds also to solution III. In the case of the upper part (point **1**) of solution III, the antenna predominantly radiates in the backward direction (at 180°, see Fig. 6a). In the case of selection of the lower part (point **2**) of solution III, the antenna radiates in the main direction (at 0°, see Fig. 6b).

If frequency further increases, two-mode solution III transforms into single-mode solution II. In the case of solution II, the metamaterial waveguide supports a backward propagating

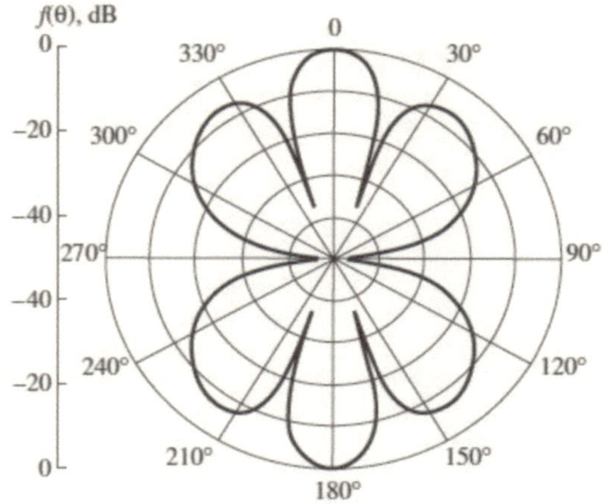

Fig. 4. Radiation pattern $f(\theta)$ on an antenna based on a planar metamaterial waveguide at a frequency of 2.5 GHz for solution I and interference of the forward and backward waves.

wave. The negative value of longitudinal wave number h should be chosen. The radiation pattern at a frequency of 3.5 GHz is shown in Fig. 7. The maximum of this radiation pattern is located at 180°.

It can be proved that this backward radiation effect in a direction of 180° is possible only for the antennas manufactured on the basis of metamaterial waveguides with negative values of the relative permittivity and relative permeability.

The condition of radiation in the backward direction (180°) is the inequality

$$f(\theta=0)-f(\theta=180)<0\,, \tag{17}$$

which can readily be solved analytically by Basharin at al. (2010). The corresponding solution is

$$\varepsilon_1 < -\varepsilon_2 \frac{k_1^2}{k_2^2}\,. \tag{18}$$

It follows from expression (18) that backward radiation is possible only for negative values of permittivity ε_1 and under the condition $\varepsilon_1, \mu_1 < 0$. Analysis shows that condition (18) is not satisfied for solution I and point 2 of solutions III. In these cases, the antenna radiates in the main direction (see Figs. 3a, 5b, 6b).

As a practical application of this effect, we can propose a two-mode scanning antenna, which on one frequency, in the case of operation in the first mode (point **1** of solution III, see Fig. 2), can radiate in the backward direction (at 180°, see Figs. 5a, 6a) and, in the case of operation (at the same frequency) in the second mode (point **2**), can radiate in the main direction (at 0°, see Figs. 5b, 6b). Switching from one mode to another can be performed, for example, by feeding the antenna from a system of electric dipoles placed parallel to the electric field lines at the feeding point of the antenna. Changing the amplitude- phase distribution in each dipole, it is possible to obtain the field distribution required for the first or the second mode.

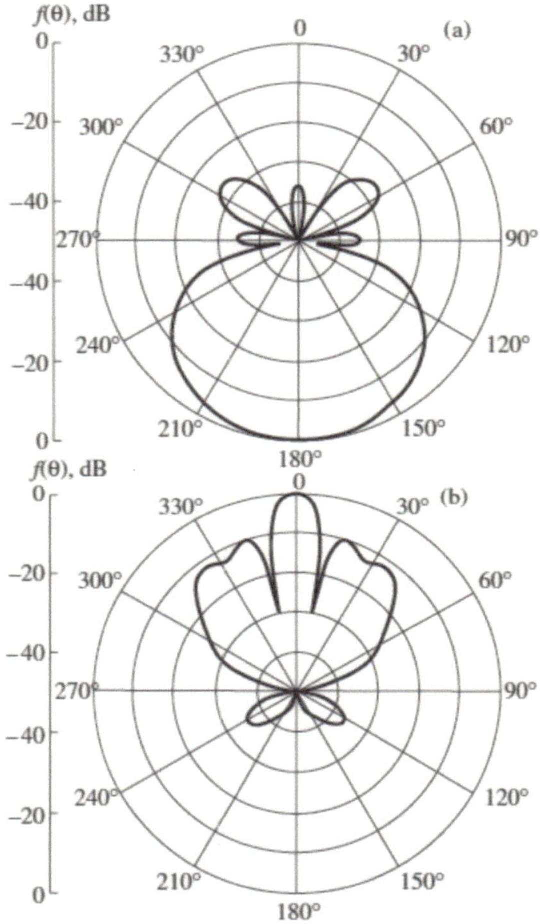

Fig. 5. Radiation patterns $f(\theta)$ on an antenna based on a planar metamaterial waveguide at a frequency of 2.8 GHz for solution III and points (a) **1** and (b) **2**.

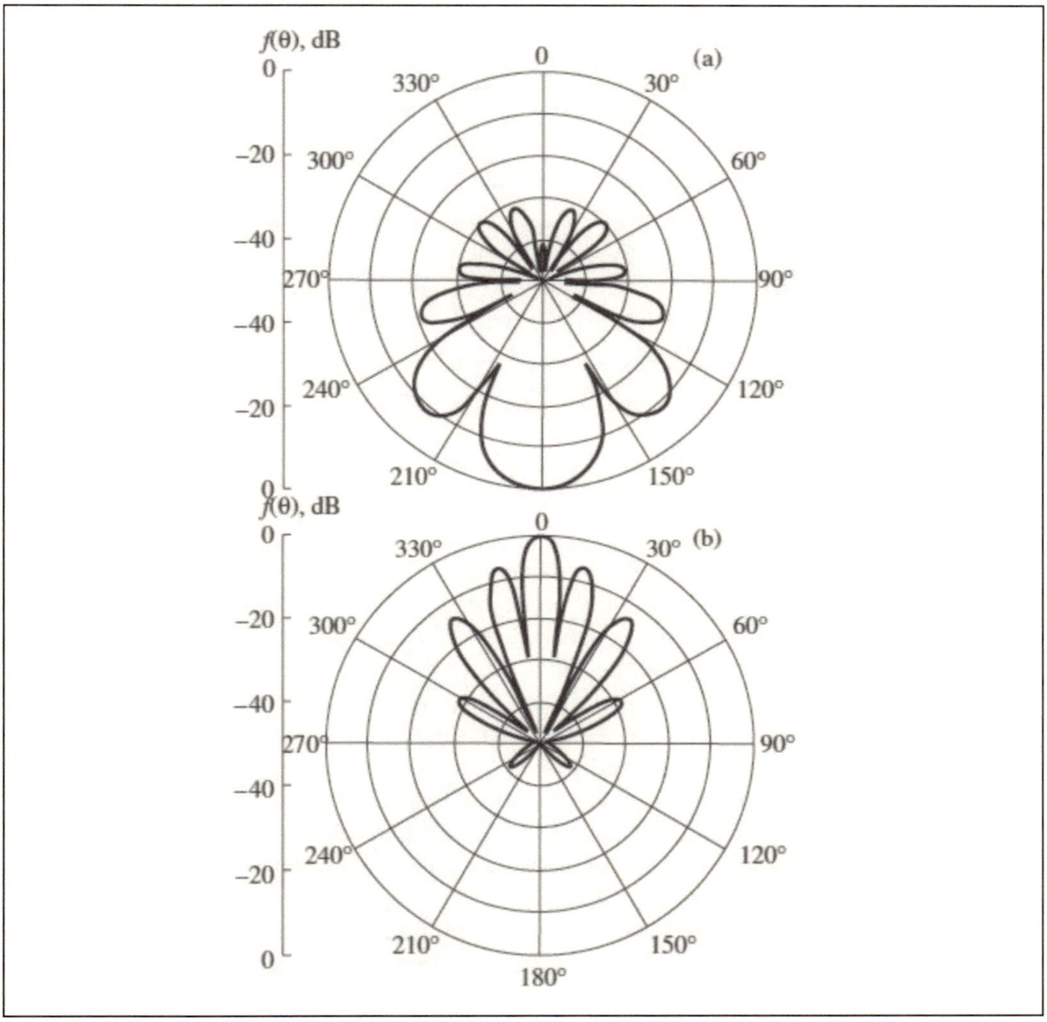

Fig. 6. Radiation patterns $f(\theta)$ on an antenna based on a planar metamaterial waveguide at a frequency of 3.0 GHz for solution III and points (a) **1** and (b) **2**.

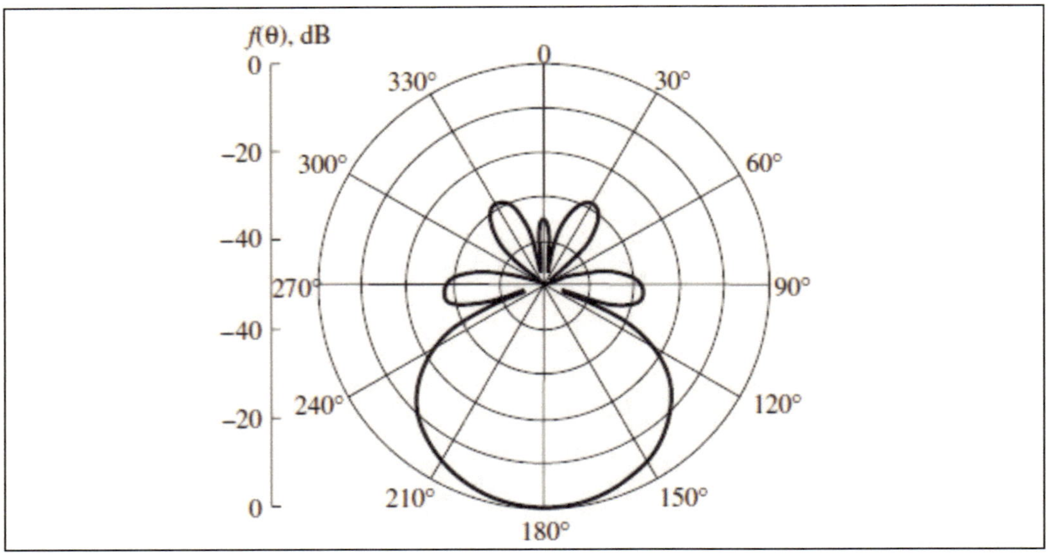

Fig. 7. Radiation pattern $f(\theta)$ on an antenna based on a planar metamaterial waveguide at a frequency of 3.5 GHz for solution II.

4. Characteristics of electromagnetic waves in a planar waveguide based on metamaterial with losses

Here we present results of calculation of dispersion characteristics for a planar waveguide made of metamaterials with losses and wave type classification in such waveguides. It demonstrates that forward and backward waves can exist in such structure, waves can permeate "without attenuation" (so the imaginary part of longitudinal wave number is near zero) in spite of presence losses in metamaterial, as well it shows presence of wave mode propagating with a constant phase velocity which does not depends on the frequency. This paper researches dispersion characteristics of the planar metamaterial waveguide with losses, analyses the waveguide characteristics and classifies wave modes.

Let's consider a planar waveguide, made of a metamaterial with losses (Fig. 1). The concerned metamaterial is not ideal and has losses, so its penetrability's are complex values and can be presented as: $\varepsilon_1=\varepsilon_1'+\varepsilon_1''$, $\mu_1=\mu_1'+\mu_1''$.

Having placed expressions (5) and (6) in equation (8) for k_1 and k_2 (continuity condition for the longitudinal wave number h) it should reduce to the following expression:

$$\cos\left[a\sqrt{k_0{}^2\varepsilon_1\mu_1-h^2}\right]\varepsilon_1\sqrt{h^2-k_0{}^2\varepsilon_2\mu_2}-\varepsilon_2\sqrt{k_0{}^2\varepsilon_1\mu_1-h^2}\sin\left[a\sqrt{k_0{}^2\varepsilon_1\mu_1-h^2}\right]=0 \quad (19)$$

Equation (19) is solved with Muller method, Muller (1965), Katin and Titarenko (2006). In the capacity of initial estimate we choose values of h within the wide range from 0 to few k_0.

For the metamaterial waveguide with parameters $\varepsilon'_1= -2$ and $\mu'_1= -1$ calculations have been done for various imaginary parts of penetrabilities. Results are presented as diagrams of h'/k_0 and h''/k_0 versus wave number of environment k_0a (Fig.8-13). Great numbers of waves, which exist simultaneously at the same frequency in the waveguide, make us give special attention on their classification

Viewing the diagrams of imaginary and real parts of longitudinal wave number we see right lines $h'=k_0$ and $h''=k_0$ allowing to divide all waves into fast ($h'<k_0$) and slow ($h'>k_0$) waves, strongly attenuation waves ($h''>k_0$) and weak attenuation ones ($h''<k_0$), Marcuvitz (1951), Vainshtein (1988), Shevchenko (1969).

Let's consider the case when the metamaterial has not got any losses, Shadrivov et al. (2003), Basharin et al. (2010). At low frequencies all waves are backward (dotted lines on the diagrams), it means that the phase velocity direction is opposite to the Poynting vector. At some frequency (points A and B in Fig.8) the wave is split into two modes: forward (continuous line on the diagrams) and backward. Fields of these waves presented on Fig. 14 and Fig. 15. Away from waveguide the fields decay on exponential law. The both modes start to propagate without any losses. The forward wave exists only within the narrow frequency range. Further increasing of frequency forces the forward wave turn into the "anti-surface wave" (improper wave) (Away from waveguide the field grows exponentially), Vainshtein (1988), Marcuvitz (1951), which exists only mathematically (dash-dotted lines).

Z component of Poynting vector is negative inside the waveguide, but it is positive outside. There is a zero density of power flow at the splitting points (points A, B in Fig.8) i.e. a

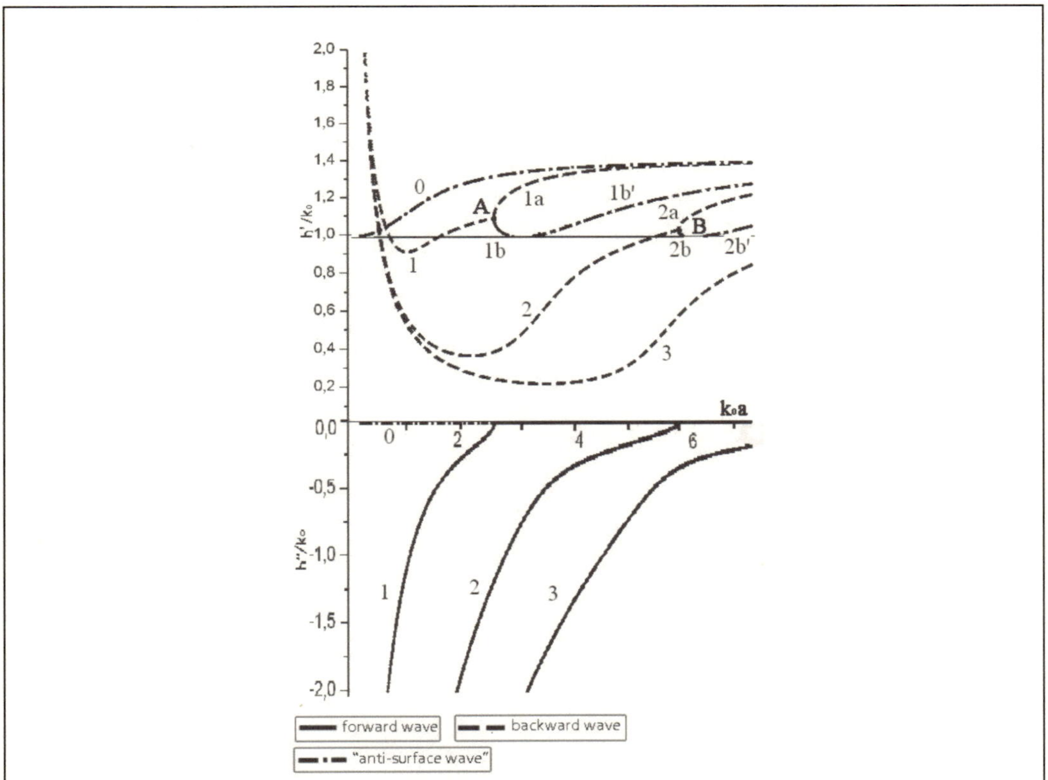

Fig. 8. Dispersion characteristic of metamaterial waveguide with ε_1=-2, μ_1=-1

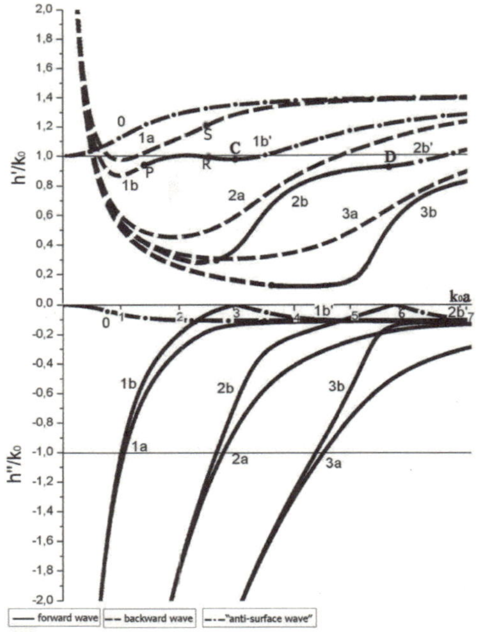

Fig. 9. Dispersion characteristic of metamaterial waveguide with ε_1=-2+i0.1, μ_1=-1+i0.1

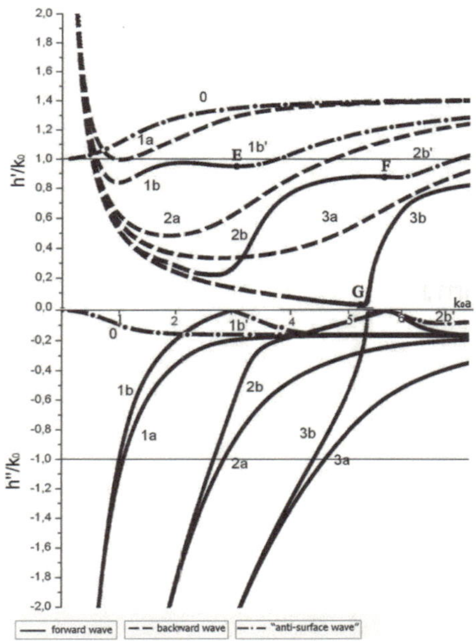

Fig. 10. Dispersion characteristic of metamaterial waveguide with ε_1=-2+i0.15, μ_1=-1+i0.15

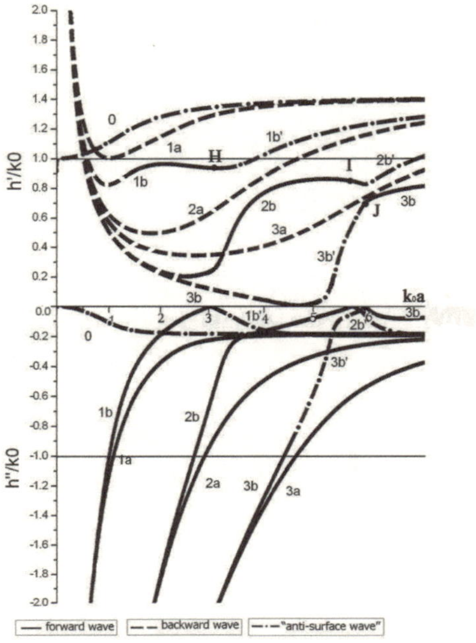

Fig. 11. Dispersion characteristic of metamaterial waveguide with ε_1=-2+i0.17, μ_1=-1+i0.17

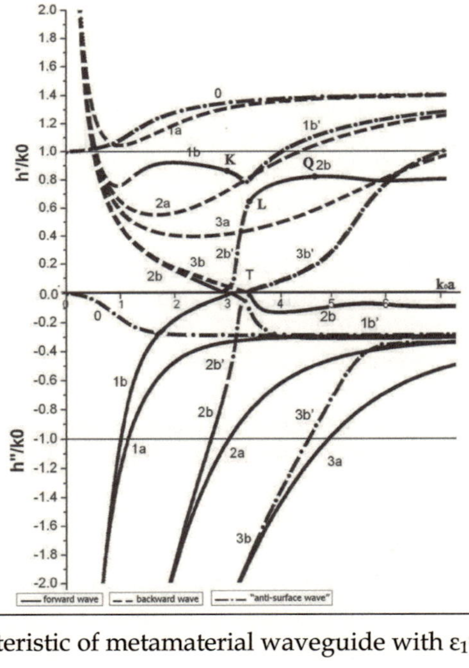

Fig. 12. Dispersion characteristic of metamaterial waveguide with ε_1=-2+i0.27, μ_1=-1+i0.27

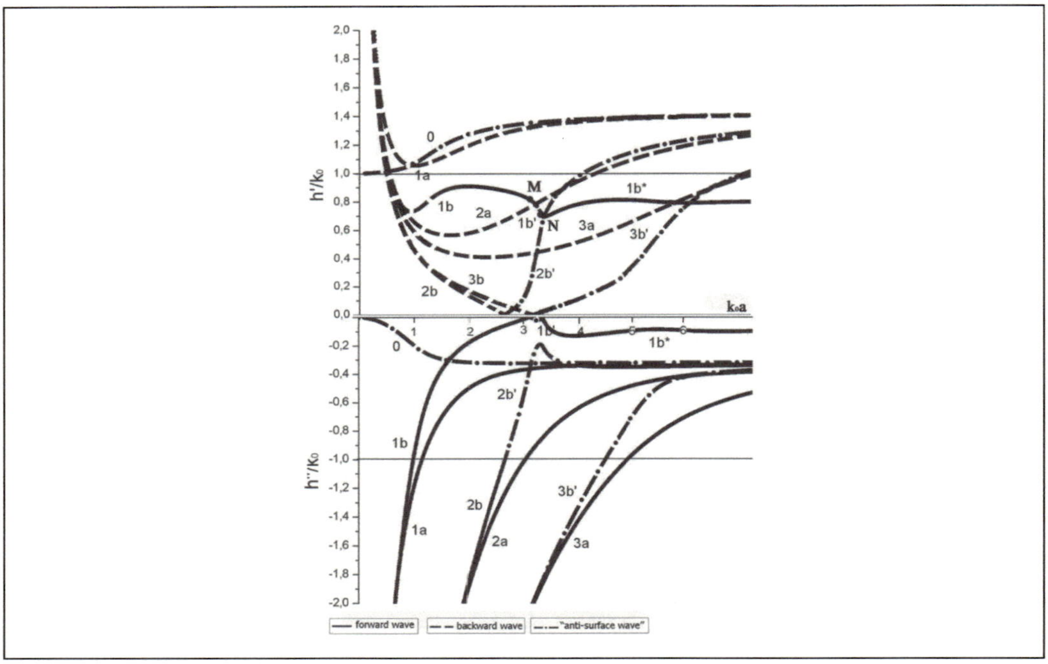

Fig. 13. Dispersion characteristic of metamaterial waveguide with ε_1=-2+i0.3, μ_1=-1+i0.3

standing wave arises as result of interference of the forward and backward wave, Basharin et al., (2010), Shatrov et al. (1974); Shevchenko at al. (2000).

When we include a loss in metamaterial, then splitting occurs at low frequencies. Both modes are supported in whole range (Fig. 9). One mode is backward and the second at a certain frequency is transformed into a forward (point P).

The range of frequencies in which there is forward wave increases as compared with the case without losses. The fields of these waves are the leaking (Fig. 16 a - forward, b - backward). These waves, as in the absence of losses, transformed in "anti-surface" ones with growing frequency. However, for sufficiently large losses wave decays before reaching the transition point in the forward wave. In this case (Fig. 12, point T) backward wave soon becomes "anti-surface".

Some results require further discussion. For sufficiently large losses (0.17 and over) for some modes there is a certain frequency range in which the forward wave is transformed into anti-surface wave. So a transition mode 3b in 3b' in Fig. 11 and fashion in 2b 2b' in Fig. 12.

Near at the change point where the surface wave turns into the "anti-surface" one and converse (excluding the point where $h'=0$) the wave field essentially concentrates outside the waveguide. Even considering the metamaterial inserts some losses, the wave can propagate without no degradation at these points, so the imaginary part of longitudinal wave number h'' is near zero (points C, D, E, F, H, G, I, J, K, L, M, N).

When the imaginary parts of penetrabilities are sufficiently great the wave modes are combined. At first two forward waves are combined with the "anti-surface wave" and when losses are being increased it disappears. It is obviously (Fig.11) the 2-nd and 3-rd modes are being combined and when the losses are great the 1-st and 2-nd modes are being combined (Figs.12, 13). These waves through exist lose in metamatirael ($\varepsilon''\neq 0$, $\mu''\neq 0$) exist points (for

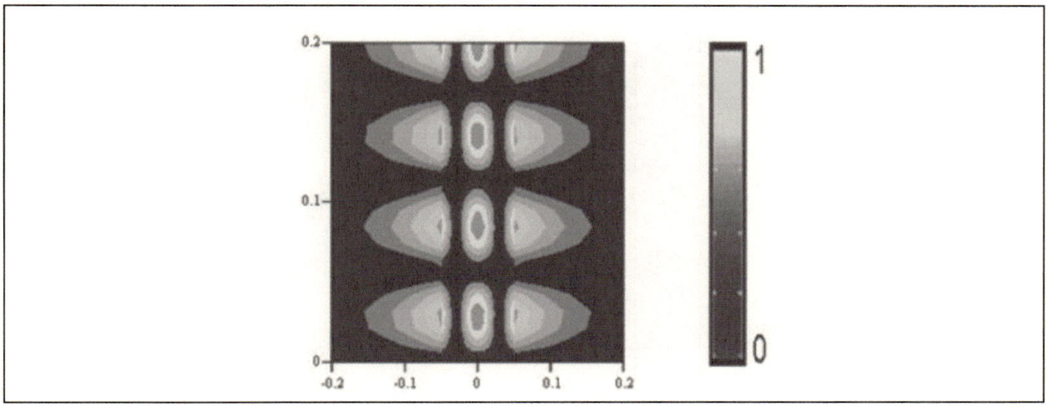

Fig. 14. Normalize magnetic field near point A (Fig. 8), forward wave.

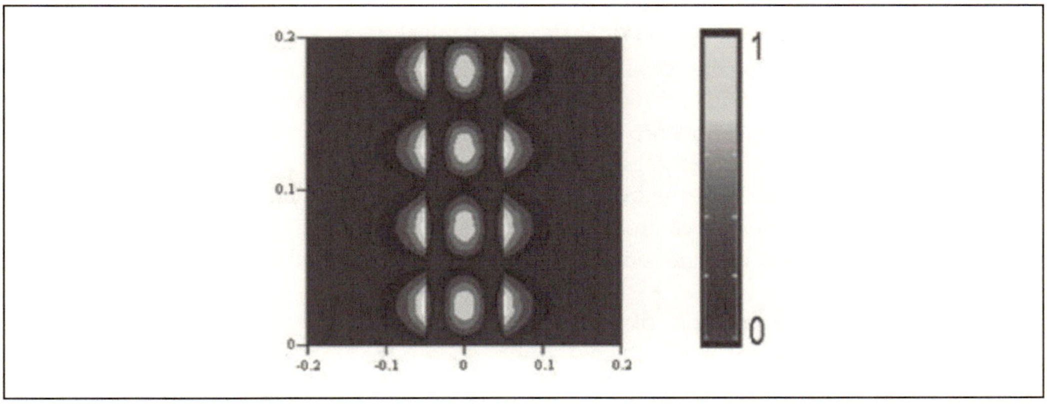

Fig. 15. Normalize magnetic field near point A (Fig. 8), backward wave

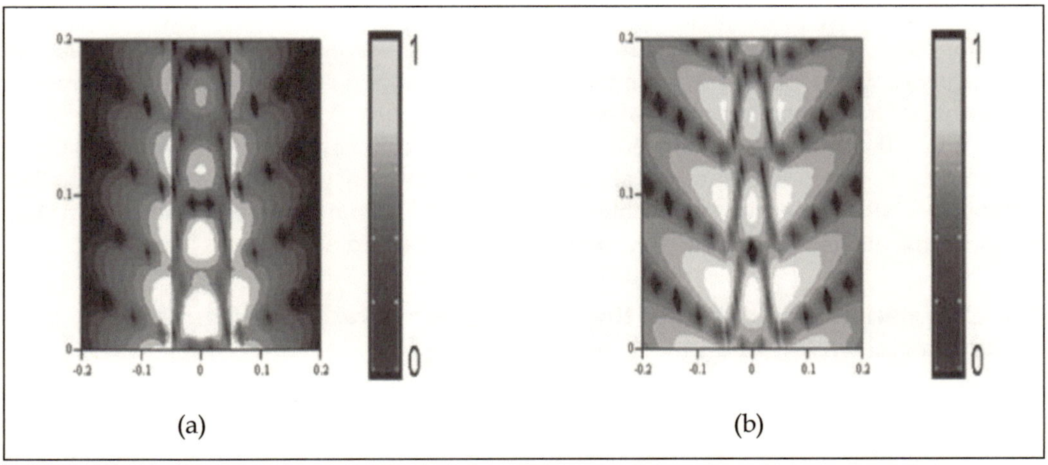

Fig. 16. Normalize magnetic field: a) backward wave, point S (Fig. 9); b) forward wave, point R (Fig. 9).

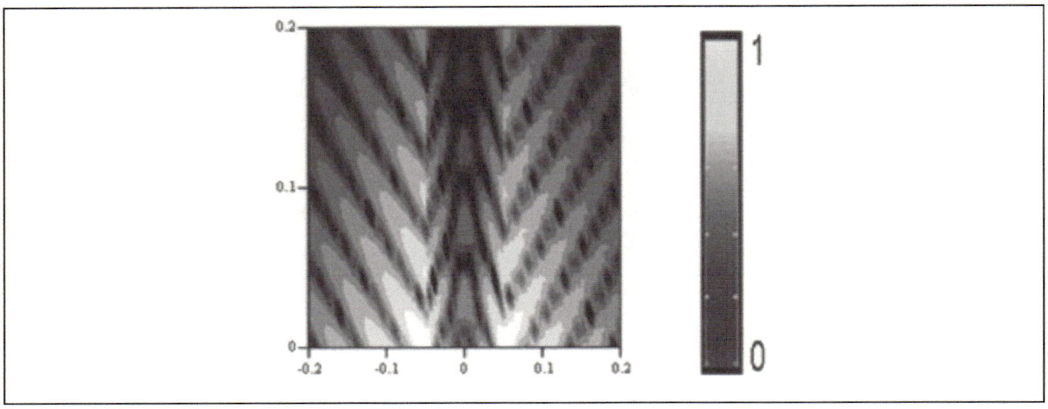

Fig. 17. Normalize magnetic field near point Q (Fig. 12).

example, point Q on Fig. 12) where imagine part of h is near zero. At this point the field is concentrated outside the waveguide (Fig. 17) and close to the free wave field.

While frequency increasing the phase velocity usually increases starting from the moment when the imaginary part *h* is getting sufficiently small to that waves can propagate. As for combined modes such relationship is not observed, the phase velocity remains almost constant while frequency increasing. Notably the velocity becomes constant with less loss, i.e. before waves combine near the modes 3b in Fig. 11 and 2b in Fig.12. Generally the fields and power flows for such waves are concentrated at the waveguide boundary. Outside the waveguide the field decays exponentially except abovementioned cases (points C, D, E, F, H, G, I, J, K, L, M, N).

Also, the mathematic models of wave types for planar waveguide made of the metamaterials with losses was researched. It was shown that forward and backward waves can propagate in such structure. It was pointed out that waves can propagate with $h''=0$ at defined frequencies despite metamaterial inserts losses. Combining of wave types is observed while losses inserted, and the combined modes propagate at a constant phase velocity which does not depend on frequency. These waves are forward, but in contrast to other forward waves they do not turn into "anti-surface waves", but they exist through all the frequency range starting from some frequency. These's fields of practice isn't percolate in waveguide and concentrate on waveguide's bound.

To be noted that the real metamaterials have strong frequency dispersion, so the material with constant permeability through all the concerned frequency range can not be made. But a metamaterial with desired permeability can be made for limited frequency ranges. In such metamaterial all above mentioned properties will be observed.

5. Experimental verification of the backward wave radiation from a metamaterial waveguide structure

This work is aimed at the experimental verification of the effect of backward radiation of electromagnetic waves by the example of an antenna in the form of a rectangular tube based on a metamaterial with the thickness *d* and negative real parts of permittivity and permeability. The essence of the effect is the preferential backward radiation of the antenna structure. The numerical simulations (with the use of the method of moments) and the measurements of the far field radiation patterns in the anechoic chamber show that such radiation is possible only at negative real parts of the permittivity and permeability of the metamaterial. However, at positive values, this antenna behaves like a usual dielectric antenna and radiates in the main direction. The geometry of the metamaterial tube waveguide antenna made is shown in Fig. 18.

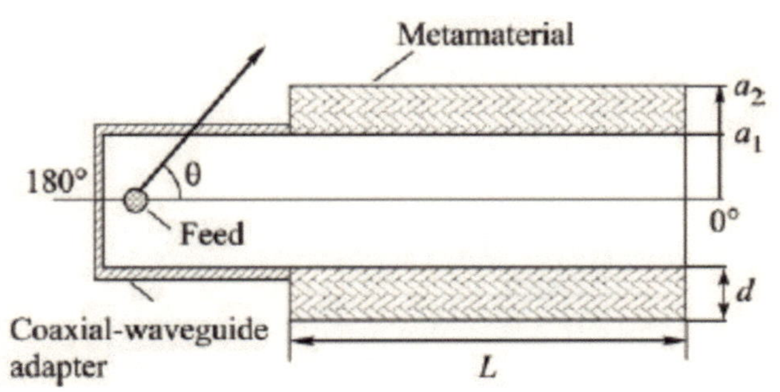

Fig. 18. Scheme of the metamaterial tube waveguide antenna.

The antenna was excited through a standard coaxial connection with a 50x25 mm waveguide for the 10 cm wavelength range. The tube length was L = 150 mm. The middle layer consisted of expanded polystyrene with a nearly unity relative permittivity. A photo of the antenna is shown in Fig. 19.

Fig. 19. Photo of the antenna under investigation.

The properties of an anisotropic metamaterial were described by Lagarkov A.N. et al. (2003). The distinctive features of the metamaterial used in this work are its spatially isotropic electric and magnetic resonance properties in the same frequency range. The metamaterial is an isotropic two dimensional array of Nichrome helixes placed on a thin 0.2 mm polyurethane substrate. To avoid the appearance of the metamaterial chirality, the right and left helixes were used in equal numbers. The helixes with an outer diameter of 5 mm consisted of three turns of a 0.4 mm Nichrome wire with a 1 mm pitch. A helixes with a nonzero pitch is simultaneously electric and magnetic dipoles, which may be effectively excited by both electric and magnetic fields polarized along the helix axis. The electric and magnetic fields of a wave incident on the metamaterial layer excite the electric and magnetic dipole moments of the helix, whose axes are parallel to the vectors **E** and **H**, respectively. At frequencies somewhat above the resonance frequency of the helixes, the phase shift between the field of the incident wave and the field induced by the currents in the metamaterial elements becomes negative; as a result, the real parts of permittivity and permeability become negative, Gulyaev et al. (2008).

In the experiments, a 5.2 mm thick sheet metamaterial was used (see Fig. 20, the coils were directed along the x, y, and z axes in equal numbers). The parameters of the sample were chosen so as to provide the resonance of permittivity and permeability at a frequency of about 3 GHz.

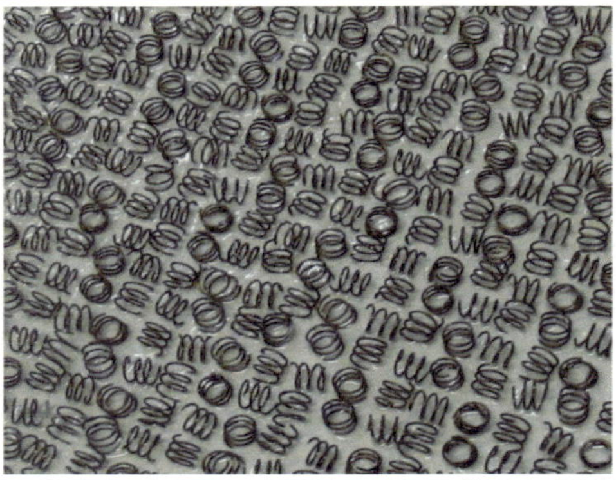

Fig. 20. Appearance of the planar metamaterial sample.

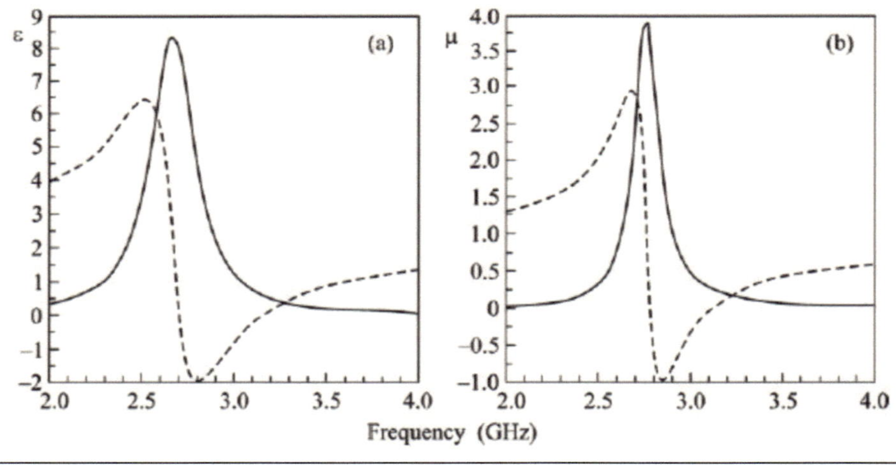

Fig. 21. Frequency dependence of the effective complex relative (a) permittivity and (b) permeability.

The experimental frequency dependence of the effective material parameters of the sample calculated from the measured values of the S parameters (complex reflection and transmission coefficients) of the planar metamaterial sample according to the Fresnel formulas is shown in Fig. 21 (the dashed and solid lines represent the real and imaginary parts, respectively). According to the studies of sheet materials by Lagarkov et al. (1990, 1997, 2003), the results of measurements for two or more layers are not considerably different from those shown in Fig. 21 and are not presented here.

The permittivity and permeability of the metamaterial are negative near a frequency of 3 GHz. The measurements were carried out with the use of the previously developed and approved technique for measuring the permittivity and permeability of sheet materials in the near field of horn antennas based on the measurement of *S* parameters. The technique was implemented with the use of a Rhode & Schwarz ZVA 24 four port Vector Network Analyzer. According to the test measurements with reference samples, the experimental error of permittivity and permeability was no more than 5%.

Let us consider the fields propagating in the waveguide part of the antenna. The calculation of the characteristics of the rectangular metamaterial waveguide may be approximately reduced to the calculation of the characteristics of the planar waveguide, Marcuvitz (1951). The fields in the waveguide part of the antenna (inside the metamaterial tube) are represented in terms of the longitudinal components of the Hertz vector (see table),

For even TE modes	For odd TE modes
$\Pi_z = A_1 \sin(k_1 x) e^{ihz}$, for $\lvert x\rvert < a_1$ (20);	$\Pi_z = A_1 \cos(k_1 x) e^{ihz}$, for $\lvert x\rvert < a_1$ (23);
$\Pi_z = (B\sin(k_2 x) + C\cos(k_2 x)) e^{ihz}$, for $a_1 < \lvert x\rvert < a_2$ (21);	$\Pi_z = (B\sin(k_2 x) + C\cos(k_2 x)) e^{ihz}$, for $a_1 < \lvert x\rvert < a_2$ (24);
$\Pi_z = De^{-k_3\lvert x\rvert} e^{ihz}$, for $\lvert x\rvert > a_2$ (22);	$\Pi_z = De^{-k_3\lvert x\rvert} e^{ihz}$, for $\lvert x\rvert > a_2$ (25),

Table 1.

where
$$k_1 = \sqrt{k_0^2 \varepsilon_0 \mu_0 - h^2} \tag{26}$$

$$k_2 = \sqrt{k_0^2 \varepsilon \mu - h^2} \tag{27}$$

$$k_3 = \sqrt{h^2 - k_0^2 \varepsilon_0 \mu_0} \tag{28}$$

h is the longitudinal wavenumber; k_0 is the wavenumber in a vacuum; ε_0 and μ_0 are the permittivity and permeability of free space, respectively; ε and μ are the relative permittivity and permeability of the metamaterial, respectively; and *A*, *B*, *C*, and *D* are constants.

Hereinafter, the time dependent factor $e^{-i\omega t}$ is omitted.

The continuity boundary conditions for the tangential components at the $x = a_1$ and $x = a_2$ interfaces allow us to obtain the characteristic equations for the waveguide modes. It is also necessary to use the continuity condition for the longitudinal wavenumber h. The characteristic equation allows us to find the longitudinal wavenumber h and the types of the modes propagating in the metamaterial waveguide. Figure 22 presents the values of h as a function of the tube thickness d for main modes of waveguide.

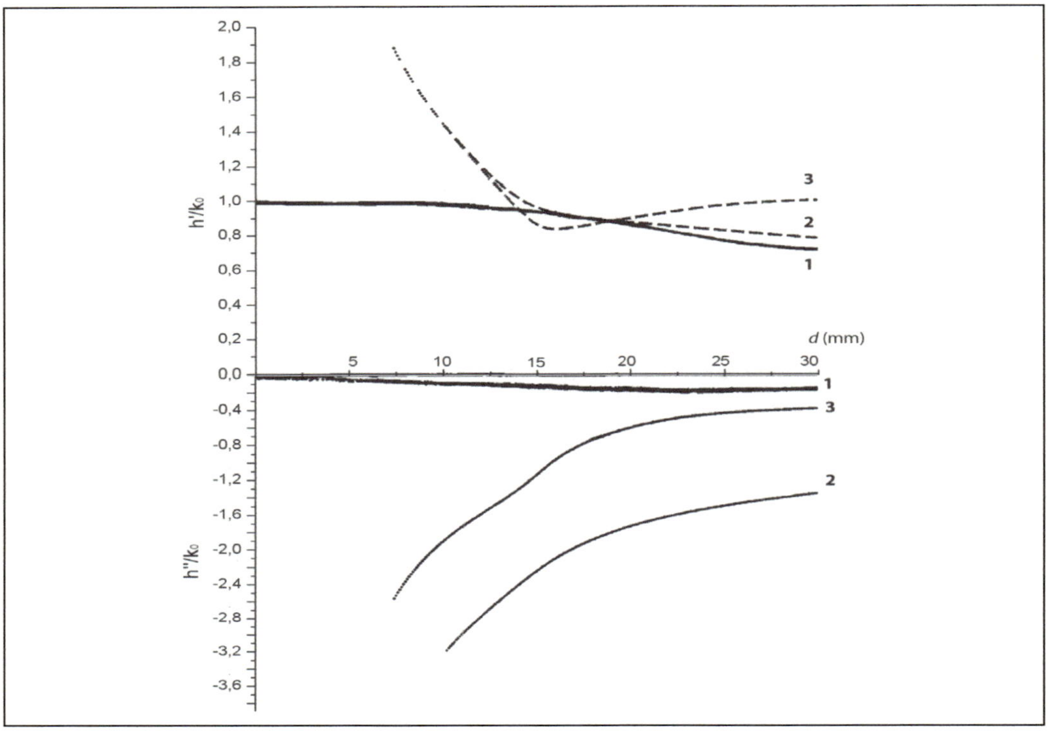

Fig. 22. Solution of the characteristic equation.

Here h` - the relative part, h`` an imaginary part of cross-section wave number h, normalized on k_0. In the waveguide supports forward, backward, and standing waves. The forward wave 1 corresponds to a wave extending in free space, waves 2 and 3- backward. In the beginning, the wave 1- dominates, and a total flux positive. Since some thickness thickness d, a wave 2 and 3 prevail also a total flux becomes negative.

The power flux in the central part of the waveguide and the surrounding waveguide free space is positive. Because of the negative values of permittivity and permeability, the flux is negative in the layer of metamaterial. Thus total power flux (in the cross section of the waveguide and the surrounding waveguide space) can be positive, negative or zero, depending on the predominance of flux in a given layer, Alu and Engheta (2005).

In other words, if the total flux is positive, then the waveguide is excited by forward wave, if negative - the backward wave, and in the case of zero flux, a standing wave.

Let us consider the behavior of the total power flux, depending on the thickness of the tube d. As a result the total power flux in such structure passes from positive in negative (Fig. 23). The normalized plot of the total power flux S_Σ as a function of the tube thickness d at a frequency of 3 GHz (the region of negative permittivity and permeability) is shown in Fig. 23. Clearly, the total power flux S_Σ may be positive ($d < 10$ mm), negative ($d > 10$ mm), or zero ($d \approx 10$ mm).

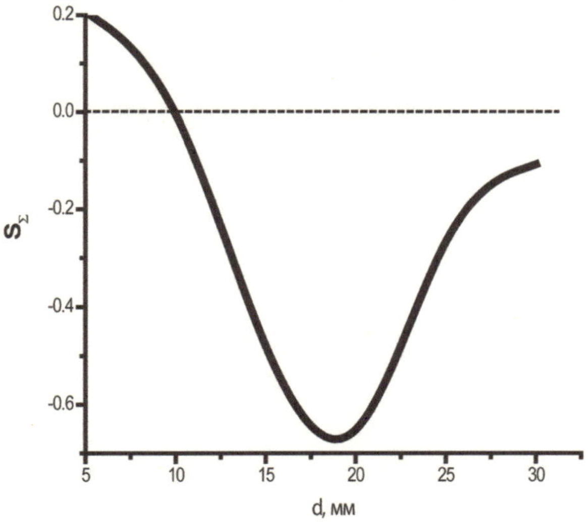

Fig. 23. Total power flux S_{Σ} versus the tube thickness d.

We have previously shown that when radiation is a single-layered planar metamaterial waveguide in the case of maintenance of the forward wave, the Radiation pattern has a maximum in the main direction. In the case of backward-wave waveguide radiates in the opposite direction, Basharin et al., 2010. Also proved, that the radiation in the opposite direction is possible only in the region of negative permittivity and permeability. It can be assumed that the existence of backward wave, two-layer waveguide radiates in the opposite direction.

6. The radiation from a metamaterial waveguide structure

Figures 24–28 present the normalized (to the maximum value in dB) radiation patterns (the field amplitudes as functions of the angular variable θ indicated in Fig. 18) of the rectangular metamaterial tube waveguide antenna excited through the coaxial waveguide adapter that were obtained by (dashed lines) numerical simulation with the use of the Method of Moments by Balanis (1997) and (solid lines) the measurements of the radiation pattern of the antenna in an anechoic chamber at frequencies close to the 3 GHz resonance frequency of the metamaterial. The difference between the measured and calculated diagrams may be as high as 5 dB.

Let us analyze the form of the radiation pattern of this waveguide antenna with various tube thicknesses d. At the tube thickness $d < 10$ mm, the total power flux is positive (see Fig. 24). According to Wu et al. (2003), Shevchenko (2005), a forward wave propagates in the waveguide in the 0° direction. Figure 24 shows the radiation pattern of the metamaterial tube antenna with the thickness d =5 mm at a frequency of 3 GHz.

The value d = 10 mm corresponds to zero total power flux S_{Σ}. A standing wave is formed in the waveguide, which results from the interference of the forward and backward waves whose longitudinal wave numbers h are equal in magnitude and opposite in sign. This corresponds to the radiation pattern of the waveguide antenna in which the front lobe is approximately equal to the back lobe (Fig. 25).

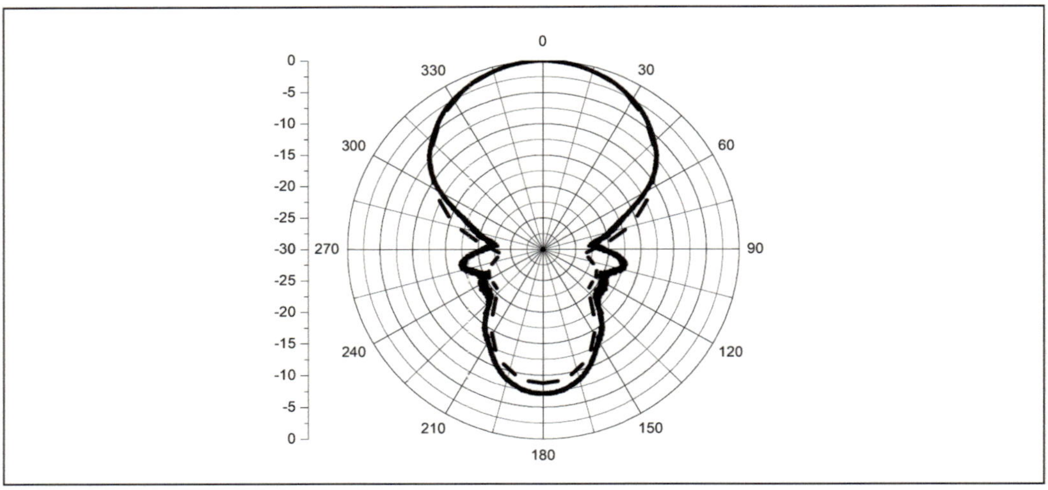

Fig. 24. Radiation pattern of the metamaterial tube waveguide antenna with the thickness d = 5 mm at a frequency of 3 GHz.

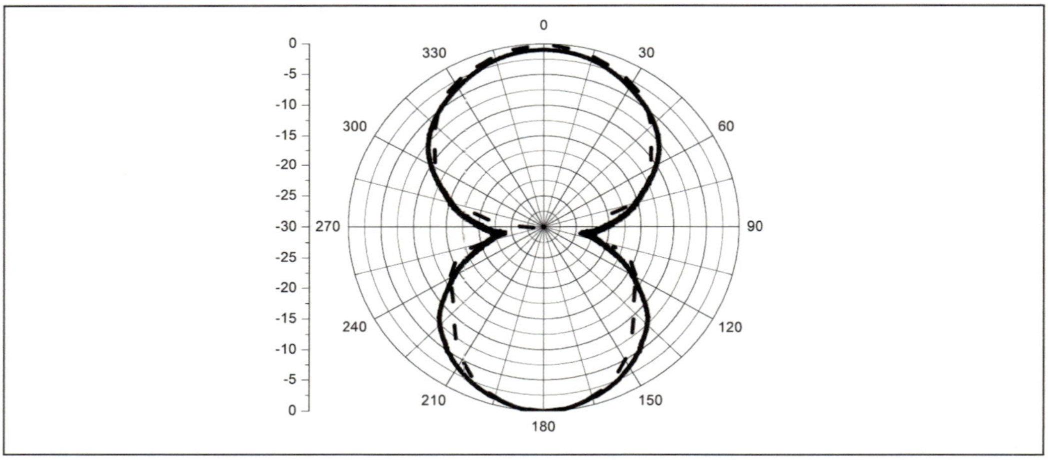

Fig. 25. Radiation pattern of the metamaterial tube waveguide antenna with the thickness d = 10 mm at a frequency of 3 GHz.

The thickness $d > 10$ mm of the metamaterial waveguide tube corresponds to the negative total power flux S_Σ; i.e., the wave propagates in the backward direction (180°). This corresponds to an increase in the amplitude of the back lobe and the "degeneration" of the main lobe. Figures 26, 27 are plotted for d = 15, 20 mm, respectively. At a thickness of 20 mm (Fig. 27), the difference between the back and front lobes of the radiation pattern is more than 15 dB and the antenna radiates predominantly in the backward direction 180°.

Thus, in the case of zero total power flux in the metamaterial waveguide, the antenna radiates equally in the directions 0° and 180°. In the presence of the forward and backward wave fields, the antenna radiates in the forward (0°) and backward (180°) directions, respectively.

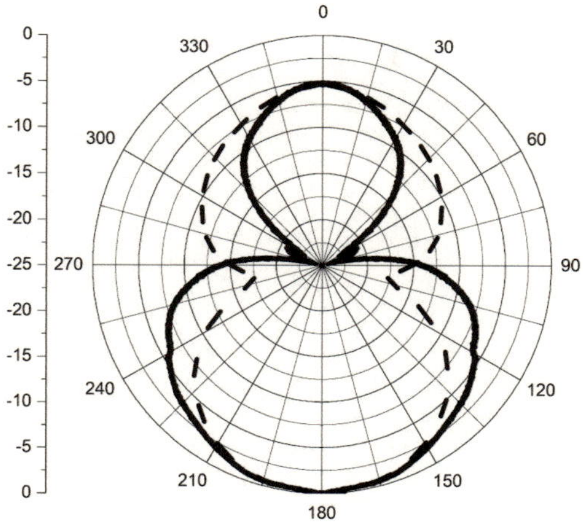

Fig. 26. Radiation pattern of the metamaterial tube waveguide antenna with the thickness *d* = 15 mm at a frequency of 3 GHz.

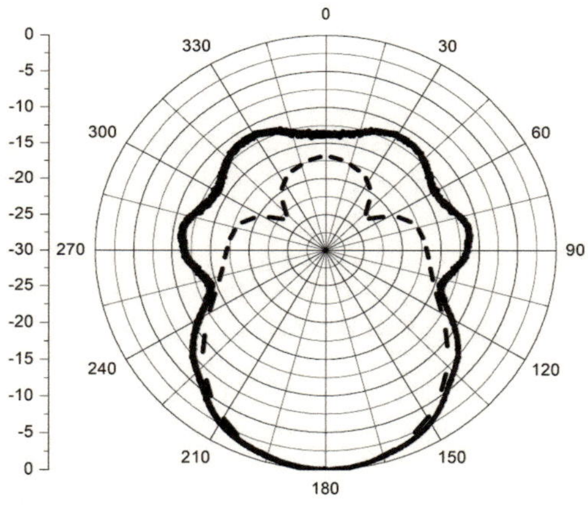

Fig. 27. Radiation pattern of the metamaterial tube waveguide antenna with the thickness *d* = 20 mm at a frequency of 3 GHz.

The backward radiation effect is observed only at the negative values of the real parts of permittivity and permeability. Figure 28 presents the radiation pattern of the metamaterial waveguide tube with the thickness d = 20 mm at a frequency of 3.5 GHz, for which permittivity and permeability are positive according to Fig. 21.

In this case, the radiation pattern is usual with the main lobe greater than the back lobe. That is, the antenna radiates mainly in the forward direction. To conclude, we have experimentally demonstrated the effect of the backward electromagnetic radiation of the metamaterial waveguide tube antenna. We have shown that the effect is observed in the presence of a backward wave field for the negative permittivity and permeability of the metamaterial, whereas the effect does not occur for the positive material parameters.

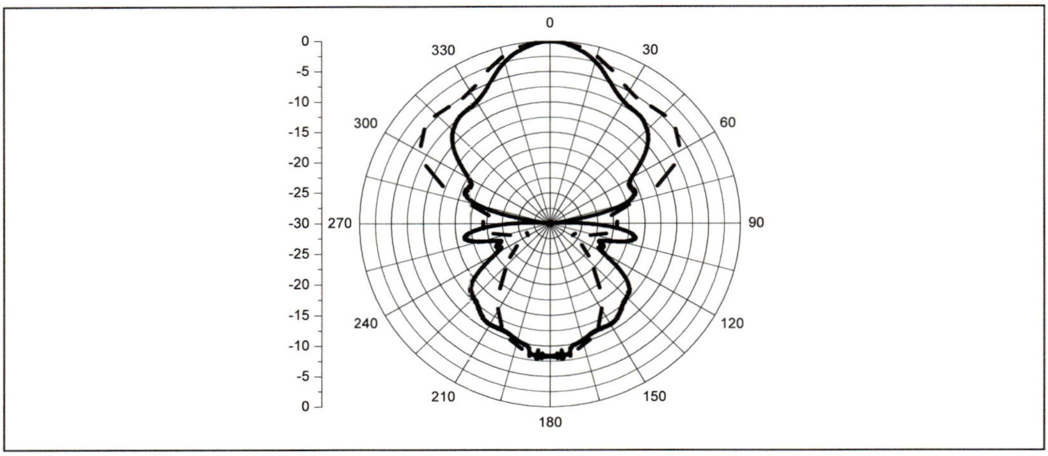

Fig. 28. Radiation pattern of the metamaterial tube waveguide antenna with the thickness $d = 20$ mm at a frequency of 3.5 GHz.

Next, we show that only the backward wave is the cause of the backward radiation effect. Of course, the radiation of the waveguide also contributes to the reflection from the open end. But the dimensions of the metamaterial waveguide are chosen so that the influence of reflected waves can be neglected, Volakis (2007), Angulo (1957).

We make a numerical experiment (MoM). Take our waveguide structure of the metamaterial (Fig. 18, 19), with the thickness of the tube $d = 20$ *mm*, and construct the E field distribution on frequency 3 GHz (ε=-1.5+i0.7, μ=-0.7+i0.4) in the three cases. 1) the open end of the waveguide (Fig. 29 a), 2) the open end of the waveguide with a reflection coefficient of zero (Perfectly matched layers) (Fig. 29 b), 3) open end of the waveguide with a reflection coefficient of one (Perfectly conducting surfaces) (Fig. 29 c). A comparison of figures (Fig. 29) shows that the structure radiates the same in the direction 180^0 for all cases. In the case 1 (Fig. 29 a) part of the wave of the leak in the direction 0^0. In case 2 (Fig. 29 b), the wave is not

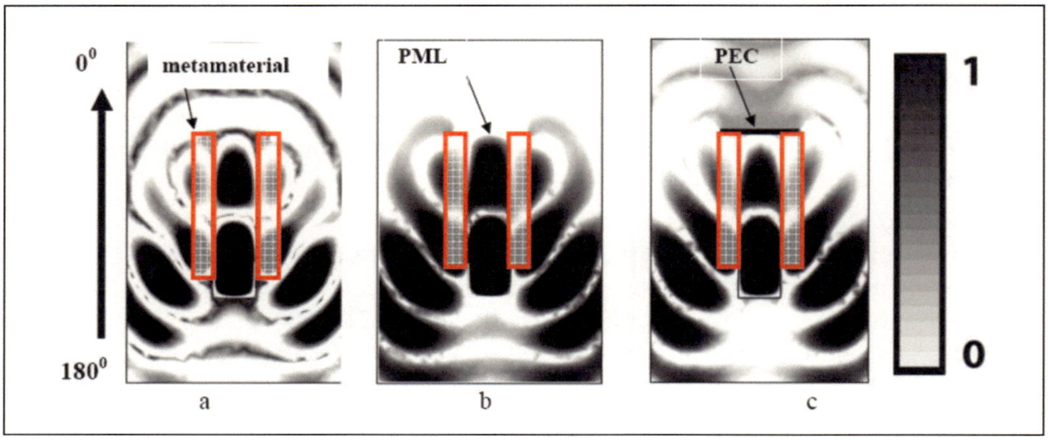

Fig. 29. Normalize electric field E. a) the open end of the waveguide, b) the open end of the waveguide with a reflection coefficient of zero (Perfectly matched layers), c) open end of the waveguide with a reflection coefficient of one (Perfectly conducting surfaces).

reflected from the end of the waveguide, and are absorbed in the PML and the radiation in the direction 0^0 of can be neglected. In case 3 (Fig. 29 c), there is diffraction on a conducting surface, which forms a small part of radiation in the direction 0^0 and in the direction 180^0.
Thus, the reflection from the open end of the metamaterial waveguide introduces a small perturbation in the total field, which does not affect the formation of reverse radiation. As a result, the main cause of back radiation is excited in the waveguide backward wave and the presence of negative permittivity and permeability.
Note that similar effect observed Sheng et al. (2009) by experimental modelling of the reversed Cherenkov radiations.
Propagation of electromagnetic waves in a metamaterial waveguide with negative values of the relative permittivity and relative permeability has been studied. It has been shown, that this waveguide can support forward, backward, and standing waves. In the case of the forward wave, an antenna manufactured on the basis of such a waveguide can radiate in a direction of 0°. In the case of the backward wave, the backward radiation effect is observed, i.e., the radiation pattern is formed in a direction of 180°. The condition for the backward radiation effect in the antennas built on the basis of metamaterial waveguides has been obtained. As follows from this condition, the backward radiation effect arises only in the antennas manufactured on the basis of waveguides consisting of metamaterials with negative values of the relative permittivity and relative permeability in the case of propagation of the backward wave.

7. Conclusion

Propagation of electromagnetic waves in a planar metamaterial waveguide with negative values of the relative permittivity and relative permeability has been studied. It has been shown, that this waveguide can support forward, backward, and standing waves. In the case of the forward wave, an antenna manufactured on the basis of such a waveguide can radiate in a direction of 0°. In the case of the backward wave, the backward radiation effect is observed, i.e., the radiation pattern is formed in a direction of 180°. The condition for the backward radiation effect in the antennas built on the basis of metamaterial waveguides has been obtained. As follows from this condition, the backward radiation effect arises only in the antennas manufactured on the basis of waveguides consisting of metamaterials with negative values of the relative permittivity and relative permeability in the case of propagation of the backward wave. The possibility of the preferential backward radiation of the structure has been shown on the basis of numerical calculations with the use of the method of moments and the measurements of the radiation pattern of the antenna in an anechoic chamber at frequencies close to the 3 GHz resonance frequency of the metamaterial.

8. Acknowledgements

Authors express their thanks to Vinogradov A.P. and Shevchenko V.V. for useful discussion.

9. References

Aizenberg G. Z. (1977), *Antennas UHF,* Svyaz, Moscow.
Alù A. and N. Engheta (2007a), Anomalies of subdiffractive guided wave propagation along metamaterial nanocomponents, *Radio Sci.*, *42*, RS6S17, doi:10.1029/2007RS003691.

Alù A. and Engheta N. (2005), "An Overview of Salient Properties of Planar Guided-Wave Structures with Double-Negative (DNG) and Single-Negative (SNG) Layers," *Negative-Refraction Metamaterials: Fundamental Principles and Applications*, G. V. Eleftheriades and K. G. Balmain (Ed.),IEEE Press, John Wiley and Sons Inc., Hoboken, NJ, Chapter 9, 339-380.

Alù A., Bilotti F., Engheta N., Vegni L.(2007b), *IEEE Trans..* V. AP-55. No 6, P. 1698.

Angulo C. M. (1957), *IRE Trans. Microwave Theory Tech.* 5, 1.

P. Baccarelli, P. Burghignoli, G. Lovat, and S. Paulotto (2003), *IEEE Antennas Wireless Propag. Letter.*, Vol. 2, no 19., pp. 269-272.

Balabukha N. P., Basharin A. A., and Semenenko V. N. (2009), *JETP Letters*, vol. 89, no. 10, pp. 500-505.

Balanis C. A. (1997), *Antenna Theory: Analysis and Design,*Wiley, New York.

Basharin A.A., Balabukha N.P., and Semenenko V.N. (2010), *J. Appl. Phys.* 107, 113301.

Caloz C., Itoh T., (2006), *Electromagnetic metamaterials: transmission line theory and microwave - applications.* N.Y.: John Wiley and Sons, Inc.

Grbic A. and Eleftheriades G. V. (2002), *J. Appl. Phys.* 92, 5930.

Gulyaev Yu. V., Lagar'kov A. N., and Nikitov S. A. (2008), *Vestn. RAN* , 78 (5).

Hanson, G., and A. Yakovlev (1999), Investigation of mode interaction on planar dielectric waveguides with loss and gain, Radio Sci., 34(6), 1349-1359.

Katin S.V., Titarenko A.A. (2006), *Antennas.* Vol 5. (108). p.24-27.

Lagarkov A. N. (1990), *Electrophysical Properties of Percolation Systems*, IVTAN, Moscow, [in Russian].

Lagarkov A.N., Semenenko V.N., Chistyaev V.A. et al (1997), *Electromagnetics.* V.17. P 213

Lagarkov A.N., Semenenko V. N., Kisel V. N., and Chistyaev V. A. (2003), *J. Magn. Magn Mater.* V 161, P 258–259.

Lagarkov A.N., Kissel V.N (2004), *Phys. Rev. Lett.* V.92. P.077401.

Marcuvitz N. (1951), *Waveguide Handbook,* MCGraw-Hill, New York.

Muller D.E. (1965).Mathematical Tables and Aids to Computation. T.10. P.208.

Nefedov, I. S. and S. A. Tretyakov (2003), Waveguide containing a backward wave slab, *Radio Sci., 38,* 1101, doi:10.1029/2003RS002900I.

Pendry J.B. (2000), *Phys. Rev. Lett..* V.85. P.3966

Shadrivov W., Sukhorukov A.A., and Kivshar Y.S. (2003), *Phys. Rev. E,* vol 67, pp.57602-1-57602-4

Shatrov A.D., Shevchenko V.V (1974), Radiophysics.. Vol. 11. C.1692.

Sheng Xi, Hongsheng Chen, Tao Jiang, Lixin Ran, Jiangtao Huangfu, Bae-Ian Wu, Jin Au Kong, and Min Chen (2009), Phys. Rev. Lett. 103, 194801.

Shevchenko V.V. (1969), *Radiotekh. Elektron.*, Vol.50, P.1768

Shevchenko V.V (2005), *Radiotekh. Elektron...* Vol. 50. p. 1363.

Shevchenko V.V. Radiotekh. Elektron. (2000), Vol. 10. p. 1157

Smith D.R., Padilla W.T., Vier D.C. et al. (2000), *Phys. Rev. Lett.* V. 84. P. 584.

Solymar L., Shamonina E. (2009), Waves in Metamaterials, OXFORD Press, New York.

Vainshtein L. A. (1988), *Electromagnetic Waves* (Radio I Svyaz', Moscow,) [in Russian].

Veselago V. G. (1967), *Usp. Fiz. Nauk*, Vol. 92, P. 517

Volakis J. A. (2007), *Antenna Engineering Handbook*, 4th ed., McGraw-Hill, New York,.

Wu B.-I., Grzegorczyk T. M., Zhang Y., and Kong J. A., *J. Appl. Phys* (2003), Vol. 93, P. 9386.

13

On the Electrodinamics of Space-Time Periodic Mediums in a Waveguide of Arbitrary Cross Section

Eduard A. Gevorkyan
Doctor of physical-mathematical science, professor
Moscow State university of Economics,Statistics and Informatics,
7 Nezhinskaya St., Moscow, 119501
Russia

1. Introduction

Theoretical investigations of parametric interaction between the electrodynamics waves and space-time periodic filling of the waveguide of arbitrary cross section are reviewed. The cases of dielectric, anisotropic and magnetodielectric periodically modulated filling are considered. The analytical method of solution of the problems of electrodynamics of space-time periodic mediums in a waveguide is given. The propagation of transverse-electric (TE) and transverse-magnetic (TM) waves in the waveguide mentioned above are investigated. Physical phenomena of electrodynamics of space-time periodic mediums in the region of "weak" and "strong" interactions between the travelling wave in the waveguide and the modulation wave are studied.

Propagation of electromagnetic waves in the medium whose permittivity and permeability are modulated in space and time with help of pump waves of various nature (electromagnetic wave, ultrasonic wave, etc.) under the harmonic law, represents one of the basic problem of the electromagnetic theory. In the scientific literature the most part of such researches concerns to electrodynamics of periodically non-stationary and non-uniform mediums in the unlimited space [1-15], while the same problems in the limited modulated mediums, for example, in the waveguides of arbitrary cross section remain still insufficiently studied and there is no strict analytical theory of the propagation of electromagnetic waves in similar systems (although in the scientific literature already appeared the articles on the problems, mentioned above [16-25].

Meanwhile the investigation of the propagation of electromagnetic waves in the waveguides with space-time periodically modulated filling represents a great interest not only from point of view of development of theory but also from point of view of possibility of practice application of similar waveguides in the ultrahigh frequency electronics. For example, the waveguides with periodically non-stationary and non-uniform filling can be applied for designing of multifrequency distributing back-coupling lasers (DBS lasers), Bragg reflection lasers (DBR lasers), mode transformers, parametric amplifiers, multifrequency generators, transformers of low and higher frequency, Bragg resonators and filters, prismatic polarizer, diffraction lattices, oscillators, mode converters, wave-channeling devices with a fine periodic structure, etc [14], [26-30].

2. Electromagnetic waves in a waveguide with space-time periodic filling

Let us consider the regular ideal waveguide of arbitrary cross section which axis coincides with the OZ axis of certain Cartesian frame. Let the permittivity and permeability of the filling of the waveguide with help of pump wave are modulated in space and time under the periodic law (Fig.1.1) [23, 25]

$$\varepsilon = \varepsilon_0\left[1+m_\varepsilon \cos k_0(z-ut)\right],\ \mu = \mu_0\left[1+m_\mu \cos k_0(z-ut)\right] \tag{1.1}$$

where m_ε и m_μ are the modulation indexes, u is the modulation wave velocity , k_0 is the modulation wave number, $k_0 u$ is the modulation wave frequency, ε_0 и μ_0 are the permittivity and permeability of the filling in the absence of modulation. The signal wave with frequency ω_0

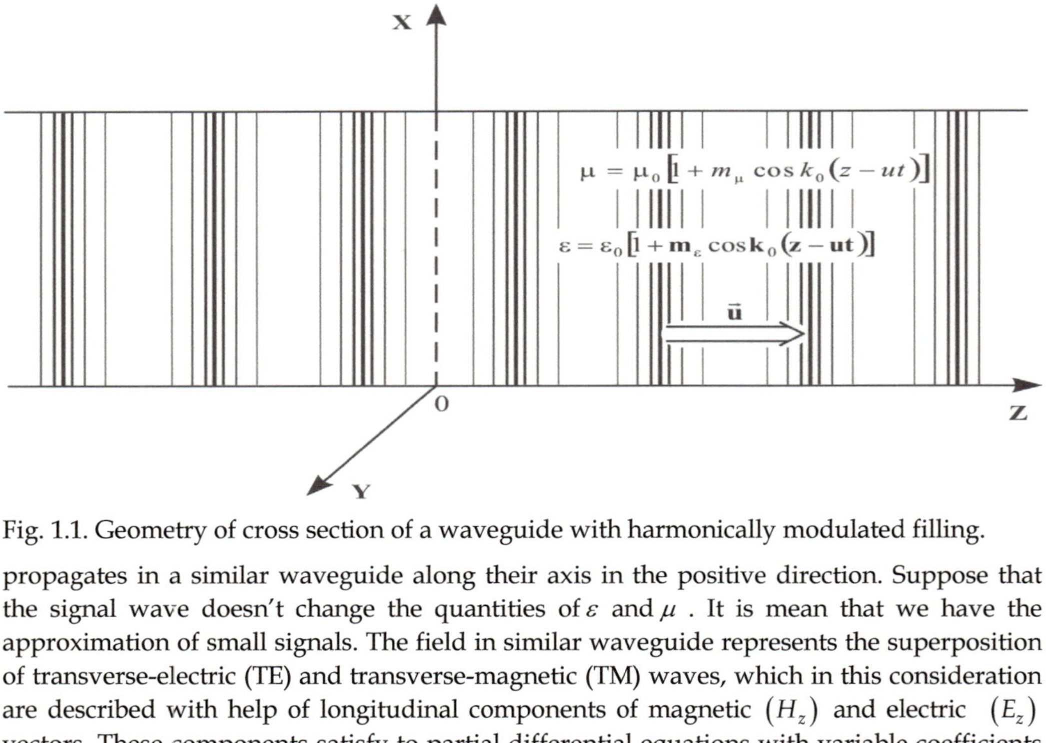

Fig. 1.1. Geometry of cross section of a waveguide with harmonically modulated filling.

propagates in a similar waveguide along their axis in the positive direction. Suppose that the signal wave doesn't change the quantities of ε and μ . It is mean that we have the approximation of small signals. The field in similar waveguide represents the superposition of transverse-electric (TE) and transverse-magnetic (TM) waves, which in this consideration are described with help of longitudinal components of magnetic (H_z) and electric (E_z) vectors. These components satisfy to partial differential equations with variable coefficients which are obtained from the Maxwell equations taking into account that the charge density and the current density are equal to zero. These wave equations have a form [23-25], [31]

TE field

$$\Delta_\perp H_z + \frac{\partial}{\partial z}\left(\frac{1}{\mu}\frac{\partial(\mu H_z)}{\partial z}\right) - \frac{1}{c^2}\frac{\partial}{\partial t}\left(\varepsilon\frac{\partial(\mu H_z)}{\partial t}\right) = 0, \tag{1.2}$$

TM field

$$\Delta_{\perp}E_z + \frac{\partial}{\partial z}\left(\frac{1}{\varepsilon}\frac{\partial(\varepsilon E_z)}{\partial z}\right) - \frac{1}{c^2}\frac{\partial}{\partial t}\left(\mu\frac{\partial(\varepsilon E_z)}{\partial t}\right) = 0\,, \tag{1.3}$$

where $\Delta_{\perp} = \partial^2/\partial x^2 + \partial^2/\partial y^2$ is the two-dimensional Laplace operator, c is the velocity of light in vacuum.

The solution of wave equations (1.2) and (1.3) we look for the form of decomposition by orthonormal eigenfunctions of the second and first boundary-value problems for the cross section of the waveguide $(\widehat{\psi}_n(x,y),\psi_n(x,y))$. These functions satisfy to the following Helmholtz equations with corresponding boundary conditions on the surface of the waveguide [23], [31]:

second boundary-value problem

$$\Delta_{\perp}\widehat{\Psi}_n(x,y) + \widehat{\lambda}_n^2\widehat{\Psi}_n(x,y) = 0, \quad \left.\frac{\partial\widehat{\Psi}_n}{\partial\vec{n}}\right|_{\Sigma} = 0\,, \tag{1.4}$$

first boundary-value problem

$$\Delta_{\perp}\Psi_n(x,y) + \lambda_n^2\Psi_n(x,y) = 0, \quad \Psi_n(x,y)\big|_{\Sigma} = 0\,, \tag{1.5}$$

where $\widehat{\lambda}_n$ and λ_n are the eigenvalues of the second and first boundary-value problems for the transverse cross section of the waveguide, Σ is the contour of the waveguide's cross section, $\vec{n}$ is the normal to Σ. From Maxwell equations the transverse components of transverse-electric (TE) and transverse-magnetic (TM) fields can be represented in terms of

$$H_z(x,y,z,t) = \sum_{n=0}^{\infty} H_n(z,t)\cdot\widehat{\Psi}_n(x,y),\; E_z(x,y,z,t) = \sum_{n=0}^{\infty} E_n(z,t)\cdot\Psi_n(x,y) \tag{1.6}$$

as follows [23]:

TE field

$$\vec{H}_{\tau}(x,y,z,t) = \frac{1}{\mu(z,t)}\sum_{n=0}^{\infty}\widehat{\lambda}_n^{-2}\frac{\partial\left[\mu(z,t)H_n(z,t)\right]}{\partial z}\nabla\widehat{\Psi}_n(x,y), \tag{1.7}$$

$$\vec{E}_{\tau}(x,y,z,t) = \frac{1}{c}\sum_{n=0}^{\infty}\widehat{\lambda}_n^{-2}\frac{\partial\left[\mu(z,t)\,H_n(z,t)\right]}{\partial t}\left[\vec{z}_0\,\nabla\widehat{\Psi}_n(x,y)\right], \tag{1.8}$$

TE field

$$\vec{H}_{\tau}(x,y,z,t) = -\frac{1}{c}\sum_{n=0}^{\infty}\lambda_n^{-2}\frac{\partial\left[\varepsilon(z.t)E_n(z,t)\right]}{\partial t}\left[\vec{z}_0\,\nabla\Psi_n(x,y)\right], \tag{1.9}$$

$$\vec{E}_{\tau}(x,y,z,t) = \frac{1}{\varepsilon(z,t)}\sum_{n=0}^{\infty}\lambda_n^{-2}\frac{\partial\left[\varepsilon(z,t)E_n(z,t)\right]}{\partial z}\nabla\Psi_n(x,y), \tag{1.10}$$

where $\nabla = \vec{i}(\partial / \partial x) + \vec{j}(\partial / \partial y)$ is the two-dimensional nabla operator.

If into the wave equations (1.2) and (1.3) of variables z and t to introduce the new quantities by the formulas

$$\tilde{H}_z = \mu H_z, \quad \tilde{E}_z = \varepsilon E_z \tag{1.11}$$

and to pass to the new variables ξ and η according to the formulas [22]

$$\xi = z - ut, \quad \eta = \frac{z}{u} - \frac{1}{u}\int_0^{\xi} \frac{d\xi}{1 - \beta^2 \frac{\varepsilon(\xi)\,\mu(\xi)}{\varepsilon_0\,\mu_0}}, \tag{1.12}$$

where $\beta^2 = u^2 \varepsilon_0 \mu_0 / c^2$ and when $u \to 0$ then $\xi \to z, \eta \to t$, and the solutions of received partial differential equations to look for the form [22]

$$\tilde{H}_z = \sum_{n=0}^{\infty} e^{i\gamma\eta} \tilde{H}_{nz}(\xi) \cdot \hat{\Psi}_n(x,y), \quad \tilde{E}_z = \sum_{n=0}^{\infty} e^{i\gamma\eta} \tilde{E}_{nz}(\xi) \cdot \Psi_n(x,y), \tag{1.13}$$

taking into account the orthonormalization of the eigenfunctions $\hat{\psi}_n(x,y)$ and $\psi_n(x,y)$ then we receive for $\tilde{H}_{nz}$ and $\tilde{E}_{nz}$ the following ordinary differential equations with variable coefficients:

$$\mu \frac{d}{d\xi}\left[\frac{1}{\mu}\left(1 - \beta^2 \frac{\varepsilon\,\mu}{\varepsilon_0\,\mu_0}\right)\frac{d\tilde{H}_{nz}}{d\xi}\right] + \frac{\hat{\chi}_n^2}{1 - \beta^2 \frac{\varepsilon\,\mu}{\varepsilon_0\,\mu_0}} \tilde{H}_{nz} = 0, \tag{1.14}$$

$$\varepsilon \frac{d}{d\xi}\left[\frac{1}{\varepsilon}\left(1 - \beta^2 \frac{\varepsilon\,\mu}{\varepsilon_0\,\mu_0}\right)\frac{d\tilde{E}_{nz}}{d\xi}\right] + \frac{\chi_n^2}{1 - \beta^2 \frac{\varepsilon\,\mu}{\varepsilon_0\,\mu_0}} \tilde{E}_{nz} = 0, \tag{1.15}$$

where

$$\hat{\chi}_n^2 = \frac{\gamma^2}{c^2}\varepsilon\,\mu - \hat{\lambda}_n^2\left(1 - \beta^2 \frac{\varepsilon\,\mu}{\varepsilon_0\,\mu_0}\right), \quad \chi_n^2 = \frac{\gamma^2}{c^2}\varepsilon\,\mu - \lambda_n^2\left(1 - \beta^2 \frac{\varepsilon\,\mu}{\varepsilon_0\,\mu_0}\right). \tag{1.16}$$

In this investigation we are limited of small quantities of modulation indexes of the waveguide filling. It is explained that in real experiment the modulation indexes are very small and they can change from 10^{-4} to $4 \cdot 10^{-2}$ (the quantity $4 \cdot 10^{-2}$ is fixed in the chrome gelatin). Note that if the velocity of modulation wave satisfies the condition $u \le 0{,}8 \cdot \upsilon_{\text{ph}}$, where $\upsilon_{ph} = c / \sqrt{\varepsilon_0\,\mu_0}$ is the phase velocity in the non-disturbance medium, then side by side of modulation indexes is small the parameter $\ell = \left(m_\varepsilon + m_\mu\right)\beta^2 / b$ ($b = 1 - \beta^2$) too, that is $l \ll 1$.

Then with help of changed of variables

$$\hat{s}=\frac{k_0\left(1-\beta^2\right)}{2\mu_0}\int_0^{\xi}\frac{\mu\,d\xi}{1-\beta^2\frac{\varepsilon\mu}{\varepsilon_0\mu_0}},\quad s=\frac{k_0\left(1-\beta^2\right)}{2\varepsilon_0}\int_0^{\xi}\frac{\varepsilon\,d\xi}{1-\beta^2\frac{\varepsilon\mu}{\varepsilon_0\mu_0}} \tag{1.17}$$

and taking into account that permittivity and permeability of the filling change by the harmonic law (1.1) the above received differential equations (1.14) and (1.15) on variables ξ and η are transformed to the differential equations with periodic coefficients of Mathie-Hill type [32]. In the first approximation on small modulation indexes they have a form [23]

$$\frac{d^2\tilde{H}_{nz}(\hat{s})}{d\hat{s}^2}+\sum_{k=-1}^{1}\hat{\theta}_k^n\,e^{2ik\hat{s}}\tilde{H}_{nz}(\hat{s})=0, \tag{1.18}$$

$$\frac{d^2\tilde{E}_{nz}(s)}{ds^2}+\sum_{k=-1}^{1}\theta_k^n\,e^{2iks}\tilde{E}_{nz}(s)=0, \tag{1.19}$$

where

$$\hat{\theta}_0^n=\frac{4}{k_0^2b^2}\left(\hat{\chi}_0^n\right)^2,\quad \theta_0^n=\frac{4}{k_0^2b^2}\left(\chi_0^n\right)^2, \tag{1.20}$$

$$\hat{\theta}_{\pm1}^n=\frac{2}{k_0^2b^2}\left(\frac{\gamma^2}{u^2}-\left(\hat{\chi}_0^n\right)^2\right)\ell-\frac{4}{k_0^2b^2}\left(\hat{\chi}_0^n\right)^2m_\mu, \tag{1.21}$$

$$\theta_{\pm1}^n=\frac{2}{k_0^2b^2}\left(\frac{\gamma^2}{u^2}-\left(\chi_0^n\right)^2\right)\ell-\frac{4}{k_0^2b^2}\left(\chi_0^n\right)^2m_\varepsilon, \tag{1.22}$$

$$\left(\hat{\chi}_0^n\right)^2=\frac{\gamma^2}{c^2}\varepsilon_0\mu_0-\hat{\lambda}_n^2\,b,\quad \left(\chi_0^n\right)^2=\frac{\gamma^2}{c^2}\varepsilon_0\mu_0-\lambda_n^2\,b. \tag{1.23}$$

The solutions of the equations (1.18) and (1.19) we look for the form

$$\tilde{H}_{nz}(\hat{s})=e^{i\hat{\mu}_n\hat{s}}\sum_{k=-1}^{1}\hat{C}_k^n\cdot e^{2ik\hat{s}},\quad \tilde{E}_{nz}(s)=e^{i\mu_n s}\sum_{k=-1}^{1}C_k^n\cdot e^{2iks}. \tag{1.24}$$

Substituting these expressions into Mathie-Hill equations (1.18) and (1.19) for determination of characteristic indexes $\hat{\mu}_n$ and μ_n we receive the following dispersion equations:

$$\hat{\mu}_n^2=\hat{\theta}_0^n+\frac{\left(\hat{\theta}_1^n\right)^2}{\left(\hat{\mu}_n-2\right)^2-\hat{\theta}_0^n}+\frac{\left(\hat{\theta}_1^n\right)^2}{\left(\hat{\mu}_n-2\right)^2-\hat{\theta}_0^n}, \tag{1.25}$$

$$\mu_n^2=\theta_0^n+\frac{\left(\theta_1^n\right)^2}{\left(\mu_n-2\right)^2-\theta_0^n}+\frac{\left(\theta_1^n\right)^2}{\left(\mu_n-2\right)^2-\theta_0^n}. \tag{1.26}$$

The analysis of these dispersion equations show that under the following conditions [33]

$$\left|1-\widehat{\theta}_0^n\right| \geq \widehat{\delta}_n \approx \ell, \quad \left|1-\theta_0^n\right| \geq \delta_n \approx \ell \tag{1.27}$$

we become to the region of weak interaction between the signal wave and the modulation wave where the characteristic indexes $\widehat{\mu}_n$ and μ_n are real and then the Mathie-Hill equations have the stable solutions. With help of obtained solutions of dispersion equations

$$(\widehat{\mu}_n)^2 = \widehat{\theta}_0^n, \quad (\mu_n)^2 = \theta_0^n \tag{1.28}$$

and the expressions for the coefficients

$$\widehat{C}_{\pm 1}^n = \frac{\widehat{\theta}_1^n \cdot \widehat{C}_0^n}{4\left(1 \pm \sqrt{\widehat{\theta}_0^n}\right)}, \quad C_{\pm 1}^n = \frac{\theta_1^n \cdot C_0^n}{4\left(1 \pm \sqrt{\theta_0^n}\right)}, \tag{1.29}$$

where $\widehat{C}_0^n$ and C_0^n are defined from the conditions of normalizing, we obtained the analytical expressions for the H_z and E_z of TE and TM fields in the waveguide in the region of weak interaction. They have a form [33]

$$H_z = \frac{1}{\mu_0} \sum_{n=0}^{\infty} \Psi_n(x,y) e^{i\left(\widehat{P}_0^n z - \omega_0 t\right)} \cdot \widehat{C}_0^n \sum_{k=-1}^{1} \widehat{V}_k^n \cdot e^{ikk_0(z-ut)}, \tag{1.30}$$

$$E_z = \frac{1}{\varepsilon_0} \sum_{n=0}^{\infty} \Psi_n(x,y) e^{i\left(P_0^n z - \omega_0 t\right)} \cdot C_0^n \sum_{k=-1}^{1} V_k^n \cdot e^{ikk_0(z-ut)}, \tag{1.31}$$

where

$$\widehat{V}_k^n = \left(k \cdot \frac{\widehat{\Delta}_0^n}{2} + \frac{\widehat{C}_k^n}{\widehat{C}_0^n} - \frac{m_\mu}{2}\right)^{|k|}, \quad V_k^n = \left(k \cdot \frac{\Delta_0^n}{2} + \frac{C_k^n}{C_0^n} - \frac{m_\varepsilon}{2}\right)^{|k|}, \tag{1.32}$$

$$\widehat{\Delta}_0^n = \frac{\widehat{\mu}_n}{2} m_\mu + \frac{\omega_0}{k_0 u}\ell, \quad \Delta_0^n = \frac{\mu_n}{2} m_\varepsilon + \frac{\omega_0}{k_0 u}\ell, \tag{1.33}$$

$$\widehat{\theta}_{\pm 1}^n = \frac{2}{k_0^2 \beta^2 b}\left[\left(\widehat{\chi}_0^n\right)^2 + \widehat{\lambda}_n^2\right]\ell - \frac{4}{k_0^2 b^2}\left(\widehat{\chi}_0^n\right)^2 m_\mu, \tag{1.34}$$

$$\theta_{\pm 1}^n = \frac{2}{k_0^2 \beta^2 b}\left[\left(\chi_0^n\right)^2 + \lambda_n^2\right]\ell - \frac{4}{k_0^2 b^2}\left(\chi_0^n\right)^2 m_\varepsilon, \tag{1.35}$$

$$\left(\widehat{P}_0^n\right)^2 = \frac{(\omega_0)^2}{c^2}\varepsilon_0\mu_0 - \widehat{\lambda}_n^2, \left(P_0^n\right)^2 = \frac{(\omega_0)^2}{c^2}\varepsilon_0\mu_0 - \lambda_n^2. \tag{1.37}$$

As is seen from the expressions (1.30) and (1.31) TE and TM fields in the waveguide with modulated filling are represented as the set of space-time harmonics with different

amplitudes. At that time the amplitude of the zero (fundamental) harmonic are independent of small modulation indexes, while the amplitudes of the plus and minus first harmonics (side harmonics) are proportional to the small modulation indexes in the first degree.
At the realization the following condition [31], [33]

$$\left|1-\widehat{\theta}_0^n\right| \leq \widehat{\delta}_n, \tag{1.38}$$

where

$$\widehat{\delta}_n = \frac{\widehat{\eta}_n^2 - \beta_\varepsilon^2}{4\sqrt{2}\beta_\varepsilon^2}\ell_\varepsilon,\ \widehat{\eta}_n = \sqrt{1+\frac{4\widehat{\lambda}_n^2}{k_0^2 b_\varepsilon}},\ \ell_\varepsilon = \frac{m_\varepsilon \beta_\varepsilon^2}{b_\varepsilon}, \tag{1.39}$$

$$\widehat{\theta}_{\varepsilon_0}^n = \frac{4\left(u\sqrt{\frac{(\omega_0)^2}{C^2}\varepsilon_0 - \widehat{\lambda}_n^2} - \beta_\varepsilon^2\,\omega_0\right)^2}{k_0^2 u^2 b_\varepsilon^2},\ \beta_\varepsilon = \frac{u}{c}\sqrt{\varepsilon_0}, b_\varepsilon = 1-\beta_\varepsilon^2 \tag{1.40}$$

(here are shown the results for the TE field when $\mu = 1$) the strong (resonance) interaction between the signal wave and the modulation wave takes place, when occurs the considerable energy exchange between them. The analytical expression for the frequency of strong interaction is found in the form

$$\omega_{0,s} - \Delta\omega_0 \leq \omega_0 \leq \omega_{0,s} + \Delta\omega_0,\ \omega_{0,s} = \frac{uk_0}{2\beta_\varepsilon}\left(\widehat{\eta}_n + \beta_\varepsilon\right) \tag{1.41}$$

and is shown that the width of strong interaction is small and proportional to the modulation index in the first degree [31], [33]

$$\Delta\omega_0 = \frac{k_0 u}{8\sqrt{2}\beta_\varepsilon}\frac{\left(1+\widehat{\eta}_n\beta_\varepsilon\right)\left(\widehat{\eta}_n^2 - \beta_\varepsilon^2\right)}{\widehat{\eta}_n}\ell_\varepsilon. \tag{1.42}$$

In the region of strong interaction the dispersion equation (1.25) has complex solutions in the following form

$$\widehat{\mu}_{\varepsilon,n} = 1 \pm i\frac{\sqrt{\left(\widehat{\theta}_{\varepsilon 1}^n\right)^2 - \widehat{\delta}_n^2}}{2},\qquad \widehat{\theta}_{\varepsilon,1}^n = 2\sqrt{2}\widehat{\delta}_n. \tag{1.43}$$

(1.43) allow to receive the analytical expressions for the amplitudes of different harmonics in the form [31], [33]

$$\left|\widehat{V}_{\varepsilon,-1}^n\right| \approx 1,\ \left|\widehat{V}_{\varepsilon,1}^n\right| \approx \frac{(\widehat{\eta}_n + \beta_\varepsilon)(\widehat{\eta}_n + 3\beta_\varepsilon)}{16\beta_\varepsilon^2}\ell_\varepsilon. \tag{1.44}$$

The analysis of these expressions shows that in the case of forward modulation, when the directions of propagation of the signal wave and the modulation wave coincide, the amplitude of minus first harmonic doesn't depend from the modulation index, while the amplitude of the plus first harmonic is proportional to the modulation index in the first degree. In other words in the region of strong interaction besides the fundamental harmonic the substantial role plays the minus first harmonic reflected from the periodic structure of the filling on the frequency

$$\omega_{-1,s} = \omega_{0,s} - k_0 u = \frac{k_0 u}{2\beta_\varepsilon}\left(\widehat{\eta}_n - \beta_\varepsilon\right), \ \widehat{\eta}_n > \beta_\varepsilon . \tag{1.45}$$

In the backward modulation case, when the directions of propagation of the signal wave and the modulation wave don't coincide, the minus first and plus first harmonics change their roles.

The results received above admit the visual physical explanation of the effect of strong interaction between the signal wave and the modulation wave. Below the physical explanation we show by example of TE field in the case of forward modulation. The zero harmonic in the modulated filling of the waveguide is incident on the density maxima of the filling at the angle $\widehat{\varphi}^n_{\varepsilon,0}$ and is reflected from them at the angle $\widehat{\varphi}^n_{\varepsilon,1}$ (Fig.2). These angles are defined from the following correlations [33]

$$\cos\widehat{\varphi}^n_{\varepsilon,0} = \frac{c}{\omega_0\sqrt{\varepsilon_0}}\sqrt{\frac{\omega_0^2}{c^2}\varepsilon_0 - \widehat{\lambda}_n^2}, \ \cos\widehat{\varphi}^n_{\varepsilon,1} = \frac{\left(1+\beta_\varepsilon^2\right)\cos\widehat{\varphi}^n_{\varepsilon_0} - 2\beta_\varepsilon}{1+\beta_\varepsilon^2 - 2\beta_\varepsilon\cos\widehat{\varphi}^n_{\varepsilon,0}}. \tag{1.46}$$

At that time the incident and reflection angles are different because of the moving of the modulation wave of the filling and the frequencies of incident and reflected waves satisfy to the following correlation [33]

$$\omega_1 \cdot \sin\widehat{\varphi}^n_{\varepsilon,1} = \omega_0 \cdot \sin\widehat{\varphi}^n_{\varepsilon,0} . \tag{1.47}$$

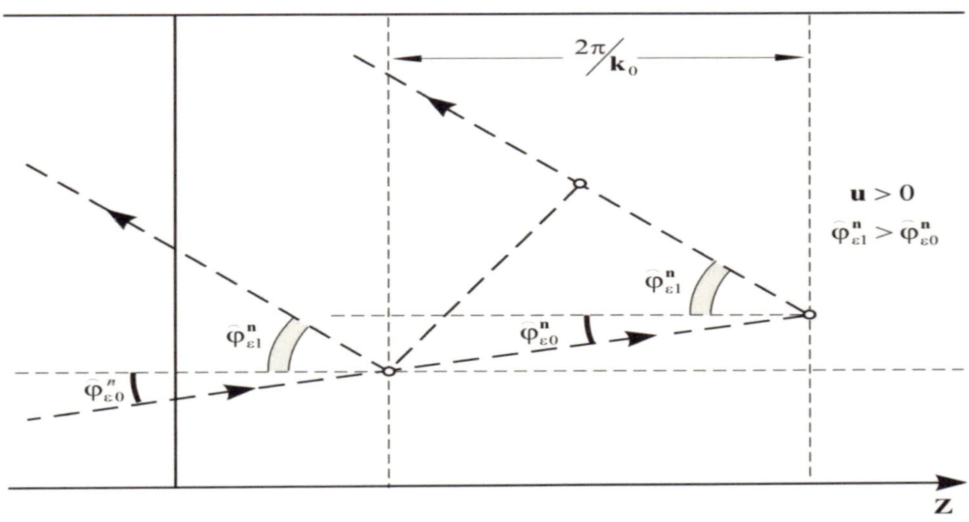

Fig. 2. The physical explanation of the effect of strong interaction.

If now we apply the first-order Wolf-Bragg condition, when the waves reflected from high-density points of the interference pattern are amplified, we obtain the following equation

$$\frac{2\sqrt{\varepsilon_0}\cdot\omega_0\left(\cos\bar{\varphi}^{n}_{\varepsilon,0}-\beta_\varepsilon\right)}{k_0 c\left(1-\beta_\varepsilon^2\right)}=1. \tag{1.48}$$

It is not difficult to note, taking into account (1.46), that the solution of the equation (1.48) precisely coincides with the expression of the frequency of strong interaction (see (1.41)), received above.

3. Propagation of electromagnetic waves in a waveguide with a periodically modulated anisotropic insert

Consider a waveguide of arbitrary cross section with an anisotropic nonmagnetic $(\mu=1)$ modulated insert (modulated uniaxial crystal) the permittivity tensor of which has the form

$$\hat{\varepsilon}=\begin{pmatrix}\varepsilon_1(z,t) & 0 & 0\\ 0 & \varepsilon_1(z,t) & 0\\ 0 & 0 & \varepsilon_2(z,t)\end{pmatrix}, \tag{2.1}$$

where components $\varepsilon_1(z,t)$ and $\varepsilon_2(z,t)$ are modulated by the pumping wave in space and time according to the harmonic law

$$\varepsilon_1(z,t)=\varepsilon_1^0[1+m_1\cos k_0(z-ut)],\quad \varepsilon_2(z,t)=\varepsilon_2^0[1+m_2\cos k_0(z-ut)]. \tag{2.2}$$

Here, $\varepsilon_1{}^0$ and $\varepsilon_2{}^0$ are the permittivities in the absence of a modulating wave; m, and m_2 are the modulation indices; and k_0 and и are, respectively, the wavenumber and velocity of the modulating wave.

Consider the propagation of a signal electromagnetic wave at frequency ω_0 in this waveguide under the assumption that the modulation indices are small $(m_1 << 1, m_2 << 1, m_1 \approx m_2)$. Note that, when the condition $\beta_1 \le 0.8$ is satisfied, where $\beta_1 = u\sqrt{\varepsilon_1^0}/c$, not only the modulation indices, but also parameter $l_1 = m_1\beta_1/\left(1-\beta_1^2\right)$ are small $(l_1 << 1)$.

As in my earlier works (see, e.g., [23], [31], [34-37]), transverse electric (TE) and transverse magnetic (TM) waves in the waveguide will be described through the longitudinal components of the magnetic (H_z) and electric (E_z) field. Then, bearing in mind that $D_x=\varepsilon_1(z,t)E_x, D_y=\varepsilon_1(z,t)E_y, D_z=\varepsilon_2(z,t)E_z$ and **B** = **H** and using the Maxwell equations, we obtain equations for H_z(x, y, z, f) and E,(x, y, z, 0; namely,

for the TE wave

$$\Delta_\perp H_z+\frac{\partial^2 H_z}{\partial z^2}-\frac{1}{c^2}\frac{\partial}{\partial t}\left[\varepsilon_1\frac{\partial H_z}{\partial t}\right]=0, \tag{2.3}$$

for the TM wave

$$\Delta_{\perp}\tilde{E}_z + \varepsilon_2 \frac{\partial}{\partial z}\left(\frac{1}{\varepsilon_1}\frac{\partial \tilde{E}_z}{\partial z}\right) - \frac{\varepsilon_2}{c^2}\frac{\partial^2 \tilde{E}_z}{\partial t^2} = 0, \tag{2.4}$$

where $\Delta_{\perp}$ is the two-dimensional Laplacian and $\tilde{E}_z = \varepsilon_2 E_z$.
It is easy to check in this case that the transverse components of the TE and TM fields can be expressed in terms of (1.6) as:

for TE wave

$$\vec{H}_{\tau} = \sum_{n=0}^{\infty} \hat{\lambda}_n^{-2} \frac{\partial H_n(z,t)}{\partial z} \nabla \hat{\Psi}_n(x,y), \tag{2.5}$$

$$\vec{E}_{\tau} = \frac{1}{c}\sum_{n=0}^{\infty} \hat{\lambda}_n^{-2} \frac{\partial H_n(z,t)}{\partial t}\left[\vec{z}_0 \nabla \hat{\Psi}_n(x,y)\right], \tag{2.6}$$

for TM wave

$$\vec{H}_{\tau} = -\frac{1}{c}\sum_{n=0}^{\infty} \lambda_n^{-2} \frac{\partial\left[\varepsilon_2 E_n(z,t)\right]}{\partial t}\left[\vec{z}_0 \nabla \Psi_n(x,y)\right], \tag{2.7}$$

$$\vec{E}_{\tau} = \frac{1}{\varepsilon_1}\sum_{n=0}^{\infty} \lambda_n^{-2} \frac{\partial\left[\varepsilon_2 E_n(z,t)\right]}{\partial z} \nabla \Psi_n(x,y). \tag{2.8}$$

Let us introduce the new variables

$$\xi = z - ut,\ \eta = \frac{z}{u} - \frac{1}{u}\int_0^{\xi} \frac{d\xi}{1 - \beta_1^2 \varepsilon_1(\xi)/\varepsilon_1^0} \tag{2.9}$$

into equations (2.3) and (2.4) and seek for solutions to the above equations in the form

$$H_z = \sum_{n=0}^{\infty} e^{i\left(\hat{P}_0^n u - \omega_0\right)\eta} H_{nz}(\xi) \hat{\Psi}_n(x,y), \tag{2.10}$$

$$\tilde{E}_z = \sum_{n=0}^{\infty} e^{i\left(P_0^n u - \omega_0\right)\eta} E_{nz}(\xi) \Psi_n(x,y). \tag{2.11}$$

Taking into account that functions $\hat{\varphi}_n(x,y)$ and $\psi_n(x,y)$ satisfy the Helmholtz equations (1.4) and (1.5), we get ordinary second-order differential equations in variable ξ to find $H_{nz}(\xi)$ and $E_{nz}(\xi)$,

$$\frac{d}{d\xi}\left[\left(1 - \beta_1^2 \frac{\varepsilon_1}{\varepsilon_1^0}\right)\frac{dH_{nz}}{d\xi}\right] + \frac{\hat{\chi}_n^2}{1 - \beta_1^2 \frac{\varepsilon_1}{\varepsilon_1^0}} H_{nz} = 0, \tag{2.12}$$

$$\varepsilon_2 \frac{d}{d\xi}\left[\frac{1}{\varepsilon_1}\left(1-\beta_1^2\frac{\varepsilon_1}{\varepsilon_1^0}\right)\frac{d\tilde{E}_{nz}}{d\xi}\right]+\frac{\chi_n^2}{1-\beta_1^2\frac{\varepsilon_1}{\varepsilon_1^0}}\tilde{E}_{nz}=0\,. \tag{2.13}$$

Here,

$$\widehat{\chi}_n^2=\frac{\left(u\widehat{p}_0^n-\omega_0\right)^2}{c^2}\varepsilon_1-\widehat{\lambda}_n^2\left(1-\beta_1^2\frac{\varepsilon_1}{\varepsilon_1^0}\right),\ \chi_n^2=\frac{\left(up_0^n-\omega_0\right)^2}{c^2}\varepsilon_1-\lambda_n^2\left(1-\beta_1^2\frac{\varepsilon_1}{\varepsilon_1^0}\right), \tag{2.14}$$

$$\left(\widehat{p}_0^n\right)^2=\frac{\omega_0^2}{c^2}\varepsilon_1^0-\widehat{\lambda}_n^2\,,\ \left(p_0^n\right)^2=\frac{\varepsilon_2^0-\varepsilon_1^0\beta_1^2}{\varepsilon_2^0-\varepsilon_1^0\beta_2^2}\frac{\omega_0^2}{c^2}\varepsilon_1^0-\lambda_n^2 b_1\,,\ b_1=1-\beta_1^2\,,\ \beta_2=\frac{u\sqrt{\varepsilon_2^0}}{c}. \tag{2.15}$$

In terms of the new variables

$$s=\frac{k_0 b_1}{2}\int_0^{\xi}\frac{d\xi}{1-\beta_1^2\frac{\varepsilon_1}{\varepsilon_1^0}},,\ s=\frac{k_0 b_1}{2\varepsilon_1^0}\int_0^{\xi}\frac{\varepsilon_1 d\xi}{1-\beta_1^2\frac{\varepsilon_1}{\varepsilon_1^0}} \tag{2.16}$$

Equations (2.12) and (2.13) take the form of the Mathieu-Hill equations

$$\frac{d^2 H_{nz}}{d\widehat{s}^2}+\frac{4\widehat{\chi}_n^2}{k_0^2 b_1^2}H_{nz}=0\,,\ \frac{d^2\tilde{E}_{nz}}{ds^2}+\frac{4\chi_n^2\left(\varepsilon_1^0\right)^2}{k_0^2 b_1^2\varepsilon_1\varepsilon_2}\tilde{E}_{nz}=0\,. \tag{2.17}$$

Note that the frequency domain described by the conditions

$$\left|1-\widehat{\theta}_0^n\right|>>\widehat{\delta}_n\cong\frac{\widehat{\theta}_1^n}{2\sqrt{2}},\ \left|1-\theta_0^n\right|>>\delta_n\cong\frac{\theta_1^n}{2\sqrt{2}}, \tag{2.18}$$

where

$$\widehat{\theta}_0^n=\frac{4}{k_0^2 b_1^2}\left(\widehat{p}_0^n-\frac{\omega_0\beta_1^2}{u}\right)^2,\ \theta_0^n=\frac{4}{k_0^2 b_1^2}\left(p_0^n-\frac{\omega_0\beta_1^2}{u}\right)^2, \tag{2.19}$$

$$\widehat{\theta}_1^n=\widehat{\theta}_{-1}^n=\frac{2}{k_0^2 b_1^2}\left[\frac{\left(u\widehat{p}_0^n-\omega_0\right)^2}{u^2}+\frac{k_0^2 b_1^2}{4}\widehat{\theta}_0^n\right]\ell_1\,,\ \theta_1^n=\theta_{-1}^n=\frac{2\varepsilon_1^0\lambda_n^2}{k_0^2 b_1^2\varepsilon_2^0}m_2-\frac{\theta_0^n}{2}m_1\,, \tag{2.20}$$

is the domain of weak interaction between the signal wave and the wave that modulates the insert. Solving (2.17) by the method developed in [23], [31], [34-37] and discarding the terms proportional to the modulation indices in the first power, we obtain the following expressions for the TE and TM field in the frequency domain defined by formulas (2.18): [38]

for the TE wave

$$H_z = \sum_{n=0}^{\infty} \widehat{\Psi}_n(x,y) e^{i\left(\widehat{P}_0^n z - \omega_0 t\right)} \widehat{c}_0^n \sum_{k=-1}^{1} \widehat{V}_k^n e^{ikk_0(z-ut)}, \tag{2.21}$$

where

$$\widehat{V}_k^n = \left(k \frac{\omega_0}{2uk_0} \ell_1 + \frac{\widehat{c}_k^n}{\widehat{c}_0^n} \right)^{|k|}, \; \widehat{c}_{\pm 1}^n = \frac{\widehat{\theta}_1^n \widehat{c}_0^n}{4\left(1 \pm \sqrt{\widehat{\theta}_0^n}\right)}, \tag{2.22}$$

for the TM wave

$$E_z = \sum_{n=0}^{\infty} \Psi_n(x,y) e^{i\left(P_0^n z - \omega_0 t\right)} c_0^n \sum_{k=-1}^{1} V_k^n e^{ikk_0(z-ut)}, \tag{2.23}$$

where

$$V_k^n = \left[k \frac{\omega_0}{2uk_0} \ell_1 + \frac{c_k^n}{c_0^n} - \frac{1}{2}(m_1 + m_2) \right]^{|k|}, \; c_{\pm 1}^n = \frac{\theta_1^n c_0^n}{4\left(1 \pm \sqrt{\theta_0^n}\right)}. \tag{2.24}$$

Note that quantities $\widehat{c}_0^n$ and c_0^n in (2.21) and (2.23) are found from the normalization condition. As follows from (2.21) and (2.23), when an electromagnetic wave propagates in a waveguide with an insert harmonically modulated in space and time, the TE and TM fields represent a superposition of space-time harmonics of different amplitudes. In the domain of weak interaction between the signal and modulation waves, the amplitudes of harmonics +1 and -1 prove to be small (they are linearly related to the modulation indices) compared with the amplitude of the fundamental harmonic (which is independent of modulation indices).
It is known [21] that, when $\widehat{\theta}_0^n$ and θ_0^n tend to unity, i.e., when the conditions

$$\left|1 - \widehat{\theta}_0^n\right| \le \widehat{\delta}_n, \quad \left|1 - \theta_0^n\right| \le \delta_n \tag{2.25}$$

are satisfied, the signal wave and the wave that modulates the insert strongly interact (the first-order Bragg condition for waves reflected from a high-density area is met) and vigorously exchange energy.
Condition (2.25) can be recast (for the TM field) as

$$\omega_{0,s} - \Delta\omega_0 \le \omega_0 \le \omega_{0,s} + \Delta\omega_0, \tag{2.26}$$

where $\omega_{0,s}$ given by

$$\omega_{0,s} = \frac{uk_0}{2\beta_1}(\beta_1 + \overline{\eta}_n), \overline{\eta}_n = \sqrt{1 + \frac{4\overline{\lambda}_n^2}{b_1 k_0^2}}, \overline{\lambda}_n^2 = \frac{\varepsilon_1^0}{\varepsilon_2^0} \lambda_n^2, \tag{2.27}$$

is the frequency near which the strong interaction takes place, and

$$\Delta\omega_0 = \frac{k_0 u\left(1+\beta_1\bar{\eta}_n\right)}{8\sqrt{2}\beta_1\bar{\eta}_n}\theta_1^n \tag{2.28}$$

is the width of the domain of strong interaction. Calculations show that

$$\left|V_{-1}^n\right| \sim 1,\ \left|V_1^n\right| \sim m_1, m_2 \tag{2.29}$$

in frequency domain (2.26). From relationships (2.29), it follows that the amplitude of reflected harmonic -1 is independent of modulation indices in the domain where the signal wave and the wave that modulates the anisotropic insert strongly interact. In other words, not only the zeroth harmonic of the signal, but also reflected harmonic -1 of frequency

$$\omega_{-1,s} = \frac{uk_0}{2\beta_1}\left(\bar{\eta}_n - \beta_1\right), \left(\bar{\eta}_n > \beta_1\right) \tag{2.30}$$

plays a significant role in this domain.

Note in conclusion that the results obtained here turn into those reported in [37] in the limit $m_1 \to 0$; in the limit $u \to 0$, one arrives at results for a waveguide with an inhomogeneous but stationary anisotropic insert.

4. Interaction of electromagnetic waves with space-time periodic anisotropic magneto-dielectric filling of a waveguide

Let the axis of a regular waveguide of an arbitrary cross section coincides with the OZ axis of a Certain Cartesian coordinate frame. Assume that the waveguide is filled with a periodically modulated anisotropic magneto- dielectric filling whose tensor permittivity and permeability are specified by the formulas

$$\hat{\varepsilon} = \begin{pmatrix} \varepsilon_1 & 0 & 0 \\ 0 & \varepsilon_1 & 0 \\ 0 & 0 & \varepsilon_2(z,t) \end{pmatrix},\ \hat{\mu} = \begin{pmatrix} \mu_1 & 0 & 0 \\ 0 & \mu_1 & 0 \\ 0 & 0 & \mu_2(z,t) \end{pmatrix}. \tag{3.1}$$

In (3.1) $\varepsilon_1 = const, \mu_1 = const$ and the $\varepsilon_2(z,t)$ and $\mu_2(z,t)$ components are harmonic functions in space and time:

$$\varepsilon_2(z,t) = \varepsilon_2^0\left[1 + m_\varepsilon \cos\left(k_0 z - k_0 u t\right)\right], \tag{3.2}$$

$$\mu_2(z,t) = \mu_2^0\left[1 + m_\mu \cos\left(k_0 z - k_0 u t\right)\right], \tag{3.3}$$

where $m_\varepsilon \ll 1$ and $m_\mu \ll 1$ are small modulation indexes, $\varepsilon_2^0 = const$ and $\mu_2^0 = const$ are, respectively, the permittivity and permeability of the filling in the absence of a modulation wave.

Let a signal wave unit amplitude with frequency ω_0 propagates in such a waveguide in a positive Direction of an axis OZ. After some algebra, the wave equations for the longitudinal components $H_z(x,y,z,t)$ and $E_z(x,y,z,t)$ of TE and TM fields can be obtained from Maxwell equations

$$curl\vec{H} = \frac{\partial \vec{D}}{\partial t}\,,\; curl\vec{E} = -\frac{\partial \vec{B}}{\partial t}\,,\; div\vec{D} = 0\,,\; div\vec{B} = 0\,, \tag{3.4}$$

$$\vec{D} = \varepsilon_0 \hat{\varepsilon} \vec{E}\,,\; \vec{B} = \mu_0 \hat{\mu} \vec{H}\,,\; \varepsilon_0 = \left(1/4\pi \cdot 9 \cdot 10^9\right) F/m\,,\; \mu_0 = 4\pi \cdot 10^{-7} H/m \tag{3.5}$$

with allowance for the equalities

$$D_x = \varepsilon_0 \varepsilon_1 E_x\,,\; D_y = \varepsilon_0 \varepsilon_1 E_y\,,\; D_z = \varepsilon_0 \varepsilon_2(z,t) E_z\,, \tag{3.6}$$

$$B_x = \mu_0 \mu_1 H_x\,,\; B_y = \mu_0 \mu_1 H_y\,,\; B_z = \mu_0 \mu_2(z,t) H_z\,. \tag{3.7}$$

We arrive at the following equations:

for TE waves

$$\Delta_\perp \tilde{H}_z + \frac{\mu_2(z,t)}{\mu_1}\frac{\partial^2 \tilde{H}_z}{\partial z^2} = \varepsilon_0 \mu_0 \varepsilon_1 \mu_2(z,t) \frac{\partial^2 \tilde{H}_z}{\partial t^2}\,, \tag{3.8}$$

for the TM waves

$$\Delta_\perp \tilde{E}_z + \frac{\varepsilon_2(z,t)}{\varepsilon_1}\frac{\partial^2 \tilde{E}_z}{\partial z^2} = \varepsilon_0 \mu_0 \mu_1 \varepsilon_2(z,t) \frac{\partial^2 \tilde{E}_z}{\partial t^2}\,, \tag{3.9}$$

where

$$\tilde{H}_z = \mu_2(z,t) H_z\,,\; \tilde{E}_z = \varepsilon_2(z,t) E_z\,. \tag{3.10}$$

With the use of Maxwell equations (3.4) and (3.5), the transverse components of TE and TM fields can be represented in terms of

$$H_z = \sum_{n=0}^{\infty} H_n(z,t) \hat{\Psi}(x,y)\,,\; E_z = \sum_{n=0}^{\infty} E_n(z,t) \Psi(x,y) \tag{3.11}$$

as follows:

for TE waves

$$\vec{H}_\tau = \frac{1}{\mu_1}\sum_{n=0}^{\infty} \hat{\lambda}_n^{-2} \frac{\partial\left[\mu_2(z,t) H_n(z,t)\right]}{\partial z} \nabla \hat{\Psi}_n(x,y)\,, \tag{3.12}$$

$$\vec{E}_\tau = \mu_0 \sum_{n=0}^{\infty} \hat{\lambda}_n^{-2} \frac{\partial\left[\mu_2(z,t) H_n(z,t)\right]}{\partial t} \left[\vec{z}_0 \nabla \hat{\Psi}_n(x,y)\right], \tag{3.13}$$

for TM waves

$$\vec{H}_\tau = -\varepsilon_0 \sum_{n=0}^{\infty} \lambda_n^{-2} \frac{\partial\left[\varepsilon_2(z,t) E_n(z,t)\right]}{\partial t} \left[\vec{z}_0 \nabla \Psi_n(x,y)\right], \tag{3.14}$$

$$\vec{E}_\tau = \frac{1}{\varepsilon_1}\sum_{n=0}^{\infty}\lambda_n^{-2}\frac{\partial\left[\varepsilon_2(z,t)E_n(z,t)\right]}{\partial z}\nabla\Psi_n(x,y). \tag{3.15}$$

With the new variables

$$\xi = z - ut,\quad \eta = \frac{z}{u} - \frac{1}{u}\frac{z-ut}{1-\beta^2} \tag{3.16}$$

where $\beta^2 = u^2\varepsilon_0\mu_0\varepsilon_1\mu_1$, wave equations (3.8) and (3.9) can be modified to obtain

$$\Delta_\perp\tilde{H}_z + \frac{\mu_2(z,t)}{\mu_1}\left(1-\beta^2\right)\frac{\partial^2\tilde{H}_z}{\partial\xi^2} - \frac{\varepsilon_0\mu_0\varepsilon_1\mu_2(z,t)}{1-\beta^2}\frac{\partial^2\tilde{H}_z}{\partial\eta^2} = 0, \tag{3.17}$$

$$\Delta_\perp\tilde{E}_z + \frac{\varepsilon_2(z,t)}{\varepsilon_1}\left(1-\beta^2\right)\frac{\partial^2\tilde{E}_z}{\partial\xi^2} - \frac{\varepsilon_0\mu_0\mu_1\varepsilon_2(z,t)}{1-\beta^2}\frac{\partial^2\tilde{E}_z}{\partial\eta^2} = 0. \tag{3.18}$$

Let us seek solutions of equations (3.17) and (3.18) in the form (1.13). Then, taking into account (1.4) and (1.5), we obtain for $\tilde{H}_{nz}(\xi)$ and $\tilde{E}_{nz}(\xi)$ the following second-order ordinary differential equations with the periodic Mathieu-Hill coefficients:

$$\frac{d^2\tilde{H}_{nz}(\xi)}{d\xi^2} + \frac{\mu_1}{\mu_2(z,t)\left(1-\beta^2\right)}\left[\frac{\varepsilon_0\mu_0\varepsilon_1\mu_2(z,t)}{1-\beta^2}\gamma^2 - \hat{\lambda}_n{}^2\right]\tilde{H}_{nz}(\xi) = 0, \tag{3.19}$$

$$\frac{d^2\tilde{E}_{nz}(\xi)}{d\xi^2} + \frac{\varepsilon_1}{\varepsilon_2(z,t)\left(1-\beta^2\right)}\left[\frac{\varepsilon_0\mu_0\mu_1\varepsilon_2(z,t)}{1-\beta^2}\gamma^2 - \lambda_n{}^2\right]\tilde{E}_{nz}(\xi) = 0. \tag{3.20}$$

With the new variable $\zeta = k_0\xi/2$ equations (3.19) and (3.20) can be modified into the form

$$\frac{d^2\tilde{H}_{nz}(\zeta)}{d\zeta^2} + \sum_{k=-1}^{1}\hat{\theta}_k^n\exp(2ik\zeta)\tilde{H}_{nz}(\zeta) = 0, \tag{3.21}$$

$$\frac{d^2\tilde{E}_{nz}(\zeta)}{d\zeta^2} + \sum_{k=-1}^{1}\theta_k^n\exp(2ik\zeta)\tilde{E}_{nz}(\zeta) = 0, \tag{3.22}$$

where quantities $\hat{\theta}_k^n$ and θ_k^n are the coefficients of the Fourier decompositions of the expressions that appear before functions $\tilde{H}_{nz}(\zeta)$ and $\tilde{E}_{nz}(\zeta)$ entering equations (3.19) and (3.20). In the first approximation for small parameters m_ε and m_μ these coefficients are expressed according to the formulas

$$\hat{\theta}_0^n = \frac{4\mu_1}{b^2\mu_2^0 k_0^2}\left(\mu_2^0\varepsilon_0\mu_0\varepsilon_1\gamma^2 - \hat{\lambda}_n^2 b\right),\ \hat{\theta}_{\pm1}^n = \frac{2\mu_1\hat{\lambda}_n^2}{b\mu_2^0 k_0^2}m_\mu, \tag{3.23}$$

$$\theta_0^n = \frac{4\varepsilon_1}{b^2 \varepsilon_2^0 k_0^2}\left(\varepsilon_2^0 \varepsilon_0 \mu_0 \mu_1 \gamma^2 - \lambda_n^2 b\right), \; \theta_{\pm 1}^n = \frac{2\varepsilon_1 \lambda_n^2}{b \varepsilon_2^0 k_0^2} m_\varepsilon , \; b = 1 - \beta^2 . \tag{3.24}$$

We seek solutions to equations (3.21) and (3.22) in the form

$$\tilde{H}_{nz}(\zeta) = e^{i\hat{\mu}_n \zeta} \sum_{k=-1}^{1} \hat{C}_k^n e^{2ik\zeta} , \; \tilde{E}_{nz}(\zeta) = e^{i\mu_n \zeta} \sum_{k=-1}^{1} C_k^n e^{2ik\zeta} . \tag{3.25}$$

It is known [33] that, under the conditions (1.27), which provide for weak interaction between the signal wave and the wave of the waveguide-filling modulation, quantities $\hat{\mu}_n, \mu_n, \hat{C}_{\pm 1}^n$ and $C_{\pm 1}^n$ have the form (1.28) and (1.29) (accurate to within small parameters m_ε and m_μ inclusively). Taking into account (3.25), (1.28), (1.29) and changing to variables z and t, we obtain from (3.10) analytic expressions for H_z and E_z of TE and TM waves. These expressions correspond to the first approximation for m_ε and m_μ, are valid in the region of weak interaction between the signal wave and the wave of the waveguide-filling modulation, and have the form [39]

$$H_z = \frac{1}{\mu_2^0} \sum_{n=0}^{\infty} \hat{\Psi}_n(x,y) e^{i\left(\hat{P}_0^n z - \omega_0 t\right)} \hat{C}_0^n \sum_{k=-1}^{1} \hat{V}_k^n e^{ikk_0(z-ut)} , \tag{3.25}$$

$$E_z = \frac{1}{\varepsilon_2^0} \sum_{n=0}^{\infty} \Psi_n(x,y) e^{i\left(P_0^n z - \omega_0 t\right)} C_0^n \sum_{k=-1}^{1} V_k^n e^{ikk_0(z-ut)} , \tag{3.26}$$

where

$$\hat{V}_k^n = \left(\frac{\hat{C}_k^n}{\hat{C}_0^n} - \frac{m_\mu}{2}\right)^{|k|}, \; V_k^n = \left(\frac{C_k^n}{C_0^n} - \frac{m_\varepsilon}{2}\right)^{|k|}, \; \hat{\theta}_0^n = \frac{4}{k_0^2 b^2}\left(\sqrt{\frac{\beta^2}{u^2}\omega_0^2 - \frac{\mu_1}{\mu_2^0}\hat{\lambda}_n^2} - \frac{\beta^2 \omega_0}{u}\right)^2 ,$$

$$\theta_0^n = \frac{4}{k_0^2 b^2}\left(\sqrt{\frac{\beta^2}{u^2}\omega_0^2 - \frac{\varepsilon_1}{\varepsilon_2^0}\lambda_n^2} - \frac{\beta^2 \omega_0}{u}\right)^2 \tag{3.27}$$

$$\hat{P}_0^n = \sqrt{\frac{\beta^2}{u^2}\omega_0^2 - \frac{\mu_1}{\mu_2^0}\hat{\lambda}_n^2} , \qquad P_0^n = \sqrt{\frac{\beta^2}{u^2}\omega_0^2 - \frac{\varepsilon_1}{\varepsilon_2^0}\lambda_n^2} . \tag{3.28}$$

Note, that for the frequency and frequency width of the strong interaction region (see [31], [33]) the following expressions can easily be obtained from (2.25):

for TE waves

$$\omega_{0,c} = \frac{k_0 u}{2\beta}\left(\beta + \hat{\tilde{\eta}}_n\right), \; \hat{\tilde{\eta}}_n = \sqrt{1 + \frac{4\mu_1 \hat{\lambda}_n^2}{\mu_2^0 b k_0^2}} , \; \Delta\omega_{0,c} = \frac{k_0 u\left(1 + \beta \hat{\eta}_n\right)}{4\beta \hat{\eta}_n}\hat{\delta}_n , \tag{3.29}$$

for TM waves

$$\omega_{0,c} = \frac{k_0 u}{2\beta}\left(\beta + \tilde{\eta}_n\right), \; \tilde{\eta}_n = \sqrt{1 + \frac{4\varepsilon_1 \lambda_n^2}{\varepsilon_2^0 b k_0^2}} , \; \Delta\omega_{0,c} = \frac{k_0 u\left(1 + \beta \eta_n\right)}{4\beta \eta_n}\delta_n . \tag{3.30}$$

For the quantities $\widehat{V}_{\pm 1}$ and $V_{\pm 1}$ in this case we obtain

$$\left|\widehat{V}_{-1}^{n}\right| \cong 1, \quad \left|\widehat{V}_{1}^{n}\right| \cong \frac{\mu_1 \widehat{\lambda}_n^2 - 2b\mu_2^0 k_0^2}{4b\mu_2^0 k_0^2} m_\mu, \tag{3.31}$$

$$\left|V_{-1}^{n}\right| \cong 1, \quad \left|V_{1}^{n}\right| \cong \frac{\varepsilon_1 \lambda_n^2 - 2b\varepsilon_2^0 k_0^2}{4b\varepsilon_2^0 k_0^2} m_\varepsilon . \tag{3.32}$$

According to (3.31) and (3.32), in the strong- interaction region a substantial role is played not only by the fundamental harmonic but also by the reflected minus-first harmonic that exists at the frequency:

for TE waves

$$\omega_{-1} = \frac{k_0 u}{2\beta}\left(\widehat{\tilde{\eta}}_n - \beta\right), \quad \widehat{\tilde{\eta}}_n > \beta, \tag{3.33}$$

for TM waves

$$\omega_{-1} = \frac{k_0 u}{2\beta}\left(\tilde{\eta}_n - \beta\right), \quad \tilde{\eta}_n > \beta . \tag{3.34}$$

Note that, in limiting case $u \to 0$ the above obtained relationships yield results for the stationary inhomogeneous anisotropic magneto-dielectric filling of a waveguide.

5. Refrences

[1] Brillouin L., Parodi M. Rasprostranenie Voln v Periodicheskikh Strukturakh. Perevod s Frantsuskovo, M.: IL, 1959.

[2] Born M., Volf E. Osnovi Optiki. Perevod s Angliyskovo, M.: Nauka, 1973.

[3] Cassedy E. S., Oliner A. A. TIIER, 51, No 10, 1330 (1963).

[4] Barsukov K. A., Bolotovskiy B. M. Izvestiya Vuzov. Seriya Radiofizika, 7, No 2, 291 (1964).

[5] Barsukov K. A., Bolotovskiy B. M. Izvestiya Vuzov. Seriya Radiofizika, 8, No 4, 760 (1965).

[6] Cassedy E. S. TIIER, 55, No 7, 37 (1967).

[7] Peng S. T., Cassedy E. S. Proceedings of the Symposium on Modern Optics. Brooklyn, N. Y.: Politecnic Press, MRI-17, 299 (1967).

[8] Barsukov K. A., Gevorkyan E. A., Zvonnikov N. A. Radiotekhnika i Elektronika, 20, No 5, 908 (1975).

[9] Tamir T., Wang H. C., Oliner A. A. IEEE, Transactions on Microwave Theory and Techniques, MTT-12, 324 (1964).

[10] Averkov S. I., Boldin V. P. Izvestiya Vuzov. Seriya Radiofizika, 23, No 9, 1060 (1980).

[11] Askne J. TIIER, 59, No 9, 244 (1968).

[12] Rao. TIIER, 65, No 9, 244 (1968).

[13] Yeh C., Cassey K. F., Kaprielian Z. A. IEEE, Transactions on Microwave Theory and Techniques, MTT-13, 297 (1965).

[14] Elachi Ch. TIIER, 64, No 12, 22 (1976).
[15] Karpov S. Yu., Stolyarov S. N. Uspekhi Fizicheskikh Nauk, 163, No 1, 63 (1993).
[16] Elachi Ch., Yeh C. Journal of Applied Physics, 44, 3146 (1973).
[17] Elachi Ch., Yeh C. Journal of Applied Physics, 45, 3494 (1974).
[18] Peng S. T., Tamir T., Bertoni H. L. IEEE, Transactions on Microwave Theory and Techniques, MTT-23, 123 (1975).
[19] Seshadri S. R. Applied Physics, 25, 211 (1981).
[20] Krekhtunov V. M., Tyulin V. A. Radiotekhnika i Elektronika, 28, 209 (1983).
[21] Simon J. C. IRE, Transactions on Microwave Theory and Techniques, MTT-8, No 1, 18 (1960).
[22] Barsukov K. A., Radiotekhnika i Elektronika, 9, No 7, 1173 (1964).
[23] Gevorkyan E. A. Proceedings of International Symposium on Electromagnetic Theory, Thessaloniki, Greece, May 25-28, 1, 69 (1998).
[24] Gevorkyan E. A. Mezhduvedomstvenniy Tematicheskiy Nauchniy Sbornik. Rasseyanie Elektromagnitnikh Voln. Taganrok, TRTU, No 12, 55 (2002).
[25] Gevorkyan E. A. Book of Abstracts of the Fifth International Congress on Mathematical Modelling, Dubna, Russia, September 30 - October 6, 1, 199 (2002).
[26] Gayduk V. I., Palatov K. I., Petrov D. M. Fizicheskie Osnovi Elektroniki SVCH, Moscow, Sovetskoe Radio (1971).
[27] Yariv A. Kvantovaya Elektronika I Nelineynaya Optika. Perevod s Angliyskovo, M.: Sovetskoe Radio (1973).
[28] Volnovodnaya Optoelektronika. Pod Redaktsiey T. Tamir. Perevod s Angliyskovo, M.: Mir (1974).
[29] Markuze D. Opticheskie Volnovodi. Perevod s Angliyskovo, M.: Mir (1974).
[30] Yariv A., Yukh P. Opticheskie Volni v Kristallakh. Perevod s Angliyskovo, M.: Mir (1987).
[31] Barsukov K. A., Gevorkyan E. A. Radiotekhnika i Elektronika, 28, No 2, 237 (1983).
[32] Mak-Lakhlan N. V. Teoriya i Prilozheniya Funktsiy Mathe. Perevod s zAngliyskovo, M.: Fizmatgiz (1963).
[33] Gevorkyan E. A. Uspekhi Sovremennoy Radioelektroniki, No 1, 3 (2006).
[34] Barsukov K. A., Gevorkyan E. A. Radiotekhnika i Elektronika, 31, 1733 (1986).
[35] Barsukov K. A., Gevorkyan E. A. Radiotekhnika i Elektronika, 39, 1170 (1994).
[36] Gevorkyan E. A. Proceedings of International Symposium on Electromagnetic Theory, Kiev, Ukraine, September 10-13, 2, 373 (2002).
[37] Gevorkyan E. A. Proceedings of International Symposium on Electromagnetic Theory, Dnepropetrovsk, Ukraine, September 14-17, 370 (2004).
[38] Gevorkyan E.A. Zhurnal Tekhnicheskoy Fiziki, 76, No 5, 134 (2006) (Technical Physics, 51, 666 (2006)).
[39] Gevorkyan E.A. Radiotekhnika i Elektronika, 53, No 5, 565 (2008) (Journal of Communications Technology and Electronics, 53, No 5, 535 (2008)).

14

The Analysis of Hybrid Reflector Antennas and Diffraction Antenna Arrays on the Basis of Surfaces with a Circular Profile

Oleg Ponomarev
Baltic Fishing Fleet State Academy
Russia

1. Introduction

Science achievements in methods of processing of the radar-tracking information define directions of development of antenna systems. These are expansion of aims and functions of antennas, achievement of optimal electric characteristics with regard to mass, dimensional and technological limits. Hybrid reflector antennas (HRA's) have the important part of the radars and modern communication systems. In HRA the high directivity is provided by system of reflectors and form of the pattern of the feed, and scanning possibility is provided by feeding antenna array. Artificial network, generic synthesis and evolution strategy algorithms of the phased antenna arrays and HRA's, numerical methods of the analysis and synthesis of HRA's on the basis of any finite-domain methods of the theory of diffraction, the wavelet analysis and other methods, have been developed for the last decades. The theory and practice of antenna systems have in impact on the ways of development of radars: radio optical systems, digital antenna arrays, synthesed aperture radars (SAR), the solid-state active phased antenna arrays (Fourikis, 1996).
However these HRA's has a disadvantage - impossibility of scanning by a beam in a wide angle range without decrease of gain and are worse than HRA's on the basis of reflector with a circular profile. In this antennas need to calibrate a phase and amplitude of a phased array feeds to yield a maximum directivity into diapason of beam scanning (Haupt, 2008).
Extremely achievable electric characteristics HRA are reached by optimization of a profile of a reflector and amplitude-phase distribution of feeding antenna array (Bucci et al., 1996). Parabolic reflectors with one focus have a simple design, but their worse then multifocuses reflectors. For example, the spherical or circular cylindrical forms are capable of electromechanical scanning the main beam. However the reflectors with a circular profile have a spherical aberration that limits their application.
The aim of this article is to elaborate the combined mathematical method of the diffraction theory for the analysis of spherical HRA's and spherical diffraction antenna arrays of any electric radius. The developed mathematical method is based on a combination of eigenfunctions/geometrical theory of diffraction (GTD) methods. All essential characteristics of physical processes give the evident description of the fields in near and far antenna areas.

2. Methods of analysis and synthesis of hybrid reflector antennas

The majority of HRA's use the parabolic and elliptic reflectors working in the range of submillimeter to decimeter ranges of wave's lengths. A designs of multibeam space basing HRA's have been developed by company Alcatel Alenia Space Italy (AAS-I) for SAR (Llombart et al., 2008). The first HRA for SAR was constructed in a Ku-range in 1997 for space born Cassini. These HRA are equipped with feeds presented as single multimode horns or based on clusters and have two orthogonal polarizations. The parabolic reflector of satellite HRA presented in (Young-Bae & Seong-Ook, 2008) is fed by horn antenna array and has a gain 37 dB in range of frequencies 30,085-30,885 GHz. A tri-band mobile HRA with operates by utilizing the geo-stationary satellite Koreasat-3 in tri-band (Ka, K, and Ku) was desined, and a pilot antenna was fabricated and tested (Eom et al., 2007).

One of the ways of development of methods of detection of sources of a signal at an interference with hindrances is the use of adaptive HRA. The effective algorithm of an estimation of a direction of arrival of signals has been developed for estimation of spectral density of a signal (Jeffs & Warnick, 2008). The adaptive beamformer is used together with HRA and consist of a parabolic reflector and a multichannel feed as a planar antenna array. A mathematical method on the basis of the GTD and physical optics (PO), a design multibeam multifrequency HRA centimeter and millimeter ranges for a satellite communication, are presented in (Jung et al., 2008). In these antennas the basic reflector have a parabolic and elliptic forms that illuminated by compound feeds are used. Use of metamaterials as a part of the feed HRA of a range of 30 GHz is discussed in (Chantalat et al., 2008). The wide range of beam scanning is provided by parabolic cylindrical reflectors (Janpugdee et al., 2008), but only in once plane of the cylinder. A novel hybrid combination of an analytical asymptotic method with a numerical PO procedure was developed to efficiently and accurately predict the far-fields of extremely long, scanning, very high gain, offset cylindrical HRA's, with large linear phased array feeds, for spaceborn application (Tap & Pathak, 2006).

Application in HRA the reflectors with a circular profile is limited due to spherical aberration and lack of methods of it correction (Love, 1962). The field analysis in spherical reflectors was carried out by a methods PO, geometrical optics (GO), GTD (Tingye, 1959), integrated equations (Elsherbeni, 1989). The field analysis was carried out within a central region of reflector in the vicinity focus $F = a/2$ (a - reflector radius), where beams are undergone unitary reflections. Because of it the central region of hemispherical reflectors was fed and their gain remained low.

However it is known that diffraction on concave bodies gives a number of effects which have not been found for improvement of electric characteristics of spherical antennas. Such effects are multireflections and effect of "whispering gallery" which can be seen when the source of a field is located near a concave wall of a reflector. By means of surface waves additional excitation of peripheral areas of hemispherical reflector can raises the gain and decreases a side lobes level (SLL) of pattern (Ponomarev, 2008).

Use of surface electromagnetic waves (EMW) together with traditional methods of correction of a spherical aberration are expedient for electronic and mechanical control of pattrn in the spherical antennas.

The existing mathematical methods based on asymptotic techniques GTD, GO and PO does not produce correct solutions near asymptotic and focal regions. Numerical methods are suitable for electrically small hemisphere. Another alternative for the analysis of HRA's is combine method eigenfunctions/GTD technique. This method allows to prove use of surface EMW for improvement of electric characteristics spherical and cylindrical HRA's and to give clear physical interpretation the phenomena's of waves diffraction.

3. Spherical hybrid reflector antennas

3.1 Surfaces electromagnetic waves

For the first time surface wave properties were investigated by Rayleigh and were named "whispering gallery" waves, which propagate on the concave surface of circumferential gallery (Rayleigh, 1945). It was determined that these waves propagate in thin layer with equal wave length. This layer covers concave surface. On the spherical surface the energy of Rayleigh waves is maximal and change on the spherical surface by value $J_m(kr)/\sqrt{r}\left(P_n(\cos\theta)+A\cdot Q_n(\cos\theta)\right)$, where $J_m(kr)$ is cylindrical Bessel function of the 1-st kind, $P_n(\cdot), Q_n(\cdot)$ are Legendre polynomials, A is the constant coefficient, $k=2\pi/\lambda$ is the number of waves of free space, (r,θ) are spherical coordinates. The same waves propagate on the solid surface of circumferential cylinder and the acoustic field potential for longitudinal and transverse waves is proportional to the values $J_\nu(k\rho)e^{i\nu\varphi}$, $J'_\nu(k\rho)e^{i\nu\varphi}$, where $J'_\nu(k\rho)$ is derivative of Bessel function about argument, $\nu \approx a$ is constant propagate of surface acoustic wave on cylinder with radius of curvature a, (ρ,φ) are cylindrical coordinates (Grase & Goodman, 1966). In electromagnetic region the surface phenomena at the bent reflectors with perfect electric conducting of the wall, were investigated (Miller & Talanov, 1956). It was showed that their energy is concentrated at the layer with approximate width $a-(\nu-\nu^{1/3})/k$. The same conclusion was made after viewing the diffraction of waves on the bent metallic list that illuminated by waveguide source and in bent waveguides (Shevchenko, 1971).
A lot of letters were aimed at investigating the properties of surface electromagnetic waves (EMW) on the reserved and unreserved isotropic and anisotropic boundaries. The generality of approaches can be clearly seen. To make mathematic model an impressed point current source as Green function. For example, for spiral-conducting parabolic reflector the feed source is a ring current, for elliptic and circumferential cylinder the feed source is an impressed thread current, for spherical perfect electric conducting surface the feed source is a twice magnetic sheet. So far surface phenomena of antenna engineering were considered to solve the problem of decreasing the SLL of pattern.

3.2 Methods of correction of spherical aberration

A process of scanning pattern and making a multibeam pattern without moving the main reflector explains the advantage of spherical antennas. On the one hand the spherical aberration makes it difficult to get a tolerable phase errors on the aperture of the hemispherical dish. On the other hand the spherical aberration allows to extend functionalities of the spherical reflector antennas (Spencer et al., 1949). As a rule the diffraction field inside spherical reflector is analyzed by means of uniform GTD based for large electrical radius of curvature ka of reflector. An interferential structure of the field along longitudinal coordinate of the hemispherical reflector z has a powerful maximum near a paraxial focus $F=ka/2$ (fig.1) (Schell, 1963). The change of parameter z from 0 to 1 is equal to the change of radial coordinate r from 0 to a. As an angle value of the reflector

has been reduced one can see the contribution of the diffraction field from the edges of the reflector.
According to the field distribution shown at fig.1 a feed of the spherical reflector can consist of set of discrete sources disposed from paraxial focus $z = 0,5$ to apex of the reflector $z = 1$. The separate discrete source illuminates a part of ring on the aperture limited by the beams unitary reflected from concave surface of the reflector.

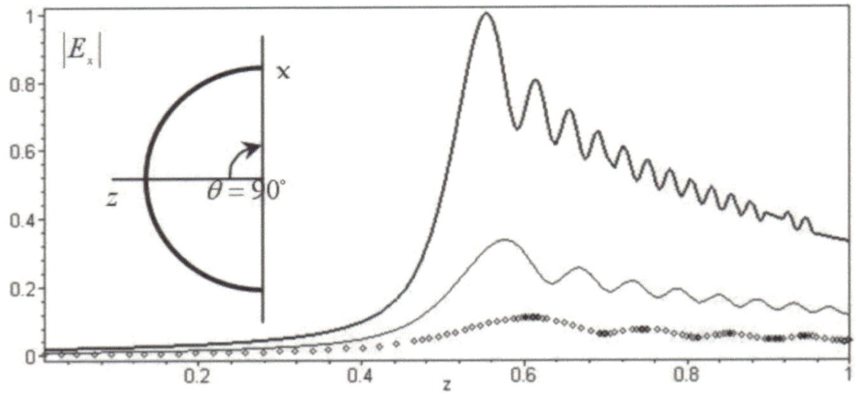

Fig. 1. Distribution of the electrical field component $|E_x|$ along axes of the spherical reflector at the wave length $\lambda = 3,14\ cm$: thick line corresponds to curvature radius of the reflector $a = 100\ cm$; thin line - $a = 50\ cm$; points - $a = 25\ cm$

The results of the investigation of hemispherical reflector antenna with radius of curvature $2a = 3\ m$ at frequency $11,2\ GHz$, are discussed (Tingye, 1959). At the excitation of the central path of the reflector with diameter $1,1\ m$ phase errors at the aperture were less than $\pi/8$, SLL of the pattern were less -25 dB.
Using the channel waveguides as a corrected line source of the spherical antennas seems to be an effective way to reduce the phase errors at the aperture of a spherical reflector (Love, 1962). This line corrected source guarantees the required distribution of the field with illuminated edges of the reflector at the sector of 70° and guarantees electromechanical scattering of the pattern over sector of 110° . An array of waveguide slot sources with dielectric elements for correcting the amplitude-phase field distribution along aperture can be used as the line feed source (Spencer et al., 1949).
Integral equations method is the effective way to solve the problem of spherical aberration correction for reflector with any electrical radius (Elsherbeni, 1989), but at the same time it has difficult physical interpretation of the results and the accuracy is not guaranteed. In theory and practice of correction of spherical aberration is considered within the limits of central area aperture where rays are unitary reflected from concave surface. The edges areas of aperture are not considered for illumination. However, when an incident wave falls on hemispherical reflector, surface EMW propagate along its concave surface. The amplitude of surface waves is concentrated in thin layer with width $\Delta r \approx \lambda$ that is bordered with the reflecting surface. Using the amplitude stability of surface EMW with respect to surface curvature is very important for spherical aberration correction for reflectors with $2\theta = 180^\circ$.

3.3 Solve of Maxwell's equations in spherical coordinates and diffraction problems

It is known that diffraction on concave bodies gives a series of effects, which are still not practically used for improving electrical characteristics of spherical reflector antennas. These are the effects of multiple reflections and the effect of "whispering gallery", that show themselves when the source of electromagnetic field is disposed near the reflector.

Solves of wave equation about electrical field in the spherical coordinates (r,θ,φ) with using a group of rotates $E_+ = -1/\sqrt{2}(E_\varphi + iE_\theta)$, $E_0 = E_r$, $E_- = 1/\sqrt{2}(E_\varphi - iE_\theta)$ will look as series of cylindrical $f_l^*(r)$ and spherical functions $T_{m,n}^l(\frac{\pi}{2}-\varphi,\theta,0)$ (Gradshteyn & Ryzhik, 2000)

$$\left.\begin{aligned} E_0(r,\theta,\varphi) &= \sum_{l=0}^{\infty} f_l^0(r) \sum_{n=-l}^{l} \alpha_{l,n} T_{0n}^l(\tfrac{\pi}{2}-\varphi,\theta,0) \\ E_+(r,\theta,\varphi) &= \sum_{l=0}^{\infty} f_l^+(r) \sum_{n=-l}^{l} \beta_{l,n} T_{1n}^l(\tfrac{\pi}{2}-\varphi,\theta,0) \\ E_-(r,\theta,\varphi) &= \sum_{l=0}^{\infty} f_l^-(r) \sum_{n=-l}^{l} \gamma_{l,n} T_{-1n}^l(\tfrac{\pi}{2}-\varphi,\theta,0) \end{aligned}\right\},$$

where $\alpha_{l,n}, \beta_{l,n}, \gamma_{l,n}$ - weighting coefficients.

Taking into consideration the generalized spherical functions over joined associated Legendre functions $P_l^1(\cos\theta)$ and polynomial Jacobi $P_{l-1}^{(0,2)}(\cos\theta)$, common solving of wave equation about components of electrical field can be written as follows

$$\left.\begin{aligned} E_r &= \sum_{l=1}^{\infty} A_l \frac{J_{l+1/2}(kr)}{(kr)^{3/2}} P_l^1(\cos\theta) e^{-i(\pi/2-\varphi)}; \\ E_\theta &= \frac{i(1+\cos\theta)}{2} \sum_{l=1}^{\infty} \frac{C_1\cos^2\frac{\pi l}{2}}{\sqrt{l(l+1)}} \left[\frac{J_{l+3/2}(kr)}{(kr)^{1/2}} - (l+1)\frac{J_{l+1/2}(kr)}{(kr)^{3/2}}\right] P_{l-1}^{0,2}(\cos\theta) e^{-i(\pi/2-\varphi)} + \\ &\quad + i\frac{1+\cos\theta}{2} \sum_{l=1}^{\infty} C_2 \sin^2\frac{\pi l}{2} \frac{J_{l+1/2}(kr)}{(kr)^{1/2}} P_{l-1}^{0,2}(\cos\theta) e^{-i(\pi/2-\varphi)}; \\ E_\varphi &= -\frac{1+\cos\theta}{2} \sum_{l=1}^{\infty} C_1 \frac{\sin^2\frac{\pi l}{2}}{\sqrt{l(l+1)}} \left[\frac{J_{l+3/2}(kr)}{(kr)^{1/2}} - (l+1)\frac{J_{l+1/2}(kr)}{(kr)^{3/2}}\right] P_{l-1}^{0,2}(\cos\theta) e^{-i(\pi/2-\varphi)} - \\ &\quad - \frac{1+\cos\theta}{2} \sum_{l=1}^{\infty} C_2 \cos^2\frac{\pi l}{2} \frac{J_{l+1/2}(kr)}{(kr)^{1/2}} P_{l-1}^{0,2}(\cos\theta) e^{-i(\pi/2-\varphi)}. \end{aligned}\right\}, \quad (1)$$

where A_l, C_1, C_2 are the constant coefficients.

The analysis of the equations (1) shows that eigenwaves propagate into hemisphere with angles $\pm\theta$. Each wave propagates with its constant of propagation $\gamma_{m(n)}$ along virtual curves with radiuses $r_{m(n)} = \gamma_{m(n)} / k$ with maintaining vectors of polarization about component E_r and E_φ (fig.2). In the area near points $r_{m(n)} = \gamma_{m(n)} / k$ the aperture of reflector is coordinated with surrounding space under conditions of eigenwaves propagation.

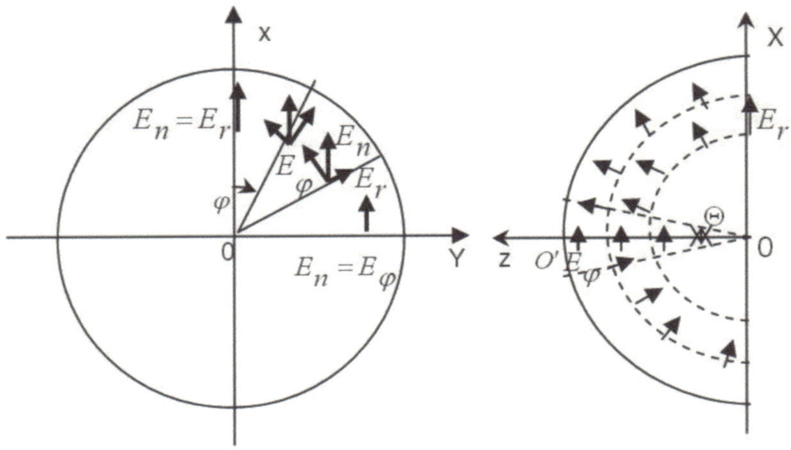

Fig. 2. Waves propagation in to hemispherical reflector

The amplitude of the electrical field strength vector $\vec{E}$ with arbitrary angle φ is defined as

$$|\vec{E}| = \sqrt{E_r^2 + E_\varphi^2},$$

where $E_r = \sqrt{(\mathrm{Re}\,E_r)^2 + (\mathrm{Im}\,E_r)^2}$; $E_\varphi = \sqrt{(\mathrm{Re}\,E_\varphi)^2 + (\mathrm{Im}\,E_\varphi)^2}$;

$$\mathrm{Re}\,E_r = (kr)^{-3/2} \sum_m A_{\gamma_m} J_{\gamma_m}(kr) \cos(\gamma_m(\theta - \pi/2)) \cdot \cos\varphi;$$

$$\mathrm{Im}\,E_r = (kr)^{-3/2} \sum_m A_{\gamma_m} J_{\gamma_m}(kr) \sin(\gamma_m(\theta - \pi/2)) \cdot \cos\varphi;$$

$$\mathrm{Re}\,E_\varphi = \sum_n A_{\gamma_n} \left[\frac{J_{\gamma_n+1}(kr)}{(kr)^{1/2}} - (\gamma_n + 1/2) \frac{J_{\gamma_n}(kr)}{(kr)^{3/2}} \right] \times \cos(\gamma_n(\theta - \pi/2)) \cdot \sin\varphi;$$

$$\mathrm{Im}\,E_\varphi = \sum_n A_{\gamma_n} \left[\frac{J_{\gamma_n+1}(kr)}{(kr)^{1/2}} - (\gamma_n + 1/2) \frac{J_{\gamma_n}(kr)}{(kr)^{3/2}} \right] \times \sin(\gamma_n(\theta - \pi/2)) \cdot \sin\varphi;$$

$$A_{\gamma_n} = \frac{1}{N_{\gamma_n}} \left[\left(1 - \frac{1}{2\gamma_n}\right) \int_0^{ka} F(z) J_{\gamma_n+1}(z) \frac{dz}{\sqrt{z}} + \left(1 + \frac{1}{2\gamma_n}\right) \int_0^{ka} F(z) J_{\gamma_n-1}(z) \frac{dz}{\sqrt{z}} \right];$$

$$A_{\gamma_m} = \frac{1}{N_{\gamma_m}} \int_0^{ka} F(z) z^{1/2} J_{\gamma_m}(z) dz;$$

$N_{\gamma_m} = \int_0^{ka} \left[J_{\gamma_m}(z) \right]^2 dz$; $N_{\gamma_n} = \left(1 - \frac{1}{2\gamma_n}\right)^2 \int_0^{ka} \left[J_{\gamma_n+1}(z) \right]^2 \frac{dz}{z} - \left(1 + \frac{1}{2\gamma_n}\right)^2 \int_0^{ka} \left[J_{\gamma_n-1}(z) \right]^2 \frac{dz}{z}$; $F(z)$ is the given distribution of the field on the aperture of a hemispherical reflector.

The distribution of electric field on the surface of virtual polarized cone inside hemispherical reflector with electrical radius of curvature $ka = 40$ and uniform distribution of incident field on the aperture $F(z) = F(kr) = 1$ has two powerful interferential maximums. First maximum is placed near paraxial focus from $kr = 10$ to $kr = 30$, second maximum is placed near reflected surface $kr = ka$ (fig.3). It is possible to observe the redistribution of interferential maximums along radial axes on a virtual cone by increasing the angle value of the cone.

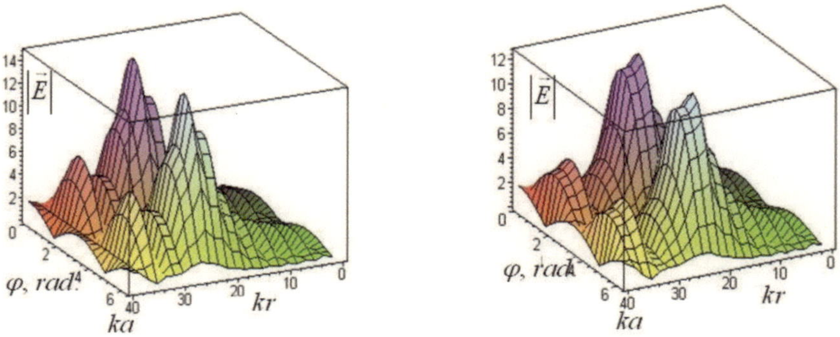

Fig. 3. Distribution of the electric field on the surface of virtual polarization cone with angle value $\Theta = 2^\circ$ (a) and $\Theta = 10^\circ$ (b)

Obviously the feed of hemispherical reflector presented as set of discrete sources that must be placed at the polarized cone according to a structure of power lines and amplitude-phase distribution of the field. According to distribution of electromagnetic field along radial axes on the arbitrary section (fig.3), the line phased feed can consist of discrete elements. It is obvious that this phased feed have an advantage about absence a shadow region of aperture. According to increase cone value Θ of polarized cone is decrease an efficiency of excitation the aperture.

Focusing properties of the edge areas of the hemispherical reflector are displayed when solving the problem of excitation a perfect electric conducting spherical surface by the symmetrical ring of electric current with coordinates (kr', θ'), where $kr' \sin\theta' >> 1$. The ring of electric current is equivalent to double magnetic sheet with thickness d with density of electrostatic charge $\pm\sigma$ for neighboring sheets (fig.4).

Let us define scalar Green function of the problem $\Gamma = \Gamma(kr, \theta, kr', \theta')$ as a function satisfactory to homogeneous wave equation anywhere, except the ring current, where electrostatic potential $\eta = \sigma \cdot d$ exposes a jump that equivalent the condition: $\partial\Gamma/\partial\theta\big|_{\theta=\theta'+0} - \partial\Gamma/\partial\theta\big|_{\theta=\theta'-0} = -4\pi\eta = -4\pi kr\delta(kr - kr')$. Let us search Green function satisfactory to radiation condition $\lim\limits_{r\to\infty}\left(r\dfrac{\partial\Gamma}{\partial r} - ikr\Gamma\right) = 0$ and boundary condition $\partial/\partial z\left(z^{3/2}\Gamma\right)_{z=ka} = 0$ as

$$\Gamma = \frac{-8\pi}{ka}\left(\frac{kr'}{kr}\right)^{3/2} \sin\theta' \times \sum_m \frac{\gamma_m}{\nu_m(\nu_m + 1)} \frac{J_{\gamma_m}(kr) J_{\gamma_m}(kr')}{J_{\gamma_m}(ka)\dfrac{\partial^2 J_{\gamma_m}(ka)}{\partial\gamma\partial(kr)}} \times \begin{Bmatrix} L_{\nu_m}(\cos\theta) P^1_{\nu_m}(\cos\theta'), \theta > \theta' \\ L_{\nu_m}(\cos\theta') P^1_{\nu_m}(\cos\theta), \theta < \theta' \end{Bmatrix}, \quad (2)$$

where $L_{\nu_m}(\cos\theta) = Q^1_{\nu_m}(\cos\theta) + i\frac{\pi}{2}P^1_{\nu_m}(\cos\theta)$; $\nu_m = \gamma_m - 1/2$; $Q^1_{\nu_m}(\cos\theta)$ - associated Legendre functions of the 2-nd kind.

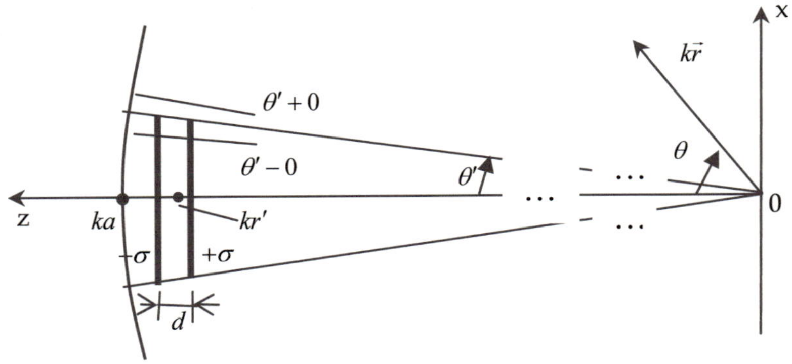

Fig. 4. Geometry of excitation of hemispherical reflector by electric current ring

Consider one-connected area D on complex surface γ and distinguish points $\gamma_1,\gamma_2,...,\gamma_m$ for this surface as solutions of Neumann boundary condition for Green function. In every point of the area D function Γ is univalent analytic function, except points $\gamma_1,\gamma_2,...,\gamma_m$ where it has simple poles. Let us place the field source on the concave hemisphere surface $(kr' = ka)$. Using Cauchy expression present (2) as a sum of waves and integral with contour that encloses part of the poles γ, satisfactory to the condition $ka < \gamma < ka - ka^{1/3}$ and describing the geometrical optic rays field. In the distance from axis of symmetry $(\gamma_m\theta >> 1)$, the contour integral is given as

$$\frac{-2\sqrt{ka}}{(kr)^{3/2}}\sqrt{\frac{\sin\theta'}{\sin\theta}}\int_{C_0}\frac{J_{\nu+1/2}(kr)}{J'_{\nu+1/2}(ka)}\left\{e^{i[(\nu+1/2)(\theta+\theta')+\pi/2]} + e^{\pm i(\nu+1/2)(\theta-\theta')}\right\}d\nu .$$

The sign "+" at index exponent corresponds for $\theta > \theta'$ and the sign "-" for $\theta < \theta'$. A factor $(\sin\theta' / \sin\theta)^{1/2}$ explains geometrical value of increasing of rays by double concave of the hemispherical reflector with comparison to cylindrical surface. First component in contour integral correspond a brighten point at the distance part of the ring with electrical current, second component - brighten point at the near part of the ring. Additional phase $\pi/2$ is interlinked with passing of rays through the axis caustic.

In the focal point area $(\gamma_m\theta \le 1)$ the contour integral can be written as

$$\frac{-4(\pi/2)^{1/2}(ka)^{1/2}}{(kr)^{3/2}}\sqrt{\frac{\theta\sin\theta'}{\sin\theta}}\int_{C_0}\frac{(\nu+1/2)^2}{\sqrt{\nu}(\nu+1)}J_1((\nu+1/2)\theta)\frac{J_{\nu+1/2}(kr)}{J'_{\nu+1/2}(ka)}e^{i[(\nu+1/2)\theta'+1/4]}d\nu .$$

In accordance with a stationary phase method the amplitude and phase structure of the field in tubes of the rays near a caustic can be investigated.

The distribution of the radial component of the electrical field E_r along axis of the hemisphere with radius of curvature $a = 22{,}5cm$ for uniform distribution field on the aperture $\theta = \pi/2$ has two powerful interferential maximums at the wave length $\lambda = 6{,}28cm$ (dotted line), $\lambda = 2{,}513cm$ (thin line) and $\lambda = 0{,}838cm$ (thick line) (fig.5).

First maximum is near paraxial focus $F = ka/2$ and is caused by diffraction of the rays, reflected from concave surface at the central area of the reflector. Second interferential maximum characterizes diffraction properties of the edges of hemispherical reflector. In this area the rays test multiple reflections from concave surface and influenced by the "whispering gallery" waves.
At decrease the wavelength λ first diffraction maximum is displaced near a paraxial focus $F = 10cm$, a field at the area $r < f$ is decreasing quickly as well as the wavelength. Second diffraction maximum is narrow at decrease of the wavelength, one can see redistribution of the interferential maximums near paraxial focus.

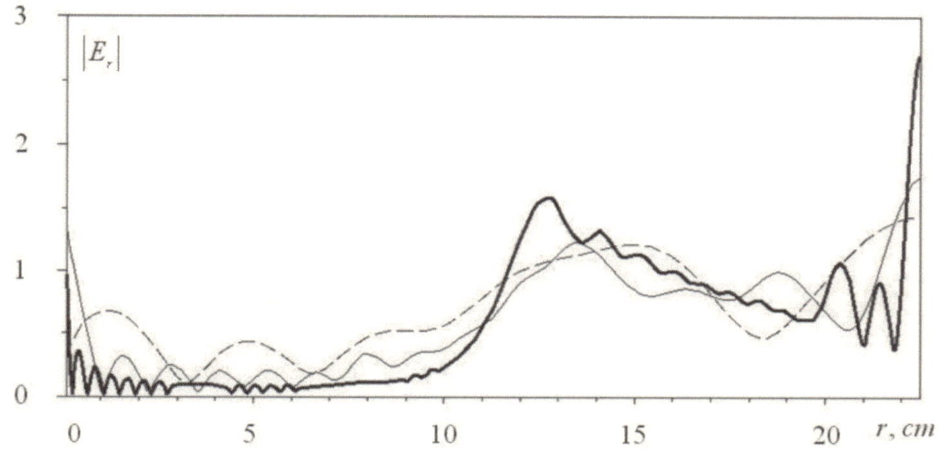

Fig. 5. Diffraction properties of the hemisphere dish with radius of curvature $a = 22,5cm$

In accordance with the distribution of the field, a spherical antenna can have a feed that consists of two elements: central feed in the area near paraxial focus and additional feed near concave surface hemisphere. This additional feed illuminates the edge areas of aperture by surface EMW. The use of additional feed can increase the gain of the spherical antennas.

3.4 Experimental investigations of spherical hybrid antenna

Accordingly a fig.5, the feed of the spherical HRA can consist of two sections. First section is ordinary feed as line source arrays that place near paraxial focus. Second section is additional feed near concave spherical surface and consists of four microstrip or waveguide sources of the surface EMW. An aperture of the additional feed must be placed as near as possible to longitudinal axis of the reflector. The direction of excitation of the additional sources is twice-opposite in two perpendicular planes. This compound feed of the spherical reflector can control the amplitude and phase distribution at the aperture of the spherical antenna.
Extended method of spherical aberration correction shows that additional sources of surface EMW must be presented as the aperture of rectangular waveguides or as microstrip sources with illumination directions along the reflector at the opposite directions. By phasing of the additional and main sources and by choosing their amplitude distribution one can control SLL of the pattern and increase the gain of the spherical antenna.

Experimental investigations of the spherical reflector antenna with diameter $2a = 31cm$ at wave length $\lambda = 3cm$ show the possibility to reduce the SLL and increase the gain by $10 - 12\%$ by means of a system control of amplitude-phase distribution between the sources (fig.6a). For correction spherical aberration at full aperture the main feed 2 (for example horn or line phase source) used for correction spherical aberration in central region of reflector 1 and placed at region near paraxial focus $F = a / 2$. Additional feed 3 consist of two (four) sources that place near reflector and radiated surface EMW in opposite directions. Therefore additional feed excite ring region at the aperture. By means of mutual control of amplitude-phase distribution between feeds by phase shifters 4, 5, 9, attenuators 6, 7, 10 and power dividers 8, 11 (waveguide tees), can be reduce SLL. There are experimental data of measurement pattern at far-field with SLL no more -36 dB (fig.7) (Ponomarev, 2008).

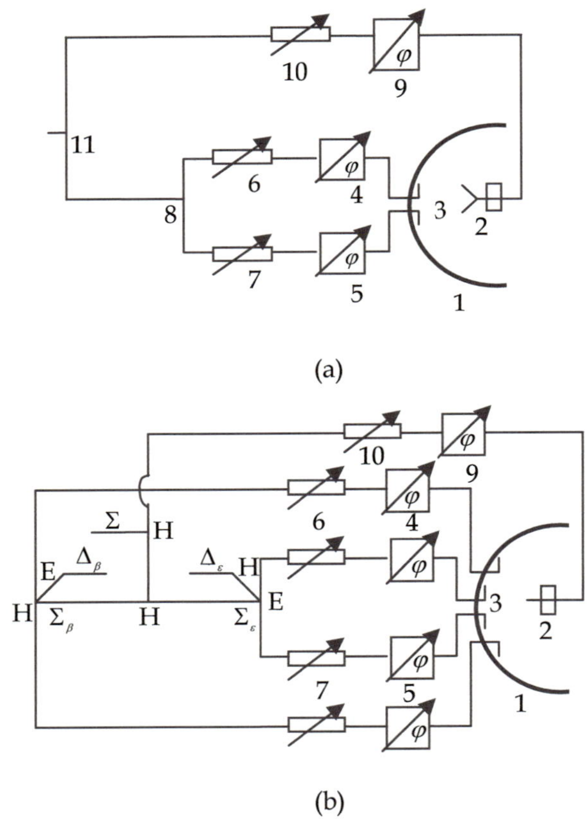

Fig. 6. Layout of spherical HRA with low SLL (a) and monopulse feed of spherical HRA (b)

For allocation of the angular information about position of the objects in two mutually perpendicular planes the monopulse feed with the basic source 2 and additional souses 3 in two mutually perpendicular areas is under construction (fig. 6b). Error signals of elevation Δ_ε and azimuth Δ_β and a sum signal Σ are allocated on the sum-difference devices (for example E- H-waveguide T-hybrid).

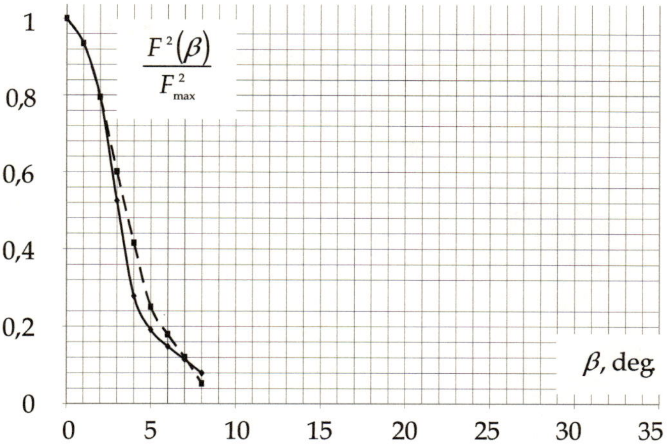

a)

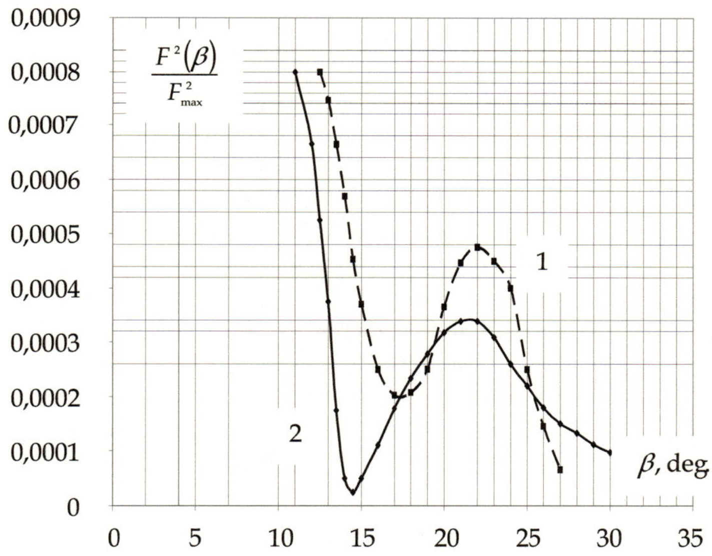

b)

Fig. 7. Pattern of spherical HRA excited with main feed (1); excited the main and additional feeds (2): a – main lobe; b – 1-st side lobe

4. Spherical diffraction antenna arrays

4.1 Analysis of spherical diffraction antenna array

Full correction of a spherical aberration is possible if to illuminate circular aperture of spherical HRA by leaky waves. For this the aperture divides into rings illuminated by separate feeds of leaky waves waveguide type. So, the spherical diffraction antenna array is forming. It consists of n hemispherical reflectors 1 (fig.8) with common axis and aperture, and $4 \cdot n$ discrete illuminators 2 near the axis of antenna array. In concordance with electrodynamics the spherical diffraction antenna array consists of diffraction elements wich are formed by two neighbouring hemispherical reflectors and illuminated by four sources. For illuminate the diffraction element between the correcting reflectors there illuminators are located in cross planes.

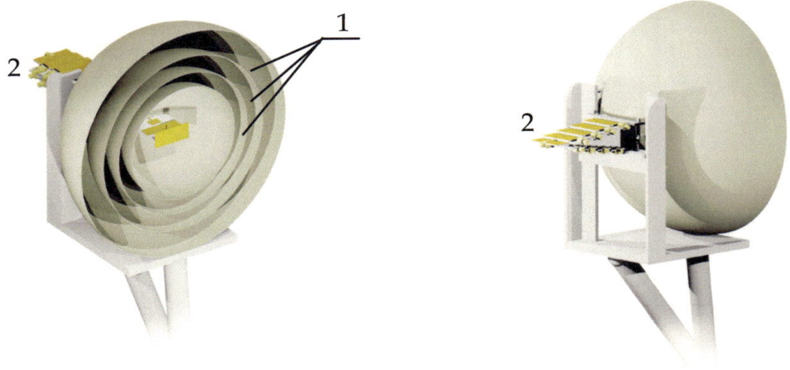

Fig. 8. Spherical diffraction antenna array: 1 – hemispherical reflectors; 2 – linear phased feed

The feed sources illuminated waves waveguide type between hemispherical reflectors which propagate along reflecting surfaces and illuminated all aperture of antenna. By means of change of amplitude-phase field distribution between feed sources the amplitude and a phase of leaky waves and amplitude-phase field distribution on aperture are controlled. For maximized of efficiency and gain of antenna the active elements should place as close as it is possible to an antenna axis. At the expense of illuminating of diffraction elements of HRA by leaky waves feeds their phase centers are "transforms" to the aperture in opposite points. Thus the realization of a phase method of direction finding in HRA is possible.

The eigenfunctions/GTD – method is selected due to its high versatility for analyzing the characteristics of diffraction antenna arrays with arbitrary electrical curvature of reflectors.

Let's assume that in diffraction element there are waves of electric and magnetic types. Let for the first diffraction element the relation of radiuses of reflectors is $\Delta = a_M / a_{M-1}$. In spherical coordinates $r \in (a_{M-1}; a_M)$, $\theta \in (0; \pi/2)$, $\varphi \in (0; \pi)$ according to a boundary problem of diffraction of EMW on ring aperture of diffraction element, an electrical potential U satisfies the homogeneous equation of Helmholtz

$$\left(\Delta_{r,\theta,\varphi} + k^2\right) U(r,\theta,\varphi) = 0$$

and boundary conditions of the 1-st kind

$$U\Big|_{r=a_M; r=a_{M-1}} = 0 \quad (3)$$

or 2-nd kind

$$\frac{\partial U}{\partial r}\Bigg|_{r=a_M; r=a_{M-1}} = 0 \quad (4)$$

The solution of the Helmholtz equation is searched in the form of a double series

$$U = \sum_{s}\sum_{m} U_{ms}(r,\theta,\varphi),$$

were transverse egenfunctions $U_{ms}(r,\theta,\varphi)$ satisfies to a condition of periodicity $U_{ms}(r,\theta,\varphi) = U_{ms}(r,\theta,\varphi + 2n\pi)$ and looks like

$$U_{ms} = -i^{m}(2m+1)\times\left[j_m(gr) - \frac{j_m(ga_{M-1})}{h_m^{(1)}(ga_{M-1})} h_m^{(1)}(gr) \right] \times P_m(\cos\theta) \begin{Bmatrix} \cos \\ \sin \end{Bmatrix} (m\varphi) \tag{5}$$

for a boundary condition (3) and

$$\tilde{U}_{ms} = -i^{m}(2m+1)\times\left[j_m(\tilde{g}r) - \frac{j'_m(\tilde{g}a_{M-1})}{h_m^{(1)'}(\tilde{g}a_{M-1})} h_m^{(1)}(\tilde{g}r) \right] \times P_m(\cos\theta) \begin{Bmatrix} \cos \\ \sin \end{Bmatrix} (m\varphi) \tag{6}$$

for a boundary condition (4).
Having substituted (5) to boundary condition (3) we will receive a following transcendental characteristic equation

$$j_m(\chi)h_m^{(1)}(\chi\cdot\Delta) - j_m(\chi\cdot\Delta)h_m^{(1)}(\chi) = 0$$

where $\chi = g \cdot a_M$; $\chi_{ms}(m = 0,1,2,...; s = 0,1,2,...)$ are roots of the equation which are eigenvalues of system of electric waves; $j_m(\cdot), h_m^{(1)}(\cdot)$ - spherical Bessel functions of 1-st and 3-rd kind, accordingly.
Similarly, having substituted expression (6) in the equation (4) we can receive the characteristic equation

$$j'_m(\tilde{\chi})h_m^{(1)'}(\tilde{\chi}\cdot\Delta) - j'_m(\tilde{\chi}\cdot\Delta)h_m^{(1)'}(\tilde{\chi}) = 0$$

where equation roots $\tilde{\chi}_{ms}(m = 0,1,2,...; s = 0,1,2,...)$ are eigenvalues of magnetic type waves.
A distributions of amplitude and a phase of a field along axis of diffraction element for a wave of the electric type limited to hemispheres in radiuses $a_M = 15{,}5\,cm$; $a_{M-1} = 13{,}423\,cm$ are presented at fig.9a,b. We can see the field maximum in the centre of a spherical waveguide.
Influence of the higher types of waves on the distribution of the field for electrical type of waves is explained by diagram's on fig. 10 where to a position (a) corresponds amplitude of interference of waves types $E_{00}, E_{01}, ..., E_{010}, E_{011}$, and to a position (b) – phase distribution of an interference of waves of the same types.
The diffracted wave in spherical waveguides of spherical diffraction antenna array according to (5), (6) can be written as

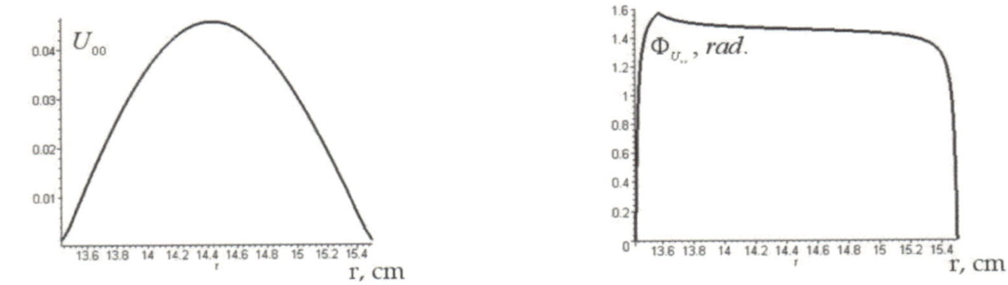

a) b)

Fig. 9. Distribution of amplitude (a) and phase (b) of fundamental mode E_{00} of spherical waveguide

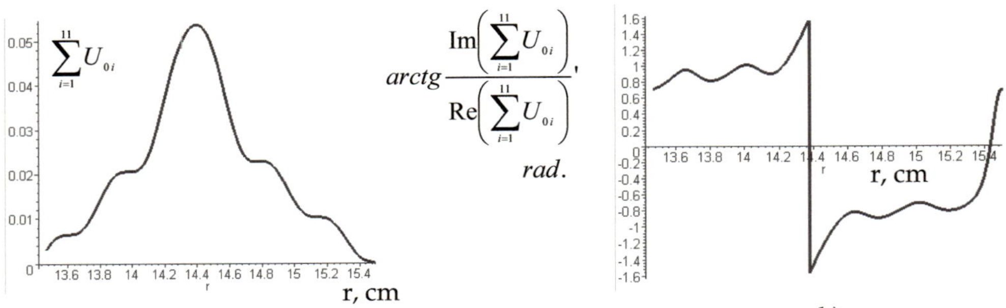

a) b)

Fig. 10. Distribution of amplitude (a) and phase (b) of eleven electrical type eigenwaves $E_{0s}\,(s = 1,2,\ldots,11)$

$$U_{ms} = \sum_{m,s} e^{-im\pi/2}(2m+1)\left[j_m(gr) - \frac{\Omega j_m(ga_{M-1})}{\Omega h_m^{(1)}(ga_{M-1})} h_m^{(1)}(gr) \right] \times P_m(\cos\theta) \begin{Bmatrix} \cos \\ \sin \end{Bmatrix} (m\varphi), \quad (7)$$

where $\Omega = \begin{Bmatrix} 1 \\ \dfrac{d}{d(gr)} \end{Bmatrix}$.

Asymptotic expression for U_{ms} at $ga_{M-1} >> 1$ looks as

$$U_{ms} \to (ka_{M-1})^{1/3} \frac{\exp\left[i \cdot \left(k\sqrt{r^2 - a^2{}_{M-1}} + \pi/12 \right) \right]}{\sqrt{2g}\left(r^2 - a^2{}_{M-1} \right)^{1/4}} \times$$
$$\times \sum_s \frac{C_s}{1 - e^{2\pi \cdot i \cdot \gamma_s}} \left\{ \exp\left[i\gamma_s \left(\theta - \arccos\frac{a_{M-1}}{r} - \frac{\pi}{2} \right) \right] + \exp\left[i\gamma_s \left(\frac{3\pi}{2} - \theta - \arccos\frac{a_{M-1}}{r} \right) \right] \right\} \quad (8)$$

where C_s - amplitudes of "creeping" waves (Keller, 1958); γ_s - poles of 1-st order for (7).

The values into (8) have a simple geometrical sense. Apparently from fig.11 values $a_{M-1}\left(\theta-\arccos\frac{a_{M-1}}{r}-\frac{\pi}{2}\right)$ and $a_{M-1}\left(\frac{3\pi}{2}-\theta-\arccos\frac{a_{M-1}}{r}\right)$ are represent lengths of arcs along which exist two rays falling on aperture of a diffraction element. These rays come off a convex surface and there are meting in point. From the point of separation to a observation point both rays pass rectilinear segment $\sqrt{r^2-a^2{}_{M-1}}$.

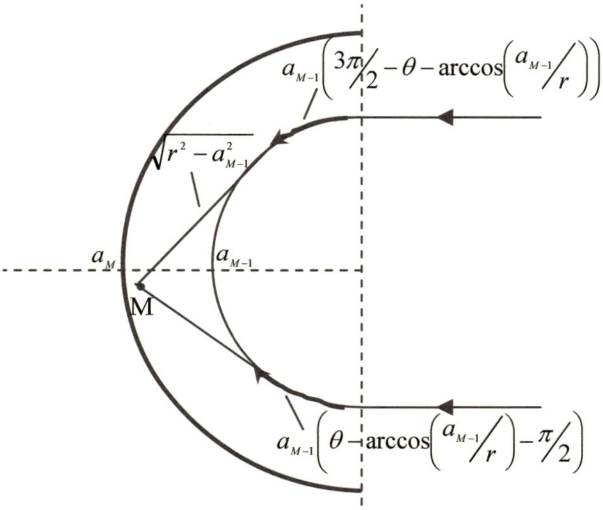

Fig. 11. The paths passed by "creeping" waves in spherical wave guide of spherical diffraction antenna array

Because of the roots γ_s have a positive imaginary part which increases with number s, each of eigenwaves attenuates along a convex spherical surface. Attenuating that faster, than it is more number s. Therefore the eigenwaves on a convex spherical surface represent “creeping” waves. Thus in diffraction element the rays of GO, leakage waves and “creeping” waves, are propagate. The leakages EMW are propagating along concave surface, the "creeping" waves are propagating along a convex surface.
At aperture the field is described by the sum of normal waves. Description of pattern provides by Huygens-Kirchhoff method. The radiation field of spherical diffraction antenna array in the main planes is defined only by x-th component of electric field in aperture and y-th component of a magnetic field.

A directivity of spherical diffraction antenna array created by electrical waves with indexes n, m is defined by expression

$$E_{nm}=\frac{D^c_{nm}\cdot\pi\cdot r(c1)\cdot i^{m-1}(\sin q')^{m-1}}{\sqrt{2}k^2 r(m)r(c2)}\times$$
$$\times\left[\frac{\cos((m-1)\varphi)}{c2}F\left(c1,c2;m;\sin^2 q'\right)-\frac{c1\cdot\sin^2 q\cdot\cos((m+1)\varphi)}{m(m+1)}F\left(c1+1,-c2+1;m+2;\sin^2 q'\right)\right], \quad (9)$$

where D_{nm}^c - weight coefficients; $F(a,b;x;z) = {}_2F_1(a,b;x;z)$ - hypergeometric functions; $c1 = (v_n + m - 0,5)/2$; $c2 = (v_n - m + 0,5)/2$; v_n - propagation constants of electrical type eigenwaves; q', φ' - observation point coordinates at far-field.
For magnetic types of waves

$$E_{sp} = \frac{\sqrt{2} \cdot W \cdot B_{sp} \cdot \pi \cdot r(c3) \cdot i^{p-1} (\sin q')^{p-1}}{w_s k^2 r(p) r(c4)} \times$$
$$\times \left[\frac{\cos((p-1)\varphi')}{c4} F(c3, -c4; p; \sin^2 q') + \frac{c3 \cdot \sin^2 q' \cdot \cos((p+1)\varphi')}{p(p+1)} F(c3+1, -c4+1; p+2; \sin^2 q') \right], \quad (10)$$

where $E_{s0} = 0$ for $p = 0$; W - free space impedance; w_s - propagation constants of magnetic type eigenwaves; $c3 = (w_s + p + 0,5)/2$; $c4 = (w_s - p - 0,5)/2$.
Generally the pattern of spherical diffraction antenna array consists of the partial characteristics enclosed each other created by electrical and magnetic types of waves that it is possible to present as follows

$$E = \sum_{n=1}^{N} \sum_{m=0}^{\infty} E_{nm} + \sum_{s=1}^{N} \sum_{p=0}^{\infty} E_{sp}, \quad (11)$$

where E_{nm}, E_{sp} - the partial patterns that defined by expressions (9), (10).

4.2 Numerical and experimental results

Numerical modeling by the (10) show that at $q = 0$ in a direction of main lobe, the far-field produced only by waves with azimuthally indexes $m = 1$, $p = 1$. At $q > 0$ the fare-field created by EMW with indexes $0 < m < \infty$, $0 < p < \infty$.
Influence of location of feeds to field distribution on aperture of spherical diffraction antenna array, is researched. At the first way feeds took places on an imaginary surface of a polarization cone (fig. 12a). On the second way feeds took places on equal distances from an axis of the main reflector (fig. 12b).
Dependences of geometric efficiency of antenna $L = S_0/S$ (S_0 - square of radiated part of antenna, S - square of aperture) versus the corner value Θ at the identical sizes of radiators for the first way (a curve 1) and the second way (a curve 2), are shown on fig.13.
Dependences of directivity versus the corner value Θ, are presented on fig.14. Growth of directivity accordingly reduction the value Θ explain the focusing properties of reflectors of diffraction elements. Most strongly these properties appear at small values of Θ. The rise of directivity is accompanied by equivalent reduction of width of main lobe on the plane yoz.
A comparison with equivalent linear phased array is shown that in spherical diffraction antenna array the increasing of directivity at 4-5 times, can be achieved.
For scanning of pattern over angle q_0 it is necessary to realize linear change of a phase on aperture by the expression $\Phi = kr \sin q_0$. As leakage waveguide modes excited between correcting reflectors, possess properties to transfer phase centers of sources for scanning of pattern over angle q_0, it is necessary that a phase of the feeds allocated between reflectors in radiuses a_n, a_{n+1}, are defines by

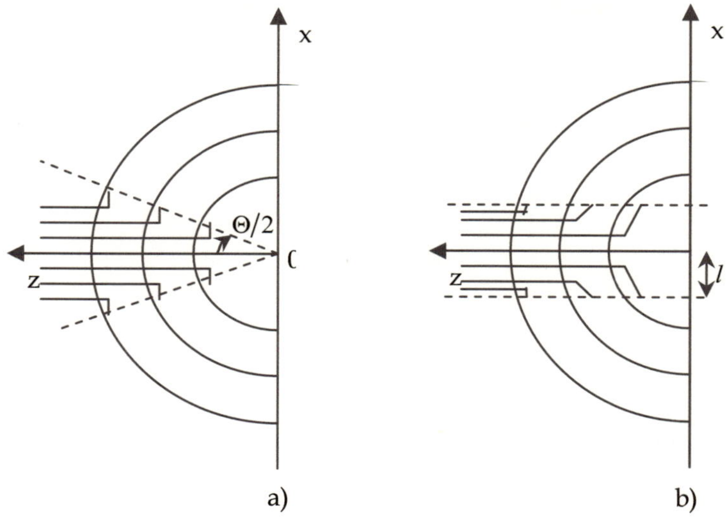

Fig. 12. Types of aperture excitation of spherical diffraction antenna array

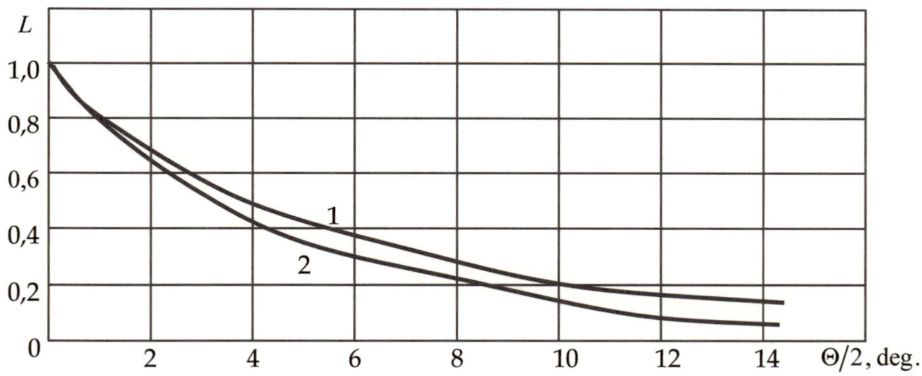

Fig. 13. Dependences of geometric efficiency of spherical diffraction antenna array versus the corner value Θ of radiator

$$\Phi_n = kr_n \sin q_0 = \nu_n \sin q_0 .$$

Accordingly the phases distribution of waves along q, we have

$$\Phi_n = \nu_n \sin q_0 + \nu_n \left(\pi/2 - h\right). \tag{12}$$

The possibility of main lobe scanning at the angle value q_0, is researched by the setting a phase of feeds according to (12). The increase of the width of a main lobe and SLL, is observed at increase q_0. Scanning of pattern is possible over angles up to 30-40 deg. (fig. 15).
Measurements of amplitude field distribution into diffraction elements of spherical diffraction antenna array are carried out for vertical and horizontal field polarization on the measurement setup (fig. 16).

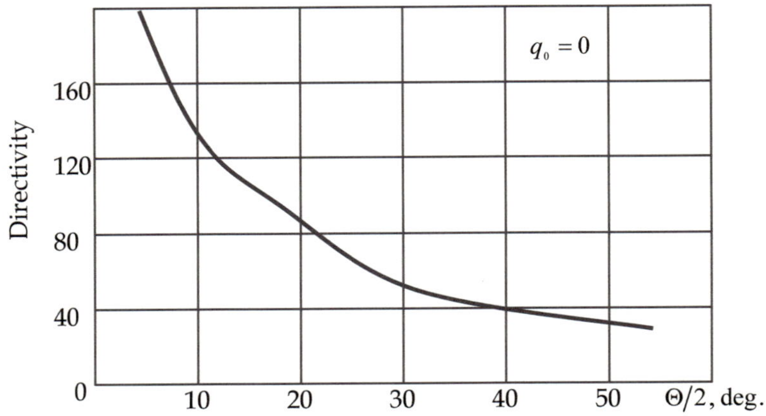

Fig. 14. Dependences of directivity versus the corner value Θ of radiator

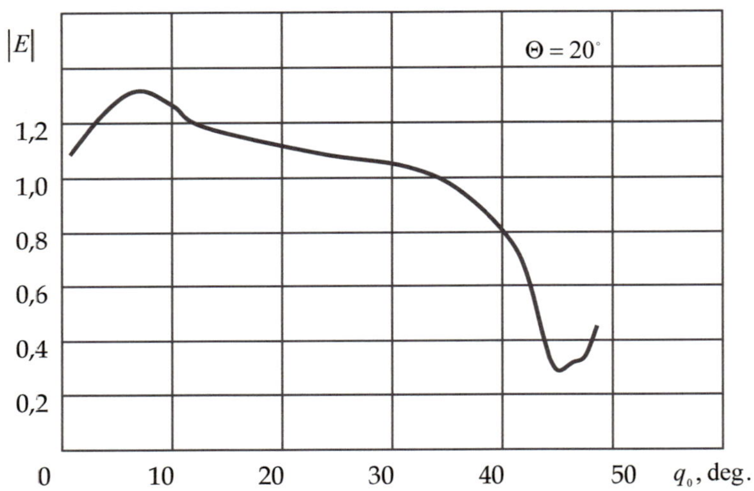

Fig. 15. Dependence of amplitude of the main lobe of spherical diffraction antenna array versus the scanning angle

Far-field and near-field properties of the aforementioned antennas were measured using the compact antenna test range facilities at the antenna laboratory of the Baltic Fishing Fleet State Academy. The experimental setting of spherical diffraction antenna array consists of two diffraction elements 1 that forms by reflectors with radiuses $a_1 = 9{,}15$ cm, $a_2 = 10{,}7$ cm and $a_3 = 12{,}6$ cm. The spherical diffraction antenna array aperture was illuminated from far zone (15 m) by vertically polarized field. The $\lambda/4$ probe which was central conductor of a coaxial cable at diameter of 2 mm, was used. It moved on the carriage 3 of positioning system QLZ 80 (BAHR Modultechnik GmbH). The output of a probe through cable assemble SM86FEP/11N/11SMA (4) with the length 50 cm (HUBER+SUHNER AG) connected with the low noise power amplifier HMC441LP3 (Hittite MW Corp.) (5) with gain equal 14 dB. The amplifier output through cable assembles SM86FEP/11N/11SMA connected to programmed detector section HMC611LP3 (6). Its output connected to digital multimeter. For selection of traveling waves into diffraction elements, the half of aperture was closed by radio absorber.

Fig. 16. Photos of experimental setting for measurement of amplitude field distribution inside the diffraction elements of spherical diffraction antenna array

During measurement of a radial component of the electrical field E_r along axis of spherical diffraction antenna array the bottom half of aperture was closed by the radio absorber (at vertical polarization of incident waves), and during measurement of a tangential component of the electrical field E_φ the left half of aperture was closed by it. The amplitude distribution of the tangential component of electric field E_φ along axis of spherical diffraction antenna array at the frequency 10 GHz is presented on fig. 17.

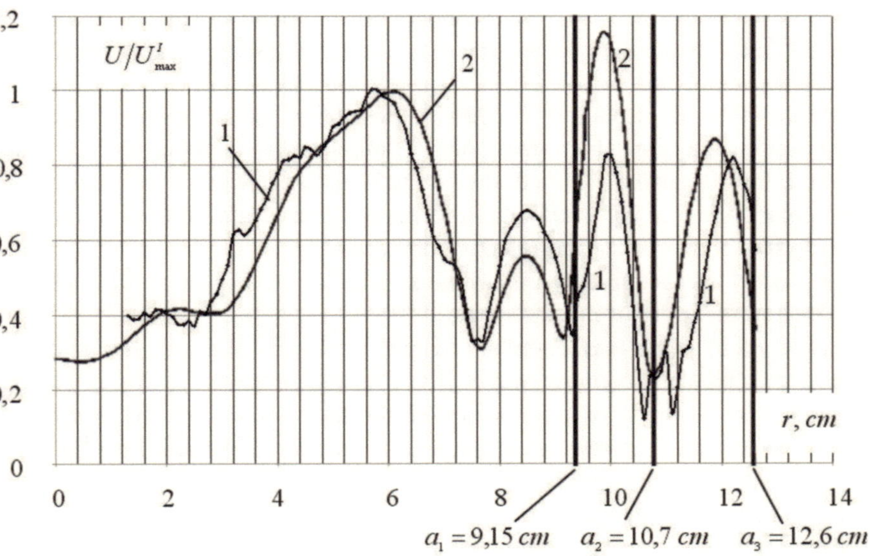

Fig. 17. Normalized amplitude distribution of the tangential component of electric field E_φ along axis of spherical diffraction antenna array at the frequency 10 GHz: 1 - measured, 2 - calculated by suggested method

The experimental measurements of partial patterns of spherical diffraction antenna array in a centimeter waves are carried out. The feed of spherical diffraction antenna array is fabricated on series 0,813 mm-thick substrates Rogers RO4003C with permittivity 3,35. The feeds consist of the packaged microstrip antennas tuned on the frequencies range 10 GHz. The technique of designing and an experimental research of packaged feeds of spherical diffraction antenna arrays include following stages: definition of geometrical characteristics of feeds (subject to radial sizes of diffraction elements and amplitude-phase distribution inside diffraction elements at the set polarization of field radiation); optimization of geometrical parameters of microstrip antennas, power dividers, feeding lines (subject to influence of metal walls of diffraction element); an experimental investigation of S-parameters of a feeds; computer optimization of a feeds geometry in Ansoft HFSS.

Experimental measurements of spherical diffraction antenna array were by a method of the rotate antenna under test in far-field zone subject to errors of measurements. At 8-11 GHz frequency range the partial patterns were measured at linear polarization of incident waves. The reflectors and the radiator are adjusted at measurement setting (fig.18).

The form and position of patterns of spherical diffraction antenna array for one side of package feed characterize electrodynamics properties of diffraction elements and edge effects of multireflector system (fig.19). The distances of partial patterns with respect to phase centre of the antenna were: for 1-st (greatest) diffraction element - 46° (curve 1), for 2-nd diffraction element - 38° (curve 2), for 3-rd diffraction element - 32° (curve 3). The radiuses of reflectors: $a_1 = 12{,}6$ cm, $a_2 = 9{,}79$ cm, $a_3 = 7{,}89$ cm and $a_4 = 6{,}28$ cm.

The comparative analysis of measurements results of spherical diffraction antenna array partial patterns shows possibility of design multibeam antenna systems with a combination of direction finding methods (amplitude and phase), of frequency ranges and of radiation (reception) field polarization. The angular rating of partial patterns can be used the spherical diffraction antenna arrays as feeds for big size HRA's or planar antenna arrays $(n \cdot 100 - n \cdot 1000)\lambda$.

The diffraction elements isolation defines a possibility of creating a multifrequency diffraction antenna arrays in which every diffraction element works on the fixed frequency in the set band. Besides, every diffraction element of spherical diffraction antenna arrays is isolated on polarization of the radiation (reception) field. Such antennas can be used as frequency-selective and polarization-selective devices.

The usage spherical diffraction antenna arrays as angular sensors of multifunctional radars of the purpose and the guidance weapon are effective.

As a rule, antenna systems surface-mounted and on board phase radars consist of four parabolic antennas with the common edges (fig. 20a) for direction finding of objects in two ortogonal planes. In the centre of antenna system the rod-shaped dielectric antenna forming wide beam of pattern can dispose. At attempt of reduction of the aperture size of antenna system D^* the distance between the phase centers of antennas becomes less then diameter of a reflector $b < D$. The principle of matching of a slope of direction finding characteristics to width of area of unequivocal direction finding is broken.

Alternative the considered antenna system of phase radar is the antenna consisting of one reflector and a radiator exciting opposite areas of aperture the surface EMW that propagating directly along a concave hemispherical reflector (fig. 20b).

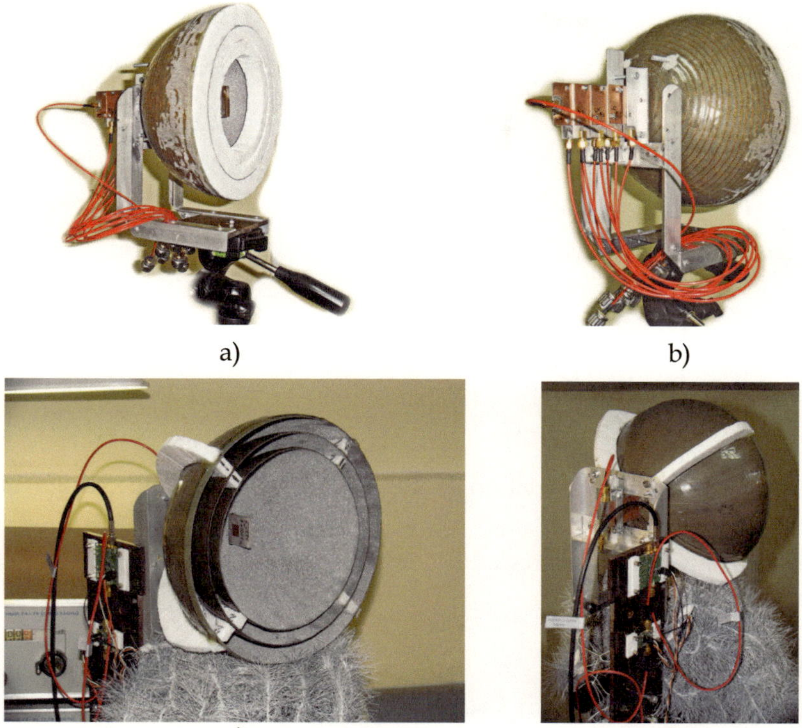

a) b)

c) d)

Fig. 18. Photos of experimental setting for measurement of spherical diffraction antenna arrays patterns of a centimeter wave: a, b – 9,5 GHz; c, d – 10,5 GHz

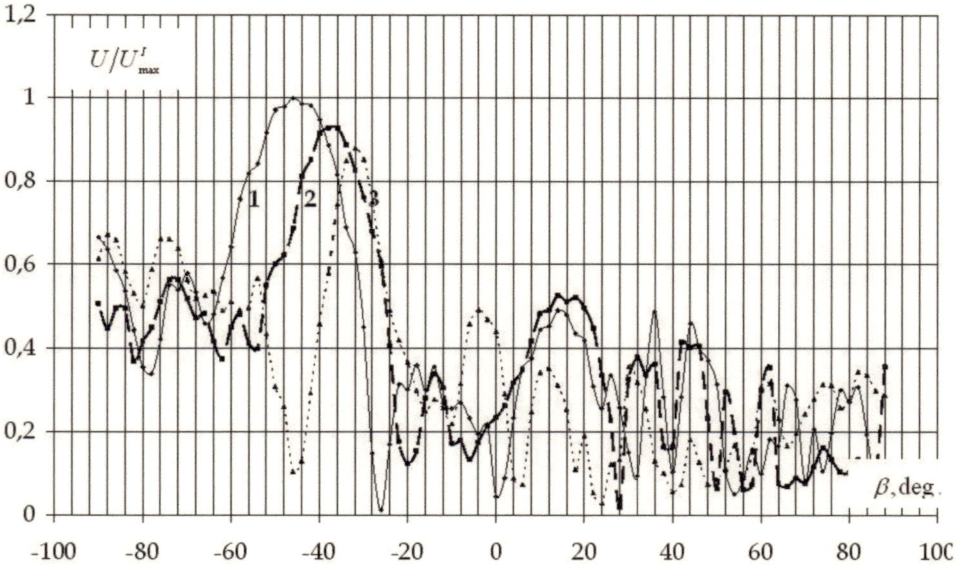

Fig. 19. The partial patterns of spherical diffraction antenna array

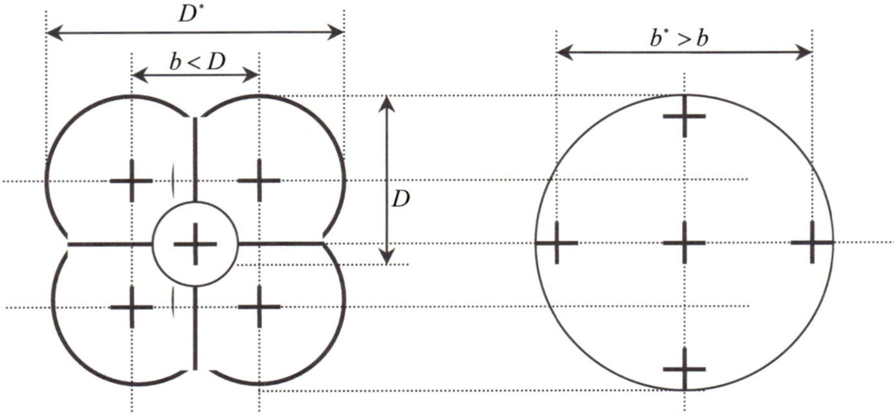

Fig. 20. Antenna system of phase radar on the basis of four parabolic reflectors (a) and one hemispherical reflector (b)

This waves transfers the phase centres of a sources on the reflector aperture in points with radial coordinates $r = \gamma_m / k \approx a$.

5. Conclusion

The review of methods of the analysis, synthesis and design features of HRA is carried out. The ways of improvement of their electrical characteristics at the expense of usage of reflectors with the circular profile and the linear phased feeds with additional sources of surface EMW or leakage waves of waveguide type is defined. Disadvantages of methods of correction of a spherical aberration on the basis of usage of a subreflector of the special form and the linear phased irradiator are revealed. The new method of correction of a spherical aberration by illuminating of edge areas of aperture of hemispherical reflector surface EMW is developed. Solves of the Maxwell equations with usage of techniques of the spherical rotates, diffractions of electromagnetic waves precisely describing and explaining a physical picture on a hemispherical surface are received. The solution of a problem of diffraction plane electromagnetic waves on a hemispherical reflector has allowed to define borders and the form interference maximums in internal area of reflectors with any electric radius. By the solve of a problem of excitation of aperture of hemispherical reflector by current ring the additional requirements to width of the pattern of sources of surface EMW are developed. Expressions for Green's function as sum of waves and beam fields according to approaches of GTD are found. The behavior of a field in special zones is found out: near caustics and focal points. The method of control of amplitude and phase fields distribution on the aperture of spherical HRA is developed. At the expense of phasing of the basic and additional feeds and control of amplitudes and phases between them it was possible to reduce the SLL no more - 36 dB and to increase the gain by 10-12 %. The method of full correction of a spherical aberration by a subdividing of circular aperture to finite number ring apertures of systems of coaxial hemispherical reflectors is developed. New type of HRA's as spherical diffraction antenna arrays is offered. The spherical HRA's and diffraction antenna arrays have a low cost and easy to fabricate. Their electrodynamics analysis by a method of eigenfunctions and in approximations of GTD is carried out. The spherical diffraction antenna array allows: to control amplitude end phase fields distribution on all aperture of HRA; to provide high efficiency because of active radiating units of feeds do not shade of aperture; to realize a combined amplitude/multibase phase method of direction finding of the objects, polarization selection of signals. The HRA's provide: increasing of range of radars operation by 8-10 %; reduce the error of measurement of coordinates at 6-8 times; reduction of probability of suppression of radar by active interferences by 20-30 %.

On the basis of such antennas use of MMIC technology of fabricate integrated feeds millimeter and centimeter waves is perspective. Embedding the micromodules into integral feeding-source antennas for HRA's and spherical diffraction antenna arrays for processing of the microwave information can be utilized for long-term evolution multifunctional radars. Future work includes a more detailed investigation the antennas for solving a problem of miniaturization of feeds for these antennas by means of MMIC technologies.

6. References

Bucci, O.M., Elia, G.D. & Romito, G. (1996). Synthesis Technique for Scanning and/or Reconfigurable Beam Reflector Antennas With Phase-only Control. *IEE Proc.-Microw. Antennas Propag.*, Vol. 143, No. 5, October, p.p. 402-412.

Chantalat, R., Menudier, C., Thevenot, M., Monediere, T., Arnaud, E. & Dumon, P. (2008). Enhanced EBG Resonator Antenna as Feed of a Reflector Antenna in the Ka Band. *IEEE Antennas and Wireless Propag.*, Vol. 7, p.p. 349-353.

Elsherbeni, A. (1989). High Gain Cylingrical Reflector Antennas with Low Sidelobes. *AEU*, Band 43, Heft 6, p.p. 362-369.

Eom, S.Y., Son, S.H., Jung, Y.B., Jeon, S.I., ganin, S.A., Shubov, A.G., Tobolev, A.K. & Shishlov, A.V. (2007). Design and Test of a Mobile Antenna System With Tri-Band Operation for Broadband Satellite Communications And DBS Reception. *IEEE Trans. on Antennas and Propag.*, Vol. 55, No. 11, November, p.p. 3123-3133.

Fourikis, N. (1996). *Phased Array-Based Systems and Applications*, John Willey & Sons., Inc.

Gradshteyn, I.S. & Ryzhik, I.M. (2000). *Table of Integrals, Series and Products*, 930, 8.533, Academic Press, New York.

Grase, O. & Goodman, R. (1966). Circumferential waves on solid cylinders. *J. Acoust. Soc. America*, Vol. 39, No. 1, p.p.173-174.

Haupt, R.L. (2008). Calibration of Cylindrical Reflector Antennas With Linear Phased Array Feeds. *IEEE Trans. on Antennas and Propag.*, Vol. 56, No. 2, February, p.p. 593-596.

Janpugdee, P., Pathak, P. & Burkholder, R. (2005). A new traveling wave expansion for the UTD analysis of the collective radiation from large finite planar arrays. *IEEE AP-S/URSI Int. Symp.*, Washington, DC, July.

Jeffs, B. & Warnick, K. (2008). Bias Corrected PSD Estimation for an Adaptive Array with Moving Interference. *IEEE Trans. on Antennas and Propag.*, Vol. 56, No. 7, July, p.p. 3108-3121.

Jung, Y.B. & Park, S.O. (2008). Ka-Band Shaped Reflector Hybrid Antenna Illuminated by Microstrip-Fed Horn Array. *IEEE Trans. on Antennas and Propag.*, Vol. 56, No. 12, December, p.p. 3863-3867.

Jung, Y.B., Shishlov, A. & Park, S.O. (2009). Cassegrain Antenna With Hybrid Beam Steering Scheme for Mobile Satellite Communications. *IEEE Trans. on Antennas and Propag.*, Vol. 57, No. 5, May, p.p. 1367-1372.

Keller, J. (1958). A geometrical theory of diffraction. Calculus of variations and its applications. *Proc. Symposia Appl. Math.*, 8, 27-52. Mc Graw-Hill., N.Y.

Llombart, N., Neto, A., Gerini, G., Bonnedal, M. & Peter De Maagt. (2008). Leaky Wave Enhanced Feed Arrays for the Improvement of the Edge of Coverage Gain in Multibeam Reflector Antennas. *IEEE Trans. on Antennas and Propag.*, Vol. 56, No. 5, May, p.p. 1280-1291.

Love, A. (1962). Spherical Reflecting Antennas with Corrected Line Sources. *IRE Trans. on Antennas and Propagation*, Vol. AP-10, September, No. 5-6, p.p.529-537.

Miller, M. & Talanov, V. (1956). Electromagnetic Surface Waves Guided by a Boundary with Small Curvature. *Zh. Tekh. Fiz*, Vol. 26, No. 12, p.p. 2755-2765.

Ponomarev, O. (2008). Diffraction of Electromagnetic Waves by Concave Circumferential Surfaces: Application for Hybrid Reflector Antennas. *Bull. of the Russian Academy of Sciences: Physics*, Vol. 72, No. 12, p.p. 1666-1670.

Rayleigh, J.W.S. (1945). *The Theory of Sound*, 2-nd ed., Vol. 2, Sec. 287, Dover Publication, ISBN 0-486-60293-1, New York.

Schell, A. (1963). The Diffraction Theory of Large-Aperture Sp-herical Reflector Antennas. *IRE Trans. on Antennas and Propagation*, July, p.p. 428-432.

Shevchenko, V. (1971). Radiation losses in bent waveguides for surface waves. *Institute of Radioengineering and Electronics, Academy of Sciences of the USSR. Translated from Izvestiya Vysshikh Uchebnykh Zavedenii, Radiofizika*, Vol. 14, No. 5, p.p. 768-777, May.

Spencer, R., Sletten, C. & Walsh, J. (1949). Correction of Spherical Aberration by a Phased Line Source. *Proc. N.E.C.*, Vol. 5, p.p. 320-333.

Tap, K. & Pathak, P.H. (2006). A Fast Hybrid Asymptotic and Numerical Physical Optics Analysis of Very Large Scanning Cylindrical Reflectors With Stacked Linear Array Feeds. *IEEE Trans. on Antennas and Propag.*, Vol. 54, No. 4, April, p.p. 1142-1151.

Tingye, L. (1959). A Study of Spherical Reflectors as Wide-Angle Scanning Antennas. *IRE Trans. on Antennas and Propagation*, July, p.p. 223-226.

Part 4

Wave Propagation in Plasmas

15

Electromagnetic Waves in Plasma

Takayuki Umeda
Solar-Terrestrial Environment Laboratory, Nagoya University
Japan

1. Introduction

No less than 99.9% of the matter in the visible Universe is in the plasma state. The plasma is a gas in which a certain portion of the particles are ionized, and is considered to be the "fourth" state of the matter. The Universe is filled with plasma particles ejected from the upper atmosphere of stars. The stream of plasma is called the stellar wind, which also carries the intrinsic magnetic field of the stars. Our solar system is filled with solar-wind-plasma particles. Neutral gases in the upper atmosphere of the Earth are also ionized by a photoelectric effect due to absorption of energy from sunlight. The number density of plasma far above the Earth's ionosphere is very low ($\sim$100cm^{-3} or much less). A typical mean-free path of solar-wind plasma is about 1AU[1] (Astronomical Unit: the distance from the Sun to the Earth). Thus plasma in Geospace can be regarded as collisionless.

Motion of plasma is affected by electromagnetic fields. The change in the motion of plasma results in an electric current, and the surrounding electromagnetic fields are then modified by the current. The plasma behaves as a dielectric media. Thus the linear dispersion relation of electromagnetic waves in plasma is strongly modified from that in vacuum, which is simply $\tilde{\omega} = kc$ where $\tilde{\omega}$, k, and c represent angular frequency, wavenumber, and the speed of light, respectively. This chapter gives an introduction to electromagnetic waves in collisionless plasma[2], because it is important to study electromagnetic waves in plasma for understanding of electromagnetic environment around the Earth.

Section 2 gives basic equations for electromagnetic waves in collisionless plasma. Then, the linear dispersion relation of plasma waves is derived. It should be noted that there are many good textbooks for linear dispersion relation of plasma waves. However, detailed derivation of the linear dispersion relation is presented only in a few textbooks (e.g., Stix, 1992; Swanson, 2003; 2008). Thus Section 2 aims to revisit the derivation of the linear dispersion relation.

Section 3 discusses excitation of plasma waves, by providing examples on the excitation of plasma waves based on the linear dispersion analysis.

Section 4 gives summary of this chapter. It is noted that the linear dispersion relation can be applied for small-amplitude plasma waves only. Large-amplitude plasma waves sometimes result in nonlinear processes. Nonlinear processes are so complex that it is difficult to provide their analytical expressions, and computer simulations play important roles in studies of nonlinear processes, which should be left as a future study.

[1] 1AU$\sim$150,000,000km

[2] This work was supported by Grant-in-Aid for Scientific Research on Innovative Areas No.21200050 from MEXT of Japan. The author is grateful to Y. Kidani for his careful reading of the manuscript.

2. Linear dispersion relation

2.1 Basic equations

The starting point is Maxwell's equations (1-4)

$$\nabla \times \vec{E} = -\frac{\partial \vec{B}}{\partial t}, \tag{1}$$

$$\nabla \times \vec{B} = \mu_0 \vec{J} + \frac{1}{c^2}\frac{\partial \vec{E}}{\partial t}, \tag{2}$$

$$\nabla \cdot \vec{E} = \frac{\rho}{\epsilon_0}, \tag{3}$$

$$\nabla \cdot \vec{B} = 0, \tag{4}$$

where $\vec{E}$, $\vec{B}$, $\vec{J}$, and ρ represent electric field, magnetic field, current density, and charge density, respectively. Here a useful relation $\epsilon_0\mu_0 = 1/c^2$ is used where ϵ_0 and μ_0 are dielectric constant and magnetic permeability in vacuum, respectively.
The motion of charged particles is described by the Newton-Lorentz equations (5,6)

$$\frac{\mathrm{d}\vec{x}}{\mathrm{d}t} = \vec{v}, \tag{5}$$

$$\frac{\mathrm{d}\vec{v}}{\mathrm{d}t} = \frac{q}{m}\left(\vec{E} + \vec{v} \times \vec{B}\right), \tag{6}$$

where $\vec{x}$ and $\vec{v}$ represent the position and velocity of a charged particle with q and m being its charge and mass. The motion of charged particles is also expressed in terms of microscopic distribution functions

$$\frac{\partial f}{\partial t} + \vec{v} \cdot \frac{\partial f}{\partial \vec{x}} + \frac{q}{m}\left(\vec{E} + \vec{v} \times \vec{B}\right) \cdot \frac{\partial f}{\partial \vec{v}} = 0, \tag{7}$$

where $f[\vec{x}, \vec{v}, t]$ represents distribution function of plasma particles in a position-velocity phase space. Equation (7) is called the Vlasov equation or the collisionless Boltzmann equation (collision terms of the Boltzmann equation in right hand side is neglected). The zeroth momentum and the first momentum of the distribution function give the charge density and the current density

$$\rho = q\int f \mathrm{d}^3\vec{v}, \tag{8}$$

$$\vec{J} = q\int \vec{v} f \mathrm{d}^3\vec{v}. \tag{9}$$

2.2 Derivation of linear dispersion equation

Let us "linearize" the Vlasov equation. That is, we divide physical quantities into an equilibrium part and a small perturbation part (for the distribution function $f = n(f_0 + f_1)$ with f_0 and f_1 being the equilibrium and the small perturbation parts normalized to unity, respectively). Then the Vlasov equation (7) becomes

$$\frac{\partial f_1}{\partial t} + \vec{v} \cdot \frac{\partial f_1}{\partial \vec{x}} + \frac{q}{m}\left(\vec{v} \times \vec{B}_0\right) \cdot \frac{\partial f_1}{\partial \vec{v}} = -\frac{q}{m}\left(\vec{E}_1 + \vec{v} \times \vec{B}_1\right) \cdot \frac{\partial f_0}{\partial \vec{v}}. \tag{10}$$

Here, the electric field has only the perturbed component ($\vec{E}_0 = 0$) and the multiplication of small perturbation parts is neglected ($f_1\vec{E}_1 \to 0$ and $f_1\vec{B}_1 \to 0$). Let us evaluate the term $\left(\vec{v} \times \vec{B}_0\right) \cdot \frac{\partial f_0}{\partial \vec{v}}$ by taking the spatial coordinate relative to the ambient magnetic field and writing the velocity in terms of its Cartesian coordinate $\vec{v} = [v_\perp \cos\phi, v_\perp \sin\phi, v_{||}]$. Here, $v_{||}$ and $v_\perp$ represent velocity components parallel and perpendicular to the ambient magnetic field, and $\phi = \Omega_c t + \phi_0$ represents the phase angle of the gyro-motion where $\Omega_c \equiv \frac{q}{m}B$ is the cyclotron angular frequency (with sign included). Then, we obtain

$$\left(\vec{v} \times \vec{B}_0\right) \cdot \frac{\partial f_0}{\partial \vec{v}} = \begin{bmatrix} v_y B_0 \\ -v_x B_0 \\ 0 \end{bmatrix} \cdot \begin{bmatrix} \frac{\partial v_\perp}{\partial v_x}\frac{\partial f_0}{\partial v_\perp} \\ \frac{\partial v_\perp}{\partial v_y}\frac{\partial f_0}{\partial v_\perp} \\ \frac{\partial f_0}{\partial v_{||}} \end{bmatrix} = \begin{bmatrix} v_y B_0 \\ -v_x B_0 \\ 0 \end{bmatrix} \cdot \begin{bmatrix} \frac{v_x}{v_\perp}\frac{\partial f_0}{\partial v_\perp} \\ \frac{v_y}{v_\perp}\frac{\partial f_0}{\partial v_\perp} \\ \frac{\partial f_0}{\partial v_{||}} \end{bmatrix} = 0.$$

This means that the distribution function must not be changed during the gyration of plasma particles around the ambient magnetic field at an equilibrium state. By using the total derivative, Eq.(10) can be rewritten as

$$\frac{\mathrm{d}f_1}{\mathrm{d}t} = -\frac{q}{m}\left(\vec{E}_1 + \vec{v} \times \vec{B}_1\right) \cdot \frac{\partial f_0}{\partial \vec{v}},$$

and the solution to which can be obtained as

$$f_1[\vec{x},\vec{v},t] = -\frac{q}{m}\int_{-\infty}^{t}\left(\vec{E}_1[\vec{x}',t'] + \vec{v}' \times \vec{B}_1[\vec{x}',t']\right) \cdot \frac{\partial f_0[\vec{v}']}{\partial \vec{v}'}\mathrm{d}t', \tag{11}$$

where $[\vec{x}',\vec{v}']$ is an unperturbed trajectory of a particle which passes through the point $[\vec{x},\vec{v}]$ when $t' = t$.

Let us Fourier analyze electromagnetic fields,

$$\begin{aligned} E_1(\vec{x},t) &\equiv E_1 \exp[i\vec{k}\cdot\vec{x} - i\tilde{\omega}t], \\ B_1(\vec{x},t) &\equiv B_1 \exp[i\vec{k}\cdot\vec{x} - i\tilde{\omega}t]. \end{aligned}$$

where $\tilde{\omega} \equiv \omega + i\gamma$ is complex frequency and $\vec{k}$ is wavenumber vector. Then Maxwell's equations yield

$$\vec{k} \times \vec{E}_1 = \tilde{\omega}\vec{B}_1, \tag{12}$$

$$\vec{k} \times \vec{B}_1 = -i\mu_0\vec{J}_1 - \frac{\tilde{\omega}}{c^2}\vec{E}_1. \tag{13}$$

Inserting Eq.(12) into Eq.(13), we obtain

$$\vec{k} \times (\vec{k} \times \vec{E}_1) = (\vec{k}\cdot\vec{E}_1)\vec{k} - |\vec{k}|^2\vec{E}_1 = -i\tilde{\omega}\mu_0\vec{J}_1 - \frac{\tilde{\omega}^2}{c^2}\vec{E}_1,$$

$$0 = \left(\vec{k}\vec{k} - |\vec{k}|^2\overleftrightarrow{I}\right)\frac{c^2}{\tilde{\omega}^2}\vec{E}_1 + \vec{E}_1 + i\frac{c^2}{\tilde{\omega}}\mu_0\vec{J}_1, \tag{14}$$

where $\overleftrightarrow{I}$ represents a unit tensor and $\vec{a}\vec{b}$ denotes a tensor such that

$$\vec{a}\vec{b} = \begin{bmatrix} a_x b_x & a_x b_y & a_x b_z \\ a_y b_x & a_y b_y & a_y b_z \\ a_z b_x & a_z b_y & a_z b_z \end{bmatrix} = \begin{bmatrix} a_x \\ a_y \\ a_z \end{bmatrix} \begin{bmatrix} b_x \\ b_y \\ b_z \end{bmatrix}^T .$$

By using Eqs.(9), (11) and (12), the last term in the right hand side of Eq.(14) yields

$$i\frac{c^2}{\tilde{\omega}}\mu_0 \vec{J}_1 = -i\frac{\Pi_p^2}{\tilde{\omega}} \int \int_{-\infty}^{t} \left(\vec{E}_1 + \vec{v}' \times \frac{\vec{k} \times \vec{E}_1}{\tilde{\omega}} \right) \cdot \frac{\partial f_0}{\partial \vec{v}'} \exp[i\vec{k}\cdot\vec{x}' - i\tilde{\omega}t'] \mathrm{d}t' \vec{v} \mathrm{d}^3\vec{v}, \tag{15}$$

where $\Pi_p \equiv \sqrt{\frac{q^2 n}{m\epsilon_0}}$ represents the plasma angular frequency. It follows that

$$\begin{aligned} & \left(\vec{E}_1 + \vec{v}' \times \frac{\vec{k} \times \vec{E}_1}{\tilde{\omega}} \right) \cdot \frac{\partial f_0}{\partial \vec{v}'} \\ = & \begin{bmatrix} E_{x1}\left(1 - \frac{v_y' k_y + v_z' k_z}{\tilde{\omega}}\right) + E_{y1}\frac{v_y' k_x}{\tilde{\omega}} + E_{z1}\frac{v_z' k_x}{\tilde{\omega}} \\ E_{x1}\frac{v_x' k_y}{\tilde{\omega}} + E_{y1}\left(1 - \frac{v_x' k_x + v_z' k_z}{\tilde{\omega}}\right) + E_{z1}\frac{v_z' k_y}{\tilde{\omega}} \\ E_{x1}\frac{v_x' k_z}{\tilde{\omega}} + E_{y1}\frac{v_y' k_z}{\tilde{\omega}} + E_{z1}\left(1 - \frac{v_x' k_x + v_y' k_y}{\tilde{\omega}}\right) \end{bmatrix} \cdot \begin{bmatrix} \frac{v_x'}{v_\perp}\frac{\partial f_0}{\partial v_\perp} \\ \frac{v_y'}{v_\perp}\frac{\partial f_0}{\partial v_\perp} \\ \frac{\partial f_0}{\partial v_{||}} \end{bmatrix} \\ = & E_{x1}h_x + E_{y1}h_y + E_{z1}h_z \end{aligned}$$

where

$$\left.\begin{aligned} h_x &= \frac{v_x'}{v_\perp}\left(1 - \frac{v_z' k_z}{\tilde{\omega}}\right)\frac{\partial f_0}{\partial v_\perp} + \frac{v_x' k_z}{\tilde{\omega}}\frac{\partial f_0}{\partial v_{||}} \\ h_y &= \frac{v_y'}{v_\perp}\left(1 - \frac{v_z' k_z}{\tilde{\omega}}\right)\frac{\partial f_0}{\partial v_\perp} + \frac{v_y' k_z}{\tilde{\omega}}\frac{\partial f_0}{\partial v_{||}} \\ h_z &= \frac{v_z'(v_x' k_x + v_y' k_y)}{\tilde{\omega} v_\perp}\frac{\partial f_0}{\partial v_\perp} + \left(1 - \frac{v_x' k_x + v_y' k_y}{\tilde{\omega}}\right)\frac{\partial f_0}{\partial v_{||}} \end{aligned}\right\}. \tag{16}$$

Now, let us consider transforming from Lagrangian coordinate along the unperturbed trajectory $[\vec{x}', \vec{v}', t']$ to Eulerian coordinate $[\vec{x}, \vec{v}, t]$ in a stationary frame. We define the velocity as

$$\left.\begin{aligned} v_x' &= v_\perp \cos[\Omega_c(t - t') + \phi_0] \\ v_y' &= v_\perp \sin[\Omega_c(t - t') + \phi_0] \\ v_z' &= v_{||} \end{aligned}\right\},$$

and integrate the velocity in the polar coordinate over time to obtain the position

$$\left.\begin{aligned} x' &= x - \tfrac{v_\perp}{\Omega_c}\{\sin[\Omega_c(t - t') + \phi_0] - \sin\phi_0\} \\ y' &= y + \tfrac{v_\perp}{\Omega_c}\{\cos[\Omega_c(t - t') + \phi_0] - \cos\phi_0\} \\ z' &= z - v_{||}(t - t') \end{aligned}\right\}.$$

Further taking the wavenumber vector $k_x = k_\perp \cos\theta, k_y = k_\perp \sin\theta, k_z = k_{||}$, we obtain

$$\begin{aligned}
\exp[i\vec{k}\cdot\vec{x}' - i\tilde{\omega}t'] &= \exp[i\vec{k}\cdot\vec{x} - i\tilde{\omega}t]\exp[i(\tilde{\omega} - v_{||}k_{||})(t-t')] \\
&\times \exp\left[-i\frac{v_\perp k_\perp}{\Omega_c}\left\{\sin[\Omega_c(t-t') + \phi_0 - \theta] - \sin[\phi_0 - \theta]\right\}\right] \\
&= \exp[i\vec{k}\cdot\vec{x} - i\tilde{\omega}t]\sum_{l,n=-\infty}^{\infty} J_l\left[\frac{v_\perp k_\perp}{\Omega_c}\right] J_n\left[\frac{v_\perp k_\perp}{\Omega_c}\right] \\
&\times \exp[i(l-n)(\phi_0 - \theta)]\exp[i(\tilde{\omega} - v_{||}k_{||} - n\Omega_c)(t-t')] \qquad (17)
\end{aligned}$$

where $J_n[x]$ is the Bessel function of the first kind of order n with

$$\exp[ia\sin\psi] = \sum_{n=-\infty}^{\infty} J_n[a]\exp[in\psi].$$

Eq.(16) also becomes

$$\left.\begin{aligned}
h_x &= \cos[\Omega_c(t-t') + \phi_0]\left\{\left(1 - \frac{v_{||}k_{||}}{\tilde{\omega}}\right)\frac{\partial f_0}{\partial v_\perp} + \frac{v_\perp k_{||}}{\tilde{\omega}}\frac{\partial f_0}{\partial v_{||}}\right\} \\
h_y &= \sin[\Omega_c(t-t') + \phi_0]\left\{\left(1 - \frac{v_{||}k_{||}}{\tilde{\omega}}\right)\frac{\partial f_0}{\partial v_\perp} + \frac{v_\perp k_{||}}{\tilde{\omega}}\frac{\partial f_0}{\partial v_{||}}\right\} \\
h_z &= \frac{v_{||}k_\perp}{\tilde{\omega}}\cos[\Omega_c(t-t') + \phi_0 - \theta]\frac{\partial f_0}{\partial v_\perp} + \left(1 - \frac{v_\perp k_\perp}{\tilde{\omega}}\cos[\Omega_c(t-t') + \phi_0 - \theta]\right)\frac{\partial f_0}{\partial v_{||}}
\end{aligned}\right\}.$$

For the time integral in Eq.(15), we use the following relationship,

$$\begin{aligned}
&\int_{-\infty}^{t}\sum_{n=-\infty}^{\infty} J_n[\lambda]\begin{bmatrix}\cos[\Omega_c(t-t') + \phi_0] \\ \sin[\Omega_c(t-t') + \phi_0] \\ 1\end{bmatrix}\exp[-in\phi_0]\exp[i(\tilde{\omega} - v_{||}k_{||} - n\Omega_c)(t-t')]dt' \\
&= \sum_{n=-\infty}^{\infty}\frac{J_n[\lambda]}{2}\begin{bmatrix}\dfrac{i\exp[-i(n-1)\phi_0]}{\tilde{\omega} - v_{||}k_{||} - (n-1)\Omega_c} + \dfrac{i\exp[-i(n+1)\phi_0]}{\tilde{\omega} - v_{||}k_{||} - (n+1)\Omega_c} \\ \dfrac{\exp[-i(n-1)\phi_0]}{\tilde{\omega} - v_{||}k_{||} - (n-1)\Omega_c} - \dfrac{\exp[-i(n+1)\phi_0]}{\tilde{\omega} - v_{||}k_{||} - (n+1)\Omega_c} \\ \dfrac{2i\exp[-in\phi_0]}{\tilde{\omega} - v_{||}k_{||} - n\Omega_c}\end{bmatrix} \\
&= \sum_{n=-\infty}^{\infty}\begin{bmatrix}\dfrac{n\Omega_c}{v_\perp k_\perp}J_n[\lambda]\dfrac{i\exp[-in\phi_0]}{\tilde{\omega} - v_{||}k_{||} - n\Omega_c} \\ -J_n'[\lambda]\dfrac{\exp[-in\phi_0]}{\tilde{\omega} - v_{||}k_{||} - n\Omega_c} \\ J_n[\lambda]\dfrac{i\exp[-in\phi_0]}{\tilde{\omega} - v_{||}k_{||} - n\Omega_c}\end{bmatrix}. \qquad (18)
\end{aligned}$$

Here,

$$\begin{aligned}
\lambda &\equiv \frac{v_\perp k_\perp}{\Omega_c}, \\
\cos\phi &= \frac{\exp[i\phi] + \exp[-i\phi]}{2}, \\
\sin\phi &= \frac{\exp[i\phi] - \exp[-i\phi]}{2i},
\end{aligned}$$

and the following Bessel identities are used,

$$\begin{aligned}
J_{n+1}[\lambda] + J_{n-1}[\lambda] &= \frac{2n}{\lambda} J_n[\lambda] \\
J_{n+1}[\lambda] - J_{n-1}[\lambda] &= -2J_n'[\lambda]
\end{aligned}$$

with

$$\sum_{n=-\infty}^{\infty} J_n[\lambda]\left(A[n-1] \pm A[n+1]\right) = \sum_{n=-\infty}^{\infty} \left(J_{n+1}[\lambda] \pm J_{n-1}[\lambda]\right) A[n]$$

By using Eq.(18), Eq.(15) is rewritten as

$$\begin{aligned}
& -i\frac{\Pi_p^2}{\tilde{\omega}} \int \int_{-\infty}^{t} (E_x h_x + E_y h_y + E_z h_z) \exp[i\vec{k} \cdot \vec{x}' - i\tilde{\omega} t'] \mathrm{d}t' \vec{v} \mathrm{d}^3\vec{v} \\
&= -i\frac{\Pi_p^2}{\tilde{\omega}} \int \sum_{l,n=-\infty}^{\infty} \left[\begin{array}{c}
i\frac{n\Omega_c}{v_\perp k_\perp} J_n[\lambda] \left\{ \left(1 - \frac{v_{||}k_{||}}{\tilde{\omega}}\right) \frac{\partial f_0}{\partial v_\perp} + \frac{v_\perp k_{||}}{\tilde{\omega}} \frac{\partial f_0}{\partial v_{||}} \right\} \\
-J_n'[\lambda] \left\{ \left(1 - \frac{v_{||}k_{||}}{\tilde{\omega}}\right) \frac{\partial f_0}{\partial v_\perp} + \frac{v_\perp k_{||}}{\tilde{\omega}} \frac{\partial f_0}{\partial v_{||}} \right\} \\
i\frac{n\Omega_c v_{||}}{\tilde{\omega} v_\perp} J_n[\lambda] \frac{\partial f_0}{\partial v_\perp} + iJ_n[\lambda] \left(1 - \frac{n\Omega_c}{\tilde{\omega}}\right) \frac{\partial f_0}{\partial v_{||}}
\end{array} \right] \\
&\qquad \times \frac{J_l[\lambda]}{\tilde{\omega} - v_{||}k_{||} - n\Omega_c} \exp[i(l-n)(\phi_0 - \theta)] \vec{E}_1 \vec{v} \mathrm{d}^3\vec{v} \\
&= \frac{\Pi_p^2}{\tilde{\omega}^2} \int \sum_{l,n=-\infty}^{\infty} \left[\begin{array}{c} v_\perp \cos\phi_0 \\ v_\perp \sin\phi_0 \\ v_{||} \end{array} \right] \left[\begin{array}{c} \frac{n\Omega_c}{v_\perp k_\perp} J_n[\lambda] U \\ iJ_n'[\lambda] U \\ J_n[\lambda] W \end{array} \right]^T \vec{E}_1 \\
&\qquad \times \frac{J_l[\lambda]}{\tilde{\omega} - v_{||}k_{||} - n\Omega_c} \exp[i(l-n)(\phi_0 - \theta)] \mathrm{d}^3\vec{v}, \qquad (19)
\end{aligned}$$

where

$$\begin{aligned}
U &\equiv v_\perp k_{||} \frac{\partial f_0}{\partial v_{||}} + (\tilde{\omega} - k_{||}v_{||}) \frac{\partial f_0}{\partial v_\perp}, \\
W &\equiv (\tilde{\omega} - n\Omega_c) \frac{\partial f_0}{\partial v_{||}} + \frac{n\Omega_c v_{||}}{v_\perp} \frac{\partial f_0}{\partial v_\perp}.
\end{aligned}$$

and

$$\begin{bmatrix} v_x(E_xh_x+E_yh_y+E_zh_z) \\ v_y(E_xh_x+E_yh_y+E_zh_z) \\ v_z(E_xh_x+E_yh_y+E_zh_z) \end{bmatrix} = \begin{bmatrix} v_xh_x & v_xh_y & v_xh_z \\ v_yh_x & v_yh_y & v_yh_z \\ v_zh_x & v_zh_y & v_zh_z \end{bmatrix} \begin{bmatrix} E_x \\ E_y \\ E_z \end{bmatrix} = \begin{bmatrix} v_x \\ v_y \\ v_z \end{bmatrix} \begin{bmatrix} h_x \\ h_y \\ h_z \end{bmatrix}^T \begin{bmatrix} E_x \\ E_y \\ E_z \end{bmatrix}$$

Let us assume that distribution functions are gyrotropic, i.e., $f_0(\vec{v}) \equiv f_0(v_{||}, v_\perp)$ ($\frac{\partial f_0}{\partial \phi_0} = 0$) and that the wavenumber vector $\vec{k}$ is taken in the $x-z$ plane, i.e., $\theta = 0$. Then, we have

$$\int \mathrm{d}^3\vec{v} = \int_0^{2\pi}\int_0^\infty\int_{-\infty}^\infty v_\perp \mathrm{d}v_{||}\mathrm{d}v_\perp \mathrm{d}\phi_0 = 2\pi\int_0^\infty\int_{-\infty}^\infty v_\perp \mathrm{d}v_{||}\mathrm{d}v_\perp$$

and

$$\begin{aligned} \int_0^{2\pi} J_l[\lambda]\cos\phi_0 \exp[i(l-n)\phi_0]\mathrm{d}\phi_0 &= \frac{2\pi}{2}\left(J_{n-1}[\lambda]+J_{n+1}[\lambda]\right) = \frac{2\pi n}{\lambda}J_n[\lambda], \\ \int_0^{2\pi} J_l[\lambda]\sin\phi_0 \exp[i(l-n)\phi_0]\mathrm{d}\phi_0 &= \frac{2\pi}{2i}\left(J_{n-1}[\lambda]-J_{n+1}[\lambda]\right) = \frac{2\pi}{i}J_n'[\lambda], \\ \int_0^{2\pi} J_l[\lambda]\exp[i(l-n)\phi_0]\mathrm{d}\phi_0 &= 2\pi J_n[\lambda]. \end{aligned}$$

Thus Eq.(14) can be rewritten as

$$0 = \left(\vec{k}\vec{k} - |\vec{k}|^2\overleftrightarrow{I}\right)\frac{c^2}{\tilde{\omega}^2} + \overleftrightarrow{\epsilon}(\tilde{\omega}, \vec{k}), \tag{20}$$

where

$$\begin{aligned} \overleftrightarrow{\epsilon}(\tilde{\omega}, \vec{k})\vec{E}_1 &= \vec{E}_1 + i\frac{c^2}{\tilde{\omega}}\mu_0\vec{J}_1, \\ \overleftrightarrow{\epsilon}(\tilde{\omega}, \vec{k}) &= \overleftrightarrow{I} + \sum_s \frac{\Pi_p^2}{\tilde{\omega}^2}\sum_{n=-\infty}^{\infty}\int \frac{1}{\tilde{\omega} - k_{||}v_{||} - n\Omega_c}\overleftrightarrow{T_n}\mathrm{d}^3\vec{v}, \end{aligned} \tag{21}$$

with

$$\overleftrightarrow{T_n} = \begin{bmatrix} \frac{n^2\Omega_c^2}{k_\perp^2 v_\perp}J_n^2[\lambda]U & i\frac{n\Omega_c}{k_\perp}J_n[\lambda]J_n'[\lambda]U & \frac{n\Omega_c}{k_\perp}J_n^2[\lambda]W \\ -i\frac{n\Omega_c}{k_\perp}J_n[\lambda]J_n'[\lambda]U & v_\perp {J_n'}^2[\lambda]U & -iv_\perp J_n[\lambda]J_n'[\lambda]W \\ \frac{n\Omega_c v_{||}}{k_\perp v_\perp}J_n^2[\lambda]U & iv_{||}J_n[\lambda]J_n'[\lambda]U & v_{||}J_n^2[\lambda]W \end{bmatrix}.$$

Here Eq.(20) is called the linear dispersion relation and Eq.(21) is called the plasma dielectric equation. Note that $\sum_s$ is added in Eq.(21) for treating multi-species (e.g., ions and electrons) plasma ($\Pi_{ps}, \Omega_{cs}, f_{0s}$).
There also exists another expression for the plasma dielectric equation,

$$\overleftrightarrow{\epsilon} = \overleftrightarrow{I} + \sum_s \frac{\Pi_p^2}{\tilde{\omega}^2} \left\{ \sum_{n=-\infty}^{\infty} \int \frac{k_{||} \dfrac{\partial f_0}{\partial v_{||}} + \dfrac{n\Omega_c}{v_\perp} \dfrac{\partial f_0}{\partial v_\perp}}{\tilde{\omega} - k_{||} v_{||} - n\Omega_c} \overleftrightarrow{S_n} d^3\vec{v} - \overleftrightarrow{I} \right\}, \tag{22}$$

where

$$\overleftrightarrow{S_n} = \begin{bmatrix} \dfrac{n^2\Omega_c^2}{k_\perp^2} J_n^2[\lambda] & i\dfrac{n\Omega_c}{k_\perp} v_\perp J_n[\lambda] J_n'[\lambda] & \dfrac{n\Omega_c}{k_\perp} v_{||} J_n^2[\lambda] \\ -i\dfrac{n\Omega_c}{k_\perp} v_\perp J_n[\lambda] J_n'[\lambda] & v_\perp^2 J_n'^2[\lambda] & -i v_{||} v_\perp J_n[\lambda] J_n'[\lambda] \\ \dfrac{n\Omega_c}{k_\perp} v_{||} J_n^2[\lambda] & i v_{||} v_\perp J_n[\lambda] J_n'[\lambda] & v_{||}^2 J_n^2[\lambda] \end{bmatrix}.$$

2.3 Linear dispersion relation for waves in Maxwellian plasma

The Maxwellian (or Maxwell-Boltzmann) distribution is usually regarded as a distribution of particle velocity at an equilibrium state. Plasma or charged particles easily move along an ambient magnetic field, while they do not move across the ambient magnetic field. Distributions of particle velocity often show anisotropy in the direction parallel and perpendicular to the ambient magnetic field. That is, the average drift velocity and the temperature in the direction parallel to the ambient magnetic field differ from those in the direction perpendicular to the ambient magnetic field. Thus the following shifted bi-Maxwellian distribution is used as a velocity distribution at an initial state or an equilibrium state,

$$\left.\begin{aligned} f(v_{||}, v_\perp) &= f_{||}(v_{||}) f_\perp(v_\perp) \\ f_{||}(v_{||}) &= \frac{1}{\sqrt{2\pi} V_{t||}} \exp\left[-\frac{(v_{||} - V_d)^2}{2V_{t||}^2} \right] \\ f_\perp(v_\perp) &= \frac{1}{2\pi V_{t\perp}^2} \exp\left[-\frac{v_\perp^2}{2V_{t\perp}^2} \right] \end{aligned}\right\}, \tag{23}$$

where V_d is the drift velocity in the direction parallel to the ambient magnetic field, and $V_{t||} \equiv \sqrt{T_{||}/m}$ and $V_{t\perp} \equiv \sqrt{T_\perp/m}$ are the thermal velocities in the direction parallel and perpendicular to the ambient magnetic field, respectively, with T being temperature of plasma particles.
When plasma has the Maxwellian velocity distribution (23), we can explicitly perform the velocity-space integral by using the following properties,

$$\begin{aligned}
\int_{-\infty}^{\infty} \frac{f_{||}\mathrm{d}v_{||}}{\tilde{\omega}-k_{||}v_{||}-n\Omega_c} &= -\frac{1}{\sqrt{2}k_{||}V_{t||}}Z_0[\zeta_n] \\
\int_{-\infty}^{\infty} \frac{v_{||}f_{||}\mathrm{d}v_{||}}{\tilde{\omega}-k_{||}v_{||}-n\Omega_c} &= -\frac{1}{\sqrt{2}k_{||}V_{t||}}Z_1[\zeta_n] = -\frac{1}{k_{||}}\left\{1+\frac{\tilde{\omega}-n\Omega_c}{\sqrt{2}k_{||}V_{t||}}Z_0[\zeta_n]\right\} \\
\int_{-\infty}^{\infty} \frac{v_{||}^2 f_{||}\mathrm{d}v_{||}}{\tilde{\omega}-k_{||}v_{||}-n\Omega_c} &= -\frac{1}{\sqrt{2}k_{||}V_{t||}}Z_2[\zeta_n] \\
&= -\frac{\sqrt{2}V_{t||}}{k_{||}}\left\{\frac{\tilde{\omega}+k_{||}V_d-n\Omega_c}{\sqrt{2}k_{||}V_{t||}}+\left(\frac{\tilde{\omega}-n\Omega_c}{\sqrt{2}k_{||}V_{t||}}\right)^2 Z_0[\zeta_n]\right\} \\
\int_{-\infty}^{\infty} \frac{\frac{\partial f_{||}}{\partial v_{||}}\mathrm{d}v_{||}}{\tilde{\omega}-k_{||}v_{||}-n\Omega_c} &= \frac{1}{\sqrt{2}k_{||}V_{t||}^3}\{Z_1[\zeta_n]-V_d Z_0[\zeta_n]\} \\
&= \frac{1}{k_{||}V_{t||}^2}\left\{1+\frac{\tilde{\omega}-k_{||}V_d-n\Omega_c}{\sqrt{2}k_{||}V_{t||}}Z_0[\zeta_n]\right\} \\
\int_{-\infty}^{\infty} \frac{v_{||}\frac{\partial f_{||}}{\partial v_{||}}\mathrm{d}v_{||}}{\tilde{\omega}-k_{||}v_{||}-n\Omega_c} &= \frac{1}{\sqrt{2}k_{||}V_{t||}^3}\{Z_2[\zeta_n]-V_d Z_1[\zeta_n]\} \\
&= \frac{\tilde{\omega}-n\Omega}{k_{||}^2 V_{t||}^2}\left\{1+\frac{\tilde{\omega}-k_{||}V_d-n\Omega_c}{\sqrt{2}k_{||}V_{t||}}Z_0[\zeta_n]\right\}
\end{aligned}$$

and

$$\begin{aligned}
\int_0^{\infty} J_n^2[\lambda]f_{\perp}2\pi v_{\perp}\mathrm{d}v_{\perp} &= \exp\left[-\frac{a^2}{2}\right]I_n\left[\frac{a^2}{2}\right] \\
\int_0^{\infty} J_n^2[\lambda]\frac{\partial f_{\perp}}{\partial v_{\perp}}2\pi\mathrm{d}v_{\perp} &= -\frac{1}{V_{t\perp}^2}\exp\left[-\frac{a^2}{2}\right]I_n\left[\frac{a^2}{2}\right]
\end{aligned}$$

$$\begin{aligned}
\int_0^{\infty} J_n[\lambda]J_n'[\lambda]f_{\perp}2\pi v_{\perp}^2\mathrm{d}v_{\perp} &= \frac{k_{\perp}V_{t\perp}^2}{\Omega_c}\exp\left[-\frac{a^2}{2}\right]\left\{I_n'\left[\frac{a^2}{2}\right]-I_n\left[\frac{a^2}{2}\right]\right\} \\
\int_0^{\infty} J_n[\lambda]J_n'[\lambda]\frac{\partial f_{\perp}}{\partial v_{\perp}}2\pi v_{\perp}\mathrm{d}v_{\perp} &= -\frac{k_{\perp}}{\Omega_c}\exp\left[-\frac{a^2}{2}\right]\left\{I_n'\left[\frac{a^2}{2}\right]-I_n\left[\frac{a^2}{2}\right]\right\}
\end{aligned}$$

$$\begin{aligned}
\int_0^{\infty} J_n'^2[\lambda]f_{\perp}2\pi v_{\perp}^3\mathrm{d}v_{\perp} &= 4V_{t\perp}^2\exp\left[-\frac{a^2}{2}\right]\left\{\frac{n^2}{2a^2}I_n\left[\frac{a^2}{2}\right]-\frac{a^2}{4}I_n'\left[\frac{a^2}{2}\right]+\frac{a^2}{4}I_n\left[\frac{a^2}{2}\right]\right\} \\
\int_0^{\infty} J_n'^2[\lambda]\frac{\partial f_{\perp}}{\partial v_{\perp}}2\pi v_{\perp}^2\mathrm{d}v_{\perp} &= -4\exp\left[-\frac{a^2}{2}\right]\left\{\frac{n^2}{2a^2}I_n\left[\frac{a^2}{2}\right]-\frac{a^2}{4}I_n'\left[\frac{a^2}{2}\right]+\frac{a^2}{4}I_n\left[\frac{a^2}{2}\right]\right\}
\end{aligned}$$

where

$$\begin{aligned}\zeta_n &\equiv \frac{\tilde{\omega} - k_{||}V_d - n\Omega_c}{\sqrt{2}k_{||}V_{t||}},\\ a &\equiv \frac{\sqrt{2}k_\perp V_{t\perp}}{\Omega_c} = \sqrt{2}\lambda\frac{V_{t\perp}}{v_\perp}.\end{aligned}$$

Here $I_n[x]$ is the modified Bessel function of the first kind of order n with the following properties,

$$\begin{aligned}\int_0^\infty xJ_n^2[tx]\exp\left[-x^2\right]\mathrm{d}x &= \frac{1}{2}\exp\left[-\frac{t^2}{2}\right]I_n\left[\frac{t^2}{2}\right],\\ \int_0^\infty x^2J_n[tx]J_n'[tx]\exp\left[-x^2\right]\mathrm{d}x &= \frac{t}{4}\exp\left[-\frac{t^2}{2}\right]\left\{I_n'\left[\frac{t^2}{2}\right]-I_n\left[\frac{t^2}{2}\right]\right\},\\ \int_0^\infty x^3J_n'^2[tx]\exp\left[-x^2\right]\mathrm{d}x &= \exp\left[-\frac{t^2}{2}\right]\left\{\frac{n^2}{2t^2}I_n\left[\frac{t^2}{2}\right]-\frac{t^2}{4}I_n'\left[\frac{t^2}{2}\right]+\frac{t^2}{4}I_n\left[\frac{t^2}{2}\right]\right\},\end{aligned}$$

and $Z_p[x]$ is the plasma dispersion function (Fried & Conte, 1961)

$$\begin{aligned}Z_0[x] &\equiv \frac{1}{\sqrt{\pi}}\int_{-\infty}^{\infty}\frac{1}{t-x}\mathrm{d}t, \qquad (24)\\ Z_p[\zeta_n] &\equiv -\frac{k_{||}}{\sqrt{\pi}}\int_{-\infty}^{\infty}\frac{v_{||}^p}{\tilde{\omega}-k_{||}v_{||}-n\Omega_c}\exp\left[-\frac{(v_{||}-V_d)^2}{2V_{t||}^2}\right]\mathrm{d}v_{||},\\ Z_1[\zeta_n] &= \sqrt{2}V_{t||}\left\{1+\frac{\tilde{\omega}-n\Omega_c}{\sqrt{2}k_{||}V_{t||}}Z_0[\zeta_n]\right\},\\ Z_2[\zeta_n] &= 2V_{t||}^2\left\{\frac{\tilde{\omega}+k_{||}V_d-n\Omega_c}{\sqrt{2}k_{||}V_{t||}}+\left(\frac{\tilde{\omega}-n\Omega_c}{\sqrt{2}k_{||}V_{t||}}\right)^2 Z_0[\zeta_n]\right\}.\end{aligned}$$

We also use the following identity of the modified Bessel function,

$$\begin{aligned}\sum_{n=-\infty}^{\infty} nI_n'[\lambda] = \sum_{n=-\infty}^{\infty}\frac{n}{2}(I_{n+1}[\lambda]+I_{n-1}[\lambda]) = \sum_{n=-\infty}^{\infty} nI_n[\lambda] = 0\\ \sum_{n=-\infty}^{\infty}\{I_n[\lambda]-I_n'[\lambda]\} = \sum_{n=-\infty}^{\infty}\left\{I_n[\lambda]-\frac{1}{2}(I_{n+1}[\lambda]+I_{n-1}[\lambda])\right\} = 0\end{aligned}$$

Then we obtain

$$\sum_{n=-\infty}^{\infty}\int\frac{1}{\tilde{\omega}-k_{||}v_{||}-n\Omega_c}\overleftrightarrow{T}_n\mathrm{d}^3\vec{v} \equiv \overleftrightarrow{K} = \begin{bmatrix} K_{1,1} & K_{1,2} & K_{1,3}\\ -K_{1,2} & K_{2,2} & K_{2,3}\\ K_{1,3} & -K_{2,3} & K_{3,3}\end{bmatrix}, \qquad (25)$$

where

$$
\begin{aligned}
K_{1,1} &= \sum_{n=-\infty}^{\infty} \frac{2n^2}{a^2} \exp\left[-\frac{a^2}{2}\right] I_n\left[\frac{a^2}{2}\right] \left\{ \left(\frac{V_{t\perp}^2}{V_{t||}^2} - 1\right) + \frac{V_{t\perp}^2}{V_{t||}^2} \xi_n Z_0[\zeta_n] \right\}, \\
K_{1,2} &= i \sum_{n=-\infty}^{\infty} n \exp\left[-\frac{a^2}{2}\right] \left\{ I_n\left[\frac{a^2}{2}\right] - I_n'\left[\frac{a^2}{2}\right] \right\} \frac{V_{t\perp}^2}{V_{t||}^2} \xi_n Z_0[\zeta_n], \\
K_{1,3} &= \sum_{n=-\infty}^{\infty} \frac{2nV_{t\perp}}{aV_{t||}} \exp\left[-\frac{a^2}{2}\right] I_n\left[\frac{a^2}{2}\right] \left\{ -\frac{n\Omega_c\left(1 - \frac{V_{t||}^2}{V_{t\perp}^2}\right)}{\sqrt{2}k_{||}V_{t||}} + \frac{\tilde{\omega} - n\Omega_c}{\sqrt{2}k_{||}V_{t||}} \xi_n Z_0[\zeta_n] \right\}, \\
K_{2,2} &= K_{1,1} - \sum_{n=-\infty}^{\infty} a^2 \exp\left[-\frac{a^2}{2}\right] \left\{ I_n'\left[\frac{a^2}{2}\right] + I_n\left[\frac{a^2}{2}\right] \right\} \frac{V_{t\perp}^2}{V_{t||}^2} \xi_n Z_0[\zeta_n], \\
K_{2,3} &= i \sum_{n=-\infty}^{\infty} \frac{aV_{t\perp}}{V_{t||}} \exp\left[-\frac{a^2}{2}\right] \left\{ I_n\left[\frac{a^2}{2}\right] - I_n'\left[\frac{a^2}{2}\right] \right\} \frac{\tilde{\omega} - n\Omega_c}{\sqrt{2}k_{||}V_{t||}} \xi_n Z_0[\zeta_n], \\
K_{3,3} &= 2 \sum_{n=-\infty}^{\infty} \exp\left[-\frac{a^2}{2}\right] I_n\left[\frac{a^2}{2}\right] \left\{ \frac{\tilde{\omega}^2 + n^2\Omega_c^2\left(1 - \frac{V_{t||}^2}{V_{t\perp}^2}\right)}{2k_{||}^2 V_{t||}^2} + \left(\frac{\tilde{\omega} - n\Omega_c}{\sqrt{2}k_{||}V_{t||}}\right)^2 \xi_n Z_0[\zeta_n] \right\},
\end{aligned}
$$

with

$$
\xi_n \equiv \frac{\tilde{\omega} - k_{||}V_d - n\Omega_c\left(1 - \frac{V_{t||}^2}{V_{t\perp}^2}\right)}{\sqrt{2}k_{||}V_{t||}}.
$$

The linear dispersion relation is obtained by solving the following equation,

$$
0 = \mathrm{Det}\left[\frac{c^2}{\tilde{\omega}^2}\left(\vec{k}\vec{k} - |\vec{k}|^2\overleftrightarrow{I}\right) + \overleftrightarrow{I} + \sum_s \frac{\Pi_p^2}{\tilde{\omega}^2}\overleftrightarrow{K}\right] \tag{26}
$$

$$
\equiv \begin{vmatrix}
\sum_s \Pi_p^2 K_{1,1} + \tilde{\omega}^2 - c^2k_{||}^2 & \sum_s \Pi_p^2 K_{1,2} & \sum_s \Pi_p^2 K_{1,3} + c^2(k_{||}k_\perp) \\
-\sum_s \Pi_p^2 K_{1,2} & \sum_s \Pi_p^2 K_{2,2} + \tilde{\omega}^2 - c^2(k_{||}^2 + k_\perp^2) & \sum_s \Pi_p^2 K_{2,3} \\
\sum_s \Pi_p^2 K_{1,3} + c^2(k_{||}k_\perp) & -\sum_s \Pi_p^2 K_{2,3} & \sum_s \Pi_p^2 K_{3,3} + \tilde{\omega}^2 - c^2k_\perp^2
\end{vmatrix}.
$$

3. Excitation of electromagnetic waves

Eq.(26) tells us what kind of plasma waves grows and damps in arbitrary Maxwellian plasma. This section gives examples on the excitation of plasma waves based on the linear dispersion analysis.

For simplicity, let us assume propagation of plasma waves in the direction parallel to the ambient magnetic field, i.e., $k_\perp = 0$. Then, we have $I_0[0] = 1$ and $I_{n\neq 0}[0] = 0$. These also gives $I'_{\pm 1}[0] = 0.5$. Thus we obtain $K_{1,1} = K_{2,2}$, $K_{1,3} = 0$[3], $K_{2,3} = 0$ and Eq.(26) becomes

$$\left\{ \left(\sum_s \Pi_p^2 K_{1,1} + \tilde{\omega}^2 - c^2 k_{||}^2 \right)^2 + \sum_s \Pi_p^2 K_{1,2}^2 \right\} \left\{ \sum_s \Pi_p^2 K_{3,3} + \tilde{\omega}^2 \right\} = 0 \tag{27}$$

The first factor is for transverse waves where $\vec{k} \perp \vec{E}$. That is, a wave propagates in the z direction while its electromagnetic fields polarize in the $x - y$ plane. The second factor is for longitudinal waves where $\vec{k} \parallel \vec{E}$. That is, a wave propagates in the z direction and only its electric fields polarize in the z direction. The longitudinal waves are also referred to as compressional waves or sound waves. Especially in the case of $\vec{k} \parallel \vec{E}$, waves are called "electrostatic" waves because these waves arise from electric charge and are expressed by the Poisson equation (3).

3.1 Transverse electromagnetic waves

The first factor of Eq.(27) becomes the following equation,

$$\sum_s \Pi_p^2 \left\{ \left(\frac{V_{t\perp}^2}{V_{t||}^2} - 1 \right) + \frac{V_{t\perp}^2}{V_{t||}^2} \frac{\tilde{\omega} - k_{||} V_d - \Omega_c \left(1 - \frac{V_{t||}^2}{V_{t\perp}^2} \right)}{\sqrt{2} k_{||} V_{t||}} Z_0 \left[\frac{\tilde{\omega} - k_{||} V_d - \Omega_c}{\sqrt{2} k_{||} V_{t||}} \right] \right\} = c^2 k_{||}^2 - \tilde{\omega}^2 \tag{28}$$

The plasma dispersion function $Z_0[x]$ (Fried & Conte, 1961) is approximated for $|x| \gg 1$ as

$$Z_0[x] \sim i\sigma\sqrt{\pi} \exp[-x^2] - \frac{1}{x} \left(1 + \frac{1}{2x^2} + \frac{3}{4x^4} + \cdots \right) \tag{29}$$

with

$$\sigma = \begin{cases} 0, & \mathrm{Im}[x] > 1/|\mathrm{Re}[x]| \\ 1, & |\mathrm{Im}[x]| < 1/|\mathrm{Re}[x]| \\ 2, & \mathrm{Im}[x] < -1/|\mathrm{Re}[x]| \end{cases}$$

Here the argument of the dispersion function x is a complex value.
Let us consider that a phase speed of waves is much faster than velocities of plasma particles. Then, the argument of the plasma dispersion function becomes a larger number. Here, the drift velocity of plasma V_d is also neglected. Equation (28) is thus rewritten by using Eq.(29) as

---

[3] $K_{1,3} = 0$ if $n = 0$, and $\lim_{a\to 0} \frac{I_n\left[\frac{a^2}{2}\right]}{a} = \lim_{a\to 0} \frac{a}{2} \frac{I_n\left[\frac{a^2}{2}\right] - I_n[0]}{\frac{a^2}{2} - 0} = \lim_{a\to 0} \frac{a}{2} I'_n \left[\frac{a^2}{2}\right] = 0$ if $n \neq 0$.

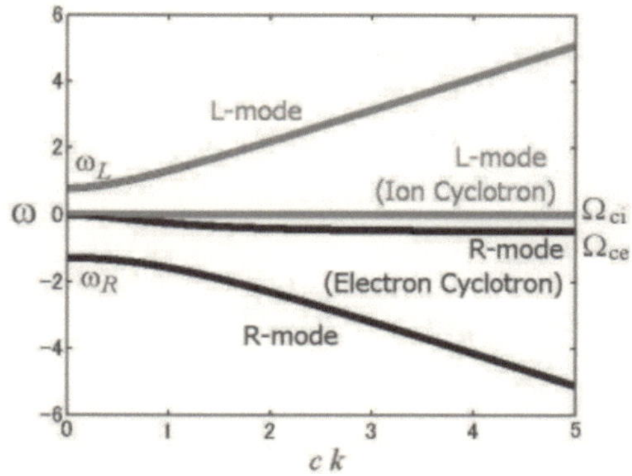

(a) Linear scale.

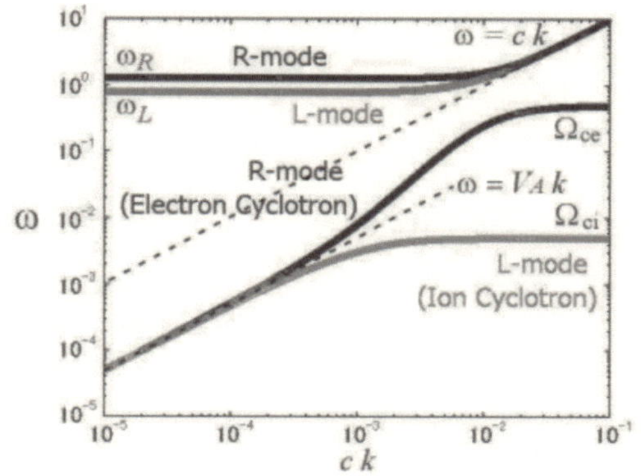

(b) Logarithmic scale.

Fig. 1. Linear dispersion relation (frequency ω versus wavenumber k) for electromagnetic waves in plasma. The quantities ω and ck are normalized by Π_{pe}.

$$\sum_s \Pi_p^2 \left\{ \left(\frac{V_{t\perp}^2}{V_{t||}^2} - 1 \right) - \frac{V_{t\perp}^2}{V_{t||}^2} \frac{\tilde{\omega} - \Omega_c \left(1 - \frac{V_{t||}^2}{V_{t\perp}^2} \right)}{\tilde{\omega} - \Omega_c} \left(1 + \frac{k_{||}^2 V_{t||}^2}{(\tilde{\omega} - \Omega_c)^2} + \cdots \right) \right\} \tag{30}$$

$$+ i\sigma\sqrt{\pi} \sum_s \Pi_p^2 \frac{V_{t\perp}^2}{V_{t||}^2} \frac{\tilde{\omega} - \Omega_c \left(1 - \frac{V_{t||}^2}{V_{t\perp}^2} \right)}{\sqrt{2} k_{||} V_{t||}} \exp\left[- \frac{(\tilde{\omega} - \Omega_c)^2}{2 k_{||}^2 V_{t||}^2} \right] - c^2 k_{||}^2 + \tilde{\omega}^2 = 0$$

The solutions to above equation are simplified when we assume that the temperature of plasma approaches to zero, i.e., $V_{t||} \to 0$ and $V_{t\perp} \to 0$. Note that this approach is known as the "cold plasma approximation." Equation (30) is rewritten by the cold plasma approximation as

$$\Pi_{pe}^2 \frac{\tilde{\omega}}{\tilde{\omega} - \Omega_{ce}} + \Pi_{pi}^2 \frac{\tilde{\omega}}{\tilde{\omega} - \Omega_{ci}} + c^2 k_{||}^2 - \tilde{\omega}^2 = 0 \tag{31}$$

Here an electron-ion pair plasma is assumed ($\Omega_{ci} > 0$ and $\Omega_{ce} < 0$). The solutions to the above equation are shown in Figure 1. Note that the imaginary part of the complex frequency (growth/damping rate) becomes zero in the present case. There exist four dispersion curves. The dispersion curves for $\omega > 0$ are called "L-mode" (left-handed circularly polarized) waves, while the dispersion curves for $\omega < 0$ are called "R-mode" (right-handed circularly polarized) waves. In the present case, a positive frequency corresponds to the direction of ion gyro-motion, which is left-handed (counter-clockwise) circularly polarized against magnetic field lines.

The dispersion curves for the high-frequency R-mode and L-mode waves approach to the following frequencies as $k_{||} \to 0$,

$$\omega_R = \frac{\sqrt{(\Omega_{ce}+\Omega_{ci})^2+4(\Pi_{pe}^2+\Pi_{pi}^2-\Omega_{ce}\Omega_{ci})}-(\Omega_{ce}+\Omega_{ci})}{2} \tag{32}$$

$$\omega_L = \frac{\sqrt{(\Omega_{ce}+\Omega_{ci})^2+4(\Pi_{pe}^2+\Pi_{pi}^2-\Omega_{ce}\Omega_{ci})}+(\Omega_{ce}+\Omega_{ci})}{2} \tag{33}$$

which are known as the R-mode and L-mode cut-off frequencies, respectively. The two high-frequency waves approach to $\omega = ck_{||}$, i.e., electromagnetic light mode waves as $k_{||} \to \infty$. On the other hand, the low-frequency wave approaches to $k_{||}V_A$ as $k_{||} \to 0$, and approaches to Ω_c as $k_{||} \to \infty$. Note that $V_A \equiv c\Omega_{ci}/\Pi_{pi}$ is called the Alfven velocity. The R-mode and L-mode low-frequency waves are called electromagnetic electron and ion cyclotron wave, respectively, or (electron and ion) whistler mode wave.
The temperature of plasma affects the growth/damping rate in the dispersion relation. Assuming $\omega \gg |\gamma|$ (where $\tilde{\omega} \equiv \omega + i\gamma$), the imaginary part of Eq.(30) gives the growth rate γ as

$$\gamma \sim -\frac{1}{2\omega - \dfrac{\Omega_{ce}\Pi_{pe}^2}{(\omega-\Omega_{ce})^2} - \dfrac{\Omega_{ci}\Pi_{pi}^2}{(\omega-\Omega_{ci})^2}} \tag{34}$$

$$\times\sigma\sqrt{\pi}\left\{\Pi_{pe}^2\frac{V_{te\perp}^2}{V_{te||}^2}\frac{\omega-\Omega_{ce}\left(1-\frac{V_{te||}^2}{V_{te\perp}^2}\right)}{\sqrt{2}k_{||}V_{te||}}\exp\left[-\frac{(\omega-\Omega_{ce})^2}{2k_{||}^2V_{te||}^2}\right]\right.$$

$$\left.+\Pi_{pi}^2\frac{V_{ti\perp}^2}{V_{ti||}^2}\frac{\omega-\Omega_{ci}\left(1-\frac{V_{ti||}^2}{V_{te\perp}^2}\right)}{\sqrt{2}k_{||}V_{ti||}}\exp\left[-\frac{(\omega-\Omega_{ci})^2}{2k_{||}^2V_{ti||}^2}\right]\right\}$$

Here the higher-order terms are neglected for simplicity. One can find that the growth rate becomes always negative $\gamma \le 0$ for light mode waves ($|\omega| > \omega_{R,L}$). On the other hand, γ becomes positive for electromagnetic electron cyclotron waves ($\Omega_{ce} < \omega < 0$) with

$$\frac{\omega}{\Omega_{ce}} > \left(1-\frac{V_{te||}^2}{V_{te\perp}^2}\right), \quad 2\frac{\omega}{\Omega_{ce}} < \frac{\Pi_{pe}^2}{(\omega-\Omega_{ce})^2} \tag{35}$$

Here ion terms are neglected by assuming $|\omega| \gg \Omega_{ci}$ and $|\omega| \gg \Pi_{pi}$. This condition is achieved when $V_{te\perp} > V_{te||}$, which is known as electron temperature anisotropy instability.
As a special case, electromagnetic electron cyclotron waves are also excited if the ion contribution (the third line in Eq.(34)) becomes larger than the electron contribution (the second line in Eq.(34)) around $|\omega| \sim \Omega_{ci}$. The growth rare becomes positive when

$$\frac{\omega}{\Omega_{ci}} > \left(1-\frac{V_{ti||}^2}{V_{ti\perp}^2}\right), \quad -\frac{\Pi_{pe}^2}{\Omega_{ce}} < \frac{\Omega_{ci}\Pi_{pi}^2}{(\omega-\Omega_{ci})^2} \tag{36}$$

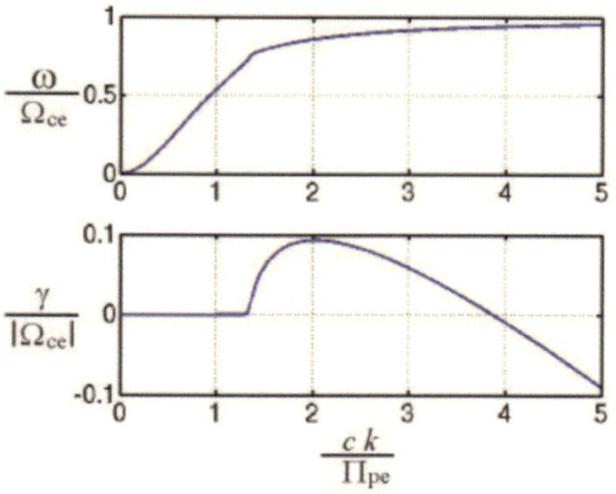

(a) Electron temperature anisotropy instability with $V_{te\perp} = 4V_{te||}$, $T_{e||} = T_{i||} = T_{i\perp}$, $\Pi_{pe} = 10|\Omega_{ce}|$, $c = 200V_{te||}$.

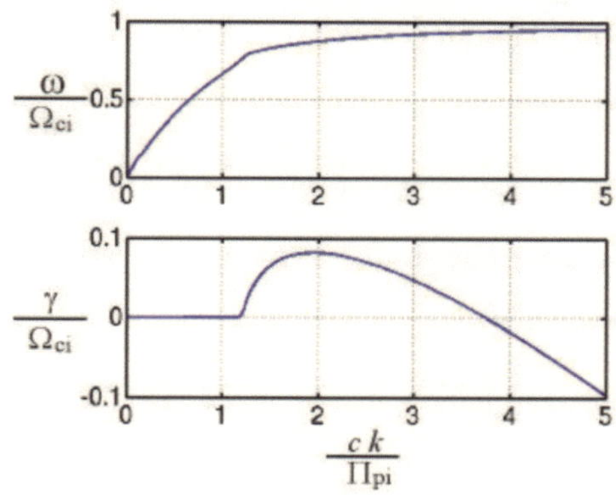

(b) Ion temperature anisotropy instability with $V_{ti\perp} = 4V_{ti||}$, $T_{i||} = T_{e||} = T_{e\perp}$, $\Pi_{pe} = 10|\Omega_{ce}|$, $c = 200V_{te||}$.

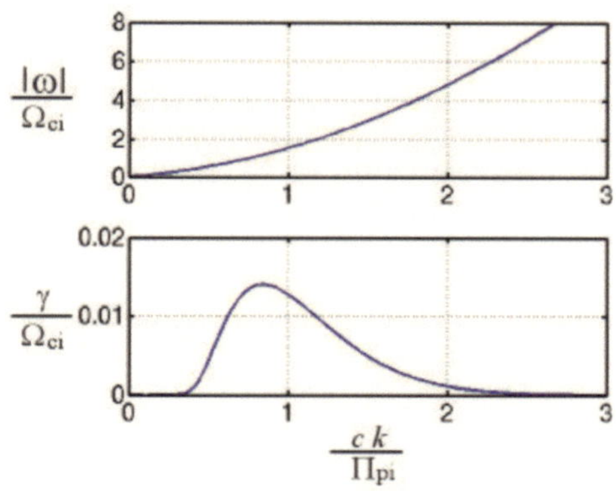

(c) Firehose instability with $V_{ti||} = 16V_{ti\perp}$, $T_{i\perp} = T_{e||} = T_{e\perp}$, $\Pi_{pe} = 10|\Omega_{ce}|$, $c = 200V_{te||}$.

Fig. 2. Linear dispersion relation for electromagnetic instabilities.

Here $|\omega| \ll |\Omega_{ce}|$ and $|\omega| \ll \Pi_{pe}$ are assumed. This condition is achieved when $V_{ti||} \gg V_{ti\perp}$, which is known as firehose instability.
For electromagnetic ion cyclotron waves $(0 < \omega < \Omega_{ci})$, the growth rare becomes positive when

$$\frac{\omega}{\Omega_{ci}} < \left(1 - \frac{V_{ti||}^2}{V_{ti\perp}^2}\right), \quad -\frac{\Pi_{pe}^2}{\Omega_{ce}} > \frac{\Omega_{ci}\Pi_{pi}^2}{(\omega - \Omega_{ci})^2} \tag{37}$$

Here $\omega \ll |\Omega_{ce}|$ and $\omega \ll \Pi_{pe}$ are used. This condition is achieved when $V_{ti\perp} > V_{ti||}$, which is known as ion temperature anisotropy instability.
Examples of these electromagnetic linear instabilities are shown in Figure 2, which are obtained by numerically solving Eq.(28).

3.2 Longitudinal electrostatic waves

The second factor of Eq.(27) becomes

$$1+\sum_s \frac{\Pi_p^2}{k_{||}^2 V_{t||}^2}\left\{1+\frac{\tilde{\omega}-k_{||}V_d}{\sqrt{2}k_{||}V_{t||}} Z_0\left[\frac{\tilde{\omega}-k_{||}V_d}{\sqrt{2}k_{||}V_{t||}}\right]\right\}=0 \tag{38}$$

By using the similar approach, the above equation is rewritten by using Eq.(29) as

$$\begin{aligned}1&-\sum_s \frac{\Pi_p^2}{(\tilde{\omega}-k_{||}V_d)^2}\left\{1+\frac{3k_{||}^2 V_{t||}^2}{(\tilde{\omega}-k_{||}V_d)^2}+\cdots\right\}\\ &+i\sigma\sqrt{\pi}\sum_s \frac{\Pi_p^2}{k_{||}^2 V_{t||}^2}\frac{\tilde{\omega}-k_{||}V_d}{\sqrt{2}k_{||}V_{t||}}\exp\left[-\frac{(\tilde{\omega}-k_{||}V_d)^2}{2k_{||}^2 V_{t||}^2}\right]=0\end{aligned} \tag{39}$$

For a simple case with $V_d=0$ and $\omega \sim \Pi_{pe}$, Eq.(39) gives

$$\omega^2=\frac{\Pi_{pe}^2+\sqrt{\Pi_{pe}^4+12\Pi_{pe}^2 k_{||}V_{te||}^2}}{2}\sim \Pi_{pe}^2+3k_{||}^2 V_{te||}^2 \tag{40}$$

which is known as the dispersion relation of Langmuir (electron plasma) waves. The imaginary part in Eq.(39) gives the damping rate as

$$\frac{\gamma}{\omega}\sim -\frac{\sigma\sqrt{\pi}}{2}\frac{\omega^5}{\sqrt{2}k_{||}^3 V_{te||}^3(\omega^2+6k_{||}^2 V_{te||}^2)}\exp\left[-\frac{\omega^2}{2k_{||}^2 V_{te||}^2}\right] \tag{41}$$

This means that the damping of the Langmuir waves becomes largest at $k_{||}\sim \Pi_{pe}/V_{te||}$, which is known as the Landau damping. Note that the second line in Eq.(39) comes from the gradient in the velocity distribution function, i.e., $\partial f_0/\partial v_{||}$. Thus electrostatic waves are known to be most unstable where the velocity distribution function has the maximum positive gradient. As an example for the growth of electrostatic waves, let us assume a two-species plasma, and one species drift against the other species at rest. Then, Eq.(39) becomes

$$\begin{aligned}0=1&-\frac{\Pi_{p1}^2}{(\tilde{\omega}-k_{||}V_{d1})^2}\left\{1+\frac{3k_{||}^2 V_{t1||}^2}{(\tilde{\omega}-k_{||}V_{d1})^2}\right\}-\frac{\Pi_{p2}^2}{\tilde{\omega}^2}\left\{1+\frac{3k_{||}^2 V_{t2||}^2}{\tilde{\omega}^2}\right\}\\ &+i\frac{\sigma\sqrt{\pi}}{k_{||}^2}\left\{\frac{\Pi_{p1}^2}{V_{t1||}^2}\frac{\tilde{\omega}-k_{||}V_{d1}}{\sqrt{2}k_{||}V_{t1||}}\exp\left[-\frac{(\tilde{\omega}-k_{||}V_{d1})^2}{2k_{||}^2 V_{t1||}^2}\right]+\frac{\Pi_{p2}^2}{V_{t2||}^2}\frac{\tilde{\omega}}{\sqrt{2}k_{||}V_{t2||}}\exp\left[-\frac{\tilde{\omega}^2}{2k_{||}^2 V_{t2||}^2}\right]\right\}\end{aligned} \tag{42}$$

By neglecting higher-order terms, the growth rate is obtained as

$$\begin{aligned}\gamma &\sim -\frac{\omega^3(\omega-k_{||}V_{d1})^3}{(\omega-k_{||}V_{d1})^3\Pi_{p2}^2+\omega^3\Pi_{p1}^2}\\ &\times\frac{\sigma\sqrt{\pi}}{2k_{||}^2}\left\{\frac{\Pi_{p1}^2}{V_{t1||}^2}\frac{\omega-k_{||}V_{d1}}{\sqrt{2}k_{||}V_{t1||}}\exp\left[-\frac{(\omega-k_{||}V_{d1})^2}{2k_{||}^2 V_{t1||}^2}\right]+\frac{\Pi_{p2}^2}{V_{t2||}^2}\frac{\omega}{\sqrt{2}k_{||}V_{t2||}}\exp\left[-\frac{\omega^2}{2k_{||}^2 V_{t2||}^2}\right]\right\}\end{aligned} \tag{43}$$

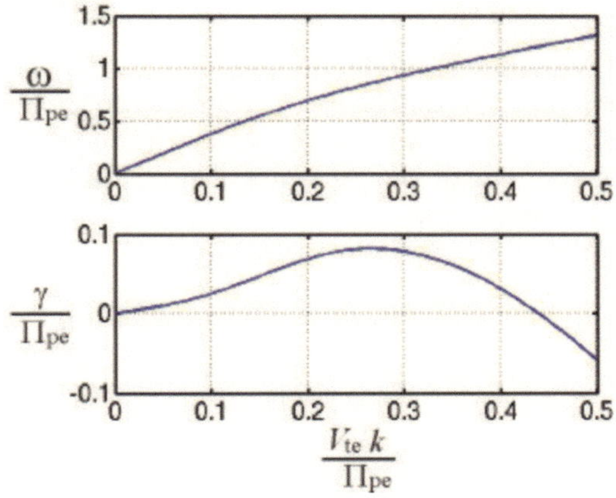

(a) Electron beam-plasma instability with $V_{te} \equiv V_{t1||} = V_{t2||}$, $V_{d1} = 4T_{te}$, $\Pi_{p2} = 9\Pi_{p1}$.

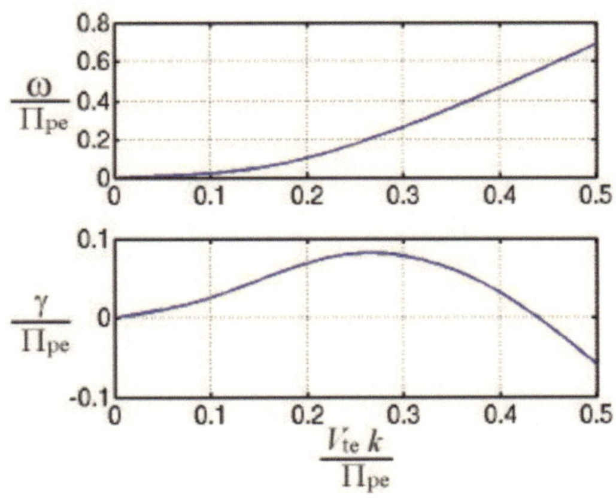

(b) Electron beam-plasma instability with $V_{te} \equiv V_{t1||} = V_{t2||}$, $V_{d1} = 4T_{te}$, $\Pi_{p1} = 9\Pi_{p2}$.

Fig. 3. Linear dispersion relation for electrostatic instabilities. The frequency is normalized by $\Pi_{pe}^2 \equiv \Pi_{p1}^2 + \Pi_{p2}^2$. Note that these two cases have the same growth rate, but the maximum growth rate is given at $\omega \sim \Pi_{p2}$.

One can find that the growth rate becomes positive when $\omega - k_{||}V_{d1} < 0$, $(\omega - k_{||}V_{d1})^3\Pi_{p2}^2 + \omega^3\Pi_{p1}^2 < 0$, and the second line in Eq.(43) is negative. If $V_{d1} \gg V_{t1} + V_{t2}$, the growth rate becomes maximum at $\omega/k_{||} \sim V_{d1} - V_{t1}$. It is again noted that the second line in Eq.(43) comes from derivative of the velocity distribution function with respect to velocity with $v = \omega/k_{||}$. Electrostatic waves are excited when the velocity distribution function has positive gradient. This condition is also called the Landau resonance. Since the positive gradient in the velocity distribution function is due to drifting plasma (or beam), the instability is known as the beam-plasma instability.

Examples of the beam-plasma instability are shown in Figure 3, which are obtained by numerically solving Eq.(38).

3.3 Cyclotron resonance

Since the Newton-Lorentz equation (6) and the Vlasov equation (7) cannot treat the relativism (such that $c \gg V_d$), plasma particles cannot interact with electromagnetic light mode waves. On the other hand, drifting plasma can interact with electromagnetic cyclotron waves when a velocity of particles is faster than the Alfven speed V_A but is slow enough such that electrostatic instabilities do not take place.

For isotropic but drifting plasma, Eq.(27) is rewritten as

$$
\begin{aligned}
0 = \tilde{\omega}^2 - c^2 k_{||}^2 - \Pi_{pe}^2 \frac{\tilde{\omega} - k_{||} V_{de}}{\tilde{\omega} - k_{||} V_{de} - \Omega_{ce}} - \Pi_{pi}^2 \frac{\tilde{\omega} - k_{||} V_{di}}{\tilde{\omega} - k_{||} V_{di} - \Omega_{ci}} \\
+ i\sigma\sqrt{\pi} \left\{ \Pi_{pe}^2 \frac{\tilde{\omega} - k_{||} V_{de}}{\sqrt{2} k_{||} V_{te||}} \exp\left[-\frac{(\tilde{\omega} - k_{||} V_{de} - \Omega_{ce})^2}{2 k_{||}^2 V_{te||}^2} \right] \right. \\
\left. + \Pi_{pi}^2 \frac{\tilde{\omega} - k_{||} V_{di}}{\sqrt{2} k_{||} V_{ti||}} \exp\left[-\frac{(\tilde{\omega} - k_{||} V_{di} - \Omega_{ci})^2}{2 k_{||}^2 V_{ti||}^2} \right] \right\}
\end{aligned}
\tag{44}
$$

Here higher-order terms are neglected. The imaginary part of the above equation gives the growth rate as

$$
\begin{aligned}
\gamma \sim -\frac{1}{2\omega - \dfrac{\Omega_{ce} \Pi_{pe}^2}{(\omega - k_{||} V_{de} - \Omega_{ce})^2} - \dfrac{\Omega_{ci} \Pi_{pi}^2}{(\omega - k_{||} V_{di} - \Omega_{ci})^2}} \\
\times \sigma\sqrt{\pi} \left\{ \Pi_{pe}^2 \frac{\omega - k_{||} V_{de}}{\sqrt{2} k_{||} V_{te||}} \exp\left[-\frac{(\omega - k_{||} V_{de} - \Omega_{ce})^2}{2 k_{||}^2 V_{te||}^2} \right] \right. \\
\left. + \Pi_{pi}^2 \frac{\omega - k_{||} V_{di}}{\sqrt{2} k_{||} V_{ti||}} \exp\left[-\frac{(\omega - k_{||} V_{di} - \Omega_{ci})^2}{2 k_{||}^2 V_{ti||}^2} \right] \right\}
\end{aligned}
\tag{45}
$$

In the case of $V_{de} \neq 0$ and $V_{di} = 0$, ion cyclotron waves have positive growth rate at $\omega \sim \Omega_{ci}$ when

$$
\frac{\omega}{k_{||}} - V_{de} < 0, \quad -\frac{\Pi_{pe}^2}{\Omega_{ce}} > \frac{\Omega_{ci} \Pi_{pi}^2}{(\omega - \Omega_{ci})^2}
\tag{46}
$$

Here $\omega \ll |\Omega_{ce}|$ and $\omega \ll \Pi_{pe}$ are assumed. The maximum growth rate is obtained at $\omega \sim k_{||}(V_{de} - V_A)$.

In the case of $V_{de} = 0$ and $V_{de} \neq 0$, electron cyclotron waves have positive growth rate at $\omega \sim -\Omega_{ci}$ when

$$
\frac{\omega}{k_{||}} - V_{di} > 0, \quad -\frac{\Pi_{pe}^2}{\Omega_{ce}} < \frac{\Omega_{ci} \Pi_{pi}^2}{(\omega - k_{||} V_{di} - \Omega_{ci})^2}
\tag{47}
$$

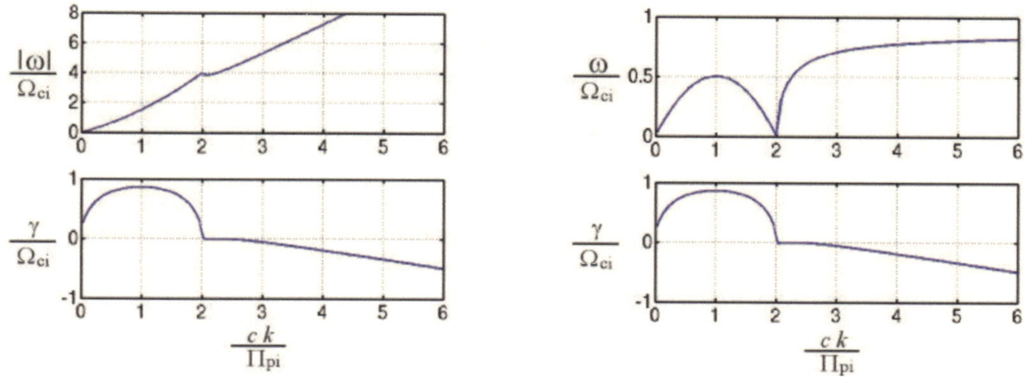

(a) Ion beam-cyclotron instability with $V_{di} = 2V_A = 0.005c$, $T_{e||} = T_{e\perp} = T_{ti||} = T_{i||}$, $\Pi_{pe} = 10|\Omega_{ce}|$.

(b) Electron beam-cyclotron instability with $V_{de} = 2V_A = 0.005c$, $T_{e||} = T_{e\perp} = T_{ti||} = T_{i||}$, $\Pi_{pe} = 10|\Omega_{ce}|$.

Fig. 4. Linear dispersion relation for beam-cyclotron instabilities.

Here $|\omega| \ll |\Omega_{ce}|$ and $|\omega| \ll \Pi_{pe}$ are assumed. The maximum growth rate is obtained at $\omega \sim k_{||}(V_{di} + V_A)$.
These conditions are called the cyclotron resonance. Note that electron cyclotron waves are excited by drifting ions while ion cyclotron waves are excited by drifting electrons. These instabilities are known as the beam-cyclotron instability.
Examples of the beam-cyclotron instability are shown in Figure 4, which are obtained by numerically solving Eq.(28).

4. Summary

In this chapter, electromagnetic waves in plasma are discussed. The basic equations for electromagnetic waves and charged particle motions are given, and linear dispersion relations of waves in plasma are derived. Then excitation of electromagnetic waves in plasma is discussed by using simplified linear dispersion relations. It is shown that electromagnetic cyclotron waves are excited when the plasma temperature in the direction perpendicular to an ambient magnetic field is not equal to the parallel temperature. Electrostatic waves are excited when a velocity distribution function in the direction parallel to an ambient magnetic field has positive gradient. Note that the former condition is called the temperature anisotropy instability. The latter condition is achieved when a high-speed charged-particle beam propagates along the ambient magnetic field, and is called the beam-plasma instability. Charged-particle beams can also interact with electromagnetic cyclotron waves, which is called the beam-cyclotron instability. These linear instabilities take place by free energy sources existing in velocity space.
It is noted that plasma is highly nonlinear media, and the linear dispersion relation can be applied for small-amplitude plasma waves only. Large-amplitude plasma waves sometimes result in nonlinear processes, which are so complex that it is difficult to provide their analytical expressions. Therefore computer simulations play essential roles in studies of nonlinear processes. One can refer to textbooks on kinetic plasma simulations (e.g., Birdsall & Langdon, 2004; Hockney & Eastwood, 1988; Omura & Matsumoto, 1994; Buneman, 1994) for further reading.

5. References

Birdsall, C. K., & Langdon, A. B. (2004), *Plasma Physics via Computer Simulation*, 479pp., Institute of Physics, ISBN 9780750310253, Bristol

Buneman, O. (1994), TRISTAN, In: *Computer Space Plasma Physics: Simulation Techniques and Software*, Matsumoto, H. & Omura,Y. (Ed.s) pp.67–84, Terra Scientific Publishing Company, ISBN 9784887041110, Tokyo

Fried, B. D. & Conte, S. D. (1961), *The Plasma Dispersion Function*, 419pp., Academic Press, New York

Hockney, R. W. & Eastwood, J. W. (1988) *Computer Simulation Using Particles*, 540pp., Institute of Physics, ISBN 9780852743928, Bristol

Omura, Y. & Matsumoto, H. (1994), KEMPO1, In: *Computer Space Plasma Physics: Simulation Techniques and Software*, Matsumoto, H. & Omura,Y. (Ed.s) pp.21–65, Terra Scientific Publishing Company, ISBN 9784887041110, Tokyo

Stix, T. H. (1992), *Waves in Plasmas*, 584pp., Springer-Verlag, ISBN 9780883188590, New York

Swanson, D. G. (2003), *Plasma Waves, Second Edition*, 400pp., Institute of Physics, ISBN 9780750309271, Bristol

Swanson, D. G. (2008), *Plasma Kinetic Theory*, 344pp., Crc Press, ISBN 9781420075809, New York

16

Propagation of Electromagnetic Waves in and around Plasmas

Osamu Sakai
Kyoto University
Japan

1. Introduction

1.1 Historical perspective

Interaction between electromagnetic waves and plasmas has been explored for several decades in various fields, such as ionosphere layers and outer space (Ginzburg, 1964), fusion plasmas (Stix, 1962; Swanson, 1989; Nishikawa & Wakatani, 1990), and plasma material-processing reactors (Lieberman & Lichtenberg, 1994). When an electromagnetic wave is launched in or near a non-magnetized plasma whose size is much larger than the wavelength, it is transmitted, reflected, or absorbed; these features are determined by a set of the electromagnetic wave frequency, the electron plasma frequency, and the electron elastic collision frequency. These three parameters determine real and imaginary parts of permittivity. In other words, plasmas equivalently behave as conductors or dielectric materials for electromagnetic waves, and these behaviours are controllable by changing complex permittivity, or electron density and gas pressure, which is associated with the electron plasma frequency and the electron elastic collision frequency; this controllability and the time-varying manner for permittivity distinguish plasmas from other electromagnetic media.

First of all, in this section, we briefly review the historical perspective of the electromagnetic waves in plasmas, and we point out the reasons why the concept of electromagnetic media composed of plasmas and their discontinuities is focused on in this chapter.

Electromagnetic waves in magnetized plasmas have been well investigated for more than half century, aiming at ultimate goals of controlled fusion plasmas for energy production and space plasmas for launching human beings using spacecrafts. In a magnetized plasma, various kinds of wave branches are present from low to high frequency ranges, and change of the external magnetic field induces "walk" on the dispersion curves in a "zoo" of plasma waves. Sometimes a branch leads to another totally-different branch; that is called mode conversion (Stix, 1962; Swanson, 1989). These waves can be launched from the outer side of the plasma, but there are also many inherent waves found as magnetohydrodynamic and micro instabilities (Swanson, 1989; Nishikawa & Wakatani, 1990). Other characteristic features of plasma waves are their nonlinearity; shock waves, solitons, and nonlinear mode conversion originate from the aspects of high-energy-state substance (Swanson 1989). The main focus in this chapter is different from such conventional scientific interests.

Before we start our description, one more comment about plasma production for industry should be addressed (Lieberman & Lichtenberg, 1994). Plasma production in fabrication

processes of thin film technology was quite successful, and several different methods have been developed for semiconductor industry, flat panel display markets, and photovoltaic cell production. Such a technology using plasmas is now somewhat mature, and several different ideas and schemes for researches on plasma science and engineering are being tested for other applicable fields.
In such a sense, we study new types of plasma-wave interactions, especially arising from discontinuities in both space and time. Progress of techniques to control shape and parameters of plasmas enables us to make discontinuities in a clear and stable state.

1.2 Emerging aspects of plasma wave propagation

Since wave propagation in a magnetized plasma is well described elsewhere (Stix, 1962; Ginzburg, 1964; Swanson 1989), we here focus on the propagation in and around a non-magnetized plasma. It is not so complicated to describe electromagnetic waves propagating in and around a bulk non-magnetized plasma (Kalluri, 1998), although the propagation in a spatially-nonuniform plasma, in which electron-density gradient is significant (Swanson, 1989; Nickel et al., 1963; Sakai et al., 2009) and/or the profiles of electron density is spatially periodic (Hojo & Mase, 2004; Sakai et al., 2005(1); Sakai et al., 2005(2); Sakai et al., 2007(1); Sakai et al., 2007(2); Sakaguchi et al., 2007; Sakai & Tachibana, 2007; Naito et al., 2008; Sakai et al., 2010(1); Sakai et al., 2010(2)), includes novel physical aspects which have not been described in usual textbooks of plasma physics.
Also, complex dielectric constant or permittivity whose imaginary part is significantly large is observed and easily controlled in a plasma as a macroscopic value (Naito et al., 2008; Sakai et al., 2010(1)). This imaginary part strongly depends on field profiles of electromagnetic waves around plasmas when their spatial discontinuities exist, and so synthesized effects with Bloch modes in periodic structure lead to not only frequency band gap but also *attenuation* gap (Naito et al., 2008; Sakai et al., 2009). In another point of view, power dissipation due to the imaginary part leads to plasma generation (Lieberman & Lichtenberg, 1994), which is a quite nonlinear phenomenon.
In this chapter, considering the spatial discontinuities and the complex permittivity, the fundamentals of theoretical understandings on electromagnetic waves in and around plasmas are described. In Section 2.1, properties of complex permittivity are generalized using equations and a newly-developed 3-dimensional (3D) drawing. In Section 2.2, starting from the momentum equation of electrons to treat plasma effects, the complex permittivity in the Drude model is derived, and the equation is compared with that for metals whose permittivity is also in the Drude model; such a description will reveal unique features in the case of discharge plasmas for control of electromagnetic waves. In Section 3.1, various methods to describe effects of periodic spatial discontinuities are demonstrated, including both analytical and numerical ones, and specific examples of band diagrams of 2D structures are shown. In Section 3.2, another aspect of the spatial discontinuity associated with surface wave propagation is described, in which propagation of surface waves on the interface with spatial electron-density gradient is clarified. Section 4 summarizes this chapter, showing emerging aspects of electromagnetic waves in and around plasmas.

2. Fundamentals of new aspects for wave propagation

In this Section, we demonstrate features and importance of complex permittivity which is usual in a low-temperature partially-ionized plasma. Section 2.1 includes general description which is also applicable to other lossy materials (Sakai et al., 2010(1)), and Section 2.2 focuses on the momentum balance equations of electrons in a discharge plasma, which is the origin of characteristic wave propagation in and around plasmas.

2.1 Complex permittivity in a plasma

To describe wave transmission and absorption as well as phase shift and reflection of the propagating waves, we here introduce a new drawing of dispersion relation in the 3D space of three coordinates consisting of wave frequency $\omega/2\pi$, real wavenumber k_r, and imaginary wavenumber k_i. A propagating wave which is launched at a spatial position $x=0$, or on the edge of a given media, is expressed as

$$\begin{aligned} A(x)\exp(\mathrm{j}\varphi(t,x)) &= A(x)\exp(\mathrm{j}(\omega t - k_r x)) \\ &= A_0\exp(k_i x)\exp(\mathrm{j}(\omega t - k_r x)) = A_0\exp(\mathrm{j}(\omega t - (k_r + \mathrm{j}k_i)x)), \end{aligned} \tag{2.1.1}$$

where $A(x)$ is the wave amplitude with the initial boundary condition of $A_0 = A(x=0)$, t is the time, and $\phi(t,x)$ is the phase of the wave with the initial condition of $\phi(0,0)=0$. The dispersion relation in a collisionless plasma is usually expressed in the $\omega - k_r$ plane, and we can also obtain a useful information about wave attenuation from k_i as a function of ω when significant loss or wave attenuation takes place.

Such a wave propagation in a bulk non-magnetized plasma is characterized by the permittivity ε_p in the Drude model in the form

$$\varepsilon_p = 1 - \frac{{\omega_{pe}}^2}{\omega^2(1 - \mathrm{j}\nu_m/\omega)}, \tag{2.1.2}$$

where ω_{pe} is the electron plasma frequency which is a function of electron density n_e, and $\nu_m/2\pi$ is the electron elastic collision frequency. We note that, since we choose a formula in equation (2.1.1) instead of $\exp(\mathrm{j}(k_r x - \omega t))$, the sign in the bracket of equation (2.1.2) becomes "-", and the imaginary part of the permittivity becomes negative in general (Pozar, 2005). The detailed derivation of equation (2.1.2) is given in Section 2.2. Figure 1 shows ε_p at a fixed wave frequency (4 GHz) as a function of n_e with various gas conditions on the complex plane (Sakai et al., 2010(1)). Here, we assume that the electron energy is 0.5 eV for a plasma in the afterglow and that the cross section for the electron elastic collisions is 5.0×10^{-16} cm^2 for He and 1.0×10^{-16} cm^2 for Ar from the literature (Raizer, 1991). At 760 Torr of He, $\mathrm{Re}(\varepsilon_p)$ is almost constant at unity for various n_e. On the other hand, at 5 Torr of Ar, $\mathrm{Im}(\varepsilon_p)$ is roughly zero while $\mathrm{Re}(\varepsilon_p)$ changes significantly in the negative polarity, and this feature almost corresponds to a collisionless plasma. This figure indicates that the change of gas species and pressure yields ε_p with $\mathrm{Im}(\varepsilon_p)/\mathrm{Re}(\varepsilon_p)$ ranging from 0 to infinity for $\mathrm{Re}(\varepsilon_p)<1$ on the complex plane.

Equation (2.1.2) gives us an understanding of dispersion relation in the 3D space (ω, k_r, k_i). Figure 2 displays a dispersion relation in a bulk non-magnetized plasma expressed by equation (2.1.2) (Sakai et al., 2010(1)). In the case at 5 Torr of Ar, which is nearly collisionless as mentioned earlier, the trajectory on the (ω, k_r) plane is well known in textbooks of plasma physics. The working point is always on the (ω, k_r) plane or on the (ω, k_i) plane,

which can be understood easily from equation (2.1.2). However, in the case at 120 Torr of He, the working point goes far away from the two planes below ω_{pe} and leaves a trajectory on the (k_r, k_i) plane; at such a point, the wave suffers attenuation as well as phase shift, as suggested in equation (2.1.1).
Drawings of dispersion relations in this 3D space reveal significant physical parameters of electromagnetic media, as shown in the following. Knowledge from microwave engineering (Pozar, 2005) shows that

$$k = \omega\sqrt{\mu\varepsilon}\sqrt{1 - j\frac{\sigma}{\omega\varepsilon}}, \tag{2.1.3}$$

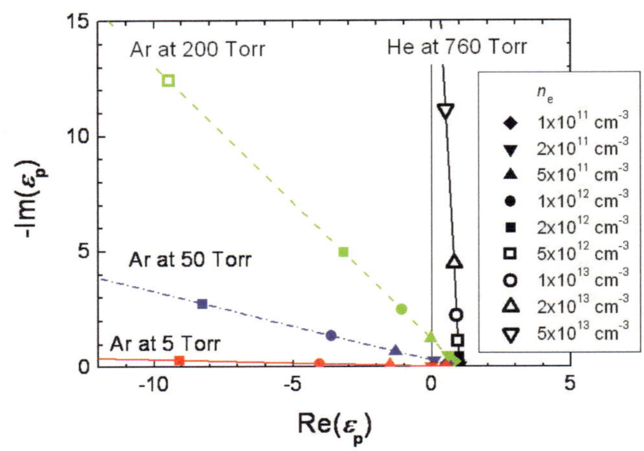

Fig. 1. Permittivity in a lossy bulk plasma with various gas condition and various n_e (Sakai et al., 2010(1)).

where ε, μ and σ is the permittivity, the permeability, and the conductivity of the media, respectively. From equation (2.1.3), the following equation is derived:

$$k_r|k_i| = \frac{\omega\mu}{2}\sigma = \frac{1}{\delta_s^{\,2}}. \tag{2.1.4}$$

Here δ_s is the skin depth of the wave into the media. $k_r|k_i|$ indicates an area on the (k_r, k_i) plane, and so a point projected on the (k_r, k_i) plane expresses conductivity of the media on the assumption that μ is constant. The inverse of the area on the (k_r, k_i) plane corresponds to square of δ_s; as the area is larger, the skin depth is shorter. Another physical parameter which is visible in this 3D drawing is the metallic/dielectric boundary. From equation (2.1.3), we also obtain

$$k_r^{\,2} - k_i^{\,2} = \omega^2\mu\varepsilon. \tag{2.1.5}$$

Comprehension of this equation gives us the following result. If $k_r > |k_i|$, ε is positive when μ is positive, leading to the fact that the media is dielectric, and if $k_r < |k_i|$, vice versa, and we can recognize that the media is metallic. The line of $k_r = k_i$ becomes the boundary between metallic and dielectric media.

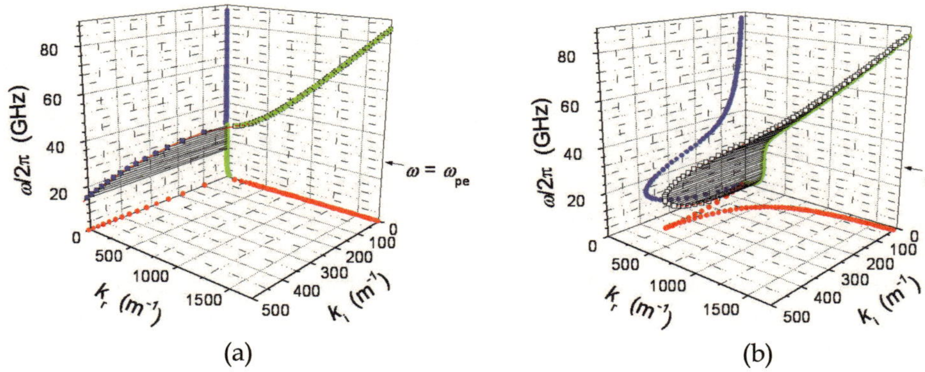

Fig. 2. Dispersion relation of electromagnetic waves in a bulk plasma with $n_e = 1\times10^{13}$ cm^{-3} in the 3D space. (a) In a plasma at 5 Torr of Ar gas. (b) In a plasma at 120 Torr of He gas. (Sakai et al., 2010(1)).

These characteristics arising from lossy plasmas are distinguishable from other electromagnetic media; unlike plasmas, any other material never has a variety of parameter sets such as complex $\varepsilon\,(=\varepsilon_r - j\varepsilon_i)$ and σ. Such a characteristic property can be enhanced by spatial periodicity; a simple periodic ε_r distribution realized in a solid material makes a photonic or electromagnetic band material which exhibits band gaps. If we introduce the effects of ε_i by a plasma array, new features can emerge with the complex-variable effects (Naito et al., 2008; Sakai et al., 2010(2)). That is, spatial periodic change in ε_r leads to formation of frequency band gaps around which propagation-permitted frequency has a certain gap. As a novel feature in plasma cases, wave attenuation due to ε_i is significantly different between two bands above and under this band gap; we call it an *attenuation* gap. This gap arises from change of field profiles of electromagnetic waves; on the upper band, fields are localized in the lower-ε_r area, and vice versa on the lower band; such properties are demonstrated in Section 3.1.1 and 3.1.4.

2.2 Electron momentum balance equation and its effects for wave propagation

When we consider wave propagation in a plasma, the rigorous starting point of description is electron momentum balance equation. Here, a non-magnetized plasma is assumed for simplicity; if an external magnetic field is present, other terms, for instance arising from the Lorentz force, might be included. The electron momentum balance equation deals with a plasma as a kind of fluids, and it keeps good matching with macroscopic parameters such as ε and μ. Another method containing plasma effects is calculation of direct particle motion, such as a particle-in-cell simulation in which momentum balance of each single test particle is treated and collective effects of charged particles are integrated by the Poisson's equation. Another comment of our treatment based on the fluid model is that ions are assumed to be immobile due to its huge relative mass compared with electrons. Dispersion relations of some electrostatic waves propagating in a plasma such as ion acoustic waves are derived with the effects of ion motions, although they are removed from our curiosity here.

Now, a general expression of the electron momentum balance is described as (Razer, 1991)

$$mn_{\mathrm{e}}(\boldsymbol{r})\frac{\mathrm{d}\boldsymbol{v}_{\mathrm{e}}(\boldsymbol{r})}{\mathrm{d}t} = -n_{\mathrm{e}}(\boldsymbol{r})e\boldsymbol{E}(\boldsymbol{r}) - \nabla p_{\mathrm{e}}(\boldsymbol{r}) - mn_{\mathrm{e}}(\boldsymbol{r})\boldsymbol{v}_{\mathrm{e}}(\boldsymbol{r})\nu_{\mathrm{m}}, \tag{2.2.1}$$

where t is the time, $\boldsymbol{r}$ the spatial position vector, $\boldsymbol{v}_{\mathrm{e}}$ the electron fluid velocity, m_{e} the mass of an electron, e the charge of an electron, and $\boldsymbol{E}$ the electric field. The left hand side contains a convective term arising from the change of electron position in time due to $\boldsymbol{v}_{\mathrm{e}}$ as $\mathrm{d}/\mathrm{d}t = \partial/\partial t + \boldsymbol{v}_{\mathrm{e}} \cdot \nabla$, although this term is neglected in the following. The first term on the right hand side is Coulomb force by $\boldsymbol{E}$. The second term expresses pressure-gradient term, where assumption of an isothermal plasma with a uniform spatial profile of electron temperature T_{e} enables to obtain $\nabla p_{\mathrm{e}}(\boldsymbol{r}) = kT_{\mathrm{e}}\nabla n_{\mathrm{e}}(\boldsymbol{r})$. The third term implies friction against neutral particles, i.e., change of momentum that is as frequent as ν_{m}; $\nu_{\mathrm{m}} = n_{\mathrm{g}}\langle\sigma v\rangle$, where n_{g} is the neutral gas density, σ, strongly depending of electron energy, is the cross section of the elastic collision, and $\langle\sigma v\rangle$ is a statistically averaged value over the velocity distribution.

First of all, complex permittivity of a plasma in the Drude model ε_{p}, as shown in equation (2.1.2), is derived in the following. In this case, we ignore the pressure term in equation (2.2.1) to see wave propagation in a bulk plasma. Then, current density by electrons J is given by

$$\frac{\partial \boldsymbol{J}}{\partial t} + \nu_{\mathrm{m}}\boldsymbol{J} = \varepsilon_0 \omega_{\mathrm{pe}}{}^2 \boldsymbol{E}. \tag{2.2.2}$$

If we express a variable $\widetilde{x}(\omega)$, as a frequency spectrum at ω, via Fourier transform from $x(t)$, equation (2.2.2) becomes for a wave with frequency $\omega/2\pi$

$$(\mathrm{j}\omega + \nu_{\mathrm{m}})\widetilde{\boldsymbol{J}} = \varepsilon_0 \omega_{\mathrm{pe}}{}^2 \widetilde{\boldsymbol{E}}. \tag{2.2.3}$$

Also, one of the Maxwell equations becomes

$$\nabla \times \widetilde{\boldsymbol{H}} = \mathrm{j}\omega\varepsilon_0 \widetilde{\boldsymbol{E}} + \widetilde{\boldsymbol{J}}. \tag{2.2.4}$$

Substituting $\widetilde{\boldsymbol{E}}$ in equation (2.2.3) into equation (2.2.4), we obtain ε_{p} in the following:

$$\nabla \times \widetilde{\boldsymbol{H}} = \mathrm{j}\omega\varepsilon_0 \widetilde{\boldsymbol{E}}\left(1 - \frac{\omega_{\mathrm{pe}}{}^2}{\omega(\omega - \mathrm{j}\nu_{\mathrm{m}})}\right) = \mathrm{j}\omega\varepsilon_{\mathrm{p}}\varepsilon_0 \widetilde{\boldsymbol{E}}, \tag{2.2.5}$$

with the expression of ε_{p} similar to equation (2.1.2), the well known form in the Drude model. We carefully note that, from the above discussion, this form is valid for sinusoidal waves.

Secondly, we take a look on the contributions of the pressure term in equation (2.2.1). The pressure term will be significant on the edge of a plasma, and so this treatment is beneficial when we consider wave propagation on the edge as a surface wave. Ignoring the friction term in equation (2.2.1), the current density in a plasma is given by

$$\frac{\partial \mathbf{J}}{\partial t} = \varepsilon_0 \omega_{\mathrm{pe}}{}^2 \mathbf{E} - \frac{ekT_e}{m_e} \nabla n_{\mathrm{e}}. \tag{2.2.6}$$

Then, we compare the effects of the pressure term with the cases of metals. In metal optics (Forstmann & Gerhardts, 1986), the theoretical framework similar to the fluid model in plasma physics is referred to as the hydrodynamic formulation. An expression of the current density is given as

$$\frac{\partial J}{\partial t} = \varepsilon_0 \omega_{\mathrm{pe}}{}^2 E - e\beta^2 \nabla n_{\mathrm{e}}. \tag{2.2.7}$$

That is, in the hydrodynamic formulation, the only different point between metals and discharge plasmas are pressure term p_{e}; in the case of metals, $p_{\mathrm{e}} = m_{\mathrm{e}}\beta^2 n_{\mathrm{e}}$ with $\beta^2 = (3/5)v_{\mathrm{F}}{}^2$, where v_{F} is the Fermi velocity, and in the case of discharge plasmas $p_{\mathrm{e}} = n_{\mathrm{e}}kT_{\mathrm{e}}$. This difference comes from the state-of-phase transition at $T_{\mathrm{e}} = T_{\mathrm{F}}$ condition where the thermal energy is equal to the Fermi energy expressed by the Fermi temperature T_{F}; in the cases of metals, $T_{\mathrm{e}} < T_{\mathrm{F}}$ and metals are in the quantum phase in the category of electron gases (Isihara, 1993). In a usual metal, parameter $m_{\mathrm{e}}\beta^2$ is by 3-4 orders larger than kT_e, which yields significant differences for dispersion relations of surface wave modes between these two cases, which is discussed in Section 3.2.

3. Emerging aspects due to spatial discontinuities

3.1 Plasma periodic structure and deformation of wave propagation

Stimulated by recent progresses in photonic crystals (Yablonovitch, 2000; Noda & Baba, 2003), a number of researches have been performed about spatially periodic plasma structures which exhibit various novel phenomena that have not been expected when we use solid materials. To the best of our knowledge, the earliest publication associated with the plasma periodic structure was by Faith and Kuo (Faith et al., 1997); in their report, deformation of the band structure by rapidly generated 1D plasma structure that was analytically calculated was used for frequency upshift, and implied presence of photonic band gaps. Kalluri (Kalluri, 1998) summarized the variety of dispersion relations in the cases of both spatial and time discontinuities, and he pointed out photonic band gaps as forbidden bands. The terminology of a "plasma photonic crystal" was first announced by Hojo and Mase with its physical and technological importance (Hojo & Mase, 2004). After that, the first experimental verification of a 2D photonic crystal was performed by Sakai and his coworkers (Sakai et al., 2005(1); Sakai et al., 2005(2)), and they continued to extend their work in both theoretical and experimental results. Several groups followed these researches mainly from the theoretical points of view, as mentioned later, although Dong and her coworkers (Dong et al., 2007) performed simulated experiments of 2D plasma self-organized pattern formation, and Lo and coworkers (Lo et al., 2010) successfully observed effects for modification of propagating electromagnetic waves using metal-plasma composites.

In this section, the theoretical fundamentals for investigation of dispersion relations of plasma photonic crystals are reviewed. 1D dispersion relations are well described by the formula of a Bloch mode, given by

$$\cos ka = \cos\frac{N\omega d_{\rm n}}{c}\cos\frac{N_{\rm d}\omega d_{\rm d}}{c} - \frac{1}{2}\left(\frac{Z}{Z_{\rm d}} + \frac{Z_{\rm d}}{Z}\right)\sin\frac{N\omega d_{\rm n}}{c}\sin\frac{N_{\rm d}\omega d_{\rm d}}{c}. \tag{3.1.1}$$

Here $d_{\rm n}$ is the thickness of the region with N where a plasma layer is present, and c is the velocity of light. Z is the wave impedance of the equivalent slab region with the plasma structure. We also assumed that dielectric layers with refractive index $N_{\rm d}$, wave impedance $Z_{\rm d}$, and a thickness $d_{\rm d}$ were set between the plasma layers. Several authors investigated 1D plasma photonic srystals using equation (3.1.1) or other fomulae (Faith et al., 1997; Kalluri, 1998; Guo, 2009; Yin et al., 2009; Qi et al., 2010; Fan & Dong, 2010). However, for derivation of dispersion relations in 2D structures, which includes more scientific understandings as well as technical applications, several different methods have been developed, and they are described in the following.

3.1.1 Plane-wave expansion method for dispersive media

The plane-wave expansion method has been widely used to derive analytically photonic band diagrams of 2D and 3D dielectric periodic structures (Ho et al., 1990; Phihal et al., 1991). A spatially periodic permittivity, which is constant over a certain frequency range, is converted into the summation of spatial Fourier coefficients, and an assumption that multiplane-waves are superposed leads to dispersion relations by solving eigenvalue problems. That is, for the first task to use the plane-wave expansion method, we have to derive the spatial Fourier coefficients in advance, which is all right if an assumed structure is quite simple, although complicated structures are hardly treated in this procedure.

When a solid dielectric array in the plane-wave expansion method is replaced by a 2D plasma structure, the permittivity depends on ω, and the normal plane-wave expansion method is ineffective. Kuzmiak and Maradudin developed a plane-wave expansion method applicable to derivation of the photonic band structures of metallic components (Kuzmiak & Maradudin, 1997), and here we use a similar technique to solve the eigenvalue problems and further investigate cases similar to the experimental conditions (Sakaguchi et al., 2007). That is, the important point of the development of this method by Kuzmiak and Maradudin is that their formulae enable us to deal with dispersive media, or frequency-dependent permittivity, one of which is observed in a plasma in the Drude model, as shown in equation (2.1.2). Here we briefly describe this modified plane-wave expansion method introduced in (Kuzmiak & Maradudin, 1997) and followed in (Sakai et al., 2007(1)).

We set the dielectric region with permittivity $\varepsilon_{\rm d}$ outside the plasma column, since in experiments we frequently confine discharge gases and plasmas in a dielectric container. Using equation (2.1.2), the position vector at the site of lattice point $\boldsymbol{x}_{||}$, and translation vector $\boldsymbol{G}_{||}(=h_1\boldsymbol{b}_1+h_2\boldsymbol{b}_2$, where $\boldsymbol{b}_1$ and $\boldsymbol{b}_2$ are the primitive translation vectors of the reciprocal lattice and h_1 and h_2 are arbitrary integers), spatially and frequency-dependent permittivity $\varepsilon(\boldsymbol{x}_{||}\,|\,\omega)$ in a square lattice is expressed by Fourier coefficients $\hat{\varepsilon}(\boldsymbol{G}_{||})$ as

$$\varepsilon(\boldsymbol{x}_{||}\,|\,\omega) = \sum_{\boldsymbol{G}_{||}}\hat{\varepsilon}(\boldsymbol{G}_{||})\exp(j\boldsymbol{G}_{||}\cdot\boldsymbol{x}_{||}) \tag{3.1.2}$$

with

$$\hat{\varepsilon}\left(\mathbf{G}_{||}\right)=\varepsilon_{\mathrm{d}}-f\left[1-\varepsilon_{\mathrm{d}}-\frac{\omega_{\mathrm{pe}}{}^{2}}{\omega\left(\omega-\mathrm{j}\nu_{\mathrm{m}}\right)}\right] \quad \mathrm{G}_{||}=0, \tag{3.1.3a}$$

$$\hat{\varepsilon}\left(\mathbf{G}_{||}\right)=f\left[1-\varepsilon_{\mathrm{d}}-\frac{\omega_{\mathrm{pe}}{}^{2}}{\omega\left(\omega-\mathrm{j}\nu_{\mathrm{m}}\right)}\right]\frac{2\mathrm{J}_{1}\left(\mathbf{G}_{||}R\right)}{\left(\mathbf{G}_{||}R\right)} \quad \mathrm{G}_{||}\neq 0, \tag{3.1.3b}$$

where $f=\pi R^{2}/a^{2}$ is the filling fraction and R is the radius of the cross section of a plasma column with lattice constant a. This frequency-dependent complex value of $\varepsilon(\boldsymbol{x}_{||}\,|\,\omega)$ complicates the eigenvalue problem.

From Maxwell equations with fields varying harmonically in time t in the form of $\exp(\mathrm{j}\omega t)$, we can expand the z component of the magnetic field in the TE (H polarization) mode, where the (x,y) plane exhibits the spatial periodicity of the permittivity as

$$\tilde{\boldsymbol{H}}_{z}\left(\boldsymbol{x}_{||}|\omega\right)=\sum_{\boldsymbol{G}_{||}}A\left(\boldsymbol{k}_{||}|\boldsymbol{G}_{||}\right)\exp\left\{\mathrm{j}\left(\boldsymbol{k}_{||}+\boldsymbol{G}_{||}\right)\cdot\boldsymbol{x}_{||}\right\}. \tag{3.1.4}$$

When we substitute equation (3.1.4) into the wave equation about $\tilde{\boldsymbol{H}}_{z}$, coefficients $\{A(\boldsymbol{k}_{||}|\boldsymbol{G}_{||})\}$ fulfill the following equation:

$$\sum_{\boldsymbol{G}_{||}'}\left(\boldsymbol{k}_{||}+\boldsymbol{G}_{||}\right)\cdot\left(\boldsymbol{k}_{||}+\boldsymbol{G}_{||}'\right)\hat{\kappa}\left(\boldsymbol{G}_{||}-\boldsymbol{G}_{||}'\right)A(\boldsymbol{k}_{||}|\boldsymbol{G}_{||})=\frac{\omega^{2}}{c^{2}}A\left(\boldsymbol{k}_{||}|\boldsymbol{G}_{||}\right), \tag{3.1.5}$$

where $\{\hat{\kappa}(\boldsymbol{G}_{||})\}$ are the Fourier coefficients of $1/\varepsilon(\boldsymbol{x}_{||}\,|\,\omega)$.

In the case of TM (E polarization) mode, the z component of electric field $\tilde{\boldsymbol{E}}_{z}$ is

$$\tilde{\mathbf{E}}_{z}\left(\mathbf{x}_{||}|\omega\right)=\sum_{\mathbf{G}_{||}}B\left(\mathbf{k}_{||}|\mathbf{G}_{||}\right)\exp\left\{\mathrm{j}\left(\mathbf{k}_{||}+\mathbf{G}_{||}\right)\cdot\mathbf{x}_{||}\right\}. \tag{3.1.6}$$

Components $\{B(\boldsymbol{k}_{||}|\boldsymbol{G}_{||})\}$ fulfill the following equation as

$$\left(\boldsymbol{k}_{||}+\boldsymbol{G}_{||}\right)^{2}B\left(\boldsymbol{k}_{||}|\boldsymbol{G}_{||}\right)=\frac{\omega^{2}}{c^{2}}\hat{\varepsilon}(\boldsymbol{0})B\left(\boldsymbol{k}_{||}|\boldsymbol{G}_{||}\right)+\frac{\omega^{2}}{c^{2}}\sum_{\boldsymbol{G}_{||}'}\hat{\varepsilon}\left(\boldsymbol{G}_{||}-\boldsymbol{G}_{||}'\right)B\left(\boldsymbol{k}_{||}|\boldsymbol{G}_{||}\right), \tag{3.1.7}$$

where $\sum_{\boldsymbol{G}_{||}'}$ denotes summation except for the case with $\boldsymbol{G}_{||}=\boldsymbol{G}_{||}'$.

In the process of solving equation (3.1.5) or (3.1.7), we obtain a polynomial formula for ω/c, which can be transferred to a linear problem using matrix forms. Practically, when we set a specific value of $\boldsymbol{k}$, which is a wavenumber vector composed of real numbers we obtain complex wave frequency, that is, $\omega=\omega_{\mathrm{r}}+\mathrm{j}\omega_{\mathrm{i}}$. We can plot them, as they are, as a function of $\boldsymbol{k}$, but if we convert ω_{i} to k_{i} via $\omega_{\mathrm{i}}=-v_{\mathrm{g}}k_{\mathrm{i}}$ with group velocity v_{g}, we can get a set of ($\omega(=\omega_{\mathrm{r}}),k_{\mathrm{r}},k_{\mathrm{i}}$).

Figure 3 shows a dispersion relation in 2D wavenumber, or a band diagram, of a plasma photonic crystal with a significant value of ν_{m} (Sakai et al., 2009), in which we use a 3D drawing in the ($\omega,k_{\mathrm{r}},k_{\mathrm{i}}$) space described in Section 2.1. In the $\omega-k_{\mathrm{r}}$ plane in Fig. 3, a band gap was clearly observed from $\omega a/2\pi c$ = 0.29 to 0.32, and the dispersion relation was divided into upper and lower bands. In the lower band, there was a gap from $\omega a/2\pi c$ = 0.17 to 0.21 arising from the crossing of one flat band, but k_r and k_i were continuous on both sides of the gap. This fact indicates that this is a simple frequency band gap caused by deformation of the flat-band crossing. However, the band gap from $\omega a/2\pi c$ = 0.29 to 0.32 yielded very large differences between the upper and lower bands. Although k_r was equal on both sides at $\omega a/2\pi c$ = 0.29 and 0.32, k_i of the upper band was one order of magnitude larger than that of the lower band. The trajectory on the k_r - k_i plane clarifies that these two bands had completely different properties, and we refer it as an *attenuation* gap. This is mainly attributed to different wave-field profiles in one lattice (Sakai & Tachibana, 2007). On the upper band, wave fields concentrate on the plasma region where ε is relatively small, but on the lower band, wave fields are localized outside the plasma region where ε is relatively large. If a band gap is located above $\omega_{pe}/2\pi$, the differences in the field profile change only the matching condition between inside and outside the array region, and the band gap shows the features of a band-stop filter. On the other hand, if a band gap is located below $\omega_{pe}/2\pi$, the field profile in the periodic structure strongly affects attenuation of the propagating waves by electron elastic collisions.

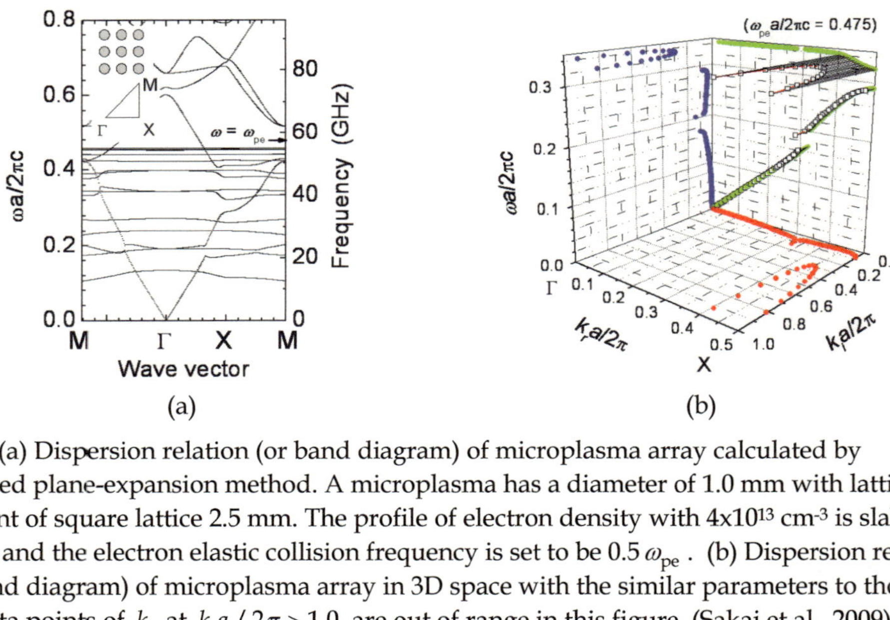

(a) (b)

Fig. 3. (a) Dispersion relation (or band diagram) of microplasma array calculated by modified plane-expansion method. A microplasma has a diameter of 1.0 mm with lattice constant of square lattice 2.5 mm. The profile of electron density with $4\text{x}10^{13}$ cm^{-3} is slab shape, and the electron elastic collision frequency is set to be 0.5 ω_{pe}. (b) Dispersion relation (or band diagram) of microplasma array in 3D space with the similar parameters to those in (a). Data points of k_i at $k_i a/2\pi > 1.0$ are out of range in this figure. (Sakai et al., 2009)

3.1.2 Direct Complex-Amplitude (DCA) method

Direct numerical analysis of Maxwell equations is usually by a finite difference time-domain (FDTD) method, and it successfully gave rise to band diagrams of a 2-dimensional plasma structures as well as those of metallic photonic crystal, which is described in Section 3.1.3. What we use here to obtain band diagrams numerically is a different method dealing with the complex values of electromagnetic fields that enables us to obtain a *static* solution of the fields in shorter central processing unit CPU time. A description of the method is found in (Sakai & Tachibana, 2006) and (Sakai et al., 2007(1)). We note that this method was originally applied to investigate wave propagation in a fusion plasma (Fukuyama et al., 1983). Here we mention its details.

From the Maxwell equations, the elimination of magnetic fields leads to a wave equation composed of electric fields $\widetilde{\boldsymbol{E}}$ and external current density $\widetilde{\boldsymbol{J}}_{\text{ext}}$, written as

$$\nabla \times \nabla \times \widetilde{\boldsymbol{E}} - \omega^2 \mu \varepsilon \varepsilon_0 \widetilde{\boldsymbol{E}} = \mathrm{j}\omega\mu \widetilde{\boldsymbol{J}}_{\text{ext}} , \tag{3.1.8}$$

which is spatially discretized (for instance, 20×20 square meshes in one lattice) based on the finite difference method. We note that ε is expressed as $\varepsilon_{\text{p}}(\omega)$ in equation (2.1.2) inside a plasma. 2D complex-value electric fields in the form expressed in equation (2.1.1) are considered, and so the TE mode in which an electric field is parallel to a 2D lattice plane can be analyzed when electric field vectors are vertical to plasma columns, in applications of the Bloch theorem to electric and magnetic (or gradient of electric) fields on the boundaries of a lattice in the form of

$$\begin{aligned} \widetilde{\boldsymbol{E}}(x_0 + a) &= \widetilde{\boldsymbol{E}}(x_0 + a)\exp(-\mathrm{j}k_x a), \\ \widetilde{\boldsymbol{E}}(y_0 + a) &= \widetilde{\boldsymbol{E}}(y_0 + a)\exp(-\mathrm{j}k_y a), \end{aligned} \tag{3.1.9}$$

and

$$\begin{aligned} \frac{\mathrm{d}\widetilde{\boldsymbol{E}}}{\mathrm{d}x}(x_0 + a) &= \frac{\mathrm{d}\widetilde{\boldsymbol{E}}}{\mathrm{d}x}(x_0 + a)\exp(-\mathrm{j}k_x a), \\ \frac{\mathrm{d}\widetilde{\boldsymbol{E}}}{\mathrm{d}x}(y_0 + a) &= \frac{\mathrm{d}\widetilde{\boldsymbol{E}}}{\mathrm{d}x}(y_0 + a)\exp(-\mathrm{j}k_y a), \end{aligned} \tag{3.1.10}$$

where (x_0, y_0) is a position vector on the boundary of lattices. In the case of the TM mode analysis, 2D magnetic field vectors will be solved numerically. From these boundary conditions, the 2D wave number of a propagating wave (k_x, k_y) is specified. The external current source is hypothetically set within a lattice. Then equation (3.1.8) is solved using the finite difference method at each discrete frequency from 0 to 150 GHz, and the local wave power density at a specific point is chosen as a cost function to detect a propagation branch. That is, when one set of frequency and wave number is matched with a propagating wave condition, the cost function represents resonance-like peaking with a very narrow frequency region (usually less than 0.1 GHz width) and with a nonlocal electric field pattern independent of the position of the hypothetical current source. On the other hand, when another set is not along the propagation branch, the electric field profile only shows a near-field pattern localized around the hypothetical current source with a very small value of the cost function value. A complex value of the electric field includes phase information, and so the wave number assumed in equations (3.1.9) and (3.1.10) can be reconfirmed by a spatially differential value of the phase $(\boldsymbol{k} \cdot \boldsymbol{x})$ at the resonance-like frequency. In this method, no derivation of a spatial Fourier coefficient is required, and so an arbitrary structure inside a lattice such as a complicated plasma shape and an arbitrary ε profile can be treated.

This method does not use time-domain discretization, unlike the FDTD method. In the case of the FDTD method, time evolution of propagating waves in media is converted into frequency spectra. To deal with the wide frequency range simultaneously, it is required to perform auxiliary calculation to reinforce the dispersive dependence of the permittivity such as shown in equation (2.2.2), and such a scheme is referred to as frequency-dependent FDTD method, which is shown in Section 3.1.3. In our method used here, a monochromatic wave at one frequency is assumed in each calculation step with a corresponding and precise value of the permittivity from equation (2.1.2). In other words, the frequency step which we set for searching wave propagation is crucial to assure the entire calculation accuracy. A narrower frequency step will yield a more accurate determination of a propagating wave, although more CPU time is required.

Using this scheme, we calculated band diagrams with $\nu_m << \omega$, that is, a collisionless case, as shown in Figs. 5-9. When ν_m is introduced as a finite value comparable to ω and ω_{pe}, note that the resonance-like frequency is searched on the complex frequency for a real wave number, like in the cases of the plane-wave expansion method described in Section 3.1.1. If we consider spatial wave damping of the static propagation in a finite region, a complex wave number is derived for a real value of frequency. This method enables us to take such a flexible approach. The relation of complex wavenumber and complex wave frequency was well investigated by Lee et al.(Lee & Mok, 2010).

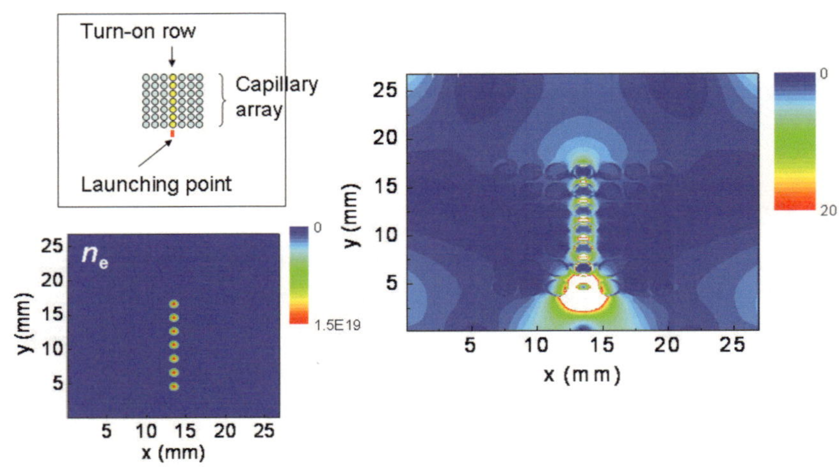

Fig. 4. 2D Wave propagation along a chain structure composed of columnar plasams at 6.2 GHz. Inset figures shows assumed configuration with assumed n_e profile in shape of 0th order Bessel function with peak density of 1.5×10^{13} cm^{-3}.

3.1.3 Finite-difference time domain method for dispersive media

In Section 3.1.2, we mentioned a different numerical method which saves computer resources, but the FDTD method is more popular and well developed if they are sufficient. We note that, even if it is possible to use the FDTD method, it provides quasi-steady solution which is difficult to be detected as a completely steady state one; human judgement will be finally required. Here, we describe the ways how the FDTD method can be applicable to analysis of wave propagation in and around plasmas.

Maxwell equations are linearized according to Yee's Algorism (Yee, 1966), as used in a conventional FDTD method. In addition, to deal with frequent-dependent permittivity equivalently, equation (2.2.2) is combined with Maxwell equations (Young, 1994) in the similar dicretization manner. Here, we ignored a pressure-gradient term from the general momentum balance equation in equation (2.2.1) because the pressure-gradient term is in the order of 10^{-7} of the right hand side of equation (2.2.2), although we have to treat it rigorously when electron temperature is quite high or when electromagnetic waves propagate with very short wavelength in the vicinity of resonance conditions. The boundary layers on the edges of the calculation area is set to be in Mur's second absorption boundary condition if the waves are assumed to be absorbed, and also we can use the Bloch or Froquet theorem described in equations (3.1.9) and (3.1.10) to assure the spatial periodic structure.

Figure 4 shows one example of the calculated results using this FDTD method. The peak n_e value in each plasma column assures the condition with $\omega < \omega_{pe}$ in which surface waves can propagate, as mentioned in Section 3.2. The launched waves from the lower side propagate along the chain structure of the isolated plasmas, and the fields are not inside plasmas but around them, similar to localized surface plasmons in the photon range along metal nanoparticles (Maier et al., 2002). That is, using this method, not only n_e profiles in one plasma but also the entire configuration surrounding plasma structures can be handled easily, although the limitation mentioned above reminds us of cross checking of the calculated results by other methods.

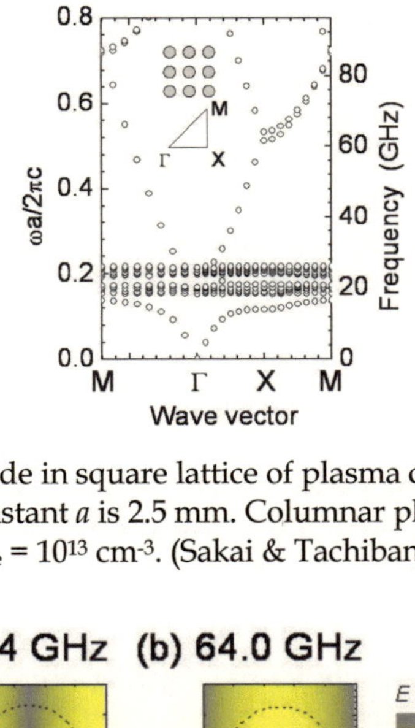

Fig. 5. Band diagram of TE mode in square lattice of plasma columns by direct complex-amplitude method. Lattice constant a is 2.5 mm. Columnar plasma with 1.75 mm in diameter is collisionless and $n_e = 10^{13}$ cm^{-3}. (Sakai & Tachibana, 2007)

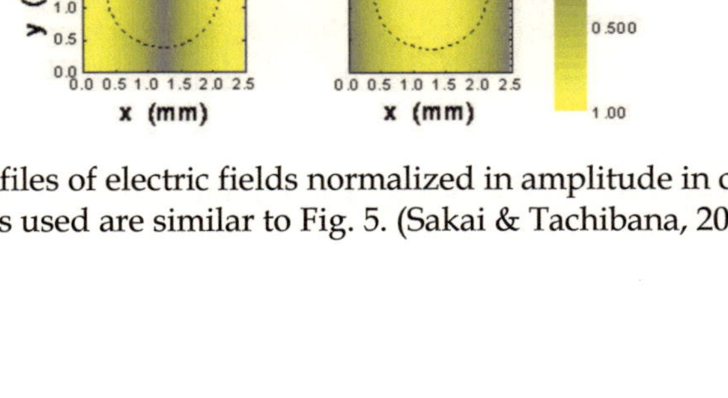

Fig. 6. Calculated profiles of electric fields normalized in amplitude in case of $k_x a/2\pi = 0.50$ and $k_y = 0$. Parameters used are similar to Fig. 5. (Sakai & Tachibana, 2007)

3.1.4 Examples of wave propagation in periodic plasma structures

Now, we demonstrate some specific examples of electromagnetic wave propagation (Sakai & Tachibana, 2007), especially with coupling of surface waves on the interface of 2-dimensional structures. The method to derive field profiles as well as band diagrams is the direct complex-amplitude (DCA) method, shown in Section 3.1.2.

Figure 5 shows a band diagram of a columnar plasma 2D array, derived by DCA method. The 2D plane was discretized into 20 × 20 meshes in one square-shaped lattice cell. This band diagram clarified typical features of 2D plasma photonic crystals, such as the band gaps, the flat bands, and the Fano mode. We also successfully obtained a case with a gradient electron density profile (Sakai et al., 2009) , in which the width of the flat band range increases due to lower density region in the periphery without changing other propagation properties; this mechanism of the wider flat band range is investigated in Section 3.2.

A unidirectional band gap in the $\Gamma - X$ direction, which lies around 61 GHz in Fig. 5, is reviewed in the following. Forbidden propagation is enhanced due to anisotropic wave propagation in the vicinity of the band gap (Sakai et al., 2007(2)). Figure 6(a) and (b) shows the electric field profiles around the plasma columns obtained as subproducts of the band calculation shown in Fig. 5 by DCA method. The electric fields showed different patterns just below or above this band gap; their amplitude was smaller in the plasma region than in the outer area just below the band gap (61.4 GHz), but their maximum region spreads over the center of the plasma just above the band gap (64.0 GHz). These structures were similar to 1D standing waves, and effects of the plasma with circular cross section were ambiguous.

Next, we focus on properties of the flat bands. As shown in Figs. 3 and 5, the flat bands with very low group velocity region are present below $\omega_{\mathrm{pe}} / 2\pi$ = 28.4 GHz. Such a wide frequency range arises from both localized surface modes and their periodicity.

Surface modes around a metal particle were well investigated in the photon frequency range (Forstmann & Gerhardts, 1986). When electromagnetic waves encounter an individual metal particle smaller than the wavelength, they are coupled with localized surface modes called "surface plasmon polalitons." Their maximum frequency spectrum is at $\omega_{\mathrm{pe}} / \sqrt{(1+\varepsilon_{\mathrm{d}})}$, where ε_{d} is the permittivity of the dielectric medium surrounding the metal particle. The localized surface modes have azimuthal (angular) mode number l around the particle, and l becomes larger as the frequency approaches $\omega_{\mathrm{pe}} / \sqrt{(1+\varepsilon_{\mathrm{d}})}$, which corresponds to ~20 GHz in Fig. 5.

In our case, however, structure periodicity complicates the problem. Recently, several reports about metallic photonic crystals (Kuzmiak & Maradudin, 1997; Ito & Sakoda, 2001; Moreno et al., 2002; Torder & John, 2004; Chern et al., 2006) have dealt with this issue. We investigated the electric field profiles calculated by DCA method along the band branches to clarify the roles of surface plasmons and their periodic effects. Figure 7 shows several amplitude profiles of electric fields in the propagating waves in the 2D columnar plasmas, with the same parameters as Figs. 5 and 6.

Electric fields in the Fano mode, present below the flat band region, are shown in Fig. 7(a). The amplitude of the electric field inside the columnar plasma was very small, and most of the wave energy was uniformly distributed and flowed outside the plasma. As we mentioned earlier, this wave branch coalesces with the flat bands at their lowest frequency as the frequency increases.

Electric fields of the waves on flat bands are shown in Fig. 7(b)–(h). A clearly different point from Fig. 7(a) is that the electric fields were localized on the boundary between the plasma and the vacuum. Another unique feature was the change of l of the standing waves around the plasma column. At lower frequency, l around the plasma column was low, and it became multiple at a higher frequency. This tendency is consistent with the general phenomena of surface plasmons around a metal particle. The highest l number (~6) was observed around 20 GHz, as shown in Fig. 7(e), and this frequency was approximately in the condition of $\omega_{pe}/\sqrt{2}$ which agrees with the predicted top frequency of the surface plasmon around a metal sphere ($\omega_{pe}/\sqrt{(1+\varepsilon_d)}$ in the case where the surrounding medium is a vacuum ($\varepsilon_d = 1$)).

However, the sequence of l along the frequency axis was not perfect for the surface waves around an individual metal sphere in the array. Above 20 GHz in Figs. 5 and 7 there are some flat bands, separated from the group below 20 GHz. In this group, however, no sequential change of l was found in Fig. 7. This might arise from the periodicity, as suggested by Ito and Sakoda (Ito & Sakoda, 2001). That is, Fig. 7(f)–(h) shows a different tendency from that below 20 GHz, and these electric field profiles imply that surface wave modes are localized in the gap region of the adjacent plasma columns and no boundary condition for standing waves around the column affects them. Note that this group of the flat bands above 20 GHz was hardly detected using the modified plane-wave method described in Section 3.1.1, as shown in Ref. (Sakai & Tachibana, 2007), where the region with no detection of flat bands ranges from 20 GHz to $\omega_{pe}/2\pi$. The structures of wave propagation are too fine to be detected in the modified plane-wave method, and therefore, an increase of assumed plane waves might be required to detect them in this method.

In summary, wave propagation on the flat bands of a 2-D columnar plasma array is mainly attributed to the dispersion of the localized surface modes around an individual columnar plasma and is modified by periodicity in the plasma array. These phenomena analogically resemble light waveguides composed of metal nanoparticle chains (Maier et al., 2002). The property observed in the aforementioned calculations will be applied to the dynamic waveguide of the electromagnetic waves composed of localized surface modes, similar to that shown in Fig. 4, since flat bands can intersect with the wave branches with various characteristic impedances.

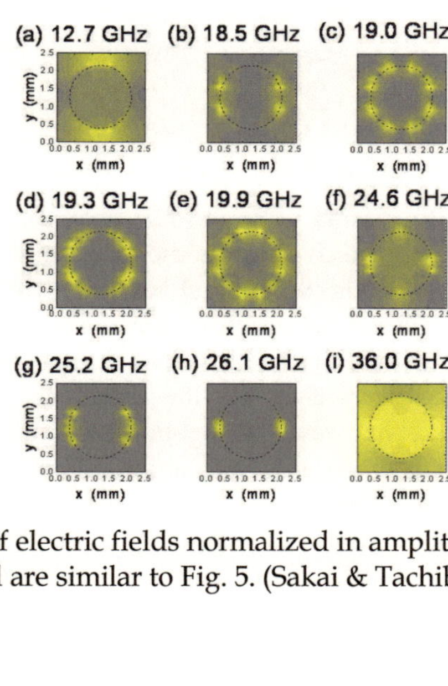

Fig. 7. Calculated profiles of electric fields normalized in amplitude in case of $k_x a/2\pi = 0.25$ and $k_y = 0$. Parameters used are similar to Fig. 5. (Sakai & Tachibana, 2007)

So far, we have investigated wave propagation in an array of plasma columns. The next target is antiparallel structure, which is an infinite-size plasma with periodic *holes*. Above $\omega_{pe}/2\pi$, periodic dielectric constant in space will contribute to form a similar band diagram. When the frequency is low enough, since there is no continuous vacuum space in this structure, wave propagation below $\omega_{pe}/2\pi$ is considered difficult from the first guess of the wave-propagation theory in a bulk plasma.

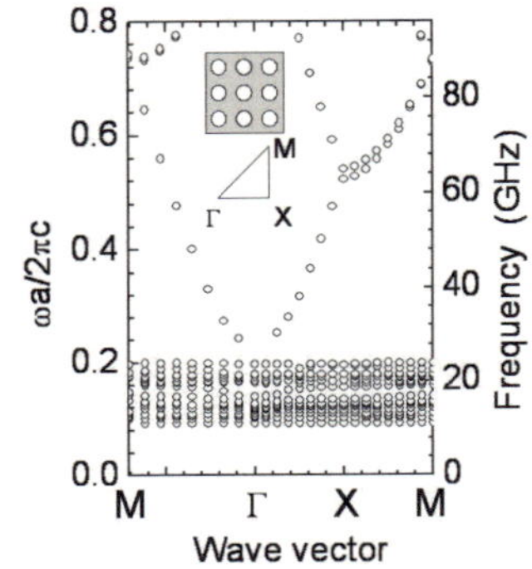

Fig. 8. Band diagram of TE mode in square lattice of plasma holes by direct complex-amplitude method. Lattice constant a is 2.5 mm. Circular holes with 1.75 mm in diameter are in a collisionless infinite plasma with $n_e = 10^{13}$ cm^{-3}. (Sakai & Tachibana, 2007)

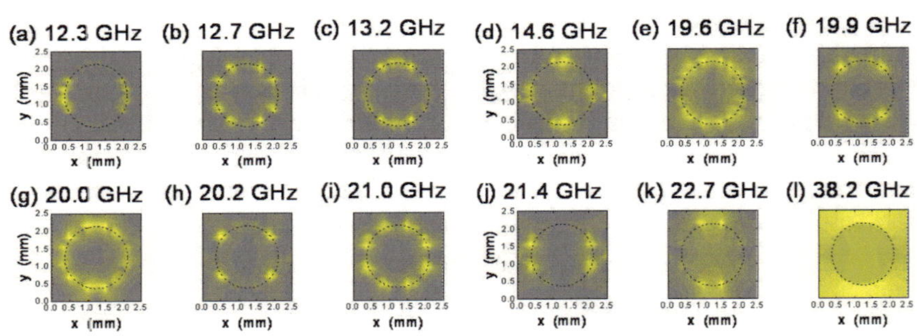

Fig. 9. Calculated profiles of electric fields normalized in amplitude in case of $k_x a/2\pi = 0.25$ and $k_y = 0$. Parameters used are similar to Fig. 8. (Sakai & Tachibana, 2007)

A band diagram of the infinite plasma with periodic holes, calculated by DCA method, is shown in Fig. 8. The basic features are common to the diagram in Fig. 5, and several different points from Fig. 5 can be found in Fig. 8. The first band-gap frequency in the Γ−X direction was slightly higher, since the filling fraction of the plasma region in one lattice cell in Fig. 8 (0.62) was larger than that in Fig. 5 (0.38) and it reduced the synthetic dielectric constant above $\omega_{pe}/2\pi$. No Fano mode was present in the low frequency region since there was no continuous vacuum region. Note that wave propagation remained below $\omega_{pe}/2\pi$ and the flat band region expanded to lower frequencies, which are examined in the following.

Figure 9 shows the electric field profiles in one lattice cell at various frequencies. In this case, no clear dependence of the azimuthal mode number l on the frequency was found; for instance, l = 1 at 12.3 GHz, l = 4 at 12.7 GHz, and l = 2 at 14.6 GHz. The path for the wave energy flow is limited to four points from the adjacent lattice cells through the short gap region between holes, and a plasma hole works as a wave cavity. Furthermore, conditions for standing eigenmodes along the inner surface of the hole are also required. In contrast, in the case of the columnar plasmas in Fig. 7, wave energy freely flows around the column, and therefore, wave patterns fulfill eigenmode conditions around the plasma columns and their periodicity. These facts yield differences between the cases of columnar plasmas and plasma holes.

It is difficult to express the penetration depth of the electromagnetic waves in surface plasmon in a simple formula (Forstmann & Gerhardts, 1986), but here, for the first approximation, we estimate usual skin depth δ_s on the plasma surface with a slab n_e profile instead. We use the well-known definition in a collisionless plasma as $\delta_s = c / \omega_{pe}$, where c is the velocity of light, and δ_s is 1.7 mm using the assumed n_e value in the aforementioned calculation as 10^{13} cm^{-3}. Since this value is comparable to the size and the gap of the plasma(s) in the aforementioned calculation, no wave propagation is expected in the normal cases in the cutoff condition. That is, wave propagation in the case of the hole array is supported not only by tunnelling effects but also by resonant field enhancement on the boundary that can amplify the local fields that strongly decay in the plasma region but couple with those in adjacent cells as near fields.

Using metals and waves in the photon range, similar phenomena will be found when holes are made in the 2-D lattice structure in the bulk metal, and waves propagate along this 2-D plane. In that case, some amount of light will pass through the metal in the usual cutoff condition; *opaque* material will become *transparent* to a certain extent, although damping by electron collisions will be present in the actual metallic materials.

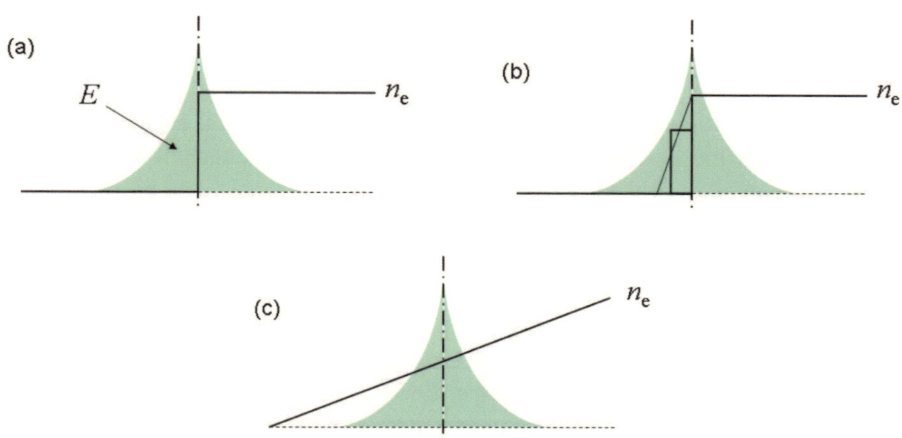

Fig. 10. Schematic view of surface waves on various models. (a) Model for ideal metal surface. (b) Bulk selvage model for metal surface. (c) Model for a discharge plasma.

3.2 Surface wave propagation in a plasma with spatially gradient electron density

In Section 3.1.4, several features of the localized surface waves in plasma periodic structures have been demonstrated. Some features are in common with the cases of light propagation on metal particles, but others are not; in this section, we clarify the different points from the surface waves or the surface plasmon polaritons on metal surfaces.

Figure 10 displays schematic views of surface waves and n_e profiles in both plasma and metal cases. In most cases of metal surfaces, since a n_e profile is almost similar to a slab shape, analysis of surface waves is rather easy, and surface plasmon polaritons have been well understood so far. On the other hand, in the plasma case, characteristic length of n_e is much larger than the presence width of the density gradient. This point is identical to plasma surface waves, although rigorous reports of these waves have been very few (Nickel et al., 1963; Trivelpiece & Gould, 1959; Cooperberg, 1998; Yasaka & Hojo, 2000). Here, we describe these waves using analytical approaches (Sakai et al., 2009).

The plasma is assumed to be infinite in the half space for the spatial coordinate $z<0$ with vacuum region for $z>0$. Since we deal with wave propagation, a variable x has two components as

$$x = x_0 + x_1, \tag{3.2.1}$$

where subscripts 0 and 1 correspond to static and fluctuating (wave-field) parts, respectively, equation (2.2.6) in the fluid or the hydrodynamic model is rewritten as

$$\frac{\partial \boldsymbol{J}_1(z)}{\partial t} = \varepsilon_0 \omega_{\mathrm{pe0}}(z)^2 \boldsymbol{E}_1(z) - \frac{ekT_e}{m} \nabla n_{\mathrm{e1}}(z). \tag{3.2.2}$$

Here, Poisson's equation given as

$$\nabla^2 \varphi_1(z) = -\rho_1 / \varepsilon_0 \tag{3.2.3}$$

with $\boldsymbol{E}_1 = -\nabla \varphi_1$ and continuity equation given as

$$\nabla \cdot \boldsymbol{J}_1 + \mathrm{d}\rho_1 / \mathrm{d}t = 0 \tag{3.2.4}$$

are coupled with equation (3.2.2), where φ is the electric potential and ρ the amount of charge. We also assume electron temperature $T_{\mathrm{e}} = 1$ eV as a constant value. To make it possible to obtain an analytical solution, a specific n_{e} profile is assumed as

$$\omega_{\mathrm{pe0}}(z)^2 = \omega_{\mathrm{pe0}}(-\infty)^2 \left(1 - \cosh^{-2}(\alpha_1 z)\right)$$

for $z<0$ and $\omega_{\mathrm{pe0}}(z)^2 = 0$ for $z \geq 0$, in the similar manner to the previous studies (Eguiluz & Quinn, 1976; Sipe, 1979), where α_1 represents density gradient factor, and solutions are derived using the similar method in Ref. (Sipe, 1979).

Here we point out common and different properties between metals and plasmas deduced from this model. Figure 11(a) shows analytical dispersion relations including two lowest-order multipole modes on a plasma half space with n_{e} gradient region characterized by $\alpha_1 = 40$ cm^{-1}, where ω_n is the eigen frequency of the multipole mode number n. There should be a number of multipole modes with every odd number n, and the two lowest cases ($n=1$ and 3) are displayed in Fig. 11(a). Higher multiple modes can exist as long as the density decay region works as a resonance cavity. A branch similar to the ordinary surface plasmon with the resonance frequency of $\omega_{\mathrm{sp}} = \omega_{\mathrm{pe}} / \sqrt{2}$ is observed, and the two multipole modes are located at much lower frequencies than ω_{sp}.

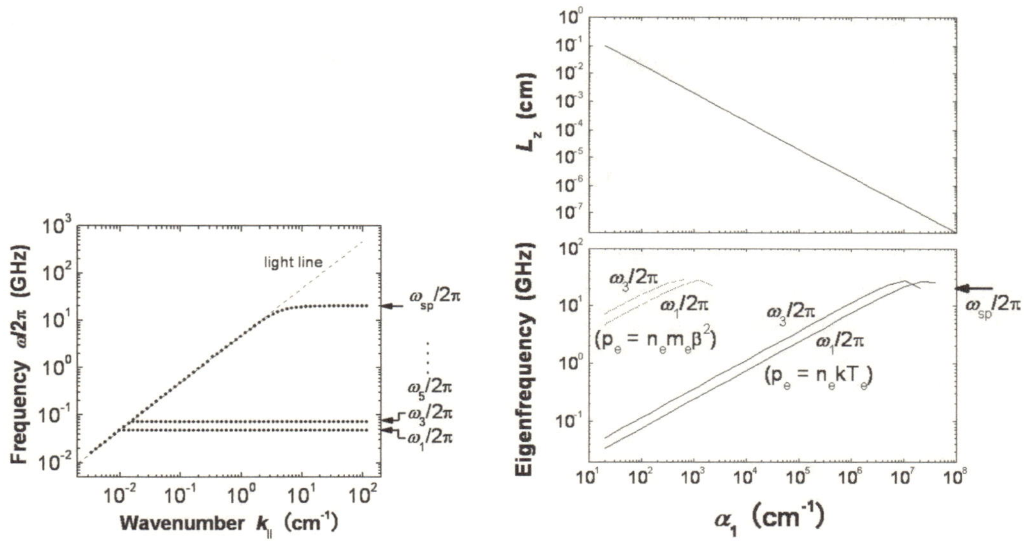

(a) (b)

Fig. 11. Analytically calculated dispersion relations of surface waves propagating along a surface of a plasma half space with a gradual electron density profile. (a) Dispersion relations with α_1 = 40 cm^{-1} and $\omega_{pe}/2\pi$ ~ 28 GHz. (b) Dependence of gradient parameter α_1 on length of density gradient L_z in the top figure and eigenfrequencies of the two lowest order in the bottom figure. Eigenfrequencies are plotted for two different pressure terms.

As previously described in Section 2.2, in a usual metal, parameter $m\beta^2$ is much larger than kT_e, which yields significant differences for dispersion relations of surface wave modes between plasma and metal cases. One of them is expressed in Fig. 11(b), which indicates the difference of frequency region of the surface wave modes. The top figure of Fig. 11(b) shows approximate length of gradient region L_z as a function of the parameter α_1. From the bottom part of Fig. 11(b), at one value of α_1, the frequency range of the surface wave modes (from ω_1 to ω_{sp}) in the case of gas-discharge plasmas is much larger than that in the case of metals with $\beta = 0.85 \times 10^8$ cm/sec. That is, not only inherent density gradient on the edge but also accelerating factor by the difference of the pressure term widens frequency region of the surface wave modes in a gas-discharge plasma.

Up to now, we have concentrated on surface waves on an infinite flat interface. Usually the excitation of surface waves on such a flat surface requires some particular methods such as ATR configuration or periodic structure like fluctuating surface. If we generate an isolated plasma from the others whose size is less than the wavelength, localized surface waves can be excited through electromagnetic waves in a free space, as shown in Section 3.1.4. In this case, we also observe similar wave propagation in comparison with the case of the flat surface (Sakai et al., 2009); the spectra of the waves propagating along the chain structure of the isolated plasmas with spatial n_e gradient are much wider than that without the density gradient. That is, using such inherent property of the density gradient with the pressure term determined by electron temperature, we expect a very wide range waveguide composed of plasma chains; an example was demonstrated in Fig. 4.

Figure 12 shows conceptual dispersion relations of surface wave modes on surfaces of isolated plasmas, as a summary of the discussion here. In a case of the gradual profile of n_e shown in Fig. 12(a), the propagating modes are on $\omega - z$ plane. Wave fields are distributed around the condition of $\varepsilon = 0$, i.e., on the layer with $\omega = \omega_{pe}$, and localized in a narrower region whose width is less than δ_s. Their frequency region is very wide and the surface modes can be present at frequencies much lower than $\omega_{pe}(0)$ by one or two orders. On the other hand, In a case of the slab profile of n_e shown in Fig. 12(b), the propagating modes are on $\omega - \varepsilon$ plane with narrow permittivity region (e.g., $-2 < \varepsilon < -1$). Wave fields are distributed around a surface of solids, i.e., $z = 0$ in Fig. 12(b), where n_e is discontinuous, and expand in a spatial range approximately equal to δ_s. Such newly-verified features of surface wave modes on small gas-discharge plasmas will open new possibilities of media for electromagnetic waves such as plasma chains demonstrated in Fig. 4 and spatially narrow waveguide on a n_e-gradient plasma surface.

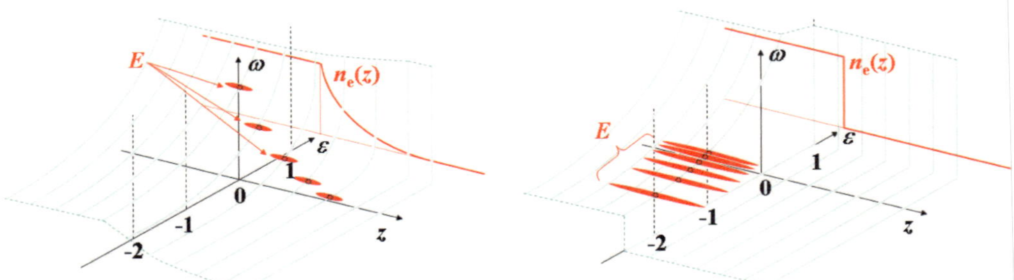

Fig. 12. Summary of dispersion relations of surface wave modes with two different electron density profiles. (a) Case of a gradual density profile. (b) Case of a slab density profile.

4. Concluding remarks

In this chapter, we investigate emerging features of electromagnetic wave propagation when we consider spatial structures of plasmas with complex permittivity. We derived the complex permittivity and introduce its drawing technique. We also obtained several methods to derive propagation of waves in two-dimensional plasma structures, and analytical solution of surface waves with the effects of significant n_e gradient. Combining these results, we verified wave propagation as localized surface modes. Clearly, the properties of wave propagation are different from those of surface waves on metals as well as those in waves propagating solid photonic crystals. These fundamentals will be applicable to various physical approaches as well as technological applications for control of electromagnetic waves.

5. References

Chern, R.L., Chang, C.C. & Chang, C.C. (2006). Analysis of surface plasmon modes and band structures for plasmonic crystals in one and two dimensions, *Phys. Rev. E*, vol. 73: 036605-1-15

Cooperberg, D. J. (1998). Electron surface waves in a nonuniform plasma lab, *Phys. Plasmas*, Vol. 5: 862-872

Dong, L., He Y., Liu, W., Gao R., Wang, H. & Zhao, H. (2007). Hexagon and square patterned air discharges, *Appl. Phys. Lett.* Vol. 90: 031504-1-3

Eguiluz, A. & Quinn, J. J. (1976). Hydrodynamic model for surface plasmons in metals and degenerate semiconductor, *Phys.Rev. B,* Vol. 14: 1347-1361

Faith, J., Kuo, S.P. & Huang, J. (1997). Frequency downshifting and trapping of an electromagnetic wave by a rapidly created spatially periodic plasma *Phys. Rev. E,* Vol. 55: 1843-1851

Fan W. & Dong L. (2010). Tunable one-dimensional plasma photonic crystals in dielectric barrier discharge, *Phys. Plasmas,* Vol. 17: 073506-1-6

Forstmann, F. & Gerhardts, R. R. (1986). *Metal Optics Near the Plasma Frequency,* Springer-Verlag, Berlin

Fukuyama, A., Goto, A., Itoh, S.I. & Itoh, K. (1983). Excitation and Propagation of ICRF Waves in INTOR Tokamak, *Jpn. J. Appl. Phys.*, Vol. 23: L613-L616

Ginzburg, V.L. (1964). *The Propagation of Electromagnetic Waves in Plasma,* Pergamon Press, Oxford

Guo, B. (2009). Photonic band gap structures of obliquely incident electromagnetic wave propagation in a one-dimension absorptive plasma photonic crystal, *Phys. Plasmas,* Vol. 16: 043508-1-6

Ho, K.M., Chan, C.T. & Soukoulis, C.M. (1990). Existence of a photonic gap in periodic dielectric structures, *Phys. Rev. Lett.* Vol. 65: 3152-3155

Hojo, H. & Mase A. (2004). Dispersion relation of electromagnetic waves in one-dimensional plasma phonic crystals, *J. Plasma Fusion Res.* Vol. 80: 89-90.

Isihara, A. (1993). *Electron Liquids,* Springer-Verlag, Berlin

Ito, T. & Sakoda, K. (2001). Photonic bands of metallic systems. II. Features of surface plasmon polaritons, *Phys. Rev. B,* vol. 64: 045117-1-8

Kalluri, D.K. (1998). *Electromagnetics of Complex Media,* CRC Press, Boca Raton

Kuzmiak, V. & Maradudin, A.A. (1997). Photonic band structures of one- and two-dimensional periodic systems with metallic components in the presence of dissipation, *Phys. Rev. B* Vol. 55: 7427-7444

Lee, H.I. & Mok, M. (2010). On the cubic zero-order solution of electromagnetic waves. I. Periodic slabs with lossy plasmas, *Phys. Plasmas,* Vol. 17: 072108-1-9

Lieberman, M.A. & Lichtenberg, A.J. (1994). *Principles of Plasma Discharges and Materials Processing,* John Wiley & Sons, New York

Lo, J., Sokoloff, J., Callegari, Th., & Boeuf, J.P. (2010). Reconfigurable electromagnetic band gap device using plasma as a localized tunable defect, *Appl. Phys. Lett.* Vol. 96: 251501-1-3

Maier, S.A., Brongersma, M.L., Kik, P.G. & Atwater, H.A. (2002). Observation of near-field coupling in metal nanoparticle chains using far-field polarization spectroscopy, *Phys. Rev. B,* vol. 65: 193408-1-4

Moreno, E., Erni, D. & Hafner, C. (2002). Band structure computations of metallic photonic crystals with the multiple multipole method, *Phys. Rev. B,* vol. 65: 155120-1-10

Naito, T., Sakai, O. & Tachibana, K. (2008). Experimental verification of complex dispersion relation in lossy photonic crystals, *Appl. Phys.Express* Vol. 1: 066003-1-3.

Nickel, J. C., Parker, J. V. & Gould, R. W. (1963). Resonance Oscillations in a Hot Nonuniform Plasma Column, *Phys. Rev. Lett.* Vol. 11: 183-185.

Nishikawa, K. & Wakatani, M. (1990). *Plasma Physics,* Aspringer-Verlag, Berlin

Noda, S. & Baba T. (ed.) (2003). *Roadmap on Photonic Crystals,* Kluwer Academic, Boston

Phihal, M., Shambrook, A., Maradudin, A.A. (1991) and P. Sheng, Two-dimensional photonic band structure, *Opt. Commun.* Vol. 80: 199-204

Pozar, D.M. (2005). *Microwave Engineering (3rd ed.),* John Wiley & Sons, Hoboken

Qi, L., Yang, Z., Lan, F., Gao, X. & Shi, Z. (2010). Properties of obliquely incident electromagnetic wave in one-dimensional magnetized plasma photonic crystals, *Phys. Plasmas,* Vol. 17: 042501-1-8

Razer, Y. P. (1991). *Gas Discharge Physics,* Springer-Verlag, Berlin

Sakaguchi, T., Sakai, O. & Tachibana, K. (2007). Photonic bands in two-dimensional microplasma arrays. II. Band gaps observed in millimeter and sub-terahertz ranges, *J. Appl. Phys.* Vol. 101: 073305-1-7

Sakai, O., Sakaguchi, T. & Tachibana, K. (2005(1)). Verification of a plasma photonic crystal for microwaves of millimeter wavelength range using two-dimensional array of columnar microplasmas, *Appl. Phys. Lett.* Vol. 87: 241505-1-3

Sakai, O., Sakaguchi, T., Ito, Y. & Tachibana, K. (2005(2)). Interaction and control of millimetre-waves with microplasma arrays, *Plasma Phys. Contr. Fusion* Vol. 47: B617-B627

Sakai, O. & Tachibana, K. (2006). Dynamic control of propagating electromagnetic waves using tailored millimeter plasmas on microstrip structures, *IEEE Trans. Plasma Sci.,* vol. 34: 80-87

Sakai, O., Sakaguchi, T. & Tachibana, K. (2007). Photonic bands in two-dimensional microplasma arrays. I. Theoretical derivation of band structures of electromagnetic waves, *J. Appl. Phys.* Vol. 101: 073304-1-9

Sakai, O., Sakaguchi T. & Tachibana K. (2007(2)). Plasma photonic crystals in two-dimensional arrays of microplasmas, *Contrib. Plasma Phys.*, vol. 47: 96-102

Sakai, O. & Tachibana, K. (2007). Properties of electromagnetic wave propagation emerging in two-dimensional periodic plasma structures, *IEEE Trans. Plasma Sci.* Vol. 35:1267-1273.

Sakai, O., Naito, T. & Tachibana K. (2009). Microplasma array serving as photonic crystals and Plasmon Chains, *Plasma Fusion Res.* Vol. 4: 052-1-8.

Sakai, O., Naito, T. & Tachibana, K. (2010(1)). Experimental and numerical verification of microplasma assembly for novel electromagnetic media, *Physics of Plasmas*, vol. 17: 057102-1-9.

Sakai, O., Shimomura, T. & Tachibana, K. (2010(2)). Microplasma array with metamaterial effects, *Thin Solid Films*, vol. 518: 3444-3448

Sipe, J. E. (1979). The ATR spectra of multipole surface plasmons, *Surf. Sci.,* Vol. 84: 75-105

Stix, T.H. (1962). *The Theory of Plasma Waves,* McGraw-Hill, New York

Swanson, D.G. (1989). *Plasma Waves,* Academic Press, Boston

Toader, O. & John, S. (2004). Photonic band gap enhancement in frequency-dependent dielectrics, *Phys. Rev. E,* vol. 70: 046605-1-15

Trivelpiece, A. W. & Gould, R. W. (1959). Surface charge waves in cylindrical plasma columns, *J. Appl. Phys.,* Vol. 30: 1784-1793

Yablonovitch, E. (2000), How to be truly photonic, *Science* Vol. 289, 557-559.

Yasaka, Y. & Hojo, H. (2000). Enhanced power absorption in planar microwave discharges, *Phys. Plasmas,* Vol. 7: 1601-1605

Yee, K. (1966). Numerical solution of initial boundary value problems involving maxwell equations in isotropic media, *IEEE Trans. Antennas Propag.,* Vol. 14: 302-307

Yin, Y., Xu, H., Yu, M.Y., Ma, Y.Y., Zhuo, H.B., Tian, C.L. & Shao F.Q. (2009). Bandgap characteristics of one-dimensional plasma photonic crystal, *Phys. Plasmas,* Vol. 16: 102103-1-5

Young, J.L. (1994). A full finite difference time domain implementation for radio wave propagation in a plasma, *Radio Sci.,* Vol. 29: 1513-1522

Part 5

Electromagnetic Waves Absorption and No Reflection Phenomena

17

Electromagnetic Wave Absorption Properties of RE-Fe Nanocomposites

Ying Liu, LiXian Lian and Jinwen Ye
Sichuan University
China

1. Introduction

Recently, the number of communication devices that utilize gigahertz range microwave radiation, such as mobile phones and LAN systems, has greatly increased. However, electromagnetic interference (EMI) has become serious. One promising technique to prevent EMI is the use of microwave absorption materials. However, the applications of conventional microwave absorption materials are limited. The reasons are that Snoek's limit of spinel-tripe ferrites is so small that the imaginary part of permeability is considerably lowered in GHz range, and metallic soft-magnet materials have high electric conductivity, which makes the high frequency permeability decreased drastically due to the eddy current loss induced by EM wave.

The $Nd_2Fe_{14}B/\alpha$-Fe composites is composed of soft magnetic α-Fe phase with high M_S and hard magnetic $Nd_2Fe_{14}B$ phase with large H_A, consequently their natural resonance frequency are at a high frequency range and permeability still remains as a large value in high frequency range. Furthermore, the electric resistivity of $Nd_2Fe_{14}B$ is higher than that of metallic soft magnetic material, which can restrain the eddy current loss. Thus, the authors have already reported that $Nd_2Fe_{14}B/\alpha$-Fe composites can fuction as a microwave absorber. In this present work, the electromagnetic and absorption properties of the $Nd_2Fe_{14}B/\alpha$-Fe nanocomposites were studied in the 0.5–18 and 26.5–40 GHz frequency ranges. Moreover, the effect of rare earth Nd content on natural resonance frequency and microwave permeability of $Nd_2Fe_{14}B/\alpha$-Fe nanocomposites was reported in this chapter. The results show that it is possible to be a good candidate for thinner microwave absorbers in the GHz range.

In order to restrain the eddy current loss of metallic soft magnetic material, Sm_2O_3 and SmN was introduced in Sm_2O_3/α-Fe and SmN/α-Fe composites as dielectric phase, and $Sm_2Fe_{17}N_x$ with high magnetocrystalline anisotropy was introduced in SmN/α-Fe/$Sm_2Fe_{17}N_x$ as hard magnetic phase. Accordingly, Sm_2O_3/α-Fe and SmN/α-Fe/$Sm_2Fe_{17}N_x$ are possible to be another good candidate for microwave absorbers in the GHz range as the authors reported in reference. Therefore, the purpose of this study is to investigate the microwave complex permeability, resonant frequency, and microwave absorption properties of nanocrystalline rare-earth magnetic composite materials Sm_2O_3/α-Fe and SmN/α-Fe/$Sm_2Fe_{17}N_x$. The absorption performance and natural resonance frequency can be controlled by adjusting phase composite proportion and optimizing the microstructure.

II. Microwave Electromagnetic Properties of $Nd_2Fe_{14}B/\alpha$-Fe

1. Experiments

The compounds NdFeB alloys were induction-melted under an argon atmosphere. The ribbons were prepared by the single-roll melt-spun at a roll surface velocity of 26 m/s, and then annealed at 923-1023K for 8-20 min in an argon atmosphere. The annealed ribbons were pulverized for 10-30h using a planetary ball milling machine. X-ray diffraction (XRD) and transmission electron microscope (TEM) were used to determine the phases and microstructure of samples. The magnetic hysteresis loops were measured using a vibrating sample magnetometer (VSM). The alloy powders were mixed with paraffin at a weight ratio of 5:1 and compacted respectively into a toroidal shape (7.00 mm outer diameter, 3.01 mm inner diameter and approximately 3 mm thickness.) and rectangular shape (L×W= 7.2×3.6: corresponding to the size of various wave guide, thickness: 0.9 mm). The vector value of reflection/transmission coefficient (scattering parameters) of samples were measured in the range of 0.5-18 GHz and 26.5-40 GHz, using an Agilent 8720ET and Agilent E8363A vector network analyzer respectively. The relative permeability (μ_r) and permittivity (ε_r) values were determined from the scattering parameters and sample thickness. Assumed the metal material was underlay of absorber, and the reflection loss (RL) curves were calculated from the relative complex permeability and permittivity with a given frequency range and a given absorber thickness (d) with the following equations:

$$RL = 20\log\left|(Z_{in} - 1) / (Z_{in} + 1)\right| \tag{1}$$

$$Z_{in} = \sqrt{\mu_r / \varepsilon_r} \tanh\left\{ j(2\pi fd / c)\sqrt{\mu_r \varepsilon_r} \right\} \tag{2}$$

, where Z_{in} is the normalized input impedance at absorber surface, f the frequency of microwave, and c the velocity of light.

2. Microwave electromagnetic properties of $Nd_{10}Fe_{78}Co_5Zr_1B_6$

In the present work, $Nd_2Fe_{14}B/\alpha$-Fe microwave electromagnetic and absorption properties of $Nd_2Fe_{14}B/\alpha$-Fe were investigated in 0.5-18 and 26.5-40GHz range.

Fig.1 (a) and Fig.1 (b) show the XRD patterns of the $Nd_{10}Fe_{84}B_6$ melt-spun ribbons after subsequent annealing and ball milling respectively. The peaks ascribed to hard magnetic phase $Nd_2Fe_{14}B$ and soft magnetic phase α-Fe can be observed clearly. After ball milling, the diffraction peaks exhibit the wider line broadening, and any other phase has not been detected on the XRD patterns. It indicates the grain size gets finer by ball-milling. The average grain size is evaluated to be about 30nm for annealed ribbons and 20nm for the ball-milling one from the line broadening of the XRD peaks, using the Scherrer's formula. Fig.2. shows TEM micrograph and electron diffraction (ED) patterns of the heat treated $Nd_{10}Fe_{78}Co_5Zr_1B_6$ melt-spun ribbons. It can be seen that the grain size is uniform and the average diameter is around 30 nm. The results are consistent with the XRD analysis. Such a microstructure of magnetic phase is effective to enhance the exchange interaction between hard and soft magnetic phases.

Magnetic hysteresis loop for $Nd_2Fe_{14}B/\alpha$-Fe nanocomposites is shown in Fig.3. The value of saturation magnetization M_s and coercivity H_{cb} is 100.03 emu/g and 2435 Oe respectively, which is rather high compared with common soft magnetic materials such as hexaferrite - FeCo nanocomposite. Furthermore, the magnetic hysteresis loops are quite smooth, which shows the characteristics of single phase hard magnetic material. This result can be explained by the effect of exchange interaction between the hard-magnetic $Nd_2Fe_{14}B$ and soft-magneticα-Fe. Comparing with conventional ferrite materials, the $Nd_2Fe_{14}B$/α-Fe permanent magnetic materials has larger saturation magnetization value and its snoek's limit is at 30-40GHz. Thus the values of relative complex permeability can still remain rather high in a higher frequency range.

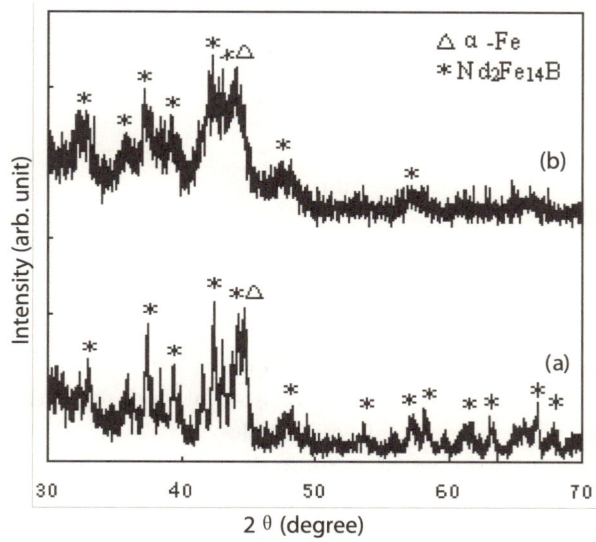

Fig. 1. XRD patterns of $Nd_{10}Fe_{78}Co_5Zr_1B_6$ composite melt-spun ribbons annealed at 973K for 8 min before (a) and after 25h milling (b)

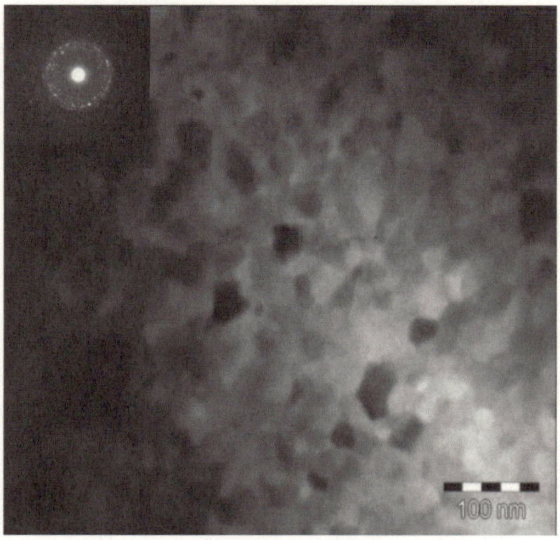

Fig. 2. TEM micrograph and diffraction patterns of the heat treated $Nd_{10}Fe_{78}Co_5Zr_1B_6$ melt-spun ribbons

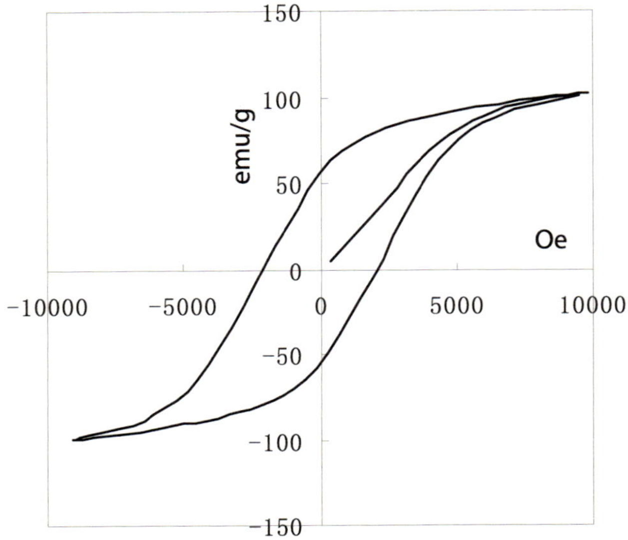

Fig. 3. Magnetic hysteresis loop for $Nd_2Fe_{14}B/\alpha$-Fe nanocomposite

Fig.4 shows the frequency dependence of the complex relative permeability and permittivity of $Nd_2Fe_{14}B/\alpha$-Fe composites. As shown in Fig.4 (a) and (b), that values of complex permittivity decrease with increasing frequency for $Nd_2Fe_{14}B/\alpha$-Fe composites in 0.5-18 GHz. However the imaginary part of permittivity ε_r'' exhibits a peak at 36 GHz. The dielectric constant of $Nd_2Fe_{14}B/\alpha$-Fe composites are higher than that of ferrites due to high electric conductivity of metal material α-Fe, and the dielectric loss plays an important role in microwave absorption property. The dielectric properties of $Nd_2Fe_{14}B/\alpha$-Fe composites arise mainly from the interfacial polarization induced by the large number of interface for nanocomposites. However low complex dielectric constant of $Nd_2Fe_{14}B/\alpha$-Fe composites is expected to satisfy the requirements of impedance matching. The permeability spectra of $Nd_2Fe_{14}B/\alpha$-Fe nanocomposites exhibits relaxation and resonance type characteristic in the 0.5-18 and 26.5-40 GHz frequency range respectively. The resonance frequency (f_r) of $Nd_{10}Fe_{78}Co_5Zr_1B_6$ nanocomposite is 30GHz due to the large anisotropy field (H_A). It is well known that the ferromagnetic resonance frequency (f_r) is related to its anisotropy fields (H_A) by the following relation:

$$2\pi f_r = \gamma H_A \tag{3}$$

,where γ is the gyromagnetic ratio. $Nd_2Fe_{14}B/\alpha$-Fe nanocomposites have a large anisotropy field H_A, and consequently their natural resonance frequency f_r is at a high frequency range. The resonance frequency of $Nd_2Fe_{14}B$ is calculated as 210GHz. However, the resonance frequency of this $Nd_2Fe_{14}B/\alpha$-Fe sample is lower than that of $Nd_2Fe_{14}B$, due to the decrease of H_A induced by the exchange interaction between hard and soft magnetic phases. Thus the observed resonance phenomena in Fig.4(c) can be attributed to the resistance to the spin rotational. And the ferromagnetic resonance plays an important role in the high frequency region.

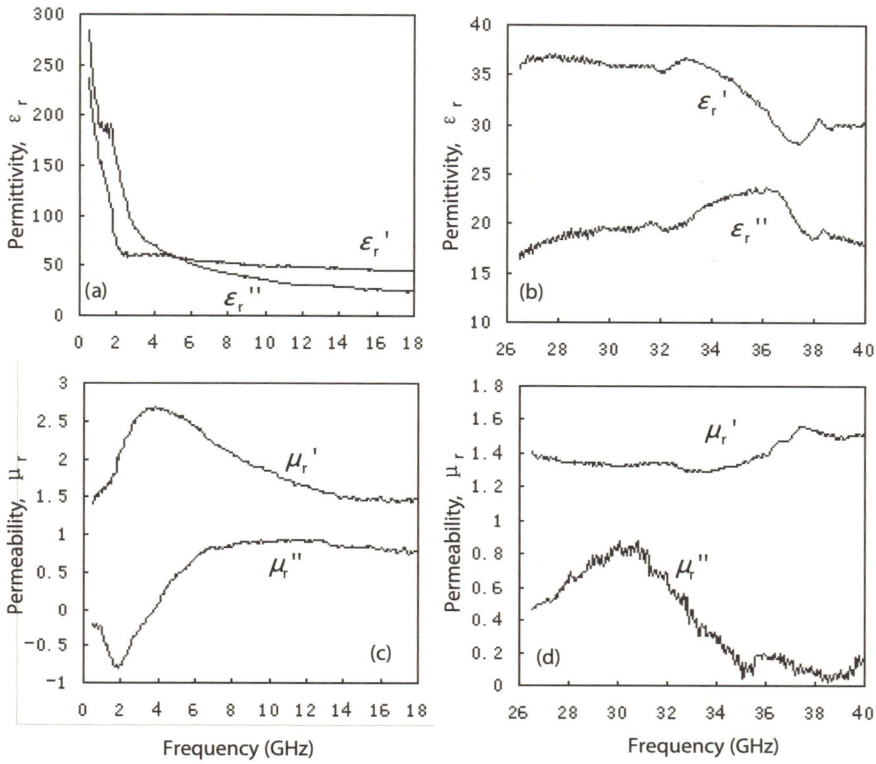

Fig. 4. The relative permittivity and permeability plotted against frequency for $Nd_2Fe_{14}B$/α-Fe composites in the 0.5-18 and 26.5-40GHz

The variation of reflection loss with frequency for composite is shown in Fig.5. This nanocomposite realized the optimum matching (reflection loss: RL < -20 dB) in 9, 17 GHz with thin matching thickness of 2, 1.2 mm respectively. Furthermore, the maximum microwave absorption -35 dB is obtained at 37 GHz with a thinner matching thickness (d_m) of 0.37 mm. Consequently, efficient EM absorption properties are observed not only in centimeter-wave band but also in millimeter-wave band.

The permeability spectra of $Nd_2Fe_{14}B$/α-Fe nanocomposites exhibits relaxation and resonance type characteristic in the 0.5-18 and 26.5-40 GHz frequency range respectively. The resonance frequency (f_r) of $Nd_{10}Fe_{78}Co_5Zr_1B_6$ nanocomposite is 30GHz. This nanocomposite also shows an excellent microwave absorption property (reflection loss: RL<-20dB) in 9, 17 GHz with thin matching thickness of 2, 1.2mm respectively, and the minimum peak of -35 dB appears at 37 GHz with a thin matching thickness (dm) of 0.37 mm.

3. Effect of Nd content on natural resonance frenquency and microwave permeability of $Nd_2Fe_{14}B$/α-Fe nanocomposites

The natural resonance frequency (f_r) is related to its anisotropy fields (H_A) by the expression (3).

$$2\pi f_r = \gamma H_A \tag{3}$$

where γ is the gyromagnetic ratio. And there is a relationship between the absorber thickness d_m and magnetic loss μ_r'' of absorbers by

$$d_m = c / 2\pi f_m \mu_r'' \tag{4}$$

where c is velocity of light and f_m is the matching frequency. Therefore, the magnetic materials which show higher μ_r'' values are suitable for the fillers of thinner microwave absorbers. However, the maximum μ_r'' value induced by natural resonance phenomenon is estimated using the saturation magnetization M_s and H_A as

$$\mu_r'' = M_s / 3\mu_0 H_A \alpha \tag{5}$$

where μ_0 is the permeability of vacuum state and α is Gilbert's damping coefficient. Consequently, d_m is inversely proportion to M_s from formulae (2) and (3), and it is effective to use a metal-based material with high M_s and adequate f_r values, such as $Nd_2Fe_{14}B$/α-Fe nanocomposites due to the high M_s of α-Fe (M_s = 2.15T) and the large H_A of $Nd_2Fe_{14}B$ (H_A =6.0MAm^{-1}). T. Maeda et al investigated the effect of exchange interaction between the hard-magnetic $Y_2Fe_{14}B$ and soft-magnetic Fe_3B on the resonance phenomenon. Kato et al. also reported a shift of the ferromagnetic resonance (FMR) frequency by changing the volume fraction of soft and hard phases in the $Nd_2Fe_{14}B$/α-Fe thin films. Therefore, it is possible to control the f_r values of $Nd_2Fe_{14}B$/α-Fe nanocomposites by changing the rare earth Nd content. Due to the effect of exchange interaction, nanocrystalline composites $Nd_2Fe_{14}B$/α-Fe magnet with high theoretical energy product $(BH)_{max}$ value attract much attention as permanent magnet.

In the present work, the effect of the rare earth Nd contents on the natural resonance frequency and microwave permeability of $Nd_2Fe_{14}B$/α-Fe nanocomposites was investigated. The $Nd_xFe_{94-x}B_6$ (x = 9.5, 10.5, 11.5) ribbons were prepared using melt-spinning and annealing method. The microwave complex permeability was measured in the 26.5-40 GHz frequency range.

Fig.6 shows the XRD patterns of the heat treated $Nd_xFe_{94-x}B_6$ melt-spun ribbons with different Nd contents. The peaks ascribed to hard magnetic phase $Nd_2Fe_{14}B$ and soft magnetic phaseα-Fe have been observed clearly. The average grain size D calculated by using Scherrer equation are about 30nm for $Nd_xFe_{94-x}B_6$ (x=9.5, 10.5, 11.5) composites. Furthermore, it is noticeable that the fraction of $Nd_2Fe_{14}B$ are gradually increased and the fraction of α-Fe are gradually decreased with the increasing of the Nd content based on checking the ratio of characteristic peaks intensity of $Nd_2Fe_{14}B$ to that of α-Fe. Thereby, the magnetic properties of $Nd_2Fe_{14}B$/α-Fe nanocomposite powder with different Nd content exhibit obvious differences as shown in Fig.7.

The values of remanent magnetization and coercivity are very high compared with soft magnetic materials, and the magnetic hysteresis loops are quite smooth. It behaves the characteristics of single hard magnetic material. This result can be explained by the effect of exchange interaction between the hard-magnetic $Nd_2Fe_{14}B$ and soft-magneticα-Fe. Fig.8. shows TEM micrograph and electron diffraction (ED) patterns of the heat treated $Nd_{9.5}Fe_{84.5}B_6$ melt-spun ribbons. It can be seen that the grain size is uniform and the average diameter is around 30 nm. The results are consistent with the XRD analysis. Such a

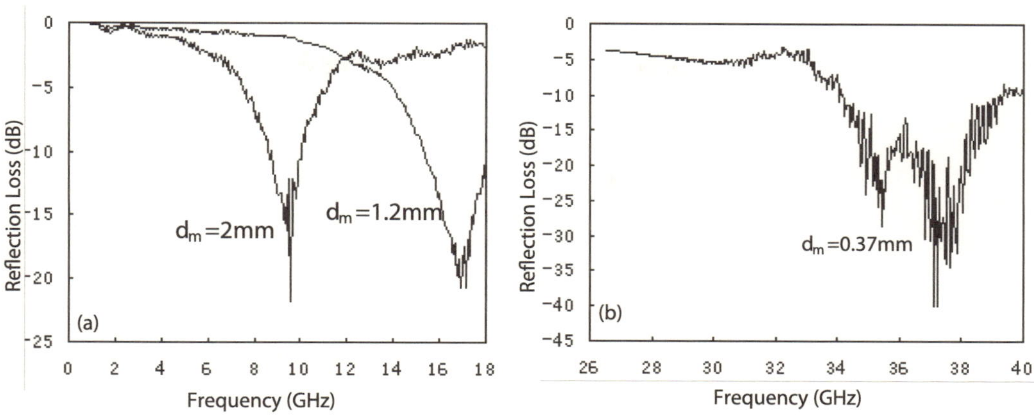

Fig. 5. Frequency dependences of RL of $Nd_2Fe_{14}B/\alpha$-Fe composite in (a) 0.5-18GHz and (b) 26.5-40GHz

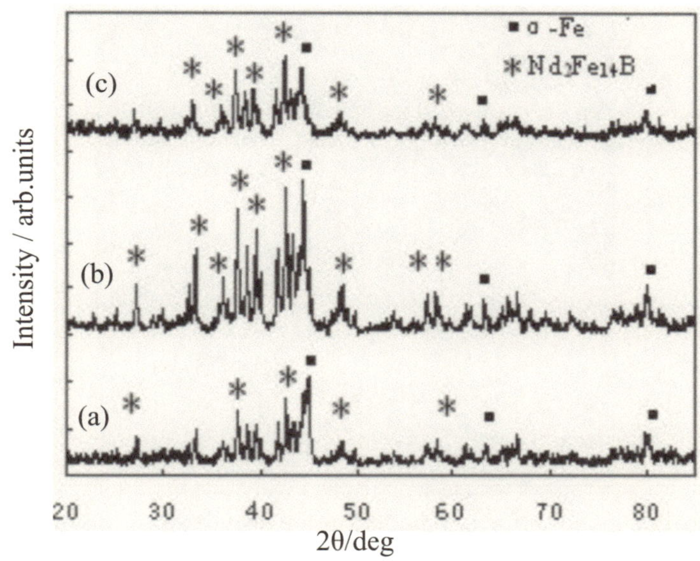

Fig. 6. XRD patterns of $Nd_xFe_{94-x}B_6$ (a) x=9.5, (b) x=10.5, and (c) x=11.5 melt-spun ribbons annealed at 973K for 8 min

microstructure of small grains of magnetic phase is available to enhance the exchange interaction between hard and soft magnetic phases. Because the exchange interaction is only efficient in surface shell, approximately within a diameter of the Block wall width δ_B, this extremely fine-grained microstructures is necessary to ensure that a considerable volume fraction of grain is affected by the exchange coupling.

At the same time, the saturation magnetization M_s is gradually decreased and the coercivity H_c is gradually increased with the increase of the Nd content, due to the decrease of volume fraction of soft-magneticα-Fe phases with high M_s and the increasing of the volume fraction of hard-magnetic $Nd_2Fe_{14}B$ phases with high H_A (see Fig.6).

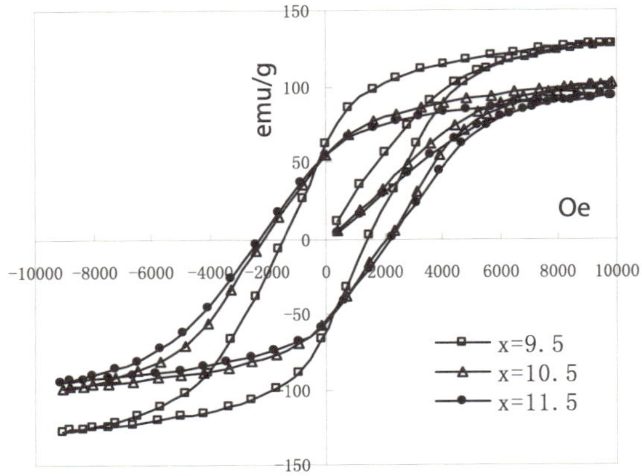

Fig. 7. Magnetic hysteresis loops for $Nd_2Fe_{14}B/\alpha$-Fe nanocomposites powder with different Nd content

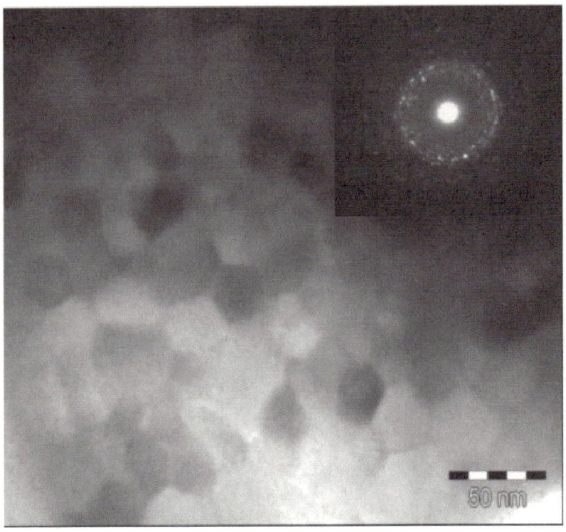

Fig. 8. TEM micrograph and diffraction patters of the heat treated $Nd_xFe_{94-x}B_6$(x=9.5) melt-spun ribbons

Because M_s and effective anisotropy constant K_{eff} are depended on the volume fraction of soft-magnetic phases f_S and that of hard-magnetic phases as shown in Eq. (6) and (7),

$$Ms = f_S M_S^S + (1 - f_S) M_S^H \tag{6}$$

$$\begin{aligned} K_{\text{eff}} &= \int \varphi(\gamma) K_1(\gamma) \varphi(\gamma) d\gamma \\ &= \langle K_1(\gamma) \rangle = f_S K_S + (1 - f_S) K_H \end{aligned} \tag{7}$$

where M_S^S and K_S are the saturation magnetization and anisotropy constant of soft-magnetic α-Fe phases respectively, M_S^H and K_H are the saturation magnetization and anisotropy constant of hard-magnetic $Nd_2Fe_{14}B$ phases respectively. Therefore microwave permeability and the resonance frequency f_r will exhibit obvious differences with different Nd content.

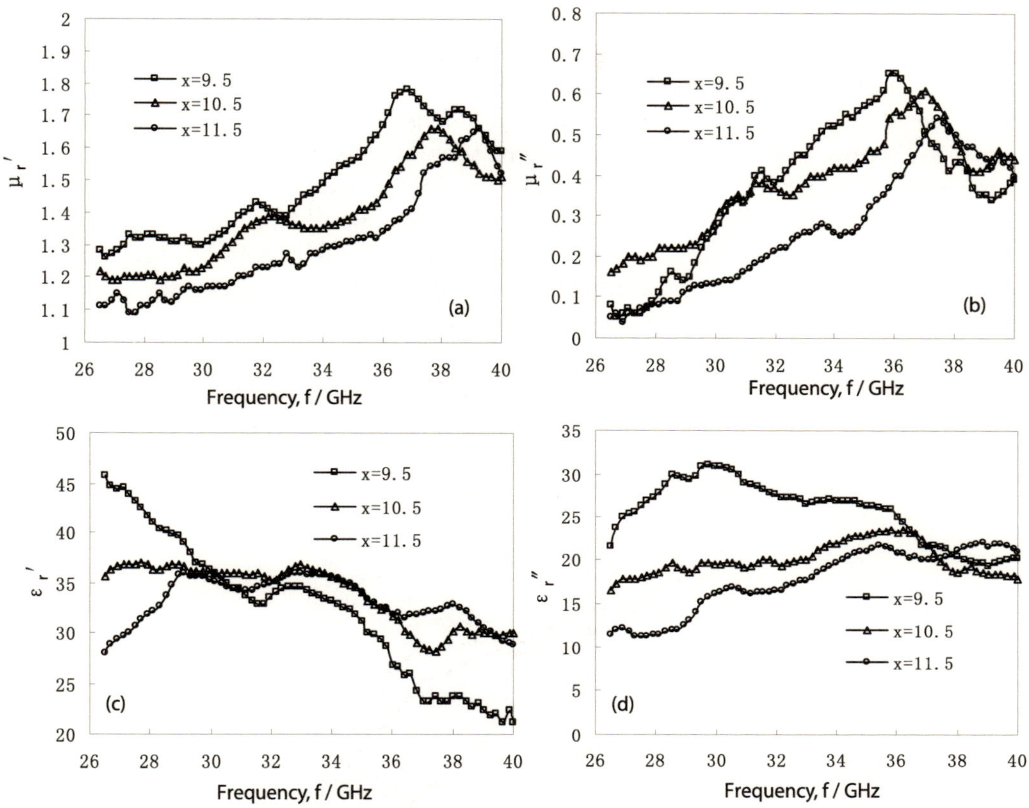

Fig. 9. The frequency dependencies of the complex relative permeability and permittivity of resin composites $Nd_xFe_{94-x}B_6$(x=9.5, 10.5, 11.5):
(a) real part μr′ of complex permeability; (b) imaginary part μ_r'' of complex permeability;

Fig.9 shows the frequency dependencies of the complex relative permeability and permittivity of resin composites $Nd_xFe_{94-x}B_6$(x=9.5, 10.5, 11.5). Compared with hexagonal W-type ferrites reported in reference, $Nd_2Fe_{14}B/\alpha$-Fe nanocomposites shows higher values in both the real $\mu_r'$and imaginary μ_r'' parts of permeability in the higher frequency region. These higher values are due to the larger magnetization of $Nd_2Fe_{14}B/\alpha$-Fe than ferrites. For Nd9.5Fe84.5B6 resin composites, the maximum of μ_r'' is 0.65 and the resonance frequency is around 36GHz. For $Nd_{10.5}Fe_{83.5}B_6$ and $Nd_{11.5}Fe_{82.5}B_6$ samples, the maxima of μ_r'' are 0.61 and 0.54, and f_r are 37GHz and 37.5GHz respectively. It shows that the μ_r'' values of $Nd_xFe_{94-x}B_6$ for x=10.5 and x=11.5 are smaller than that of $Nd_{9.5}Fe_{84.5}B_6$ sample, whereas the resonance frequency f_r of these two composites are higher than that of the $Nd_{9.5}Fe_{84.5}B_6$ sample. This interesting resonance phenomenon could be explained as follows. On the one hand, the volume fractions of soft-magneticα-Fe phase with higher M_s decreases with the increase of rare earth Nd content (shown in Fig.6), resulting in the decrease of μ_r'' based on

Eq. (5). On the other hand, the volume fractions of hard-magnetic $Nd_2Fe_{14}B$ phase with higher HA increases gradually. As a result, resonance frequency f_r shifts to a higher frequency range with the increase of rare earth Nd content, according to Eq. (3)..

$Nd_2Fe_{14}B/\alpha$-Fe composites have much larger H_A than common absorber materials such as ferrites and metal soft magnetic materials, and behave the characteristics of hard magnetic material. Therefore magnetic spectrum of $Nd_2Fe_{14}B/\alpha$-Fe composites shows some difference with other absorber materials, and the real μ_r' doesn't decrease with frequency in the resonance region as general rule. The detailed reasons are expected to be investigated.

It can be seen from Fig.9 (c) and Fig.9 (d), that values of the real part of complex permittivity ε_r' are found to decrease with increasing frequency for $Nd_xFe_{94-x}B_6$ composites. The imaginary part of permittivity ε_r'' exhibits a peak at 30GHz ,35GHz and 38GHz for $Nd_xFe_{94-x}B_6$(x=9.5, 10.5, 11.5) respectively. It can be seen that the dielectric constant are higher than ferrites, the dielectric loss play an important role in microwave absorption property. Thus, microwave absorption properties of $Nd_2Fe_{14}B/\alpha$-Fe composites depend on cooperate effect of magnetic loss and dielectric loss. The dielectric properties of $Nd_2Fe_{14}B/\alpha$-Fe composites arise mainly due to the interfacial polarization. It also shows that the complex dielectric constant is composition dependent. However low complex dielectric constant of $Nd_2Fe_{14}B/\alpha$-Fe composites is expected to satisfy the requirements of impedance matching.

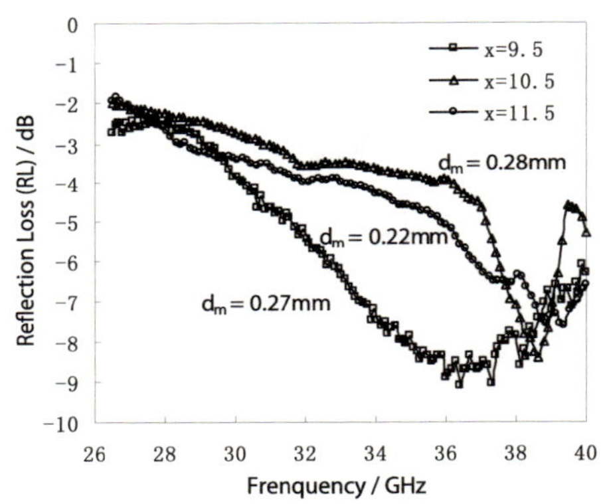

Fig. 10. Frequency dependence of RL for the resin composites NdxFe94-xB6 (x=9.5, 10.5, 11.5)

Finally, the RL of the resin composites $Nd_xFe_{94-x}B_6$ (x=9.5, 10.5, 11.5) are calculated from the microwave complex permeability and permittivity, and absorber thickness. Their frequency dependence is shown in Fig.10.The optimum matching condition is realized when absorber thickness is 0.27mm and a minimum RL value of -8.9dB is obtained at the f_m of 36GHz for the $Nd_{9.5}Fe_{84.5}B_6$ sample. For $Nd_{10.5}Fe_{83.5}B_6$ and $Nd_{11.5}Fe_{82.5}B_6$ composites, the f_m are 38.6GHz and 39.4GHz respectively, and higher than that of the $Nd_{9.5}Fe_{84.5}B_6$ sample. This is attribute to the increase in H_A due to the increase of hard-magnetic $Nd_2Fe_{14}B$ phases. This result is in good agreenment with the results from Fig.9. Further more, $Nd_2Fe_{14}B/\alpha$-Fe composites have a thinner matching thickness than ferrites absorber materials demonstrated by Y. J.

The microwave permeability and the frequency range of microwave absorption of $Nd_2Fe_{14}B/\alpha$-Fe nanocomposites can be controlled effectively by adjusting rare earth Nd content. Microwave permeability reduces and natural resonance frequency f_r shifts to a higher frequency with the increase of Nd content. $Nd_{9.5}Fe_{84.5}B_6$ resin composites shows the maximum μ_r'' of 0.65 at 36GHz and the maximum microwave absorption(RL=-8.9dB) is obtained at 36GHz with the matching thickness of 0.27mm. $Nd_2Fe_{14}B/\alpha$-Fe nanocomposites are promising microwave absorbers in the 26.5-40GHz frequency range.

4. Electromagnetic wave absorption properties of NdFeB alloys with low Nd content

In this section, the electromagnetic and absorption properties of NdFeB alloys with low Nd content comprised withα-Fe /$Nd_2Fe_{14}B$ nanocomposites were studied in the 0.5–18 GHz frequency ranges.Fig.11 shows the XRD patterns of $Nd_6Fe_{91}B_3$ melt-spun ribbons after annealing at 1073K for 15 min. The peaks ascribed to soft magnetic phaseα-Fe and hard magnetic phase $Nd_2Fe_{14}B$ have been observed clearly. The average grain size D calculated by using Scherrer equation are about 35nm forα-Fe phase.

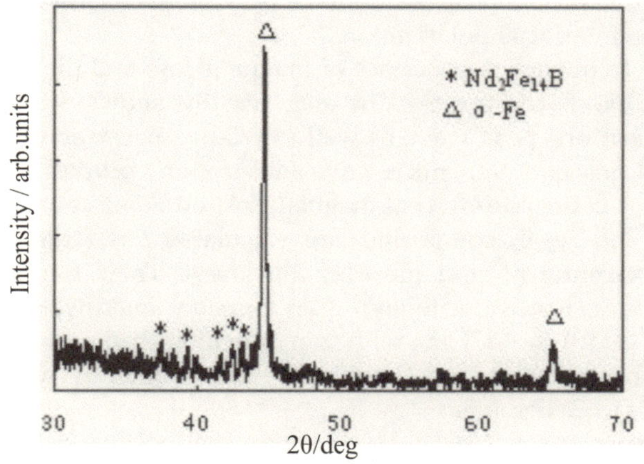

Fig. 11. XRD patterns of $Nd_6Fe_{91}B_3$ melt-spun ribbons annealed at1073K for 15 min

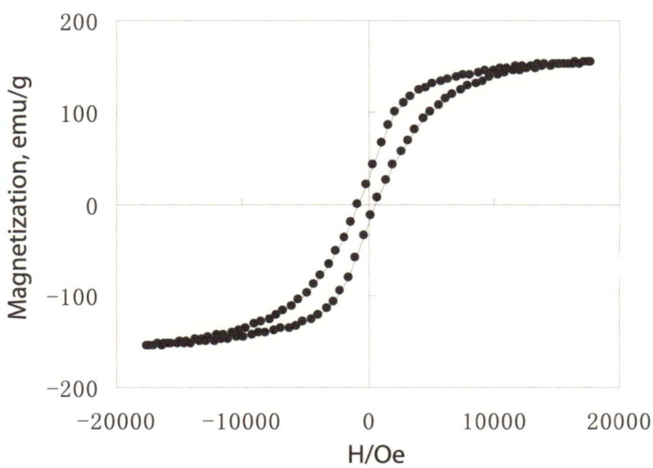

Fig. 12. Magnetic hysteresis loops for $Nd_6Fe_{91}B_3$ compositions after annealing at 1073K

Magnetic hysteresis loop for $Nd_2Fe_{14}B/\alpha$-Fe nanocomposites is shown in Fig.12. The value of saturation magnetization M_s and coercivity H_{cb} is 154.35 emu/g and 667 Oe respectively. The values of remanent magnetization and coercivity are very high compared with common soft magnetic materials such as hexaferrite - FeCo nanocomposite. Furthermore, the magnetic hysteresis loops are quite smooth and it behaves the characteristics of single hard magnetic material. This result can be explained by the effect of exchange interaction between the hard-magnetic $Nd_2Fe_{14}B$ and soft-magnetic α-Fe.

Fig.13 shows the frequency dependence of relative permeability, permittivity. For $Nd_6Fe_{91}B_3$ composites, the real part μ' and imaginary part μ_r'' of relative permeability shows two dispersion peaks, and the first peak may induced by the size resonance due to the inhomogeneous of composites. The permeability spectra of $Nd_2Fe_{14}B/\alpha$-Fe nanocomposites varies very slowly and exhibits relaxation type characteristic in the 0.5-18 GHz frequency range, and the second peak value of imaginary part μ_r'' obtains at 13GHz. The imaginary part of permittivity for the composites exhibits a peak at 9GHz. The dielectric properties of arise mainly from the interfacial polarization.

Fig.14 illustrates the frequency dependence of magnetic loss and dielectric loss in the range of 0.5-18GHz for $Nd_6Fe_{91}B_3$ composites. The magnetic loss shows two dispersion peaks and the dielectric loss show one peak value as well as relative permeability, permittivity in this frequency range. Consequently, microwave absorption properties of α-Fe/$Nd_2Fe_{14}B$ composites depend on cooperate effect of magnetic loss and dielectric loss.

Finally, the RL of $Nd_6Fe_{91}B_3$ composites are calculated from the microwave complex permeability and permittivity, and absorber thickness. Their frequency dependence is shown in Fig.15. This nanocomposite shows an excellent microwave absorption property (reflection loss: RL<-20dB) in 9-17GHz with thin matching thickness 1.6-2.5 mm. Therefore, α-Fe/$Nd_2Fe_{14}B$ nanocomposites are thought to be a potential candidate for thinner microwave absorbers in GHz range.

The permeability spectra of $Nd_2Fe_{14}B/\alpha$-Fe nanocomposites exhibits relaxation type characteristic in the 0.5-18 GHz frequency range. Microwave absorption properties of this composites depend on cooperate effect of magnetic loss and dielectric loss. A minimum RL of -37dB is observed at 16GHz with an absorber thickness of 1.6 mm. $Nd_2Fe_{14}B/\alpha$-Fe nanocomposites are promising microwave absorbers in GHz frequency range.

5. Effect of microstructure on microwave complex permeability of $Nd_2Fe_{14}B/\alpha$-Fe nanocomposites

The effect of ball milling process on the microstructure, morphology and microwave complex permeability of $Nd_2Fe_{14}B/\alpha$-Fe nanocomposites have been investigated. The mechanical ball milling can reduce the grain sizes and the particle sizes of $Nd_2Fe_{14}B/\alpha$-Fe nanocomposite material as shown in Fig.16 and Table 1, and the particle of powder becomes fine and thin, the grain size reduces during the process of milling. and enhance the microwave complex permeability and the complex permittivity. The effect of exchange interaction between the hard-magnetic $Nd_2Fe_{14}B$ and soft-magneticα-Fe enhance, its anisotropy fields H_A and saturation magnetization M_s increase as shown in Fig.17, and the microwave complex permeability increase and the resonance frequencies f_r shift to high frequency with proper ball milling time as shown in Fig.18. The optimal complex

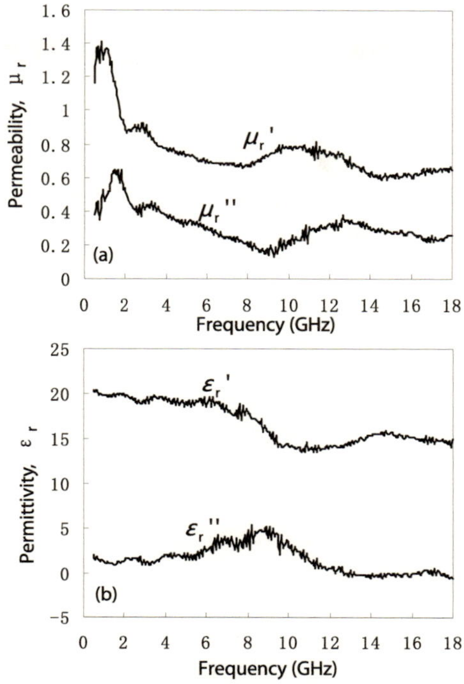

Fig. 13. Frequency dependence of relative complex permeability (a), permittivity (b) of $Nd_6Fe_{91}B_3$ compositions

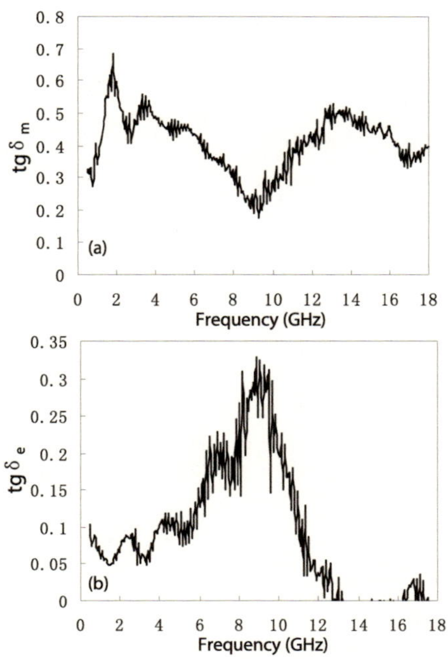

Fig. 14. Frequency dependence of dielectric and magnetic loss of $Nd_6Fe_{91}B_3$ compositions

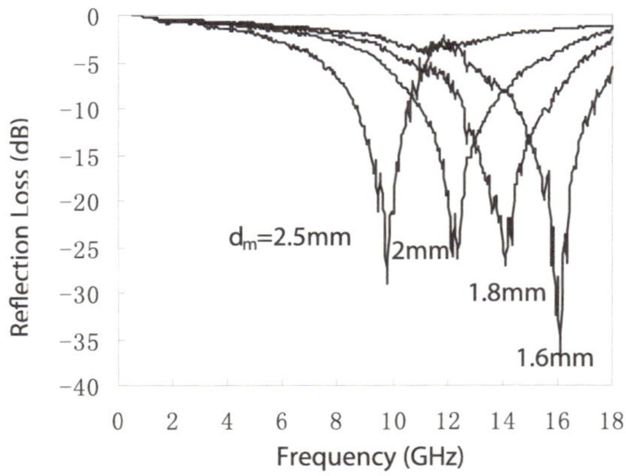

Fig. 15. Frequency dependence of reflection loss for $Nd_6Fe_{91}B_3$ compositions with different matching thickness

permeability are obtained after ball milling for 25h, and the maximum values of μ' is 2.6 at 3GHz, the maximum values of μ'' is 1.2 at 12GHz, and furthermore , the maximum values of μ'' is 0.85 at 30.5GHz. Thus the resonance frequencies f_r of $Nd_2Fe_{14}B/\alpha$-Fe nanocomposites can be controlled, and these novel materials can be used for microwaveabsorbers operating in both centimeter wave and millimeter wave. The ball milling process is an efficient way to optimize the microstructure and improve microwave electromagnetic properties of $Nd_2Fe_{14}B/\alpha$-Fe nanocomposites.

Fig. 16. SEM micrographs of $Nd_{10}Fe_{84}B_6$ composite with various milling time

	0h	10h	20h	30h
Powder size / μm	400	4.88	3.81	3.2
Grain size / nm	37	25	20	18

Table 1. Even size of powder and grain for sample after various milling time

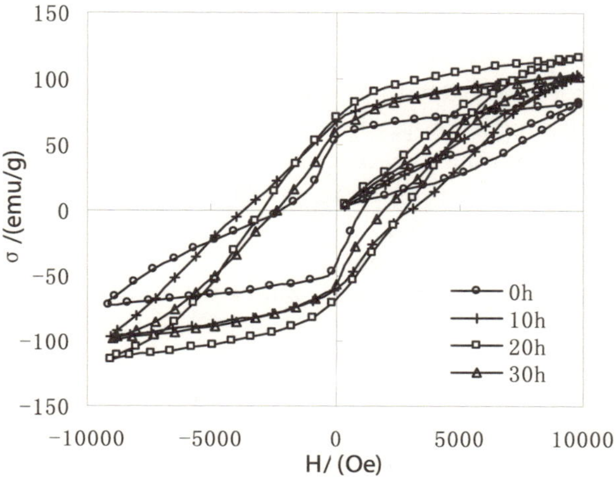

Fig. 17. Magnetic hysteresis loops for $Nd_{10}Fe_{84}B_6$ composite with various milling time

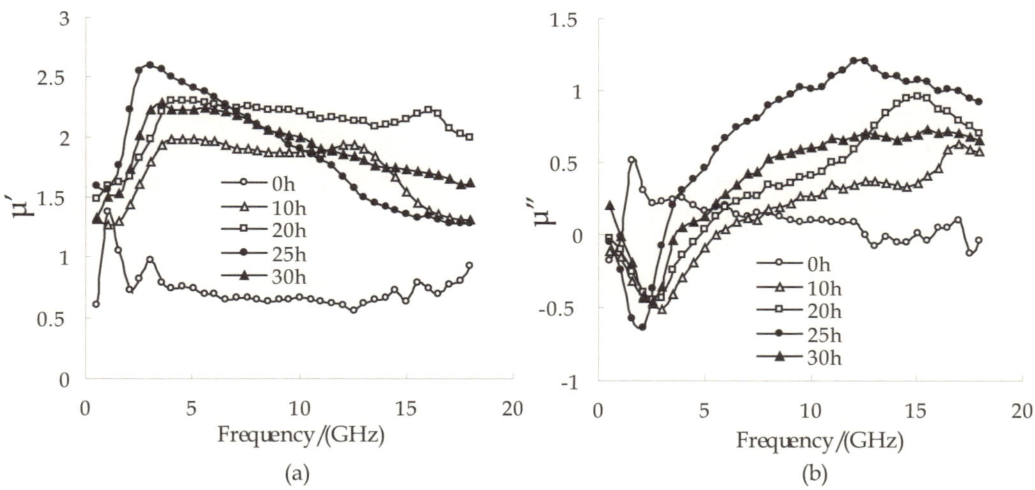

Fig. 18. The real part μ_r'(a) and imaginary part μ_r'' (b) of complex relative permeability against frequency for $Nd_{10}Fe_{84}B_6$ composite after various milling time in the 0.5-18GHz

III. Electromagnetic wave absorption properties of nano-crystalline SmN/α-Fe and SmO/α-Fe composite

In our early study on $Sm_2Fe_{17}N_x$ magnetic material, it showed that the Sm_2Fe_{17} compound disproportionates into a two-phase, α-Fe/SmH_2 microstructure after heating under a hydrogen atmosphere (hydrogen-disproportionation) at temperatures from 873 to 1173 K. A nanometer scale lamellar structure composed of SmH_2 and α-Fe is obtained after a heat treatment at temperatures close to 873 K. However, at higher temperatures, the SmH_2 lamellae grow to form spheres several hundred micrometers in diameter, embedded within an α-Fe phase. If the disproportionated microstructure heated in nitrogen and then oxidation treatment in oxygen or air at lower temperature, a similar disproportionated microstructure of α-Fe/SmN and α-Fe/SmO finally can be obtained, due to the difference in free energy change for the formation of oxide between rare earth elements and iron. In addition, the SmO compounds exhibit a lager resistivity than RH_2 and SmN phase. Therefore, a similar disproportionated microstructure of α-Fe/SmO can be a EMI material with high microwave absorption properties. However, there are hardly reports relevant to the application of this effect for microwave absorbers. Therefore, in this chapter, the effect of the microstructure and preparation processes on EM wave absorption properties in GHz-range microwave absorption is investigated.

1. Preparation process and measurement

Fig.19 shows the preparation process.Sm_2Fe_{17} alloys were produced by industrial melting method with initial materials whose purities were 99.9% or above in argon atmosphere(see Fig.20). The ingot was annealed in argon atmosphere at 1323 K for 24h(see Fig.21). Then the homogenized ingot was crushed into powders of less than 100μm in size. The crushed Sm_2Fe_{17} powders were ball-milled into powders of about 10 μm for 20 minute. The Hydrogenation-Disproportion (HD)and nitrogen process were conducted as following: the crushed powders were placed in a furnace and heated to 875K at a rate of 5K/min in a high purity hydrogen atmosphere of 0.1MPa , held for an hour, Subsequently the furnace was vacuumized to $\times10^{-4}$Pa and then cooled to 773K in a high purity nitrogen atmosphere of 0.3MPa, held for 5h.The nitrified powder was heated in oxygen or air (oxygen-disproportionation) at 573K for 2h.

X-ray diffraction (XRD) and transmission electron microscope (TEM) were used to determine the phases and microstructure of samples. The magnetic hysteresis loops were measured using a vibrating sample magnetometer (VSM)(see Fig.22). The alloy powders were mixed with paraffin at a weight ratio of 5:1 and compacted respectively into a toroidal shape (7.00 mm outer diameter, 3.01 mm inner diameter and approximately 3 mm thickness)The vector value of reflection/transmission coefficient (scattering parameters) of samples were measured in the range of 0.5-18 GHz, using an Agilent 8720ET vector network analyzer respectively. The relative permeability (μ_r) and permittivity (ε_r) values were determined from the scattering parameters and sample thickness. Assumed the metal material was underlay of absorber, and the reflection loss (RL) curves were calculated from the relative complex permeability and permittivity with a given frequency range and a given absorber thickness (d) with the following equations :

$$RL = 20\lg\left|(Z_{in}-1)/(Z_{in}+1)\right| \tag{1}$$

$$Z_{in} = \sqrt{\mu_r/\varepsilon_r}\tanh\left\{j(2\pi fd/c)\sqrt{\mu_r\varepsilon_r}\right\} \tag{2}$$

where Z_{in} is the normalized input impedance at absorber surface, f the frequency of microwave, and c the velocity of light.

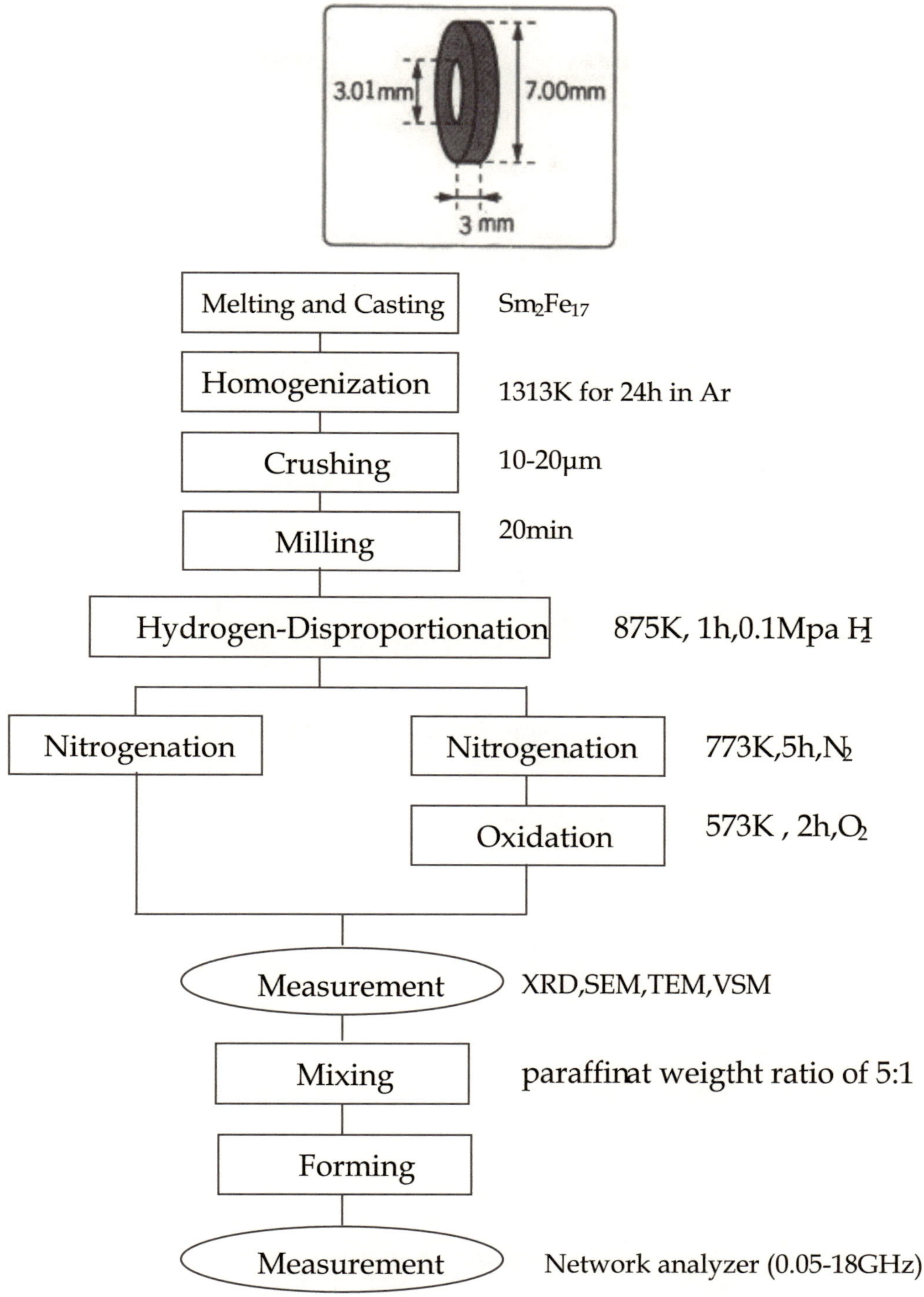

Fig. 19. Preparation procedure

Fig. 20. Medium frequence induction melting furnace

Fig. 21. Vacuum heat treatment furnace

Fig. 22. LakeShore7410 VSM

2. Microstructure of SmN / α-Fe and SmO / α-Fe phase coupling absorbing material

Fig. 23 shows X-ray diffraction patterns of the homogenized Sm_2Fe_{17} powders (a) , (b)after hydrogen-disproportionation at 873 K for 1 h , (c) followed by nitrified at 773 K for 5 h in0.3Mpa N_2, and (d) after oxygen-disproportionation at 573 K for 1 h. .It can be seen from Fig.23(a) that the main phase in as-homogenized ingot is Sm_2Fe_{17} with rhombohedra Th_2Zn_{17}-type structure ,co-existing with small fraction of Sm-rich and α-Fe phases. It is confirmed from Fig.23 b that after hydrogenation at 873K for 1h, the Sm_2Fe_{17} (113) peak almost can not be detected, and the alloy is almost composed of SmHx and α-Fe phase. It suggests that disproportation completed according to the reaction:

$$Sm_2Fe_{17}+H_2 \rightarrow SmH_x + \alpha\text{-}Fe \qquad (3)$$

Fig.24 shows TEM image and diffraction patterns of SmH_x/α-Fe. It can be seen that a nano-meter scale lamellar structure composed of SmH_2 and α-Fe is obtained after a heat treatment at temperatures close to 873 K according to diffraction pattern. Also, as local high temperature in the sample, the SmH_2 lamellae grow to form spheres several hundred micrometers in diameter, embedded within an α-Fe phase. The top left corner of the picture in Fig.24 is to enlarge the box office chart, from which can be drawn that the growth of the alloy grains and the dimensions of the grains are about 10nm. These correspond to the result of XRD analysis. After nitrified at 773 K for 5 h， the SmH_x phase is completely transformed into SmN phase, while the α-Fe phase content is essentially the same(see Fig.23c).Thus the electromagnetic wave absorption materials with SmN/α-Fe phase coupling structure were obtained. When the powder with SmN/α-Fe two-phase structure heated in the air after 1 h, it can be seen from Fig.23(d), that the alloy is composite of SmO and α-Fe phase, oxide of iron is not detected ,namely SmN phase has transformed into SmO. The difference of diffraction peak between SmN (PDF 30-1104) and SmO (PDF 65-2915) is not obvious. The characteristic diffraction peaks were enlarged in the diagram at the upper left corner of Fig.23, it illustrate that the position of two diffraction peak are different at enlarged picture. Combined with the color change of oxidized powder before and after, all of these confirmed again that SmN phase has been changed into SmO phase.

Fig.25 is SEM picture of nitrified and oxidized powder of SmFe alloy. It Shows that each particle size is between 1 ~ 5 μm. Small particles can reduce the eddy current, and then helps to reduce magnetic permeability induced by eddy current. Compared with SmN phase, the resistivity of SmO phase is lager, which can increase skin depth of eddy current, and further reduce the eddy current. Otherwise, this structure can generate an exchange coupling effect, which is conducive to increase the natural vibration frequency of materials.

3. Electromagnetic wave absorption properties of SmN / α-Fe and SmO / α-Fe composites

Fig.26 shows the frequency dependence of the complex relative permeability and permittivity of SmN / α-Fe and SmO / α-Fe composites. As shown in Fig.26(a), that the real part of complex permittivity ε_r (= ε_r - $j\varepsilon_r$) decre ε_r ase with increasing frequency for SmN / α-Fe and SmO / α-Fe composites in 0.5-18 GHz, and that of the nitrified and oxidized sample remains almost constant in the frequency range of about 9HGz. But the relative

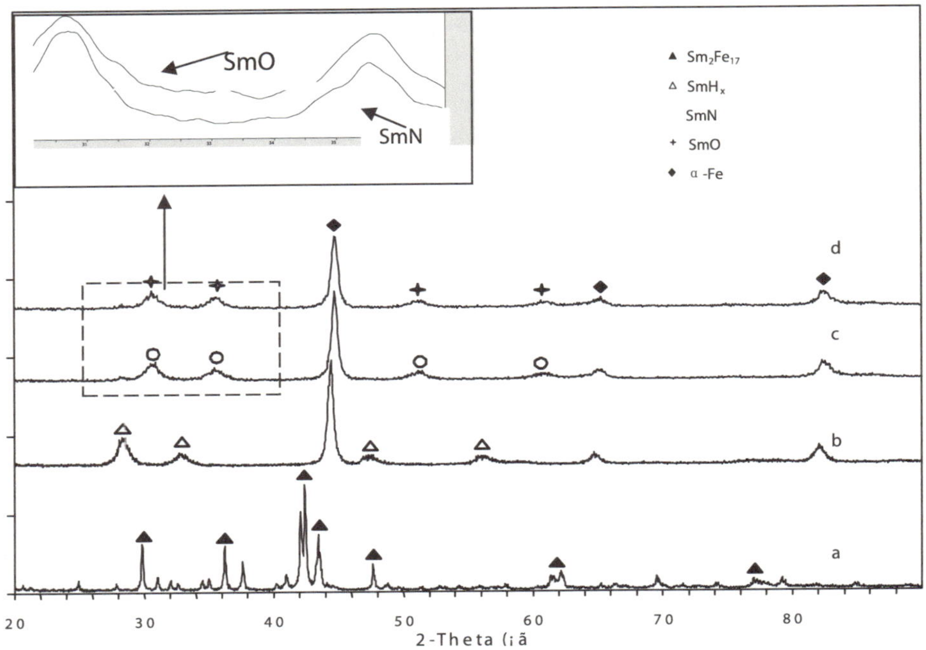

Fig. 23. The XRD patterns of Sm_2Fe_{17}: (a) homogenized powders, (b) after hydrogenation-disproportionation at 873 K for 1 h in H_2, (c) after nitriding the sample (b) at 773 K for 5 h, and (d) after oxidizing the sample (c) in O2 at 573 K for 1 h

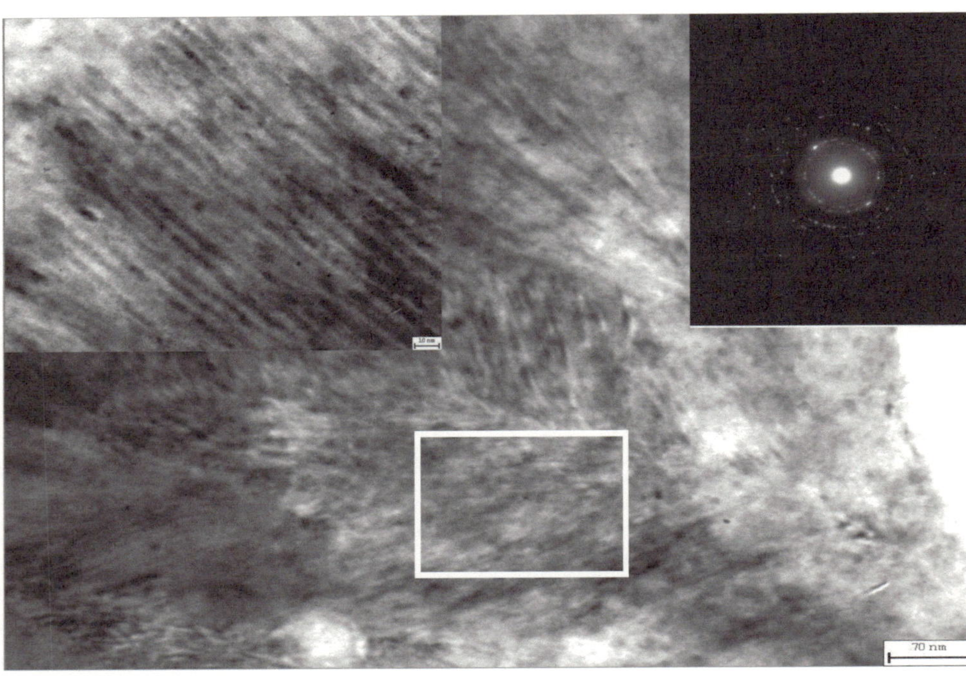

Fig. 24. TEM image , diffraction patterns of SmHx/α-Fe and enlarge picture

(a) (b)

Fig. 25. SEM image of SmN/α-Fe and SmO/α-Fe composites (a) SmN/α-Fe composite;(b) SmO/α-Fe composite

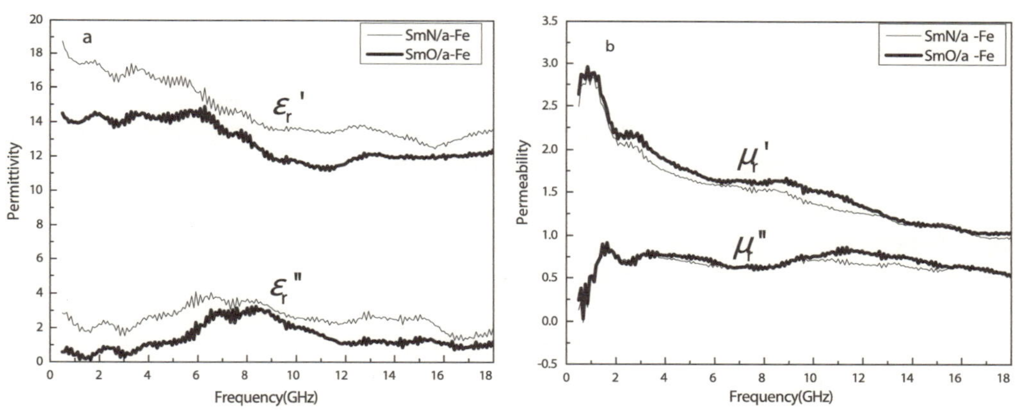

Fig. 26. (a)The relative permittivity plotted against frequency for SmN/α-Fe and SmN/α-Fe composites in the 0.5-18GHz. (b) The relative permeability plotted against frequency for SmN/α-Fe and SmO/α-Fe composites in the 0.5-18GHz

permittivity ε_r'' exhibits a peak at the frequency nearly 8HGz. As can be seen from Fig. 26 (b) ,that complex permeability of SmN/α-Fe and SmO/α-Fe composites are similar, the real part of complex permeability μ_r' show maximum of value 3.2 at a frequency of 1GHz.While the relative permeability μ_r'' have two peaks, and remain a high value in a wide frequency band. It found that it shows the phenomenon of multiple resonances for Fe-Co-Ni bonded composite elastic material with the Fe-Co-Ni powder particles decreased from micron to sub-micron. As to SmN/α-Fe and SmO /α-Fe composites, because that its powder particle is uniform and size distribution is between the 1-5μm (see Fig.25), the relative permeability exhibits two peaks in the 0.5-18GHz, which is the characteristics of the multiple resonance.

In addition, it can be found from Fig.8 that the real part and relative permittivity of SmO/α-Fe composite are lower than that of SmN/α-Fe composite. This is ascribed to the lower resistivity of SmN than SmO phase's, due to the permittivity of SmN/α-Fe composite being lager, and more fluctuations with the frequency. As we all know, the relative permeability can express by the formula (4):

$$\mu_r'' = \frac{M_s}{3\mu_0 H_A \alpha} \tag{4}$$

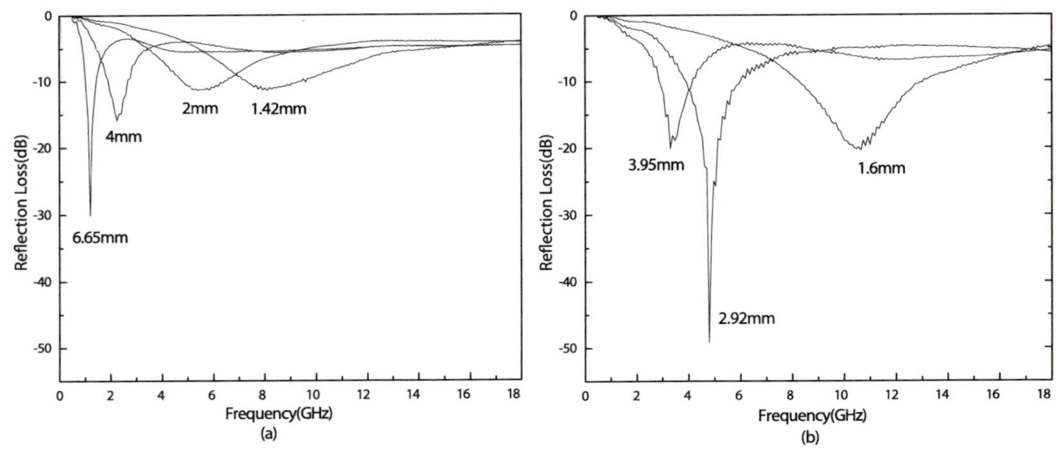

Fig. 27. Frequency dependences of RL of SmN/α-Fe and SmO/α-Fe composites at different thicknesses.(a) SmN/α-Fe composite;(b) SmO/α-Fe composite

where M_s is saturation magnetization, μ_0 is vacuum permeability, H_A is anisotropy magnetic field,αis constant. Also, their complex permeability are similar according to Fig.26, which are main contribute to α-Fe in the two materials. It can be calculated by XRD patterns (Fig.23)that the content of α-Fe content in the two materials are almost identical, which is confirm by VSM measurement with saturation magnetization value of 120 emu/g for SmN/α-Fe and SmO/α-Fe composites. As can be seen from Fig.26(b) , the permeability of SmO/α-Fe composite is slightly higher in fact. The reason is that the resistivity increases during the SmN phase transform to SmO phase, due to reduce the eddy current and decreased slowly the permeability with increase of frequency. It can be drawn from the formula (1) that in order to achieve full non-reflective, that is, R = 0, the value of complex permittivity and complex permeability for the absorbing materials must be equal, that is to say , the normalized input impedance at absorber surface Z_{in}= 1. Hence, high-performance absorber material is characterized by as possible to keeping both permeability and permittivity approximately equal at a wide frequency range. It can be deduced that SmO/α-Fe composite have better absorbing performance from the permeability and permittivity in Fig.26.
Figure 27 shows frequency dependences of RL of SmO/α-Fe composite at different thickness. It can be seen that the two materials in the 0.5-18 GHz have good absorbing properties, with the absorbing coating thickness increases, the minimum reflection loss of absorber moves to the low frequency. Because the match between the permeability and the dielectric constant for SmO/α-Fe is better than that for SmN/α-Fe, the former has better EM wave absorption property and its RL (reflection loss) is less than -20 dB with absorber

match thickness of 1.60~3.95 mm in the frequency range of 3.30~10.65 GHz. The minimum RL is -50 dB, absorber match thickness is 2.92 mm at 4.8 GHz.

4. Acknowledgements

This work is supported by National Natureal Science Foundation of China (No.10804079) and Doctoral Fund for New Teachers No. 15205025 from the Ministry of Education of China.

5. References

[1] J.L. Snoek, Physica 14, 207(1948).
[2] D. Rousselle, A. Berthault, O. Acher, et al. [J] J. Appl. Phys.,74,475(1993).
[3] E.P. Wohlfarth, K.H.J. Buschow, in: Ferromagnetic Materials, Vol.4, Elsevier Science Publishers B.V., Amsterdam, 1988.
[4] H. Ono, T. Tayu, N. Waki, et al. [J] J. Appl. Phys.,93,4060(2003).
[5] L.X. Lian, L.J. Deng, M. Han, et al. [J] J. Alloys Compd., 441, 301 (2007).
[6] Li-Xian Lian, L. J. Deng, and M. Han. [J] J. Appl. Phys., 101, 09M520 (2007).
[7] Chen Xianfu, Ye Jinwen, Liu Ying, et al. [J] Rare Metal Materials and Engineering,38, 726(2009).
[8] Y. Naito, K. Suetake, IEEE Trans. Microwave Theory Tech. MTT-19, 65(1971).
[9] C.Sudakar, G.N.Subbanna, and T.R.N.Kutty. [J] J. Appl. Phys., 94, 6030(2003).
[10] E. F. Kneller, R. Hawig, IEEE Trans. Magn. 27, 3588 (1991).
[11] R. Skomski, J. M.D. Coey, Phys. Rev. B., 48, 15812(1993).
[12] Y. J. Kim and S. S. Kim, IEEE Trans. Mag., 38, 3108(2002).
[13] T. Maeda, S. Sugimoto, T. Kagotani, Nobuki Tezuka, Koichiro, J. Magn. Magn. Mater.281, 195 (2004).
[14] R. Gerber, C.D. Wright, G. Asti (Eds.), Applied Magnetism, Kluwer Academic Publisher, Doldrecht, pp.457(1992).
[15] E.P. Wohlfarth, K.H.J. Buschow, in: Ferromagnetic Materials, Vol.4, Elsevier Science Publishers B.V., Amsterdam, (1988).
[16] H. Kato, M. Ishizone, T. Miyazaki, K. Koyame, H. Nojiri, M. Motokawa, IEEE Trans. Magn. 37, 2567(2001).
[17] J. Jakubowicz, M.Giersig, J Alloys Compounds, 349, 311(2003).
[18] L.X. Lian, Y. Liu, S.J. Gao, M.J. Tu, J. Rare Earths, 23(2), 203(2005).
[19] I. Panagiotopoulos, L. Withanawasam, A.S. Murthy, G.C. Hadjipanayis, J.Appl. Phys. 79, 4827(1996).
[20] J. Jakubowicz, M. Jurczyk, J. Magn. Magn. Mater., 208, 163(2000).
[21] J. Bauer, M. Seegrer, A .Zern, H. Kronmüller, J. Appl. Phys., 80 , 1667(1996).
[22] T. Schrefl, J. Fidler, H. Kronmüller, Phys.Rev.B., 49, 6100 (1994).
[23] T. Leineweber, H. Kronmüller, J. Magn. Magn. Mater., 176, 145(1997).
[24] C.Sudakar, G.N.Subbanna, and T.R.N.Kutty. [J] J. Appl. Phys., 94, 6030(2003).
[25] Koji Miura, Masahiro Masuda, etc. [J].Journal of Alloys and Compounds 408–412, 1391(2006).
[26] Zhou Shouzeng,Dong Qingfei. Super Permanent Magnets [M]. BeiJing: Metallurgy Industry Publishing House, 406,(1999).
[27] YE Jinwen, Liu Ying, Gao Shengji, Tu Ming-jing. [J].Journal of theChinese Rare Earth Socity. 23(3):303(2005).

[28] YE Jinwen, Liu Ying, Gao Shengji, Tu Ming-jing. [J].Rare Metal Materials and Engineering, 34(12): 2002(2005).
[29] Coey J M D, Sun H, Otani Y. Proceeding of 11th Int. Workshop on RE Magnets and their applications, 36~40(1990).
[30] Li Fang, Liu Yingetc. [J]. Metaltlc Functional Materlals. 11(3)(2004).
[31] S.Sugimoto,T.Kagotani etc. [J] Journal of Alloys and Compounds 330-332, 301 (2002).
[32] Jiu Rong Liu, Masahiro Itoh, Jianzhuang Jiang, Ken-ichi Machida. [J].Journal of Magnetism and Magnetic Materials 271, L147 (2004).

18

Electromagnetic Wave Absorption Properties of Nanoscaled ZnO

Yue Zhang, Yunhua Huang and Huifeng Li
University of Science and Technology Beijing
China

1. Introduction

Microwave absorbing material (MAM) is a kind of functional material that can absorb electromagnetic wave effectively and convert electromagnetic energy into heat or make electromagnetic wave disappear by interference (Kimura et al., 2007). MAM is currently gaining much attention in the field of civil and military applications. For example, the materials have been widely applied to minimize the reflection of microwave darkrooms, airplanes, steamboats, tanks and so on (Zou et al., 2008). Generally, the electromagnetic absorbing performance of any MAM is linked to its intrinsic electromagnetic properties (i.e. conductivity, complex permittivity and permeability) as well as to extrinsic properties such as the thickness and working frequencies. It is clear that the microwave absorption properties can be improved by changing the above parameters. However, the traditional MAMs or novel nanomaterials still have some disadvantages such as high density, narrow band, and low absorptivity (Zou et al., 2006). Therefore, demands for developing more economical MAMs with "low density, wide band, thin thickness, and high absorptivity" are ever increasing.

Wurtzite-structured ZnO is of great importance for its versatile applications in optoelectronics, piezoelectricity, electromagnetic wave absorption, laser, acous-optical divices, sensors, and so on (Wang et al., 2007). One-dimensional nanostructures of ZnO, such as nanowires, nanobelts, and nanotetrapods, have been a hot research topic in nanotechnology for their unique properties and potential applications. Moreover, several types of three-dimensional ZnO nanostructures have been synthesized. Because of the high surface/volume ratio and integrated platform, three-dimensional oxide networks have been demonstrated for building ultrasensitive and highly selective gas sensors and optoelectronics applications (Zhu et al., 2007). It is worth mentioning that the ZnO nanostructures have shown great attraction for microwave radiation absorption and shielding material in the high-frequency range due to their many unique chemical and physical properties (Zhuo et al., 2008). Some research works focused on nanoscaled ZnO as a vivid microwave absorption material due to their light weight, high surface/volume ratio, and semiconductive and piezoelectric properties (Wang & Song, 2008).

On the other hand, carbon nanotubes (CNTs) as conductive filler have been widely studied in MAMs due to the unique spiral and tubular structure since the discovery of CNTs by Iijima in 1991 (Iijima, 1991). CNTs/polymer composites exhibit a strong microwave absorption in the frequency range of 2-18 GHz and have the potential application as broad frequency radar absorbing materials (Fan et al., 2006). For example, Zhao et al. demonstrate that carbon nanocoils are chiral microwave absorbing materials and exhibit superior microwave absorption (Zhao & Shen, 2008). However, there are few reports concerning electromagnetic wave absorption properties of ZnO and CNTs nanostructures composites. Furthermore, the nanocrystalline structure of tetraleg ZnO (T-ZnO) is constituted of a central part and four needle-like legs, and exhibits super high strength, wear resistance, vibration insulation and can be widely applied as MAMs (Dai et al, 2002). So, it is necessary to study the absorption properties of T-ZnO and CNTs nanostructures composites.

In this chapter, we will report the synthesis methods of T-ZnO nanomatrials and ZnO micro-/nanorod networks, the fabrication methods of wave absorption coatings using T-ZnO and T-ZnO plus multi-walled CNTs as absorbent respectively, the measurement of wave absorption properties of coatings, the effects of absorbent contents, thickness of coatings on the properties, the measurement of electromagnetic parameters and the calculated properties of T-ZnO and ZnO networks, and the wave absorption mechanisms.

2. Preparation and structure of microwave absorbing materials

The nanoscaled ZnO used for microwave absorbing samples, including tetraleg ZnO nanorods and three-dimensional ZnO micro-/nanorod networks were synthesized in our laboratory. The other materials, multi-walled carbon nanotubes, were purchased from commercial company. The fabrication methods and structures of nanoscaled ZnO, and the structural characterization of CNTs will be presented as follows.

2.1 Fabrication of tetraleg ZnO

The tetraleg ZnO nanorods are the one type of the easy synthesized morphologies through thermal evaporation method. Tetraleg ZnO nanostructures were fabricated by the following procedure. The metal zinc powders (99.9%) with thickness of 1~3 mm were placed in an alumina ceramics boat in a tubular furnace under a constant flow of argon and oxygen, and the fraction of oxygen was 5 ~ 10%. The furnace was kept to 700 ~ 800°C, i. e. the reaction temperature, for 20~30 minutes. No catalyzer was utilized in all the deposition processes. White fluffy products were obtained. The materials for the wave-absorbing coatings were synthesized and accumulated as above process. The synthesized products were characterized using X-ray diffraction (XRD) (D/MAX-RB) with Cu-Ka radiation, field-emission scanning electron microscopy (FE-SEM) (LEO1530), and transmission electron microscopy (TEM) (HP-800). Figure 1 shows the SEM images of the morphologies of the tetraleg ZnO nanorods. The obtained ZnO nanostructures are of a tetrapod shape having four legs. The image at low magnification shows that uniform T-ZnO nanorods form in high yield (Fig. 1a). No particles are produced. The high-magnified image (Fig. 1b) indicates that the surfaces of nanorods are smooth. The length of legs of T-ZnO nanorods is 2–4μm. Very little secondary growth components are observed. T-ZnO nanorods we obtained are uniform nanorods.

XRD measurements were made on the mass nanorods to assess the overall structure and phase purity. A typical XRD pattern of the T-ZnO nanorods is shown in Fig. 2. The diffraction peaks can be indexed to a wurtzite structure of ZnO with cell constants of a = 0.324 nm and c = 0.519 nm. No diffraction peaks from Zn or other impurities were found in any of the samples.

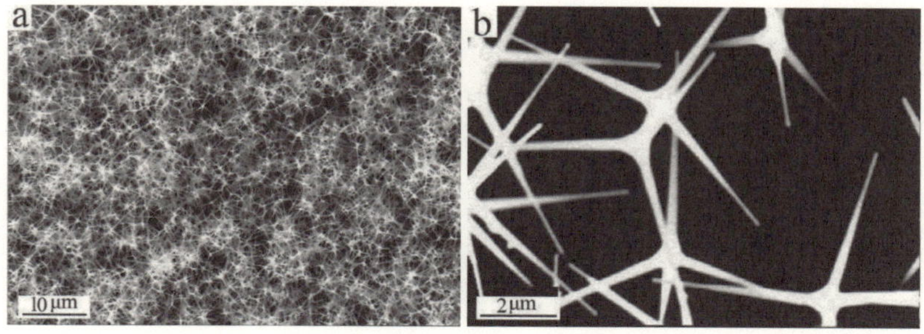

Fig. 1. SEM images of tetraleg ZnO nanorods, (a) Low-magnified, and (b) high-magnified

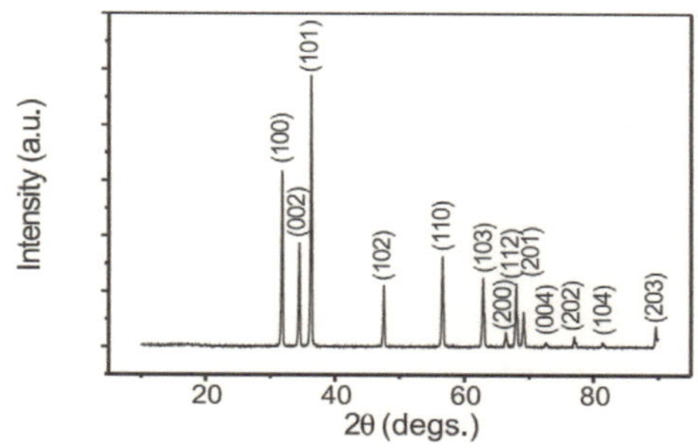

Fig. 2. XRD pattern obtained from a bulk sample of T-ZnO nanorods

HRTEM observation of the T-ZnO nanostructure is shown in Fig. 3. A low-magnification image given in Fig. 3(a) shows the projected four-fold twin structure at the central region. Fig. 3(b) is a corresponding HRTEM image from the central region. It reveals the structure of the twin boundaries between the element crystals. From the HRTEM image, it can be seen that the interfaces are sharp and show no amorphous layer. These twins are smoothly conjugated fairly coherently at the boundaries with little lattice distortion. A Fourier transform of Fig. 3(b) is given in Fig. 3(c), based on which the twin planes can be determined to be the $\{11\bar{2}2\}$ family. The index corresponding the grain at the bottom-left corner of Fig. 3(b) is labeled in Fig. 3(c). The twin plane is indicated by an arrowhead. The incident beam direction is $[\bar{2}4\bar{2}3]$, along which the four twin boundaries are imaged edge-on.

As for the growth mechanism, Iwanaga proposed the octahedral multiple twin (octa-twin) nucleus models (Fujii et al., 1993), and Dai et al. directly revealed the structure of the T-ZnO nanostructures by HRTEM for the first time (Dai et al., 2003). According to the octa-twin nucleus model, ZnO nuclei form in an atmosphere containing oxygen are octa-twins nuclei which consist of eight tetrahedral-shape crystals, each consisting of three $\{11\bar{2}2\}$ pyramidal facets and one (0001) basal facet (Fig. 4(a)). The eight tetrahedral crystals are connected together by making the pyramidal faces contacting one with another to form an octahedron. The surfaces of the octa-twin are all basal planes. An important additional condition is that every twin is of the inversion type, i.e. the polarities of the twinned crystals are not mirror-symmetric with respect to the contact plane but antisymmetric. Thus the eight basal surfaces of the octa-twin are alternately the plus (0001) surface (+c) and the minus surface $(000\bar{1})$ (-c), as shown in Fig. 4(b). The formation of the tetraleg structure has to do with the following two factors based on the octa-twin nucleus. It is known through the study of ZnO nanowires and nanobelts, [0001] is the fastest growth direction in the formation of

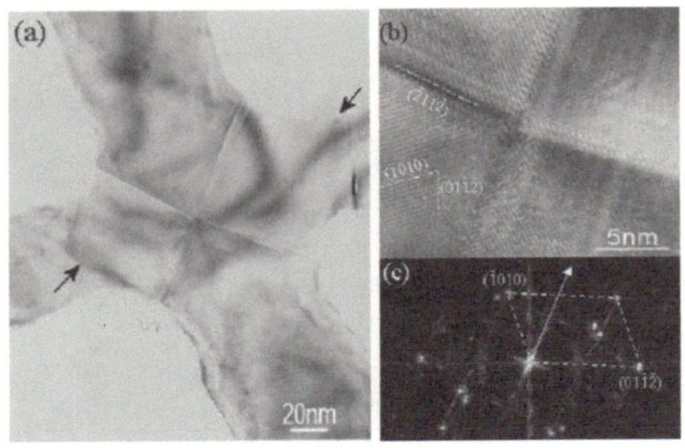

Fig. 3. (a) Low magnification TEM image of a tetraleg ZnO nanostructure. (b) A high-resolution TEM image recorded from the center of the tetraleg structure. (c) A Fourier transform of the image given in (b) and the indexes corresponding to one of the bottom-left grain in (b). The incident beam direction is [$\bar{2}4\ \bar{2}3$]

nanostructure. The octa-twin has four positively charged (0001) surfaces and four negatively charged (0001) surfaces. The positively charged surfaces are likely to be terminated with Zn, which may be the favorable sites to attracting vapor species, resulting in the growth of whiskers along four [0001] directions that have a geometrical configuration analogous to the diamond bonds in diamond. The growth mechanism is believed to be a solid–vapor process.

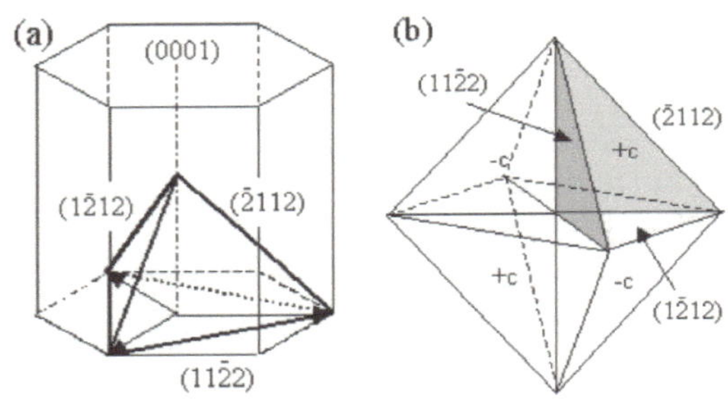

Fig. 4. (a) A pyramid formed by three {11 $\bar{2}$ 2} and one (0001) facets. (b) The octa-twin model composed of eight pyramidal inversion twin crystals

2.2 Fabrication of three-dimensional ZnO networks

The three-dimensional ZnO netlike micro-/nanostructures were fabricated by the following procedure. First, high pure Zn (99.99%) and graphite powders with molar ratio of 10:1 were ground fully into a mixture before being loaded into a quartz boat. The Si substrate with the polished side facing the powder was fixed upon the boat, and the boat with the mixture was placed at the center of the furnace. The vertical distance between the zinc source and the substrate was about 4-6 mm. And then the alumina ceramics boat was inserted into a quartz tube (30 mm inside diameter) of a tubular furnace under a constant flow of argon and oxygen. The flow rate of Argon was 100 standard cubic centimeters per minute (sccm) and the fraction of oxygen was 4 sccm. The quartz tube was heated up to 910 °C, and retain reaction temperature for 30 minutes. After the evaporation finished, a layer of woollike product was formed on the walls of the boat and the surface of the substrate.

The SEM images in Figure 5 show the morphologies of ZnO netlike microstructures. It can be clearly seen that these ZnO micro/nanorods form crossed network, and the rods have the diameter in the range of 0.2-2μm and the length of 50-100 μm (Fig. 5a). The high magnified image of partial network is shown in Figure1b and the ZnO rods have the diameter of 1-2μm. Also, the Fig. 5b indicates the ZnO microrods with a rough surface, possibly due to the competition between surface energy and strain energy (Li et al., 2010).

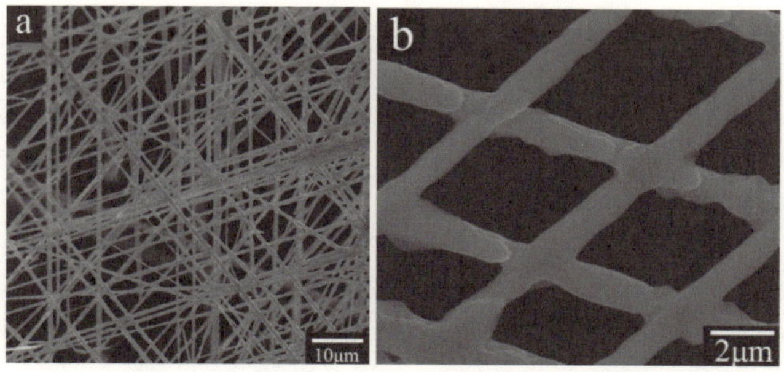

Fig. 5. SEM images of ZnO netlike microstructures, (a) Low-magnified, and (b) high-magnified

In order to obtain more detailed structural information of the ZnO products, typical transmission electron microscopy (TEM) and high-resolution transmission electron microscopy (HRTEM) images were recorded, as shown in Figure 6. Figure 6a reveals a ZnO micro-/nanorods bundle in the 3D networks. The individual ZnO nanorods have the diameters in the range of 200-500 nm and length of several microns. Figure 6b shows the HRTEM image and corresponding SAED pattern taken from the nanorod. The HRTEM image of the fraction in Figure 6b clearly shows the lattice fringes with the d-space of 0.52 nm, which matches that of (0001) planes of the wurtzite structural ZnO. The inset of Figure

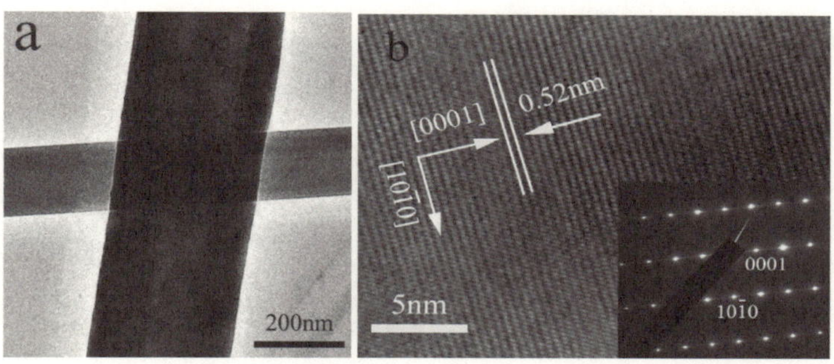

Fig. 6. (a) The low-magnified TEM image of ZnO netlike micro-/nanostructures, (b) the HRTEM image of ZnO netlike micro-/nanostructures, inset: SEAD image of ZnO netlike micro-/nanostructures

6b shows the corresponding SAED pattern taken from the nanorod. Combined HRTEM images with the corresponding SAED pattern, the growth direction of the fraction can be determined along [0001] and [10 $\bar{1}$0]. It is noteworthy that the netlike structures, such as the TEM samples, are sufficiently stable, which cannot be destroyed even after ultrasonication for a long time. Therefore, these electron microscopy characterizations reveal the formation mechanism of ZnO netlike structure is following the V-S mode presented in the literature (Wang et al., 2003).

2.3 Characterization of carbon nanotubes

The multi-walled CNTs were purchased from Beijing Nachen Corporation (Beijing, China), and were observed by a field emission scanning electric microscopy (FE-SEM) (Zeiss, SUPRA-55). The low and high magnified morphologies of the CNTs are shown respectively in figure 7.

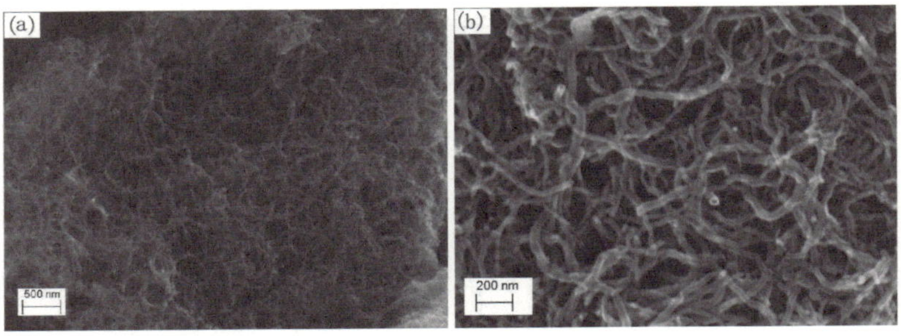

Fig. 7. SEM images of CNTs (a) low-magnified image; (b) high-magnified image

3. Absorption properties of T-ZnO / EP coatings

3.1 Fabrication of T-ZnO / EP coatings

T-ZnO/Epoxy resin (EP) wave-absorbing coatings were fabricated with nanosized T-ZnO as the absorbent and epoxy resin as the binder as follows. The nano T-ZnO was added into the EP resin which was diluted by absolute ethyl alcohol, vibrated by ultrasonic wave for about 1h, and then the curing agent was put into the composite, stirred gently. The mixture was sprayed layer by layer onto aluminum plate with a square of 180mm X 180mm and cured at 25-30°C for at least 2h. The images of the surface and cross-section of the wave-absorbing coating are shown in figure 8 (Cao et al., 2008).

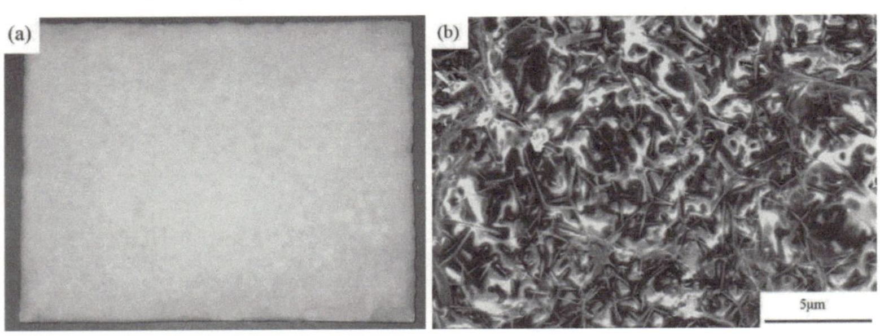

Fig. 8. Images of the cross-section of nano tetraleg ZnO/EP resin coating

The reflectivity of the composites were measured by a reflectivity scanning measurement system (HP 83751B) integrated a signal source (HP 8757E) working at the 2-18 GHz band. The linear scanning frequency was used, and the testing accuracy was better than 0.1 dB. Both the real and imaginary parts of the complex permittivity and permeability of samples were measured by a vector network analyzer system (HP8722ES) in the frequency range of 2-18 GHz. The sample obtained by mixing nanoscaled T-ZnO with molten paraffin was made into a ring of 7.00 / 3.00 × 2.00 mm (outer diameter / inner diameter × thickness) for electromagnetic parameters measurement. The paraffin is transparent for microwave. The details of the measurement system for the microwave absorption properties are shown in Schematic 1.

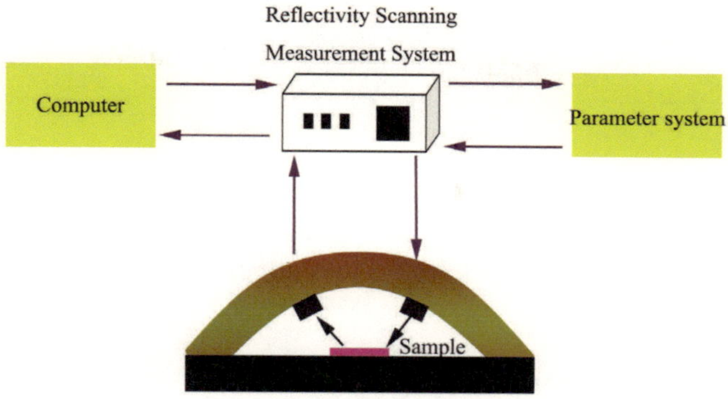

Schem. 1. The schematic of automatic parameter sweep vector network measurement system for measurements the microwave absorption properties

3.2 Microwave absorption properties of T-ZnO / EP coatings

1. **Impacts of concentration of T-ZnO plus CNTs on microwave absorption properties**

The microwave absorption properties of the nano T-ZnO/EP resin coatings with different ZnO concentration and thickness of 1.5 mm are summarized in Table 1. The measurement results, as shown in Fig. 9, reveal that the absorption properties improve as the concentration of nano T-ZnO increases. The minimum reflection loss is -1.74dB when the concentration of nano T-ZnO is 11%, and reduces to -3.23dB, when the content of nano T-ZnO is 16%. The sample A3 with the concentration of 20% shows the minimum reflection loss of -3.89dB at 17.4GHz. The difference on minimum reflection loss of the coatings is associated with the concentration of nano T-ZnO in the coating, which attenuates the electromagnetic wave energy mainly by forming conductive networks.

Sample number	*T-ZnO concentration (wt%)*	*Thickness (mm)*	*Minimum reflection loss (dB)*	*Corresponding frequency (GHz)*
A1	11	1.5	-1.74	15.7
A2	16	1.5	-3.23	18.0
A3	20	1.5	-3.89	17.4

Table 1. Absorption properties of ZnO /EP resin coatings with different ZnO concentration

2. **Impacts of the coating thickness on microwave absorption properties**

In other research, the absorption properties of ZnO /EP resin coatings with different thickness were measured. A list of the microwave absorption properties of the samples is presented in Table 2 and Fig. 10. The results indicate that the absorption properties improve as the coating thickness increases. The minimum reflection loss is -0.38dB when the thickness is 11%, and reduce to -5.30dB, as the thickness is -2.5 dB. When the thickness increases to 3.5mm, the minimum reflection loss reaches to -9.11dB. On the other hand, it can also be seen that the maximum absorbing peak shifts towards a lower frequency as the thickness increases. When the coating thickness enlarges from 1.5mm to 2.5 and 3.5 mm, the peak frequency is 14.0, 12.9 and 8.8 GHz respectively.

Sample number	T-ZnO concentration (wt%)	Thickness (mm)	Minimum reflection loss (dB)	Corresponding frequency (GHz)
B1	28.6	1.5	-0.38	14.0
B2	28.6	2.5	-5.30	12.9
B3	28.6	3.5	-9.11	8.8

Table 2. Absorption properties of ZnO /EP resin coatings with different thickness

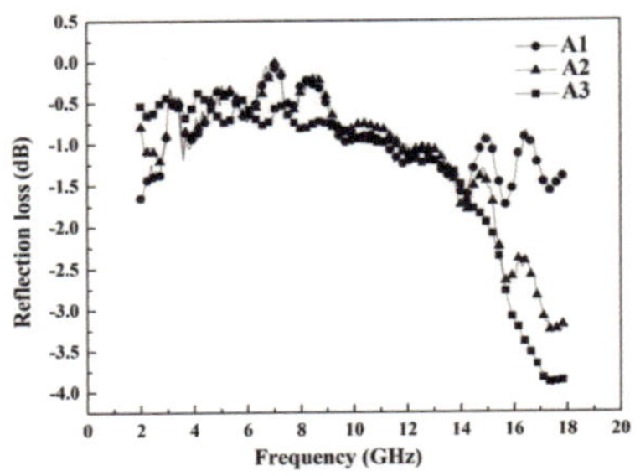

Fig. 9. Absorption characteristics of ZnO/EP resin coatings with different ZnO concentration

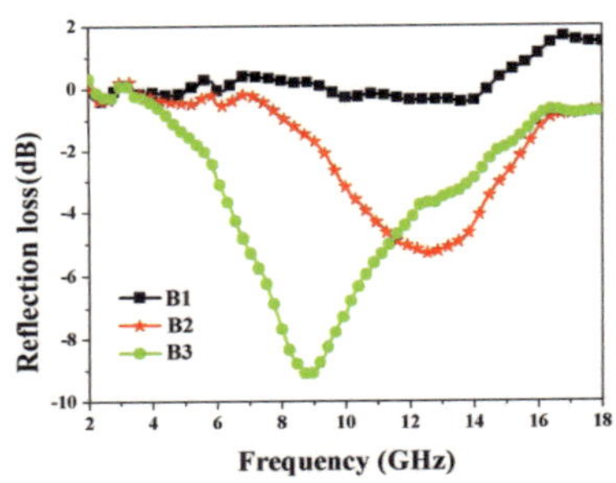

Fig. 10. Absorption characteristics of ZnO/EP resin coatings with different thickness

3. Intrinsic reasons of T-ZnO plus CNTs for microwave absorption

Microwave absorption may result from dielectric loss and/or magnetic loss. They are characterized with the complex relative permittivity ε_r ($\varepsilon_r = \varepsilon' - j\varepsilon''$, where ε' is the real part, ε'' the imaginary part) and the complex relative permeability μ_r ($\mu_r = \mu' - j\mu''$, where $\mu'$is the real part, μ'' the imaginary part) (Zhang et al., 2008).

In order to investigate the intrinsic reasons for microwave absorption of the coating, the complex permittivity ε and permeability μ of the nano T-ZnO were measured. Fig.11a and Fig.11b show the frequency dependence of the permittivity and permeability of nano T-ZnO, respectively. From the figures, it is found that the values of imaginary part of permittivity of T-ZnO nanorods are larger than that of permeability of T-ZnO nanorods, the value of imaginary part of permittivity and permeability are getting close to 3.0 and 1.0, respectively. The results revealed that the value of dielectric loss $\tan\delta_E$ ($\varepsilon''/\varepsilon'$) is larger than that of magnetic loss $\tan\delta_M$ (μ''/μ'). Thus, the electromagnetic wave absorptions of T-ZnO nanorods are mainly caused by dielectric loss rather than magnetic loss.

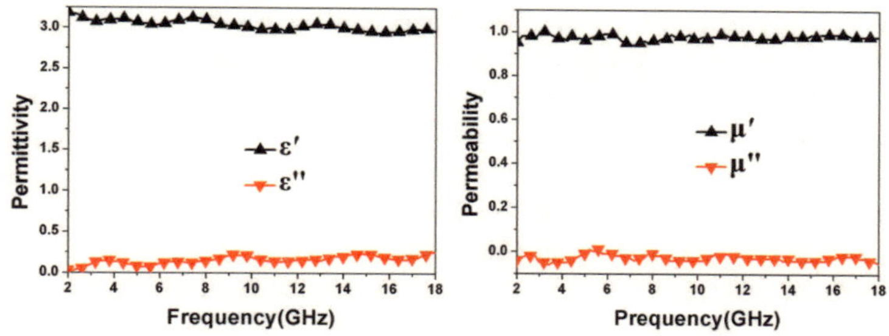

Fig. 11. Frequency dependence of the permittivity (a) and permeability (b) of tetraleg ZnO

From the above analysis, the wave-absorbing mechanism can be derivated. Firstly, the diameter of needle-body of nano T-ZnO belongs to the nanoscale range, so the quantum confine effect makes the wave-absorbing properties of nano T-ZnO change greatly. According to the Kubo theory, the energy levels in nano T-ZnO are not continuous but split because of the quantum confine effect. When an energy level is in the range of microwave energy, the electron will absorb a photon to hop from a low energy level to a higher one. Also, the defects and suspending band can cause multiple scattering and interface polarization, which result in the electromagnetic wave absorption. Secondly, the wavelength of 2~18GHz electromagnetic wave is lager than the size of nano T-ZnO, which can reduce the electromagnetic wave reflection. It can easily lead to Rayleigh scattering when the incident electromagnetic wave reacts with the nano T-ZnO, which results in the electromagnetic wave absorption in all direction. Furthermore, it can be found that the coating is constituted of networks resulted from the tetraleg-shaped structure of nano T-ZnO, and nano T-ZnO have good conductive property in comparison with the common oxides, so it is available for the electromagnetic wave to penetrate the cellular material formed by the numerous conductive networks of nano T-ZnO and the energy will be induced into dissipative current, which leads to the energy attenuation. The earlier analysis of the related system indicates that the charge concentration at the needles' tip of the T-ZnO is distinct when the material is under an electric field because of the larger aspect ratio and the limited conductivity of the nano T-ZnO. So, it is reasonable that the concentrated tips will act as multipoles that will be tuned with the incident electromagnetic waves and contributes to strong absorption (Zhou et al., 2003). Besides above, the piezoelectric character of nano T-ZnO is also a factor of damaging the entered energy of microwave and reducing the reflectivity.

4. Absorption properties of T-ZnO plus CNTs / EP coatings

4.1 Fabrication of T-ZnO plus CNTs / EP coatings

As the multi-walled CNTs and T-ZnO nanostructures were prepared, the typical fabrication process of CNT/T-ZnO/EP composites is as follows. The calculated amount of mixed raw CNTs and T-ZnO nanostructures were sufficiently dispersed by ultrasonication for about 30 minutes. Then the mixture was added into EP, which was diluted by absolute ethyl alcohol, and dispersed by ultrasonication for about 30 minutes again, and then the curing agent was put into the composites, stirred gently. The mixture was sprayed layer by layer onto an aluminum plate with a square of 180 mm × 180 mm and cured at room temperature for at least 24 hours. Considering the preparing conditions, the thickness error of the epoxy composites was controlled to ± 0.1 mm. The measurement results are same within the range of the system measurement error range. A list of the microwave absorption properties of all manufactured samples is presented in Table 3. There are three samples of the same lot which have been prepared and tested, and we have obtained the average data of three samples to examine their absorption properties.

The morphologies of absorbents and CNTs/T-ZnO/EP composites were also observed by a FE-SEM (Zeiss, SUPRA-55), as shown in figure 12.

The measurements of reflectivity of the composites, complex permittivity and permeability of samples, and the measurement system refer to part 3.1. The sample was made by mixing nanoscaled T-ZnO/CNTs with molten paraffin into a ring for electromagnetic parameters measurement.

Sample Number	*CNT concentration (wt%)*	*T-ZnO concentration (wt%)*	*Thickness (mm)*	*Minimum reflection loss (dB)*	*Corresponding frequency (GHz)*	*Absorption bandwidth (<10GHz) (GHz)*
1 #	0	20	1.2±0.1	-3.48	10.24	0
2#	8	0	1.2±0.1	-7.83	18.00	0
3#	12	0	1.2±0.1	-9.35	18.00	0
4 #	20	0	1.2±0.1	-8.48	17.78	0
5 #	8	12	1.2±0.1	-11.21	16.16	1.5
6 #	12	8	1.2±0.1	-13.36	14.24	2.8
7 #	12	8	1.5±0.1	-23.07	12.16	5
8 #	12	8	2.2±0.1	-23.23	12.8	4.4
9 #	12	8	2.7±0.1	-19.95	8.16	2.56

Table 3. Microwave absorption properties of prepared CNTs/T-ZnO/EP composites

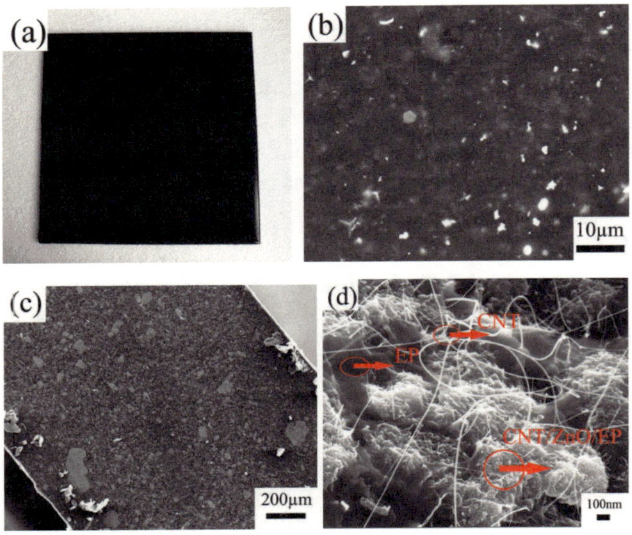

Fig. 12. Typical SEM images of CNTs/T-ZnO /EP composites: (a) and (b) coating surface; (c) and (d) fractured cross-section.

4.2 Microwave absorption properties of T-ZnO plus CNTs / EP coatings

1. **Impacts of concentration of T-ZnO plus CNTs on microwave absorption properties**

In order to investigate the impacts of concentration of CNTs and T-ZnO nanostructures on microwave absorption properties, the absorption properties of CNTs/T-ZnO/EP composites with thickness of 1.2 mm were measured as shown in Fig. 13 (Li et al., 2010). It can be seen that CNTs and T-ZnO nanostructures concentration has an obvious effect on microwave absorption properties. T-ZnO/EP and CNTs/EP composites have weak absorption performance. The value of the minimum reflection loss for T-ZnO/EP composite corresponding to sample 1# is -3.48 dB at 10.24 GHz, and for CNTs/EP composite corresponding to sample 2#, 3# and 4#, the value of the minimum reflection loss are -7.83 dB at 18.00 GHz, -9.35 dB at 18.00 GHz, and -8.48 dB at 17.78 GHz, respectively. CNTs/T-ZnO/EP composites corresponding to sample 5# achieve a maximum absorbing value of -11.21 dB at 16.16 GHz , and reflection loss is over 10 dB (90% absorption) between 15.52 GHz and 17.04 GHz, when the content of CNTs and T-ZnO nanostructures are 8 wt% and 12 wt%, respectively. The maximum absorption for CNTs/T-ZnO/EP composite corresponding to sample 6# reaches 13.36 dB at 14.24 GHz, and the reflection loss is over 10 dB between 13.28 GHz and 16 GHz, when the content of CNTs increases to 12 wt% and T-ZnO nanostructures decreases to 8 wt%, respectively. The curves indicate that the CNTs mixed with an appropriate amount of T-ZnO nanostructures can optimize the absorbent impedance matching and attenuation characteristics (Yusoff et al., 2002). The microwave absorption properties of CNTs/T-ZnO/EP composites are improved significantly with the content of CNTs and T-ZnO nanostructures being 12 wt% and 8 wt%, respectively. It is clear that the positions of microwave absorption peaks move towards the lower frequencies due to the increase of T-ZnO nanostructures amount. The result is similar to the previous report on CNTs/Ag-NWs coatings [19], which shows that the positions of microwave absorption peaks move towards the lower frequencies by filling the Ag nanowires into multi-walled CNTs. This indicates that the absorption peak frequency of the CNTs/T-ZnO/EP composites can be modulated easily by changing the amount of CNTs and T-ZnO nanostructures (Fan et al., 2006).

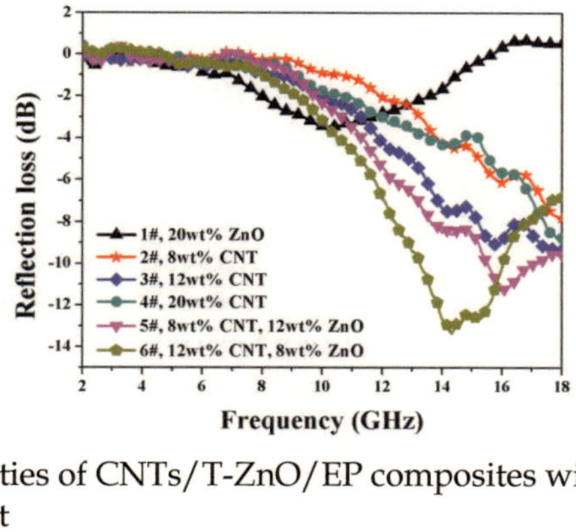

Fig. 13. Absorption properties of CNTs/T-ZnO/EP composites with different CNT and T-ZnO nanostructure content

About the mechanism of the CNTs/T-ZnO/EP, we think that the nanostructures of T-ZnO possess isotropic crystal symmetry play an important role in the process of microwave absorption. It can form isotropic quasiantennas and some incontinuous networks in the composites. Then, it is available for the electromagnetic wave to penetrate the nanocomposites formed by the numerous antenna-like semiconductive T-ZnO nanostructures and the energy will be induced into a dissipative current, and then the current will be consumed in the incontinuous networks, which lead to the energy attenuation [14, 15]. On the other hand, there are many interfaces between the epoxy matrix and CNTs outer surfaces. Compared with the CNTs, there are more interfaces between the T-ZnO and CNTs inner surfaces in the composites. Therefore, interfacial multipoles contribute to the absorption of the CNT/T-ZnO/EP composites (Zhao et al., 2008). These results were confirmed the Cao et al. by the theoretical calculation (Fang et al., 2010). Furthermore, the size, defects, and impurities also have effects on the microwave absorption property of the T-ZnO.

2. **Impacts of the coating thickness on microwave absorption properties**

To further study the influence of the composite thickness on microwave absorption properties, CNTs/T-ZnO/EP composites with various thickness were prepared by fixing CNTs and T-ZnO content of 12 wt% and 8 wt%, respectively. Figure 14(a) shows that the value of the minimum reflection loss for CNTs/T-ZnO/EP composite corresponding to sample 6# is -13.36 dB at 14.24 GHz with a thickness of 1.2 mm, and the bandwidth corresponding to reflection loss below -10 dB is 2.8 GHz. When the thickness of CNTs/T-ZnO/EP composite corresponding to sample 7# is 1.5 mm, the microwave absorption properties have been improved obviously. The value of the minimum reflection loss for CNTs/T-ZnO/EP composite is -23.00 dB at 12.16 GHz, and the bandwidth corresponding to reflection loss below -10 dB is 5 GHz (from 10.60 GHz to 15.60 GHz). Continuing to increase the thickness to 2.2 mm, the microwave absorption properties have little change. The value of the minimum reflection loss for CNTs/T-ZnO/EP composite corresponding to sample 8# is -23.24 dB at 12.71 GHz, and the bandwidth corresponding to the reflection loss below -10 dB is 4.4 GHz. When composite thickness increases to 2.7 mm, the value of the minimum reflection loss for CNTs/T-ZnO/EP composite corresponding to sample 9# is -19.95 dB at 8.16 GHz, the bandwidth correspondingly is 2.56 GHz. It can also be seen that the maximum absorbing peaks shift towards a lower frequency with the increase of composite thickness, which is due to the interference resonance vibration caused by electromagnetic wave and the coating (Zhang et al., 2008).

To clarify the impact of composite thickness on microwave absorption properties, we plotted a curve about the frequency dependence of absorption bandwidth and thickness of CNTs/T-ZnO/EP composites, as shown in Fig. 14(b). The frequency bandwidth of CNTs/T-ZnO/EP composites initially increases, and then decreases with the increase of the composite thickness. When the thickness is up to 1.5mm, the frequency bandwidth reaches the maximum value of 5 GHz. This indicates that the CNTs mixed with T-ZnO nanostructures have potential application as the broad frequency absorbing materials.

3. **Intrinsic reasons of T-ZnO plus CNTs for microwave absorption**

In order to investigate the intrinsic reasons for microwave absorption of CNTs and T-ZnO nanostructures composites, the complex permittivity and permeability of the studied samples were measured. The preparation of the samples can be seen in the experimental part. As CNTs and T-ZnO nanostructures are dielectric absorbents, the real and imaginary parts of the complex permittivity are shown in Fig. 15. Compared with nanoscaled CNTs and T-ZnO/CNTs composites, it is apparent that both the real and imaginary parts of the complex permittivity of T-ZnO nanostructures are greatly smaller. The tangent loss of permittivity of nanoscaled T-ZnO/CNTs are shown in Fig. 16. It can be seen that the values of tangent loss of permittivity of nanoscaled T-ZnO/CNTs are sensitive to the content of

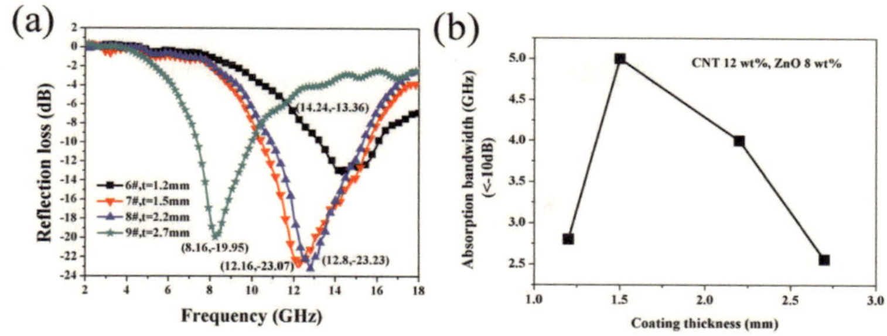

Fig. 14. (a) Absorption properties of CNTs/T-ZnO/EP composites with different thickness; (b) Frequency dependence of absorption bandwidth and coating thickness of CNTs/T-ZnO/EP composites

CNTs. The larger the CNTs mass fraction, the higher are the tangent loss constant. The sample with 20 wt% CNTs has the largest values. There is an obvious peak at 13 GHz for the sample with 12 wt% CNTs and 8 wt% T-ZnO nanostructures. So we calculated the reflection loss of 20% CNTs and 12% CNTs mixed with 8% T-ZnO nanostructures at different thickness as shown in Fig. 17. It is apparent that 12% CNTs mixed with 8% T-ZnO nanostructures gives the optimum microwave absorption. The minimum reflection loss is -12.20 dB at 17.2 GHz with 1.2 mm thickness. The maximum bandwidth under -5 dB (68% absorption) is 6.7 GHz with 1.5 mm thickness, which is consistent with the experimental results as shown in Fig. 14. The experiment and calculation results indicate that CNTs mixed T-ZnO nanostructures have excellent absorption property.

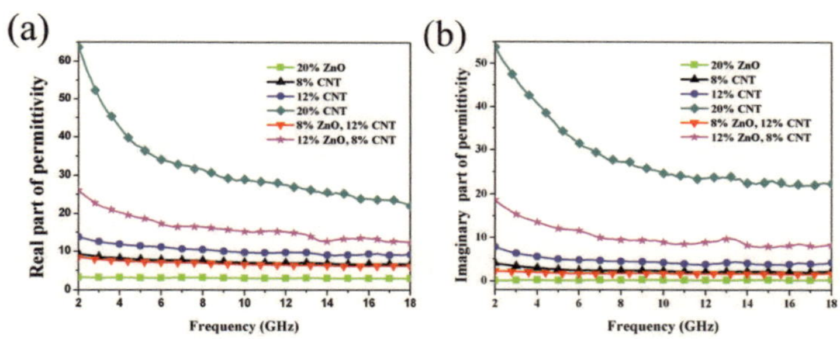

Fig. 15. Frequency dependence of complex permittivity: (a) the real part of permittivity; (b) the imaginary part of permittivity

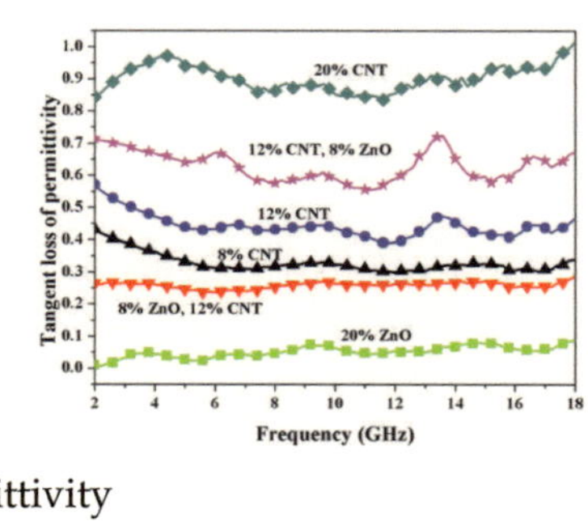

Fig. 16. Tangent loss of permittivity

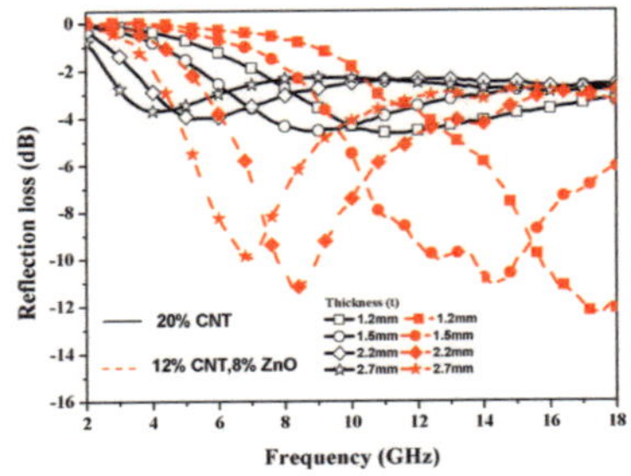

Fig. 17. Calculation of reflection loss of 20% and 12% CNTs mixed with 8% T-ZnO at different thickness

According to the results mentioned above, CNTs/T-ZnO/EP composites exhibit excellent microwave absorption properties compared with that of CNTs/EP and T-ZnO/EP composites. A continuous network formed by a small number of CNTs and some discontinuous networks formed by T-ZnO nanostructures coexist together with CNTs and EP in the composites. This special morphology is available for the electromagnetic wave to penetrate the composites and the energy will be induced into a dissipative current, and then the current will be consumed in the discontinuous networks, which leads to energy attenuation (Zhuo et al., 2008). More importantly, the microwave absorption of CNTs/T-ZnO/EP composites is attributed to interfacial electric polarization. The multi-interfaces between CNTs, T-ZnO nanostructures, EP and lots of agglomerates can be benefit for the microwave absorption because of the interactions of electromagnetic radiation with charge multipoles at the interface (Fang et al., 2010).

5. Absorption properties of three-dimensional ZnO micro-/nanorod networks

Both three-dimensional netlikes ZnO and T-ZnO composite samples used for microwave absorption measurement were prepared by mixing the ZnO netlike micro-/nanostructures and the T-ZnO with paraffin wax with 50 vol % of the ZnO, respectively. The composite samples were then pressed into cylindrical-shaped compact (φ_{out}: 7.00 mm; φ_{in}: 3.04 mm) for the tests of the complex permittivity ε and permeability μ. The measurement system and method are the same as part 3.1.

To investigate the microwave absorption property of netlike ZnO micro-/nanostructures, the complex permittivity ε of the netlike micro-/nanostructures was measured. For comparison, T-ZnO structures were corresponding studied. Figure 18 shows the plots of the frequency versus the complex permittivity of the primary ZnO netlike micro-/nanostructures composites with 50 vol % ZnO netlike nanostructures (SEM image shown in the insert of Fig. 18a) and T-ZnO composites with 50 vol % T-ZnO (See the insert in Fig. 18b). The real permittivity of T-ZnO composite is about 3.1, and the imaginary permittivity is about 0.2 (corresponding green curve). However, for the ZnO netlike micro-/nanostructures composite, the ε' and ε'' values show a complex variation (corresponding red curve). The real part of relative permittivity (ε') declines from 15 to 6 in the frequency range of 2-18 GHz (Fig. 18a). The imaginary part of permittivity (ε'') decreases from 4.4 to 3.6 and the curve exhibits two broad peaks in the 7-9 and 12-15 GHz ranges (Fig. 18b). It is worthy to notice that the peaks of the ε'' curve appear at 8 and 14 GHz, suggesting a resonance behavior, which is expected when the composite is highly conductive and skin effect becomes significant. The imaginary part ε'' of ZnO netlike micro-/nanostructures is relatively higher in contrast to that of T-ZnO composites, which implies the distinct dielectric loss properties arising from the morphology variation. It is reasonable that the dielectric loss is attributed to the lags of polarization between the 3D frame interfaces as the frequency is varied. ZnO netlike micro-/nanostructures possess more complicated interfaces than T-ZnO, resulting in better dielectric loss properties (Zhuo et al., 2008).

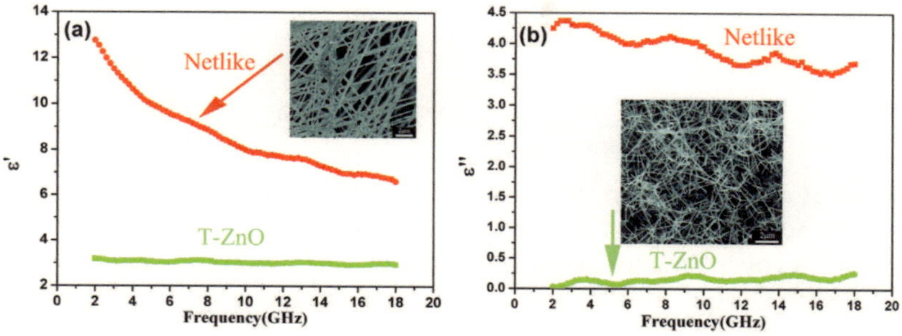

Fig. 18. The frequency dependence of (a) real ε' and (b) imaginary ε'' parts of relative complex permittivity for ZnO nanotetrapod composite and netlike structures composite. The SEM image of ZnO netlike micro-/nanostructures and T-ZnO nanostructures insert in Fig.4a and Fig.4b, respectively

To explain the microwave absorption properties of the samples measured above, the reflection loss (RL) of the netlike micro-/nanostructures and T-ZnO were calculated, respectively, using the relative complex permeability and permittivity at a given frequency and thickness layer according to the transmit-line theory, which follows the equations below (Sun et al., 2009):

$$Z_{in} = Z_0(\mu_r/\varepsilon)^{1/2}\tanh[j(2\pi fd/c)(\mu_r\varepsilon)^{1/2}] \quad (1)$$

$$RL(dB) = 20 \log |(Z_{in}-Z_0) / (Z_{in}+Z_0)| \quad (2)$$

where, f is the microwave frequency, d is the thickness of the absorber, c is the velocity of light, Z_0 is the impedance of air and Z_{in} is the input impedance of the absorber. The relative complex permeability and permittivity were tested on a network analyzer in the range 2-18 GHz. The simulations of the reflection loss of the two composites with a thickness of 2.0 mm are shown in Figure 19a. The ZnO netlike micro-/nanostructures composite possesses a strong microwave absorption property, the value of the minimum reflection loss for the composite is -30 dB at 14.4 GHz. However, the T-ZnO composite almost has no absorption. Figure 19b shows simulations of reflection loss of ZnO netlike micro-/nanostructures composite with different thicknesses. The value of the minimum reflection loss for the ZnO netlike micro-/nanostructures composite is -37 dB at 6.2 GHz with a thickness of 4.0 mm. Compared with the previous report, (Zhuo et al., 2008) in which the value of minimum reflection loss for the composite with 50 vol % ZnO dendritic nanostructures is -25dB at 4.2 GHz with a thickness of 4.0 mm, the ZnO netlike micro-/nanostructures have more excellent properties.

About the mechanism of microwave absorption, P. X. Yan et al. explained the ZnO nanotrees microwaves absorption performances using isotropic antenna mechanism. The random distribution of the isotropic quasi-antenna ZnO semiconductive crystals not only leads to diffuse scattering of the incident microwaves, which results in the attenuation of electromagnetic (EM) energy, but also acts as receivers of microwaves, which can produce

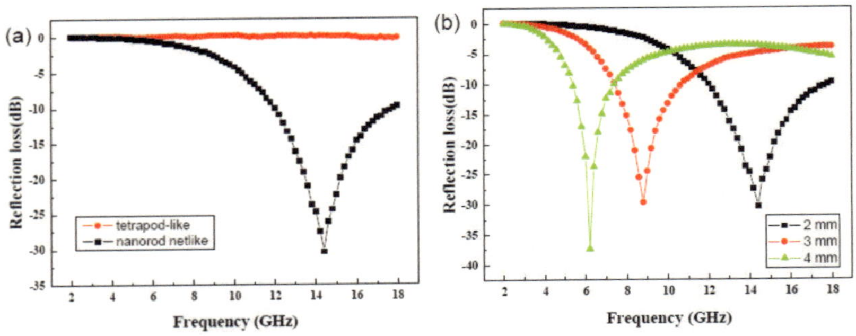

Fig. 19. (a) Simulation of reflection loss of 50 vol % ZnO nanotetrapod and 50 vol % ZnO netlike micro-/nanostructures composites with a thickness of 2.0 mm. (b) Simulation of reflection loss of 50 vol % ZnO netlikes micro-/nanostructures composite with different thickness

vibrating microcurrent in the local networks. Here, in our experiment, we noted that the ZnO netlike structures have special geometrical morphology. Such isotropic crystal symmetry can form continuous isotropic antennas networks in the composites. Moreover, it is available for the EM wave to penetrate the nanocomposites formed by the numerous conductive ZnO networks, and the energy will be induced into dissipative current by random distributed isotropic antennas, and then part of the current will generate EM radiation and the rest will be consumed in the discontinuous networks, which lead to the energy attenuation. Thus, the tanglesome network frame of ZnO nanostructures in the composite will induce a certain extent of conductive loss. Compared with ZnO netlike micro-/nanostructures, no complex frame exists in T-ZnO with the quasi-One-dimensional nanostructures as same as nanowire. A large part of EM radiation will counteract with each other when the orientation of these isotropic quasi-antennas distributes randomly. In a word, ZnO netlike nanostructures, acting as receiving antenna, can receive EM energy and transform it into dissipative current. And they also act as sending antenna transforming the vibrating current into EM radiation. Besides these, the interfacial electric polarization should also be considered. The multi-interfaces between isotropic antenna frame, and air bubbles can benefit for the microwave absorption because of the interactions of EM radiation with charged multipoles at the interfaces (Chen et al., 2004 & Zhuo et al., 2008).

6. Conclusion

1. The effective absorption of T-ZnO/ EP coatings is distributed in the wave band from 15GHz to 18GHz when the content of ZnO is 11-20 wt% and the coating thickness is 1.5mm. As the content of ZnO and the coating thickness increase to 28.6 wt% and 3.5mm, the absorption properties of the coatings improve significantly, especially in the wave band from 6GHz to 12GHz. The minimum reflection loss value is -9.11 dB at 8.80 GHz, and the bandwidth is 4.6 GHz as the reflection loss reaches below -5 dB.
2. For CNTs/ T-ZnO/ EP composite coatings, when the content of CNTs and T-ZnO nanostructures are 12 wt% and 8 wt%, respectively, and the coating thickness is 1.5 mm, the minimum reflection loss of CNTs/ T-ZnO/ EP coatings is -23.00 dB at 12.16 GHz, and the bandwidth corresponding to reflection loss below -5 dB is 7.8 GHz.
3. Both T-ZnO/ EP coatings and CNTs/ T-ZnO/ EP composite coatings have the potential applications in electromagnetic wave shielding, but the CNTs/ T-ZnO composites used

as the microwave absorbents are more effective in the conditions of the thin and light shielding coatings.

4. On the other hand, according to the complex permittivity and the reflection loss simulations of T-ZnO and ZnO networks used as the absorbent in the coatings with the ZnO proportion of 50 vol % and the thickness of 2.0 mm respectively, ZnO networks possess preferable microwave absorption property, and the minimum reflection loss value is -30 dB at 14.4 GHz.

7. Acknowledgements

This work was supported by the National Basic Research Program of China (Grant No. 2007CB936201), the Funds for International Cooperation and Exchange (Grant Nos. 50620120439, 2006DFB51000), the Science Foundation (Grant Nos. 50772011, NCET-07-0066) and the Fundamental Research Funds for the Central Universities.

8. References

Cao, J. W.; Huang, Y. H.; Zhang, Y.; Liao, Q. L.; Deng, Z. Q. Research on electromagnetic wave absorbing properties of nano tetraleg ZnO. (2008). *Acta Physica Sinica*, 6, 3641-3645.

Chen, Y. J.; Cao, M. S.; Wang, T. H.; Wan, Q. (2004). Microwave absorption properties of the ZnO nanowire-polyester composites. *Appl. Phys. Lett.*, 84, 3367-3369.

Dai, Y.; Zhang, Y.; Li, Q. K.; Nan, C. W. (2002). Synthesis and optical properties of tetrapod-like zinc oxide nanorods. *Chem. Phy. Lett.*, 358, 83-86.

Dai, Y.; Zhang, Y.; Wang, Z. L. (2003). The octa-twin tetraleg ZnO nanostructures, *Solid State Commun.*, 126, 629–633

Fan, Z. J.; Luo, G. H.; Zhang, Z. F.; Zhou, L.; Wei, F. (2006). Electromagnetic and microwave absorbing properties of multi-walled carbon nanotubes/polymer composites. *Mater. Sci. Eng. B*, 132, 85-89.

Fang, X. Y.; Cao, M. S.; Shi, X. L.; Hou, Z. L.; Song, W. L.; Yuan, J.; (2010). Microwave responses and general model of nanotetraneedle ZnO: Integration of interface scattering, microcurrent, dielectric relaxation, and microantenna. *J. Appl. Phys.*, 107, 054304.

Fujii, M.; Iwanaga, H.; Ichihara, M.; Takeuchi, S. (1993). Structure of tetrapod-like ZnO crystals. *J. Cryst. Growth*, 128, 1095-1098.

Iijima, S. (1991). Helical microtubules of graphitic carbon. *Nature*, 354, 56-58.

Kimura, S.; Kato, T.; Hyodo, T.; Shimizu, Y.; Egashira, M. (2007). Electromagnetic wave absorption properties of carbonyl iron-ferrite/PMMA composites fabricated by hybridization method. *J. Magn. Magn. Mater.*, 312, 181-186.

Li, H. F.; Huang, Y. H.; Sun, G. B.; Yan, X. Q.; Yang, Y.; Wang, J.; Zhang, Y. (2010). Directed Growth and Microwave Absorption Property of Crossed ZnO Netlike Micro-/ Nanostructures. *J. Phys. Chem. C*, 114, 10088–10091.

Li, H. F.; Wang, J,; Huang, Y. H.; Yan, X. Q.; Qi, J. J.; Liu, J.; Zhang, Y. (2010). Microwave absorption properties of carbon nanotubes and tetrapod-shaped ZnO nanostructures composites. *Mater. Sci. Eng. B*, 175, 81-85

Sun, G. B.; Zhang, X. Q.; Cao, M. H.; Wei, B. Q.; Hu, C. W. (2009). Facile Synthesis, Characterization, and Microwave Absorbability of CoO Nanobelts and Submicrometer Spheres. *J. Phys. Chem. C,* 113, 6948-6954.

Wang, X. D.; Song, J. H.; Wang, Z. L. (2007). Nanowire and nanobelt arrays of zinc oxide from synthesis to properties and to novel devices. *J. Mater. Chem.,* 17, 711-720.

Wang, Z. L.; Kong, X. Y.; Zuo, J. M. (2003). Induced Growth of Asymmetric Nanocantilever Arrays on Polar Surfaces. *Phys. Rev. Lett.,* 91, 185502.

Wang, Z.L.; Song, J.H. (2006). Piezoelectric Nanogenerators Based on Zinc Oxide Nanowire Arrays. *Science,* 312, 242-246.

Yusoff, A. N.; Abdullah, M. H.; Ahmad, S. H.; Jusoh, S. F.; Mansor, A. A.; Hamid, S. A. A. (2002). Electromagnetic and absorption properties of some microwave absorbers. *J. Appl. Phys.,* 92, 876-882.

Zhang, L.; Zhu, H.; Song, Y.; Zhang, Y. M.; Huang, Y. (2008). The electromagnetic characteristics and absorbing properties of multi-walled carbon nanotubes filled with Er_2O_3 nanoparticles as microwave absorbers. *Mater.Sci.Eng.B,* 153. 78–82.

Zhao, D. L.; Li, X.; Shen, Z. M. (2008). Electromagnetic and microwave absorbing properties of multi-walled carbon nanotubes filled with Ag nanowires. *Mater. Sci. Eng. B,* 150, 105-110.

Zhao, D. L.; Shen, Z. M. (2008). Preparation and microwave absorption properties of carbon nanocoils. *Mater. Lett.,* 62, 3704-3706.

Zhou, Z.; Chu, L.; Tang, W.; Gu, L. (2003). Studies on the antistatic mechanism of tetrapod-shaped zinc oxide whisker. *J. Electrostatics,* 57, 347-354.

Zhu, J.; Peng, H. L.; Chan, C. K.; Jarausch, K.; Zhang, X. F.; Cui, Y. (2007). Hyperbranched Lead Selenide Nanowire Networks. *Nano Lett.,* 7, 1095-1099.

Zhuo, R. F.; Feng, H. T.; Chen, J. T.; Yan, D.; Feng, J. J.; Li, H. J.; Geng, B. S.; Cheng, S.; Xu; X. Y.; Yan, P. X. (2008). Multistep Synthesis, Growth Mechanism, Optical, and Microwave Absorption Properties of ZnO Dendritic Nanostructures. *J. Phys. Chem. C.,* 112, 11767.

Zhuo, R. F.; Qiao, L.; Feng, H. T.; Chen, J. T.; Yan, D.; Wu, Z. G.; Yan, P. X. (2008). Microwave absorption properties and the isotropic antenna mechanism of ZnO nanotrees. *J. Appl. Phys.,* 104, 094101.

Zou, Y. H.; Jiang, L. Y.; Wen, S. C.; Shu, W. X.; Qing, Y. J.; Tang, Z. X.; Luo, H. L.; Fan, D. Y. (2008). Enhancing and tuning absorption properties of microwave absorbing materials using metamaterials. *Appl. Phys. Lett.,* 93, 261115.

Zou, Y. H.; Liu, H. B.; Yang, L.; Chen, Z. Z. (2006). The influence of temperature on magnetic and microwave absorption properties of Fe/graphite oxide nanocomposites. *J. Magn. Magn. Mater.,* 302, 343-347.

19

Composite Electromagnetic Wave Absorber Made of Soft Magnetic Material Particle and Metal Particle Dispersed in Polystyrene Resin

Kenji Sakai, Norizumi Asano, Yoichi Wada,
Yang Guan, Yuuki Sato and Shinzo Yoshikado
Doshisha University
Japan

1. Introduction

The development of an electromagnetic wave absorber suitable for frequencies higher than 1 GHz is required with the increasing use of wireless telecommunication systems. Moreover, new wireless telecommunication devices will be developed owing to the advances in the information and telecommunication field in the future. For this reason, electromagnetic wave absorber suitable for these new devices is required, especially at frequencies above 10 GHz. In this chapter, we deal with a metal-backed single-layer absorber that has a low cost and is easy to fabricate. To design a metal-backed single-layer absorber, the control of the frequency dependences of the relative complex permeability μ_r^* and the relative complex permittivity ε_r^* is important because the absorption of an electromagnetic wave is determined by both μ_r^* and ε_r^*. In particular, μ_r', the real part of μ_r^*, must be less than unity to satisfy the non-reflective condition of electromagnetic wave for a metal-backed single-layer absorber at frequencies above 10 GHz. Therefore, the frequency dependences of μ_r^*, ε_r^*, and the absorption characteristics of a composite made of a soft magnetic material dispersed in an insulating matrix have been investigated ([a]Kasagi et.al., 1999; Lim et. al., 2005; Song et. al., 2005). However, the frequency dependences of μ_r' and μ_r'', the imaginary part of μ_r^*, has not been investigated at frequencies above 10 GHz.

Thus, for the purpose of designing electromagnetic wave absorbers with good absorption properties at frequencies above 10 GHz, the frequency dependences of μ_r^*, ε_r^*, and the return loss were investigated for the composite made of particles of soft magnetic material dispersed in polystyrene resin in the frequency range from 100 MHz to 40 GHz. Soft magnetic material used in this study is sendust (an alloy of Al 5%, Si 10%, and Fe 85%), because sendust is a low-cost material and does not contain any rare metals and sendust is suitable for use in a practical absorber. In addition, the frequency dependences of μ_r^*, ε_r^*, and the return loss of the composite made of permalloy (an alloy of Ni 45%, Fe 55%) were also investigated for the comparison. Both sendust and permalloy have high permeability values in the frequency range above 1 GHz. This characteristic makes it possible to fabricate an electromagnetic wave absorber suitable for this frequency band. Moreover, the values of μ_r' for the composite made of sendust or permalloy dispersed in polystyrene resin are expected to be less than unity because of the natural magnetic resonance and magnetic moment

	Sendust [vol%]	Aluminum [vol%]
Composite A	25	25
Composite B	12.5	37.5
Composite C	10	40

Table 1. The volume mixture ratios of the composites made of sendust and aluminum particles.

generated by the induced eddy current ([b]Kasagi et.al., 2006). This characteristic allows electromagnetic wave absorption at frequencies exceeding 10 GHz because μ_r' must be less than unity for a constant relative dielectric constant ε_r' and the composite made of soft magnetic material dispersed in polystyrene resin has constant value of ε_r'.

Furthermore, a composite electromagnetic wave absorber that can adjust the absorption characteristics by controlling μ_r' and μ_r'' are discussed in this chapter. The frequency dependence of μ_r^* for the composite made of soft magnetic material and polystyrene resin is similar to that required to satisfy the non-reflective condition. Therefore, the absorption with a wide frequency is expected if an optimum frequency dependence of μ_r^* can be obtained. However, the frequency dependence of μ_r^* is determined by the magnetic property, such as magnetic anisotropy. Thus, flexible control of the frequency dependence of μ_r^* is difficult using the composite made of a soft magnetic material and polystyrene resin.

Meanwhile, it has been reported that the composite made of metal particles, such as aluminum particles, dispersed in polystyrene resin can control the values of both μ_r' and μ_r'' by the volume mixture ratio and the size of metal particle, and can be used as an electromagnetic wave absorber (Nishikata, 2002; Sakai et. al., 2008). In particular, the value of μ_r' for the composite made of aluminum and polystyrene becomes less than unity. Thus, the frequency where the absorption of electromagnetic waves occurs can be controlled by adjusting μ_r' and μ_r''. However, the values of μ_r' and μ_r'' for this composite are almost independent of frequency, hence the non-reflective condition is satisfied in a narrow frequency range and the bandwidth of absorption is narrow. From above results, if both soft magnetic material and metal particles are dispersed in polystyrene resin, the frequency dependence of μ_r^* is expected to be controlled flexibly so that the non-reflective condition is satisfied. Therefore, the frequency dependences of μ_r^*, ε_r^*, and the absorption characteristics of a composite made of both sendust and aluminum particles dispersed in polystyrene resin were also evaluated and the flexible design of an electromagnetic wave absorber is discussed.

2. Experiment

Commercially available sendust (Al 5%, Si 10%, Fe 85%) particles and permalloy (Ni 45%, Fe 55%) particles were used. The sendust and permalloy particles were granular. The compositions of sendust and permalloy were confirmed using a scanning electron microscope (SEM) and energy-dispersive X-ray spectrometry (EDX). The average particle size (diameter) of the permalloy particles was approximately 10 μm and that of sendust particles was approximately 5 μm. For the composite made of both sendust and aluminum particles dispersed in polystyrene resin, commercially available aluminum particles of approximately 8 μm were used. The volume mixture ratios of sendust and aluminum particles for the composite made of both sendust and aluminum particles are shown in Table 1.

Chips of polystyrene resin were dissolved in acetone. The dissolved polystyrene resin and sendust or permalloy particles were mixed to uniformly disperse and isolate the particles. After mixing, the mixture was heated to melt the polystyrene resin and was then hot-pressed at a pressure of 5 MPa into a pellet. This was allowed to cool naturally to room temperature and was processed into a toroidal-core shape (outer diameter of approximately 7 mm, inner diameter of approximately 3 mm) for use in a 7 mm coaxial line in the frequency range 100 MHz to 12.4 GHz, or into a rectangular shape (P-band: 12.4-18 GHz, 15.80 mm × 7.90 mm, K-band: 18-26.5 GHz, 10.67 mm × 4.32 mm, R-band: 26.5-40 GHz, 7.11 mm × 3.56 mm) for use in a waveguide. The sample was mounted inside the coaxial line or waveguide using silver past to ensure that no gap existed between the sample and the walls of the line/waveguide.

The complex scattering matrix elements S_{11}^* (reflection coefficient) and S_{21}^* (transmission coefficient) for the TEM mode (coaxial line) or TE_{10} mode (rectangular waveguide) were measured using a vector network analyzer (Agilent Technology, 8722ES) by the full-two-port method in the frequency range from 100 MHz to 40 GHz. The values of μ_r^* ($\mu_r^* = \mu_r' - j\mu_r''$, $j = \sqrt{-1}$) and ε_r^* ($\varepsilon_r^* = \varepsilon_r' - j\varepsilon_r''$) were calculated from the data of both S_{11}^* and S_{21}^*. The complex reflection coefficient Γ^* for a metal backed single layer absorber was then determined from the values of μ_r^* and ε_r^*. The return loss R for each sample thickness was calculated from Γ^* using the relation $R = 20\log_{10}|\Gamma^*|$. R was calculated at 0.1 mm intervals in the sample thickness range 0.1 to 30 mm.

3. Results and discussion

3.1 Frequency dependences of μ_r^* and ε_r^* for the composite made of soft magnetic material dispersed in polystyrene resin

Figure 1 shows the frequency dependences of μ_r' and μ_r'' for the composite made of sendust or permalloy particles dispersed in polystyrene resin. The values of μ_r' decreased gradually with increasing frequency and became less than unity at frequencies around 10 GHz. The values of μ_r'' increased with increasing frequency, had maximum at frequencies of several GHz and decreased. As seen in Figs. 1(a) and (b), frequency dependences of μ_r' and μ_r'' for the composite made of sendust were similar to those for the composite made of permalloy. This result indicates that the magnetic property of the composite made of sendust or permalloy in the GHz range was almost the same. On the other hand, the values of both μ_r' and μ_r'' increased and the shape of frequency dependences of μ_r' and μ_r'' were different when the volume mixture ratio of sendust increased. This is because the amount of magnetic material increased and the response of the incident magnetic field changed. The variation of the frequency dependences of μ_r' and μ_r'' by the volume mixture ratio of magnetic material suggests that the absorption characteristics can be adjusted for the desired frequency range by simply changing the volume mixture ratio of magnetic material. However, when the amount of magnetic material increased or decreased, the values of ε_r' and ε_r'' increased or decreased, as shown in Fig. 2.

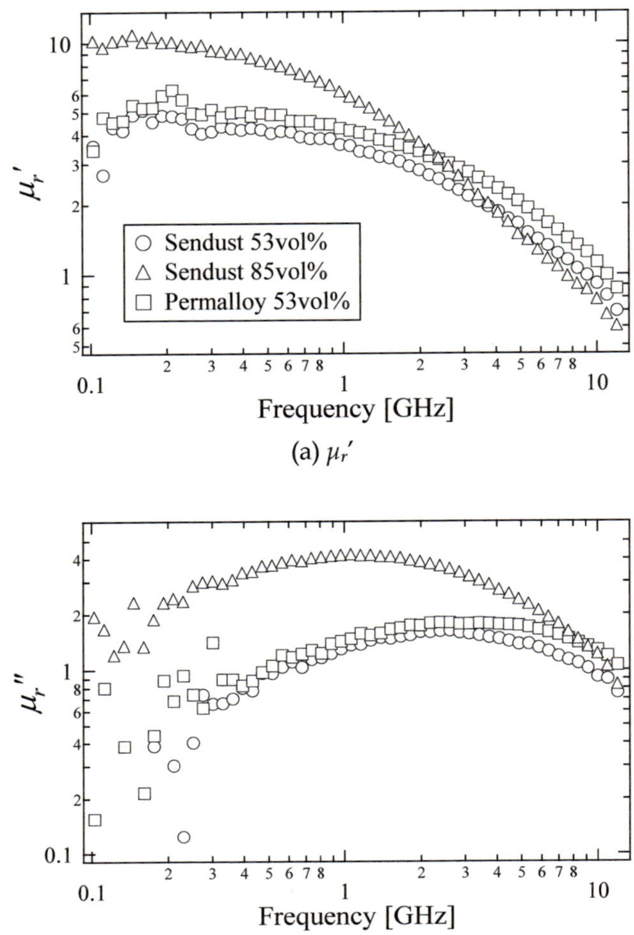

Fig. 1. Frequency dependences of μ_r' and μ_r'' for composites made of sendust or permalloy particles dispersed in polystyrene resin.

Figure 2 shows the frequency dependences of ε_r' and ε_r'' for the composite made of sendust or permalloy particles dispersed in polystyrene resin. Although, the values of ε_r' and ε_r'' for the composite made of sendust were similar to those for the composite made of permalloy when the volume mixture ratio of sendust and permalloy is the same, both ε_r' and ε_r'' increased markedly for the composite made of 85 vol%-sendust particles. Therefore, the increase or decrease in ε_r' and ε_r'' should be considered to design an absorber when the volume mixture ratio of magnetic material is varied.

To investigate the difference of frequency dependences of μ_r^* and ε_r^* for various magnetic materials, nickel and ferrite particles were dispersed in polystyrene resin and the frequency dependences of μ_r^* and ε_r^* were evaluated in the frequency range from 100 MHz to 40 GHz.

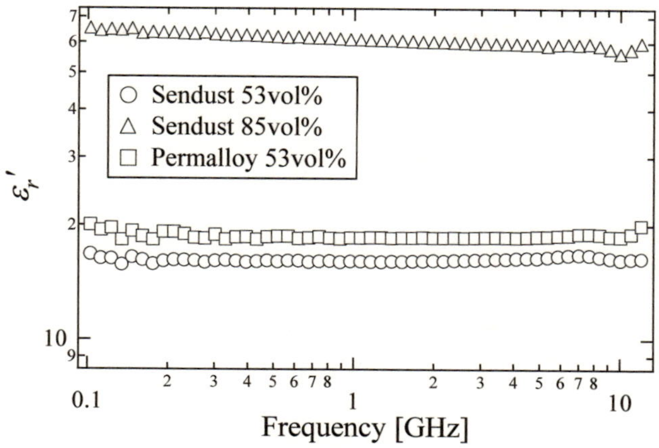

(a) ε_r'

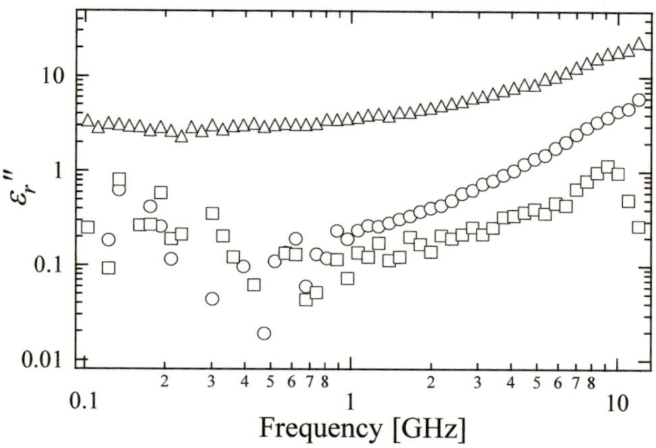

(b) ε_r''

Fig. 2. Frequency dependences of ε_r' and ε_r'' for composites made of sendust or permalloy particles dispersed in polystyrene resin.

The average particle size of nickel was between 10 and 20 μm and that of ferrite was 5 μm and 10 nm. The volume mixture ratio of nickel particles and ferrite particles of 5 μm was 50 vol%. Ferrite particles of 10 nm were difficult to disperse uniformly in polystyrene resin with increasing the amount of ferrite. Thus, the volume mixture ratio of ferrite particles of 10 nm was 40 vol% where the ferrite particles could be dispersed uniformly. Figures 3 and 4 show the frequency dependences of μ_r' and μ_r'' for the composite made of various magnetic materials. As shown in Fig. 3, the frequency dependences of μ_r' and μ_r'' for the composite made of ferrite particles of 10 nm were considerably different from those of other composite. It is speculated that the particle size of magnetic material affects the frequency dependences of μ_r' and μ_r'' because ferrite particles of 10 nm is much smaller than other magnetic material particles. The frequency dependences of μ_r' and μ_r'' for the composite made of sendust, nickel, or ferrite of 5 μm were similar. However, for the composite made of sendust, the

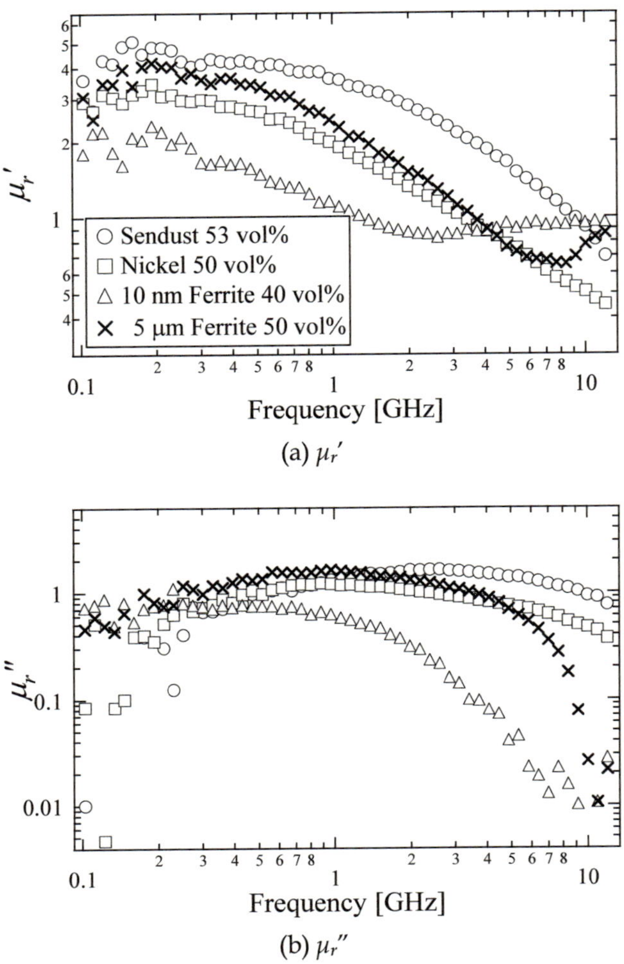

Fig. 3. Frequency dependences of μ_r' and μ_r'' for composites made of sendust, nickel or ferrite particles dispersed in polystyrene resin. The frequency range is from 100 MHz to 12.4 GHz.

frequency where μ_r' begins to decrease and μ_r'' is maximum was higher than that of other composites. This result is speculated to be the property that sendust and permalloy have high permeability values in the high frequency range.

On the other hand, in the high frequency range, the composite made of sendust, permalloy, nickel, or 10 nm-ferrite showed different frequency dependences of μ_r' and μ_r'', as shown in Figs. 4(a) and (b). For the composite made of 10 nm-ferrite, the values of μ_r' was almost 1 and μ_r'' was almost zero. This result suggests that the composite made of 10 nm-ferrite has no magnetic response to the incident electromagnetic wave of high frequency. However, the values of μ_r' for the composite made of nickel was minimum near 15 GHz and increased up to 20 GHz. Then, μ_r' decreased with increasing frequency. Moreover, the values of μ_r'' decreased up to 20 GHz and increased with increasing frequency. These frequency dependences of μ_r' and μ_r'' are different from those for the composite made of ferrite

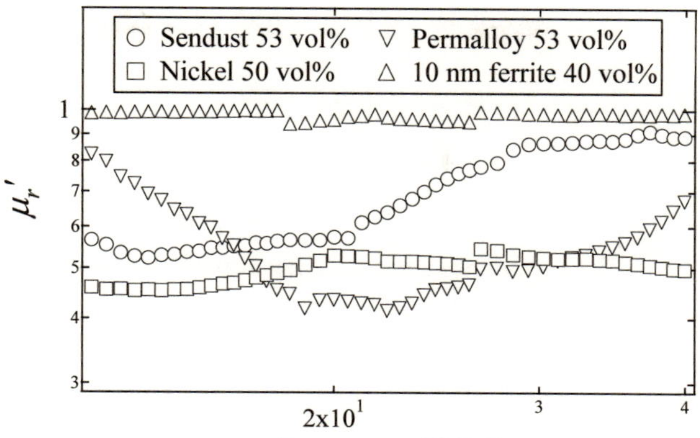

(a) μ_r'

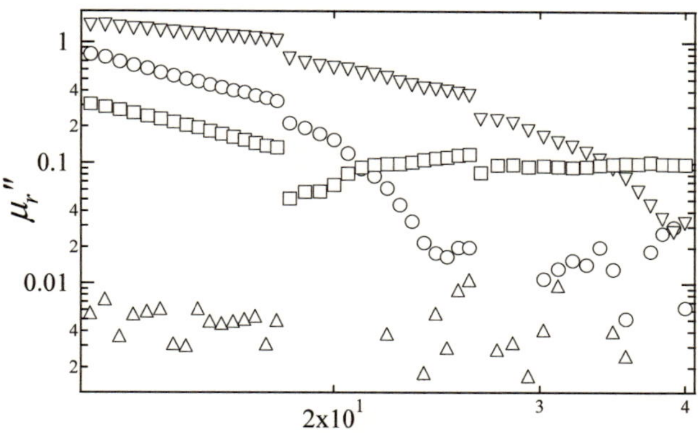

(b) μ_r''

Fig. 4. Frequency dependences of μ_r' and μ_r'' for composites made of sendust, nickel or ferrite particles dispersed in polystyrene resin. The frequency range is from 12.4 to 40 GHz.

although nickel and ferrite are magnetic material. These differences are speculated to be as follows.

Nickel is a magnetic material and is conductive. The resistivity of nickel was 6.84×10^{-8} Ωm and the skin depth δ of nickel is estimated to be approximately 1.3 μm at 1 GHz. Thus, the eddy current flows on the surface of nickel particles when an electromagnetic wave of high frequency enters inside nickel particle. We have reported that this phenomenon was observed in the composite made of aluminum particles dispersed in polystyrene resin (Sakai et. al., 2008). Thus, the effects of both the natural magnetic resonance caused by the magnetism and the magnetic moments generated by the eddy current are observed in the composite made of nickel. In particular, the effect of natural magnetic resonance on μ_r' and μ_r'' caused by the magnetism of nickel decreases as the frequency increases far from the resonance frequency and the effect of magnetic moment is dominant in the high frequency

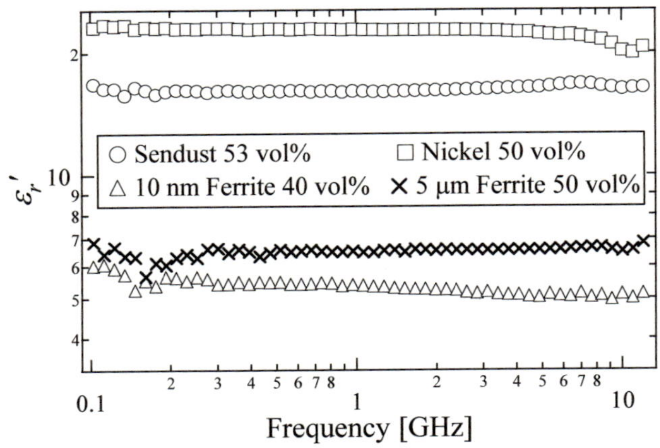

(a) ε_r'

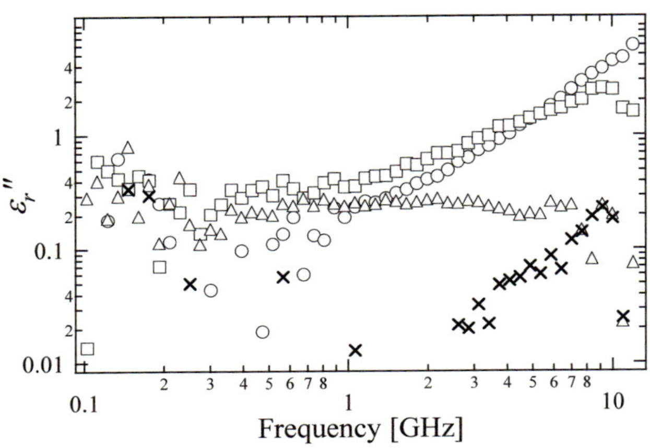

(b) ε_r''

Fig. 5. Frequency dependences of ε_r' and ε_r'' for composites made of sendust, nickel or ferrite particles dispersed in polystyrene resin. The frequency range is from 100 MHz to 12.4 GHz.

range. Therefore, the composite made of nickel had different frequency dependences of μ_r' and μ_r'' in the high frequency range.

The values of μ_r' for the composite made of sendust were minimum near 13 GHz, increased gradually, and were almost a constant value of 0.9. Moreover, μ_r'' decreased with increasing frequency and was almost constant or fractional increase in the high frequency range, as shown in Fig. 4(b). A similar frequency dependences of μ_r' and μ_r'' were obtained for the composite made of permalloy. These frequency dependences for the composite made of sendust or permalloy are similar to those for the composite made of nickel and the mechanism of the frequency dependences of μ_r' and μ_r'' is speculated to be explained by the same reason as that of nickel composite because sendust and permalloy are conductive. The above results indicate that an absorber which can operate at frequencies above 10 GHz is

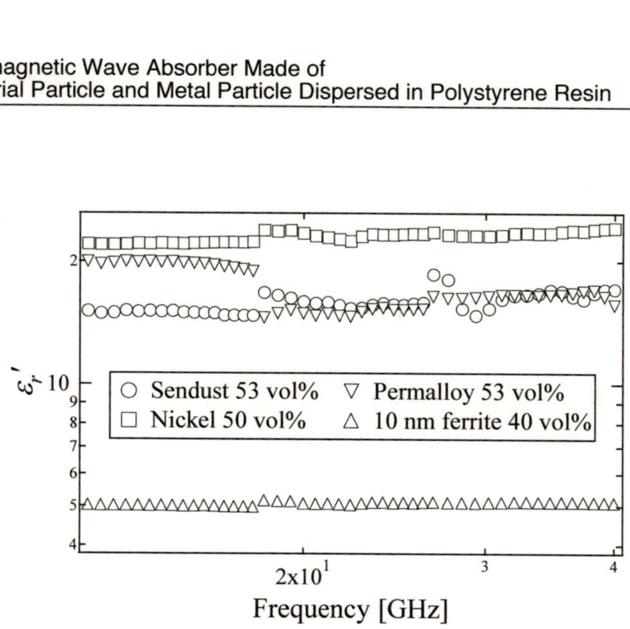

(a) ε_r'

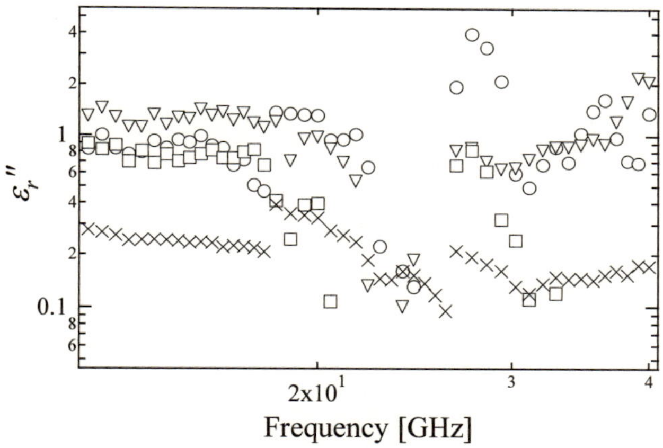

(b) ε_r''

Fig. 6. Frequency dependences of ε_r' and ε_r'' for composites made of sendust, nickel or ferrite particles dispersed in polystyrene resin. The frequency range is from 12.4 to 40 GHz.

possible using a composite made of soft magnetic material particles dispersed in polystyrene resin because the values of μ_r' and μ_r'' has frequency dispersion in the high frequency range and the non-reflective condition can be satisfied in this frequency range.
Figures 5 and 6 show the frequency dependences of ε_r' and ε_r'' for the composite made of various magnetic materials. The values of ε_r' for all composite were almost constant to the frequency in spite of the difference of the magnetic material. However, the values of ε_r' and ε_r'' for the composite made of ferrite were low whereas those for the composite made of nickel were high. This is caused by the difference of the resistivity; ferrite has a high value of resistivity.

3.2 Frequency dependences of return loss for the composite made of soft magnetic material dispersed in polystyrene resin

The frequency dependence of the return loss in free space was calculated from the measured values of μ_r^* and ε_r^* to investigate the absorption characteristics of the composite. The absorber used for the calculation was a metal-backed single layer absorber and the incident electromagnetic wave was perpendicular to the surface. Figures 7 and 8 show the frequency dependences of return loss for the composite made of sendust and that made of permalloy. The percentages shown in the graphs represent the normalized −20 dB bandwidth (the bandwidth Δf corresponding to a return loss of less than −20 dB divided by the absorption center frequency f_0). A value of −20 dB corresponds to the absorption of 99% of the electromagnetic wave power.

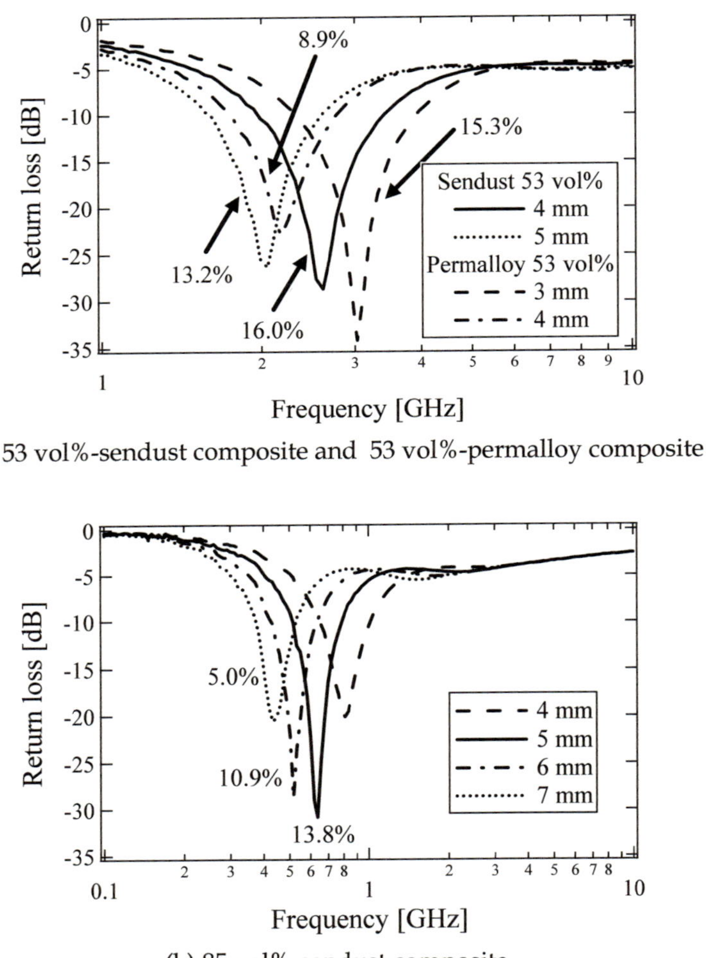

Fig. 7. Frequency dependences of return loss for composites made of sendust or permalloy particles dispersed in polystyrene resin.

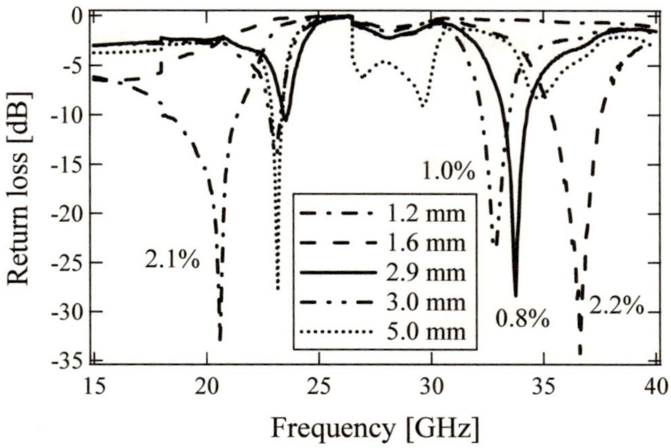

(a) 53 vol%-sendust composite

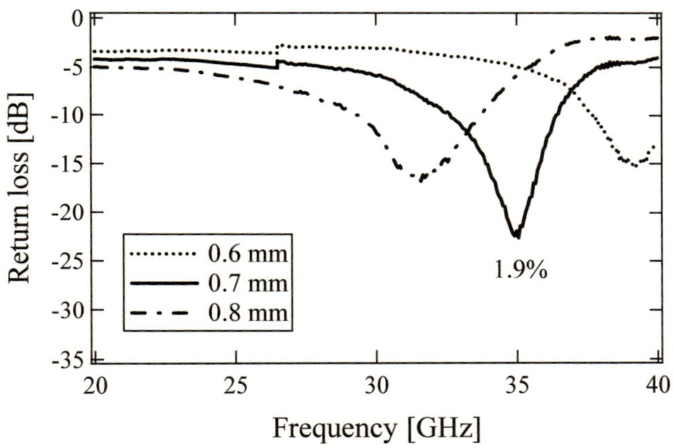

(b) 53 vol%-permalloy composite

Fig. 8. Frequency dependences of return loss for composites made of sendust or permalloy particles dispersed in polystyrene resin.

The composite made of sendust and that made of permalloy with the volume mixture ratio of 53 vol% exhibited a return loss of less than −20 dB in the frequency range of several GHz, as shown in Fig. 7(a). The sample thickness where the return loss is less than −20 dB was thin; several millimeters, and the normalized −20 dB bandwidth was over 10% (the normalized −20 dB bandwidth of commercially available absorber in the frequency range of several GHz ranges from 10 to 30%). Thus, it was found that the composite made of sendust particles dispersed in polystyrene resin can be used as a practical absorber. Moreover, as shown in Fig. 8(a), the return loss for the composite made of sendust was less than −20 dB near 20 and 36 GHz although the value of normalized −20 dB bandwidth was small. It is speculated that the absorption characteristics for the composite made of sendust can be improved in the high frequency range by selecting a suitable volume mixture ratio of sendust, because μ_r' and μ_r'' varied when the amount of sendust was changed, as discussed in section 3.1.

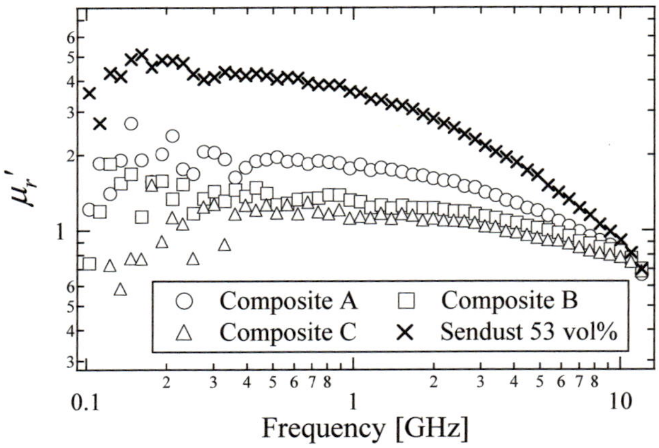

(a) μ_r'

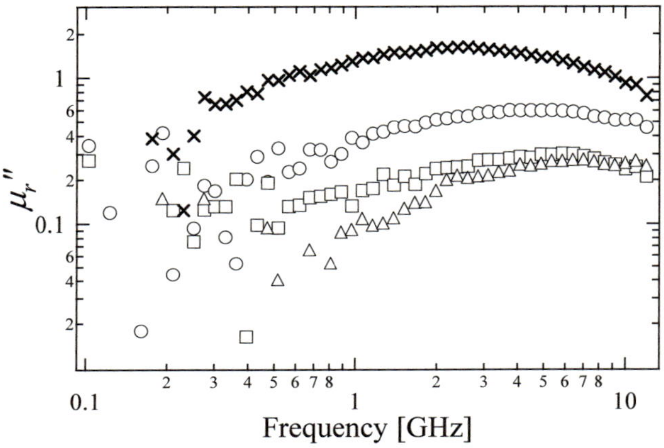

(b) μ_r''

Fig. 9. Frequency dependences of μ_r' and μ_r'' for composites made of both sendust and aluminum particles dispersed in polystyrene resin. The frequency range is from 100 MHz to 12.4 GHz.

The composite made of sendust with the volume mixture ratio of 85 vol% did not have a return loss of less than −20 dB at frequencies above 1 GHz although the return loss was less than −20 dB in the frequency range of several hundred MHz. This is because the values of ε_r' and ε_r'' increased markedly with a high volume mixture ratio of sendust, as shown in section 3.1. Thus, it is speculated that good absorption characteristic can be obtained with a low volume mixture ratio of sendust.

3.3 Frequency dependences of μ_r^* and ε_r^* for the composite made of both sendust and aluminum particles dispersed in polystyrene resin

Figures 9 and 10 show the frequency dependences of μ_r' and μ_r'' for the composite made of both sendust and aluminum particles dispersed in polystyrene resin. At frequencies of below 10 GHz, the values of both μ_r' and μ_r'' were low and the shape of the frequency

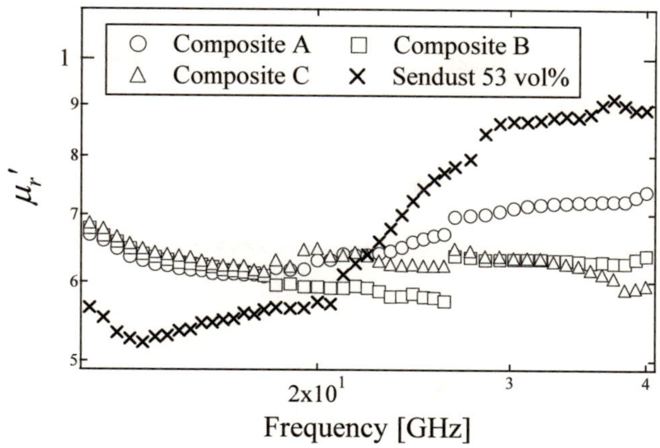

(a) μ_r'

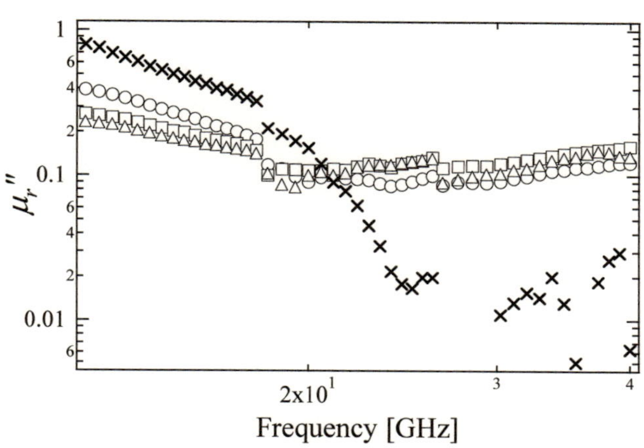

(b) μ_r''

Fig. 10. Frequency dependences of μ_r' and μ_r'' for composites made of both sendust and aluminum particles dispersed in polystyrene resin. The frequency range is from 12.4 to 40 GHz.

dependence of μ_r' became flat with increasing the amount of aluminum particle. This result indicates that the magnetic response to the incident electromagnetic wave reduced because the magnetic material of sendust dcreased and the non-magnetic material of aluminum increased. Meanwhile, in the high frequency range, the values of μ_r' for the composite made of both sendust and aluminum were less than unity and the magnitude of μ_r' was smaller than that for the composite made of only sendust. Moreover, the values of μ_r'' for the composite made of both sendust and aluminum increased gradually at frequencies above 20 GHz. These phnomena observed in the high frequency range is considered to be that the eddy current flowed the surface of aluminum particle and the magnetic moment and magnetic loss caused by the eddy current affected the frequency dependences of μ_r' and μ_r''. From these results, it is found that the frequency dependences of μ_r' and μ_r'' that is not obtained by the composite made of only sendust can realize using the composite made of

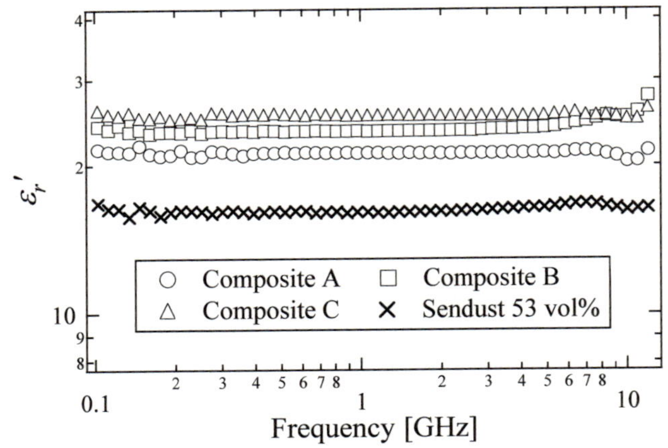

(a) ε_r'

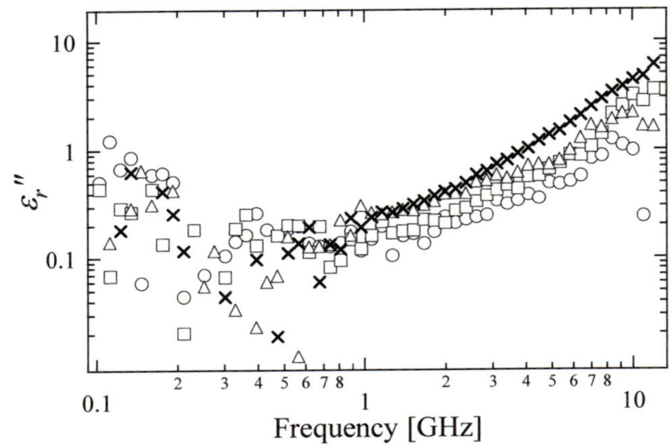

(b) ε_r''

Fig. 11. Frequency dependences of ε_r' and ε_r'' for composites made of both sendust and aluminum particles dispersed in polystyrene resin. The frequency range is from 100 MHz to 12.4 GHz.

both sendust and aluminum, and the values of μ_r' and μ_r'' is changed to satisfy satisfy the non-reflective condition by simply adjusting the total volume and ratio of sendust and aluminum. Furthermore, when changing the ratio of sendust and aluminum with the same total volume of sendust and aluminum, the variation of ε_r' and ε_r'' was a little although the frequency dependences of μ_r' and μ_r'' changed considerably, as shown in Figs. 11 and 12. Figures 11 and 12 show the frequency dependences of ε_r' and ε_r'' for the composite made of both sendust and aluminum particles dispersed in polystyrene resin. Therfore, the composite made of both sendust and aluminum can control the values of μ_r' and μ_r'' reducing the increase or decrease in ε_r' and ε_r''. This leads to a flexible design of an electromagnetic wave absorber compared with the conventional method of designing an absorber.

3.4 Frequency dependences of return loss for the composite made of both sendust and aluminum particles dispersed in polystyrene resin

Figure 13 show the frequency dependences of return loss for the composite made of both sendust and aluminum in the freqeuncy rage from 1 to 10 GHz. Composites A, B, and C had a return loss of less than −20 dB at a sample thickness of several millimeter. However, the value of normalized −20 dB bandwidth was small compared with that for the composite made of only sendust as shown in Fig. 7. This is because the frequency dependences of μ_r' and μ_r'' became flat and the frequency range where the non-reflective condition is satisfied was narrow.

Figure 14 show the the frequency dependences of return loss for composites B and C in the freqeuncy rage from 15 to 40 GHz. The return loss of composite A was not less than −20 dB in this frequency range. As shown in Fig. 14, composites B and C exhibited a return loss of

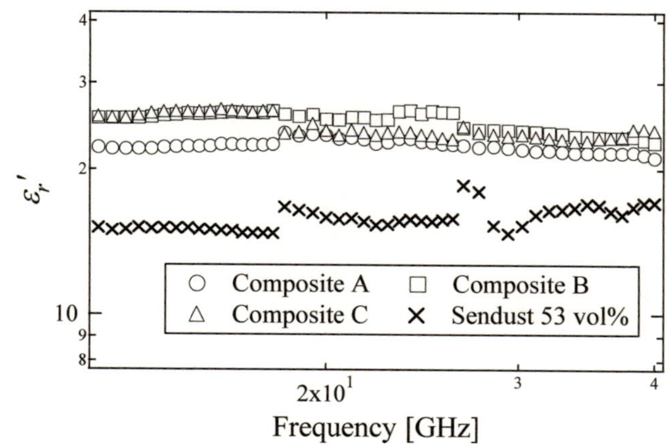

(a) ε_r'

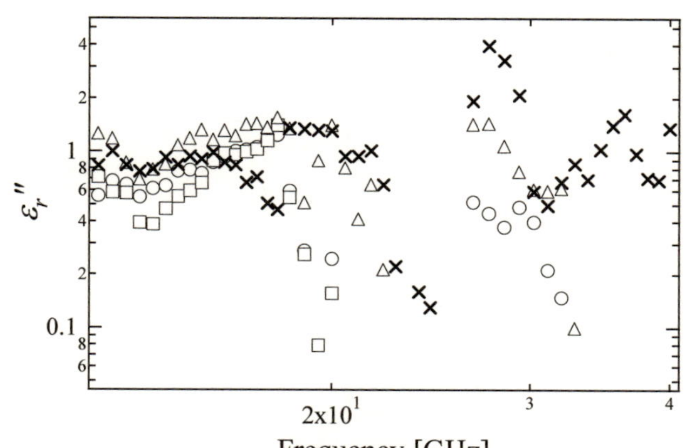

(b) ε_r''

Fig. 12. Frequency dependences of ε_r' and ε_r'' for composites made of both sendust and aluminum particles dispersed in polystyrene resin. The frequency range is from 12.4 to 40 GHz.

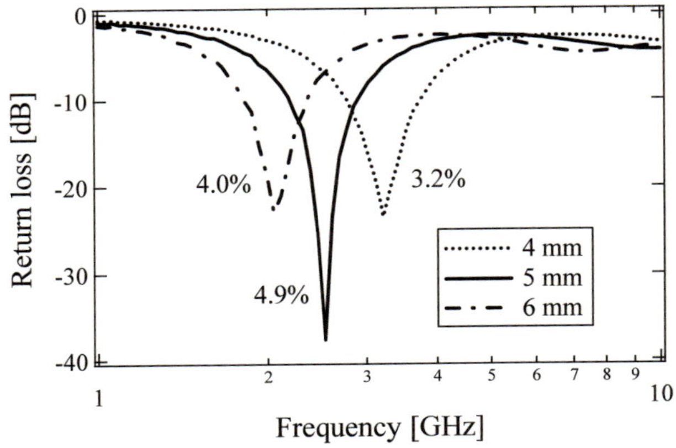

(a) Composite A (Sendust 25 vol%, Al 25 vol%)

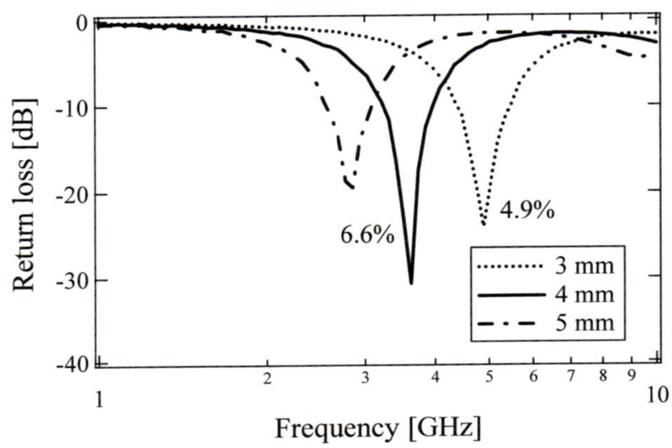

(b) Composite B (Sendust 12.5 vol%, Al 37.5 vol%)

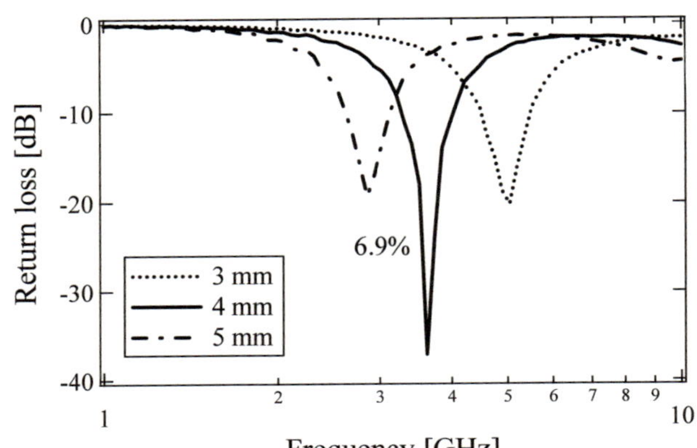

(c) Composite C (Sendust 10 vol%, Al 40 vol%)

Fig. 13. Frequency dependences of return loss for composites made of both sendust and aluminum particles dispersed in polystyrene resin. The frequency range is from 1 to 10 GHz.

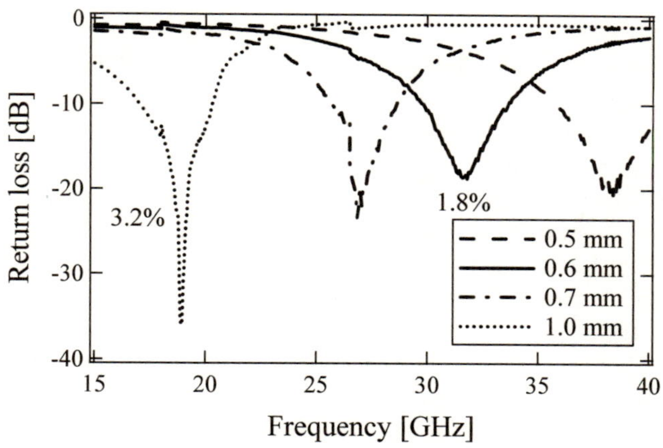

(a) Composite B (Sendust 12.5 vol%, Al 37.5 vol%)

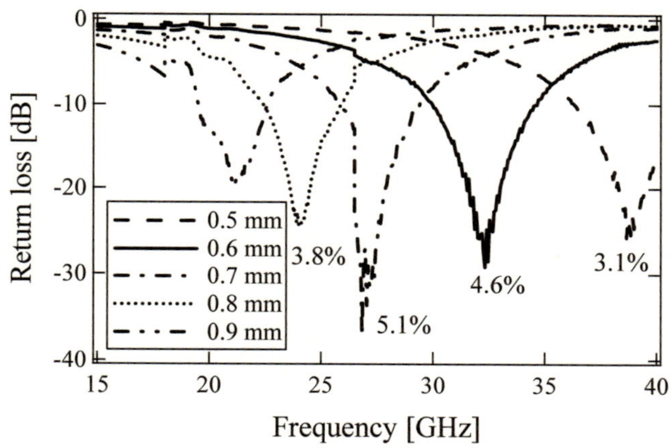

(b) Composite C (Sendust 10 vol%, Al 40 vol%)

Fig. 14. Frequency dependences of return loss for composites made of both sendust and aluminum particles dispersed in polystyrene resin. The frequency range is from 15 to 40 GHz.

less than −20 dB at frequencies above 15 GHz and the normalized −20 dB bandwidth of these two composites was broader than that for the composite made of only sendust. In addition, the sample thicknesses where the return loss is less than −20 dB were very thin. In particular, composite C had high values of normalized −20 dB bandwidth in spite of the high frequency range and the frequency where the return loss is less than −20 dB can be selected by changing the sample thickness in the range from 0.5 to 0.7 mm.

It is concluded from these results that the absorption characteristics of the composite made of sendust in the high frequency range could be improved by adding aluminum particles. Moreover, aluminum is low cost, abundant chemical element, and light wight. For example, the mass densities of composites A, B and C were approximately 2.4, 2.1, and 2.0 g / cm^3, respectively, while the mass density of the composite made of 53 vol%-sendust is approximately 3.5 g / cm^3. Therefore, a light-weight absorber can be fabricated by incorporating both sendust and aluminum.

4. Conclusion

Frequency dependences of μ_r^*, ε_r^*, and absorption characteristics for the composite made of sendust were investigated in the frequency range from 100 MHz to 40 GHz. By the comparison with the composite made of various magnetic material, the mechanism of the frequency dependences of μ_r' and μ_r'' was found to be the magnetic resonance in the low frequency range and the magnetic moment in the high frequency range. The values of μ_r' for the composite made of sendust were less than unity and the values of μ_r'' was not zero like the composite made of ferrite particles dispersed in polystyrene resin. Thus, the composite made of sendust had a return loss of less than −20 dB at frequencies above 10 GHz in addition to the absorption at frequencies of several GHz. From these result, it is concluded that a practical absorber suitable for frequencies above 10 GHz is possible using a composite made of sendust. The values of μ_r' and μ_r'' for the composite made of both sendust and aluminum particles dispersed in polystyrene resin could be controlled by changing the volume mixture ratio of sendust and aluminum. Thus, the absorption characteristics at frequencies above 10 GHz for the composite made of only sendust could be improved by using the composite made of both sendust and aluminum by selecting a suitable volume mixture ratio of sendust and aluminum, and a flexible design of an absorber was proposed.

5. Acknowledgement

This research was supported by the Japan Society for the Promotion of Science (JSPS) and Grant-in-Aid for JSPS Fellows.

6. References

[a]Kasagi, T., Tsutaoka, T., & Hatakeyama, K. (1999). Particle size effect on the complex permeability for permalloy composite materials. *IEEE Trans. Magn.*, Vol. 35, No. 5, pp. 3424-3426, 0018-9464

[b]Kasagi, T., Tsutaoka, T., & Hatakeyama, K. (2006). Negative Permeability Spectra in Permalloy Granular Composite Materials. *Appl. Phys. Lett.*, Vol. 88, 17502, 0003-6951

Lim, K. M., Lee, K. A., Kim, M. C., & Park, C. G. (2005). Complex permeability and electromagnetic wave absorption properties of amorphous alloy-epoxy composites. *J. Non-Cryst. Solids.*, Vol. 351, pp. 75-83, 0022-3093

Nishikata, A. (2002). New Radiowave Absorbers Using Magnetic Loss Caused by Metal Particles' Internal Eddy Current. *Proceedings of EMC EUROPE 2002 International Symposium on Electromagnetic Compatibility*, pp. 697-702, Sorrento Italy, September 2002

Song, J. M., Kim, D. I., Choi, J. H., & Jeung, J. H., (2005). EM Wave Absorbers Prepared with Sendust. *APMC2005 Proceedings*, 0-7803-9433, Suzhou China, December 2005

Wada, Y., Asano, N., Sakai, K., & Yoshikado, S. (2008). Preparation and Evaluation of Composite Electromagnetic Wave Absorbers Made of Fine Aluminum Particles Dispersed in Polystyrene Medium. *PIERS Online*, Vol. 4, pp. 838-845, 1931-7360

20

No-Reflection Phenomena for Chiral Media

Yasuhiro Tamayama, Toshihiro Nakanishi,
Kazuhiko Sugiyama, and Masao Kitano
Department of Electronic Science and Engineering, Kyoto University
Japan

1. Introduction

Interest continues to grow in controlling the propagation of electromagnetic waves by utilizing periodically or randomly arranged artificial structures made of metal, dielectric, and other materials. When the size of the constituent structures and the separation between the neighboring structures are much smaller than the wavelength of the electromagnetic waves, the structure arrays behave as a continuous medium for the electromagnetic waves. That is, macroscopic medium parameters such as effective permittivity and permeability can be defined for the array. The artificial continuous medium is called a "metamaterial."

In the frequency region below the microwave frequency, the use of metallic structures as artificial media has been studied since the late 1940's (Collin, 1990). At first, only control of the permittivity was studied and not that of the permeability. However, Pendry et al. (1999) proposed methods for fabricating artificial magnetic media, namely, magnetic metamaterials, which were built from nonmagnetic conductors. It was shown that not only can relative permeability be changed from unity but it also can have a negative value. Although the relative permeabilities of naturally occurring media are almost unity in such high frequency regions as microwave, terahertz, and optical regions, the restriction that the relative permeability is almost unity can be removed using the metamaterial. Moreover, the magnetic metamaterial enabled us to fabricate media with simultaneous negative permittivity and permeability, or negative refractive index media that were predicted by Veselago (1968). In fact, Shelby et al. (2001) made the first experimental verification of a negative refractive index metamaterial in the microwave region. This increased researcher interest in metamaterials.

It was not possible to independently control the wavenumber and the wave impedance in a medium until magnetic metamaterials were developed. The wavenumber is related to the propagation and refraction of electromagnetic waves, and the wave impedance is connected with the reflection. Phenomena about electromagnetic waves are described by these two quantities. In dielectric media, both of the wavenumber and wave impedance change with a change of the permittivity, and we cannot set these parameters independently. However, the wavenumber and wave impedance can be changed independently in metamaterials because we can control the permeability as well as the permittivity with metamaterials. By utilizing the flexibility of the wavenumber and wave impedance in metamaterials, such novel phenomena as a perfect lens (superlens) (Pendry, 2000; Lagarkov & Kissel, 2004), a hyperlens (Jacob et al., 2006; Liu et al., 2007), and an invisibility cloak (Pendry et al., 2006; Leonhardt, 2006; Schurig et al., 2006) have been proposed and verified experimentally.

In this chapter, we focus on Brewster's no-reflection effect in metamaterials. The Brewster condition is one of the laws of reflection and refraction of electromagnetic waves at a boundary between two distinct media (Saleh & Teich, 2007). For a particular angle of incidence, known as the Brewster angle, the reflected wave vanishes. The Brewster effect is applied in optical instruments, for example, to generate completely polarized waves from unpolarized waves only with a glass plate and to suppress the insertion losses of intracavity elements.

The Brewster effect arises for transverse-magnetic (TM) waves [transverse-electric (TE) waves] at an interface between two distinct dielectric (magnetic) media. Hence, this phenomenon can only be observed for TM waves and not for TE waves in naturally occurring media that do not respond to high-frequency magnetic fields. However, since we can fabricate magnetic media in high frequency regions with a metamaterial technique (Pendry et al., 1999; Holloway et al., 2003; Zhang et al., 2005), the Brewster condition for TE waves can be satisfied (Doyle, 1980; Futterman, 1995; Fu et al., 2005). In fact, the TE Brewster effect has been experimentally observed in the microwave region (Tamayama et al., 2006) and also in the optical region (Watanabe et al., 2008).

In addition to permittivity and permeability, chirality parameter and non-reciprocity parameter can be controlled using metamaterials. It is also possible to control the anisotropy in electromagnetic responses. Therefore, investigating the no-reflection condition for generalized media is important. Brewster's condition has been studied for anisotropic media (Grzegorczyk et al., 2005; Tanaka et al., 2006; Shen et al., 2006; Shu et al., 2007), chiral media (bi-isotropic media) (Bassiri et al., 1988; Lindell et al., 1994), and bi-anisotropic media (Lakhtakia, 1992). However, thus far, the explicit relations among the medium parameters for achieving non-reflectivity in chiral and bi-anisotropic media have not been determined. The purpose of this chapter is to derive the explicit relation among the permittivity, permeability, and chirality parameter of the chiral medium that satisfy the no-reflection condition for a planar interface between a vacuum and the chiral medium.

The no-reflection condition is derived from the vanishing eigenvalue condition of the reflection Jones matrix. The analysis can be largely simplified by decomposing the reflection Jones matrix into the unit and Pauli matrices (Tamayama et al., 2008).

We find that in general chiral media, the no-reflection condition is satisfied by elliptically polarized incident waves for at most one particular angle of incidence. This is merely a natural extension of the usual Brewster effect for achiral (nonchiral) media. When the wave impedance and the absolute value of the wavenumber in the chiral medium equal those in a vacuum for one of the circularly polarized (CP) waves, the corresponding CP wave is transmitted to the medium without reflection for all angles of incidence. The no-reflection effect for chiral nihility media resembles that for achiral media.

We provide a finite-difference time-domain (FDTD) analysis (Taflove & Hagness, 2005) of the no-reflection effect for CP waves. We analyze the scatterings of electromagnetic waves by a cylinder and a triangular prism made of a chiral medium whose medium parameters satisfy the no-reflection condition for one of the CP waves. The simulation demonstrates that the corresponding CP wave is not scattered and the other CP wave is largely scattered. We show that a circular polarizing beam splitter can be achieved by utilizing the no-reflection effect.

2. Propagation of electromagnetic waves in chiral media

We calculate the wavenumber and wave impedance in chiral media. The constitutive equations for chiral media have several types of expressions. The Post and Tellegen representations are mainly used as the constitutive equations. The Post representation is written as

$$D = \varepsilon_{\rm P} E - {\rm i}\xi_{\rm P} B, \quad H = \mu_{\rm P}^{-1} B - {\rm i}\xi_{\rm P} E, \tag{1}$$

and the Tellegen representation is written as

$$D = \varepsilon_{\rm T} E - {\rm i}\kappa_{\rm T} H, \quad B = \mu_{\rm T} H + {\rm i}\kappa_{\rm T} E, \tag{2}$$

where $\varepsilon_{\rm P,T}$ is the permittivity, $\mu_{\rm P,T}$ is the permeability, and $\xi_{\rm P}$ and $\kappa_{\rm T}$ are the chirality parameters. The subscript P (T) stands for the Post (Tellegen) representation. These representations are equivalent and interchangeable with the following transformation (Lakhtakia, 1992):

$$\varepsilon_{\rm T} = \varepsilon_{\rm P} + \mu_{\rm P}\xi_{\rm P}^2, \quad \mu_{\rm T} = \mu_{\rm P}, \quad \kappa_{\rm T} = \mu_{\rm P}\xi_{\rm P}. \tag{3}$$

In this chapter, we consistently use the Post representation and omit the subscript P for simplicity.

Maxwell's equation for a monochromatic plane electromagnetic wave is given by

$$\boldsymbol{k} \times \boldsymbol{E} = \omega \boldsymbol{B}, \quad \boldsymbol{k} \times \boldsymbol{H} = -\omega \boldsymbol{D}, \tag{4}$$

where $\boldsymbol{k}$ is the wavenumber vector and ω is the angular frequency. Substituting Eq. (1) into Eq. (4) and assuming $\boldsymbol{k} = k\boldsymbol{e}_z$ ($\boldsymbol{e}_z$ is the unit vector in the z-direction), we obtain

$$k \begin{bmatrix} 1 & & & O \\ & 1 & & \\ & & 1 & \\ O & & & 1 \end{bmatrix} \begin{bmatrix} E_x \\ E_y \\ H_x \\ H_y \end{bmatrix} = \omega \begin{bmatrix} 0 & {\rm i}\mu\xi & 0 & \mu \\ -{\rm i}\mu\xi & 0 & -\mu & 0 \\ 0 & -(\varepsilon + \mu\xi^2) & 0 & {\rm i}\mu\xi \\ \varepsilon + \mu\xi^2 & 0 & -{\rm i}\mu\xi & 0 \end{bmatrix} \begin{bmatrix} E_x \\ E_y \\ H_x \\ H_y \end{bmatrix}. \tag{5}$$

From the condition for a non-trivial solution of Eq. (5), we have

$$k = \pm\omega(\sqrt{\varepsilon\mu + \mu^2\xi^2} + \mu\xi), \quad \pm\omega(\sqrt{\varepsilon\mu + \mu^2\xi^2} - \mu\xi). \tag{6}$$

After substitution of the derived wavenumber into Eq. (5), the relation among the wavenumber and the electromagnetic fields is obtained and summarized in Table 1. Here we define the wave impedance $Z_{\rm C}$ of the chiral medium as

$$Z_{\rm C} = \sqrt{\frac{\mu}{\varepsilon + \mu\xi^2}}. \tag{7}$$

The eigenpolarizations in chiral media are found to be CP waves because $E_y/E_x = \pm{\rm i}$ is satisfied.

Equations (6) and (7) contain double-valued square root functions. Thus, the wavenumber and wave impedance cannot be calculated without ambiguity. To choose the correct branch, we diagonalize Eq. (5). By using the transformation matrix

$$U = \frac{1}{2} \begin{bmatrix} 1 & 1 & 1 & 1 \\ -{\rm i} & {\rm i} & {\rm i} & -{\rm i} \\ {\rm i}Z_{\rm C}^{-1} & {\rm i}Z_{\rm C}^{-1} & -{\rm i}Z_{\rm C}^{-1} & -{\rm i}Z_{\rm C}^{-1} \\ Z_{\rm C}^{-1} & -Z_{\rm C}^{-1} & Z_{\rm C}^{-1} & -Z_{\rm C}^{-1} \end{bmatrix}, \quad U^{-1} = \frac{1}{2} \begin{bmatrix} 1 & {\rm i} & -{\rm i}Z_{\rm C} & Z_{\rm C} \\ 1 & -{\rm i} & -{\rm i}Z_{\rm C} & -Z_{\rm C} \\ 1 & -{\rm i} & {\rm i}Z_{\rm C} & Z_{\rm C} \\ 1 & {\rm i} & {\rm i}Z_{\rm C} & -Z_{\rm C} \end{bmatrix}, \tag{8}$$

k	k_+	$-k_+$	k_-	$-k_-$
E_x	1	1	1	1
E_y	$-\mathrm{i}$	i	i	$-\mathrm{i}$
H_x	$\mathrm{i}Z_\mathrm{c}^{-1}$	$\mathrm{i}Z_\mathrm{c}^{-1}$	$-\mathrm{i}Z_\mathrm{c}^{-1}$	$-\mathrm{i}Z_\mathrm{c}^{-1}$
H_y	Z_c^{-1}	$-Z_\mathrm{c}^{-1}$	Z_c^{-1}	$-Z_\mathrm{c}^{-1}$

Table 1. Relation among wavenumber and electromagnetic fields in chiral media. Ratio of each electromagnetic field component to E_x is written for each eigenmode. Here $k_\pm = \omega(\sqrt{\varepsilon\mu + \mu^2\xi^2} \pm \mu\xi) = \omega[Z_\mathrm{c}(\varepsilon + \mu\xi^2) \pm \mu\xi]$ and $Z_\mathrm{c} = \sqrt{\mu/(\varepsilon + \mu\xi^2)}$ [$\mathrm{Re}(Z_\mathrm{c}) > 0$].

Eq. (5) is diagonalized as follows:

$$k \begin{bmatrix} E_x + \mathrm{i}E_y - \mathrm{i}Z_\mathrm{c}H_x + Z_\mathrm{c}H_y \\ E_x - \mathrm{i}E_y - \mathrm{i}Z_\mathrm{c}H_x - Z_\mathrm{c}H_y \\ E_x - \mathrm{i}E_y + \mathrm{i}Z_\mathrm{c}H_x + Z_\mathrm{c}H_y \\ E_x + \mathrm{i}E_y + \mathrm{i}Z_\mathrm{c}H_x - Z_\mathrm{c}H_y \end{bmatrix} = \begin{bmatrix} k_+ & & & O \\ & -k_+ & & \\ & & k_- & \\ O & & & -k_- \end{bmatrix} \begin{bmatrix} E_x + \mathrm{i}E_y - \mathrm{i}Z_\mathrm{c}H_x + Z_\mathrm{c}H_y \\ E_x - \mathrm{i}E_y - \mathrm{i}Z_\mathrm{c}H_x - Z_\mathrm{c}H_y \\ E_x - \mathrm{i}E_y + \mathrm{i}Z_\mathrm{c}H_x + Z_\mathrm{c}H_y \\ E_x + \mathrm{i}E_y + \mathrm{i}Z_\mathrm{c}H_x - Z_\mathrm{c}H_y \end{bmatrix}, \quad (9)$$

where we use the relation $Z_\mathrm{c}(\varepsilon + \mu\xi^2) = \mu Z_\mathrm{c}^{-1}$ [$Z_\mathrm{c}^2 = \mu/(\varepsilon + \mu\xi^2)$] and set

$$k_\pm = \omega[Z_\mathrm{c}(\varepsilon + \mu\xi^2) \pm \mu\xi]. \quad (10)$$

If Z_c is determined, $k_\pm$ can be calculated unambiguously. The real part of the wave impedance $\mathrm{Re}(Z_\mathrm{c})$ is related to the time-averaged Poynting vector, which governs the power flow of electromagnetic waves. When a branch of Z_c is chosen so that $\mathrm{Re}(Z_\mathrm{c}) > 0$ is satisfied, the power flows of eigenmodes represented by the first and third (second and fourth) rows of Eq. (9) are directed to the positive (negative) z-direction. Thus, k_+ and k_- ($-k_+$ and $-k_-$) are the wavenumbers for the eigenmodes whose power flows are directed to the positive (negative) z-direction. Even if we choose a branch of Z_c that satisfies $\mathrm{Re}(Z_\mathrm{c}) < 0$, we can obtain the same result by regarding $-Z_\mathrm{c}$ as the wave impedance. Therefore, there is no loss of generality in supposing that the real part of Z_c is positive. The wavenumber and wave impedance can be calculated from Eqs. (7) and (10) and the condition $\mathrm{Re}(Z_\mathrm{c}) > 0$ without ambiguity. We define the eigenmodes represented by the first and second (third and fourth) rows of Eq. (9) as left circularly polarized (LCP) [right circularly polarized (RCP)] waves.

3. Reflectivity and transmissivity for chiral media

We derive the reflectivity and transmissivity at the boundary between a vacuum and an isotropic chiral medium (Bassiri et al., 1988). As shown in Fig. 1, suppose that a monochromatic plane electromagnetic wave is incident from the vacuum (permittivity ε_0, permeability μ_0) on the chiral medium at an incident angle of θ. The electromagnetic fields of the incident (i), reflected (r), and transmitted (t) waves are written as follows:

$$\boldsymbol{E}_\mathrm{i} = \boldsymbol{E}_1 \exp[\mathrm{i}k_0(x\cos\theta - y\sin\theta)], \quad (11)$$
$$\boldsymbol{H}_\mathrm{i} = \boldsymbol{H}_1 \exp[\mathrm{i}k_0(x\cos\theta - y\sin\theta)], \quad (12)$$
$$\boldsymbol{E}_\mathrm{r} = \boldsymbol{E}_2 \exp[\mathrm{i}k_0(-x\cos\theta - y\sin\theta)], \quad (13)$$
$$\boldsymbol{H}_\mathrm{r} = \boldsymbol{H}_2 \exp[\mathrm{i}k_0(-x\cos\theta - y\sin\theta)], \quad (14)$$
$$\boldsymbol{E}_\mathrm{t} = \boldsymbol{E}_3 \exp[\mathrm{i}k_+(x\cos\theta_+ - y\sin\theta_+)] + \boldsymbol{E}_4 \exp[\mathrm{i}k_-(x\cos\theta_- - y\sin\theta_-)], \quad (15)$$
$$\boldsymbol{H}_\mathrm{t} = \boldsymbol{H}_3 \exp[\mathrm{i}k_+(x\cos\theta_+ - y\sin\theta_+)] + \boldsymbol{H}_4 \exp[\mathrm{i}k_-(x\cos\theta_- - y\sin\theta_-)], \quad (16)$$

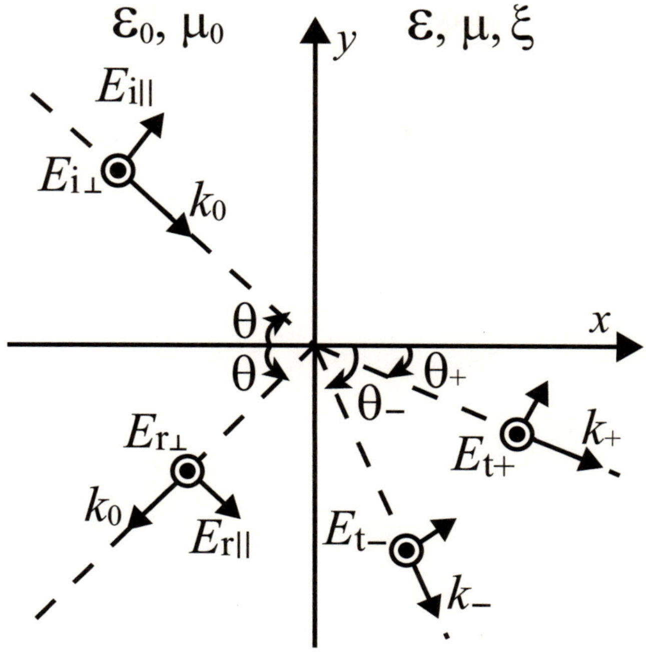

Fig. 1. Geometry of coordinate system. Incident, reflected, and transmitted waves are denoted by subscripts i, r, and t. Region $x < 0$ represents vacuum, and region $x \geq 0$ represents the chiral medium.

where

$$\boldsymbol{E}_1 = E_{\mathrm{i}\perp}\boldsymbol{e}_z + E_{\mathrm{i}\parallel}(\cos\theta\boldsymbol{e}_y + \sin\theta\boldsymbol{e}_x), \tag{17}$$

$$\boldsymbol{H}_1 = Z_0^{-1}[E_{\mathrm{i}\parallel}\boldsymbol{e}_z - E_{\mathrm{i}\perp}(\cos\theta\boldsymbol{e}_y + \sin\theta\boldsymbol{e}_x)], \tag{18}$$

$$\boldsymbol{E}_2 = E_{\mathrm{r}\perp}\boldsymbol{e}_z + E_{\mathrm{r}\parallel}(-\cos\theta\boldsymbol{e}_y + \sin\theta\boldsymbol{e}_x), \tag{19}$$

$$\boldsymbol{H}_2 = Z_0^{-1}[E_{\mathrm{r}\parallel}\boldsymbol{e}_z + E_{\mathrm{r}\perp}(\cos\theta\boldsymbol{e}_y - \sin\theta\boldsymbol{e}_x)], \tag{20}$$

$$\boldsymbol{E}_3 = E_{\mathrm{t}+}[\mathrm{i}(\cos\theta_+\boldsymbol{e}_y + \sin\theta_+\boldsymbol{e}_x) + \boldsymbol{e}_z], \tag{21}$$

$$\boldsymbol{H}_3 = E_{\mathrm{t}+}Z_{\mathrm{c}}^{-1}[-(\cos\theta_+\boldsymbol{e}_y + \sin\theta_+\boldsymbol{e}_x) + \mathrm{i}\boldsymbol{e}_z], \tag{22}$$

$$\boldsymbol{E}_4 = E_{\mathrm{t}-}[-\mathrm{i}(\cos\theta_-\boldsymbol{e}_y + \sin\theta_-\boldsymbol{e}_x) + \boldsymbol{e}_z], \tag{23}$$

$$\boldsymbol{H}_4 = E_{\mathrm{t}-}Z_{\mathrm{c}}^{-1}[-(\cos\theta_-\boldsymbol{e}_y + \sin\theta_-\boldsymbol{e}_x) - \mathrm{i}\boldsymbol{e}_z]. \tag{24}$$

In the above equations, $k_0 = \omega\sqrt{\varepsilon_0\mu_0}$ is the wavenumber in the vacuum, $Z_0 = \sqrt{\mu_0/\varepsilon_0}$ is the wave impedance of the vacuum, and $\boldsymbol{e}_x$, $\boldsymbol{e}_y$, and $\boldsymbol{e}_z$ are respectively the unit vectors in the x-, y-, and z-directions. Due to the translational invariance of the interface, Snell's equations

$$k_0 \sin\theta = k_+ \sin\theta_+ = k_- \sin\theta_-, \tag{25}$$

are satisfied. From the continuity of the tangential components of the electromagnetic fields across the boundary, we obtain

$$E_{\mathrm{i}\perp} + E_{\mathrm{r}\perp} = E_{\mathrm{t}+} + E_{\mathrm{t}-}, \tag{26}$$
$$E_{\mathrm{i}\|}\cos\theta - E_{\mathrm{r}\|}\cos\theta = \mathrm{i}E_{\mathrm{t}+}\cos\theta_{+} - \mathrm{i}E_{\mathrm{t}-}\cos\theta_{-}, \tag{27}$$
$$Z_0^{-1}(E_{\mathrm{i}\|} + E_{\mathrm{r}\|}) = \mathrm{i}Z_{\mathrm{c}}^{-1}(E_{\mathrm{t}+} - E_{\mathrm{t}-}), \tag{28}$$
$$Z_0^{-1}(-E_{\mathrm{i}\perp}\cos\theta + E_{\mathrm{r}\perp}\cos\theta) = -Z_{\mathrm{c}}^{-1}(E_{\mathrm{t}+}\cos\theta_{+} + E_{\mathrm{t}-}\cos\theta_{-}). \tag{29}$$

The reflection and transmission matrices are derived from Eqs. (26)-(29) and written as

$$\begin{bmatrix} E_{\mathrm{r}\perp} \\ E_{\mathrm{r}\|} \end{bmatrix} = \begin{bmatrix} R_{11} & R_{12} \\ R_{21} & R_{22} \end{bmatrix} \begin{bmatrix} E_{\mathrm{i}\perp} \\ E_{\mathrm{i}\|} \end{bmatrix}, \quad \begin{bmatrix} E_{\mathrm{t}+} \\ E_{\mathrm{t}-} \end{bmatrix} = \begin{bmatrix} T_{++} & T_{+-} \\ T_{-+} & T_{--} \end{bmatrix} \begin{bmatrix} E_{\mathrm{i}+} \\ E_{\mathrm{i}-} \end{bmatrix}, \tag{30}$$

where

$$R_{11} = \frac{(Z_{\mathrm{c}}^2 - Z_0^2)\cos\theta(\cos\theta_{+} + \cos\theta_{-}) + 2Z_0Z_{\mathrm{c}}(\cos^2\theta - \cos\theta_{+}\cos\theta_{-})}{\Delta}, \tag{31}$$
$$R_{12} = \frac{\mathrm{i}2Z_0Z_{\mathrm{c}}\cos\theta(\cos\theta_{+} - \cos\theta_{-})}{\Delta}, \tag{32}$$
$$R_{21} = \frac{-\mathrm{i}2Z_0Z_{\mathrm{c}}\cos\theta(\cos\theta_{+} - \cos\theta_{-})}{\Delta}, \tag{33}$$
$$R_{22} = \frac{-(Z_{\mathrm{c}}^2 - Z_0^2)\cos\theta(\cos\theta_{+} + \cos\theta_{-}) + 2Z_0Z_{\mathrm{c}}(\cos^2\theta - \cos\theta_{+}\cos\theta_{-})}{\Delta}, \tag{34}$$
$$T_{++} = \frac{2Z_{\mathrm{c}}(Z_{\mathrm{c}} + Z_0)\cos\theta(\cos\theta + \cos\theta_{-})}{\Delta}, \tag{35}$$
$$T_{+-} = \frac{-2Z_{\mathrm{c}}(Z_{\mathrm{c}} - Z_0)\cos\theta(\cos\theta - \cos\theta_{-})}{\Delta}, \tag{36}$$
$$T_{-+} = \frac{-2Z_{\mathrm{c}}(Z_{\mathrm{c}} - Z_0)\cos\theta(\cos\theta - \cos\theta_{+})}{\Delta}, \tag{37}$$
$$T_{--} = \frac{2Z_{\mathrm{c}}(Z_{\mathrm{c}} + Z_0)\cos\theta(\cos\theta + \cos\theta_{+})}{\Delta}, \tag{38}$$
$$\Delta = (Z_{\mathrm{c}}^2 + Z_0^2)\cos\theta(\cos\theta_{+} + \cos\theta_{-}) + 2Z_0Z_{\mathrm{c}}(\cos^2\theta + \cos\theta_{+}\cos\theta_{-}), \tag{39}$$
$$E_{\mathrm{i}\pm} = \frac{E_{\mathrm{i}\perp} \mp \mathrm{i}E_{\mathrm{i}\|}}{2}. \tag{40}$$

4. No-reflection conditions for chiral media

We find from Eqs. (30)-(34) that the relation between the electric field of the incident wave and that of the reflected wave is written as (Tamayama et al., 2008)

$$\begin{bmatrix} E_{\mathrm{r}\perp} \\ E_{\mathrm{r}\|} \end{bmatrix} = \frac{1}{\Delta} M_{\mathrm{R}} \begin{bmatrix} E_{\mathrm{i}\perp} \\ E_{\mathrm{i}\|} \end{bmatrix}, \quad M_{\mathrm{R}} = c_{\mathrm{u}}I + c_2\sigma_2 + c_3\sigma_3, \tag{41}$$
$$c_{\mathrm{u}} = 2Z_0Z_{\mathrm{c}}(\cos^2\theta - \cos\theta_{+}\cos\theta_{-}), \tag{42}$$
$$c_2 = -2Z_0Z_{\mathrm{c}}\cos\theta(\cos\theta_{+} - \cos\theta_{-}), \tag{43}$$
$$c_3 = (Z_{\mathrm{c}}^2 - Z_0^2)\cos\theta(\cos\theta_{+} + \cos\theta_{-}), \tag{44}$$

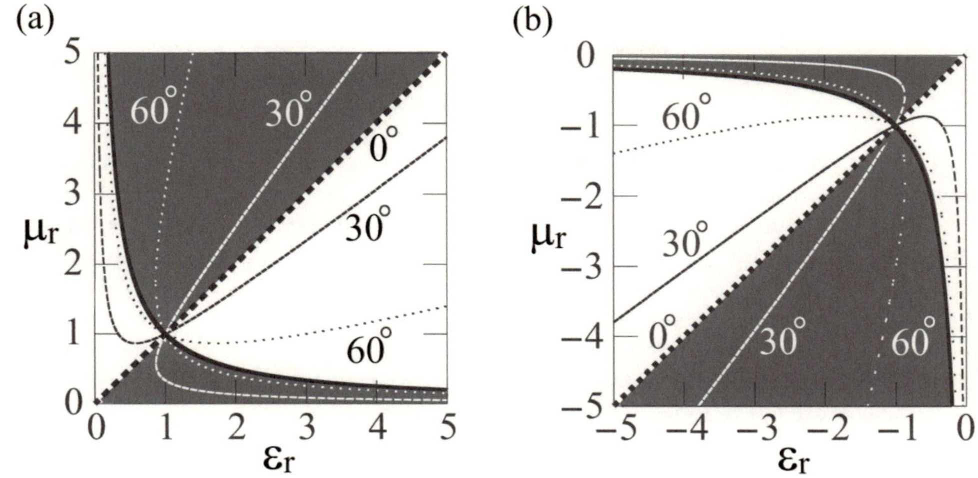

Fig. 2. Contour lines of no-reflection angles for TM and TE waves (a) in first quadrant and (b) in third quadrant of (ε_r, μ_r)-plane. No-reflection condition exists for TM (TE) waves in white (gray) region.

where we introduce the unit matrix I and the Pauli matrices (Sakurai, 1994):

$$\sigma_2 = \begin{bmatrix} 0 & -\mathrm{i} \\ \mathrm{i} & 0 \end{bmatrix}, \quad \sigma_3 = \begin{bmatrix} 1 & 0 \\ 0 & -1 \end{bmatrix}. \tag{45}$$

The reflection matrix M_R can be rewritten as

$$M_\mathrm{R} = c_\mathrm{u} I + c_\varphi \sigma_\varphi, \tag{46}$$

where $c_\varphi = \sqrt{c_2^2 + c_3^2}$, $\sigma_\varphi = \sigma_2 \sin\varphi + \sigma_3 \cos\varphi$, $\sin\varphi = c_2/c_\varphi$, and $\cos\varphi = c_3/c_\varphi$. The no-reflection condition is satisfied when M_R has at least one vanishing eigenvalue, namely, $\det(M_\mathrm{R}) = 0$ or $\mathrm{rank}(M_\mathrm{R}) \leq 1$. For the incident wave with the corresponding eigenpolarization, the reflection is nullified. From Eq. (46), we observe that the eigenvalue problem for M_R is reduced to that for σ_φ. The eigenvalues of σ_φ are ± 1, and their corresponding eigenpolarizations are $\boldsymbol{e}_{\varphi+} = \cos(\varphi/2)\boldsymbol{e}_z + \mathrm{i}\sin(\varphi/2)(\boldsymbol{e}_x \sin\theta + \boldsymbol{e}_y \cos\theta)$ and $\boldsymbol{e}_{\varphi-} = \sin(\varphi/2)\boldsymbol{e}_z - \mathrm{i}\cos(\varphi/2)(\boldsymbol{e}_x \sin\theta + \boldsymbol{e}_y \cos\theta)$. Therefore, M_R has one vanishing eigenvalue when $c_\mathrm{u} = c_\varphi \neq 0$ ($c_\mathrm{u} = -c_\varphi \neq 0$) is satisfied, and no-reflection is achieved for the incident wave with polarization $\boldsymbol{e}_{\varphi-}$ ($\boldsymbol{e}_{\varphi+}$). When $c_\mathrm{u} = c_\varphi = 0$, M_R becomes a zero matrix; no-reflection is achieved for arbitrary polarized incident waves.

4.1 In case of $\xi = 0$ (achiral media)

The reflection matrix is written as $M_\mathrm{R} = c_\mathrm{u} I + c_3 \sigma_3$. The eigenpolarizations are $\boldsymbol{e}_x \sin\theta + \boldsymbol{e}_y \cos\theta$ and $\boldsymbol{e}_z$; therefore, the no-reflection condition can only be satisfied for linearly polarized waves. The no-reflection effect is observed at a particular incident angle that satisfies $c_\mathrm{u} = \pm c_3$. The condition $c_\mathrm{u} = c_3$ ($c_\mathrm{u} = -c_3$) yields a no-reflection angle, called the Brewster angle, for TM (TE) waves in isotropic achiral media.

From $c_{\mathrm{u}} = \pm c_3$, the no-reflection angles θ_{TM} and θ_{TE} for TM and TE waves are derived as follows:

$$\theta_{\mathrm{TM}} = \arcsin\sqrt{\frac{\varepsilon_{\mathrm{r}}^2 - \varepsilon_{\mathrm{r}}\mu_{\mathrm{r}}}{\varepsilon_{\mathrm{r}}^2 - 1}}, \quad \theta_{\mathrm{TE}} = \arcsin\sqrt{\frac{\mu_{\mathrm{r}}^2 - \varepsilon_{\mathrm{r}}\mu_{\mathrm{r}}}{\mu_{\mathrm{r}}^2 - 1}}, \tag{47}$$

where $\varepsilon_{\mathrm{r}} = \varepsilon/\varepsilon_0$ is the relative permittivity and $\mu_{\mathrm{r}} = \mu/\mu_0$ is the relative permeability. Figure 2 shows the contour lines of the no-reflection angles. The no-reflection effect can be observed for TM (TE) waves in the white (gray) region. The no-reflection condition exists in the whole region of the first and third quadrants of the $(\varepsilon_{\mathrm{r}}, \mu_{\mathrm{r}})$-plane except $\varepsilon_{\mathrm{r}} = \mu_{\mathrm{r}}^{-1} \neq \pm 1$. The intersection of the contour lines of the no-reflection angles in the first quadrant of the $(\varepsilon_{\mathrm{r}}, \mu_{\mathrm{r}})$-plane corresponds to a vacuum $(\varepsilon_{\mathrm{r}}, \mu_{\mathrm{r}}) = (1,1)$ and that in the third quadrant corresponds to an anti-vacuum $(\varepsilon_{\mathrm{r}}, \mu_{\mathrm{r}}) = (-1,-1)$. For the medium with parameters that correspond to these intersections, M_{R} becomes a zero matrix for any incident angle; arbitrary polarized waves are not reflected for all angles of incidence.

4.2 In case of $\xi \neq 0$, $k_+ \neq -k_-$, and $Z_{\mathrm{c}} \neq Z_0$ (impedance unmatched chiral media)

The conditions $\xi \neq 0$, $k_+ \neq -k_-$, and $Z_{\mathrm{c}} \neq Z_0$ give $\varphi \neq n\pi/2$ with integer n. The eigenpolarizations are $\boldsymbol{e}_{\varphi\pm}$; hence, the no-reflection condition can only be satisfied for elliptically polarized (EP) waves. The no-reflection effect is observed at a particular incident angle satisfying $c_{\mathrm{u}} = \pm c_{\varphi}$, which is a natural extension of the usually observed no-reflection effect, or the Brewster effect in achiral media.

The no-reflection angles are derived from the zero eigenvalue condition $c_{\mathrm{u}} = \pm c_{\varphi}$. The contour lines of the no-reflection angles are shown in the left panels of Fig. 3. The no-reflection condition in the case of $\xi = 0$ exists in the whole region of the first and third quadrants of the $(\varepsilon_{\mathrm{r}}, \mu_{\mathrm{r}})$-plane except $\varepsilon_{\mathrm{r}} = \mu_{\mathrm{r}}^{-1} \neq \pm 1$, as shown in Fig. 2, while in the case of $\xi \neq 0$, there is a region where the no-reflection condition does not exist, which is represented as the gray region in Fig. 3. In addition, the no-reflection condition also exists in the second and fourth quadrants, which correspond to strong chiral media ($k_+k_- < 0$). The right panels of Fig. 3 show the incident polarization for which the no-reflection condition is satisfied. The polarization is described in terms of the ellipticity $\alpha = \arctan(E_{\parallel}/\mathrm{i}E_{\perp})$. The condition $\alpha > 0$ ($\alpha < 0$) denotes left (right) elliptically polarized wave and $|\alpha| = 90^\circ$ ($|\alpha| = 0$) corresponds to TM (TE) wave. When $|\alpha| > 45^\circ$ ($|\alpha| < 45^\circ$), the major axis of the polarization ellipse is perpendicular (parallel) to $\boldsymbol{e}_z$ and the minor axis is parallel (perpendicular) to $\boldsymbol{e}_z$, namely, the no-reflection condition is satisfied for TM-like (TE-like) EP waves.

4.3 In case of $\xi \neq 0$, $k_+ \neq -k_-$, and $Z_{\mathrm{c}} = Z_0$ (impedance matched chiral media)

The reflection matrix becomes $M_{\mathrm{R}} = c_{\mathrm{u}}I + c_2\sigma_2$. The eigenpolarizations are $[\boldsymbol{e}_z \pm \mathrm{i}(\boldsymbol{e}_x \sin\theta + \boldsymbol{e}_y \cos\theta)]/\sqrt{2}$; hence, the no-reflection condition can only be satisfied for CP waves. The condition $\cos\theta_+ = \cos\theta$ ($\cos\theta_- = \cos\theta$) is required to satisfy $c_{\mathrm{u}} = -c_2$ ($c_{\mathrm{u}} = c_2$), which is the no-reflection condition for LCP (RCP) waves. Note that once $|k_+| = k_0$ ($|k_-| = k_0$) is satisfied by selecting the constants of medium, $c_{\mathrm{u}} = -c_2$ ($c_{\mathrm{u}} = c_2$) is satisfied for any incident angle. That is, the no-reflection condition is satisfied for all angles of incidence (Tamayama et al., 2008).

We derive the explicit relations among ε, μ, and ξ for the no-reflection condition for CP waves. From the above discussion, both $Z_{\mathrm{c}} = Z_0$ and $|k_+| = k_0$ ($|k_-| = k_0$) are necessary and yield the

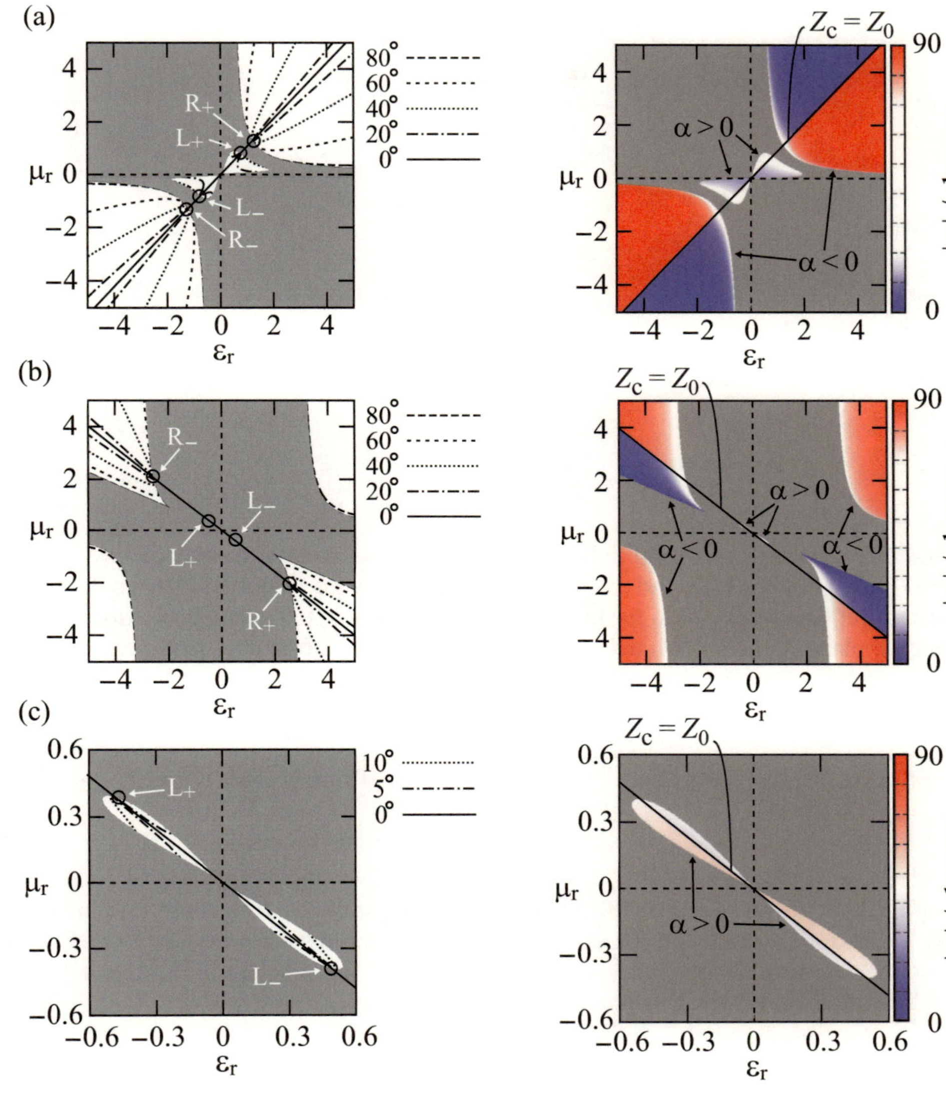

Fig. 3. No-reflection conditions (a) for $\xi_r = \xi Z_0 = 0.2$ and (b), (c) for $\xi_r = 1.5$. [Note that the scale of (c) is different from other figures.] (left panels); contour lines of no-reflection angles. L_+ and R_+ (L_- and R_-) represent the no-reflection conditions for LCP and RCP waves written in Eq. (48) [Eq. (49)], respectively. (right panels); ellipticity α of incident polarization for which no-reflection condition is satisfied. Sign of α is reversed if sign of ξ_r is reversed. No-reflection conditions do not exist in gray region.

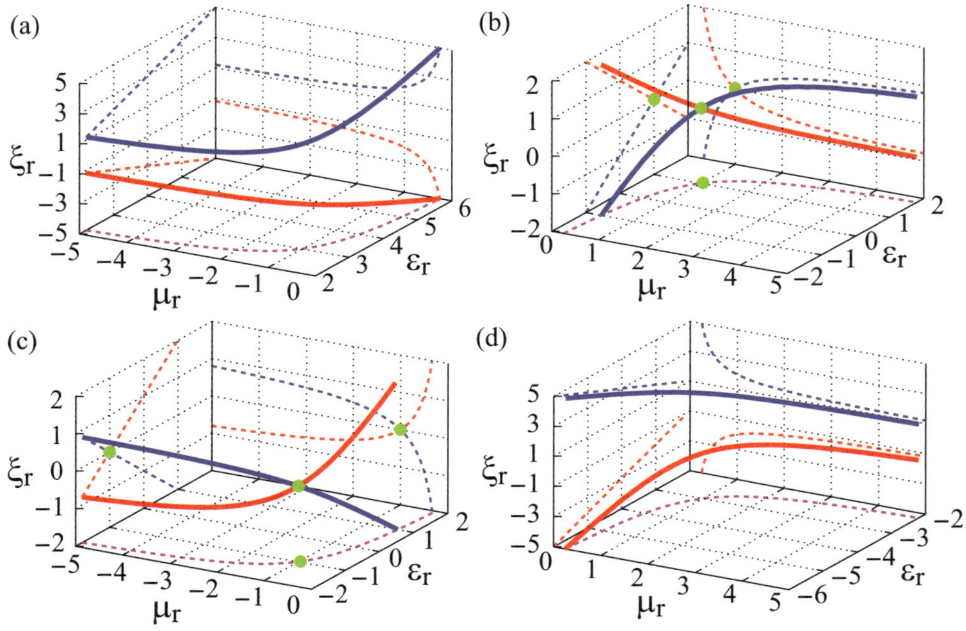

Fig. 4. Relations among ε_r, μ_r, and ξ_r that satisfy no-reflection condition for CP waves: (a), (b) relations given by Eq. (48), and (c), (d) relations given by Eq. (49). Red (blue) solid lines represent no-reflection condition for LCP (RCP) waves. Dashed lines are projections of solid lines on each plane. Green solid circles correspond to (b) vacuum and (c) anti-vacuum.

following relations:

$$\varepsilon_r = -(\pm\xi_r - 1), \quad \mu_r = \frac{\pm 1}{\xi_r \pm 1} \quad (\text{from } k_\pm = k_0), \tag{48}$$

$$\varepsilon_r = \pm\xi_r - 1, \quad \mu_r = -\frac{\pm 1}{\xi_r \pm 1} \quad (\text{from } k_\pm = -k_0), \tag{49}$$

where $\xi_r = \xi Z_0$ is the normalized chirality parameter. The positive (negative) sign in Eqs. (48) and (49) indicates the condition for LCP (RCP) waves. Figures 4(a) and 4(b) [4(c) and 4(d)] represent the relations among ε_r, μ_r, and ξ_r shown in Eq. (48) [Eq. (49)]. Note that the no-reflection conditions for CP waves correspond to the intersections of the contour lines of the no-reflection angles in Fig. 3. By using the electric susceptibility $\chi_e = \varepsilon_r - 1$ and magnetic susceptibility $\chi_m = 1 - \mu_r^{-1}$, Eqs. (48) and (49) are reduced to simpler forms:

$$\chi_e = \chi_m = \mp\xi_r, \tag{50}$$

$$\chi_e + 2 = \chi_m - 2 = \pm\xi_r, \tag{51}$$

respectively, where the upper (lower) sign corresponds to the condition for LCP (RCP) waves. We clarify the physical meaning of the no-reflection effect for CP waves by considering the medium polarization $\boldsymbol{P}$ and magnetization $\boldsymbol{M}$ induced by $\boldsymbol{E}$ and $\boldsymbol{B}$ in CP waves. For simplicity, assume that the no-reflection condition is satisfied for LCP waves. $\boldsymbol{P}$ and $\boldsymbol{M}$ are given by $\boldsymbol{P} = \boldsymbol{P}_E + \boldsymbol{P}_B$ and $\boldsymbol{M} = \boldsymbol{M}_B + \boldsymbol{M}_E$, where $\boldsymbol{P}_E = (\varepsilon - \varepsilon_0)\boldsymbol{E}$, $\boldsymbol{P}_B = -i\xi\boldsymbol{B}$, $\boldsymbol{M}_B = -(\mu^{-1} - \mu_0^{-1})\boldsymbol{B}$, and $\boldsymbol{M}_E = i\xi\boldsymbol{E}$ (Serdyukov et al., 2001). First, we calculate $\boldsymbol{P}$ and $\boldsymbol{M}$ when Eq. (48) is satisfied. From the relation $\boldsymbol{H} = (\mathrm{i}/Z_c)\boldsymbol{E}$ that is satisfied for LCP waves (see Table 1) and Eqs. (1) and (48), it is not difficult to confirm that $\boldsymbol{P} = 0$ and $\boldsymbol{M} = 0$ are satisfied regardless of the propagation direction. Due to the electromagnetic mixing attributed to ξ, the polarization $\boldsymbol{P}_\mathrm{B}$, which is induced by the magnetic flux density, completely cancels out the polarization $\boldsymbol{P}_\mathrm{E}$, which is induced by the electric field. Similarly, $\boldsymbol{M}_\mathrm{E}$ cancels out $\boldsymbol{M}_\mathrm{B}$. As a result of the destructive interference of electric and magnetic responses, net polarization and magnetization vanish in the case of LCP waves in the chiral medium. This implies that the chiral medium is identical to the vacuum for LCP waves. Next, we calculate $\boldsymbol{P}$ and $\boldsymbol{M}$ when Eq. (49) is satisfied. By applying a similar procedure, we obtain $\boldsymbol{P} = -2\varepsilon_0\boldsymbol{E}$ and $\boldsymbol{M} = -2\boldsymbol{H}$, which equal the corresponding value of the anti-vacuum. Therefore, the chiral medium behaves as an anti-vacuum for LCP waves.

4.4 In case of $k_+ = -k_-$ (chiral nihility media)

In the case of chiral nihility media ($k_+ = -k_-$) (Tretyakov et al., 2003), we obtain $M_\mathrm{R} = c_\mathrm{u}I + c_3\sigma_3$, which is the same representation as that in the achiral case. The no-reflection angles for TM and TE waves are written as

$$\theta_\mathrm{TM} = \arcsin\sqrt{\frac{Z_\mathrm{r}^{-2}-1}{Z_\mathrm{r}^{-2}-n^{-2}}}, \quad \theta_\mathrm{TE} = \arcsin\sqrt{\frac{Z_\mathrm{r}^{2}-1}{Z_\mathrm{r}^{2}-n^{-2}}}, \tag{52}$$

respectively, where $Z_\mathrm{r} = Z_\mathrm{c}/Z_0$ and $n = k_+/k_0 = -k_-/k_0$. The no-reflection effect in this case resembles but is different from that in the achiral case. While the transmitted wave is a linearly polarized wave in the achiral case, LCP and RCP waves that satisfy $\theta_+ = -\theta_-$ are transmitted in this case. Equations (30) and (35)-(38) show that the intensities of the transmitted LCP and RCP waves are equal.

The medium parameters satisfying the no-reflection condition are derived from

$$(\varepsilon_\mathrm{r} + \mu_\mathrm{r}\xi_\mathrm{r}^2)\mu_\mathrm{r} = 0, \quad \mu_\mathrm{r}/(\varepsilon_\mathrm{r} + \mu_\mathrm{r}\xi_\mathrm{r}^2) = Z_\mathrm{r}^2, \quad \mu_\mathrm{r}\xi_\mathrm{r} = \pm n. \tag{53}$$

The first is the condition for chiral nihility ($k_+ = -k_-$), and the second and third are the conditions for the wave impedance and wavenumber. To satisfy these equations at a particular angular frequency ω_0, the medium parameters need to be written as the following equations around ω_0:

$$\varepsilon_\mathrm{r} = \frac{n}{Z_\mathrm{r}}\left[f(\omega-\omega_0) - \frac{1}{f(\omega-\omega_0)}\right], \quad \mu_\mathrm{r} = nZ_\mathrm{r}f(\omega-\omega_0), \quad \xi_\mathrm{r} = \pm\frac{1}{Z_\mathrm{r}f(\omega-\omega_0)}, \tag{54}$$

where the function f satisfies $\lim_{\omega\to\omega_0} f(\omega-\omega_0) = 0$.

When $Z_\mathrm{r} = 1$ and $n = \pm 1$, namely, $Z_\mathrm{c} = Z_0$ and $|k_\pm| = k_0$ are satisfied, the reflection matrix M_R becomes a zero matrix. Since the conditions $Z_\mathrm{c} = Z_0$ and $|k_\pm| = k_0$ are independent of the incident angle, M_R becomes a zero matrix for all angles of incidence; arbitrary polarized waves are not reflected for any incident angle. This phenomenon has been confirmed by numerically calculating the reflectivity when $Z_\mathrm{c} = Z_0$ and $|k_\pm| \approx k_0$ are satisfied (Qiu et al., 2008).

We consider the physical meaning of Eq. (54) when both $Z_\mathrm{r} = 1$ and $n = \pm 1$ are satisfied. For simplicity, suppose that $(\varepsilon_\mathrm{r}, \mu_\mathrm{r}, \xi_\mathrm{r}) = [f(\omega-\omega_0) - f(\omega-\omega_0)^{-1}, f(\omega-\omega_0), f(\omega-\omega_0)^{-1}]$ are satisfied in this paragraph. The medium polarization and magnetization are found to be $\boldsymbol{P} = \varepsilon_0 f(\omega-\omega_0)\boldsymbol{E} \to 0$ and $\boldsymbol{M} = f(\omega-\omega_0)\boldsymbol{H} \to 0$ for LCP waves and $\boldsymbol{P} = \varepsilon_0[f(\omega-\omega_0) - 2]\boldsymbol{E} \to -2\varepsilon_0\boldsymbol{E}$ and $\boldsymbol{M} = [f(\omega-\omega_0) - 2]\boldsymbol{H} \to -2\boldsymbol{H}$ for RCP waves when $\omega \to \omega_0$. This implies that the medium behaves as a vacuum for LCP waves and as an anti-vacuum for RCP waves.

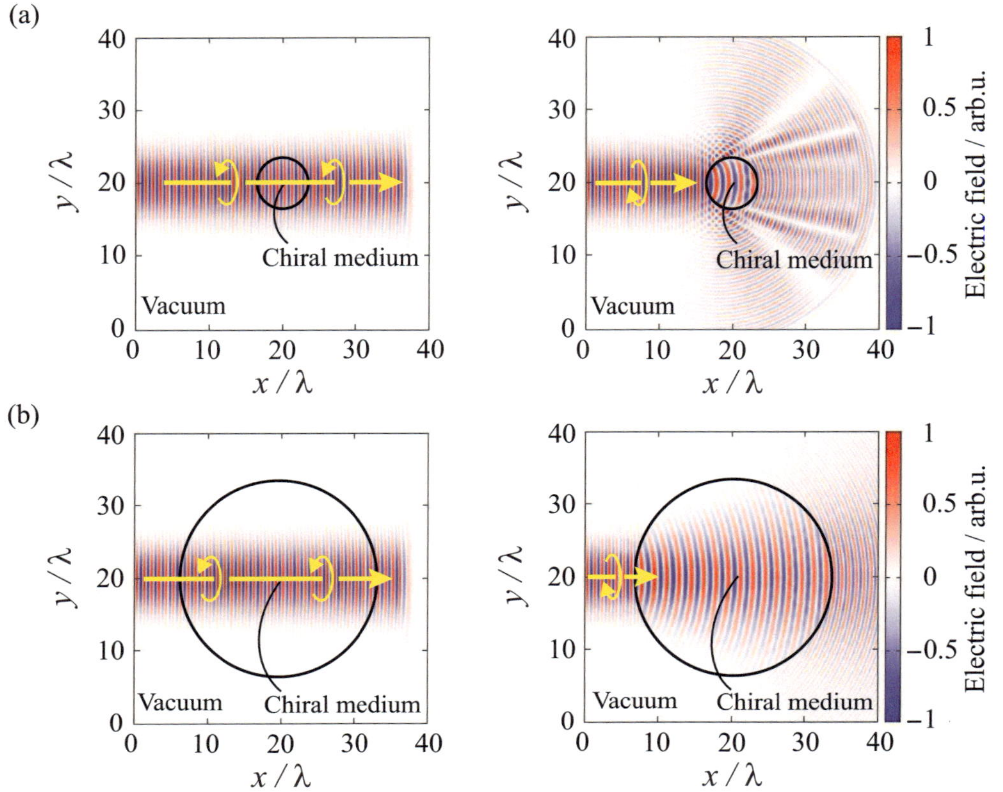

Fig. 5. Scattering of electromagnetic waves by cylindrical chiral medium when diameter of chiral medium is (a) smaller and (b) larger than beam width of electromagnetic wave. Left panels show propagations of LCP waves and right panels show those of RCP waves.

The no-reflection condition in the case of $Z_c = Z_0$ and $|k_\pm| = k_0$ can be regarded as the case of $|\xi_r| \to \infty$ in Eqs. (48) and (49). From Eqs. (48) and (49), we find that the vacuum condition for LCP (RCP) waves and the anti-vacuum condition for RCP (LCP) waves can be simultaneously satisfied when $|\xi_r| \to \infty$. In other words, the points represented by L_+ and R_- (L_- and R_+) in Fig. 3 approach each other with increasing $|\xi_r|$ and the two points become identical in the case of $|\xi_r| \to \infty$.

5. FDTD analysis of no-reflection effect for CP waves

From now, we focus on the no-reflection effect for CP waves and analyze the no-reflection effect by an FDTD method (Tamayama et al., 2008). Here, the parameters of the chiral medium are set as $\varepsilon_r = 0.75$, $\mu_r = 0.8$, and $\xi_r = 0.25$, which give $Z_c = Z_0$, $k_+ = k_0$, and $k_- = 0.6k_0$. That is, the no-reflection condition (vacuum condition) is satisfied for LCP waves.

First, we analyze the scattering of electromagnetic waves by a cylinder made of the chiral medium. To adopt the two-dimensional FDTD method, Maxwell's equations for CP waves

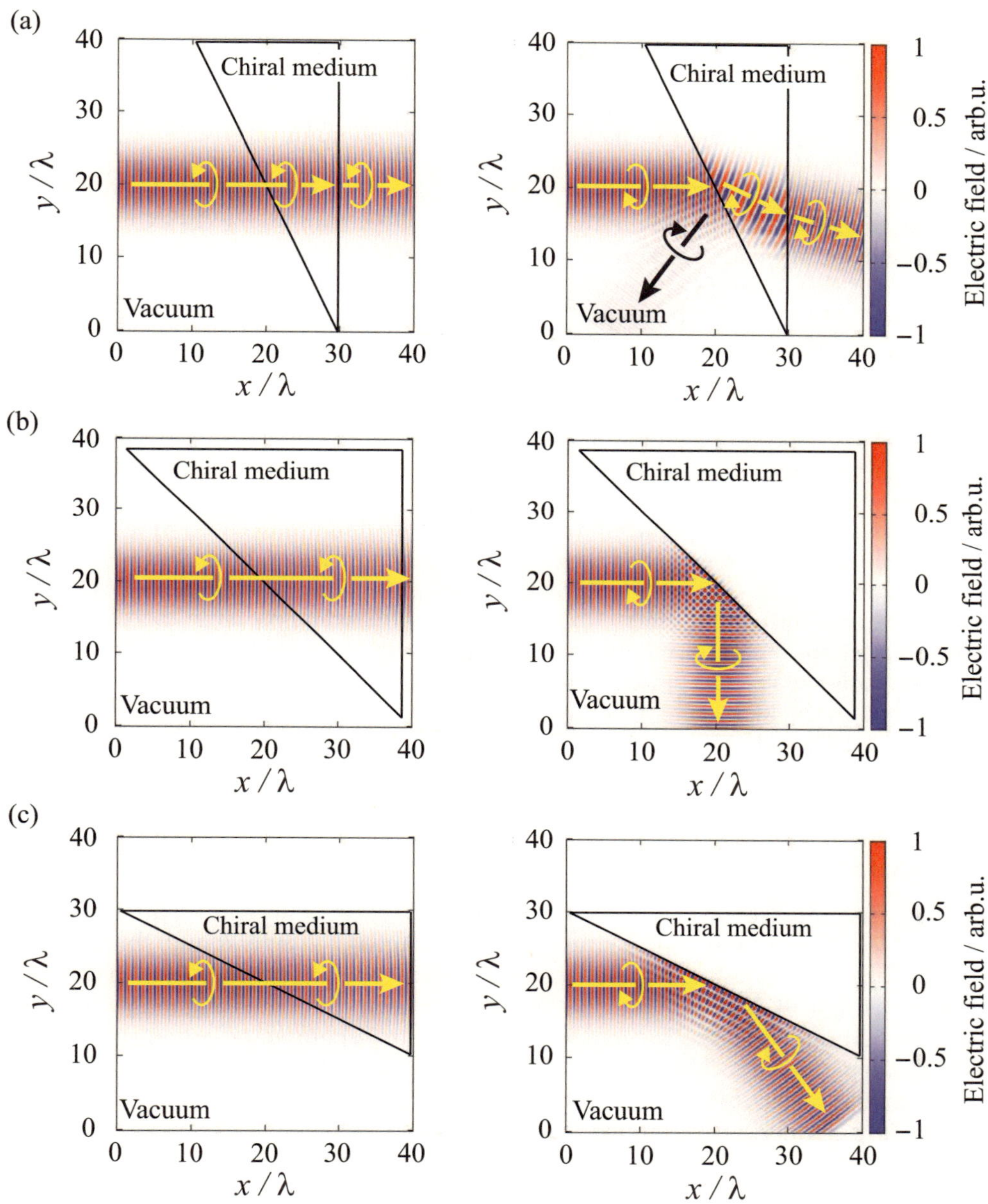

Fig. 6. Circular polarizing beam splitter. Incident angles are (a) 26.5°, (b) 45°, and (c) 63.5°. (left panels) Propagations of LCP waves and (right panels) of RCP waves.

are rearranged as follows:

$$\frac{\partial E_{z\pm}}{\partial y} = \mathrm{i}\omega(\mu \pm \mu\xi Z_\mathrm{c})H_{x\pm}, \tag{55}$$

$$-\frac{\partial E_{z\pm}}{\partial x} = \mathrm{i}\omega(\mu \pm \mu\xi Z_\mathrm{c})H_{y\pm}, \tag{56}$$

$$\frac{\partial H_{y\pm}}{\partial x} - \frac{\partial H_{x\pm}}{\partial y} = -\mathrm{i}\omega\left[(\varepsilon + \mu\xi^2) \pm \frac{\mu\xi}{Z_\mathrm{c}}\right]E_{z\pm}, \tag{57}$$

where the relation $\boldsymbol{H} = \pm(\mathrm{i}/Z_\mathrm{c})\boldsymbol{E}$ is used and the positive (negative) sign corresponds to LCP (RCP) waves. Section 4.3 and Eqs. (30) and (35)-(38) show that the incident CP wave is not converted into the other CP wave on the reflection and refraction because the wave impedance matching condition $Z_\mathrm{c} = Z_0$ is satisfied in this case. Thus, we may separately analyze the propagations of LCP and RCP waves. Figure 5 shows the propagations of electromagnetic waves when the diameter of the cylindrical chiral medium is smaller and larger than the beam width of the electromagnetic waves. One sees that LCP waves propagate with no scattering, and RCP waves are largely scattered.

Next, we analyze the propagation of electromagnetic waves when they are incident on a triangular prism made of the chiral medium. For $k_- = 0.6k_0$, Snell's equation for RCP waves is expressed as $\sin\theta = 0.6\sin\theta_-$; hence, the critical angle for RCP waves is $\theta_\mathrm{c} = \arcsin(0.6) \simeq 37°$. Therefore, LCP waves are completely transmitted without any reflection, while RCP waves are totally reflected with the incident angle greater than 37°. This implies that we can divide the incident waves into LCP and RCP waves. That is, the prism can be utilized as a circular polarizing beam splitter. The left (right) panels of Fig. 6 show the propagations of LCP (RCP) waves. Simulations are performed for three incident angles: 26.5°, 45°, and 63.5°. The LCP wave is transmitted straight through the chiral medium without reflection for any incident angle. Although the RCP wave is partially reflected and partially transmitted in the case of $\theta < \theta_\mathrm{c}$, it is totally reflected at the surface of the chiral medium when $\theta > \theta_\mathrm{c}$. This result confirms that the incident wave can be split into LCP and RCP waves, and the circular polarizing beam splitter is achieved when the incident angle exceeds θ_c.

6. Conclusion

We studied the no-reflection conditions for a planar boundary between a vacuum and a chiral medium. The comparison of the no-reflection conditions for achiral and chiral media is shown in Table 2. While the no-reflection effect arises for TM and TE waves in the case of achiral media ($\xi = 0$), it arises for EP waves in the case of chiral media ($\xi \neq 0$) whose wave impedances do not equal the vacuum wave impedance. These no-reflection conditions are satisfied for a particular incident angle. When the wave impedance and the absolute value of the wavenumber in the chiral medium equal those in the vacuum for one of the CP waves, the corresponding CP wave is transmitted with no-reflection for all angles of incidence. Although the no-reflection effect for chiral nihility media resembles that for achiral media, the two cases of the no-reflection effect are different from each other in the transmitted waves.

We analyzed the no-reflection effect for CP waves by an FDTD method. The simulation results showed that a chiral medium, whose medium parameters satisfy the no-reflection effect for one of the CP waves, does not scatter the corresponding CP wave and it largely scatters the other CP wave. The FDTD simulation also demonstrated that a circular polarizing beam splitter can be achieved by a triangular prism made of the chiral medium.

$\xi = 0$

Wave impedance	$Z_c \neq Z_0$		$Z_c = Z_0$
Medium parameters, Incident angle	$c_u = c_3$	$c_u = -c_3$	$k \equiv k_\pm = \pm k_0$
Polarization	TM	TE	—
No-reflection angle	${}^\exists\theta$	${}^\exists\theta$	${}^\forall\theta$
Refraction	yes	yes	no (when $k = k_0$) yes (when $k = -k_0$)

$\xi \neq 0$

Wave impedance	$Z_c \neq Z_0$		$Z_c = Z_0$	
Medium parameters, Incident angle	$c_u = c_\varphi, Z_c > Z_0$ or $c_u = -c_\varphi, Z_c < Z_0$	$c_u = -c_\varphi, Z_c > Z_0$ or $c_u = c_\varphi, Z_c < Z_0$	$k_+ = \pm k_0$	$k_- = \pm k_0$
Polarization	TM-like EP	TE-like EP	LCP	RCP
No-reflection angle	${}^\exists\theta$	${}^\exists\theta$	${}^\forall\theta$	${}^\forall\theta$
Refraction	yes	yes	no (when $k_+ = k_0$) yes (when $k_+ = -k_0$)	no (when $k_- = k_0$) yes (when $k_- = -k_0$)

Table 2. Classification of no-reflection conditions for achiral ($\xi = 0$) and chiral ($\xi \neq 0$) media.

For future studies, we must prepare metamaterials whose ε_r, μ_r, and ξ_r satisfy the no-reflection conditions for EP and CP waves. Such metamaterials can be realized by employing chiral structures (Kuwata-Gonokami et al., 2005; Zhang et al., 2009; Rockstuhl et al., 2009; Wang et al., 2009) and by electromagnetically induced chirality in atomic systems (Sautenkov et al., 2005; Kästel et al., 2007).

7. Acknowledgments

This research was supported by a Grant-in-Aid for Scientific Research on Innovative Areas (No. 22109004) and the Global COE program "Photonics and Electronics Science and Engineering" at Kyoto University. One of the authors (Y.T.) would like to acknowledge the support of a Research Fellowship from the Japan Society for the Promotion of Science.

8. References

Bassiri, S., Papas, C. H. & Engheta, N. (1988). Electromagnetic wave propagation through a dielectric-chiral interface and through a chiral slab, *J. Opt. Soc. Am. A* 5(9): 1450–1459.

Collin, R. E. (1990). *Field Theory of Guided Waves*, 2nd edn, IEEE Press, Piscataway, NJ.

Doyle, W. T. (1980). Graphical approach to Fresnel's equations for reflection and refraction of light, *Am. J. Phys.* 48(8): 643–647.

Fu, C., Zhang, Z. M. & First, P. N. (2005). Brewster angle with a negative-index material, *Appl. Opt.* 44(18): 3716–3724.

Futterman, J. (1995). Magnetic Brewster angle, *Am. J. Phys.* 63(5): 471.

Grzegorczyk, T. M., Thomas, Z. M. & Kong, J. A. (2005). Inversion of critical angle and brewster angle in anisotropic left-handed metamaterials, *Appl. Phys. Lett.* 86(25): 251909.

Holloway, C. L., Kuester, E. F., Baker-Jarvis, J. & Kabos, P. (2003). A Double Degative (DNG) Composite Medium Composed of Magnetodielectric Spherical Particles Embedded in a Matrix, *IEEE Trans. Antennas Propag.* 51(10): 2596–2603.

Jacob, Z., Alekseyev, L. V. & Narimanov, E. (2006). Optical Hyperlens: Far-field imaging beyond the diffraction limit, *Opt. Express* 14(18): 8247–8256.

Kästel, J., Fleischhauer, M., Yelin, S. F. & Walsworth, R. L. (2007). Tunable Negative Refraction without Absorption via Electromagnetically Induced Chirality, *Phys. Rev. Lett.* 99(7): 073602.

Kuwata-Gonokami, M., Saito, N., Ino, Y., Kauranen, M., Jefimovs, K., Vallius, T., Turunen, J. & Svirko, Y. (2005). Giant Optical Activity in Quasi-Two-Dimensional Planar Nanostructures, *Phys. Rev. Lett.* 95(22): 227401.

Lagarkov, A. N. & Kissel, V. N. (2004). Near-Perfect Imaging in a Focusing System Based on a Left-Handed-Material Plate, *Phys. Rev. Lett.* 92(7): 077401.

Lakhtakia, A. (1992). General schema for the Brewster conditions, *Optik (Stuttgart)* 90(4): 184–186.

Leonhardt, U. (2006). Optical Conformal Mapping, *Science* 312(5781): 1777–1780.

Lindell, I. V., Sihvola, A. H., Tretyakov, S. A. & Viitanen, A. J. (1994). *Electromagnetic Waves in Chiral and Bi-Isotropic Media*, Artech House, Boston, MA.

Liu, Z., Lee, H., Xiong, Y., Sun, C. & Zhang, X. (2007). Far-Field Optical Hyperlens Magnifying Sub-Diffraction-Limited Objects, *Science* 315(5819): 1686.

Pendry, J. B. (2000). Negative Refraction Makes a Perfect Lens, *Phys. Rev. Lett.* 85(18): 3966–3969.

Pendry, J. B., Holden, A. J., Robbins, D. J. & Stewart, W. J. (1999). Magnetism from Conductors and Enhanced Nonlinear Phenomena, *IEEE Trans. Microwave Theory Tech.* 47(11): 2075–2084.

Pendry, J. B., Schurig, D. & Smith, D. R. (2006). Controlling Electromagnetic Fields, *Science* 312(5781): 1780–1782.

Qiu, C.-W., Burokur, N., Zouhd, S. & Li, L.-W. (2008). Chiral nihility effects on energy flow in chiral materials, *J. Opt. Soc. Am. A* 25(1): 55–63.

Rockstuhl, C., Menzel, C., Paul, T. & Lederer, F. (2009). Optical activity in chiral media composed of three-dimensional metallic meta-atoms, *Phys. Rev. B* 79(3): 035321.

Sakurai, J. J. (1994). *Modern Quantum Mechanics*, revised edn, Addison-Wesley, Reading, MA.

Saleh, B. E. A. & Teich, M. C. (2007). *Fundamentals of Photonics*, 2nd edn, Wiley-Interscience, Hoboken, NJ.

Sautenkov, V. A., Rostovtsev, Y. V., Chen, H., Hsu, P., Agarwal, G. S. & Scully, M. O. (2005). Electromagnetically Induced Magnetochiral Anisotropy in a Resonant Medium, *Phys. Rev. Lett.* 94(23): 233601.

Schurig, D., Mock, J. J., Justice, B. J., Cummer, S. A., Pendry, J. B., Starr, A. F. & Smith, D. R. (2006). Metamaterial Electromagnetic Cloak at Microwave Frequencies, *Science* 314(5801): 977–980.

Serdyukov, A., Semchenko, I., Tretyakov, S. & Sihvola, A. (2001). *Electromagnetics of Bi-anisotropic Materials: Theory and Applications*, Gordon and Breach Science Publishers, LH, Amsterdam.

Shelby, R. A., Smith, D. R. & Schultz, S. (2001). Experimental verification of a negative index of refraction, *Science* 292(5514): 77–79.

Shen, N.-H., Wang, Q., Chen, J., Fan, Y.-X., Ding, J. & Wang, H.-T. (2006). Total transmission of electromagnetic waves at interface associated with an indefinite medium, *J. Opt. Soc. Am B* 23(5): 904–912.

Shu, W., Ren, Z., Luo, H. & Li, F. (2007). Brewster angle for anisotropic materials from the extinction theorem, *Appl. Phys. A* 87(2): 297–303.

Taflove, A. & Hagness, S. C. (2005). *Computational Electrodynamics: The Finite-Difference Time-Domain Method*, 3rd edn, Artech House, Norweed, MA.

Tamayama, Y., Nakanishi, T., Sugiyama, K. & Kitano, M. (2006). Observation of Brewster's effect for transverse-electric electromagnetic waves in metamaterials: Experiment and theory, *Phys. Rev. B* 73(19): 193104.

Tamayama, Y., Nakanishi, T., Sugiyama, K. & Kitano, M. (2008). An invisible medium for circularly polarized electromagnetic waves, *Opt. Express* 16(25): 20869–20875.

Tanaka, T., Ishikawa, A. & Kawata, S. (2006). Unattenuated light transmission through the interface between two materials with different indices of refraction using magnetic metamaterials, *Phys. Rev. B* 73(12): 125423.

Tretyakov, S., Nefedov, I., Sihvola, A., Maslovski, S. & Simovsli, C. (2003). Waves and energy in chiral nihility, *J. of Electromagn. Waves and Appl.* 17(5): 695–706.

Veselago, V. G. (1968). The electrodynamics of substances with simultaneously negative values of ε and μ, *Sov. Phys. Usp.* 10(4): 509–514.

Wang, B., Zhou, J., Koschny, T. & Soukoulis, C. M. (2009). Nonplanar chiral metamaterials with negative index, *Appl. Phys. Lett.* 94(15): 151112.

Watanabe, R., Iwanaga, M. & Ishihara, T. (2008). s-polarization Brewster's angle of stratified metal-dielectric metamaterial in optical regime, *Phys. Stat. Sol. (b)* 245(12): 2696–2701.

Zhang, S., Fan, W., Malloy, K. J., Brueck, S. R. J., Panoiu, N. C. & Osgood, R. M. (2005).

Near-infrared double negative metamaterials, *Opt. Express* 13(13): 4922–4930.
Zhang, S., Park, Y.-S., Li, J., Lu, X., Zhang, W. & Zhang, X. (2009). Negative Refractive Index in Chiral Metamaterials, *Phys. Rev. Lett.* 102(2): 023901.

Part 6

Nonlinear Phenomena and Electromagnetic Wave Generation

21

Manipulating the Electromagnetic Wave with a Magnetic Field

Shiyang Liu[1], Zhifang Lin[2] and S. T. Chui[3]
[1]*Institute of Information Optics, Zhejiang Normal University*
[2]*Surface Physics Laboratory, Department of Physics, Fudan University*
[3]*Department of Physics and Astronomy, University of Delaware*
[1,2]*China*
[3]*USA*

1. Introduction

Metamaterials are composite, artificial materials, composed of sub-wavelength building blocks acting like man-made atoms, which exhibit novel and unique electromagnetic (EM) properties, not occurring in natural materials (Pendry, 2000; Shelby et al., 2001). With the advent of metamaterials the molding of the flow of light becomes more and more flexible. Exhilaratingly, through the state-of-art design of metamaterial, nearly arbitrary profile of EM constitutive parameters can be engineered, enabling the manifestation of an unprecedented variety of exotic behaviors from microwave region to optical region such as negative refraction (Shalaev, 2007; Valentine et al., 2008), cloaking (Leonhardt, 2006; Pendry et al, 2006; Schurig et al., 2006), and subwavelength propagation (Maier, 2002; Maier et al., 2003) of EM wave. Metamaterials are attracting more and more theorist, experimentalist, and engineers from many research fields. With the further development, we can expect a revolutionary influence that metamaterials will offer in the field of microwave engineering, opto-electronic information technology, even photonics and optical communication.
The building blocks are usually made of metallic material (Pendry et al., 1996; 1999), accompanied with the drawbacks of intrinsic loss and tunability limitation. Here we present a class of metamaterials composed of building blocks made of magnetic material, which are accordingly called magnetic metamaterials (MM). Due to the dependence of permeability on external magnetic field (EMF), the EM properties of MM can be manipulated by an EMF, providing us with more degrees of tunability. For this reason, MM can be a promising candidate for the design of optical devices. In addition, different from dielectric or metallic materials, the permeability of magnetic material is a second rank tensor with nonzero off-diagonal elements. The time reversal symmetry is thus broken in MM systems (Wang et al., 2008), based on which we can observe some other phenomena and design special optical devices. Besides, we also examine the EM properties of a kind of magnetic photonic crystal (MPC) and magnetic graded PC (MGPC), made of an array of ferrite rods periodically arranged in air under uniform and nonuniform EMFs. They can also offer us the manipulability on the flow of EM wave with the tunability by an EMF.
The research content of the present chapter consists of four parts. In the fist part, we examine the mechanism governing the photonic band gaps (PBGs) of a two dimensional (2D) MPC,

based on the simulations on the photonic band diagrams and the transmission spectra, by which we can identify three different types of PBGs. In particular, it is shown that "magnetic surface plasmon" (MSP) resonance induced PBGs can be completely tunable by an EMF and robust against position and size disorder of the ferrite rods. In the second part, we develop an effective-medium theory (EMT) within the coherent potential approximation, which is specially suitable to retrieve the effective constitutive parameters of the anisotropic MM. In addition, we demonstrate a construction of negative index metamaterial (NIM) based on MM, which possesses effective constitutive parameters $\varepsilon_{\text{eff}} = \mu_{\text{eff}} = -1$ and the magnetically tunable working frequency. In the third part, we propose and conceptualize an alternative type of graded PC, MGPC. With the MGPC, we can also mold the flow of EM wave, resulting in the focusing effect and the mirage effect for an TM Gaussian beam. In fourth part, we examine the reflection behavior of an EM waves from an MM slab. It is demonstrated that at some frequency the reflected wave is found to exhibit a giant circulation in that locally the angular momenta of the components are all of the same sign due to the MSP resonance and the broken of time reversal symmetry (TRS). As a result of this finite circulation, a dramatic change is exhibited for beams incident from different directions.

2. Formation of the robust and magnetically tunable photonic band gap

PCs or PBG materials have attracted more and more attentions since the conception was brought out in the 1980s (John, 1987; Yablonovitch, 1987) and became the subject of intensive theoretical and experimental research due to their promising applications in micro- and optoelectronics (Joannopoulos et al., 1995). One unique characteristic of PCs is the existence of PBGs, the frequency ranges over which all EM modes are forbidden, which offers much more flexibilities in manipulating the EM waves. To make it more applicable, tunable PCs have been proposed and investigated, where the PBGs can be modulated extrinsically by some external parameters such as the electric field (Busch & John, 1999; Kang et al., 2001), the magnetic field (Figotin et al., 1998; Golosovsky et al., 1999; Kee et al., 2000), the temperature (Halevi & Ramos-Mendieta, 2000; Kee & Lim, 2001), or the strain (Kim & Gopalan, 2001). Here, we shall concentrate on the effect of EMF on the PBGs of MPCs. As a typical tunable PC, MPCs have attracted a lot of interest (Chen et al., 2007; Inoue et al., 2006; Lin & Chui, 2007; Liu et al., 2008; Lyubchanskii et al., 2003; Sigalas et al., 1997) due to the fast switching time of magnetic systems and the potential tunability of the PBGs by the EMF.

2.1 Band structure calculations of the PBG materials

To illustrate the typical behavior of the photonic band diagram of an MPC, we consider the example of a 2D hexagonal lattice of ferrite rods in the nonmagnetic plexiglas background, with the lattice constant a and rod axes along z direction. When fully magnetized, the magnetic permeability in form (Pozar, 2004)

$$\widehat{\mu} = \begin{pmatrix} \mu_r & -i\mu_\kappa & 0 \\ i\mu_\kappa & \mu_r & 0 \\ 0 & 0 & 1 \end{pmatrix}, \quad \widehat{\mu}^{-1} = \begin{pmatrix} \mu_r' & -i\mu_\kappa' & 0 \\ i\mu_\kappa' & \mu_r' & 0 \\ 0 & 0 & 1 \end{pmatrix}, \tag{1}$$

with

$$\mu_r = 1 + \frac{\omega_m(\omega_0 - i\alpha\omega)}{(\omega_0 - i\alpha\omega)^2 - \omega^2}, \quad \mu_\kappa = \frac{\omega_m\omega}{(\omega_0 - i\alpha\omega)^2 - \omega^2}, \quad \mu_r' = \frac{\mu_r}{\mu_r^2 - \mu_\kappa^2}, \quad \mu_r' = \frac{-\mu_\kappa}{\mu_r^2 - \mu_\kappa^2},$$

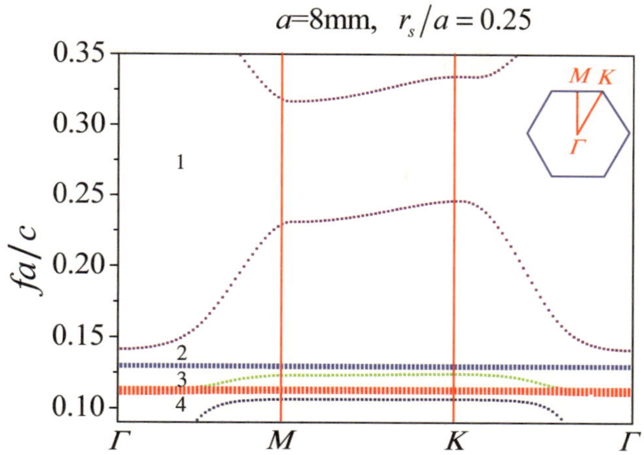

Fig. 1. The photonic band diagram of a 2D hexagonal MPC, with $a = 8$ mm, $r = 2$ mm, and $H_0 = 900$ Oe. The reduced Brillouin zone is shown in the inset together with the high symmetry points. Four PBGs marked with 1, 2, 3, and 4, exist from high to low frequency.

where $\omega_0 = \gamma H_0$ is the resonance frequency with $\gamma = 2.8$ MHz/Oe the gyromagnetic ratio; H_0 the sum of the EMF applied in z direction and the shape anisotropy field Pozar (2004), $\omega_m = 4\pi\gamma M_s$ is the characteristic frequency with $4\pi M_s$ the saturation magnetization, and α is the damping coefficient of the ferrite. In the calculation of photonic band diagram, we set $\alpha = 0$ (Wang et al., 2008) and consider the transverse magnetic (TM) mode only.

Most photonic band diagram calculations have been based on the plane-wave expansion method (Ho et al., 1990) or the finite difference time domain method (Chan et al., 1995). However, for dispersive systems with gyromagnetic scatterer, it is rather difficult to achieve a reliable convergence of the solutions. In our work, the multiple scattering method (MSM) is employed, which is proved to be most powerful for the PCs consisting of non-overlap spheres or circular cylinders (Li & Zhang, 1998; Lin & Chui, 2004; 2007; Liu & Lin, 2006; Moroz, 2002; Stefanou et al., 1998).

We have shown a typical photonic band diagram in Fig. 1 for $H_0 = 900$ Oe, where the lattice constant is $a = 8$ mm, and the radius of the ferrite rod is $r_s = \frac{1}{4}a = 2$ mm. The relative permittivity and permeability of the nonmagnetic plexiglas are $\varepsilon_m = 2.6$ and $\mu_m = 1$. For the ferrite rod $\varepsilon_s = 12.3$ and the saturation magnetization $4\pi M_s = 1700$ gauss. The reduced Brillouin zone is illustrated as well in the inset. The corresponding high symmetry points are $\Gamma = \frac{2\pi}{a}(0,0)$, $M = \frac{2\pi}{a}(0, \frac{1}{\sqrt{3}})$, and $K = \frac{2\pi}{a}(\frac{1}{3}, \frac{1}{\sqrt{3}})$ with c the speed of light in vacuum. It can be seen that there exist four PBGs centered at $fa/c = 0.275$, 0.138, 0.130, and 0.108, labeled by 1, 2, 3, and 4, respectively, in Fig. 1. Later on, we will examine the origin and the tunability of these PBGs by an EMF.

Firstly, let us focus on the 1st PBG centered at $fa/c = 0.275$. To understand the origin of this PBG, we present in Fig. 2 the amplified part of the band diagram around this PBG, together with the band diagram for a scaled system with $a = 5$ mm. It can be found that both PBGs are centered at almost the same position in reduced frequency of c/a. In addition, the photonic band diagram around the PBG are also similar, showing a pretty good scale-length invariant behavior. It is the characteristic of the PBG originating from the Bragg scattering. To illustrate the effect of an EMF on this PBG, we also present the band diagram in Fig. 3 for $H_0 = 700$ Oe (a) and $H_0 = 900$ Oe (b), while keeping all the other parameters the same as in Fig. 1. It can

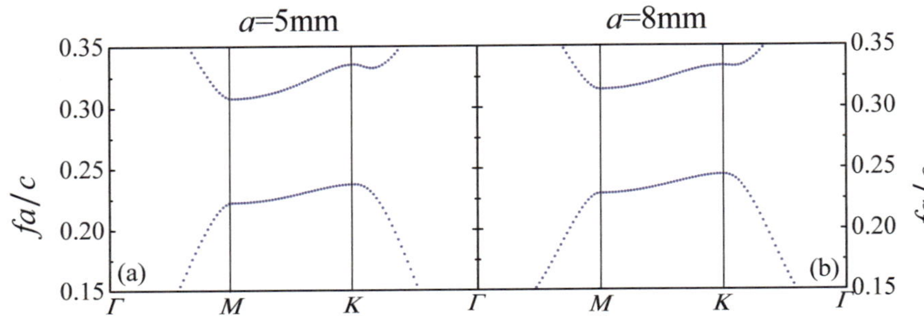

Fig. 2. The photonic band diagram near the 1st PBG of a 2D MPC with $a = 5$ mm (a) and $a = 8$ mm (b), while keeping all the other parameters the same as those in Fig. 1.

be found that the two band diagrams are nearly the same, suggesting that the EMF has a very weak effect on the 1st PBG. This explains the recent experimental results that show a tiny shift of the PBGs in response to the change of the EMF (Chen et al., 2007).

Next, let us examine the 2nd PBG centered at $fc/a = 0.138$. To demonstrate the effect of the EMF, we have calculated the photonic band diagram near the 2nd PBG as shown in Fig. 4(a) where $H_0 = 700$ Oe and all the other parameters are the same as those in Fig. 1. For comparison, we have also shown in Fig. 4(b) the corresponding part of Fig. 1. Different from the 1st PBG, a significant change can be observed in Fig. 4. The increase of the EMF shifts the 2nd PBG upwards substantially, so do the 3rd and 4th PBGs, suggesting that the 2nd PBG are intimately dependent on the magnetic permeability of the ferrite rod. An important characteristic is the appearance of a dense sets of flat bands between the 2nd and the 3rd PBGs in Fig. 4, signifying the occurrence of some kind of resonance.

In the study of plasmonics (Zayats et al., 2005), flat photonic bands have been observed near the surface plamon frequency (McGurn & Maradudin, 1993; Kuzmiak et al., 1994) with $\varepsilon = -1$ in 2D. In our case the flat bands occur when the effective magnetic susceptibility $\mu_r + \mu_\kappa = -1$ at frequency $f_s = \frac{1}{2\pi}\left(\omega_0 + \frac{1}{2}\omega_m\right)$, which can be considered as the magnetic analog of the surface plasmon in metal. For the MPC under investigation, with $4\pi M_s = 1700$ gauss and $a = 8$ mm, we have $f_s a/c = 0.1158$ and 0.1308 for $H_0 = 700$ Oe and $H_0 = 900$ Oe, respectively. It can be seen clearly from Fig. 4 that f_s is near the top of the flat bands, above and below which open up the 2nd and 3rd PBGs, respectively. As a result, the 2nd and 3rd PBGs are governed by the resonance at frequency f_s. We have also studied the photonic band diagram

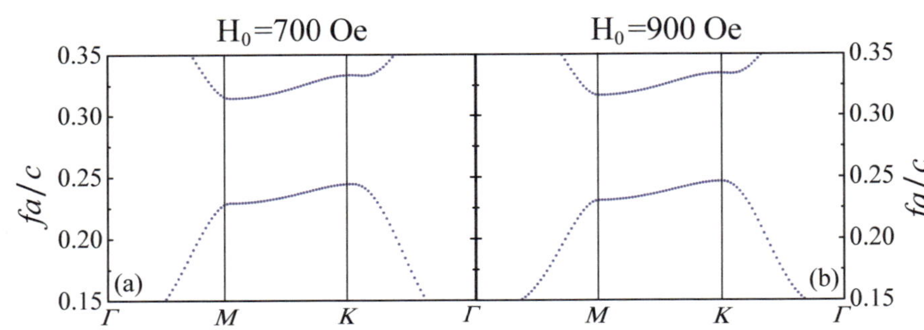

Fig. 3. (a) The photonic band diagram near the 1st PBG of a 2D MPC with $H_0 = 700$ Oe and all the other parameters the same as those in Fig. 1. (b) The same as that in Fig. 2(b).

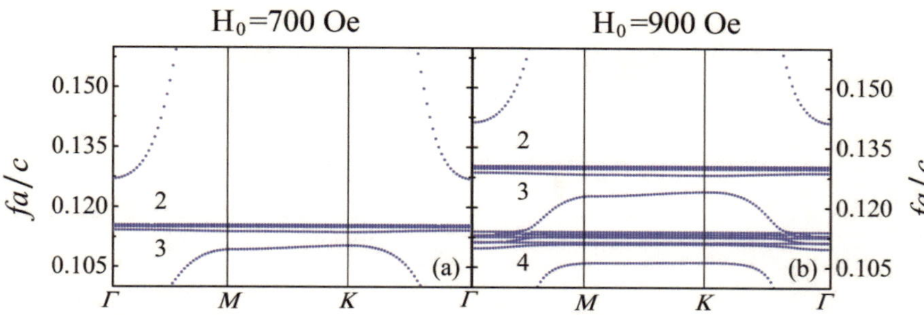

Fig. 4. (a) The photonic band diagram near the 2nd and 3rd PBGs of a 2D hexagonal MPC with $H_0 = 700$ Oe and all the other parameters the same as those in Fig. 1. (b) The same as that in Fig. 2(b). Numbers in the figure label the indices of the PBGs as shown in Fig. 1.

for systems with the lattice constants scaled by a variety of factors. Typical results are shown in Fig. 5, where we fixed $H_0 = 700$ Oe and scaled system from $a = 5$ mm to $a = 8$ mm. It can be found that the flat bands between the 2nd and the 3rd PBGs are located at the same frequency, indicating the independence of the flat bands on the lattice constant and rules out the possibility of the Bragg scattering mechanism. The flat bands is solely determined by f_s and therefore depends on the EMF and the saturation magnetization only.

In the end, we examine the 4th PBG in Fig. 1, above which there also appears a dense set of flat bands, implying once again some kind of resonance. From Eq. (1), it can be found that $\mu_r = \mu_r' = 0$ at a frequency $f_m = \frac{1}{2\pi}\sqrt{\omega_0(\omega_0 + \omega_m)}$. Accordingly, the wave vector inside the ferrite cylinder (proportional to $1/\sqrt{\mu_r'}$) tends to infinity at frequency f_m, corresponding to the occurrence of the "spin wave resonance". For our system with $H_0 = 900$ Oe, we have $f_m = 0.114c/a$, corresponding to the top of the flat bands, as shown in Fig. 4(b). The frequency of the 4th PBG can therefore be tuned by the EMF in a similar way as the case for the 2nd and 3rd PBGs.

2.2 Simulations on the transmission coefficients

To examine the PBGs further, we have also performed simulations on the transmission coefficients, which can be expressed as the forward scattering amplitude of the MPC (Li & Zhang, 1998). Different from the calculation of photonic bands, in this section we adopt the practical material parameters and take into account the absorption of the sample.

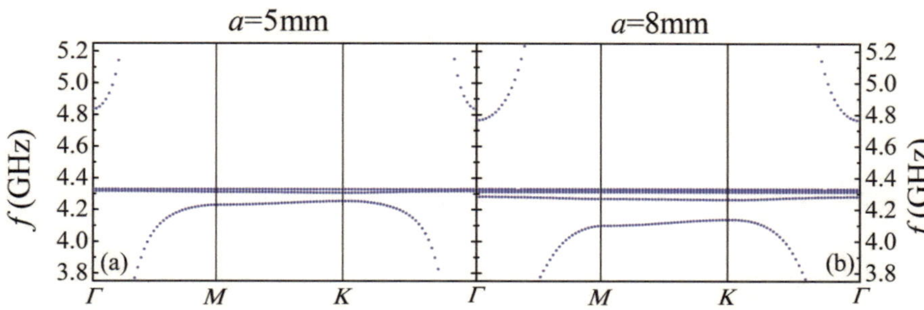

Fig. 5. The photonic band diagram near the 2nd and 3rd PBGs of a 2D hexagonal MPC with $a = 5$ mm (a) and $a = 8$ mm (b), respectively. The applied EMF is such that $H_0 = 700$ Oe, while all the other parameters are the same as those in Fig. 1.

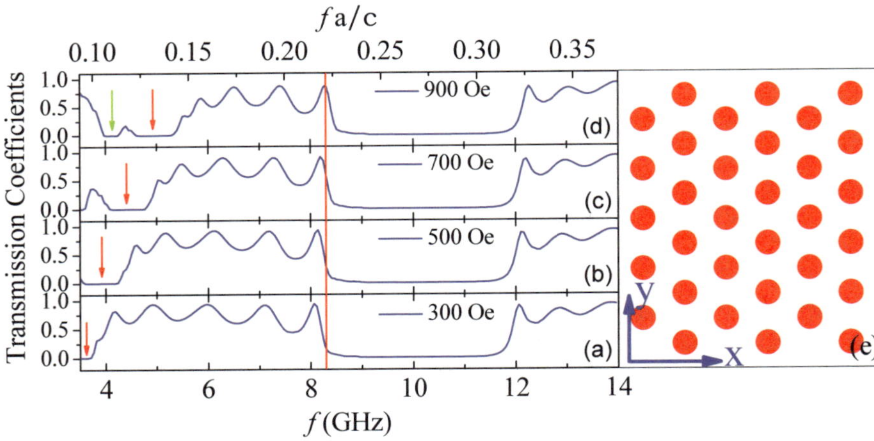

Fig. 6. Transmission coefficients of a 7-layer MPC slab under EMF $H_0 = 300$ Oe (a), 500 Oe (b), 700 Oe (c), and 900 Oe (d). The lattice constant is $a = 8$ mm and the radius of the ferrite rod is $r_s = \frac{1}{4}a = 2$ mm. The right panel (e) is the sample configuration.

Concretely, $\varepsilon_b = 2.6 + i5 \times 10^{-3}$, $\varepsilon_s = 12.3 + i6 \times 10^{-4}$, $\mu_b = 1.0$, and μ is given by Eq. (1) with damping coefficient taken as $\alpha = 7 \times 10^{-3}$ (Chen et al., 2007; Pozar, 2004).

Typical results for different H_0 are shown in Fig. 6 where the sample is a 7-layer slab with $a = 8$ mm and $r_s = \frac{1}{4}a$, as schematically illustrated in Fig. 6(e). It can be seen from Fig. 6 that with the increase of the EMF the 2nd, the 3rd (denoted by the red arrows), and the 4th PBGs (denoted by the green arrow) shift significantly to higher frequencies. However, the 1st PBG shows only a tiny shift when H_0 is varied from 300 Oe to 900 Oe. The transmission coefficients are in quantitative agreement with the photonic band diagram calculations. To be specific, the middle frequency of the 1st transmission gap in Fig. 6(d) is about $0.275c/a$; the middle frequency of the 2nd transmission gap (denoted by the red arrows) is located around $0.135c/a$; the 3rd transmission gap (denoted by the green arrow) has the middle frequency $0.125c/a$. They correspond exactly to the corresponding part of the photonic band diagram.

From the analysis of the photonic band diagrams, we know that the 2nd, the 3rd, and the 4th PBGs arise from the resonance of the individual ferrite rod. Therefore, we can expect that the

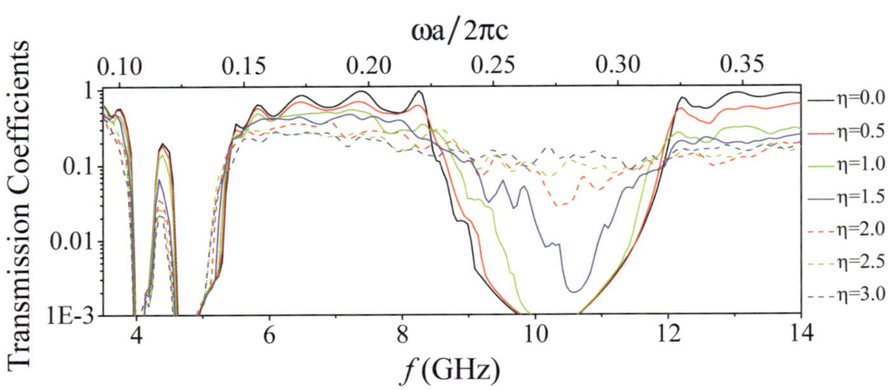

Fig. 7. Transmission coefficients of a 7-layer MPC slab under EMF $H_0 = 900$ Oe and with the position disorder characterized by η. Lattice constant $a = 8$ mm and radius $r_s = 2$ mm.

PBGs are robust against the position disorder. This is corroborated by the simulation on the transmission, where the disorder is introduced so that the position of each rod is displaced by $d_i = \xi \cdot \eta \cdot d_{i0}$ with $i = x$ or y. Here, ξ is a random number uniformly distributed between -0.5 and 0.5, $d_{x0} = \frac{\sqrt{3}}{2}a - 2r$ and $d_{y0} = a - 2r$ are the maximum position variations along x and y directions, η is a parameter introduced to measure the degree of disorder. In Fig. 7, we present the transmission coefficients with different randomness η. The corresponding parameters are the same as in Fig. 5 except that H_0 is fixed at 900 Oe. The results are obtained by averaging 20 realizations of the disordered system. It can be seen that with the increase of disorder the 1st PBG becomes narrower and narrower and disappears eventually due to the breakdown of the Bragg scattering. However, the resonance-induced PBGs remain essentially unchanged. Accordingly, it can be concluded that the introduction of the positional disorder affects only the Bragg type of PBG significantly, while the other two types of PBGs arising from resonance are quite robust against the position disorder. The similar effect has also been demonstrated in the systems consisting of electric active entities (Modinos et al., 2000; Yannopapas & Vitanov, 2006).

3. Negative index material manipulable with an external magnetic field

A particularly important class of metamaterials is the NIM (Pendry, 2000). Its corresponding properties were theoretically proposed by Veselago as early as in 1968 (Veselago, 1968). However, it is only fairly recently that the NIMs were experimentally realized based on Pendry's scheme, and then the interest in such material surged immediately. Various schemes are proposed and proved to possess the negative refractive index (Dolling et al., 2006; Liu et al., 2008; Peng et al., 2007; Rachford et al., 2007; Zhang et al., 2005; 2009). To investigate and characterize the optical properties of the metamaterials, it is crucial to retrieve the effective constitutive parameters (Chen et al., 2004; Jin et al., 2009; Koschny et al., 2003; 2005; Sarychev et al., 2000; Smith et al., 2002; 2005; Wu et al., 2006). In this section, we will present the EMT developed specially for the anisotropic MM. In particular, a ferrite-based NIM with $\mu_{\text{eff}} = \varepsilon_{\text{eff}} = -1$ is designed, which exhibits the negative refraction and the superlens effect.

3.1 Effective-medium theory for anisotropic magnetic metamaterials

We present in this section the formulation of the EMT for anisotropic MM in the 2D case, generalization to three dimensions is straightforward. For convenience, we consider a system made of ferrite rods with permittivity ε_s and magnetic permeability $\widehat{\mu}$ given in Eq. (1) arranged periodically as a square lattice in an isotropic homogeneous medium with permittivity ε_0 and permeability μ_0. The geometry of the system is shown in Fig. 8 where the rod axes are along z direction and the radii of the ferrite rods are r_s. We consider the TM mode with an anisotropic and tunable permeability.

The scheme of the EMT is as follows: (i) Transform the discrete periodic system in figure 8(a) into the effective medium with effective constitutive parameters ε_{eff} and μ_{eff} in figure 8(b); (ii) Take the unit cell of the MM as an equal-area coated rod with ferrite rod as its inner core and the background medium as the coated layer with radius $r_0 = a/\sqrt{\pi}$; (iii) The effective constitutive parameters ε_{eff} and μ_{eff} are determined by the condition that the scattering cross section of the coated rod in the effective medium vanishes in the long wave limit $k_{\text{eff}} r_0 \ll 1$ where $k_{\text{eff}} = k_0\sqrt{\varepsilon_{\text{eff}}}\sqrt{\mu_{\text{eff}}}$ with k_0 the wave-number in the vacuum, $\mu_{\text{eff}} = (\mu_{re}^2 - \mu_{\kappa e})^2/\mu_{re}$. μ_{re} and $\mu_{\kappa e}$ are the components of the effective magnetic permeability tensor. The concept of our theory is the so-called coherent potential approximation (CPA) (Sheng et al., 1995), which has

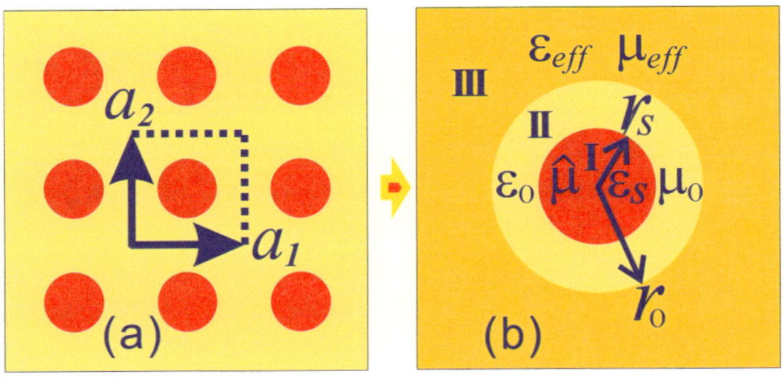

Fig. 8. (a) A 2D anisotropic MM consisting of the ferrite rods arranged as square lattice in the air background, and (b) its corresponding geometry in the effective-medium theory. The red circles, light yellow region, and dark yellow region correspond to the ferrite rods, the isotropic homogeneous background, and effective medium, respectively.

been used to retrieve the effective isotropic parameters for the EM (Hu et al., 2006; Wu et al., 2006), acoustic (Hu et al., 2008), and even elastic metamaterials (Wu et al., 2007).
As the procedure of the Mie theory, we expand the EM waves in regions I, II, and III of figure 8(b) into the summation of the vector cylindrical wave functions (VCWFs)

$$\mathbf{L}_n^{(J)}(k,\mathbf{r}) = \left[\frac{dz_n^{(J)}(\rho)}{d\rho}\mathbf{e}_r + \frac{in}{\rho}z_n^{(J)}(\rho)\mathbf{e}_\phi\right]e^{in\phi}, \tag{2a}$$

$$\mathbf{M}_n^{(J)}(k,\mathbf{r}) = \left[\frac{in}{\rho}z_n^{(J)}(\rho)\mathbf{e}_r - \frac{dz_n^{(J)}(\rho)}{d\rho}\mathbf{e}_\phi\right]e^{in\phi}, \tag{2b}$$

$$\mathbf{N}_n^{(J)}(k,\mathbf{r}) = z_n^{(J)}(\rho)e^{in\phi}\mathbf{e}_z, \tag{2c}$$

with k the wave vector, $\mathbf{r} = r\cos\phi\mathbf{e}_x + r\sin\phi\mathbf{e}_y$ the position vector, $\rho = kr$, $z_n^{(J)}(\rho)$ the Bessel function $J_n(\rho)$ and the first kind Hankel function $H_n^{(1)}(\rho)$ for $J = 1$ and 3, respectively. For the EM field in the region I

$$\mathbf{E}_s(k_s,\mathbf{r}) = \sum_n q_n^s\mathbf{N}_n^{s(1)}, \qquad \mathbf{H}_s(k_s,\mathbf{r}) = \sum_n \frac{k_s}{i\omega}q_n^s\left[i\mu'_{\kappa s}\mathbf{L}_n^{s(1)} + \mu'_{rs}\mathbf{M}_n^{s(1)}\right], \tag{3}$$

where $k_s = k_0\sqrt{\varepsilon_s}\sqrt{\mu_s}$ with $\mu_s = (\mu_{rs}^2 - \mu_{\kappa s}^2)/\mu_{rs} = 1/\mu'_{rs}$, μ_{rs}, and $\mu_{\kappa s}$ are given in equation (1), the superscript s in q_n^s means k_s is involved in VCWFs. In the coated layer area, *i.e.*, in region II the corresponding EM field components are

$$\mathbf{E}_b(k_0,\mathbf{r}) = \sum_n\left[q_n^0\mathbf{N}_n^{0(1)} - b_n^0\mathbf{N}_n^{0(3)}\right], \quad \mathbf{H}_b(k_0,\mathbf{r}) = \sum_n\frac{k_0}{i\omega\mu_0}\left[q_n^0\mathbf{M}_n^{0(1)} - b_n^0\mathbf{M}_n^{0(3)}\right], \tag{4}$$

where the superscript 0 in q_n^0 means k_0 is involved in VCWFs. Finally, in region III, we have

$$\begin{aligned}\mathbf{E}_e(k_{\text{eff}},\mathbf{r}) &= \sum_n\left[q_n^e\mathbf{N}_n^{e(1)} - b_n^e\mathbf{N}_n^{e(3)}\right],\\ \mathbf{H}_e(k_{\text{eff}},\mathbf{r}) &= \sum_n\frac{k_{\text{eff}}}{i\omega}\left\{q_n^e\left[i\mu'_{\kappa e}\mathbf{L}_n^{e(1)} + \mu'_{re}\mathbf{M}_n^{e(1)}\right] - b_n^e\left[i\mu'_{\kappa e}\mathbf{L}_n^{e(3)} + \mu'_{re}\mathbf{M}_n^{e(3)}\right]\right\},\end{aligned} \tag{5}$$

where the superscript e in q_n^e means k_{eff} is involved in VCWFs, μ'_{re} and $\mu'_{\kappa e}$ are defined in Eq. (1).
By matching the standard boundary conditions at $r = r_0$, it can be obtained that

$$\begin{pmatrix} q_n^e \\ b_n^e \end{pmatrix} = \mathcal{F} \begin{pmatrix} \mathcal{A}_{11} & \mathcal{A}_{12} \\ \mathcal{A}_{21} & \mathcal{A}_{22} \end{pmatrix} \begin{pmatrix} q_n^0 \\ b_n^0 \end{pmatrix}, \quad (6)$$

with the prefactor $\mathcal{F} = \pi/(2i\mu_0\mu'_{re})$, and the matrix elements

$$\mathcal{A}_{11} = -x_0 J'_n(x_0) H_n^{(1)}(x_{\text{eff}}) + n\mu_0\mu'_{\kappa e} J_n(x_0) H_n^{(1)}(x_{\text{eff}}) + x_{\text{eff}}\mu_0\mu'_{re} J_n(x_0) H_n^{(1)\prime}(x_{\text{eff}}), \quad (7a)$$

$$\mathcal{A}_{12} = x_0 H_n^{(1)\prime}(x_0) H_n^{(1)}(x_{\text{eff}}) - n\mu_0\mu'_{\kappa e} H_n^{(1)}(x_0) H_n^{(1)}(x_{\text{eff}}) - x_{\text{eff}}\mu_0\mu'_{re} H_n^{(1)}(x_0) H_n^{(1)\prime}(x_{\text{eff}}), \quad (7b)$$

$$\mathcal{A}_{21} = -x_0 J'_n(x_0) J_n(x_{\text{eff}}) + n\mu_0\mu'_{\kappa e} J_n(x_0) J_n(x_{\text{eff}}) + x_{\text{eff}}\mu_0\mu'_{re} J_n(x_0) J'_n(x_{\text{eff}}), \quad (7c)$$

$$\mathcal{A}_{22} = x_0 H_n^{(1)\prime}(x_0) J_n(x_{\text{eff}}) - n\mu_0\mu'_{\kappa e} H_n^{(1)}(x_0) J_n(x_{\text{eff}}) - x_{\text{eff}}\mu_0\mu'_{re} H_n^{(1)}(x_0) J'_n(x_{\text{eff}}), \quad (7d)$$

with $x_0 = k_0 r_0$ and $x_{\text{eff}} = k_{\text{eff}} r_0$.
According to rule (iii) of the EMT, the coated rod located in the effective medium subjects to null scattering. According to the Mie theory, the total scattering cross section is $C_{sca} = 4\sum_n |D_n^e|^2 / k_{\text{eff}}$ with $D_n^e = b_n^e / q_n^e$ the Mie scattering coefficients of the coated rod in the effective medium. In the limit $k_{\text{eff}} r_0 \ll 1$, only the dominated terms of $n = 0, \pm 1$ need to be considered. Taking account of $D_n^e = b_n^e = 0$, it is straightforward from equation (6) to get

$$\frac{\mathcal{A}_{21}}{\mathcal{A}_{22}} = -\frac{b_n^0}{q_n^0} = -D_n^s, \quad n = 0, \pm 1, \quad (8)$$

where D_n^s are the Mie coefficients of the inner core ferrite rod, which are given by

$$D_n^s = \frac{k_0 r_s J_n(k_s r_s) J'_n(k_0 r_s) - \mu_0 J_n(k_0 r_s) j}{k_0 r_s H_n^{(1)\prime}(k_0 r_s) J_n(k_s r_s) - \mu_0 H_n^{(1)}(k_0 r_s) j}, \quad (9)$$

with $j = n\mu'_{\kappa s} J_n(k_s r_s) + k_s r_s \mu'_{rs} J'_n(k_s r_s)$ where r_s is the radius of the inner core ferrite rod, μ'_{rs} and $\mu'_{\kappa s}$ are given in Eq. (1). Suppose $k_{\text{eff}} r_0 \ll 1$ and $k_0 r_0 \ll 1$, we can make approximations on $J_n(x)$, $J'_n(x)$, $H_n^{(1)}(x)$, and $H_n^{(1)\prime}(x)$ with $x = k_{\text{eff}} r_0$ or $k_0 r_0$ for convenience.
Then we can obtain the simplified equations determining ε_{eff} and μ_{eff}.

$$\varepsilon_{\text{eff}} = (1-f)\varepsilon_0 + f\tilde{\varepsilon}_s, \qquad \frac{\mu_{\text{eff}} - \mu_0}{\mu_{\text{eff}} + \mu_0} = f\frac{\tilde{\mu}_s - \mu_0 - \xi}{\tilde{\mu}_s + \mu_0 + \xi}, \quad (10)$$

where f is the filling fraction with $f = r_s^2 / r_0^2$, and

$$\tilde{\varepsilon}_s = 2\varepsilon_s F_2(x_s), \quad \tilde{\mu}_s = \mu_s G_2(x_s), \quad \xi = -\frac{(1-f)\mu_0^2(\mu_\kappa/\mu_r)^2(\tilde{\mu}_s/\mu_s)^2}{(1-f)\mu_0 + (1+f)\tilde{\mu}_s}, \quad (11)$$
$$F_2(x_s) = J_1(x_s)/[x_s J_0(x_s)], \quad G_2(x_s) = J_1(x_s)/\left[x_s J'_1(x_s)\right],$$

with $x_s = k_s r_s$. It is noted that if μ_κ is set to be zero, equation (10) can be recovered to that for the isotropic case (Wu et al., 2006).

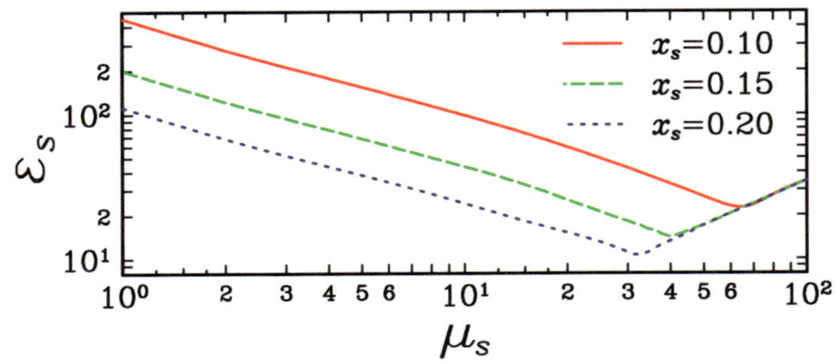

Fig. 9. Phase diagram for the 2D metamaterials composed of dielectric rods of radius r_s with respect to the permittivity ε_s and permeability μ_s. Above each curve the system may possess simultaneously negative magnetic and electric responses.

3.2 Simultaneous negative effective constitutive parameters obtained with EMT

We exemplify the ferrite-only-based designs with the 2D case, it is straightforward to generalize to the 3D case. We present in Fig. 9 the phase diagram of the system as a function of the permittivity ε_s and the permeability μ_s for the dielectric rods at a typical filling ratio $f = 0.385$. Above each curve is the region where the system may simultaneously display negative electric and magnetic responses within a certain frequency range that satisfies $x = (\omega/c)r_s \leq x_s$. Here ω is the angular frequency, r_s is the rod radius, and c is the speed of light in vacuum. As can be seen from the phase diagram, by employing ferrites with permeability over 10 as building block, the requisite permittivity can be lowered by an order of magnitude. In particular, the magnetically tunable permeability can serve to achieve manipulability of the operational frequency range and the wave impedance.

As a candidate of the ferrite materials, single-crystal yttrium-iron-garnet (YIG) can be the first choice for its negligible low damping $\alpha = 5 \times 10^{-4}$ (Pozar, 2004). The permeability of single-crystal YIG bears the same form as that given in Eq. (1) with the saturation magnetization $M_s = 1750$ Oe. In our proof-of-principle demonstration we set $\alpha = 0$ for convenience.

Based on the newly developed EMT, we have calculated the effective constitutive parameters of the MM. In Fig. 10(b) and (d) we present ε_{eff} and μ_{eff} as the functions of the angular frequency ω for an MM composed of ferrite rods arranged as a square lattice in the air with $a = 10$ mm, $r_s = 3.5$ mm, and $\varepsilon_s = 25$. The exerted EMFs satisfy $H_0 = 500$ Oe and $H_0 = 475$ Oe, for the results shown in Fig. 10(b) and (d), respectively. It can be seen from Fig. 10(b) that within the frequency range from 15.5 GHz to 16.2 GHz, the system exhibits simultaneous negative ε_{eff} and μ_{eff}, implying a working frequency range for the NIM. To confirm the result, the corresponding photonic band diagrams are calculated by using the rigorous frequency-domain multiple scattering theory as shown in Fig. 10 (a) and (c). Excellent agreement between two independent results is obtained. In particular, the working frequency of NIM corresponds to the second photonic band denoted by the blue line in Fig 10 (a). At about 16.2 GHz, one gets a so-called epsilon-near-zero (ENZ) medium (Silveirinha & Engheta, 2006), which can induce a longitudinal EM mode in the system, corresponding to the nearly flat third band labeled by the black line in Fig. 10(a). The comparison between the band diagram and ε_{eff} and μ_{eff} confirms the validity of the CPA for the MM. A particularly important results obtained with the EMT is that at $\omega = 16$ GHz, the effective constitutive parameters $\varepsilon_{\text{eff}} = \mu_{\text{eff}} = -1$ is realized as marked by the horizontal solid green line in Fig. 10(a) and (b).

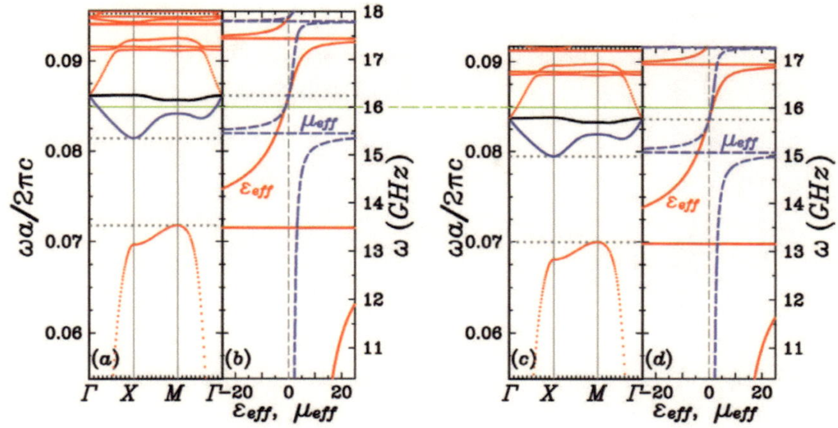

Fig. 10. (a) Photonic band diagram for an MM of square lattice with lattice constant $a = 10$ mm, rod radius $r_s = 3.5$ mm, and $H_0 = 500$ Oe. (b) Effective permittivity ε_{eff} and effective permeability μ_{eff} as the functions of circular frequency ω for the system with the same parameters as in (a). (c) and (d) are the same as (a) and (b) except $H_0 = 475$ Oe.

To demonstrate the manipulability of the negative refractive index by the EMF, we present in Fig. 10(d) ε_{eff} and μ_{eff} as the functions of ω at a smaller EMF such that $H_0 = 475$ Oe. The corresponding photonic band diagram is also given in Fig. 10(c). Good agreement can still be found for Fig. 10(c) and (d), which confirms once again the validity of the EMT. In addition, by comparing Fig. 10 (a) with (b) it can be found that the photonic band diagram is shifted to a lower frequency range, so does the operational frequency bandwidth of the NIM. At $\omega = 16$ GHz, the refractive index is no longer negative as can be seen from Fig. 10(d). Instead, it corresponds to the refractive index given by $n_{\text{eff}} = 1.13$ with $\varepsilon_{\text{eff}} = 0.92$ and $\mu_{\text{eff}} = 1.4$. The operating frequency for the NIM is shifted downwards and located at about 15.5 GHz. Therefore, by appropriately adjusting the EMF, we can manipulate the working frequency for the NIM, which adds considerably to the flexibility in practical applications.

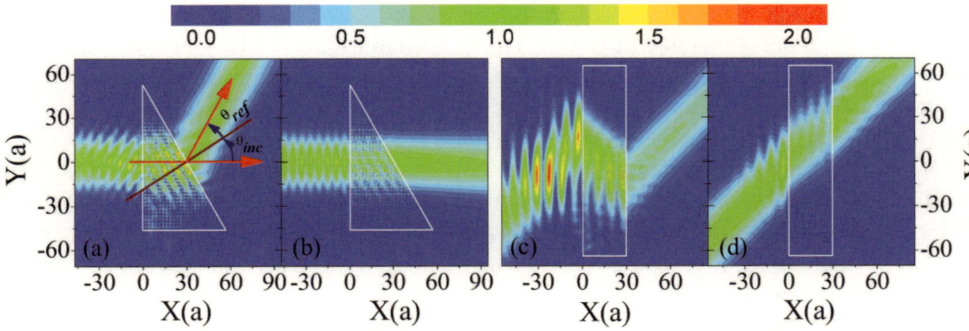

Fig. 11. A Gaussian beam incident on the MM with wedge and slab geometry experiences negative and positive refractions under the EMFs of $H_0 = 500$ Oe (a) and (c), and $H_0 = 475$ Oe (b) and (d), showing magnetic tunability from negative to positive refraction.

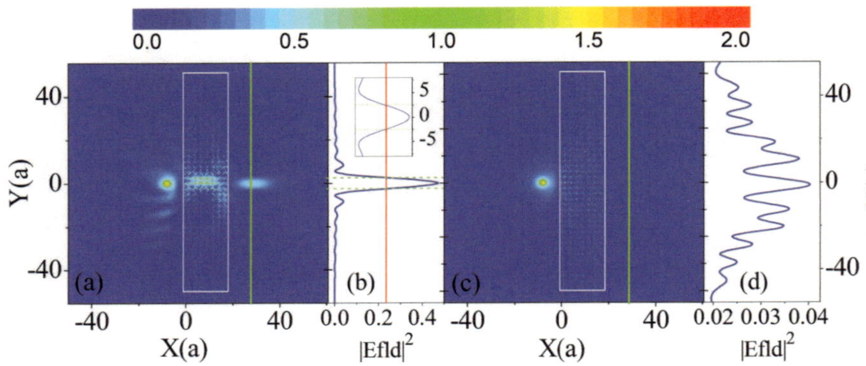

Fig. 12. Simulations on superlensing effect of an MM slab under the EMFs of $H_0 = 500$ Oe (a) and $H_0 = 475$ Oe (c). (b) and (d) demonstrate the intensity profiles along the solid green lines in (a) and (c), respectively. The inset in (b) is the amplified view of the image width at half maximum intensity, showing that the transverse image size is less than $\lambda/2$ for $H_0 = 500$ Oe.

3.3 Negative refraction controllable with a magnetic field

To corroborate the above theoretical design, we have performed simulations to examine the propagation of a TM Gaussian beam through a wedge-shaped MM under an EMF of $H_0 = 500$ Oe at the working frequency $\omega = 16$ GHz. The results are presented in Fig. 11(a) and (b). The beam shines on the interface normally from the left hand side and transmits through the wedge, as shown in Fig. 11(a). The incident angle at the wedge surface is $\theta_{\rm inc} = \theta_0$ and the refraction angle is labeled as $\theta_{\rm ref}$. A negative refraction is clearly demonstrated in Fig. 11(a) with the refraction angle $\theta_{\rm ref} \approx \theta_{\rm inc}$, suggesting the effective refractive index $n_{\rm eff} \approx -1$. Note in passing that only a very weak field appears at the bottom surface of the sample, suggesting that the Gaussian beam is subjected to a tiny reflection at the wedge surface, providing a persuasive evidence for the wave impedance match of the wedge-air interfaces. When H_0 is decreased to 475 Oe, the refractive index becomes positive from the CPA. The corresponding simulation result is shown in Fig. 11(b). The beam is seen to bend downwards a little bit, indicating a positive refraction, in agreement with the effective refractive index $n_{\rm eff} = 1.13$ from the EMT.

The negative refraction behavior of the MM and its manipulability by the EMF can also be manifested by observing the propagation of a Gaussian beam through a flat slab. The results are typically presented in Fig. 11(c) and (d) for $H_0 = 500$ Oe and 475 Oe, respectively. A Gaussian beam, incident obliquely from the left hand side of the slab with incident angle $\theta_{\rm inc} = 45°$, is seen to refract negatively into the slab, with the refractive angle $\theta_{\rm ref} \approx \theta_{\rm inc}$. And then, it is subjected to another negative refraction at the right surface of the slab, as shown in Fig. 11(c), indicating once again the negative refractive behavior with $n_{\rm eff} \approx -1$. One important thing to be noted is that at each interface of the slab we can hardly find the reflected beam, suggesting once again that effective permittivity $\varepsilon_{\rm eff}$ and effective permeability $\mu_{\rm eff}$ are both nearly equal to -1, resulting in the wave impedance match at the interface. The tunability of the negative refraction behavior is shown in Fig. 11(d), with $H_0 = 475$ Oe, where the beam transmits through the slab without obvious change of its direction, consistent with positive effective index $n_{\rm eff} = 1.13$ from the CPA.

3.4 Subwavelength imaging controllable with a magnetic field

One of the most unique characteristics for the NIM is the slab superlensing effect (Pendry, 2000), which enables many potential applications. Typical slab imaging phenomena, together with its magnetic manipulability, based on our design of NIM is shown in Fig. 12, where the thickness of the slab is $t_s = 18a$. A monochromatic line source radiating EM waves at $\omega = 16$ GHz is placed at a distance $d_p = 8a$ from the left surface of the slab. When $H_0 = 500$ Oe, an image is formed on the opposite side of the slab with the image centered at a distance $d_i = 10.2a$ from the right surface of the slab, as shown in Fig. 12(a). The profile of the field intensity along the green line that goes through the image center in Fig. 12(a) is presented in Fig. 12(b), which corresponds to a transverse image size $w \lesssim 5a \approx 0.42\lambda$, demonstrating a possible subwavelength resolution below the conventional diffraction limit $\frac{1}{2}\lambda$. The separation d between the line source and the image is $d = d_p + d_i + t_s = 36.2a \approx 2t_s$, consistent with negative refractive index $n_{\text{eff}} = -1$ from EMT calculation. The manipulability of the EMF on the negative refractive index is exhibited in Fig. 12(c) and (d), where all the parameters are the same as those in Fig. 12(a) and (b) except that the EMF is take as $H_0 = 475$ Oe instead of 500 Oe. As analyzed above, the refractive index is tuned from $n_{\text{eff}} = -1$ to a positive one with $n_{\text{eff}} = 1.13$. In this case, the slab shows no negative refraction behavior. For this reason, no image is formed on the opposite side of the slab as demonstrated in Fig. 12(c).

4. Molding the flow of EM wave with magnetic graded PC

PCs are composite materials with periodic optical index and characterized by anisotropic photonic band diagram and even PBG (Joannopoulos et al., 1995; John, 1987; Yablonovitch, 1987), enabling the manipulation of EM waves in novel and unique manners, paving the way to many promising applications. To achieve more degree of tunability, MPC with EM properties controllable by EMF has been proposed and investigated, which has ranged from photonic Hall effect (Merzlikin et al., 2005; Rikken & Tiggelen, 1996), extrinsic PCs (Xu et al., 2007), and giant magnetoreflectivity (Lin & Chui, 2007) to magnetically tunable negative refraction (Liu et al., 2008; Rachford et al., 2007), magnetically created mirage (Chen et al., 2008), magnetically tunable focusing (Chen et al., 2008), and unidirectional waveguides (Haldane & Raghu, 2008; Wang et al., 2008; 2009; Yu et al., 2008). In previous research on PCs, most efforts are devoted to the PBG-relevant effects and its potential applications. Actually, the richness of the photonic bands of the PCs supplies to us more manipulability on the control of the EM wave. Of particular import paradigms are the negative refraction in PCs (Luo et al., 2002) and the superlensing effect based on it (Decoopman et al., 2006).
Graded PC is a kind of structured material constructed by introducing appropriate gradual modifications of some PC parameters such as the lattice periodicity (Centeno & Cassagne, 2005; Centeno et al., 2006), the filling factor (Chien & Chen, 2006), or the optical index. It can further modify the photonic dispersion bands or isofrequency curves of the PCs, and thus leads to some new ways of manipulability on the EM waves. In this section, we will propose and conceptualize an alternative type of graded PC: magnetic graded PC (MGPC). The photonic dispersion bands are tuned by applying a nonuniform EMF, instead of the graduate modification of the intrinsic parameters such as lattice periodicity or filling factor. To exemplify the idea of the MGPC and its applications, we present two proof-of-principle demonstrations in the following: one is the focusing effect by taking advantage of the MGPC, the other one is the mirage effect created by MGPC.

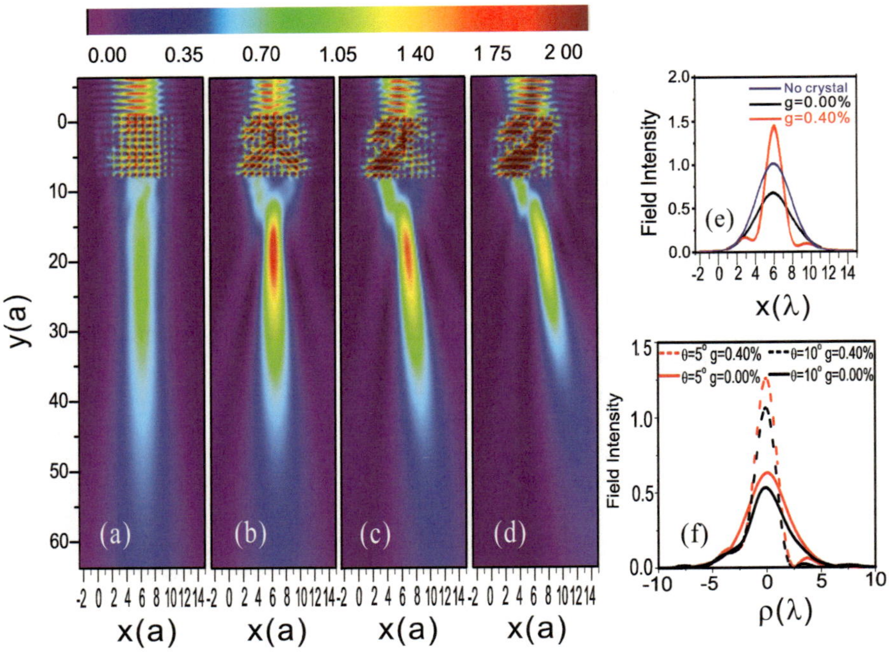

Fig. 13. The field pattern for an MGPC illuminated by a TM Gaussian incident from the top normally (a), (b) and obliquely (c), (d) with the gradient of the EMFs equal to $g = 0.00\%$ for (a) and $g = 0.4\%$ for (b), (c), and (d). The incident angle are $\theta = 5^\circ$ and $\theta = 10^\circ$ for (c) and (d), respectively. (e) and (f) are the field intensity at the focal plane as the functions of the abscissa x and distance from the focus ρ, respectively.

4.1 Subwavelength focusing effect based on magnetic graded PC

The MGPC proposed is composed of 117 (13 columns $\times$ 9 rows) ferrite rods of radius $r_c = 6$ mm arranged periodically in the air as a square lattice with lattice constant $a = 48$ mm. For the EMF exerted along the z (rod axis) direction, the magnetic permeability of the ferrite rods is given in Eq. (1). With a slightly nonuniform EMF applied to the MPC, the permeability is gradually tuned, resulting in the modification of the refractive index, a graded PC is therefore obtained. The relative permittivity of the ferrite rod is $\varepsilon_s = 12.3 + i3 \times 10^{-3}$, the saturation magnetization is $M_s = 1786$ Oe, and the damping coefficients is taken as 5×10^{-4}, typical for single-crystal YIG ferrite. We fix the Cartesian coordinates of the ferrite rods by $(x, y) = [(i-1)a, (j-1)a]$, with $i = 1, 2, ..., 13$ and $j = 1, 2, ..., 9$ the column and the row indices in x and y directions, respectively. The magnitude of the nonuniform EMF varies along the x direction, such that the EMF applied at the center region is weaker than that applied close to the edge of the MGPC sample. To be specific, the ferrite rod at the j-th row and the i-th column inside the MGPC is subjected to $H_0 = h_0[1 + (7-i)g]$ for $i \leq 7$ and $H_0 = h_0[1 + (i-7)g)]$ for $i > 7$, where g is a parameter measuring the gradient of the EMF in x direction.

4.1.1 Focusing effect for a normally/obliquely incident TM Gaussian beam

Firstly, we consider the focusing effect of the MGPC on a collimated EM beam. In Fig. 13 (a) and (b), we present the field intensity pattern for the MGPC illuminated by a TM Gaussian beam. Figure 13(a) corresponds to the case when a uniform EMF in z direction is exerted

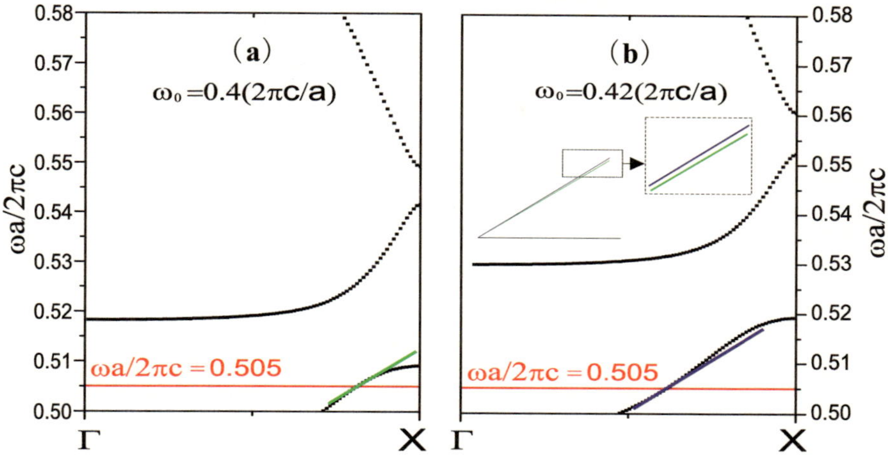

Fig. 14. Photonic band diagrams of the PC under two different uniform EMFs with $\omega_0 = 0.4(2\pi c/a)$ (a) and $\omega_0 = 0.42(2\pi c/a)$ (b). The green and blue lines are the tangents of the dispersion band at incident frequency $\omega = 0.505(2\pi c/a)$, shown also in the inset in Fig. 14(b).

so that one has a conventional MPC. It can be seen that the beam just transmits through the MPC without significant change of the beam waist radius, as can be seen from Fig. 13(e). When a slightly nonuniform EMF is exerted so that an MGPC is formed, the Gaussian beam is focused after passing through the MGPC, with the waist radius reduced to about half of the incoming beam, as demonstrated in Fig. 13(b) and (e), where the gradient g of the EMF is $g = 0.4\%$ and the Gaussian beam of waist radius $w_0 = 2\lambda$ is illuminated normally from the top of the MGPC with the wavelength λ ($a = 0.505\lambda$). We also present the results when the beam is obliquely incident on the MGPC, as shown in Fig. 13 (c) and (d), corresponding to the incident angles of $\theta_i = 5°$ and $10°$, respectively. It can be found that the beam can still be focused. However, the intensity at the focus decreases with the increase of the incident angle as can seen from Fig. 13(f), mainly due to the stronger reflection for the the larger incident angle at the interface. It should be pointed out that the weak gradient (with $g < 0.7\%$ in all cases) of the EMF and the small ferrite rod radius ($r_c = \frac{1}{8}a$) allow us to assume that each rod is subjected to a uniform EMF. Within this approximation, the simulations can be performed by using the multiple scattering method (Liu & Lin, 2006; Liu et al., 2008).

4.1.2 Physical understanding of the effect from the aspect of photonic band diagram

The focusing effects observed above can be understood using the concept of the local photonic band diagram as in the case of the conventional graded PC. In Fig. 14, we plot the photonic band diagram for the PC subjected to two different uniform EMFs where ω is the circular frequency of the incident EM beam. It can be seen that the photonic band diagram exhibits a noticeable difference when the magnitude of the EMF is slightly changed. At $\omega a/2\pi c = 0.505$, it can be seen from the inset in Fig. 14(b) that the slope of the photonic band is larger under the greater EMF. With the knowledge that $d\omega/dk \propto 1/n$ (n the effective optical index), the greater EMF produces the smaller optical index. Therefore, it can be understood that the gradient of the EMF yields a gradient optical index, leading to the formation of the MGPC.

We now further examine the magnetic tunability of the MGPC on the EM Gaussian beam. Firstly, we consider the effect of the number of rows m_r on the light focusing. In Fig. 15(a)

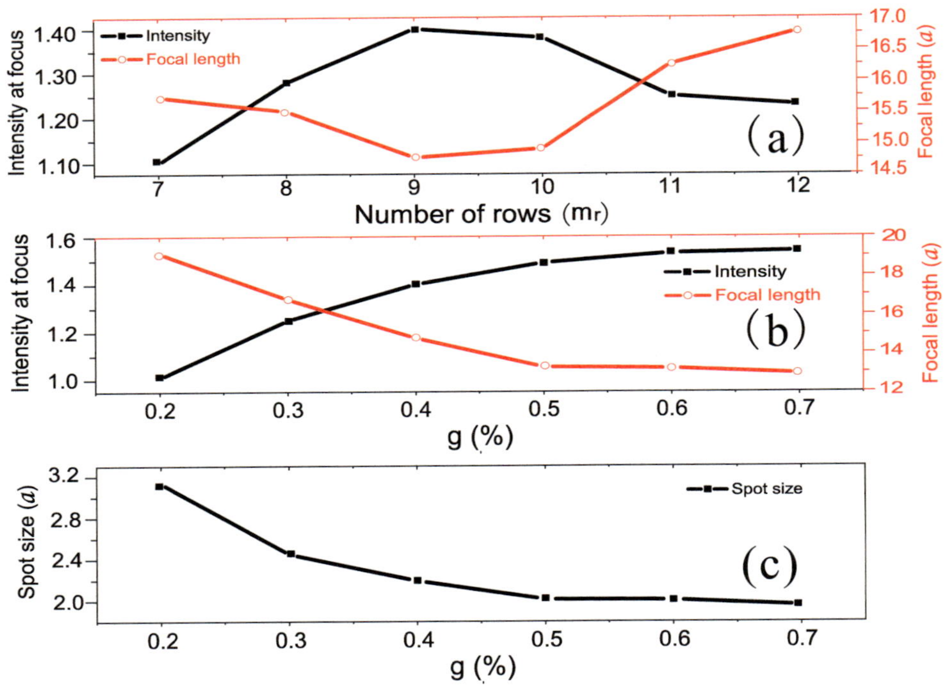

Fig. 15. The field intensity at the focus and the focal length as the functions of the number m_r of rows (a) and the gradient g of the EMF (b). (c) The waist radius of the focused beam versus the gradient g of the EMF. The number of columns is $m_c = 13$ and $g = 0.40\%$ for (a).

we present the focal length and the intensity at the focus versus m_r. It can be seen that the intensity at the focus increases at first with the increase of m_r, and reaches its maximum at $m_r = 9$. This is because the more rows the light beam goes across, the more focusing effect it will experience. As m_r increase further, the intensity at the focus decreases, due to the damping occurring when the light propagates through the MGPC. We also examine the effect of the gradient g of the EMF on the focusing property. It can be seen from Fig. 15(b) that the intensity at the focus increases with the increase of g. At $g = 0.7\%$, the intensity is twice as that for the MGPC under the uniform EMF. In addition, the focal length decreases as g increases, ranging from $13a$ to $19a$. From Fig. 15(c), it can be seen that the spot size decreases as g increases, and shrinks even to $1.5a$ at $g = 0.70\%$, less than the wavelength of the incident wave. The effect of tuning the gradient g bears a close similarity to the case of modifying the curvature or the central thickness of the conventional lens in classical optics, demonstrating the magnetical tunability of the MGPC on the focusing properties.

4.2 Tunable mirage effect based on magnetic graded PC

The MGPC considered is composed of 20 rows and 80 columns of 1600 (80 columns $\times$ 20 rows) ferrite rods arranged periodically in the air as a square lattice. The ferrite rod has the same parameters as in the last section. The lattice constant is still $a = 48$ mm$=8r_c$. The Cartesian coordinates of the rods are given by $(x,y) = [(i-1)a,(j-1)a]$, with $i = 1,2,\cdots,80$ and $j = 1,2,\cdots,20$ labeling the column and row indices in x and y directions, respectively. An EMF oriented along z with the gradient in y direction is exerted upon the MGPC such that the

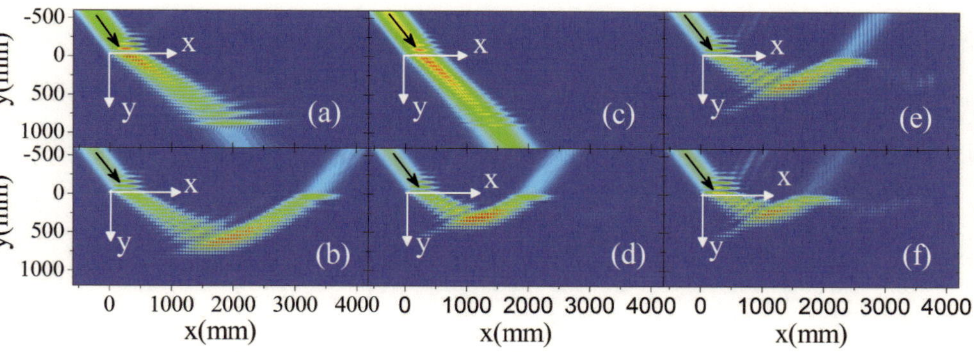

Fig. 16. Field intensity patterns for an MGPC illuminated by a TM Gaussian beam with wavelength $\lambda = 91.427$ mm for panel (f) and $\lambda = 90.564$ mm for other panels. The black arrows denote the direction of the incident beam. The MGPC-air interfaces are located at $y = -24$ mm and $y = 936$ mm. The applied EMFs exerted satisfy that (a) $h_0 = 893$ Oe, $g = 0.00\%$; (b) $h_0 = 893$ Oe, $g = 0.23\%$; (c) $h_0 = 848$ Oe, $g = 0.00\%$; (d) $h_0 = 893$ Oe, $g = 0.40\%$; (e) $h_0 = 904$ Oe, $g = 0.23\%$; and (f) $h_0 = 893$ Oe, $g = 0.23\%$.

ferrite rod in the j-th row and i-th column is subjected to magnetic field $H_0 = h_0[1 + (j-1)g]$ with g the quantity measuring the amplitude of the gradient. The incident EM wave is the TM Gaussian beam with the waist radius $w_0 = 3\lambda$, where λ is the wavelength of the beam in vacuum.

4.2.1 Creating a mirage effect for a TM Gaussian beam based on an MGPC

In Fig. 16, we present the electric field intensity patterns for an MGPC under different EMFs illuminated by an incident TM Gaussian beam with the incident angle $\theta_{inc} = 45°$. It can be seen that the beam is deflected in different manners with different EMFs exerted upon the MGPC. When a uniform EMF with gradient $g = 0$ is applied, the MGPC is actually an ordinary MPC. It can be seen from Fig. 16(a) that in this case the beam enters the MGPC at $y = -24$ mm with a refraction angle greater than 45° and finally transmits across the crystal. Very differently, when a slight gradient is introduced to the EMF such that $g = 0.23\%$, the beam is deflected layer by layer during its propagation in the MGPC and eventually reflected back off the MGPC, leading to the appearance of a mirage effect as shown in Fig. 16(b).

4.2.2 Physical understanding of the effect from the aspect of isofrequency curve

In nature, a mirage is an optical phenomenon occurring when light rays bend and go along a curved path. The reason lies in the gradual variation of the optical index of air, arising from the change of the atmosphere temperature with the height. Roughly speaking, in our case the nonuniform EMF produces a similar effect on the effective refraction index of the MGPC as the temperature does on the atmosphere, so that an mirage effect is created. More exact analysis relies on the isofrequency (IF) curves of the MPC (Kong, 1990). Due to the weak gradient of the EMF (less than 0.5%) in all our simulations, the propagation of the EM wave can be interpreted according to the local dispersion band or the IF curve (Centeno et al., 2006). In Fig. 17(a), we present three IF curves for the operating wavelength $\lambda = 90.564$ mm, corresponding to the MPC under three different uniform EMFs (with $g = 0$). The blue vertical line in Fig. 17(a) denotes parallel component of the wavevector for the incident EM wave. According to the conservation of tangential wavevector at the MPC-air interface, the parallel

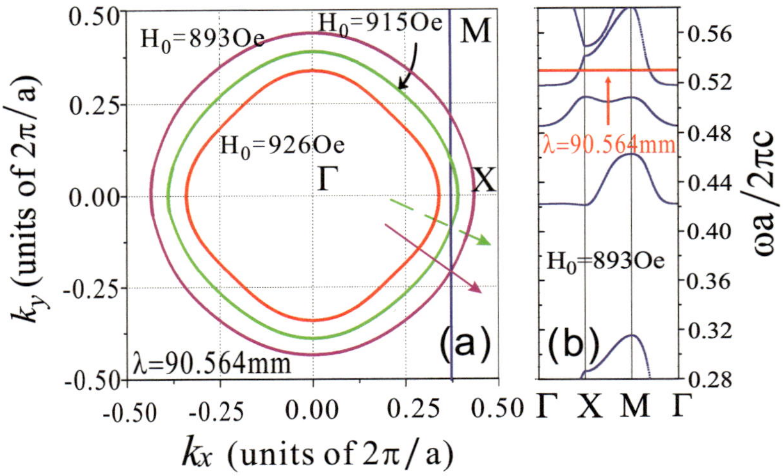

Fig. 17. (a) Isofrequency curves corresponding to the wavelength $\lambda = 90.564$ mm of the MPC under three different EMFs. The straight arrows denote the direction of the group velocity. The blue vertical line in Fig. 17 (a) is the construction line corresponding to an incident angle of 45°, obtained by the conservation of tangential momentum at the interface. (b) Photonic band diagram of the MPC under $H_0 = 893$ Oe.

wavevector of the refracted EM wave in the MPC can be obtained. Then the refractive angle can be determined by the surface normal at the intersection point of the IF curve and the blue vertical line, as marked by the blue solid and green dashed arrows in Fig. 17(a), corresponding to the direction of the group velocities $V_g = \nabla_k \omega(k)$ in the MPC under $H_0 = 893$ and 915 Oe, respectively. For convenience, we also present in Fig. 17(b) the photonic band diagram of an MPC under $H_0 = 893$ Oe, where the red solid line marks the operating frequency in Fig. 17(a). As we have shown in the previous section, the increase of the EMF will result in the shift of the photonic bands to higher frequency. Correspondingly, the IF curves will shrink with the increase of the EMF as can be seen from Fig. 17(a), resulting in the increase of the refraction angle by comparing the blue solid arrow with green dashed arrow.

The above analysis based on the IF curves and the phtonic band diagram can be corroborated by comparing Figs. 16(a) ($H_0 = 893$ Oe) with 16(c) ($H_0 = 848$ Oe), where a stronger EMF corresponds to a larger refraction angle. Accordingly, a gradient EMF enables a continuous change of refraction angle of the beam propagating in the MGPC. A typical result is shown in Fig. 16(b) where an EMF with gradient $g = 0.23\%$ is applied. It can be seen that as the beam goes deeper and deeper into the MGPC, the refraction angle will increase layer by layer in the MGPC due to the increase of the EMF along the y direction. Physically, with the increases of EMF along y, the local IF shrinks little by little until it becomes tangent to the vertical construction line. This occurs when H_0 lies between 915 Oe and 926 Oe, as illustrated in Fig. 17(a), leading to a total internal reflection. The beam is therefore reflected back, resulting in a mirage effect.

To examine the sensitivity of the mirage effect to the gradient g of the EMF and the wavelength of the incident beam, we present in Fig. 19 the separation d_s between the incoming and outgoing beam as their functions. The separation d_s characterizes the degree of bending for the mirage effect. In Fig. 19(a) it can be found that d_s decrease as the gradient increases, which can also be observed by comparing Figs. 16(b) and (d), indicating a stronger bending by a

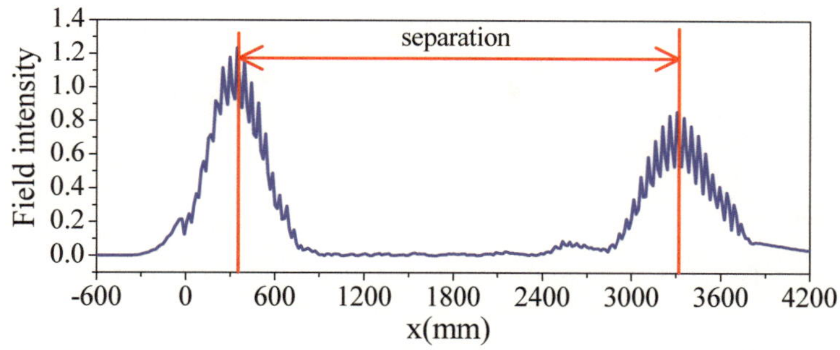

Fig. 18. The field intensity at the top of the MGPC for the case shown in Fig. 16(b).

higher gradient EMF. In addition, from Fig. 19(b) it can be seen that the separation exhibits a sensitive dependence on the incident wavelength, which is also demonstrated by comparing Figs. 16(b) and (f), suggesting a possible application in multiplexer and demultiplexer.

5. Unidirectional reflection behavior on magnetic metamaterial surface

Magnetic materials are irreplaceable ingredients in optical devices such as isolators and circulators. Different from dielectric or metallic materials, the permeability of magnetic material is a second rank tensor with nonzero off-diagonal elements as given in Eq. (1). Accordingly, the time reversal symmetry is broken in a MM system (Wang et al., 2008), based on which some very interesting phenomena can be realized. A particular one is the one-way edge state which has been investigated theoretically (Chui et al., 2010; Haldane & Raghu, 2008; Wang et al., 2008; Yu et al., 2008) and experimentally (Wang et al., 2009) recently. Besides, MSP resonance occurs in MPC when effective permeability equals to -1 in 2D case, in the vicinity of which the behavior of MM is very different, it is therefore the frequency region we will focus on.

In magnetic systems with inversion symmetry, even though the dispersion is symmetric, the wave functions for opposite propagating directions can become asymmetric. For this reason, the reflected wave develops a finite circulation, specially near the MSP resonance frequency, it can be substantially amplified. This effect can be exploited to construct one-way subwavelength waveguides that exhibit a superflow behavior. In this section, our work

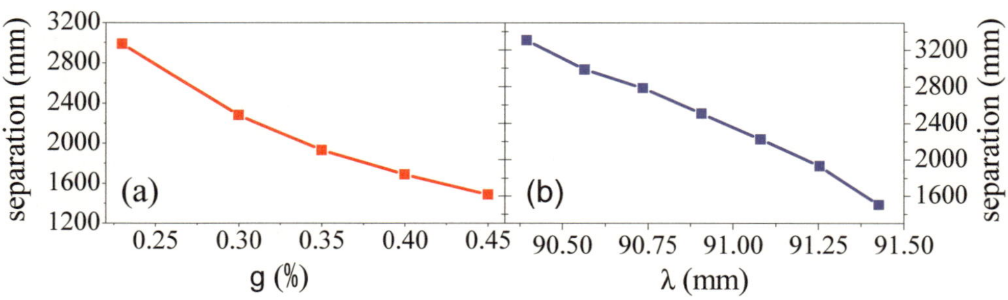

Fig. 19. The peak separation is plotted as the function of the field gradient g (a) and the incident wavelength λ (b). The other parameters are the same as those in Fig. 16(b). The separation d_s can be determined from the field intensity distribution along the x axis at $y = 0$, as is illustrated in Fig. 18 for parameters corresponding to those in Fig. 16(b).

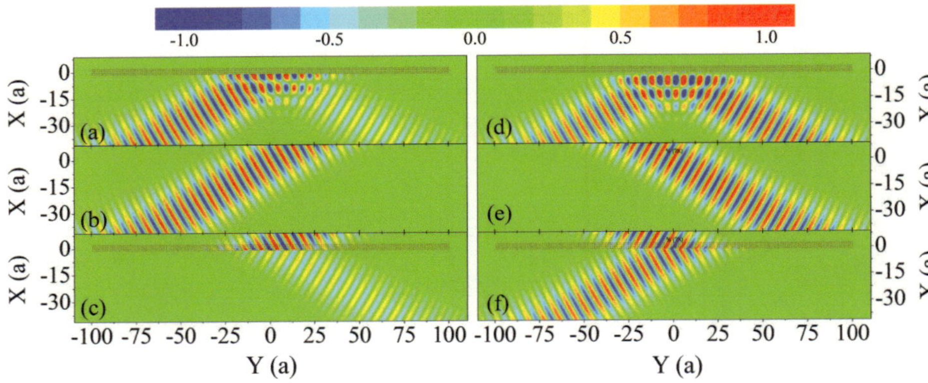

Fig. 20. The electric field pattern of the total field (a), (d), the incoming field (b), (e), and the scattered field (c), (f) for a TM Gaussian beam incident from left hand side (incident angle $\theta_{inc} = 60°$) and right hand side (incident angle $\theta_{inc} = -60°$) upon a four-layer MM slab. The positions of the ferrite rods are marked by the black dot.

is devoted to understanding the mechanism of this "finite circulation" and its physical consequence.

5.1 Unidirectional reflection of an EM Gaussian beam from an MM surface

The MM considered is composed of an array of ferrite rods arranged periodically in the air as a square lattice with the lattice constant $a = 8$ mm. The ferrite rod has the radius $r = 0.25a = 2$ mm. The permittivity of the ferrite rod $\varepsilon_s = 12.6 + i7 \times 10^{-3}$. The magnetic susceptibility tensor is of the same form as that given by Eq. (1). Here, the saturation magnetization is $4\pi M_s = 1700$ Oe and the EMF is fixed so that $H_0 = 900$ Oe, corresponding to the MSP resonance frequency $f_s = \frac{1}{2\pi}\gamma(H_0 + 2\pi M_s) = 4.9$ GHz with γ the gyromagnetic ratio. The damping damping coefficient is $\alpha = 7 \times 10^{-3}$, typical for the NiZn ferrite.

By use of the multiple scattering method, we demonstrate the reflection behavior of a TM Gaussian beam reflected from a finite four-layer MM slab with each layer consisting of 200 ferrite rods. We have examined the cases of the incoming Gaussian beams with incident angles of $\pm 60°$. The beam center is focused on the middle (100-th) ferrite rod in the first layer. The working frequency is fixed as $f_w = 4.84$ GHz, located in the vicinity of the MSP resonance. The results are illustrated in Fig. 20 where we present the electric field patterns of the Gaussian beams with opposite components of wavevector parallel to the MM slab. For the Gaussian beam incident from the left hand side, the reflected wave is very weak as shown in Fig 20(a). However, for the Gaussian beam incident from the right hand side, the intensity of the reflected wave remains substantial as shown in Fig 20(d). It is evident that there exists remarkable difference for the reflected Gaussian beams at different directions. The similar behavior can also be observed for a line source close to the MM slab where the reflection nearly disappears on one side of the line source.

5.2 Scattering amplitude corresponding to different angular momenta

The MM slab considered in our calculation is a geometrically left-right symmetric sample and the bulk photonic band are also the same for the above two incoming directions. Our effect arises from the characteristic of the wave functions at the working frequency. To gain a deeper understanding of our results, we calculate the scattering amplitudes corresponding

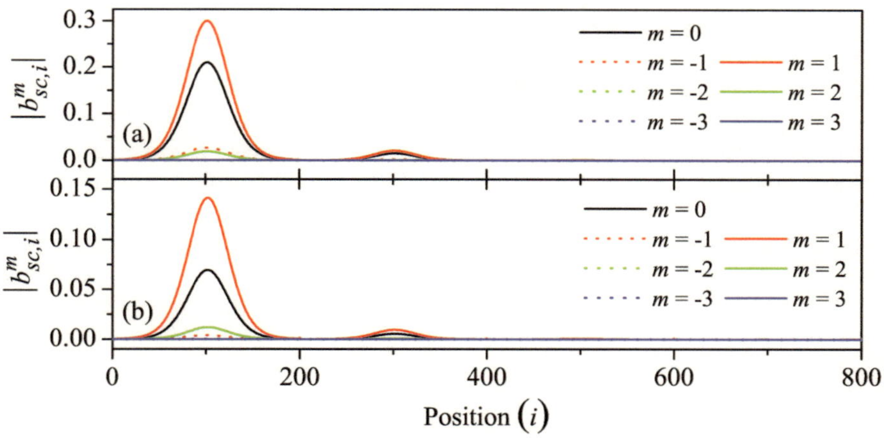

Fig. 21. The scattering amplitude $|\boldsymbol{b}^m_{sc,i}|$ of different angular momenta at rod i for an incoming Gaussian beam with incident angle of $\theta_{inc} = 60°$ (a) and $-60°$ (b). Labels 1-200, 201-400, 401-600, and 601-800 correspond to the first, second, third, and fourth layers.

to different angular momenta $|\boldsymbol{b}^m_{sc,i}|$ at the sites of different rods i in the MM slab. The results are shown in Fig. 21(a) and (b), corresponding to the cases of incident angles equal to 60° and $-60°$, respectively. As can be found from Fig. 21(a) for $\theta_{inc} = 60°$, only the components of positive angular momenta m (0 and 1) are dominant, while all the other components are nearly suppressed. The result is absolutely different from the usual case that the scattering amplitudes of the opposite angular momenta are equal and that corresponding to angular momentum 0 is the largest. The similar behavior also exist for the opposite incoming direction as shown in Fig. 21(b), however, the amplitude is relatively weaker. Actually, the behavior originates from the breaking of the time-reversal symmetry, which support the energy flow only in one direction. In the vicinity of the MSP resonance the effect is intensified so that we can observe the sharply asymmetric reflection demonstrated in Fig. 20.

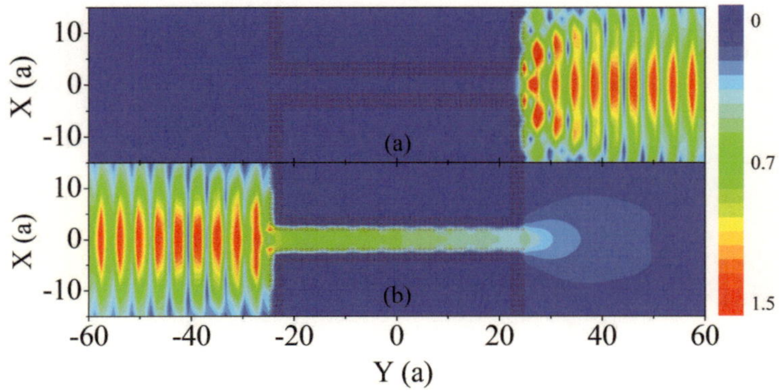

Fig. 22. The electric field patterns of a Gaussian beam incident along the channel with the width equal to $4a$. The Gaussian beam can pass the channel for one direction (b), while for the opposite direction it is completely suppressed (a).

5.3 Design of a possible EM device based on the effect

Finally, we demonstrate a potential application of our effect by constructing a unidirectional waveguide composed of two MM slabs of opposite magnetizations. A typical result is illustrated in Fig. 22 where the electric field patterns of a Gaussian beam incident along an interconnect/waveguide with the width of the channel equal to $4a$ are simulated. Due to the impedance mismatch at the interface, the field inside is less than that outside. Once inside, the Gaussian beam can pass the channel for one direction, while for the opposite direction it is completely suppressed. It should be pointed out that the damping coefficient in our design is twenty times larger than that used in the previous design by Wang and coworkers (Wang et al., 2008). While the transmission field still has an adequate intensity, which make it more applicable.

6. Conclusion

In summary, we have considered a kind of structured material which can be used to manipulate the flow of EM wave with much more flexibilities than conventional dielectrics. Firstly, a PBG material with robust and completely tunable photonic gap is designed and analyzed. Then, we design a ferrite-based negative index metamaterial with effective constitutive parameters $\varepsilon_{\text{eff}} = \mu_{\text{eff}} = -1$. In addition, the corresponding EM property can be manipulated by an external magnetic field. After that, we propose an alternative type of graded PC, a magnetic graded PC, formed by a nonuniform external magnetic field exerted on an MPC, based on which a focusing effect and a mirage effect can be created and tuned. In the end, we demonstrate an exotic reflection behavior of a Gaussian beam from an MM slab, arising from the breaking of time reversal symmetry and the MSP resonance.

Based on the fast switching effect of the phenomena observed above, we can expect many promising applications such as a band filter, a multiplexer/demultiplexer, slab superlens, unidirectional waveguide, beam bender, and beam splitter with manipulability by an external magnetic field. Besides, we have also developed a set of theory for the calculations of band diagram, effective constitutive EM parameters, and electric field pattern. However, what we have considered is just for the two dimensional case. Generalization to three dimension will make the theory and the designed electromagnetic devices more applicable. Another matter should be mentioned is that the working frequency of the magnetic metamaterials is in the microwave region. In our future work, we will try to extend our design to higher frequency.

7. Acknowledgements

This work is supported by the China 973 program (2006CB921706), NNSFC (10774028, 10904020), MOE of China (B06011), SSTC (08dj1400302), China postdoctoral science foundation (200902211), and Zhejing Normal University initiative foundation. STC is partly supported by the DOE.

8. References

Busch, K. & John, S. (1999). Liquid-crystal photonic-band-gap materials: The tunable electromagnetic vacuum. *Phys. Rev. Lett.*, Vol. 83, No. 05, 967-970

Centeno, E. & Cassagne, D. (2005). Graded photonic crystals. *Opt. Lett.*, Vol. 30, No. 17, 2278-2280

Centeno, E.; Cassagne, D. & Albert, J. P. (2006). Mirage and superbending effect in two-dimensional graded photonic crystals. *Phys. Rev. B.*, Vol. 73, No. 23, 235119

Chan, C. T.; Yu, Q. L. & Ho, K. M. (1995). Order-N spectral method for electromagnetic waves. *Phys. Rev. B*, Vol. 51, No. 23, 16635-16642

Chen, X.; Grzegorczyk, T. M.; Wu, B. I.; Pacheco, J. & Kong, J. A. (2004). Robust method to retrieve the constitutive effective parameters of metamaterials. *Phys. Rev. E*, Vol. 70, No. 1, 016608

Chen, P.; Wu, R. X.; Xu, J.; Jiang, A. M. & Ji, X. Y. (2007). Effects of magnetic anisotropy on the stop band of ferromagnetic electromagnetic band gap materials. *J. Phys.: Condens. Matter*, Vol. 19, No. 10, 106205

Chen, S. W.; Du, J. J.; Liu, S. Y.; Lin, Z. F. & Chui, S. T. (2008). Molding the flow of electromagnetic wave and creating mirage with a magnetic field. *Phys. Rev. A*, Vol. 78, No. 4, 043803

Chen, S. W.; Du, J. J.; Liu, S. Y.; Lin, Z. F. & Chui, S. T. (2008). Focusing the electromagnetic wave with a magnetic field. *Opt. Lett.*, Vol. 33, No. 21, 2476-2478

Chien, H. T. & Chen, C. C. (2006). Focusing of electromagnetic waves by periodic arrays of air holes with gradually varying radii. *Opt. Express*, Vol. 14, No. 22, 10759-10764

Chui, S. T.; Liu, S. Y. & Lin Z. F. (2010). Reflected wave of finite circulation from magnetic photonic crystals. *J. Phys.: Condens. Matter*, Vol. 22, No. 18, 182201

Decoopman, T.; Tayeb, G.; Enoch, S.; Maystre, D. & Gralak B. (2006). Photonic crystal lens: From negative refraction and negative index to negative permittivity and permeability. *Phys. Rev. Lett.*, Vol. 97, No. 7, 073905

Dolling, G.; Enkrich, C.; Wegener, M.; Soukoulis, C. M. & Linden, S. (2006). Simultaneous Negative Phase and Group Velocity of Light in a Metamaterial. *Science*, Vol. 312, No. 5775, 892-894

Figotin, A.; Godin, Y. A. & Vitebsky, I. (1998). Two-dimensional tunable photonic crystals. *Phys. Rev. B*, Vol. 57, No. 5, 2841-2848

Golosovsky, M.; Saado, Y. & Davidov, D. (1999). Self-assembly of floating magnetic particles into ordered structures: A promising route for the fabrication of tunable photonic band gap materials. *Appl. Phys. Lett.*, Vol. 75, No. 26, 4168-4170

Haldane, F. D. M. & Raghu, S. (2008). Possible realization of directional optical waveguides in photonic crystals with broken time-reversal symmetry. *Phys. Rev. Lett.*, Vol. 100, No. 1, 013904

Halevi, P. & Ramos-Mendieta, F. (2000). Tunable photonic crystals with semiconducting constituents. *Phys. Rev. Lett.*, Vol. 85, No. 9, 1875-1878

Ho, K. M.; Chan, C. T. & Soukoulis, C. M. (1990). Existence of a photonic gap in periodic dielectric structures. *Phys. Rev. Lett.*, Vol. 65, No. 25, 3152-3155

Hu, X. H.; Chan, C. T.; Zi, J.; Li, M. & Ho, K. M. (2006). Diamagnetic response of metallic photonic crystals at infrared and visible frequencies. *Phys. Rev. Lett.*, Vol. 96, No. 22, 223901

Hu, X. H; Ho, K. M.; Chan, C. T. & Zi, J. (2008). Homogenization of acoustic metamaterials of Helmholtz resonators in fluid. *Phys. Rev. B*. Vol. 77, No. 17, 172301

Inoue, M.; Fujikawa, R.; Baryshev, A.; Khanikaev, A.; Lim, P. B.; Uchida, H.; Aktsipetrov, O.; Fedyanin, A.; Murzina, T. & Granovsky, A. (2006). Magnetophotonic crystals. *J. Phys D*, Vol. 39, No. 8, R151-R161

Jin, J. J.; Liu, S. Y.; Lin, Z. F. & Chui, S. T. (2009). Effective-medium theory for anisotropic magnetic metamaterials. *Phys. Rev. B*, Vol. 80, No. 11, 115101

Joannopoulos, J. D.; Meade, R. D. & Winn, J. N. (1995). *Photonic Crystals*, NJ: Princeton University Press, Princeton.

John, S. (1987). Strong localization of photons in certain disordered dielectric superlattices. *Phys. Rev. Lett.*, Vol. 58, No. 23, 2486-2489

Kang, D.; Maclennan, J. E.; Clar, N. A.; Zakhidov, A. A. & Baughman, R. H. (2001). *Phys. Rev. Lett.*, Vol. 86, No. 18, 4052-4055

Kee, C. S.; Kim, J. E.; Park, H. Y. & Lim, H. (2000). Two-dimensional tunable magnetic photonic crystals. *Phys. Rev. B*, Vol. 61, No.23, 15523-15525

Kee, C. S. & Lim, H. (2001). Tunable complete photonic band gaps of two-dimensional photonic crystals with intrinsic semiconductor rods. *Phys. Rev. B*, Vol. 64, No. 12, 121103

Kim, S. & Gopalan, V. (2001). Strain-tunable photonic band gap crystals. *Appl. Phys. Lett.*, Vol. 78, No. 20, 3015-3017

Kong, J. A. (1990). *Electromagnetic Wave Theory*, Wiley, New York.

Koschny, T.; Markoš, P.; Smith, D. R. & Soukoulis, C. M. (2003). Resonant and antiresonant frequency dependence of the effective parameters of metamaterials. *Phys. Rev. E*, Vol. 68, No. 6, R065602

Koschny, T.; Markoš, P.; Economou, E. N.; Smith, D. R.; Vier, D. C. & Soukoulis, C. M. (2005). Impact of inherent periodic structure on effective medium description of left-handed and related metamaterials. *Phys. Rev. B*, Vol. 71, No. 24, 245105

Kuzmiak, V.; Maradudin, A. A. & Pincemin, F. (1994). Photonic band structures of two dimensional systems containing metalic components. *Phys. Rev. B*, Vol. 50, No. 23, 16835-16844

Leonhardt, U. (2006). Optical conformal mapping. *Science*, Vol. 312, No. 5781, 1777-1780

Li, L. M. & Zhang, Z. Q. (1998). Multiple-scattering approach to finite-sized photonic band-gap materials. *Phys. Rev. B*, Vol. 58, No. 15, 9587-9590

Lin, Z. F. & Chui, S. T. (2004). Electromagnetic scattering by optically anisotropic magnetic particle. *Phys. Rev. E*, Vol. 69, No. 5, 056614

Lin, Z. F. & Chui, S. T. (2007). Manipulating electromagnetic radiation with magnetic photonic crystals. *Opt. Lett.*, Vol. 32, No. 16, 2288-2290

Liu S. Y. & Lin, Z. F. (2006). Opening up complete photonic bandgaps in three-dimensional photonic crystals consisting of biaxial dielectric spheres. *Phys. Rev. E*, Vol. 73, No. 6, 066609

Liu, S. Y.; Du, J. J.; Lin, Z. F.; Wu, R. X. & Chui S. T. (2008). Formation of robust and completely tunable resonant photonic band gaps. *Phys. Rev. B*, Vol. 78, No. 15, 155101

Liu, S. Y.; Chen, W. K.; Du, J. J.; Lin, Z. F.; Chui, S. T. & Chan, C. T. (2008). Manipulating negative-refractive behavior with a magnetic field. *Phys. Rev. Lett.*, Vol. 101, No. 15, 157407

Luo, C.; Johnson, S. G.; Joannopoulos, J. D. & Pendry, J. B. (2002). All-angle negative refraction without negative effective index. *Phys. Rev. B*, Vol. 65, No. 20, R201104

Lyubchanskii, I. L.; Dadoenkova, N. N.; Lyubchanskii, M. I.; Shapovalov, E. A. & Rasing, Th. (2003). Magnetic photonic crystals. *J. Phys. D*, Vol. 36, No. 18, R277-R287

Maier, S. A.; Kik, P. G. & Atwater H. A. (2002). Observation of coupled plasmon-polariton modes in Au nanoparticle chain waveguides of different lengths: Estimation of waveguide loss. *Appl. Phys. Lett.*, Vol. 81, No. 9, 1714-1716

Maier, S. A.; Kik, P. G.; Atwater, H. A.; Meltzer, S.; Harel, E.; Koel, B. E. & Requicha, A. A. G. (2003). Local detection of electromagnetic energy transport below the diffraction limit in metal nanoparticle plasmon waveguides. *Nature Mater.*, Vol. 2, No.4, 229-232

McGurn, A. R. & Maradudin, A. A. (1993). Photonic band strucutures of two and three dimensional periodic metal or semiconductor arrays. *Phys. Rev. B*, Vol. 48, No. 23, 17576-17579

Merzlikin, A. M.; Vinogradov, A. P.; Inoue, M. & Granovsky, A. B. (2005). Giant photonic Hall effect in magnetophotonic crystals. *Phys. Rev. E*, Vol. 72, No. 4, 046603

Modinos, A.; Yannopapas, V. & Stefanou, N. (2000). Scattering of electromagnetic waves by nearly periodic structures. *Phys. Rev. B*, Vol. 61, No. 12, 8099-8107

Moroz, A. (2002). Metallo-dielectric diamond and zinc-blende photonic crystals. *Phys. Rev. B*, Vol. 66, No. 11, 115109

Pendry, J. B.; Holden,A. J.; Stewart, W. J. & Youngs, I. (1996). Extremely low frequency plasmons in metallic mesostructures. *Phys. Rev. Lett.*, Vol. 76, No. 25, 4773-4776

Pendry, J. B.; Holden, A. J.; Holden, D. J. & Stewart, W. J. (1999). Magnetism from conductors and enhanced nonlinear phenomena. *IEEE Tran. Microwave Theory Tech.*, Vol. 47, No. 11, 2075-2084

Pendry, J. B. (2000). Negative refraction makes a perfect lens. *Phys. Rev. Lett.*, Vol. 85, No. 18, 3966-3969

Pendry, J. B.; Schurig, D. & Smith, D. R. (2006). Controlling electromagnetic fields. *Science*, Vol. 312, No. 5781, 1780-1782

Peng, L.; Ran, L.; Chen, H.; Zhang, H.; Kong, J. A. & Grzegorczyk, T. M. (2007). Experimental observation of left-handed behavior in an array of standard dielectric resonators. *Phys. Rev. Lett.*, Vol. 98, No. 15, 157403

Pozar, D. M. (2004). *Microwave Engineering*, Wiley, New York.

Rachford, F. J.; Armstead, D. N.; Harris, V. G. & Vittoria, C. (2007). Simulations of ferrite-dielectric-wire composite negative index materials. *Phys. Rev. Lett.*, Vol. 99, No. 05, 057202

Rikken, G. L. J. A. & van Tiggelen, B. A. (1996). Observation of magnetically induced transverse diffusion of light. *Nature*, Vol. 381, No. 6577, 54-55

Sarychev, A. K.; McPhedran, R. C. & Shalaev, V. M. (2000). Electrodynamics of metal-dielectric composites and electromagnetic crystals. *Phys. Rev. B*, Vol. 62, No. 12, 8531-8539

Schurig, D.; Mock, J. J.; Justice, B. J.; Cummer, S. A.; Pendry, J. B.; Starr, A. F. & Smith, D. R. (2006). Metamaterial electromagnetic cloak at microwave frequencies. *Science*, Vol. 314, No. 5801, 977-980

Shalaev, V. M. (2007). Optical negative-index metamaterials. *Nature Photon.*, Vol. 1, No. 1, 41-48

Shelby, R. A.; Smith, D. R. & Schultz S. (2001). Experimental verification of a negative index of reflection. *Science*, Vol. 292, No. 5514, 77-79

Sheng, P. (1995). *Introduction to Wave Scattering, Localization, and Mesoscopic Phenomena*, Academic Press, San Diego.

Sigalas, M. M.; Soukoulis, C. M.; Biswas, R. & Ho, K. M. (1997). Effect of the magnetic permeability on photonic band gaps. *Phys. Rev. B*, Vol. 56, No. 3, 959-962

Silveirinha, M. & Engheta, N. (2006). Tunneling of electromagnetic energy through subwavelength channels and bends using ε-near-zero materials. *Phys. Rev. Lett.*, Vol. 97, No. 15, 157403

Smith, D. R.; Schultz, S.; Markoš, P. & Soukoulis, C. M. (2002). Determination of effective permittivity and permeability of metamaterials from reflection and transmission coefficients. *Phys. Rev. B*, Vol. 65, No. 19, 195104

Smith, D. R.; Vier, D. C.; Koschny, T. & Soukoulis, C. M. (2005). Electromagnetic parameter retrieval from inhomogeneous metamaterials. *Phys. Rev. E*, Vol. 71, No. 3, 036617

Stefanou, N.; Yannopapas, V. & Modinos, A. (1998). Heterostructures of photonic crystals:

frequency bands and transmission coefficients. *Compu. Phys. Commun.*, Vol. 113, No. 1, 49-77

Valentine, J.; Zhang, S.; Zentgraf, T.; Ulin-Avila, E.; Genov, D. A.; Bartal, G. & Zhang, X. (2008). Three-dimensional optical metamaterial with a negative refractive index. *Nature*, Vol. 455, No. 7211, 376-379

Veselago, V. C. (1968). The electrodynamics of substrates whith simultaneously negative values of ε and μ. *Sov. Phys. Usp.*, Vol. 10, No. 4, 509-514

Wang, Z.; Chong, Y. D.; Joannopoulos, J. D. & Soljačić, M. (2008). Reflection-free one-way edge modes in a gyromagnetic photonic crystal. *Phys. Rev. Lett.*, Vol. 100, No. 1, 013905

Wang, Z.; Chong, Y. D.; Joannopoulos, J. D. & Soljačić, M. (2009). Observation of unidirectional backscattering-immune topological electromagnetic states. *Nature*, Vol. 461, No. 7265, 772-775

Wu, Y.; Li, J.; Zhang, Z. Q. & Chan, C. T. (2006). Effective medium theory for magnetodielectric composites: Beyond the long-wavelength limit. *Phys. Rev. B*, Vol. 74, No. 8, 085111

Wu, Y.; Lai, Y. & Zhang, Z. Q. (2007). Effective medium theory for elastic metamaterials in two dimensions. *Phys. Rev. B*, Vol. 76, No. 20, 205313

Xu, C.; Han, D.; Wang, X.; Liu, X. & Zi, J. (2007). Extrinsic photonic crystals: Photonic band structure calculations of a doped semiconductor under a magnetic field. *Appl. Phys. Lett.*, Vol. 90, No. 6, 061112

Yablonovitch, E. (1987). Inhibited spontaeous emission in solid state physics and electronics. *Phys. Rev. Lett.*, Vol. 58, No. 20, 2059-2062

Yannopapas, V. & Vitanov, N. V. (2006). Photoexcitation-induced magnetism in arrays of semiconductor nanoparticles with a strong excitonic oscillator strength. *Phys. Rev. B*, Vol. 74, No. 19, 193304

Yu, Z. F.; Veronis, G.; Wang, Z. & Fan, S. H. (2008). One-Way electromagnetic waveguide formed at the interface between a plasmonic metal under a static magnetic field and a photonic crystal. *Phys. Rev. Lett.*, Vol 100, No. 2, 023902

Zhang, S.; Fan, W.; Panoiu, N. C.; Malloy, K. J.; Osgood, R. M. & Brueck, S. R. J. (2005). Experimental demonstration of near-infrared negative-index metamaterials. *Phys. Rev. Lett.*, Vol. 95, No. 13, 137404

Zhang, S.; Park, Y. S.; Li, J.; Lu, X.; Zhang, W. & Zhang, X. (2009). Negative refractive index in chiral metamaterials. *Phys. Rev. Lett.*, Vol. 102, No. 2, 023901

Zayats, A. V.; Smolyaninov, I. I. & Maradudin, A. A. (2005). Nano-optics of surface plasmon polaritons. *Phys. Rep.*, Vol. 408, No. 3-4, 131-314

22

The Nonlinear Absorption of a Strong Electromagnetic Wave in Low-dimensional Systems

Nguyen Quang Bau and Hoang Dinh Trien
Hanoi University of Science, Vietnam National University
Vietnam

1. Introduction

It is well known that in low-dimensional systems, the motion of electrons is restricted. The confinement of electron in these systems has changed the electron mobility remarkably. This has resulted in a number of new phenomena, which concern a reduction of sample dimensions. These effects differ from those in bulk semiconductors, for example, electron-phonon interaction effects in two-dimensional electron gases (Mori & Ando, 1989; Rucker et al., 1992; Butscher & Knorr, 2006), electron-phonon interaction and scattering rates in one-dimensional systems (Antonyuk et al., 2004; Kim et al., 1991) and dc electrical conductivity (Vasilopoulos et al., 1987; Suzuki, 1992), the electronic structure (Gaggero-Sager et al., 2007), the wave function distribution (Samuel & Patil, 2008) and electron subband structure and mobility trends in quantum wells (Ariza-Flores & Rodriguez-Vargas, 2008). The absorption of electromagnetic wave in bulk semiconductors, as well as low dimensional systems has also been investigated (Shmelev et al., 1978; Bau & Phong, 1998; Bau et al., 2002; 2007). However, in these articles, the author was only interested in linear absorption, namely the linear absorption of a weak electromagnetic wave has been considered in normal bulk semiconductors (Shmelev et al., 1978), the absorption coefficient of a weak electromagnetic wave by free carriers for the case of electron-optical phonon scattering in quantum wells are calculated by the Kubo-Mori method in quantum wells (Bau & Phong, 1998) and in doped superlattices (Bau et al., 2002), and the quantum theory of the absorption of weak electromagnetic waves caused by confined electrons in quantum wires has been studied based on Kubo's linear response theory and Mori's projection operator method (Bau et al., 2007); the nonlinear absorption of a strong electromagnetic wave by free electrons in the normal bulk semiconductors has been studied by using the quantum kinetic equation method (Pavlovich & Epshtein, 1977). However, the nonlinear absorption problem of an electromagnetic wave, which has strong intensity and high frequency, in low dimensional systems is still open for study.

In this book chapter, we study the nonlinear absorption of a strong electromagnetic wave in low dimensional systems (quantum wells, doped superlattices, cylindrical quantum wires and rectangular quantum wires) by using the quantum kinetic equation method. Starting from the kinetic equation for electrons, we calculate to obtain the electron distribution functions in low dimensional systems. Then we find the expression for current density vector and the nonlinear absorption coefficient of a strong electromagnetic wave in low dimensional

systems. The problem is considered in two cases: electron-optical phonon scattering and electron-acoustic phonon scattering. Numerical calculations are carried out with a AlAs/GaAs/AlAs quantum well, a compensated n-p n-GaAs/p-GaAs doped superlattices, a specific GaAs/GaAsAl quantum wire.
This book chapter is organized as follows: In section 2, we study the nonlinear absorption of a strong electromagnetic wave by confined electrons in a quantum well. Section 3 presents the nonlinear absorption of a strong electromagnetic wave by confined electrons in a doped superlattice. The nonlinear absorption of a strong electromagnetic wave by confined electrons in a cylindrical quantum wire and in a rectangular quantum wire is presented in section 4 and section 5. Conclusions are given in the section 6.

2. The nonlinear absorption of a strong electromagnetic wave by confined electrons in a quantum well

2.1 The electron distribution function in a quantum well

It is well known that in quantum wells, the motion of electrons is restricted in one dimension, so that they can flow freely in two dimension. The Hamiltonian of the electron - phonon system in quantum wells in the second quantization representation can be written as (in this chapter, we we select $\hbar$=1)

$$H = H_0 = \sum_{n,\vec{p}_\perp} \varepsilon_n(\vec{p}_\perp - \frac{e}{c}\vec{A}(t))a^+_{n,\vec{p}_\perp}a_{n,\vec{p}_\perp} + \sum_{\vec{q}}\omega_{\vec{q}}b^+_{\vec{q}}b_{\vec{q}} + \\ + \sum_{n,n',\vec{p}_\perp,\vec{q}} C_{\vec{q}}I_{n,n'}(q_z)a^+_{n',\vec{p}_\perp+\vec{q}_\perp}a_{n,\vec{p}_\perp}(b_{\vec{q}} + b^+_{-\vec{q}}), \quad (1)$$

where e is the electron charge, c is the velocity of light, n denotes the quantization of the energy spectrum in the z direction (n = 1,2,...), (n, $\vec{p}_\perp$) and (n', $\vec{p}_\perp + \vec{q}_\perp$) are electron states before and after scattering, respectively. $\vec{p}_\perp(\vec{q}_\perp)$ is the in plane (x,y) wave vector of the electron (phonon), $a^+_{n,\vec{p}_\perp}$ and $a_{n,\vec{p}_\perp}$ ($b^+_{\vec{q}}$ and $b_{\vec{q}}$) are the creation and the annihilation operators of electron (phonon), respectively. $\vec{q}=(\vec{q}_\perp,q_z)$, $\vec{A}(t) = \frac{c}{\Omega}\vec{E}_0cos(\Omega t)$ is the vector potential, $\vec{E}_0$ and Ω are the intensity and the frequency of the EMW, $\omega_{\vec{q}}$ is the frequency of a phonon, $C_{\vec{q}}$ is the electron-phonon interaction constants, $I_{n',n}(q_z)$ is the electron form factor in quantum wells.
In order to establish the quantum kinetic equations for electrons in a quantum well, we use the general quantum equation for the particle number operator (or electron distribution function) $n_{n,\vec{p}_\perp}(t) = \left\langle a^+_{n,\vec{p}_\perp}a_{n,\vec{p}_\perp}\right\rangle_t$

$$i\frac{\partial n_{n,\vec{p}_\perp}(t)}{\partial t} = \left\langle [a^+_{n,\vec{p}_\perp}a_{n,\vec{p}_\perp}, H]\right\rangle_t, \quad (2)$$

where $\left\langle\psi\right\rangle_t$ denotes a statistical average value at the moment t, and $\left\langle\psi\right\rangle_t = Tr\left(\widehat{W}\widehat{\psi}\right)$($\widehat{W}$ being the density matrix operator). Starting from the Hamiltonian Eq. (1) and using the commutative relations of the creation and the annihilation operators, we obtain the quantum kinetic equation for electrons in quantum wells:

$$\frac{\partial n_{n,\vec{p}_\perp}(t)}{\partial t} = -\sum_{\vec{q},n'}|C_{\vec{q}}|^2|I_{n,n'}|^2 \sum_{k,s=-\infty}^{\infty} J_k(\frac{e\vec{E}_0\vec{q}_\perp}{m\Omega^2})J_s(\frac{e\vec{E}_0\vec{q}_\perp}{m\Omega^2})exp[-i(s-k)\Omega]\int_{-\infty}^{t} dt'$$

$$\times\Big\{[n_{n,\vec{p}_\perp}(t')N_{\vec{q}} - n_{n',\vec{p}_\perp+\vec{q}_\perp}(t')(N_{\vec{q}}+1)]exp[i(\varepsilon_{n',\vec{p}_\perp+\vec{q}_\perp} - \varepsilon_{n,\vec{p}_\perp} - \omega_{\vec{q}} - k\Omega + i\delta)(t-t')]$$
$$+ [n_{n,\vec{p}_\perp}(t')(N_{\vec{q}}+1) - n_{n',\vec{p}_\perp+\vec{q}_\perp}(t')N_{\vec{q}}]exp[i(\varepsilon_{n',\vec{p}_\perp+\vec{q}_\perp} - \varepsilon_{n,\vec{p}_\perp} + \omega_{\vec{q}} - k\Omega + i\delta)(t-t')]$$
$$- [n_{n',\vec{p}_\perp-\vec{q}_\perp}(t')N_{\vec{q}} - n_{n,\vec{p}_\perp}(t')(N_{\vec{q}}+1)]exp[i(\varepsilon_{n,\vec{p}_\perp} - \varepsilon_{n',\vec{p}_\perp-\vec{q}_\perp} - \omega_{\vec{q}} - k\Omega + i\delta)(t-t')]$$
$$- [n_{n',\vec{p}_\perp-\vec{q}_\perp}(t')(N_{\vec{q}}+1) - n_{n,\vec{p}_\perp}(t')N_{\vec{q}}]exp[i(\varepsilon_{n,\vec{p}_\perp} - \varepsilon_{n',\vec{p}_\perp-\vec{q}_\perp} + \omega_{\vec{q}} - k\Omega + i\delta)(t-t')]\Big\}. \quad (3)$$

where $J_k(x)$ is the Bessel function, m is the effective mass of the electron, $N_{\vec{q}}$ is the time - independent component of the phonon distribution function, and the quantity δ is infinitesimal and appears due to the assumption of an adiabatic interaction of the electromagnetic wave.

It is well known that to obtain the explicit solutions from Eq. (3) is very difficult. In this paper, we use the first - order tautology approximation method (Pavlovich & Epshtein, 1977; Malevich & Epstein, 1974; Epstein, 1975) to solve this equation. In detail, in Eq. (3), we use the approximation:

$n_{n,\vec{p}_\perp}(t') \approx \bar{n}_{n,\vec{p}_\perp}$, $n_{n,\vec{p}_\perp+\vec{q}_\perp}(t') \approx \bar{n}_{n,\vec{p}_\perp+\vec{q}_\perp}$, $n_{n,\vec{p}_\perp-\vec{q}_\perp}(t') \approx \bar{n}_{n,\vec{p}_\perp-\vec{q}_\perp}$.

where $\bar{n}_{n,\vec{p}_\perp}$ is the time - independent component of the electron distribution function. The approximation is also applied for a similar exercise in bulk semiconductors (Pavlovich & Epshtein, 1977; Malevich & Epstein, 1974). We perform the integral with respect to t. Next, we perform the integral with respect to t of Eq. (3). The expression of electron distribution function can be written as

$$n_{n,\vec{p}_\perp}(t) = -\sum_{\vec{q},n'}|C_{\vec{q}}|^2|I_{n,n'}|^2\sum_{k,l=-\infty}^{\infty}J_k\Big(\frac{e\vec{E}_0\vec{q}_\perp}{m\Omega^2}\Big)J_{k+l}\Big(\frac{e\vec{E}_0\vec{q}_\perp}{m\Omega^2}\Big)\frac{1}{l\Omega}e^{-il\Omega t}$$
$$\times\Big\{-\frac{\bar{n}_{n,\vec{p}_\perp}N_{\vec{q}} - \bar{n}_{n',\vec{p}_\perp+\vec{q}_\perp}(N_{\vec{q}}+1)}{\varepsilon_{n',\vec{p}_\perp+\vec{q}_\perp} - \varepsilon_{n,\vec{p}_\perp} - \omega_{\vec{q}} - k\Omega + i\delta} - \frac{\bar{n}_{n,\vec{p}_\perp}(N_{\vec{q}}+1) - \bar{n}_{n',\vec{p}_\perp+\vec{q}_\perp}N_{\vec{q}}}{\varepsilon_{n',\vec{p}_\perp+\vec{q}_\perp} - \varepsilon_{n,\vec{p}_\perp} + \omega_{\vec{q}} - k\Omega + i\delta}$$
$$+\frac{\bar{n}_{n',\vec{p}_\perp-\vec{q}_\perp}N_{\vec{q}} - \bar{n}_{n,\vec{p}_\perp}(N_{\vec{q}}+1)}{\varepsilon_{n,\vec{p}_\perp} - \varepsilon_{n',\vec{p}_\perp-\vec{q}_\perp} - \omega_{\vec{q}} - k\Omega + i\delta} + \frac{\bar{n}_{n',\vec{p}_\perp-\vec{q}_\perp}(N_{\vec{q}}+1) - \bar{n}_{n,\vec{p}_\perp}N_{\vec{q}}}{\varepsilon_{n,\vec{p}_\perp} - \varepsilon_{n',\vec{p}_\perp-\vec{q}_\perp} + \omega_{\vec{q}} - k\Omega + i\delta}\Big\}. \quad (4)$$

From Eq.(4) we see that the electron distribution function depends on the constant in the case of electron - phonon interaction, the electron form factor and the electron energy spectrum in quantum wells. Eq.(4) also can be considered a general expression of the electron distribution function in two dimensional systems with the electron form factor and the electron energy spectrum of each systems.

2.2 Calculations of the nonlinear absorption coefficient of a strong electromagnetic wave by confined electrons in a quantum well

In a quantum well, the motion of electrons is confined and that energy spectrum of electron is quantized into discrete levels. We assume that the quantization direction is the z direction. The total wave function of electrons can be written as

$$\psi_n(\vec{r}) = \psi_0 e^{i\vec{p}_\perp\vec{r}_\perp}sin(p_z^n z), \quad (5)$$

where ψ_0 is normalization constant, the electron energy spectrum takes the simple form:

$$\varepsilon_{n,\vec{p}_\perp} = \frac{1}{2m}(p_\perp^2 + p_z^{n2}). \tag{6}$$

where p_z^n takes discrete values: $p_z^n = n\pi/L$, L is width of a quantum well.
The electron form factor can be written as

$$I_{n',n}(q_z) = \frac{2}{L}\int_0^L sin(p_z^{n'}z)sin(p_z^n z)e^{iq_z z}dz \tag{7}$$

The carrier current density formula in quantum wells takes the form (Pavlovich & Epshtein, 1977)

$$\vec{j}_\perp(t) = \frac{e}{m}\sum_{n,\vec{p}_\perp}(\vec{p}_\perp - \frac{e}{c}\vec{A}(t))n_{n,\vec{p}_\perp}(t). \tag{8}$$

Because the motion of electrons is confined along the z direction in a quantum well, we only consider the in - plane (x,y) current density vector of electrons, $\vec{j}_\perp(t)$. Using Eq. (4), we find the expression for current density vector:

$$\vec{j}_\perp(t) = -\frac{e^2}{mc}\sum_{n,\vec{p}_\perp}\vec{A}(t)n_{n,\vec{p}_\perp}(t) + \sum_{l=1}^{\infty}\vec{j}_l sin(l\Omega t). \tag{9}$$

Here,

$$\vec{j}_l = 2\pi\frac{e}{m}\frac{1}{l\Omega}\sum_{n,n'}\sum_{\vec{p},\vec{p}_\perp}|C_{\vec{q}}|^2|I_{n,n'}|^2\sum_{k=-\infty}^{\infty}\vec{q}_\perp J_k\Big(\frac{e\vec{E}_0\vec{q}_\perp}{m\Omega^2}\Big)\Big[J_{k+l}\Big(\frac{e\vec{E}_0\vec{q}_\perp}{m\Omega^2}\Big) + J_{k-l}\Big(\frac{e\vec{E}_0\vec{q}_\perp}{m\Omega^2}\Big)\Big]$$
$$\times N_{\vec{q}}(\bar{n}_{n,\vec{p}_\perp} - \bar{n}_{n',\vec{p}_\perp+\vec{q}_\perp})\{\delta(\varepsilon_{n',\vec{p}_\perp+\vec{q}_\perp} - \varepsilon_{n,\vec{p}_\perp} + \omega_{\vec{q}} - k\Omega) + [\omega_{\vec{q}} \to -\omega_{\vec{q}}]\}. \tag{10}$$

Using the expression of the nonlinear absorption coefficient of a strong electromagnetic wave (Pavlovich & Epshtein, 1977)

$$\alpha = \frac{8\pi}{c\sqrt{\chi_\infty}E_0^2}\Big\langle \vec{j}_\perp(t)\vec{E}_0 sin\Omega t\Big\rangle_t, \tag{11}$$

and properties of Bessel function, we obtain the nonlinear absorption coefficient of a strong electromagnetic wave by confined electrons in quantum well

$$\alpha = \frac{8\pi^2\Omega}{c\sqrt{\chi_\infty}E_0^2}\sum_{n,n'}|I_{n,n'}|^2\sum_{\vec{q},\vec{p}}|C_{\vec{q}}|^2 N_{\vec{q}}\sum_{k=-\infty}^{\infty}\big[\bar{n}_{n,\vec{p}} - \bar{n}_{n',\vec{p}+\vec{q}}\big]\times$$
$$\times kJ_k^2\Big(\frac{eE_0\vec{q}}{m\Omega^2}\Big)\{\delta(\varepsilon_{n',\vec{p}+\vec{q}} - \varepsilon_{n,\vec{p}} + \omega_{\vec{q}} - k\Omega) + [\omega_{\vec{q}} \to -\omega_{\vec{q}}]\} \tag{12}$$

In the following, we study the problem with different electron-phonon scattering mechanisms. We only consider the absorption close to its threshold because in the rest case (the absorption far away from its threshold) α is very smaller. In the case, the condition $|k\Omega - \omega_{\vec{q}}| \ll \bar{\varepsilon}$ must be satisfied (Pavlovich & Epshtein, 1977). We restrict the problem to the case of one photon absorption and consider the electron gas to be non-degenerate:

$$\bar{n}_{n,\vec{p}} = n_0^* exp(-\frac{\varepsilon_{n,\vec{p}}}{k_b T}), \text{ with } n_0^* = \frac{n_0 (e\pi)^{\frac{3}{2}}}{V(m_0 k_b T)^{\frac{3}{2}}} \quad (13)$$

where, V is the normalization volume, n_0 is the electron density in quantum well, m_0 is the mass of free electron, k_b is Boltzmann constant.

2.2.1 Electron - optical phonon scattering

In this case, The electron-optical phonon interaction constants can be taken as (Shmelev et al., 1978; Pavlovich & Epshtein, 1977) $|C_{\vec{q}}|^2 \equiv |C_{\vec{q}}^{op}|^2 = 2\pi e^2 \omega_0 (1/\chi_\infty - 1/\chi_0) / \epsilon_0 (q_\perp^2 + q_z^2) V$, here V is the volume, ϵ_0 is the permittivity of free space, χ_∞ and χ_0 are the high and low-frequency dielectric constants, respectively. $\omega_{\vec{q}} \equiv \omega_0$ is the frequency of the optical phonon in the equilibrium state. By using the electron - optical phonon interaction factor $C_{\vec{q}}^{op}$, the Bessel function and the electron distribution function $n_{n,\vec{p}_\perp}$, from the general expression for the nonlinear absorption coefficient of a strong electromagnetic wave in a quantum well Eq.(12), we obtain the explicit expression of the nonlinear absorption coefficient α in quantum well for the case electron-optical phonon scattering:

$$\alpha = \frac{\alpha_0}{\pi L} \sum_{nn'} exp\Big(- \frac{\pi^2 n'^2}{2mk_B T L^2} \Big) \Big\{ \Big[\Big\{ exp\Big(\frac{\Omega - \omega_0}{k_B T} \Big) - 1 \Big\}$$

$$\Big\{ 1 + \frac{e^2 E_0^2}{m\Omega^3} \frac{3k_B T}{8\Omega} \Big[1 + \frac{1}{2k_B T} \Big((\omega_0 - \Omega) + \frac{\pi^2 (n'^2 - n^2)}{2mL^2} \Big) \Big] \Big\} \Big] + [\omega_0 \to -\omega_0] \Big\}, \quad (14)$$

where

$$\alpha_0 = \frac{\pi e^4 n_0^* (k_B T)^2}{2\epsilon_0 c \sqrt{\chi_\infty} \Omega^3} \Big(\frac{1}{\chi_\infty} - \frac{1}{\chi_0} \Big). \quad (15)$$

In bulk materials, there is a strong dispersion when the phonon energy is close to the optical phonon energy. However, in a quantum well, we will see an increase in the absorption coefficient of Electromagnetic Wave (see the numerical calculation and the discussion sections). This is due to the surprising changes in the electron spectrum and the wave function in quantum system. This also results in an significant property for low - dimensional materials.

2.2.2 Elctron - acoustic phonon scattering

In the case, $\omega_{\vec{q}} \ll \Omega$ ($\omega_{\vec{q}}$ is the frequency of acoustic phonons), so we let it pass. The electron-acoustic phonon interaction constants can be taken as (Mori & Ando, 1989; Shmelev et al., 1978) $|C_{\vec{q}}|^2 \equiv |C_{\vec{q}}^{ac}|^2 = \xi^2 q / 2\rho v_s V$, here V, ρ, v_s, and ξ are the volume, the density, the acoustic velocity and the deformation potential constant, respectively. In this case, we obtain the explicit expression of α in quantum well for the case electron-acoustic phonon scattering:

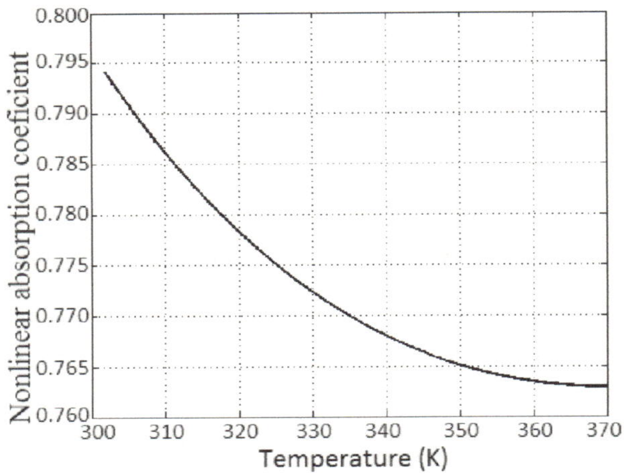

Fig. 1. The dependence of α on T in a quantum well

$$\alpha = \frac{(k_BT)^3 e^2 m n_0 \xi^2}{c\sqrt{\chi_\infty} v_s^2 \rho \Omega^3 L} \sum_{n,n'} exp\Big[-\frac{1}{k_BT}\Big(\frac{\pi^2(n'^2-n^2)}{2mL^2}\Big)\Big]$$

$$\times \Big[exp(\frac{\Omega}{k_BT})-1\Big]\Big\{1+\frac{\beta}{2k_BT}+\frac{3}{32}\frac{e^2E_0^2}{m\Omega^4 k_BT}(\beta^2+3\beta k_BT+12(k_BT)^2)\Big\} \quad (16)$$

$$with \quad \beta = \frac{\pi^2(n'^2-n^2)}{2mL^2} - \Omega. \quad (17)$$

From Eqs. 14-17 we see that the nonlinear absorption coefficient are complex and has difference from those obtained in normal bulk semiconductors. the nonlinear absorption coefficient has the sum over the quantum number n. In addition, when the term in proportion to a quadratic in the intensity of the electromagnetic wave (E_0^2) tend toward zero, the nonlinear result will turn back to the linear case which was calculated by another method-the Kubo - Mori (Bau & Phong, 1998).

2.3 Numerical results and discussion

In order to clarify the mechanism for the nonlinear absorption of a strong electromagnetic wave in a quantum well, we will evaluate, plot, and discuss the expression of the nonlinear absorption coefficient for the case of a specific quantum well: AlAs/GaAs/AlAs. The parameters used in the calculations are as follows (Bau et al., 2002; Pavlovich & Epshtein, 1977): $\chi_\infty = 10.9, \chi_0 = 12.9, n_0 = 10^{23} \quad m^{-3}, L = 100 \quad A^o, m = 0.067m_0$, m_0 being the mass of a free electron, $\hbar\omega_0 = 36.25 \quad meV, and \quad \Omega = 2.10^{14} \quad s^{-1}$.

2.3.1 Electron - optical phonon scattering

Figure 1 show the nonlinear absorption in quantum wells. When the temperature T of the system rises, its absorption coefficient decreased. However, for the case of bulk semiconductors, the absorption coefficient increases following its temperature. In addition, the absorption coefficient in bulk semiconductors is smaller than in quantum wells. The fact

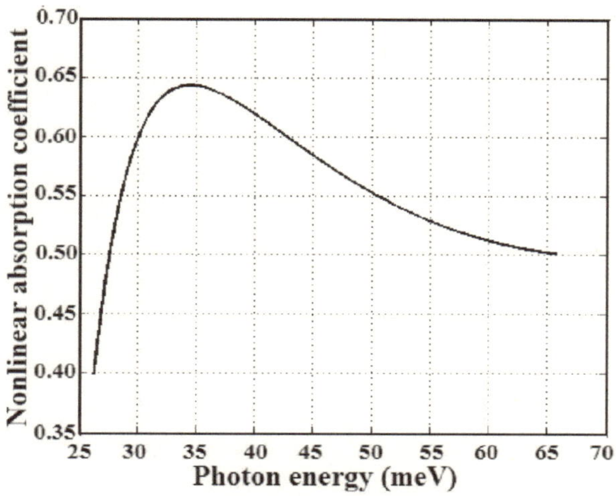

Fig. 2. The dependence of α on photon energy in a quantum well

proves that confined electrons in quantum wells have enhanced electromagnetic absorption ability.

Figure 2 shows the nonlinear absorption coefficient as a function of the electromagnetic wave energy (photon energy) for the case electron - optical phonon scattering. This figure shows that the curve has a maximum where $\Omega = \omega_0$.

Figure 3 shows the dependence of the nonlinear absorption coefficient depends on well's width L at different values of the electromagnetic wave energy, each curve has one maximum peak. The resonance peak only appears when $20\ nm < L < 40\ nm$, and it will be sharper if the frequency of the electromagnetic wave is close to the frequency of the optical phonon $\Omega = \omega_0$. This suggests that when external parameters are not changed, we can change the width of quantum well to get the absorption of a strong electromagnetic wave the best.

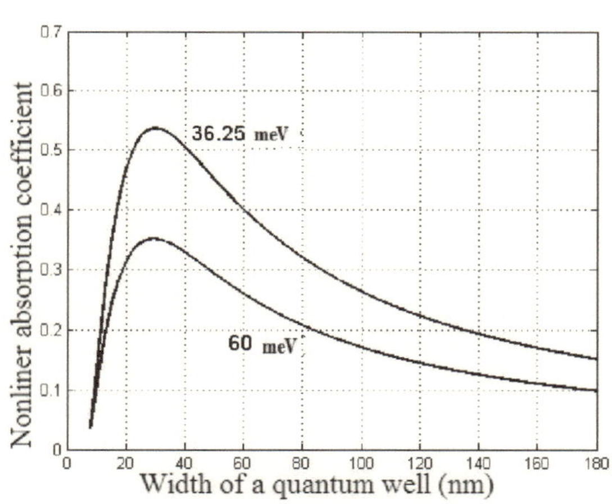

Fig. 3. The dependence of α on L (well's width) in a quantum well

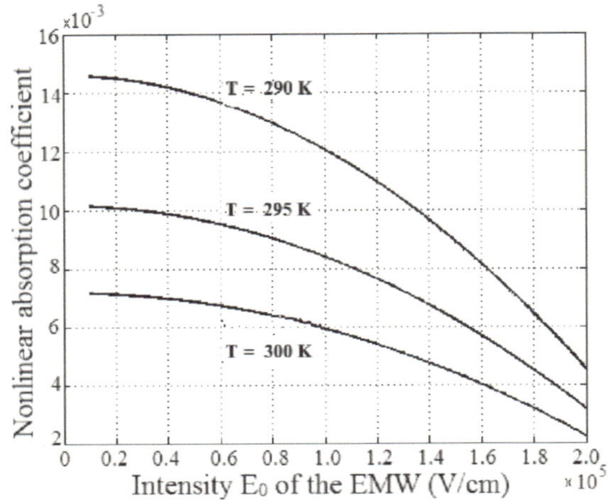

Fig. 4. The dependence of α on E_0 in a quantum well

2.3.2 Electron - acoustic phonon scattering

The parameters used in the calculations are as follow (Bau et al., 2002; Pavlovich & Epshtein, 1977): $\chi_\infty = 10{,}9, \xi = 13.5 \ eV, v_s = 5370 \ m/s, \rho = 5320 \ kg/m^3, n_0 = 10^{23} \ m^{-3}, L = 100 \ A^o, m = 0{,}067m_0, m_0$ being the mass of free electron.

Figure 4 and figure 5 shows the nonlinear absorption coefficient in quantum wells for the case electron - acoustic phonon scattering. The most important point is that in this case, the absorption coefficient is very small. Figure. 5 shows the nonlinear absorption coefficient dependence on the electromagnetic wave energy for the case electron - acoustic phonon scattering. Different from the case electron - optical phonon scattering, the nonlinear absorption coefficient α in this case has not maximum values (peaks).

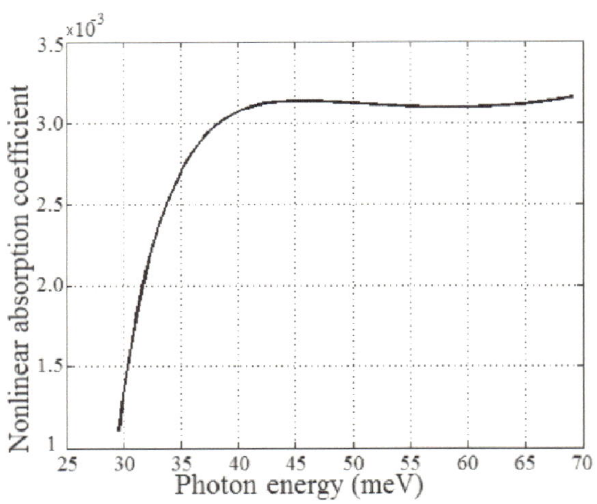

Fig. 5. The dependence of α on photon energy in a quantum well

3. The nonlinear absorption of a strong electromagnetic wave by confined electrons in a doped superlattice

3.1 Calculations of the nonlinear absorption coefficient of a strong electromagnetic wave by confined electrons in a doped superlattice

The total wave function of electrons and the electron energy spectrum in a doped superlattice can be written as (Bau et al., 2002)

$$\psi_{n,p_z}(z) = \sum_{j=0}^{s_0} e^{ip_z jz} \psi_n(z - jd), \tag{18}$$

$$\varepsilon_n(\vec{p}_\perp) = \frac{\vec{p}_\perp^2}{2m} + \omega_p\left(n + \frac{1}{2}\right), \tag{19}$$

Here, $\psi_n(z)$ is the wave function of the n-th state in one of the one-dimensional potential wells which compose the doped superlattice potential, d is the doped superlattice period, s_0 is the number of doped superlattice period, $\omega_p = \left(\frac{4\pi e^2 n_D}{\chi_0 m}\right)^{1/2}$ is the frequency plasma caused by donor doping concentration, n_D is the doping concentration. The electron form factor, $I_{n,n'}(q_z)$, is written as

$$I_{n,n'}(q_z) = \sum_{j=1}^{s_0} \int_0^d e^{iq_z d} \psi_n(z - jd)\, \psi_{n'}(z - jd)\, dz. \tag{20}$$

In order to establish analytical expressions for the nonlinear absorption coefficient of a strong electromagnetic wave by confined electrons in a doped superlattice, We insert the expression for $n_{n,\ell,\vec{p}}(t)$ into the expression for $\vec{j}(t)$ and then insert the expression for $\vec{j}(t)$ into the expression for α in Eq.(11). Using the properties of Bessel function and realizing the calculations, we obtain the nonlinear absorption coefficient of a strong electromagnetic wave by confined electrons in a doped superlattice as

$$\alpha = \frac{8\pi^2\Omega}{c\sqrt{\chi_\infty}E_0^2} \sum_{n,n'} |I_{n,n'}|^2 \sum_{\vec{q},\vec{p}} |C_{\vec{q}}|^2 N_{\vec{q}} \sum_{k=-\infty}^{\infty} [\bar{n}_{n,\vec{p}} - \bar{n}_{n',\vec{p}+\vec{q}}] k J_k^2\left(\frac{eE_0\vec{q}}{m\Omega^2}\right) \times$$
$$\times\, \delta\left(\frac{(\vec{p}+\vec{q})^2}{2m} + \omega_p(n' + 1/2) - \frac{\vec{p}_\perp^2}{2m} - \omega_p(n + 1/2) + \omega_{\vec{q}} - k\Omega\right) \tag{21}$$

Using the time - independent component of the electron distribution function, the Bessel function and the electron-optical phonon interaction constants, we can calculate to obtain expression of the carrier current density and the nonlinear absorption coefficient of a strong electromagnetic wave by confined electrons in a doped superlattice. we obtain the explicit expression of α in a doped superlattice for the case electron-optical phonon scattering:

$$\alpha = \frac{\sqrt{2}\pi n_0^* (k_B T)^2 e^4}{8c\sqrt{m\chi_\infty}\hbar^3\Omega^3} \left(\frac{1}{\chi_\infty} - \frac{1}{\chi_0}\right) \sum_{n,n'} |I_{n,n'}(q_z)|^2 \exp\left(-\frac{\hbar\omega_p\left(n + \frac{1}{2}\right) + \frac{\xi}{2}}{k_B T}\right)$$
$$\times\, e^{-2\sqrt{\rho\sigma}} \left(\frac{\rho}{|\xi|\sigma}\right)^{\frac{1}{2}} \left\{1 + \frac{3}{16\sqrt{\rho\sigma}} + \frac{3e^2E_0^2}{32m^2\Omega^4}\left(\frac{\rho}{\sigma}\right)^{\frac{1}{2}}\left[1 + \frac{1}{\sqrt{\rho\sigma}} + \frac{1}{16\rho\sigma}\right]\right\} \tag{22}$$

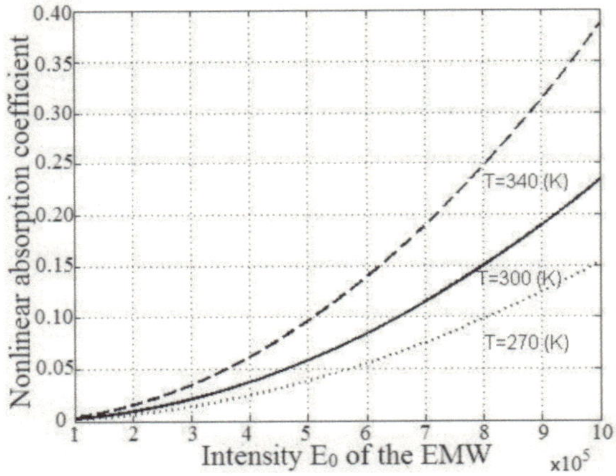

Fig. 6. The dependence of α on the E_0 in a doped superlattice

Here: $\zeta = \hbar\omega_p (n' - n) + \hbar\omega_0 - \hbar\Omega; a = \frac{3}{8}\left(\frac{eE_0}{2m\Omega^2}\right)^2; \rho = \frac{m\zeta^2}{2\hbar^2 k_B T}; \sigma = \frac{\hbar^2}{8mk_B T}$.
In a doped superlattice, the nonlinear absorption coefficient is more complex those obtained in quantum well. The term in proportion to quadratic intensity of a strong electromagnetic wave tend toward zero, the nonlinear result will turn back to the linear case which was calculated by another method-the Kubo - Mori (Bau et al., 2002)

3.2 Numerical results and discussion

In order to clarify the mechanism for the nonlinear absorption of a strong electromagnetic wave in a doped superlattice, we will evaluate, plot, and discuss the expression of the nonlinear absorption coefficient for the case of a specific doped superlattice: n-GaAs/p-GaAs. Figure 6 shows the dependence of the nonlinear absorption coefficient on intensity E_0 of electromagnetic wave in a doped superlattice. When intensity E_0 of the electromagnetic wave rises up, its absorption coefficient speeds up too. The absorption coefficient in bulk semiconductors is smaller than it is in a doped superlattice. Otherwise, the absorption

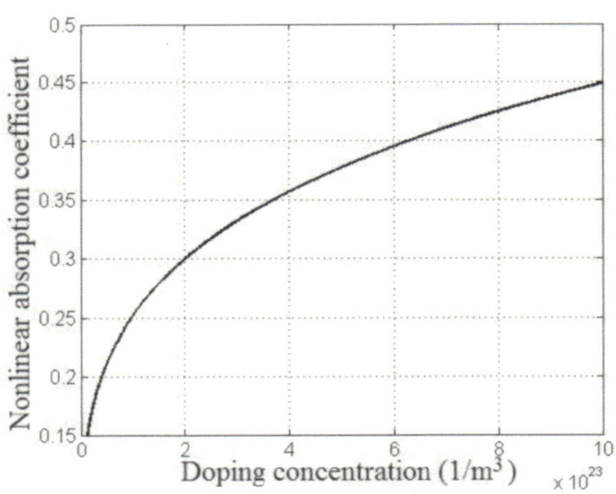

Fig. 7. The dependence of α on the n_D in a doped superlattice

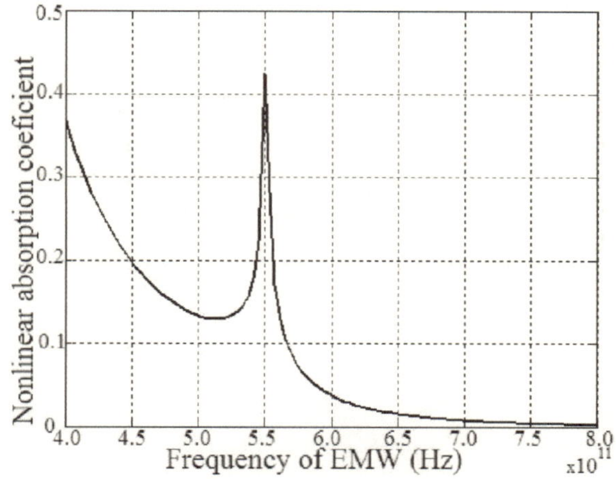

Fig. 8. The dependence of α on the Ω in a doped superlattice

coefficient changes insignificantly in bulk semiconductor. Figure 7 shows that the nonlinear absorption coefficient in a doped superlattice depends strongly on the doping concentration n_D. When the doping concentration of the system rises up, its absorption coefficient speeds up too.

Figure 8 presents the dependence of the nonlinear absorption coefficient on the frequency of the electromagnetic wave. This figure shows that the curve has a maximum coincide with the case $\Omega = \omega_0$. That is, appear a resonance peak at $\Omega = \omega_0$. However, compared with quantum well, these absorption peaks are sharper.

4. The nonlinear absorption of a strong electromagnetic wave by confined electrons in a cylindrical quantum wire

4.1 The electron distribution function in a cylindrical quantum wire

The Hamiltonian of the electron-phonon system in quantum wires. in the presence of a laser field $\vec{E}(t) = \vec{E}_0 sin(\Omega t)$, can be written as

$$H = \sum_{n,l,\vec{p}} \varepsilon_{n,l}(\vec{p} - \frac{e}{c}\vec{A}(t))a^+_{n,l,\vec{p}}\, a_{n,l,\vec{p}} + \sum_{\vec{q}} \omega_{\vec{q}}\, b^+_{\vec{q}} b_{\vec{q}}$$
$$+ \sum_{n,l,n',l',\vec{p},\vec{q}} C_{\vec{q}} I_{n,l,n',l'}(\vec{q}) a^+_{n,l,\vec{p}+\vec{q}}\, a_{n',l',\vec{p}}\, (b_{\vec{q}} + b^+_{-\vec{q}}), \quad (23)$$

where e is the electron charge, c is the velocity of light, $\vec{A}(t) = \frac{c}{\Omega}\vec{E}_0 cos(\Omega t)$ is the vector potential, $\vec{E}_0$ and Ω are the intensity and the frequency of the electromagnetic wave , $a^+_{n,\ell,\vec{p}}$ $(a_{n,\ell,\vec{p}})$ is the creation (annihilation) operator of an electron, $b^+_{\vec{q}}$ $(b_{\vec{q}})$ is the creation (annihilation) operator of a phonon for a state having wave vector $\vec{q}$, $\omega_{\vec{q}}$ is the frequency of a phonon, $C_{\vec{q}}$ is the electron-phonon interaction constants. $I_{n,l,n',l'}(\vec{q})$ is the electron form factor. In order to establish expressions for the electron distribution function in quantum wires, we use the quantum kinetic equation for particle number operator of an electron, $n_{n,\ell,\vec{p}}(t) = \langle a^+_{n,\ell,\vec{p}} a_{n,\ell,\vec{p}} \rangle_t$:

$$i\frac{\partial n_{n,\ell,\vec{p}}(t)}{\partial t} = \langle [a^{+}_{n,\ell,\vec{p}} a_{n,\ell,\vec{p}}, H] \rangle_t. \tag{24}$$

From Eq. (24), using the Hamiltonian in Eq. (23) and realizing the calculations, we obtain the quantum kinetic equation for the confined electrons in a cylindrical quantum wire. Using the first-order tautology approximation method to solve this equation, we obtain the expression of electron distribution function in cylindrical quantum wires, $n_{n,\ell,\vec{p}}(t)$:

$$n_{n,\ell,\vec{p}}(t) = -\sum_{\vec{q},n',\ell'} |C_{\vec{q}}|^2 |I_{n,\ell,n',\ell'}|^2 \sum_{k,l=-\infty}^{\infty} J_k\Big(\frac{e\vec{E}_0,\vec{q}}{m\Omega^2}\Big) J_{k+l}\Big(\frac{e\vec{E}_0,\vec{q}}{m\Omega^2}\Big) \frac{1}{l\Omega} e^{-il\Omega t} \times$$
$$\times \Bigg\{ -\frac{\bar{n}_{n,\ell,\vec{p}}(N_{\vec{q}}+1) - \bar{n}_{n',\ell',\vec{p}+\vec{q}} N_{\vec{q}}}{\varepsilon_{n',\ell',\vec{p}+\vec{q}} - \varepsilon_{n,\ell,\vec{p}} + \omega_{\vec{q}} - k\Omega + i\delta} - \frac{\bar{n}_{n,\ell,\vec{p}} N_{\vec{q}} - \bar{n}_{n',\ell',\vec{p}+\vec{q}}(N_{\vec{q}}+1)}{\varepsilon_{n',\ell',\vec{p}+\vec{q}} - \varepsilon_{n,\ell,\vec{p}} - \omega_{\vec{q}} - k\Omega + i\delta} +$$
$$+ \frac{\bar{n}_{n',\ell',\vec{p}-\vec{q}}(N_{\vec{q}}+1) - \bar{n}_{n,\ell,\vec{p}} N_{\vec{q}}}{\varepsilon_{n,\ell,\vec{p}} - \varepsilon_{n',\ell',\vec{p}-\vec{q}} + \omega_{\vec{q}} - k\Omega + i\delta} + \frac{\bar{n}_{n',\ell',\vec{p}-\vec{q}} N_{\vec{q}} - \bar{n}_{n,\ell,\vec{p}}(N_{\vec{q}}+1)}{\varepsilon_{n,\ell,\vec{p}} - \varepsilon_{n',\ell',\vec{p}-\vec{q}} - \omega_{\vec{q}} - k\Omega + i\delta} \Bigg\}, \tag{25}$$

where $N_{\vec{q}}$ ($\bar{n}_{n,\vec{p}}$) is the time-independent component of the phonon (electron) distribution function, $J_k(x)$ is the Bessel function, and the quantity δ is infinitesimal and appears due to the assumption of an adiabatic interaction of the electromagnetic wave. Eq.(25) also can be considered a general expression of the electron distribution function in quantum wires.

4.2 Calculations of the nonlinear absorption coefficient of a strong electromagnetic wave by confined electrons in a cylindrical quantum wire

We consider a wire of GaAs with a circular cross section with a radius R and a length L_z embedded in AlAs. The carriers (electrons) are assumed to be confined by infinite potential barriers and to be free along the wire's axis (Oz). It is noted that a cylindrical quantum wire with radius $R \sim 160$ Å has already been fabricated experimentally. In this case, the total wave function of electrons in cylindrical coordinates (r,ϕ,z) takes the form (Zakhleniuk et al., 1996)

$$\psi_{n,\ell,\vec{p}}(r,\phi,z) = \frac{1}{\sqrt{V_0}} e^{in\phi} e^{ip_z z} \psi_{n,\ell}(r), \qquad r < R, \tag{26}$$

where $V_0 = \pi R^2 L_z$ is the wire volume, $n = 0, \pm 1 \pm 2, ...$ is the azimuthal quantum number, $\ell = 1, 2, 3, ...$ is the radial quantum number, $\vec{p} = (0, 0, p_z)$ is the electron wave vector (along the wire's z axis), and $\psi_{n,\ell}(r)$ is the wave function of electron moving in the (x,y) plane and takes the form

$$\psi_{n,\ell}(r) = \frac{1}{J_{n+1}(B_{n,\ell})} J_n(B_{n,\ell}\frac{r}{R}), \tag{27}$$

with $B_{n,\ell}$ being the ℓ-th root of the n-th order Bessel function, corresponding to the equation $J_n(B_{n,\ell}) = 0$, for example, $B_{01} = 2.405$ and $B_{11} = 3.832$. The electron energy spectrum takes the form [18]

$$\varepsilon_{n,\ell}(\vec{p}) = \varepsilon(p_z) + \varepsilon_{n,\ell}, \tag{28}$$

where $\varepsilon(p_z) = p_z^2/2m$ is the electron kinetic energy in the z-direction and $\varepsilon_{n,\ell} = B^2_{n,\ell}/2mR^2$ is the quantized energy in the other directions, m is the effective mass of the electron.

The electron form factor can be written as (Wang & Lei, 1994)

$$I_{n,\ell,n'\ell'}(q_\perp) = \frac{2}{R^2}\int_0^R J_{|n-n'|}(q_\perp R)\psi^*_{n',\ell'}(r)\psi_{n,\ell}(q_\perp R)rdr. \quad (29)$$

Due to the complexity of the expression for the radial function in Eq. (26), the integral in Eq. (29) cannot be calculated analytically. However, according to (Gold & Ghazali, 1990), it can be calculated for ground states of electrons by applying the approximate expression for the wave function and for the energies of states: namely,

$$\psi_{0,1} \approx \sqrt{3}(1-x^2), \qquad I_{01,0,1}(q_\perp) = 24\frac{J_3(q_\perp R)}{(q_\perp R)^3} \quad (30)$$

$$\psi_{\pm 1,1} \approx \sqrt{12}(x-x^2), \qquad I_{\pm 1,1,0,1}(q_\perp) = 48\frac{J_4(q_\perp R)}{(q_\perp R)^3}. \quad (31)$$

The carrier current density $\vec{j}(t)$ and the nonlinear absorption coefficient of a strong electromagnetic wave α take the form

$$\vec{j}(t) = \frac{e}{m}\sum_{n,\ell,\vec{p}}(\vec{p} - \frac{e}{c}\vec{A}(t))n_{n,\ell,\vec{p}}(t);\ \alpha = \frac{8\pi}{c\sqrt{\chi_\infty}E_0^2}\langle\vec{j}(t)\vec{E}_0 sin\Omega t\rangle_t, \quad (32)$$

In order to establish analytical expressions for the nonlinear absorption coefficient of a strong electromagnetic wave by confined electrons in a cylindrical quantum wire, We insert the expression for $n_{n,\ell,\vec{p}}(t)$ into the expression for $\vec{j}(t)$ and then insert the expression for $\vec{j}(t)$ into the expression for α in Eq.(32). Using the properties of Bessel function and realizing the calculations, we obtain the nonlinear absorption coefficient of a strong electromagnetic wave by confined electrons in a cylindrical quantum wire as

$$\alpha = \frac{8\pi^2\Omega}{c\sqrt{\chi_\infty}E_0^2}\sum_{n,\ell,n',\ell'}|I_{n,\ell,n',\ell'}|^2\sum_{\vec{q},\vec{p}}|C_{\vec{q}}|^2 N_{\vec{q}}\sum_{k=-\infty}^{\infty}\left[\bar{n}_{n,\ell,\vec{p}} - \bar{n}_{n',\ell',\vec{p}+\vec{q}}\right]\times$$
$$\times kJ_k^2\left(\frac{eE_0\vec{q}}{m\Omega^2}\right)\{\delta(\frac{(\vec{p}+\vec{q})^2}{2m} + \frac{B^2_{n',\ell'}}{2mR^2} - \frac{\vec{p}^2}{2m} + \frac{B^2_{n,\ell}}{2mR^2} + \omega_{\vec{q}} - k\Omega) + [\omega_{\vec{q}} \to -\omega_{\vec{q}}]\}, \quad (33)$$

where $\delta(x)$ is the Dirac delta function.

4.2.1 Electron-acoustic phonon scattering

Using the electron - acoustic phonon interaction factor $C^{ac}_{\vec{q}}$, the time-independent component of the electron distribution function $n_{n,\vec{p}_\perp}$ the Bessel function and the energy spectrum of an electron in a cylindrical quantum wire, we obtain an explicit expression for α in a cylindrical quantum wire for the case of electron-acoustic phonon scattering:

$$\alpha = \frac{\sqrt{2m\pi}e^2 n_0^*\xi^2(k_bT)^{5/2}}{4c\sqrt{\chi_\infty}\rho v_s^2\Omega^3 V}\sum_{n,\ell,n',\ell'}|I_{n,\ell,n',\ell'}|^2 exp\left\{\frac{1}{2k_bT}D_1\right\}\left[1 - exp\{\frac{\Omega}{k_bT}\}\right]\times$$
$$\times\frac{D_1}{2k_bT}\left[1 + \frac{3e^2E_0^2(k_bT)^2}{4m\Omega^4 D_1}\left(\frac{D_1^2}{4(k_bT)^2} + \frac{3D_1}{4k_bT} + 3\right)\right], \quad (34)$$

where $D_1 = (B^2_{n'\ell'} - B^2_{n,\ell})/2mR^2 - \Omega$.

4.2.2 Electron-optical Phonon Scattering

By using the electron - optical phonon interaction factor $C_{\vec{q}}^{op}$, the Bessel function and the time-independent component of the electron distribution function $n_{n,\vec{p}_\perp}$, from the general expression for the nonlinear absorption coefficient of a strong electromagnetic wave in a quantum well (Eq.33), we obtain the explicit expression for α in a cylindrical quantum wire for the case of electron-optical phonon scattering:

$$\alpha = \frac{\sqrt{2\pi} e^4 n_0^*(k_bT)^{3/2}}{4c\epsilon_0\sqrt{m\chi_\infty}\Omega^3 V}\Big(\frac{1}{\chi_\infty}-\frac{1}{\chi_0}\Big)\sum_{n\ell,\acute{n},\acute{\ell}}|I_{n\ell,\acute{n},\acute{\ell}}|^2\Big\{\Big[\Big[exp\Big\{\frac{1}{k_bT}(\omega_0-\Omega)\Big\}-1\Big]\times$$

$$\times\, exp\Big\{\frac{1}{k_bT}B_1\Big\}\Big[1+\frac{3e^2E_0^2k_bT}{8m\Omega^4}(1+\frac{B_1}{2k_bT})\Big]\Big]+[\omega_0\to-\omega_0]\Big\}, \quad (35)$$

where $B_1 = (B^2_{n'\ell'} - B^2_{n,\ell})/2mR^2 + \omega_0 - \Omega$.

From the analytic expressions for the nonlinear absorption coefficient of a strong electromagnetic wave caused by confined electrons in a cylindrical quantum wire with an infinite potential (Eq. 34 and Eq. 35), we can see that when the term proportional to the quadratic intensity of the electromagnetic wave (E_0^2) tends to zero, the nonlinear result will become a linear result (Bau et al., 2007).

4.3 Numerical results and discussions

In order to clarify the results that have been obtained, in this section, we numerically calculate the nonlinear absorption coefficient of a strong electromagnetic wave for a $GaAs/GaAsAl$ cylindrical quantum wire. The nonlinear absorption coefficient is considered as a function of the intensity E_0 and energy of strong electromagnetic wave, the temperature T of the system, the radius R of cylindrical quantum wire. The parameters used in the numerical calculations (Ariza-Flores & Rodriguez-Vargas, 2008; Bau et al., 2002) are $\xi = 13.5eV$, $\rho = 5.32gcm^{-3}$, $v_s = 5378ms^{-1}$, ϵ_0=12.5, $\chi_\infty = 10.9$, $\chi_0 = 13.1$, $m = 0.066m_0$, m_0 being the mass of free electron, $\hbar\omega_0 = 36.25meV$, $k_b = 1.3807\times10^{-23}j/K$, $n_0 = 10^{23}m^{-3}$, $e = 1.60219\times10^{-19}C$, $\hbar = 1.05459\times 10^{-34}j.s.$

4.3.1 Electron-acoustic phonon scattering

Figure 9 shows the dependence of the nonlinear absorption coefficient of a strong electromagnetic wave on the wire's radius at different values of the intensity, E0, of the electromagnetic wave. It can be seen from this figure that the absorption coefficient depends strongly and nonlinearly on the radius R of the wire. The absorption has the same maximum values (peaks), but with different values of the radius of the wire. For example, at $E0 = 1{,}6\times10^6(V/m)$ and $E0 = 3.6\times10^6(V/m)$, the peaks correspond to $R\sim 23nm$ and $R\sim 28nm$ respectively. The absorption coefficient has negative values, which was seen in the case linear absorption (Bau et al., 2007) and is the difference between quantum wires and bulk semiconductors (Pavlovich & Epshtein, 1977) as well as quantum wells and doped duperlattices. α was changed strongly by the confinement of electron in a cylindrical quantum wire.

The Figure 10 presents the dependence of the nonlinear absorption coefficient α on the electromagnetic wave energy at different values of the wire's radius R. It is seen that different from the normal bulk semiconductors(Pavlovich & Epshtein, 1977) and two-dimensional systems, the nonlinear absorption coefficient α in quantum wire has the maximum values

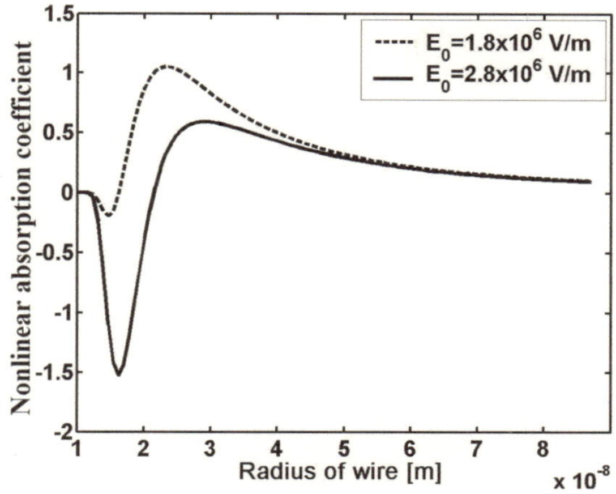

Fig. 9. The dependence of α on R in a cylindrical quantum wire (electron-acoustic phonon scattering)

(peaks). The electromagnetic wave energy at which α has a maximum are changed as the radius R of wire is varied.

Figure 11 shows the dependence of the nonlinear absorption coefficient α on the temperature T of the system at different values of the wire's radius R. It can be seen from this figure that the nonlinear absorption coefficient α has depends strongly and nonlinear on T. The nonlinear absorption coefficient α has the same maximum value, but with different values of T. For example, at $R = 15nn$ and $R = 25nn$, the peaks correspond to $T \sim 135K$ and $120K$, respectively, it is also a difference compared to the normal bulk semiconductors (Pavlovich & Epshtein, 1977), quantum wells and doped superlattices . To start from the maximum value, the nonlinear absorption coefficient α decreases when the temperature T rises.

Figure 12 presents the dependence of the nonlinear absorption coefficient α on the intensity E_0 of electromagnetic wave. This dependence shows that the nonlinear absorption coefficient α is descending when the intensity E_0 of electromagnetic wave increases. Different from

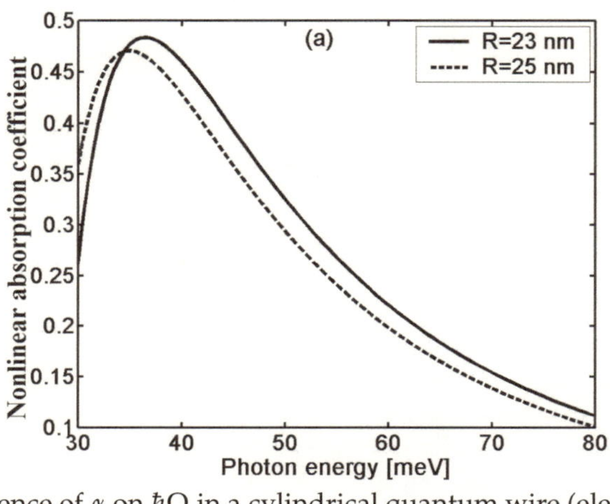

Fig. 10. The dependence of α on $\hbar\Omega$ in a cylindrical quantum wire (electron-acoustic phonon scattering)

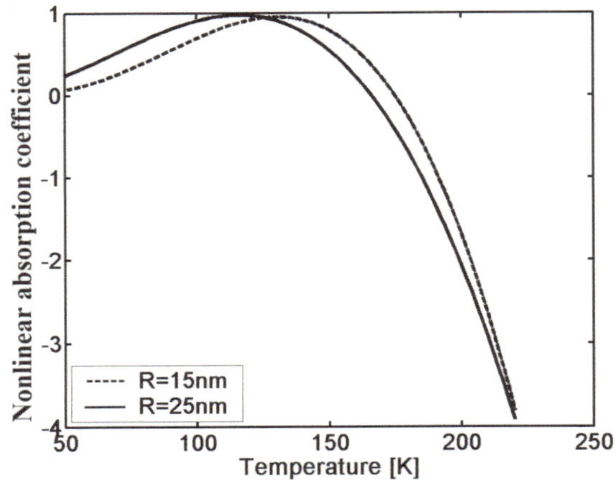

Fig. 11. The dependence of α on T in a cylindrical quantum wire (electron-acoustic phonon scattering)

normal bulk semiconduction (Pavlovich & Epshtein, 1977) and two-dimensional systems, the nonlinear absorption coefficient α in quantum wire is bigger. This is explained that when electrons are confined in quantum wire, the electron energy spectrum continue to be quantized. So the absorption of a strong electromagnetic wave is better. This fact is also reflected in the expressions of the nonlinear absorption coefficient (Eqs 34-35). Besides the sum over quantum n (as in quantum well), the expressions of the nonlinear absorption coefficient in quantum wire have the sum over the quantum number ℓ.

4.3.2 Electron-optical phonon scattering

Figures 13 shows the dependence of α on the radius R of wires in the case electron- optical phonon scattering. It can be seen from this figure that like in the case electron- acoustic phonon scattering, the nonlinear absorption coefficient α has the peak. But the absorption coefficient

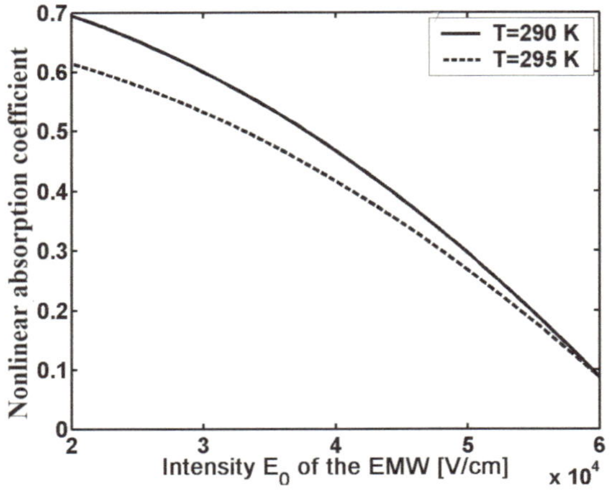

Fig. 12. The ependence of α on E_0 in a cylindrical quantum wire (electron-acoustic phonon scattering)

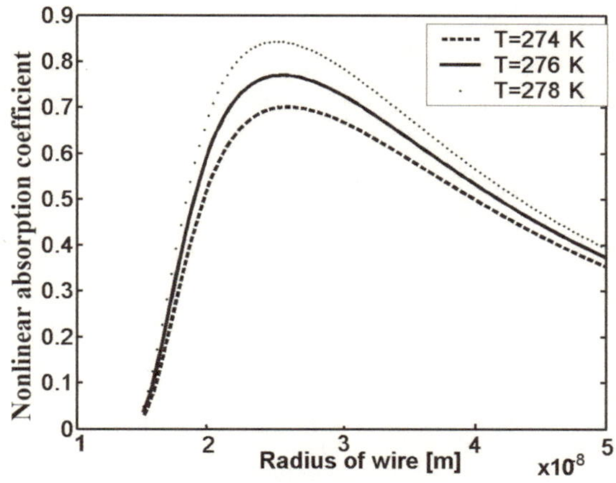

Fig. 13. The dependence of α on radius R in a cylindrical quantum wire (electron-optical phonon scattering)

does not have not negative values. Figure 14 presents the dependence of α on the intensity E_0 of electromagnetic wave. Different from the case electron - acoustic phonon scattering, in this case, α increases when the intensity E_0 of electromagnetic wave increases. Figure 15 presents the dependence of α on the electromagnetic wave energy at different values of the radius of wire. It is seen that α has the same maximum values (peaks) at $\Omega \equiv \omega$. The electromagnetic wave energy at which α has a maximum are not changed as the radius of wire is varied. This means that α depends strongly on the frequency Ω of the electromagnetic wave and resonance conditions are determined by the electromagnetic wave energy.

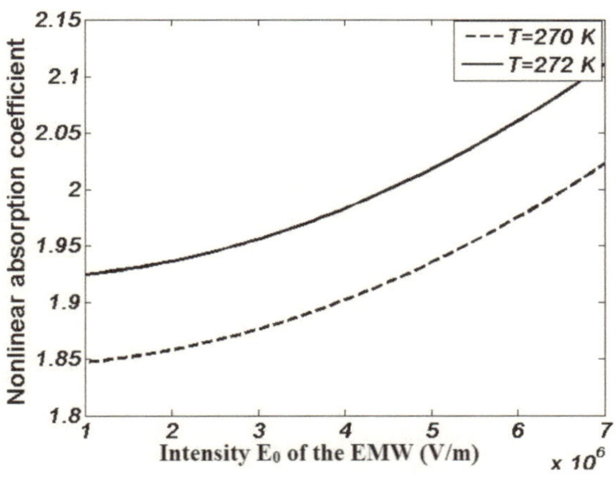

Fig. 14. The dependence of α on the intensity E_0 in a cylindrical quantum wire (electron-optical phonon scattering)

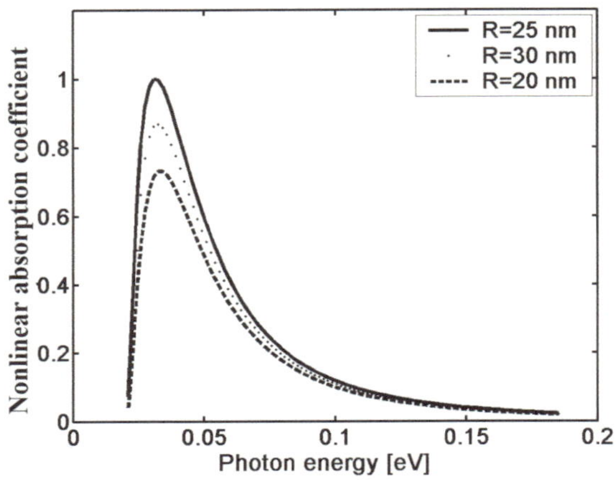

Fig. 15. The dependence of α on $\hbar\Omega$ in a cylindrical quantum wire (electron-optical phonon scattering)

5. The nonlinear absorption of a strong electromagnetic wave by confined electrons in a rectangular quantum wire

5.1 Calculations of the nonlinear absorption coefficient of a strong electromagnetic wave by confined electrons in a rectangular quantum wire

In our model, we consider a wire of GaAs with rectangular cross section ($Lx \times Ly$) and length Lz, embedded in GaAlAs. The carriers (electron gas) are assumed to be confined by an infinite potential in the (x,y) plane and are free in the z direction in Cartesian coordinates (x,y,z). The laser field propagates along the x direction. In this case, the state and the electron energy spectra have the form (Mickevicius & Mitin, 1993)

$$|n,\ell,\vec{p}\rangle = \frac{2e^{ip_z z}}{\sqrt{L_z L_x L_y}} sin\Big(\frac{\pi n x}{L_x}\Big) sin\Big(\frac{\pi \ell y}{L_y}\Big);\ \varepsilon_{n,\ell}(\vec{p}) = \frac{p^2}{2m} + \frac{\pi^2}{2m}\Big(\frac{n^2}{L_x^2} + \frac{\ell^2}{L_y^2}\Big) \quad (36)$$

where n and ℓ (n, ℓ=1, 2, 3, ...) denote the quantization of the energy spectrum in the x and y direction, $\vec{p} = (0,0,p_z)$ is the electron wave vector (along the wire's z axis), m is the effective mass of electron. The electron form factor, it is written as (Mickevicius & Mitin, 1993)

$$I_{n,l,\acute{n},\acute{l}}(\vec{q}) = \frac{32\pi^4(q_x L_x n\acute{n})^2(1-(-1)^{n+\acute{n}}cos(q_x L_x))}{[(q_x L_x)^4 - 2\pi^2(q_x L_x)^2(n^2+\acute{n}^2) + \pi^4(n^2-\acute{n}^2)^2]^2} \times \frac{32\pi^4(q_y L_y \ell\acute{\ell})^2(1-(-1)^{\ell+\acute{\ell}}cos(q_y L_y))}{[(q_y L_y)^4 - 2\pi^2(q_y L_y)^2(\ell^2+\acute{\ell}^2) + \pi^4(\ell^2-\acute{\ell}^2)^2]^2} \quad (37)$$

In order to establish analytical expressions for the nonlinear absorption coefficient of a strong electromagnetic wave by confined electrons in a rectangular quantum wire, we insert the expression of $n_{n,\ell,\vec{p}}(t)$ into the expression of $\vec{j}(t)$ and then insert the expression of $\vec{j}(t)$ into the expression of α. Using properties of Bessel function and realizing calculations, we obtain the nonlinear absorption coefficient of a strong electromagnetic wave by confined electrons in a rectangular quantum wire.

$$\alpha = \frac{8\pi^2\Omega}{c\sqrt{\chi_\infty}E_0^2}\sum_{n,\ell,n',\ell'}|I_{n,\ell,n',\ell'}|^2\sum_{\vec{q},\vec{p}}|C_{\vec{q}}|^2 N_{\vec{q}}\sum_{k=-\infty}^{\infty}\left[\bar{n}_{n,\ell,\vec{p}} - \bar{n}_{n',\ell',\vec{p}+\vec{q}}\right]\times$$
$$\times kJ_k^2\Big(\frac{eE_0\vec{q}}{m\Omega^2}\Big)\{\delta(\frac{(\vec{p}+\vec{q})^2-\vec{p}^2}{2m}+\frac{\pi^2}{2m}\Big(\frac{n'^2-n^2}{L_x^2}+\frac{\ell'^2-\ell^2}{L_y^2}\Big)+\omega_{\vec{q}}-k\Omega)+[\omega_{\vec{q}}\to-\omega_{\vec{q}}]\} \tag{38}$$

In the following, we study the problem with different electron-phonon scattering mechanisms.

5.1.1 Electron-optical phonon scattering

In this case, $\omega_{\vec{q}} \equiv \omega_0$ is the frequency of the optical phonon in the equilibrium state. Using he electron-optical phonon interaction constants $C_{\vec{q}}^{op}$, Bessel function and Fermi-Dirac distribution function for electron, we obtain the explicit expression of α in a rectangular quantum wire for the case electron-optical phonon scattering

$$\alpha = \frac{\sqrt{2\pi}e^4 n_0^*(k_bT)^{3/2}}{4c\epsilon_0\sqrt{m\chi_\infty}\Omega^3 V}\Big(\frac{1}{\chi_\infty}-\frac{1}{\chi_0}\Big)\sum_{n\ell,\acute{n},\acute{\ell}}|I_{n\ell,\acute{n},\acute{\ell}}|^2\Big\{\Big[\Big[exp\Big\{\frac{1}{k_bT}(\omega_0-\Omega)\Big\}-1\Big]\times$$
$$\times exp\Big\{\frac{1}{k_bT}\frac{\pi^2}{2m}\Big(\frac{\acute{n}^2}{L_x^2}+\frac{\acute{\ell}^2}{L_y^2}\Big)\Big\}\Big[1+\frac{3e^2E_0^2k_bT}{8m\Omega^4}(1+\frac{B}{2k_bT})\Big]\Big]+[\omega_o\to-\omega_0]\Big\} \tag{39}$$

where $B = \pi^2[(\acute{n}^2-n^2)/L_x^2+(\acute{\ell}^2-\ell^2)/L_y^2]/2m+\omega_0-\Omega$.

5.1.2 Electron- acoustic phonon scattering

In the case, $\omega_{\vec{q}} \ll \Omega$ ($\omega_{\vec{q}}$ is the frequency of acoustic phonon), so we let it pass. Using electron-acoustic phonon interaction constants $C_{\vec{q}}^{ac}$, we obtain the explicit expression of α in a rectangular quantum wire for the case electron-acoustic phonon scattering

$$\alpha = \frac{\sqrt{2m\pi}e^2 n_0^*\xi^2(k_bT)^{5/2}}{4c\sqrt{\chi_\infty}\rho v_s^2\Omega^3 V}\sum_{n,\ell,n',\ell'}|I_{n,\ell,n',\ell'}|^2 exp\Big\{\frac{1}{k_bT}\frac{\pi^2}{2m}\Big(\frac{\acute{n}^2}{L_x^2}+\frac{\acute{\ell}^2}{L_y^2}\Big)\Big\}\times$$
$$\times\Big[exp\{\frac{\Omega}{k_bT}\}-1\Big]\Big[1+\frac{D}{2k_bT}\Big[1+\frac{3e^2E_0^2(k_bT)^2}{4m\Omega^4 D}\Big(\frac{D^2}{4(k_bT)^2}+\frac{3D}{4k_bT}+3\Big)\Big]\Big] \tag{40}$$

where $D = \pi^2[(\acute{n}^2-n^2)/L_x^2+(\acute{\ell}^2-\ell^2)/L_y^2]-\Omega$

5.2 Numerical results and discussions

In order to clarify the results that have been obtained, in this section, we numerically calculate the nonlinear absorption coefficient of a strong electromagnetic wave for a $GaAs/GaAsAl$ rectangular quantum wire. The nonlinear absorption coefficient is considered as a function of the intensity E_0 and energy of strong electromagnetic wave, the temperature T of the system , and the parameters of a rectangular quantum wire.

Figure 16 shows the dependence of α of a strong electromagnetic wave on the size L (L_y and L_x) of wire. It can be seen from this figure that α depends strongly and nonlinear on size L of

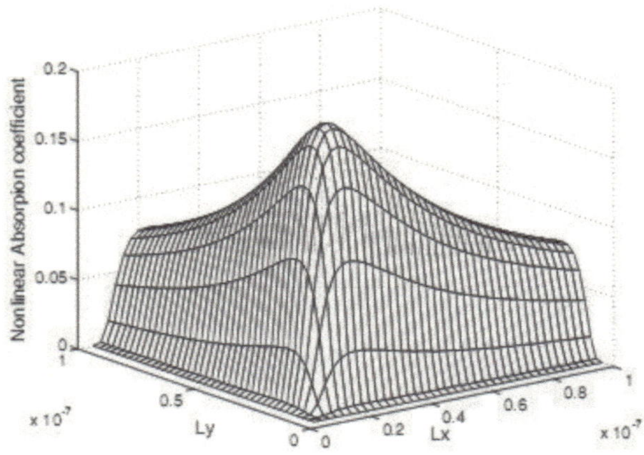

Fig. 16. The dependence of α on L_y and L_x in a rectangular quantum wire (electron-acoustic phonon scattering)

wire. When L decreases, the nonlinear absorption coefficient will increases until its maximum at L_x and $L_y \sim 24nm$ then started to decrease.

Figure 17 presents the dependence of the nonlinear absorption coefficient α on the temperature T of the system at different values of the intensity E_0 of the external strong electromagnetic wave. It can be seen from this figure that the nonlinear absorption coefficient α has depends strongly and nonlinearly on the temperature T and it has the same maximum value but with different values of T. For example, at $E_0 = 2.6 \times 10^6 V/m$ and $E_0 = 2 \times 10^6 V/m$, the peaks correspond to $T \sim 170K$ and $190K$, respectively, this fact was not seen in bulk semiconductors (Pavlovich & Epshtein, 1977) as well as quantum wells and doped superlattices, but it fit the case of linear absorption (Bau et al., 2007)

Figure 18 presents the dependence of α on the electromagnetic wave energy at different values of the radius of wire. It is seen that α has the same maximum values (peaks) at $\Omega \equiv \omega_0$. The electromagnetic wave energy at which α has a maximum are not changed as the radius of wire

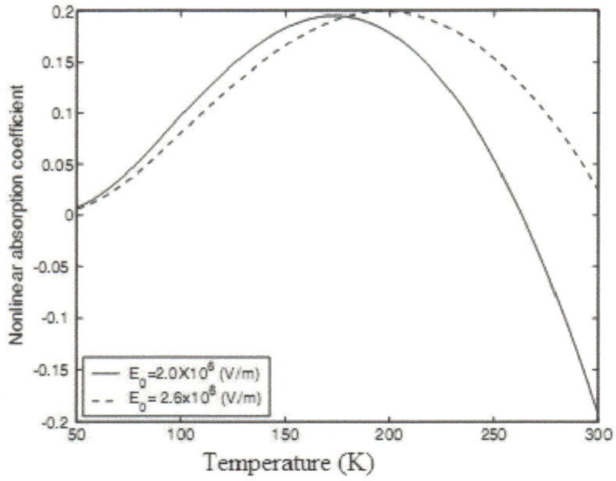

Fig. 17. The dependence of α on T in a rectangular quantum wire (electron-acoustic phonon scattering)

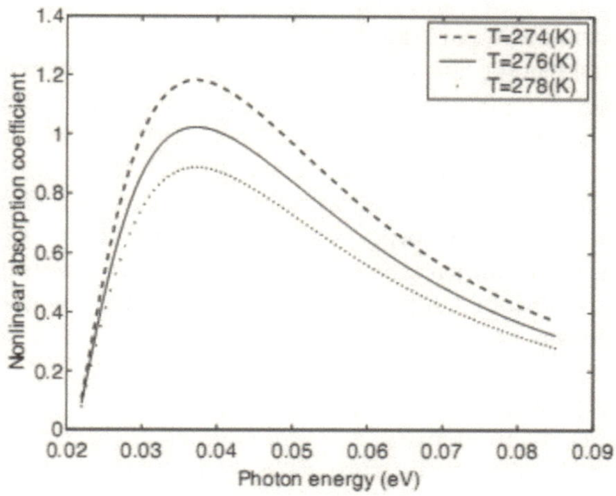

Fig. 18. The dependence of α on $\hbar\Omega$ in a rectangular quantum wire (electron-optical phonon scattering)

is varied. This means that α depends strongly on the frequency Ω of the electromagnetic wave and resonance conditions are determined by the electromagnetic wave energy.

6. Conclusion

In this chapter, the nonlinear absorption of a strong electromagnetic wave by confined electrons in low-dimensional systems is investigated. By using the method of the quantum kinetic equation for electrons, the expressions for the electron distribution function and the nonlinear absorption coefficient in quantum wells, doped superlattics, cylindrical quantum wires and rectangular quantum wires are obtained. The analytic results show that the nonlinear absorption coefficient depends on the intensity E_0 and the frequency Ω of the external strong electromagnetic wave, the temperature T of the system and the parameters of the low-dimensional systems as the width L of quantum well, the doping concentration n_D in doped superlattices, the radius R of cylindrical quantum wires, size L_x and L_y of rectangular quantum wires. This dependence are complex and has difference from those obtained in normal bulk semiconductors (Pavlovich & Epshtein, 1977), the expressions for the nonlinear absorption coefficient has the sum over the quantum number n (in quantum wells and doped superlattices) or the sum over two quantum numbers n and ℓ (in quantum wires). It shows that the electron confinement in low dimensional systems has changed significantly the nonlinear absorption coefficient. In addition, from the analytic results, we see that when the term in proportion to a quadratic in the intensity of the electromagnetic wave (E_0^2)(in the expressions for the nonlinear absorption coefficient of a strong electromagnetic wave) tend toward zero, the nonlinear result will turn back to a linear result (Bau & Phong, 1998; Bau et al., 2002; 2007). The numerical results obtained for a AlAs/GaAs/AlAs quantum well, a n-GaAs/p-GaAs doped superlattice, a $GaAs/GaAsAl$ cylindrical quantum wire and a a $GaAs/GaAsAl$ rectangular quantum wire show that α depends strongly and nonlinearly on the intensity E_0 and the frequency Ω of the external strong electromagnetic wave, the temperature T of the system, the parameters of the low-dimensional systems. In particular, there are differences between the nonlinear absorption of a strong electromagnetic wave in low-dimensional systems and the nonlinear absorption of a strong electromagnetic wave in normal bulk semiconductors (Pavlovich & Epshtein, 1977), the nonlinear absorption coefficient in a low-dimensional systems has the same maximum values (peaks) at $\Omega \equiv \omega_0$, the electromagnetic wave energies at which α has maxima are not changed as other parameters is varied, the nonlinear absorption coefficient in a low-dimensional systems is bigger. The results show a geometrical dependence of α due to the confinement of electrons in low-dimensional systems. The nonlinear absorption in each low-dimensional systems is also different, for example, these absorption peaks in doped superlattices are sharper than those in quantum wells, the nonlinear absorption coefficient in quantum wires is bigger than those in quantum wells and doped superlattices,... It shows that the nonlinear absorption of a strong electromagnetic wave by confined electrons depends significantly on the structure of each low-dimensional systems.

However in this study we have not considered the effect of confined phonon in low-dimensional systems, the influence of external magnetic field (or a weak electromagnetic wave) on the nonlinear absorption of a strong electromagnetic wave. This is still open for further studying.

7. Acknowledgments

This work is completed with financial support from the Vietnam National Foundation for Science and Technology Development (NAFOSTED 103.01.18.09).

8. References

Antonyuk, V. B., Malshukov, A. G., Larsson, M. & Chao, K. A. (2004). *Phys. Rev. B* 69: 155308.
Ariza-Flores, A. D. & Rodriguez-Vargas, I. (2008). *PIER letters* 1: 159.
Bau, N. Q., Dinh, L. & Phong, T. C. (2007). *J. Korean. Phys. Soc* 51: 1325.
Bau, N. Q., Nhan, N. V. & Phong, T. C. (2002). *J. Korean. Phys. Soc* 41: 149.
Bau, N. Q. & Phong, T. C. (1998). *J. Phys. Soc. Japan* 67: 3875.
Butscher, S. & Knorr, A. (2006). *Phys. Rev. L* 97: 197401.
Epstein, E. M. (1975). *Sov. Communicacattion of HEE of USSR, Ser. Radio* 18: 785.
Gaggero-Sager, M. L., Moreno-Martinez, N., Rodriguez-Vargas, I., Perez-Alvarez, R., Grimalskyand, V. V. & Mora-Ramos, M. E. (2007). *PIERS Online* 3: 851.
Gold, A. & Ghazali, A. (1990). *Phys. Rev. B* 41: 7626.
Kim, K. W., Stroscio, M. A., Bhatt, A., Mickevicius, R. & Mitin, V. V. (1991). *Appl. Phys.* 70: 319.
Malevich, V. L. & Epstein, E. M. (1974). *Sov. Quantum Electronic* 1: 1468.
Mickevicius, R. & Mitin, V. (1993). *Phys. Rev. B* 48: 17194.
Mori, N. & Ando, T. (1989). *Phys. Rev. B* 40: 6175.
Pavlovich, V. V. & Epshtein, E. M. (1977). *Sov. Phys. Solid State* 19: 1760.
Rucker, H., Molinari, E. & Lugli, P. (1992). *Phys. Rev. B* 45: 6747.
Samuel, E. P. & Patil, D. S. (2008). *PIER letters* 1: 119.
Shmelev, G. M., Chaikovskii, L. A. & Bau, N. Q. (1978). *Sov. Phys. Tech. Semicond* 12: 1932.
Suzuki, A. (1992). *Phys. Rev. B* 45: 6731.
Vasilopoulos, P., Charbonneau, M. & Vlier, C. N. V. (1987). *Phys. Rev. B* 35: 1334.
Wang, X. F. & Lei, X. L. (1994). *Phys. Rev. B* 94: 4780.
Zakhleniuk, N. A., Bennett, C. R., Constantinou, N. C., Ridley, B. K. & Babiker, M. (1996). *Phys. Rev. B* 54: 17838.

23

Electromagnetic Waves Generated by Line Current Pulses

Andrei B. Utkin
INOV - INESC Inovação
Portugal

1. Introduction

Solving electromagnetic problems in which both the source current and the emanated wave have complicated, essentially nonsinusoidal structure is of paramount interest for many real-word applications including weaponry, communications, energy transportation, radar, and medicine (Harmuth, 1986; Fowler at al., 1990; Harmuth et al., 1999; Hernández-Figueroa et al., 2008). In this chapter we will focus on electromagnetic fields produced by source-current pulses moving along a straight line. The explicit space-time representation of these fields is important for investigation of man-made (Chen, 1988; Zhan & Qin, 1989) and natural (Master & Uman, 1984) travelling-wave radiators, such as line antennas and lightning strokes.

Traditional methods of solving the electromagnetic problems imply passing to the frequency domain via the temporal Fourier (Laplace) transform or introducing retarded potentials. However, the resulted spectra do not provide adequate description of the essentially finite-energy, space-time limited source-current pulses and radiated transient waves. Distributing jumps and singularities over the entire frequency domain, the spectral representations cannot depict explicitly the propagation of leading/trailing edges of the pulses and designate the electromagnetic-pulse support (the spatiotemporal region in which the wavefunction is nonzero). Using the retarded potentials is not an easy and straightforward technique even for the extremely simple cases, such as the wave generation by the rectangular current pulse – see, e.g., the analysis by Master & Uman (1983), re-examined by Rubinstein & Uman (1991). In the general case of the sources of non-trivial space-time structure, the integrand characterizing the entire field via retarded inputs can be derived relatively easily. In contrast, the definition of the limits of integration is intricate for any moving source: one must obtain these limits as solutions of a set of simultaneous inequalities, in which the observation time is bounded with the space coordinates and the radiator's parameters. The explicit solutions are thus difficult to obtain.

In the present analysis, another approach, named *incomplete separation of variables in the wave equation*, is introduced. It can be generally characterized by the following stages:

- The system of Maxwell's equations is reduced to a second-order partial differential equation (PDE) for the electric/magnetic field components, or potentials, or their derivatives.
- Then one or two spatial variables are separated using the expansions in terms of eigenfunctions or integral transforms, while one spatial variable and the temporal variable remain bounded, resulting in a second-order PDE of the hyperbolic type, which, in its turn, is solved using the Riemann method.

- Sometimes these solutions, being multiplied by known functions of the previously separated variables, result in the expressions of a clear physical meaning (nonsteady-state modes), and for these cases we have explicit description of the field in the space-time representation. When it is possible, we find the explicit solution harnessing the procedure that is inverse with respect to the separation of variables, summing up the expansions or doing the inverse integral transform. In this case the solution yields the space-time structure of the entire transient field rather than its modal expansion or integral representation.

2. Electromagnetic problem

As far as the line of the current motion is the axis of symmetry, it is convenient to consider the problem of wave generation in the cylindrical coordinate system ρ,φ,z, for which the direction of the z-axis coincides with the direction of the current-density vector, $\mathbf{j} = j\mathbf{e}_z$. Following the concept discussed above, we suppose that the space-time structure of the source corresponds to a finite-energy pulse turned on in some fixed moment of time. Introduction of the time variable in the form $\tau = ct$, where t is time reckoned from this moment and c is the speed of light, results in the conditions

$$j_z \equiv 0, \quad \mathbf{E} \equiv 0, \quad \mathbf{B} \equiv 0 \quad \text{for} \quad \tau < 0. \tag{1}$$

Here $\mathbf{E}$ and $\mathbf{B}$ conventionally denote the force-related electromagnetic field vectors – the electric field intensity and the magnetic induction. The current pulse is supposed to be generated at one of the radiator's ends, $z = 0$, to travel with constant front and back velocity $v = \beta c$ $(0 < \beta \le 1)$ along the radiator and to be completely absorbed at the other end, $z = l$, as illustrated in Fig. 1.

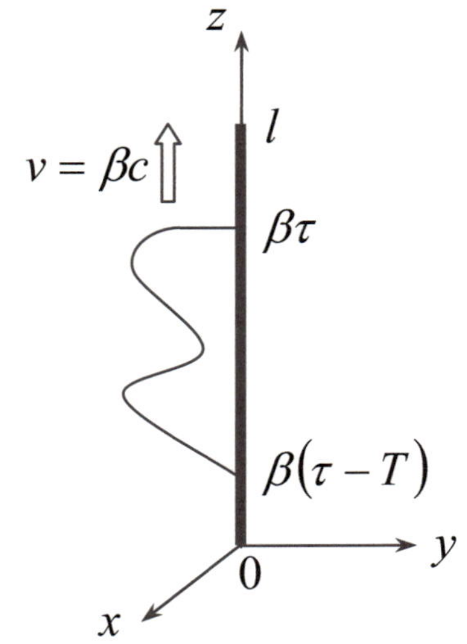

Fig. 1. Space-time structure of the source current.

Introducing, along with the finite radiator length l, the finite current pulse duration T, one can express the current density using the Dirac delta function $\delta(\rho)$ and the Heaviside step function $h(z)=\begin{cases}1 & \text{for} \quad z>0\\ 0 & \text{for} \quad z<0\end{cases}$ as

$$j(\rho,z,\tau)=\frac{\delta(\rho)}{2\pi\rho}J(z,\tau)h\left(\tau-\frac{z}{\beta}\right)h\left(\frac{z}{\beta}-\tau+T\right)h(z)h(l-z), \tag{2}$$

where $J(z,\tau)$ is an arbitrary continuous function describing the current distribution.
Bearing in mind the axial symmetry of the problem, let us seek the solution in the form of a TM wave whose components can be expressed via the Borgnis-Bromwich potential W (Whittaker, 1904; Bromwich, 1919) as

$$E_\rho=\frac{1}{\varepsilon_0 c}\frac{\partial^2 W}{\partial\rho\partial z},\quad E_z=\frac{1}{\varepsilon_0 c}\left(-\frac{\partial^2 W}{\partial\tau^2}+\frac{\partial^2 W}{\partial z^2}\right),\quad B_\phi=-\mu_0\frac{\partial^2 W}{\partial\rho\partial\tau}, \tag{3}$$

where ε_0 and μ_0 are the electric and magnetic constants. Substitution of representation (3) into the system of Maxwell's equations yields the scalar problem

$$\begin{gathered}\left(\frac{\partial^2}{\partial\tau^2}-\frac{\partial^2}{\partial z^2}-\frac{1}{\rho}\frac{\partial}{\partial\rho}\left(\rho\frac{\partial}{\partial\rho}\right)\right)\Psi(\rho,z,\tau)=j(\rho,z,\tau),\\ \Psi\equiv 0 \quad \text{for} \quad \tau<0\end{gathered} \tag{4}$$

with respect to the function $\Psi(\rho,z,\tau)=\dfrac{\partial W}{\partial\tau}$.

3. Solving algorithm

3.1 Transverse coordinate separation

Let us separate ρ by the Fourier-Bessel transform

$$\begin{pmatrix}\tilde{\Psi}(s,z,\tau)\\ \tilde{j}(s,z,\tau)\end{pmatrix}=\int_0^\infty\begin{pmatrix}\Psi(\rho,z,\tau)\\ j(\rho,z,\tau)\end{pmatrix}J_0(s\rho)\rho\,\mathrm{d}\rho,\quad \begin{pmatrix}\Psi(\rho,z,\tau)\\ j(\rho,z,\tau)\end{pmatrix}=\int_0^\infty\begin{pmatrix}\tilde{\Psi}(s,z,\tau)\\ \tilde{j}(s,z,\tau)\end{pmatrix}J_0(s\rho)s\,\mathrm{d}s, \tag{5}$$

(J_0 is the Bessel function of the fist kind of order zero) which turns problem (4) into one for the 1D Klein-Gordon equation

$$\left(\frac{\partial^2}{\partial\tau^2}-\frac{\partial^2}{\partial z^2}+s^2\right)\tilde{\Psi}(s,z,\tau)=\tilde{j}(s,z,\tau), \tag{6}$$

with the initial conditions

$$\tilde{\Psi}\equiv 0 \quad \text{for} \quad \tau<0, \tag{7}$$

where, in accordance with representation (5),

$$\tilde{j}(s,z,\tau)=\tilde{j}(z,\tau)=\frac{1}{2\pi}J(z,\tau)h\left(\tau-\frac{z}{\beta}\right)h\left(\frac{z}{\beta}-\tau+T\right)h(z)h(l-z). \tag{8}$$

3.2 Riemann (Riemann–Volterra) method

Problem (6) can easily be solved for arbitrary source function by the Riemann (also known as Riemann–Volterra) method. Although being very powerful, this method is scarcely discussed in the textbooks; a few considerations (see, for example, Courant & Hilbert, 1989) treat one and the same case related to the first canonical form of a more general equation

$$\hat{L}(a,b,c;u)=\left(\frac{\partial^2}{\partial\xi\partial\eta}+a(\xi,\eta)\frac{\partial}{\partial\xi}+b(\xi,\eta)\frac{\partial}{\partial\eta}+c(\xi,\eta)\right)u(\xi,\eta)=f(\xi,\eta),\quad a,b,c\in C^1, \tag{9}$$

aiming to represent the solution at a point $P(\xi_0,\eta_0)$ in terms of f and the values of u and its normal derivative $\frac{\partial u}{\partial n}$ on the initial-data curve Σ as depicted in Fig. 2(a).

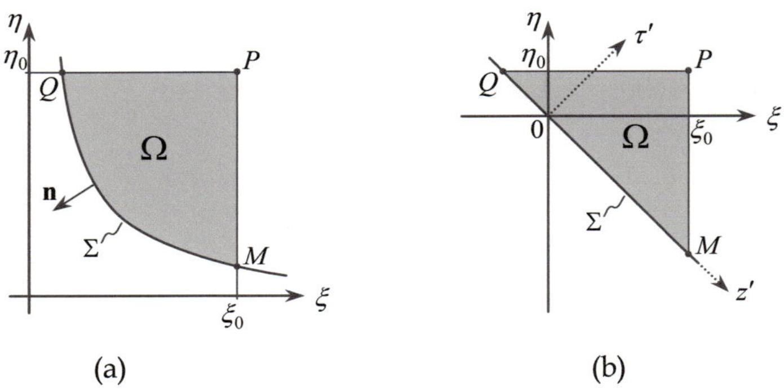

Fig. 2. Characteristic ξ,η diagrams representing the initial-data curve Σ and the integration domain Ω for the standard (a) and *ad hoc* (b) Riemann-method procedures.

As far as our objectives are limited to solving problem (6), (7), we will consider simplified *ad hoc* Riemann-method procedure involving the differential operator

$$\hat{L}(u)=\left(\frac{\partial^2}{\partial\xi\partial\eta}+s^2\right)u(\xi,\eta),\quad s=const\,, \tag{10}$$

and the extension of this procedure to the case of the second canonical form of the 1D Klein-Gordon equation (6). Corresponding diagram on the ξ,η plane is represented in Fig. 2(b); the initial data are defined on the straight line $\eta=-\xi$. The procedure is based on the fact that for any two functions u and R the difference $R\hat{L}(u)-u\hat{L}(R)$ is a divergence expression

$$\begin{gathered}R\hat{L}(u)-u\hat{L}(R)=R\frac{\partial^2 u}{\partial\xi\partial\eta}-u\frac{\partial^2 R}{\partial\xi\partial\eta}=\frac{1}{2}\left(\frac{\partial A_\eta}{\partial\xi}-\frac{\partial A_\xi}{\partial\eta}\right),\\ A_\xi=u\frac{\partial R}{\partial\xi}-R\frac{\partial u}{\partial\xi},\quad A_\eta=R\frac{\partial u}{\partial\eta}-u\frac{\partial R}{\partial\eta}.\end{gathered} \tag{11}$$

Thus, integrating over the domain Ω with boundary $\partial\Omega$, one obtains by the Gauss-Ostrogradski formula

$$I_\Omega \overset{def}{=} \iint_\Omega \left[R\hat{L}(u) - u\hat{L}(R)\right] \mathrm{d}\xi\, \mathrm{d}\eta = \frac{1}{2}\oint_{\partial\Omega} \left(A_\xi\, \mathrm{d}\xi + A_\eta\, \mathrm{d}\eta\right), \tag{12}$$

where the contour integration must be performed counterclockwise. Applying formula (12) to the particular case in which:

a. the integration domain Ω corresponds to that of Fig. 2(b);
b. the function u is the desired solution of the inhomogeneous equation

$$\hat{L}(u) = \left(\frac{\partial^2}{\partial\xi\partial\eta} + s^2\right) u(\xi,\eta) = f(\xi,\eta); \tag{13}$$

c. the function R is the Riemann function corresponding to the linear differential operator (10) and the observation point $P(\xi_0,\eta_0)$, that is

$$\hat{L}(R) = 0, \quad R\big|_{QP} = R\big|_{MP} = 1; \tag{14}$$

we have

$$I_\Omega = \iint_\Omega \left[R\hat{L}(u) - u\hat{L}(R)\right] \mathrm{d}\xi \mathrm{d}\eta = \iint_\Omega Rf\, \mathrm{d}\xi\, \mathrm{d}\eta\,. \tag{15}$$

On the other hand

$$\begin{aligned} I_\Omega &= \frac{1}{2}\oint_{\partial\Omega}\left[\left(u\frac{\partial R}{\partial\xi} - R\frac{\partial u}{\partial\xi}\right)\mathrm{d}\xi + \left(R\frac{\partial u}{\partial\eta} - u\frac{\partial R}{\partial\eta}\right)\mathrm{d}\eta\right] \\ &= \frac{1}{2}\int_{QM+MP+PQ}\left[\left(u\frac{\partial R}{\partial\xi} - R\frac{\partial u}{\partial\xi}\right)\mathrm{d}\xi + \left(R\frac{\partial u}{\partial\eta} - u\frac{\partial R}{\partial\eta}\right)\mathrm{d}\eta\right]. \end{aligned} \tag{16}$$

For the contour $\partial\Omega$ of Fig. 2(b) $\mathrm{d}\xi = 0$ on *MP* while $\mathrm{d}\eta = 0$ on *PQ* and $\mathrm{d}\eta = -\mathrm{d}\xi$ on *QM*, which reduces the integral to

$$\begin{aligned} I_\Omega &= \frac{1}{2}\int_{QM}\left(u\frac{\partial R}{\partial\xi} - R\frac{\partial u}{\partial\xi} - R\frac{\partial u}{\partial\eta} + u\frac{\partial R}{\partial\eta}\right)\mathrm{d}\xi \\ &+ \frac{1}{2}\int_{MP}\left(R\frac{\partial u}{\partial\eta} - u\frac{\partial R}{\partial\eta}\right)\mathrm{d}\eta + \frac{1}{2}\int_{PQ}\left(u\frac{\partial R}{\partial\xi} - R\frac{\partial u}{\partial\xi}\right)\mathrm{d}\xi. \end{aligned} \tag{17}$$

Noticing that

$$\frac{1}{2}\int_{MP}\left(R\frac{\partial u}{\partial\eta} - u\frac{\partial R}{\partial\eta}\right)\mathrm{d}\eta = \frac{1}{2}\int_{MP}\left[\frac{\partial}{\partial\eta}(Ru) - 2u\frac{\partial R}{\partial\eta}\right]\mathrm{d}\eta = \frac{1}{2}Ru\Big|_M^P - \int_{MP} u\frac{\partial R}{\partial\eta}\mathrm{d}\eta\,, \tag{18}$$

$$\frac{1}{2}\int_{PQ}\left(u\frac{\partial R}{\partial \xi}-R\frac{\partial u}{\partial \xi}\right)\mathrm{d}\xi=\frac{1}{2}\int_{QP}\left(R\frac{\partial u}{\partial \xi}-u\frac{\partial R}{\partial \xi}\right)\mathrm{d}\xi$$
$$=\frac{1}{2}\int_{QP}\left[\frac{\partial}{\partial \xi}(Ru)-2u\frac{\partial R}{\partial \xi}\right]\mathrm{d}\xi=\frac{1}{2}Ru\Big|_Q^P-\int_{QP}u\frac{\partial R}{\partial \xi}\mathrm{d}\xi \tag{19}$$

and, due to the second of properties (14),

$$\left.\frac{\partial R}{\partial \eta}\right|_{MP}=\left.\frac{\partial R}{\partial \xi}\right|_{QP}=0,\quad R|_P=1, \tag{20}$$

one has

$$I_\Omega=-\frac{1}{2}\int_{QM}\left(R\frac{\partial u}{\partial \xi}+R\frac{\partial u}{\partial \eta}-u\frac{\partial R}{\partial \xi}-u\frac{\partial R}{\partial \eta}\right)\mathrm{d}\xi+u|_P-\frac{1}{2}\left(Ru|_Q+Ru|_M\right). \tag{21}$$

Substituting the LHS of Eq. (21) by the RHS of Eq. (15) and solving the resulting equation with respect to $u|_P$ yield the Riemann formula corresponding to operator (10)

$$u|_P=\frac{1}{2}\left(Ru|_Q+Ru|_M\right)+\frac{1}{2}\int_{QM}\left[R\left(\frac{\partial u}{\partial \xi}+\frac{\partial u}{\partial \eta}\right)-u\left(\frac{\partial R}{\partial \xi}+\frac{\partial R}{\partial \eta}\right)\right]\mathrm{d}\xi+\iint_\Omega Rf\,\mathrm{d}\xi\,\mathrm{d}\eta, \tag{22}$$

with the Riemann function (Courant & Hilbert, 1989)

$$R(\xi_0,\eta_0;\xi,\eta)=J_0\left(s\sqrt{4(\xi_0-\xi)(\eta_0-\eta)}\right). \tag{23}$$

To apply this result to problem (4), let us postulate that the variables ξ,ξ_0 and η,η_0 are related to the longitudinal-coordinate z,z' and time τ,τ' variables via the expressions

$$\xi=\frac{1}{2}(\tau'+z'),\quad \xi_0=\frac{1}{2}(\tau+z),\quad \eta=\frac{1}{2}(\tau'-z'),\quad \eta_0=\frac{1}{2}(\tau-z). \tag{24}$$

Axes corresponding to the variables z' and τ' are shown in Fig. 2(b) as dotted lines while the entire z',τ' diagram of the Riemann-method procedure is represented in Fig. 3.
In the new variables

$$R(z,\tau;z',\tau')=J_0\left(s\sqrt{(\tau-\tau')^2-(z-z')^2}\right), \tag{25}$$

$$\mathrm{d}\xi\,\mathrm{d}\eta=\frac{\partial(\xi,\eta)}{\partial(z',\tau')}\mathrm{d}z'\,\mathrm{d}\tau'=\frac{1}{2}\mathrm{d}z'\,\mathrm{d}\tau', \tag{26}$$

$$\frac{\partial}{\partial \xi}=\frac{\partial}{\partial z'}+\frac{\partial}{\partial \tau'},\quad \frac{\partial}{\partial \eta}=-\frac{\partial}{\partial z'}+\frac{\partial}{\partial \tau'}\quad\Rightarrow\quad \frac{\partial}{\partial \xi}+\frac{\partial}{\partial \eta}=2\frac{\partial}{\partial \tau'} \tag{27}$$

and the differential operator (10) takes the second canonical form

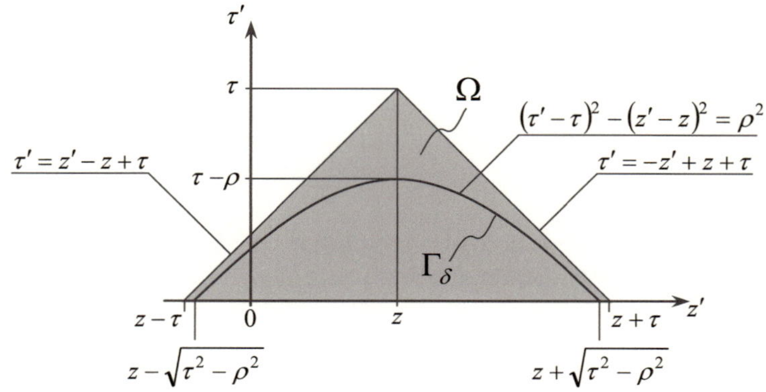

Fig. 3. A z',τ' plane diagram representing the initial 2D integration domain Ω eventually reduced to the segment of the hyperbola $(\tau'-\tau)^2-(z'-z)^2=\rho^2$, the support of kernel (36).

while on the integration segment QM

$$\hat{L}(\tilde{u})=\left(\frac{\partial^2}{\partial\tau^2}-\frac{\partial^2}{\partial z^2}+s^2\right)\tilde{u}(z,\tau),\quad \tilde{u}(z,\tau)=u(\xi_0,\tau_0)\Big|_{\xi_0=\frac{1}{2}(\tau+z),\,\eta_0=\frac{1}{2}(\tau-z)}, \tag{28}$$

$$\mathrm{d}\xi=\frac{1}{2}\mathrm{d}z'. \tag{29}$$

In view of (28)-(29), the Riemann formula for the second canonical form of the 1D Klein-Gordon equation reduces to

$$\begin{aligned}\Psi(z,\tau)=&\frac{1}{2}\left[R(z,\tau;z-\tau,0)\tilde{u}(z-\tau,0)+R(z,\tau;z+\tau,0)\tilde{u}(z+\tau,0)\right]\\&+\frac{1}{2}\int_{QM}\left[2R(z,\tau;z',0)\left.\frac{\partial\tilde{u}(z',\tau')}{\partial\tau'}\right|_{\tau'=0}-2\tilde{u}(z',0)\left.\frac{\partial R(z,\tau;z',\tau')}{\partial\tau'}\right|_{\tau'=0}\right]\frac{1}{2}\mathrm{d}z'\\&+\frac{1}{2}\iint_{\Omega}R(z,\tau;z',\tau')f(z',\tau')\mathrm{d}z'\mathrm{d}\tau',\end{aligned} \tag{30}$$

whose explicit representation for the problem

$$\left(\frac{\partial^2}{\partial\tau^2}-\frac{\partial^2}{\partial z^2}+s^2\right)\tilde{u}(z,\tau)=f(z,\tau),\quad \tilde{u}(z,0)=u_0(z),\quad \frac{\partial\tilde{u}}{\partial\tau}(z,0)=u_\tau(z) \tag{31}$$

is

$$\begin{aligned}\tilde{u}(z,\tau)=&\frac{1}{2}\left[R(z,\tau;z-\tau,0)u_0(z-\tau)+R(z,\tau;z+\tau,0)u_0(z+\tau)\right]\\&+\frac{1}{2}\int_{z-\tau}^{z+\tau}\left[R(z,\tau;z',0)u_\tau(z')-u_0(z')\frac{\partial R}{\partial\tau'}(z,\tau;z',0)\right]\mathrm{d}z'+\frac{1}{2}\int_{z-\tau}^{z+\tau}\int_{0}^{\tau-|z-z'|}R(z,\tau;z',\tau')f(z',\tau')\mathrm{d}\tau'\mathrm{d}z',\end{aligned} \tag{32}$$

where the Riemann function is defined by Eq. (25).

3.3 Space-time domain solution

In the particular case of problem (6)-(8) with the homogeneous initial conditions, the Riemann method yields

$$\tilde{\Psi}(s,z,\tau)=\frac{1}{2}\int_{z-\tau}^{z+\tau}\int_{0}^{\tau-|z-z'|} J_0\left(s\sqrt{(\tau-\tau')^2-(z-z')^2}\right)\tilde{j}(z',\tau')\mathrm{d}\tau'\mathrm{d}z'. \tag{33}$$

To obtain the explicit representation of the solution to the original problem (4), let us perform the inverse Fourier-Bessel transform (5)

$$\begin{aligned}\Psi(\rho,z,\tau)&=\int_0^\infty \tilde{\Psi}(s,z,\tau)J_0(s\rho)s\mathrm{d}s\\ &=\frac{1}{2}\int_0^\infty\left(\int_{z-\tau}^{z+\tau}\int_{0}^{\tau-|z-z'|} J_0\left(s\sqrt{(\tau-\tau')^2-(z-z')^2}\right)\tilde{j}(z',\tau')\mathrm{d}\tau'\mathrm{d}z'\right)J_0(s\rho)s\mathrm{d}s.\end{aligned} \tag{34}$$

Changing the order of integration, one gets the source-to-wave integral transform

$$\Psi(\rho,z,\tau)=\frac{1}{2}\int_{z-\tau}^{z+\tau}\int_{0}^{\tau-|z-z'|} K(\rho,z,\tau,z',\tau')\tilde{j}(z',\tau')\mathrm{d}\tau'\mathrm{d}z', \tag{35}$$

where

$$K(\rho,z,\tau,z',\tau')\overset{def}{=}\int_0^\infty J_0\left(s\sqrt{(\tau-\tau')^2-(z-z')^2}\right)J_0(s\rho)s\mathrm{d}s. \tag{36}$$

Crucial reduction of the integral wavefunction representation (35) can be achieved using the closure equation (Arfken & Weber, 2001, p. 691)

$$\int_0^\infty J_0(s\rho)J_0(s\rho')s\mathrm{d}s=\frac{1}{\rho}\delta(\rho-\rho'), \tag{37}$$

which enables kernel (36) to be represented in the form

$$K(\rho,z,\tau,z',\tau')=\frac{1}{\rho}\delta\left(\rho-\sqrt{(\tau-\tau')^2-(z-z')^2}\right). \tag{38}$$

A more explicit relationship can be obtained treating the kernel as a function of τ' and using the representation of the delta function with simple zeros $\{\tau_i\}$ on the real axis (Arfken & Weber, 2001, p. 87)

$$\delta(g(\tau'))=\sum_i\frac{\delta(\tau'-\tau_i)}{\left|\frac{\partial g}{\partial\tau'}(\tau_i)\right|}. \tag{39}$$

Two zeros must be taken into account

$$\tau_{1,2}=\tau\mp\sqrt{\rho^2+(z-z')^2}, \tag{40}$$

but $\tau_2 = \tau + \sqrt{\rho^2 + (z - z')^2} > \tau$ results in the delta function whose support always lies outside the integration domain and therefore corresponds to zero input. Thus we can write

$$K(\rho, z, \tau, z', \tau') = \frac{\delta\left(\tau' - \tau + \sqrt{\rho^2 + (z - z')^2}\right)}{\sqrt{\rho^2 + (z - z')^2}}, \tag{41}$$

which, together with Eq. (8), yields

$$\Psi(\rho, z, \tau) = \frac{1}{4\pi} \int_{z-\tau}^{z+\tau} \int_{0}^{\tau - |z - z'|} \frac{\delta(\tau' - \tau + r')}{r'} J(z', \tau') h\left(\tau' - \frac{z'}{\beta}\right) h\left(\frac{z'}{\beta} - \tau' + T\right) h(z') h(l - z') \mathrm{d}\tau' \mathrm{d}z', \tag{42}$$

where

$$r' = \sqrt{\rho^2 + (z - z')^2} \tag{43}$$

denotes the distance between the observation point ρ, z and the source location $0, z'$. The integration domain, now reduced to the inlying support of the delta function, a segment of the hyperbolic curve

$$\begin{aligned} \Gamma_\delta &\stackrel{def}{=} \Omega \cap \left\{ z', \tau' : (\tau' - \tau)^2 - (z' - z)^2 = \rho^2 \right\} \\ &= \left\{ z', \tau' : (\tau' - \tau)^2 - (z' - z)^2 = \rho^2,\ (\tau' - \tau) + (z' - z) < 0,\ \tau' > 0 \right\}, \end{aligned} \tag{44}$$

is shown in Fig. 3.

4. Explicit representations

4.1 Preliminary considerations

Formula (42) requires further examination in order to resolve inequalities implicitly introduced by the step functions in the integrand and obtain analyzable expressions.

Although it is possible to consider one-dimensional inequalities that bound only the longitudinal variable z' – just using the property of the delta function while performing integration with respect to τ' and passing to the single-integral relation

$$\Psi(\rho, z, \tau) = \frac{1}{4\pi} \int_{z - \sqrt{\tau^2 - \rho^2}}^{z + \sqrt{\tau^2 - \rho^2}} \frac{J(z', \tau - r')}{r'} h\left(\tau - r' - \frac{z'}{\beta}\right) h\left(\frac{z'}{\beta} - \tau + r' + T\right) h(z') h(l - z') \,\mathrm{d}z' \tag{45}$$

– a more convenient study can be done using basic expression (42) and the two-dimensional z', τ' plane diagrams, in which the inequalities bound both z' and τ', have a linear form, and admit illustrative graphical representation.

This study results in a set of particular expressions for certain interrelations between the spatiotemporal coordinates ρ, z, τ, the radiator length l, and the current pulse duration T.

For each observation point ρ, z, the set of expression may have one of two distinct forms, depending on what information reaches the observer first: one concerning the finiteness of the current pulse or one about the radiator finiteness. The finiteness of the current pulse comes into the scene at the spatiotemporal point $\rho = 0, z = 0, \tau = T$ (see Fig. 4). Corresponding information is carried by the back of the electromagnetic pulse with the speed of light and arrives at the point ρ, z

$$r = \sqrt{\rho^2 + z^2} \tag{46}$$

time units after, that is, at the moment $T + r$. The stopping of the source-current motion along the z axis due to the finiteness of the radiator is first manifested at $\rho = 0, z = l, \tau = l / \beta$. Related information is propagated through the distance

$$r_l = \sqrt{\rho^2 + (z - l)^2} \tag{47}$$

with the speed of light and reaches the point ρ, z at the moment $l / \beta + r_l$.
From here on the source current pulse will be called *short* provided that

$$r + T < r_l + \frac{l}{\beta} \tag{48}$$

and *long* in the opposite case. This definition depends on ρ, z, so a current pulse considered to be short for one observation point may appear as long for another, and vice versa.

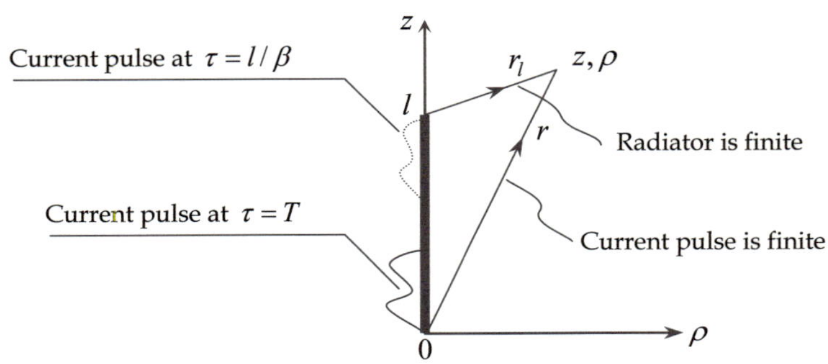

Fig. 4. On definition of the short and long pulse types.

4.2 Definition of the integration limits

The z', τ' plane diagrams for the case of a short source-current pulse are shown in Fig. 5. The step-function factors in formula (42) define the parallelogram area Ω_h within which the integrand differs from zero, and the eventual integration domain is the intersection of Ω_h and the segment of hyperbola Γ_δ defined by Eq. (44). Progression of the observation time τ unfolds the following concretization of the general formula:

- Case aS: $-\infty < \tau < r$, Fig. 5(a).

$\Gamma_\delta = \varnothing$ (for $\tau < \rho$ the hyperbola branch resides below $\tau' = 0$) or there is no intersection between Ω_h and Γ_δ, $\Omega_h \cap \Gamma_\delta = \varnothing$, so

$$\Psi(\rho,z,\tau)=\Psi_{aS}(\rho,z,\tau)=0. \tag{49}$$

This is in a complete accord with the casualty principle, as any effect of the light-speed-limited process initiated at the spatiotemporal point $\rho=0, z=0, \tau=0$ cannot reach the point ρ, z prior to $\tau=r$.

- Case bS: $r<\tau<r+T$, Fig. 5(b).

$\Omega_h \cap \Gamma_\delta$ is a segment of Γ_δ limited by $z=0$ and $z=z_0$, where

$$z_0=\beta\frac{\tau^2-r^2}{\tau-\beta z+\sqrt{(1-\beta^2)\rho^2+(z-\beta\tau)^2}} \tag{50}$$

is defined by the intersection of Γ_δ and the line $\tau'-z'/\beta=0$.

- Case cS: $r+T<\tau<r_l+l/\beta$, Fig. 5(c).

$\Omega_h \cap \Gamma_\delta$ is a segment of Γ_δ limited by $z=z_T$ and $z=z_0$, where

$$z_T=z_0\big|_{\tau\mapsto\tau-T}=\beta\frac{(\tau-T)^2-r^2}{\tau-T-\beta z+\sqrt{(1-\beta^2)\rho^2+\left[z-\beta(\tau-T)\right]^2}} \tag{51}$$

is defined by the intersection of Γ_δ and the line $z'/\beta-\tau'+T=0$.

- Case dS: $r_l+l/\beta<\tau<r_l+l/\beta+T$, Fig. 5(d).

$\Omega_h \cap \Gamma_\delta$ is a segment of Γ_δ limited by $z=z_T$ and $z=l$.

- Case eS: $r_l+l/\beta+T<\tau<\infty$, Fig. 5(e).

The hyperbola branch resides above Ω_h, $\Omega_h \cap \Gamma_\delta=\varnothing$, and as in Case aS

$$\Psi(\rho,z,\tau)=\Psi_{eS}(\rho,z,\tau)=0. \tag{52}$$

This situation relates to the epoch after passing of the electromagnetic-pulse back, corresponding to complete disappearance of the source current pulse at the spatiotemporal point $\rho=0, z=l, \tau=l/\beta+T$, which manifests itself r_l units of time later, at $\tau=r_l+l/\beta+T$.

Diagrams for a long source-current pulse are shown in Fig. 6. They correspond to the following set of cases:

- Case aL: $-\infty<\tau<r$, Fig. 6(a).

This case is identical to Case aS: $\Omega_h \cap \Gamma_\delta=\varnothing$, and

$$\Psi(\rho,z,\tau)=\Psi_{aL}(\rho,z,\tau)=0. \tag{53}$$

- Case bL: $r<\tau<r_l+l/\beta$, Fig. 6(b).

Apart from the condition imposed on τ, this case is identical to Case bS; the limits are $z=0$ and $z=z_0$.

- Case cL: $r_l+l/\beta<\tau<r+T$, Fig. 6(c).

$\Omega_h \cap \Gamma_\delta$ is a segment of Γ_δ limited by $z=0$ and $z=l$.

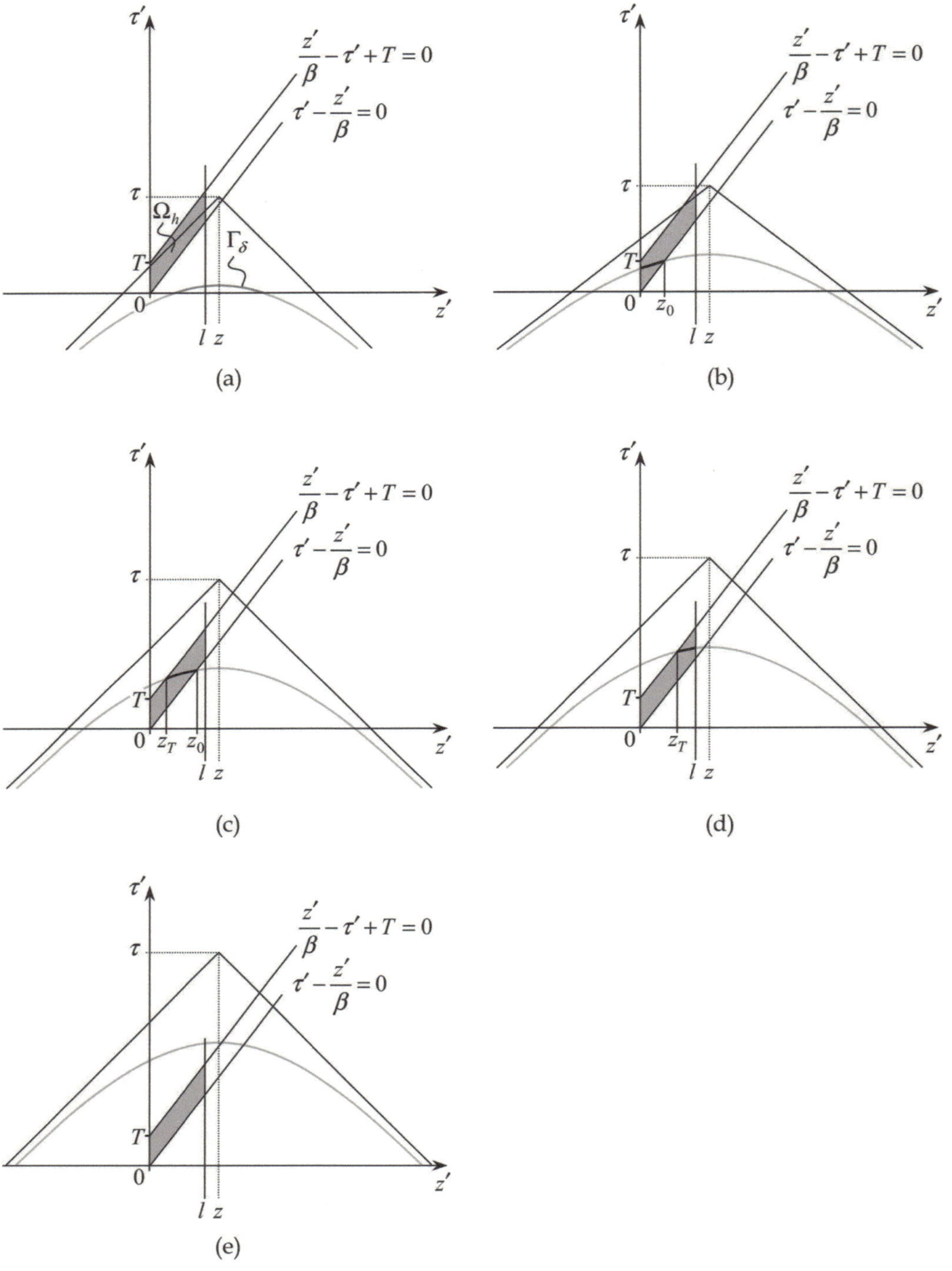

Fig. 5. Definition of the integration limits for a short source-current pulse.

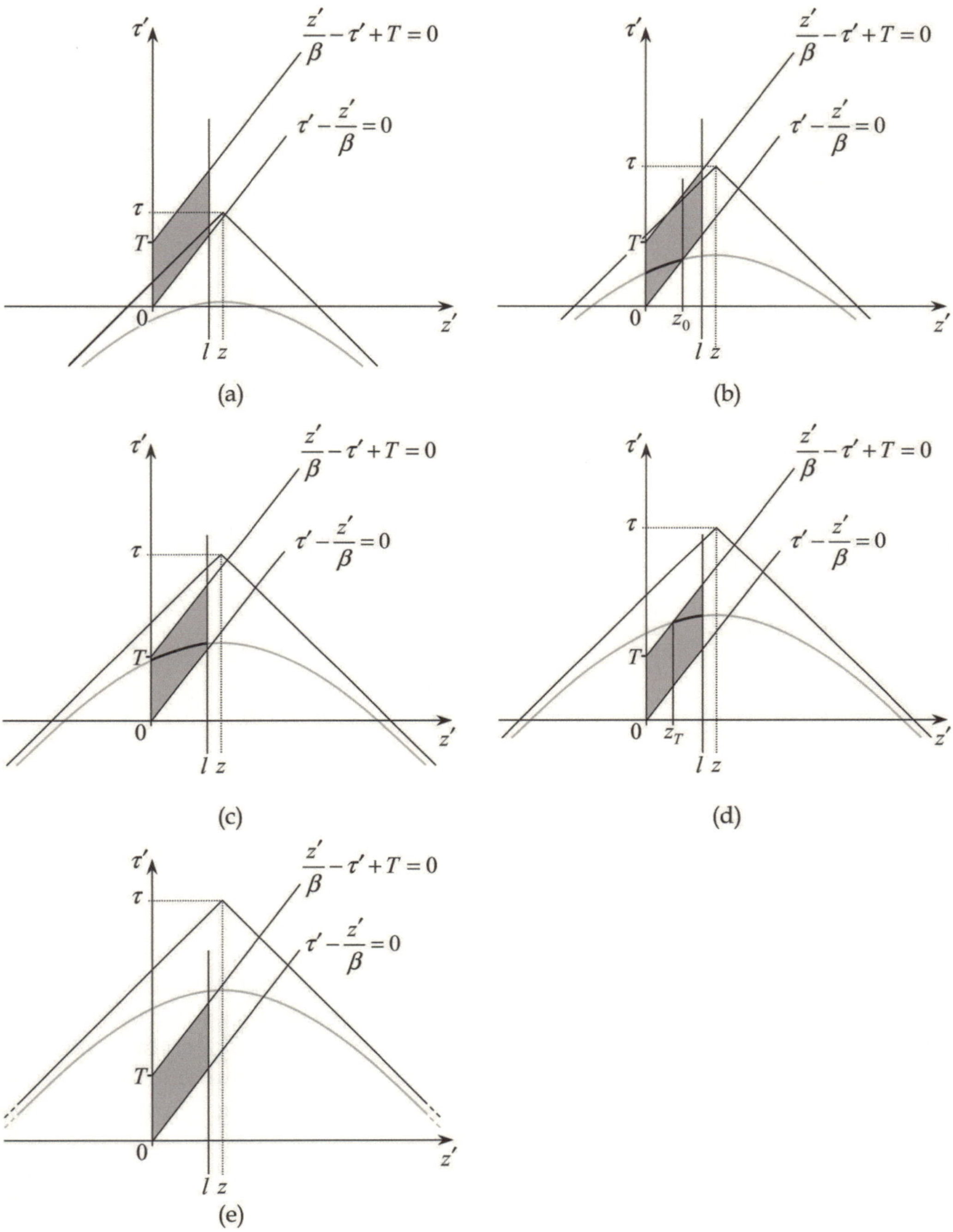

Fig. 6. Definition of the integration limits a long source-current pulse.

- Case dL: $r+T<\tau<r_l+l/\beta+T$, Fig. 6(d).

$\Omega_h \cap \Gamma_\delta$ is a segment of Γ_δ limited by $z=z_T$ and $z=l$. Apart from the condition imposed on τ, this case is identical to Case dS.

- Case eL: $r_l+l/\beta+T<\tau<\infty$, Fig. 6(e).

This case is identical to Case eS: $\Omega_h \cap \Gamma_\delta = \varnothing$ and

$$\Psi(\rho,z,\tau)=\Psi_{eL}(\rho,z,\tau)=0. \tag{54}$$

4.3 General solutions

The results obtained for the integration limits are summarized in Table 1. With all the integration limits defined, for cases corresponding to nonvanishing solution, the explicit representation of the wavefunction takes the general form akin to (45)

$$\Psi(\rho,z,\tau)=\left\{\begin{array}{ll}0 & \text{case } \lambda=a,e \\ \Psi_{\lambda C}(\rho,z,\tau)=\dfrac{1}{4\pi}\displaystyle\int_{z_{1\lambda C}}^{z_{2\lambda C}}\frac{J(z',\tau-r')}{r'}\mathrm{d}z' & \text{case } \lambda=b,c,d\end{array}\right\},\quad C=\begin{cases}S & \text{for short pulse}\\ L & \text{for long pulse}\end{cases} \tag{55}$$

where the case choice λ depends on the current value of time variable, see the column $\mathrm{Cond}(\lambda C)$ of Table 1.

One can notice from the diagrams or proof by direct calculations that the piecewise solution (55) provides continuous joining, that is,

$$\begin{gathered}\Psi_{bS,L}\big|_{\tau=r}=\Psi_{aS,L}\big|_{\tau=r}=0,\quad \Psi_{cS}\big|_{\tau=r+T}=\Psi_{bS}\big|_{\tau=r+T},\quad \Psi_{cL}\big|_{\tau=r_l+\frac{l}{\beta}}=\Psi_{bL}\big|_{\tau=r_l+\frac{l}{\beta}},\\ \Psi_{dS}\big|_{\tau=r_l+\frac{l}{\beta}}=\Psi_{cS}\big|_{\tau=r_l+\frac{l}{\beta}},\quad \Psi_{dL}\big|_{\tau=r+T}=\Psi_{cL}\big|_{\tau=r+T},\quad \Psi_{eS,L}\big|_{\tau=r_l+\frac{l}{\beta}}=\Psi_{dS,L}\big|_{\tau=r_l+\frac{l}{\beta}}=0.\end{gathered} \tag{56}$$

Relation (55) represents the solution of the scalar problem (4). With this solution constructed, one can readily find the magnetic induction using relation (3)

$$\mathbf{B}=B\mathbf{e}_\phi,\quad B=B_\phi=-\mu_0\frac{\partial^2 W}{\partial\rho\partial\tau}=-\mu_0\frac{\partial\Psi}{\partial\rho} \tag{57}$$

while the definition of the electric field components in the near zone requires calculation of the Borgnis–Bromwich potential itself, which leads to integration with respect to the time variable. Due to the initial conditions for which the charge distribution must be specified, such a procedure requires consideration that is specific to physical realization of the model (wire antenna, lightning, macroscopic current pulse accompanying absorption of hard radiation by a medium, etc.) and will not be discussed in the scope of the present work. Notably, $\mathbf{E}$ (and, consequently, the entire electromagnetic field and the electromagnetic energy density) in the far field, $r>>l$, can be found from the known magnetic induction $\mathbf{B}$ (Dlugosz & Trzaska 2010; Stratton, 2007).

Solutions (55), (57) describe emanation of finite transient electromagnetic pulses by line source-current pulses of arbitrary shape $J(z,\tau)$. They constitute the most practical and illustrative concretization of general solution (35), (41) for the pulsed sources whose front and back propagate with the same constant velocity v. The z',τ' plane diagrams admit definition of the actual integration limits in (35) for arbitrary temporal dependence of the velocities of the current-pulse front and back. In this case the limiting straight lines $\tau' - z'/\beta = 0$ and $z'/\beta - \tau' + T = 0$ must be replaced by curves $z' = z_f(\tau')$ and $z' = z_b(\tau')$ characterizing the front/back motion.

Models based on infinitely long source-current pulses, $T \to \infty$, results into the set of cases aL, bL, and cL. Electromagnetic problems describing waves generated by exponentially decaying current pulses are discussed in (Utkin 2007, 2008).

Case	Condition	Limits of integration
λC	Cond(λC)	$z_{1\lambda C}$, $z_{2\lambda C}$
aS,L	$-\infty < \tau < r$	not applicable, $\Psi_{aS,L}(\rho,z,\tau) = 0$
bS	$r < \tau < r + T$	$z_{1bS} = z_{1bL} = 0$ $z_{2bS} = z_{2bL} = z_0 = \beta \dfrac{\tau^2 - r^2}{\tau - \beta z + \sqrt{(1-\beta^2)\rho^2 + (z - \beta\tau)^2}}$
bL	$r < \tau < r_l + \dfrac{l}{\beta}$	
cS	$r + T < \tau < r_l + \dfrac{l}{\beta}$	$z_{1cS} = z_T = \beta \dfrac{(\tau - T)^2 - r^2}{\tau - T - \beta z + \sqrt{(1-\beta^2)\rho^2 + [z - \beta(\tau - T)]^2}}$ $z_{2cS} = z_0 = \beta \dfrac{\tau^2 - r^2}{\tau - \beta z + \sqrt{(1-\beta^2)\rho^2 + (z - \beta\tau)^2}}$
cL	$r_l + \dfrac{l}{\beta} < \tau < r + T$	$z_{1cL} = 0$ $z_{2cL} = l$
dS	$r_l + \dfrac{l}{\beta} < \tau < r_l + \dfrac{l}{\beta} + T$	$z_{1dS} = z_{1dL} = z_T = \beta \dfrac{(\tau - T)^2 - r^2}{\tau - T - \beta z + \sqrt{(1-\beta^2)\rho^2 + [z - \beta(\tau - T)]^2}}$ $z_{2dS} = z_{2dL} = l$
dL	$r + T < \tau < r_l + \dfrac{l}{\beta} + T$	
eS,L	$r_l + \dfrac{l}{\beta} + T < \tau < \infty$	not applicable, $\Psi_{eS,L}(\rho,z,\tau) = 0$

Table 1. Conditions and parameters of the wavefunction representation via explicit formula (55).

5. Current pulse with high-frequency filling

Of special interest is investigation of waves launched by a pulse with high-frequency filling, which was stimulated by the problem of launching directional scalar and electromagnetic waves (missiles) as well as by results of experimental investigation of superradiation waveforms (Egorov et al., 1986). The model in question can roughly describe a number of traditional artificial as well as natural line radiators and, being characterized by two different velocities -- the phase velocity of the carrier wave and the source-pulse velocity, – explains characteristic features observed in the laboratory and natural conditions for waves emanated by sources with high-frequency filling: their directionality, frequency transform, and beats.

5.1 Specific solutions

The problem of wave generation by the current pulse with high-frequency filling corresponds to a particular case of the electromagnetic problem discussed in Section 2, for which the continuous function describing the current distribution can be expressed in the form

$$J(z,\tau)=\tilde{U}(z,\tau)M_{\pm}(z,\tau), \tag{58}$$

where a differentiable function $\tilde{U}(z,\tau)$ represents the source-current envelope and

$$M_{\pm}=\exp\left(\mathrm{i}k\left(v_{ph}t\pm z\right)\right)=\exp\left(\mathrm{i}k\left(\beta_{ph}\tau\pm z\right)\right) \tag{59}$$

the factor corresponding to the cosinusoidal (the real part) and sinusoidal (the imaginary part) modulating wave with the spatial period $2\pi/k$ and the phase velocity $v_{ph}=\beta_{ph}c$. The minus sign corresponds to propagation of the modulating wave in the positive z direction (in the same direction as the source-current pulse front and back, case of copropagation) while the plus sign describes the situation in which the modulating wave propagates in the negative z direction, opposite to the direction of propagation of the pulse front and back (case of counterpropagation).

As far as superluminal phase velocity v_{ph} is readily admissible for a much wider range of real-world models than the superluminal front velocity v, the values of β_{ph} are supposed to vary from 0 to infinity. To be able to pass to spatiotemporal coordinates in the frame moving with velocity v, we will assume that $v<c$, that is, $\beta<1$. All the final results obtained in this chapter are easy to extend to the case of luminal source-current pulse taking the limit $\beta\to 1$.

To make the solutions easier to analyze, let us express $\tilde{U}(z,\tau)$ as a function of $\tau\pm z$

$$U(\tau-z,\tau+z)\overset{def}{=}\tilde{U}(z,\tau). \tag{60}$$

Then, substituting (58) into general solution (55), one has

$$\Psi_{\lambda C}^{(\pm)}(\rho,z,\tau)=\frac{1}{4\pi}\int\limits_{z_{1\lambda C}}^{z_{2\lambda C}}\frac{U(\tau-r'-z',\tau-r'+z')}{r'}\exp\left(\mathrm{i}k\left[\beta_{ph}(\tau-r')\pm z'\right]\right)\mathrm{d}z'. \tag{61}$$

In the case of copropagation of the modulating wave (M_{-}), changing the integration variable in representation (61) to $\zeta=\tau-r'-z'$ yields

$$\begin{aligned}\Psi_{\lambda C}^{(-)}&=\frac{1}{4\pi}\int\limits_{\zeta_{1\lambda C}}^{\zeta_{2\lambda C}}\frac{U(\zeta,S)}{\tau-z-\zeta}\exp\left(\mathrm{i}K\left[\zeta+\varepsilon S(\zeta)\right]\right)\mathrm{d}\zeta\\&=\frac{1}{4\pi\mathrm{i}K}\int\limits_{\zeta_{1\lambda C}}^{\zeta_{2\lambda C}}U(\zeta,S)\frac{\tau-z-\zeta}{(\tau-z-\zeta)^2-\varepsilon\rho^2}\frac{\partial}{\partial\zeta}\exp\left(\mathrm{i}K\left[\zeta+\varepsilon S(\zeta)\right]\right)\mathrm{d}\zeta,\end{aligned} \tag{62}$$

where

$$\zeta_{1\lambda C} = \zeta\big|_{z'=z_{2\lambda C}}, \quad \zeta_{2\lambda C} = \zeta\big|_{z'=z_{1\lambda C}},$$
$$S(\zeta) = \tau - r' + z' = \tau + z - \frac{\rho^2}{\tau - z - \zeta}, \quad \varepsilon = \frac{\beta_{ph} - 1}{\beta_{ph} + 1}, \quad K = \frac{\beta_{ph} + 1}{2} k. \tag{63}$$

Assuming that $U(\zeta, S(\zeta))$ is a slowly varying function of ζ, $q(\zeta) = U(\zeta, S) \dfrac{\tau - z - \zeta}{(\tau - z - \zeta)^2 - \varepsilon\rho^2}$ is the continuous function and $q'(\zeta)$ is the absolutely integrable function, one gets by integration by parts the following estimation

$$\Psi_{\lambda C}^{(-)} = \frac{1}{4\pi \mathrm{i}K} \Bigg\{ U(\zeta_{2\lambda C}, S(\zeta_{2\lambda C})) \frac{\tau - z - \zeta_{2\lambda C}}{(\tau - z - \zeta_{2\lambda C})^2 - \varepsilon\rho^2} \exp\left(\mathrm{i}K\left[\zeta_{2\lambda C} + \varepsilon S(\zeta_{2\lambda C})\right]\right)$$
$$-U(\zeta_{1\lambda C}, S(\zeta_{1\lambda C})) \frac{\tau - z - \zeta_{1\lambda C}}{(\tau - z - \zeta_{1\lambda C})^2 - \varepsilon\rho^2} \exp\left(\mathrm{i}K\left[\zeta_{1\lambda C} + \varepsilon S(\zeta_{1\lambda C})\right]\right) \Bigg\} + \mathrm{o}\left(\frac{1}{Kl}\right). \tag{64}$$

Neglecting terms of order $(Kl)^{-1}$ and higher, we readily get the following magnetic induction approximation

$$B_{\lambda C}^{(-)} = -\mu_0 \frac{\partial \Psi_{\lambda C}^{(-)}}{\partial \rho}$$
$$\cong \frac{\mu_0}{4\pi} \Bigg\{ -\frac{U(\zeta_{2\lambda C}, S(\zeta_{2\lambda C}))(\tau - z - \zeta_{2\lambda C})}{(\tau - z - \zeta_{2\lambda C})^2 - \varepsilon\rho^2} \left[\frac{\partial \zeta_{2\lambda C}}{\partial \rho} + \varepsilon \frac{\partial}{\partial \rho}(S(\zeta_{2\lambda C})) \right] \exp\left(\mathrm{i}K\left[\zeta_{2\lambda C} + \varepsilon S(\zeta_{2\lambda C})\right]\right) \tag{65}$$
$$+ \frac{U(\zeta_{1\lambda C}, S(\zeta_{1\lambda C}))(\tau - z - \zeta_{1\lambda C})}{(\tau - z - \zeta_{1\lambda C})^2 - \varepsilon\rho^2} \left[\frac{\partial \zeta_{1\lambda C}}{\partial \rho} + \varepsilon \frac{\partial}{\partial \rho}(S(\zeta_{1\lambda C})) \right] \exp\left(\mathrm{i}K\left[\zeta_{1\lambda C} + \varepsilon S(\zeta_{1\lambda C})\right]\right) \Bigg\}.$$

Finally, using the explicit representation of $S(\zeta_{1,2\lambda C})$ and making the differentiation, Eq. (65) can be reduced to

$$B_{\lambda C}^{(-)} \cong B^{(-)}(\zeta_{2\lambda C}) - B^{(-)}(\zeta_{1\lambda C}) \tag{66}$$

where

$$B^{(-)}(\Phi) \overset{def}{=} \frac{\mu_0}{4\pi} U(\Phi, S(\Phi)) \left[\frac{1}{\tau - z - \Phi} \left(-\frac{\partial \Phi}{\partial \rho} \right) + \varepsilon \frac{2\rho}{(\tau - z - \Phi)^2 - \varepsilon\rho^2} \right] \exp\left(\mathrm{i}K\left[\Phi + \varepsilon S(\Phi)\right]\right). \tag{67}$$

Application of a similar procedure in the case of counterpropagation of the modulating wave (M_+), which harness a new integration variable $\rho^2/(z' - z - r')$, yield approximations for the wavefunction $\Psi_{\lambda C}^{(+)}$ and the magnetic induction $B_{\lambda C}^{(+)}$.
Making routine calculations for each case, one can express the final result in the form

$$B(\rho,z,\tau)=\begin{cases}\text{Case bS, bL:} & B_{bS,L}^{(\pm)}=B_0^{(\pm)}-B_\beta^{(\pm)}\\ \text{Case cS:} & B_{cS}^{(\pm)}=B_T^{(\pm)}-B_\beta^{(\pm)}\\ \text{Case cL:} & B_{cL}^{(\pm)}=B_0^{(\pm)}-B_l^{(\pm)}\\ \text{Case dS, dL:} & B_{dS,L}^{(\pm)}=B_T^{(\pm)}-B_l^{(\pm)}\\ \text{otherwise:} & 0\end{cases} \tag{68}$$

where all RHS terms are of the structure

$$B_\alpha^{(\pm)}=\frac{\mu_0}{4\pi}\frac{U_\alpha}{r_\alpha}\chi_\alpha^{(\pm)}R_\alpha^{(\pm)},\quad \alpha=0,\beta,T,l. \tag{69}$$

The part depending on the source-current shape is represented by the factors U_α and r_α^{-1}, which are the same for both modulation types:

$$\begin{aligned}&U_0=U_\beta\big|_{\beta=0}=U(\tau-r,\tau-r),\\ &U_\beta=U\left(\sqrt{\frac{1-\beta}{1+\beta}}(\tau_\beta-r_\beta),\sqrt{\frac{1+\beta}{1-\beta}}(\tau_\beta-r_\beta)\right),\\ &U_T=U\left(\sqrt{\frac{1-\beta}{1+\beta}}(\tau_T-r_T)+T,\sqrt{\frac{1+\beta}{1-\beta}}(\tau_T-r_T)+T\right),\\ &U_l=U(\tau-r_l-l,\tau-r_l+l),\end{aligned} \tag{70}$$

$$\begin{aligned}&r_0=r=\sqrt{\rho^2+z^2},\\ &\tau_\beta=\frac{\tau-\beta z}{\sqrt{1-\beta^2}},\quad r_\beta=\sqrt{\rho^2+z_\beta^2},\quad z_\beta=\frac{z-\beta\tau}{\sqrt{1-\beta^2}},\\ &\tau_T=\frac{\tau-T-\beta z}{\sqrt{1-\beta^2}},\quad r_T=\sqrt{\rho^2+z_T^2},\quad z_T=\frac{z-\beta(\tau-T)}{\sqrt{1-\beta^2}},\\ &r_l=\sqrt{\rho^2+z_l^2},\quad z_l=z-l.\end{aligned} \tag{71}$$

The factor $R_\alpha^{(\pm)}$ defines the characteristic angular structure of the emanated wave due to given parameters of the wave excitation β,β_{ph},T and l

$$R_\lambda^{(\pm)}=\begin{cases}\dfrac{\beta_{ph}\sin\theta_\alpha}{1\pm\beta_{ph}\cos\theta_\alpha} & \alpha=0,l\\[2ex] \dfrac{(\beta_{ph}\pm\beta)\sin\theta_\alpha}{1\pm\beta_{ph}\beta\pm(\beta_{ph}\pm\beta)\cos\theta_\alpha} & \alpha=\beta,T,\end{cases} \tag{72}$$

where the angle θ_α is a part of spheric-coordinate representation of ρ,z_α via r_α:

$$\rho=r_\alpha\sin\theta_\alpha,\quad z_\alpha=r_\alpha\cos\theta_\alpha. \tag{73}$$

The modulation factor $\chi_\alpha^{(\pm)}$,

$$\begin{aligned}
\chi_0^{(\pm)} &= \exp\left(\frac{\mathrm{i}}{c}\omega_0(\tau - r)\right),\\
\chi_\beta^{(\pm)} &= \exp\left(\frac{\mathrm{i}}{c}\left[\omega_0\left(1 \pm \frac{\beta}{\beta_{ph}}\right)\right]\frac{\tau_\beta - r_\beta}{\sqrt{1-\beta^2}}\right),\\
\chi_T^{(\pm)} &= \exp\left(\frac{\mathrm{i}}{c}\omega_0 T\right)\exp\left(\frac{\mathrm{i}}{c}\left[\omega_0\left(1 \pm \frac{\beta}{\beta_{ph}}\right)\right]\frac{\tau_T - r_T}{\sqrt{1-\beta^2}}\right),\\
\chi_l^{(\pm)} &= \exp(\pm \mathrm{i}kl)\exp\left(\frac{\mathrm{i}}{c}\omega_0(\tau - r_l)\right),
\end{aligned} \tag{74}$$

where $\omega_0 = kv_{ph} = k\beta_{ph}c$ is the source modulation frequency, characterize the local high-frequency modulation, so the product $\frac{\mu_0}{4\pi}\frac{U_\alpha}{r_\alpha}R_\alpha^{(\pm)}$ may be treated as the wave envelope.

5.2 Directionality of the emanated waves

Inherent directionality of the waves produced by the current pulse with high-frequency filling in question is defined by the terms of the type $\dfrac{\beta_{ph}\sin\theta_\alpha}{1 \pm \beta_{ph}\cos\theta_\alpha}$ in the case of $B_0^{(\pm)}$, $B_l^{(\pm)}$ and those of the type $\dfrac{(\beta_{ph} \pm \beta)\sin\theta_\alpha}{1 \pm \beta_{ph}\beta \pm (\beta_{ph} \pm \beta)\cos\theta_\alpha}$ in the case of $B_\beta^{(\pm)}$, $B_T^{(\pm)}$.

In the case of the luminal phase velocity, $v_{ph} = c$, $\beta_{ph} = 1$ the factors $R_\alpha^{(\pm)}$ get much simpler forms (Borisov et al., 2005)

$$R_\alpha^{(-)}\Big|_{\beta_{ph}=1} = \frac{\sin\theta_\alpha}{1-\cos\theta_\alpha} = \cot\frac{\theta_\alpha}{2}, \quad R_\alpha^{(+)}\Big|_{\beta_{ph}=1} = \frac{\sin\theta_\alpha}{1+\cos\theta_\alpha} = \tan\frac{\theta_\alpha}{2}, \tag{75}$$

indicating that in the case of M_- modulation the electromagnetic wave is predominantly emanated along the direction of the source-current pulse propagation, $\theta_\alpha \cong 0$, while the case of M_+ is characterized by the opposite wave directionality, $\theta_\alpha \cong \pi$. That is, it is the direction of propagation of the modulating wave, rather than the carrier, that defines the angular localization of the emanated radiation. In spite of the apparent divergence of the terms due to the presence of the tangent/cotangent factors, their sums composing the magnetic induction remains finite everywhere except for the source domain. For example, in the case of $r + T < \tau < r_l + l/\beta$ the radiation intensity for a short source-current pulse $I_{cS}^{(\pm)}$ in the far field is given by

$$\begin{aligned}
I_{cS}^{(-)} &\cong I_0\left(\frac{U_\beta}{r_\beta}\right)^2 \cot^2\frac{\theta_\beta}{2}\,\sin^2\left(kT\frac{\beta}{1+\beta}\sin^2\frac{\theta_\beta}{2}\right),\\
I_{cS}^{(+)} &\cong I_0\left(\frac{U_\beta}{r_\beta}\right)^2 \tan^2\frac{\theta_\beta}{2}\,\sin^2\left(kT\frac{\beta}{1+\beta}\cos^2\frac{\theta_\beta}{2}\right),
\end{aligned} \tag{76}$$

while for a long pulse we have

$$I_{cL}^{(-)} \cong I_0 \left(\frac{U_0}{r}\right)^2 \cot^2\frac{\theta_0}{2}\ \sin^2\left(kl\sin^2\frac{\theta_0}{2}\right),$$
$$I_{cL}^{(+)} \cong I_0 \left(\frac{U_0}{r}\right)^2 \tan^2\frac{\theta_0}{2}\ \sin^2\left(kl\cos^2\frac{\theta_0}{2}\right), \tag{77}$$

where $I_0 = \left(8\pi^2\varepsilon_m c\right)^{-1}$, ε_m is the permittivity of the medium.

For subliminal and superluminal phase velocities, $\beta_{ph} \neq 1$, the angular factors of the field components $R_\alpha^{(\pm)}$ have more complicated form (72); their dependence on β_{ph} and $\theta_{0,l}$ is illustrated in Fig. 7.

As seen from the figure, for subluminal phase velocity the tendency of $B_{0,l}^{(-)}$ to be directed at $\theta_{0,l} = 0$ and $B_{0,l}^{(+)}$ at $\theta_{0,l} = \pi$, clearly manifested at $\beta_{ph} = 1$, holds down to $\beta_{ph} \approx 0.7$. For superluminal phase velocities the angular factors demonstrate lateral (towards $\theta_{0,l} = \pi/2$) shift of the propagation directionality, which is observed in the vicinity of the singularity curves, $\theta_{0,l} = \arccos\beta_{ph}^{-1}$ for $R_{0,l}^{(-)}$ and $\theta_{0,l} = \pi - \arccos\beta_{ph}^{-1}$ for $R_{0,l}^{(+)}$.

For analysis of more complicated factors $R_{\beta,T}^{(\pm)}$ let us represent them in a two-parameter form

$$R_{\beta,T}^{(\pm)} = \frac{\sin\theta_{\beta,T}}{\Theta^{(\pm)} \pm \cos\theta_{\beta,T}}, \quad \Theta^{(\pm)} \overset{def}{=} \frac{1 \pm \beta_{ph}\beta}{\beta_{ph} \pm \beta}. \tag{78}$$

As functions of β and β_{ph}, the parameters $\Theta^{(-)}$ and $\Theta^{(+)}$ demonstrate different behaviour: $\Theta^{(-)}$ varies from $-\infty$ to ∞ and has the area of singularity $\beta_{ph} = \beta$ while the sign of $\Theta^{(+)}$ is always positive and the area of singularity is limited to the point $\beta_{ph} = \beta = 0$, see Fig. 8. Note that $\lim\limits_{\beta\to 1}\Theta^{(\pm)} = \pm 1$, $\lim\limits_{\beta_{ph}\to +\infty}\Theta^{(\pm)} = \pm\beta$, and $\Theta^{(\pm)}\Big|_{\beta_{ph}=1} = 1$. The angular factors $R_{\beta,T}^{(\pm)}$ are finally plotted as functions of $\Theta^{(\pm)}$ and $\theta_{\beta,T}$ in Fig. 9.

5.3 Frequency transform

The modulation frequencies corresponding to the terms $B_{0,\beta,T,l}^{(\pm)}$ composing the magnetic field strength are defined by the arguments of the modulation factors $\chi_{0,\beta,T,l}^{(\pm)}$. As $\chi_0^{(\pm)} = \exp\left(\frac{i}{c}\omega_0(\tau - r)\right)$ and $\chi_l^{(\pm)} = \exp(\pm ikl)\exp\left(\frac{i}{c}\omega_0(\tau - r_l)\right)$ always oscillate with the initial modulation frequency ω_0, the frequency transform is observed only in $B_{\beta,T}^{(\pm)}$. The frequency transform range can be found considering wave propagation in the two limiting directions: parallel ($\theta_{\beta,T} = 0$) and anti-parallel ($\theta_{\beta,T} = \pi$) to the direction of propagation of the source current. In the case $\theta_{\beta,T} = 0$ one has $\tau_{\beta,T} - r_{\beta,T} = \tau_{\beta,T} - z_{\beta,T}$ and, coming back to the initial frame of reference τ, z, one can express the modulation factors as

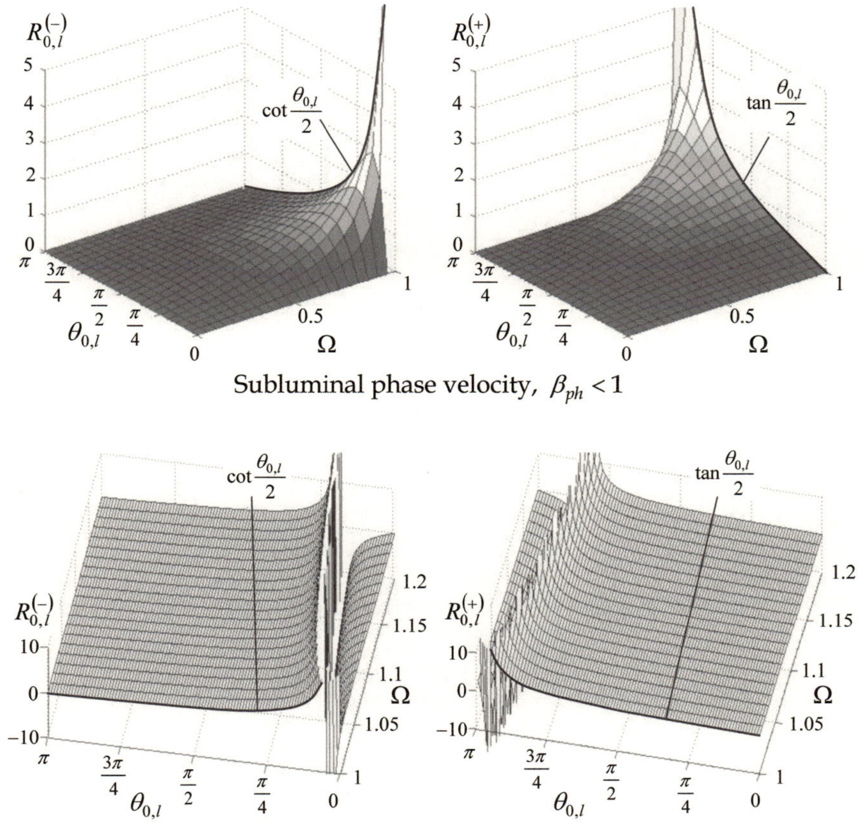

Fig. 7. Dependence of the factors $R_{0,l}^{(\pm)}$ on the dimensionless modulation phase velocity β_{ph} and the angular parameter $\theta_{0,l}$.

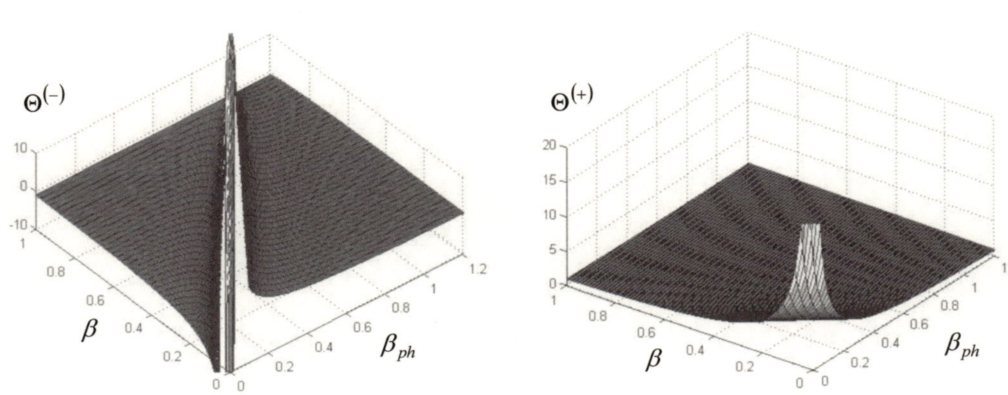

Fig. 8. Parameters $\Theta^{(\pm)}$ plotted versus the dimensionless velocities β_{ph} and β.

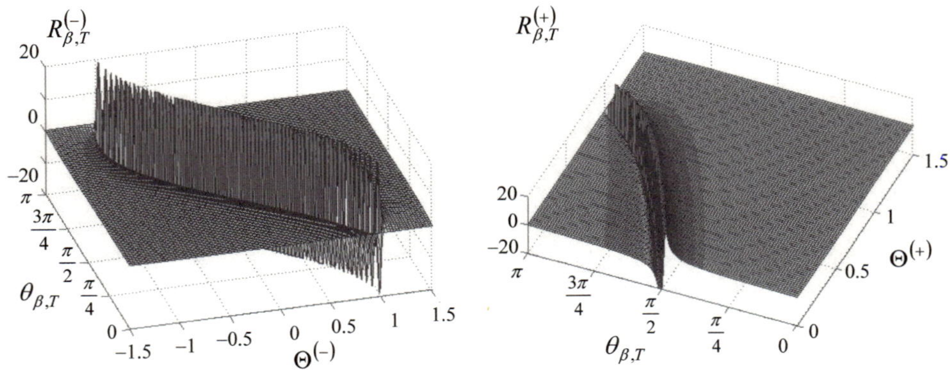

Fig. 9. The angular factors $R^{(\pm)}_{\beta,T}$ plotted as functions of $\Theta^{(\pm)}$ and $\theta_{\beta,T}$.

$$\chi^{(\pm)}_{\beta,T} = const \times \exp\left(\frac{\mathrm{i}}{c}\left[\omega_0 \frac{\left|1 \pm \beta / \beta_{ph}\right|}{1-\beta}\right](\tau - r)\right), \quad \cos\theta_{\beta,T} = 1. \tag{79}$$

The other case $\theta_{\beta,T} = \pi$, corresponding to $\tau_{\beta,T} - r_{\beta,T} = \tau_{\beta,T} + z_{\beta,T}$, results in

$$\chi^{(\pm)}_{\beta,T} = const \times \exp\left(\frac{\mathrm{i}}{c}\left[\omega_0 \frac{\left|1 \pm \beta / \beta_{ph}\right|}{1+\beta}\right](\tau - r)\right), \quad \cos\theta_{\beta,T} = -1. \tag{80}$$

This shows that for different directions of wave propagation θ the magnetic-induction components $B^{(\pm)}_{\beta,T}$ are subjected to the modulation-frequency transform with respect to the initial modulation frequency of the source current ω_0: $\omega_0 \to \omega^{(\pm)}_{\beta,T}(\theta)$. The range of this transformation is defined by the inequality

$$\frac{\left|1 \pm \beta / \beta_{ph}\right|}{1+\beta}\omega_0 = \omega^{(\pm)}_{\beta,T}(\pi) \;\le\; \omega^{(\pm)}_{\beta,T}(\theta) \;\le\; \omega^{(\pm)}_{\beta,T}(0) = \frac{\left|1 \pm \beta / \beta_{ph}\right|}{1-\beta}\omega_0 . \tag{81}$$

In the particular case of a nearly luminal or luminal phase velocity,

$$\beta_{ph} = 1 - \varepsilon_{ph}, \quad 0 \le \varepsilon_{ph} << 1, \tag{82}$$

inequality (81) takes the form

$$\left(\frac{1-\beta}{1+\beta} - \varepsilon_{ph}\frac{\beta}{1+\beta}\right)\omega_0 \;\le\; \omega^{(-)}_{\beta,T}(\theta) \;\le\; \left(1 - \varepsilon_{ph}\frac{\beta}{1-\beta}\right)\omega_0 \;\le\; \omega_0 \tag{83}$$

for $\omega^{(-)}_{\beta,T}$, describing a *red shift* at copropagation of the modulating wave, and the form

$$\omega_0 \;\le\; \omega_0\left(1 + \varepsilon_{ph}\frac{\beta}{1+\beta}\right) \;\le\; \omega^{(+)}_{\beta,T}(\theta) \;\le\; \omega_0\left(\frac{1+\beta}{1-\beta} + \varepsilon_{ph}\frac{\beta}{1-\beta}\right), \tag{84}$$

demonstrating a *blue shift* at counterpropagation of the modulating wave, phenomena described for the particular case of luminal phase velocity ($\varepsilon_{ph} = 0$) by Borisov et al. (2005).

5.4 Beats

One more phenomenon discussed by Borisov et al. (2005) for the case of luminal phase velocity is the appearance of low-frequency modulated envelopes (beats) due to the excitation of two wave components modulated with slightly different frequencies. As seen from Eqs. (68), (69) and (74) this situation may occur in Cases bS, bL and dS, dL and results in frequency subtraction $\Delta\beta_{ph} \sim \left|\omega_0 - \omega_{\beta,T}^{(\pm)}\right|$ at mixing the wave components $B_0^{(\pm)}$ and $B_\beta^{(\pm)}$ (Cases bS, bL) or $B_l^{(\pm)}$ and $B_T^{(\pm)}$ (Cases dS, dL). Remarkably, for $\beta_{ph} < 1$ one more type of beats can be observed in Cases bS, bL, cS, and dS, dL for $\beta_{ph} \sim \beta$ in a single component, $B_\beta^{(-)}$ or $B_T^{(-)}$, due to interference of the carrier and modulating waves propagating with nearly the same speed in the same direction: as seen from Eq. (74), $\lim_{\beta_{ph} \to \beta} \omega_{\beta,T}^{(-)}(\theta) = 0$.

6. Conclusion

The theoretical basics of incomplete separation of variables in the wave equation discussed in this chapter can be applied for a wide range of problems involving scalar and electromagnetic wave generation, propagation and diffraction. The use of Riemann method and the z',τ' plane diagrams provides rigorously substantiated and easy-to-follow procedures resulting in construction of analyzable *signal solutions* (Harmuth et al., 1999) of essentially nonsinusoidal nature. Concretization of the general solutions for source currents of particular shape often leads to analytical expressions. Practically important analytical solutions describing waves generated by a linear combination of exponentially decaying current pulses propagating in lossy media are constructed in (Utkin, 2008).

Although the present discussion is constrained to the line sources, its extension to the more complicated source configurations is straightforward: for instance, the multipole expansion and introduction of the Debye potential result, in the spherical coordinate system, to the Euler–Poisson–Darboux equation of known Riemann function with respect to the transient spherical-harmonic expansion coefficients of the desired wavefunction (Borisov et al., 1996). Less complex solutions were obtained by Borisov and Simonenko for sources located on moving and expanding circles (Borisov & Simonenko, 1994, 1997, 2000). An example of application of the method in the case of superluminal source pulses can be found in (Borisov, 2001).

The procedure of constructing the Riemann function $R(z,\tau;z',\tau')$ to the equation

$$\left[\frac{\partial^2}{\partial z^2} - \frac{\partial^2}{\partial \tau^2} + p_1(z) + p_2(\tau)\right]\Psi(z,\tau) = 0 \tag{85}$$

on the basis of known Riemann functions $R_{1,2}(z,\tau;z',\tau')$ for the reduced equations

$$\left[\frac{\partial^2}{\partial z^2}-\frac{\partial^2}{\partial \tau^2}+p_1(z)\right]\Psi(z,\tau)=0 \text{ and } \left[\frac{\partial^2}{\partial z^2}-\frac{\partial^2}{\partial \tau^2}+p_2(\tau)\right]\Psi(z,\tau)=0 \quad (86)$$

via the integral formula

$$R(z,\tau;z',\tau')=R_1(z,\tau;z',\tau')+\int_{\tau-\tau'}^{z-z'} R_1(z,\zeta;z',0)\frac{\partial}{\partial\zeta}R_2(\zeta,\tau;0,\tau')\mathrm{d}\zeta \quad (87)$$

proposed by Olevskii (1952) enables the applicability of the method of incomplete separation of variables to be extended to more complicated conditions. In particular, some processes of wave generation in dispersive media can be described via integral formulas (Borisov, 2002, 2003, 2008).

More specific results obtained for the current pulses with high-frequency filling demonstrate how some non-stationary phenomena can be investigated even within the framework of the high-frequency asymptotic approach. Abandoning the immediate separation of the time variable, that is, the enforcement of the temporal dependence $\exp(-i\omega_0 t)$ – a well grounded cornerstone of the physical theory of diffraction and stealth technology (Ufimtsev, 2007), but not at all a universal technique – results in direct spatiotemporal representation of the emanated electromagnetic pulses. Using this representation, one can analytically describe transformation of the frequency of the electromagnetic wave carrier with respect to the initial frequency of the source current modulation, which is manifested as the red or ultraviolet shift for the modulation factors M_- and M_+ correspondingly. In certain space-time domains, the waves of two different frequencies, fundamental and shifted, are excited, which leads to low-frequency modulated envelopes (beats). One more type of beats can be observed in the case of M_- modulation when the modulating-wave velocity comes close to the source-current pulse propagation velocity.

Spatiotemporal description of transients also gains increasing importance for localized wave generation, anti-stealth radar applications, electronic warfare and radio-frequency weaponization (Fowleret al., 1990). Formation of localized waves by source pulses with Gaussian transverse variation is discussed in (Borisov & Utkin, 1994). Having the space-time structure akin to Brittingham's focus wave modes (Brittingham, 1993), such waves can be expressed via Lommel's functions of two variables.

The first steps towards creating the theory of diffraction and guided propagation of transient waves were made in monographs by Harmuth (1986), Harmuth et al. (1999) and Borisov (1987). The practical needs of the ultra-wideband technology are believed to give rise to further advances in this area.

7. References

Arfken, G. B. & Weber, H. J. (2001). *Mathematical Methods for Physicists*, 5th ed., Academic Press, ISBN 0-12-059825-6, New York

Borisov, V. V. (1987). *Nonsteady-State Electromagnetic Waves*, Leningrad State University Press, Leningrad (in Russian)

Borisov, V. V. & Utkin, A. B. (1994). On formation of focus wave modes. *J. Phys. A: Math. Gen.*, 27, 7, (Abril 1994) 2587-2591, ISSN 0305-4470

Borisov, V. V. & Simonenko, I. I. (1994). Transient waves generated by a source on a circle. *J. Phys. A: Math. Gen.*, 27, 18, (September 1994) 6243-6252, ISSN 0305-4470

Borisov, V. V.; Manankova, A. V. & Utkin, A. B. (1996). Spherical harmonic representation of the electromagnetic field produced by a moving pulse of current density. *J. Phys. A: Math. Gen.*, 29, 15, (August 1996) 4493-4514, ISSN 0305-4470

Borisov, V. V. & Simonenko I. I. (1997). Consruction of Bessel-Gauss type solutions for the telegraph equation. *Journal de Physique I France*, 7, 8, (August 1997) 923-930, ISSN 1155-4304

Borisov, V. V. & Simonenko, I. I. (2000). Electromagnetic fields produced by sources on a spherical wavefront. *Eur. Phys. J. B*, 18, 1, (November 2000) 85-93, ISSN 1434-6028

Borisov, V. V. (2001). Spherical-harmonic representation of transient waves produced by sources on a superluminal expanding sphere, *Proc. of International Conference "Days on Diffraction - 2001"*, pp. 38-43, ISBN 5-7997-0366-9, St. Petersburg, May 2001, St. Petersburg State University, St. Petersburg

Borisov, V. V. (2002). On transient waves in dispersive media produced by moving point sources. *J. Phys. A: Math. Gen.*, 35, 26, (July 2002) 5403–5409, ISSN 0305-4470

Borisov, V. V. (2003). On spherical harmonic representation of transient waves in dispersive media. *J. Phys. A: Math. Gen.*, 36, 39, (October 2003) 10131–10140, ISSN 0305-4470

Borisov, V. V.; Reutova, N. M. & Utkin, A.B. (2005). Electromagnetic waves produced by a travelling current pulse with high-frequency filling. *J. Phys. A: Math. Gen.*, 38, 10, (March 2005) 2225-2240, ISSN 0305-4470

Borisov, V. V. (2008). Transient waves produced by Gaussian's transverse sources in a dispersive medium, *Proc. of International Conference "Days on Diffraction - 2008"*, pp. 20-22, ISBN 978-5-9651-0277-8, St. Petersburg, June 2008, IEEE, St. Petersburg

Brittingham, J. N. (1993). Focus waves modes in homogeneous Maxwell's equations: Transverse electric mode. *J. Appl. Phys.*, 54, 3, (March 1993) 1179-1189, ISSN 0021-8979

Bromwich, T. J., (1919). Electromagnetic waves. *Philosophical Magazine*, 38, 223, (July 1919) 143-164, ISSN 1478-6435

Chen, Z. (1988). Theoretical solutions of transient radiation from traveling-wave linear antennas. *IEEE Transactions on Electromagnetic Compatibility*, 30, 1, (February 1988) 80-83, ISSN 0018-9375

Courant, R. & Hilbert, D. (1989). *Methods of Mathematical Physics*, Vol. 2, John Wiley & Sons, ISBN 0-471-50439-4, New York

Dlugosz, T. & Trzaska, H. (2010). How to measure in the near field and in the far field. *Communication and Network*, 2, 1, (February 2010), 65-68, ISSN 1949-2421

Egorov, V. S.; Laptev, V. D.; Reutova, N. M. & Sokolov, I. V. (1986). The asymmetry of superradiation under the retarded excitation. *Sov. J. Quantum Electronics* 13, 4, (April 1986), 729-733, ISSN 0368-7147

Fowler, C.; Entzminger, J. & Corum, J. (1990). Assessment of ultra-wideband (UWB) technology. *Aerospace and Electronic Systems Magazine, IEEE*, 5, 11, (November 1990) 45-49, ISSN 0885-8985

Harmuth, H. F. (1986). *Propagation of Nonsinusoidal Electromagnetic Waves*. Advances in Electronics and Electron Physics Supplement, Academic Press, ISBN 0-120-14580-4, New York

Harmuth, H. F.; Boules, R. N. & Hussain, M. G. M. (1999). *Electromagnetic signals: reflection, focusing, distortion, and their practical applications*. Springer, ISBN 0-306-46054-8, New York

Hernández-Figueroa, H. E.; Zamboni-Rached, M. & Recami, E., Eds. (2008). *Localized Waves*. Wiley Series in Microwave and Optical Engineering, John Wiley & Sons, ISBN 978-0-470-10885-7, New Jersey

Master, M. J. & Uman, M. A. (1983). Transient electric and magnetic fields associated with establishing a finite electrostatic dipole. *American Journal of Physics*, 51, 2, (February 1983) 118-126, ISSN 0002-9505

Master, M. J. & Uman, M. A. (1984). Lightning induced voltages on power lines: Theory. *IEEE Transactions on Power Apparatus and Systems*, 103, 9, (September 1984) 2502-2518, ISSN 0018-9510

Olevskii, M. N. (1952). On the Riemann function for the differential equation $\partial^2 u / \partial x^2 - \partial^2 u / \partial \tau^2 + (p_1(x) + p_2(\tau))u = 0$. *Doklady Akademii Nauk SSSR*, 87, 3, (March 1952) 337-340, ISSN 0869-5652

Rubinstein, M. & Uman, M. A. (1991). Transient electric and magnetic fields associated with establishing a finite electrostatic dipole, revisited. *IEEE Transactions on Electromagnetic Compatibility*, 33, 4, (November 1991) 312-320, ISSN 0018-9375

Stratton, J. A. (2007). *Electromagnetic Theory* (IEEE Press Series on Electromagnetic Wave Theory). John Wiley & Sons, ISBN 0-470-13153-5, Hoboken, NJ

Ufimtsev P. Y. (2007). *Fundamentals of the Physical Theory of Diffraction*. John Wiley & Sons, ISBN 978-0-470-09771-7, Hoboken, NJ

Utkin, A. B. (2007). Pulsed radiation produced by a travelling exponentially decaying bipolar current pulse with high-frequency filling. *Proc. of International Conference "Days on Diffraction - 2007"*, pp. 137-142, ISBN 5-9651-0118-X, St. Petersburg, May 2007, IEEE, St. Petersburg

Utkin, A. B. (2008). Electromagnetic waves generated by line exponentially decaying current pulses propagating in lossy media, *Proc. of International Conference "Days on Diffraction - 2008"*, pp. 181-185, ISBN 978-5-9651-0277-8, St. Petersburg, June 2008, IEEE, St. Petersburg

Whittaker, E. T. (1904). On an expression of the electromagnetic field due to electrons by means of two scalar potential functions. *Proceedings of the London Mathematical Society*, s2-1, 1, (January 1904) 367-372, ISSN 0024-6115

Zhan, J. & Qin, Q. (1989). Analytic solutions of travelling-wave antennas excited by nonsinusoidal currents. *IEEE Transactions on Electromagnetic Compatibility*, 31, 3, (August 1989) 328-330, ISSN 0018-9375

Part 7

Radar Investigations

24

A Statistical Theory of the Electromagnetic Field Polarization Parameters at the Scattering by Distributed Radar Objects

Victor Tatarinov and Sergey Tatarinov
Tomsk State University of Control Systems and Radioelectronics
Russian Federation

1. Introduction

The development of coherent radars leads to discovery of coherent images specific properties at the electromagnetic waves scattering random distributed radar objects (RDRO). It was fixed that these images are having a stochastic - spotty structure. This structure is stipulated by interference of waves scattered by a random collection of RDRO secondary radiation centers. The similar phenomenon in optics has named as "speckles". So far all existent investigations both in the optic area and in RDRO coherent images for microwaves area (speckles theory) were made only for the case of electromagnetic waves linear polarization. However, the results of scattered wave's polarization parameters extensive experimental investigations at the scattering by man - made RDRO have demonstrated the existence of speckle - effect for scattered field polarization parameters.

One from main aspects of a developed polarization parameters statistical theory appears an establishment of connection between RCRO coherent images polarization parameters with these objects space spectra. It is demonstrated that a formation process of RDRO coherent images polarization - energetically parameters can be considered as the interference process. A validity of emergence principle for polarization - energetically parameters at the scattering by RDRO has demonstrated. A polarization coherence notion has introduced.

A generalization of Fresnel - Arago interference laws has demonstrated an emergence of a new physical effect, which demonstrates that at the scattering by RDRO a scattered field polarization - energetically parameters are not defined by an union of RDRO separate scatterers polarization parameters. How it is demonstrated, in the reality these parameters are defined be relations between RDRO separate scatterers properties.

As far as polarization - energetically parameters of RDRO coherent images are having the intensity dimension, then RCRO random polarization - energetically responses autocorrelation functions (ACF) will be as correlations of the 4-th order. It is demonstrated for the first time that ACF of the 4-th degree and a distribution of polarization proximity (distance) along a RDRO space spectra are related by Fourier transformations pair. Thus, a connection between scattered field polarization - energetically parameters distribution and polarization parameters distribution along a RDRO space spectrum can be found.

2. Electromagnetic waves scattering by random distributed object

2.1 Scattering geometry for random distributed object

A considerable number of articles are devoted to radar objects (RO) scattering properties. So, the scattering properties of RO having the simplest form were analyzed in «Proceedings of the IEEE» (1965. Vol. 53, № 8) and in the book (Ufimtsev, 1963). The special issues of «Proc. of the IEEE» (1985, Vol. 77, № 5), and «IEEE Trans. on Antennas and Propagation» (1989, № 5) were devoted to complex radar objects scattering properties. At this case complex radar objects were defined as non-regular bodies in contrast with simplest objects, which are usual body of rotation. However, all these investigations did not take into account a polarization of electromagnetic field. Complex (distributed) RO can be presented also as a combination of "point" scatterers (point RO) (Ostrovityanov, 1982), (Shtager, 1994). It is necessary to point out into attention that the notion of a "brilliant" point (or secondary center of radiation), which is defined by Fresnel the first zone is well known an it is used in radar theory at the definition of a radar object cross section ((RCS). For the correct definition of radiation secondary center we will use the expression for a radar object cross section in the form

$$\sigma = \frac{\pi}{\lambda^2}\left|\int \dot{I}(z)\exp\left\{j2k_0 z\cos\frac{\beta}{2}\right\}dz\right|, \qquad (1)$$

which is valid both two-position and one-position radar (Kell, 1965). The expression (1) is one from the forms of Stratton-Chu integral (Stratton & Chu, 1939) for the electromagnetic wave scattering by object having an arbitrary form . Here value β is two-positional scattering angle. The OZ axis is coinciding with this angle bisector. A vectorial function $\dot{I}(z)$ is connected with electrical and magnetic vectors of an incident wave. The integral (1) can be decomposed to an union of integrals. Every from these integrals is integrated into an area z, where integrand function is continuous. All these integrals can be interpreted as a "simplest" center of secondary radiation (Kell, 1965), which is stipulated by a stationary phase area . A contribution of every secondary radiation center in full diffraction integral (1) is connected with a stationary phase area size, i.e. an area near considered radiation center. Into the limits of this area a summarized phase of integrand function $\dot{I}(z)\exp\{j2k_0 z\cos\beta/2\}$ is differing not more than $\pi/2$ from its value in the central point of secondary radiation area.

If a radar object will have only one area (center) of secondary radiation, it will be an one-point radar object. In this case an unique are of secondary radiation must have a fixed phase center. An object area, which is defined a scattered field can be as ideal conducting surface or it can be having an electric anisotropy (Kell, 1965).

An electric anisotropy property of a secondary radiation area is defined the connection between of incident and scattered waves electric vectors $\dot{\vec{E}}_I$ in the form of matrix equation, which for the case of one-position radar has the form [Tatarinov et al, 2006]

$$\begin{Vmatrix}\dot{E}_{1S}\\ E_{2S}\end{Vmatrix} = \begin{Vmatrix}\dot{S}_{11} & \dot{S}_{12}\\ \dot{S}_{21} & \dot{S}_{22}\end{Vmatrix}\begin{Vmatrix}\dot{E}_{1I}\\ \dot{E}_{2I}\end{Vmatrix} \qquad (2)$$

where the matrix $\left\|\dot{S}_{jl}\right\|$ $(j,\ l=1,\ 2)$ is so-called back-scattering matrix (or scattering matrix) of a point scattering radar object.

Let us consider that a random complex radar object (RDRO) is a system of N rigidly connected scattering centers T_M, $(M=1, \dots N)$ each having local scattering matrix $\left\|\dot{S}_{jl}^{m}\right\|$. All scattering centers are contained in area T. The outline of this area is the outline of the RDRO (see fig. 1).

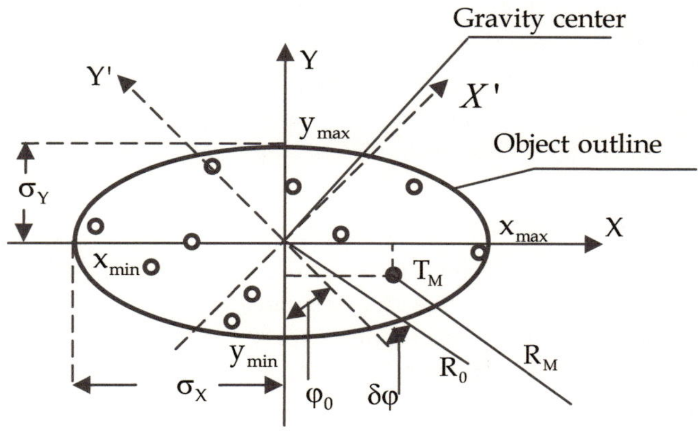

Fig. 1. The scattering geometry

Every realization of such a random system is a collection of a large number N of points reflectors (scattering centers). The $M-th$ scattering center has a position x_M, y_M in the coordinate system XOY connected with the object. The origin of coordinates is the "center of gravity" of the reflecting system. The conditions $\sum_{M=1}^{N} x_M = 0$; $\sum_{M=1}^{N} y_M = 0$ are existing in this connection. The values σ_X, σ_Y are essential mean square of the random reflecting system sizes and these values can be written as $\sigma_X = \sqrt{(1/N-1)\sum_{M=1}^{N} x_M^2}$; $\sigma_Y = \sqrt{(1/N-1)\sum_{M=1}^{N} y_M^2}$. For the case $N >> 1$ the approximation

$$\sigma_X = \sqrt{(1/N)\sum_{M=1}^{N} x_M^2}\ ; \quad \sigma_Y = \sqrt{(1/N)\sum_{M=1}^{N} y_M^2} \tag{3}$$

will be correct. Then mean square sizes of the distributed reflecting system along essential directions OX, OY can be defined in the form

$$L_X = 2\sqrt{(1/N)\sum_{M=1}^{N} x_M^2}\ ; \quad L_Y = 2\sqrt{(1/N)\sum_{M=1}^{N} y_M^2}\ . \tag{4}$$

If the probability density of reflectors coordinates is uniform, we can write (Kobak, 1975)

$$L_X = (x_{\max} - x_{\min})/\sqrt{3}\ ; \quad L_Y = (y_{\max} - y_{\min})/\sqrt{3}. \tag{5}$$

The distributed (complex) radar object representation in the form of scattered centers collection can be confirmed by the experimental results.

Fig. 2. The sea ship picture

So, on fig. 2 the picture of a sea ship is shown and on fig. 3 the optical simulation results of electromagnetic waves scattering by this ship are placed (Shtager, 1994). On this picture a lot of brilliant points are represented. These points are secondary scattering centers. A random collection of brilliant points is characterizing an electromagnetic field scattered both by complex (distributed) radar object and by the sea surface.

Fig. 3. The optical simulation results of electromagnetic waves scattering by the ship

Modern methods of complex radar objects experimental investigations allows us to have the sizes of the resolution cell equal to 1×1 m. In this case the resolution along the range is provided using ultra short pulses (5-7 nanosecond) and angular resolution is provided by aperture synthesis. This experimental radar has been designed in USA (Steinberg, 1989).. The fig. 4 presents two-dimensional radar map of the plane Locheed L - 1011 that was obtained with the use radar mentioned and its picture. It follows from these pictures that scattering centers distribution correctly represents the plane structure.

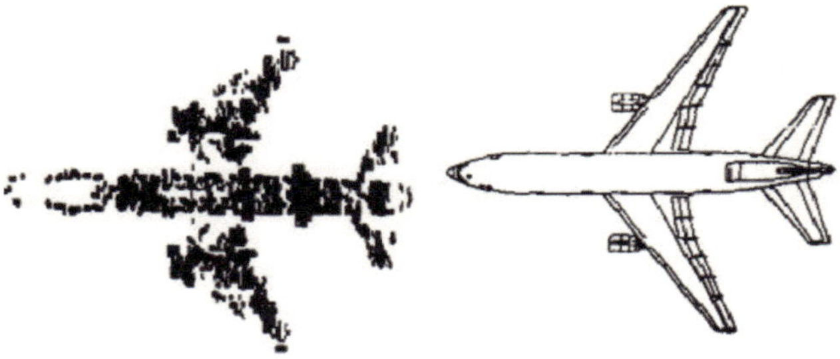

Fig. 4. The secondary scattering centers of the plane Locheed L - 1011 and its picture

Let us now to find a distance R_M between a scatterer T_M and radar, taking into account that a positional angle φ of object has defined in the form $\varphi = \varphi_0 \pm \delta\varphi$, where φ_0 is a mean positional angle. In what follows we will define a scattered field into some angular interval $\pm\delta\varphi$ relative to angle φ_0. In far zone a distance R_M can be found in the form

$$R_M = R_0 - \left[(x_M \cos\varphi_0 - y_M \sin\varphi_0)\sin\delta\varphi + (y_M \cos\varphi_0 + x_M \sin\varphi_0)\cos\delta\varphi \right]. \quad (6)$$

Using the approximation $\sin\delta\varphi \approx \delta\varphi$, $\cos\delta\varphi \approx 1$ we see that the expression (6) will have the form $R_M = R_0 - [x'_M \delta\varphi + y'_M]$, where $x'_M = x_M \cos\varphi_0 - y_M \sin\varphi_0$, $y'_M = y_M \cos\varphi_0 + x_M \sin\varphi_0$ are Cartesian coordinates of M - th scattering center into new coordinate system $X'0Y'$ that is rotated on an angle φ_0 relative to initial coordinate system XOY (see fig. 1). With regard to statistical independence of values x_M, y_m we can find mean square sizes of distributed radar object into new coordinate system $X'0Y'$ in the form

$$L'_X = \sqrt{0{,}5\left(L_X^2 + L_Y^2\right)}\sqrt{1+\gamma_T C_{2\varphi 0}} \; ; \; L'_Y = \sqrt{0{,}5\left(L_X^2 + L_Y^2\right)}\sqrt{1-\gamma_T C_{2\varphi 0}} \quad (7)$$

Here and later on the designation $C_\varphi \equiv \cos\varphi$; $S_\varphi \equiv \sin\varphi$ for the writing shortening are used. Into the expressions (7) the designation $\gamma_T = \left(L_X^2 - L_Y^2\right)/\left(L_X^2 + L_Y^2\right)$ was introduced, where values L_X, L_Y are mean square sizes of distributed radar object into its initial coordinate system XOY. So, a value γ_T is dimensionless factor for characteristic of distributed radar object on the average. We will name this factor as "distributed radar object geometric anisotropy degree".

2.2 Polarization invariants of distributed object elementary scatterer and their geometric representation

For the definition of an electromagnetic field at the scattering by random distributed object we must determine of polarization parameters system for description of scattering process by elementary (point) scatterer in the best way.

Let us to write an electromagnetic field, scattered by point scattering center T_M in the form

$$\dot{E}_S^M(\omega) = \frac{\exp(-j2kR_M)}{R_M\sqrt{4\pi}} \left\| S_{jl}^M \right\| \dot{E}_0 , \quad (8a)$$

taking into account the coherent nature of scattering process. Here a value R_M is the distance between scatterer T_m and radar; matrix $\left\| \dot{S}_{jl}^M \right\|$ is a scattering matrix of distributed radar object elementary scatterer; vectors $\dot{E}_0$ and $\dot{E}_S$ are complex Jones vectors of radiated and scattered waves respectively. Now we will analyze most important invariant of elementary scatterer scattering matrix - a complex degree of point radar object polarization anisotropy. For this parameter introduction we will use the writing of radar object scattering matrix in the Cartesian polarization basis under the condition of arbitrary angle of object eigencoordinates system and radar coordinates system mutual orientation

$$\left\|\dot{S}_{jl}\right\| = \begin{Vmatrix} C_\beta & -S_\beta \\ S_\beta & C_\beta \end{Vmatrix} \begin{Vmatrix} \dot{\lambda}_1 & 0 \\ 0 & \dot{\lambda}_2 \end{Vmatrix} \begin{Vmatrix} C_\beta & S_\beta \\ -S_\beta & C_\beta \end{Vmatrix} = 0{,}5 \begin{Vmatrix} (\dot{\lambda}_1 + \dot{\lambda}_2) + (\dot{\lambda}_1 - \dot{\lambda}_2)C_{2\beta} & (\dot{\lambda}_1 - \dot{\lambda}_2)S_{2\beta} \\ (\dot{\lambda}_1 - \dot{\lambda}_2)S_{2\beta} & (\dot{\lambda}_1 + \dot{\lambda}_2) - (\dot{\lambda}_1 - \dot{\lambda}_2)C_{2\beta} \end{Vmatrix}. \quad (8b)$$

Here β is mutual orientation angle, $\dot{\lambda}_1 = \lambda_1 \exp\{i\varphi/2\}$, $\dot{\lambda}_2 = \lambda_2 \exp\{-i\varphi/2\}$ are scattering matrix complex eigenvalues, λ_1, λ_2 are their modules and φ is arguments difference. Let us decompose this matrix using Pauli matrix system:

$$\left\|\dot{S}_{jl}\right\| = 0,5\left\{S_0^{SM}\begin{Vmatrix}1 & 0\\ 0 & 1\end{Vmatrix} + S_1^{SM}\begin{Vmatrix}1 & 0\\ 0 & -1\end{Vmatrix} + S_2^{SM}\begin{Vmatrix}0 & 1\\ 1 & 0\end{Vmatrix} + S_0^{SM}\begin{Vmatrix}0 & i\\ -i & 0\end{Vmatrix}\right\},$$

where $S_j^{SM} = Sp\left\{\left\|\sigma_j\right\|\left\|\dot{S}_{jl}\right\|\right\}$, (j=0, 1, 2, 3) are developing coefficients. So,we can rewrite the expression (8b) in the form

$$\left\|S_{jl}\right\| = 0,5\left(\dot{\lambda}_1 + \dot{\lambda}_2\right)\left\{\begin{Vmatrix}1 & 0\\ 0 & 1\end{Vmatrix} + \dot{\mu}\begin{Vmatrix}\cos 2\beta & \sin 2\beta\\ \sin 2\beta & -\cos 2\beta\end{Vmatrix}\right\}, \tag{9}$$

where the complex value

$$\dot{\mu} = \left(\dot{\lambda}_1 - \dot{\lambda}_2\right)/\left(\dot{\lambda}_1 + \dot{\lambda}_2\right) \tag{10}$$

is the "complex degree of the polarization anisotropy" (CDPA). It follows from this expression that a scattering matrix $\left\|S_{jl}\right\|$ can be presented in the weighted sum of an isotropic radar object, i.e. trihedral corner reflector and dihedral corner reflector, a weight of which is defined by $\dot{\mu}$ value. Let us write the CDPA module as follows

$$|\dot{\mu}| = \sqrt{\dot{\mu}\dot{\mu}^*} = \sqrt{\left(\lambda_1^2 + \lambda_2^2 - 2\lambda_1\lambda_2 C_\varphi\right)/\lambda_1^2 + \lambda_2^2 + 2\lambda_1\lambda_2 C_\varphi}\ . \tag{11}$$

The real part, imaginary part and argument of the CDPA can be written in the form

$$\operatorname{Re}\{\dot{\mu}\} = \frac{\lambda_1^2 - \lambda_2^2}{\lambda_1^2 + \lambda_2^2 + 2\lambda_1\lambda_2 C_\varphi};\ \operatorname{Im}\{\dot{\mu}\} = \frac{2\lambda_1\lambda_2 S_\varphi}{\lambda_1^2 + \lambda_2^2 + 2\lambda_1\lambda_2 C_\varphi};\ \arg\{\dot{\mu}\} = \arctan\left(\frac{2\lambda_1\lambda_2 S_\varphi}{\lambda_1^2 - \lambda_2^2}\right), \tag{12}$$

where λ_1 and λ_2 are the modules of the eigen values; $\varphi = \arg\{\dot{\lambda}_1\} - \arg\{\dot{\lambda}_2\}$ is the arguments' difference corresponding to the phase shift caused by the object' electric properties. It is obvious that the values $|\dot{\mu}|$, $\operatorname{Re}\{\dot{\mu}\}$, $\operatorname{Im}\{\dot{\mu}\}$, $\arg\{\dot{\mu}\}$ are also the invariants of the scattering matrix. Thus, the CDPA fully describes the polarization properties of the object. The introduction of the CDPA notion allows presenting on the complex plane the objects' polarization properties how it was demonstrated in (Tatarinov at al, 2006).

Let us consider this aspect more explicitly. It can be shown with the use of expressions (11) and (12) that $\dot{\mu}$-value can be presented on the complex plane $\dot{\mu} = \operatorname{Re}\{\dot{\mu}\} + j\operatorname{Im}\{\dot{\mu}\}$ (fig.5). The points of this plane are connected with the polarization properties of the radar objects as far as $\dot{\mu}$-value is a radar object scattering matrix polarization invariant. The origin of the coordinates $(\dot{\mu} = 0)$ corresponds to the object characterizing by values $\lambda_1 = \lambda_2$, $\varphi = 0$ (a phase shift between eigenvalues is zero). It follows from the necessity to simultaneously satisfy the conditions $\operatorname{Re}\{\dot{\mu}\} = 0$, $\operatorname{Im}\{\dot{\mu}\} = 0$. The physical analog of this point is polarization isotropic radar object (trihedral corner reflector, sphere, metal plate). The point at infinity corresponds to the object having the following polarization properties:

$\lambda_1 = \lambda_2$, $\varphi = \pm\pi$. The physical analog of the point at infinity is the dihedral corner reflector. All points of the imaginary axis correspond to the objects with $\mathrm{Re}\{\dot{\mu}\} = 0$, i.e. $\lambda_1 = \lambda_2$. In this case, the points laying on the positive imaginary semi-axis, present radar objects which are characterized by the phase shift $\varphi > 0$, while the negative imaginary semi-axis

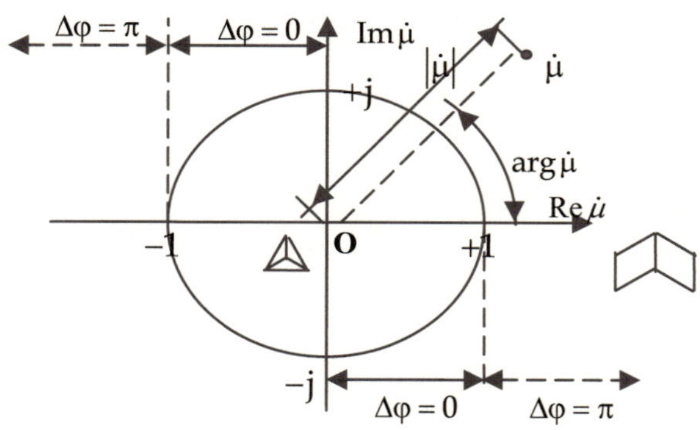

Fig. 5. The complex plane of radar objects

($\mathrm{Im}\{\dot{\mu}\} < 0$) depicts the objects with $\varphi < 0$. The points $\pm j$ present the objects having $\varphi = \pm(\pi / 2)$ phase shift. All real axis's points of the complex $\dot{\mu}$ –plane correspond to the objects with zero phase shift $\varphi = 0$; (i. e. $\mathrm{Im}\{\dot{\mu}\} = 0$). However, the given case is complicated by the fact that the object, which corresponds to the point at infinity, is the dihedral reflector. This contradiction can be solved, considering the equality $\sin\varphi = 0$ both for $\varphi = 0$; and $\varphi = \pi$; cases. Then the points of the real axis of the complex $\dot{\mu}$ –plane must be determined with the use of the conditions ($\cos 0 = 1,\ \cos\pi = -1$) as $\mathrm{Re}\{\dot{\mu}\} = \left(\lambda_1^2 - \lambda_2^2\right) / \left(\lambda_1^2 + \lambda_2^2 + 2\lambda_1\lambda_2 C_\varphi\right)$. Thus, into the interval $\mathrm{Re}\{\dot{\mu}\} = 0$; $\mathrm{Re}\{\dot{\mu}\} = 1$ the value λ_2 reduces from $\lambda_1 = \lambda_2$ in the origin up to $\lambda_2 = 0$ in the point $\mathrm{Re}\{\dot{\mu}\} = 1$ (horizontal oriented object). This point depicts the "degenerated" radar object (long linear object, dipole, i.e. polarizer). The phase shift in the point has an undefined value. It changes spasmodically by π when passing the point $\mathrm{Re}\{\dot{\mu}\} = 1$. Then, the value λ_2 increases from $\lambda_2 = 0$ (in the point $\mathrm{Re}\{\dot{\mu}\} = 1$) up to $\lambda_1 = \lambda_2$ (the point at infinity). In this case, the phase shift along the ray $\mathrm{Re}\{\dot{\mu}\} = 1$; $\mathrm{Re}\{\dot{\mu}\} = \infty$ equals to π . The similar analysis can be made with respect to the negative semi-axis $\mathrm{Re}\{\dot{\mu}\}$. Thus, the complex $\dot{\mu}$ –plane has the properties equivalent to the properties of the circular complex plane . However, at that time when the circular complex plane presents the polarization properties of electromagnetic waves, the complex $\dot{\mu}$ –plane is intended for presentation of the invariant polarization parameters of the radar objects scattering matrix.

Analyzing the similarity, which exists between the $\dot{\mu}$ –plane and the circular complex plane, we can conclude that it is expedient to choose the circular basis as the basis for presenting the radiated and scattered waves. The scattering matrix (8) in the circular basis can be found in the form (Tatarinov at al, 2006)

$$\left\|S_{jl}^{RL}\right\| = 0,5j \left\| \begin{matrix} (\dot{\lambda}_1 - \dot{\lambda}_2)\exp\{j(2\beta - \pi/2)\} & (\dot{\lambda}_1 + \dot{\lambda}_2) \\ (\dot{\lambda}_1 + \dot{\lambda}_2) & (\dot{\lambda}_1 - \dot{\lambda}_2)\exp\{-j(2\beta - \pi/2)\} \end{matrix} \right\|. \quad (13)$$

Change of the rotation direction under backscattering is also considered in this expression. Let us present the radiated wave in the circular basis $(\dot{\vec{e}}_R, \dot{\vec{e}}_L)$. The circular polarization ratio for this wave can be written as $\dot{P}_R^{RL} = \dot{E}_R / \dot{E}_L$. Then, the circular polarization ratio for the scattered wave will have the form

$$\dot{P}_S^{RL} = \left\{1 + \dot{\mu}\exp[-j(2\beta - \pi/2)]\dot{P}_R^{RL}\right\} / \left\{\dot{\mu}\exp[-j(2\beta - \pi/2)] + \dot{P}_R^{RL}\right\}. \quad (14)$$

It is possible to set the specific polarization state of the radiated wave, when polarization ratio of the scattered wave will have an unique form. So, if $\dot{P}_R^{RL} = \infty$, (right circular polarized wave) then we can rewrite the expression (14) in the form

$$\dot{P}_S^{RL} = \lim_{\dot{P}_R^{RL} \to \infty} \frac{1 + \dot{\mu}\exp\{-j(2\beta - \pi/2)\}\dot{P}_R^{RL}}{\dot{\mu}\exp\{-j(2\beta - \pi/2)\} + \dot{P}_R^{RL}} \exp\{-j[2\beta - \arg(\dot{\mu}) - \pi/2]\}. \quad (15)$$

If $\theta = 0$, then we get

$$\dot{P}_S^{RL} = |\dot{\mu}|\exp\{-j[\arg(\dot{\mu}) + \pi/2]\}. \quad (16a)$$

Using the Jones vector $\dot{\vec{E}}_{RL}$ we can find the circular polarization ratio in the form

$$\dot{P}^{RL} = \tan(\alpha + \pi/4)\exp\{-i(2\beta - \pi/2)\}. \quad (16b)$$

Here α is an ellipticity angle and β is an orientation angle of polarization ellipse.
The comparison of expressions (16a, b)shows that the measured module of the circular polarization ratio of the scattered wave (when the radiated wave has right circular polarization) is equal to the complex degree polarization anisotropy (CDPA) module

$$\left|\dot{P}_S^{RL}\right| = \tan(\alpha + \pi/4) = |\dot{\mu}|. \quad (17)$$

The argument of the $\dot{P}_S^{RL}$ (for the case $\theta = 0$) can be presented as $\arg\{\dot{\mu}\} + \pi/2 = -2\beta + \pi/2$ or $\arg\{\dot{\mu}\} = -2\beta$. The last expression demonstrates that the value of CDPA argument determines the orientation of the polarization ellipse in the eigencoordinates system of the scattering object. If $\theta \neq 0$, then the polarization ellipse will be rotated additionally an angle of 2β. The correspondence between the circular complex plane and the Riemann sphere, having unit diameter, was analyzed in details in (Tatarinov at al, 2006) with the use of the stereographic projection equations, which are connecting on-to-one the circular complex plane points $\dot{P}^{RL} = \mathrm{Re}\,\dot{P}^{RL} + j\,\mathrm{Im}\,\dot{P}^{RL}$ with Cartesian coordinates X_1, X_2, X_3 of the point S, laying on the Riemann sphere surface. The transition from the circular complex plane to the Poincare sphere , having unit radius, can be realized with the use the modified stereographic projection equations

$$X = 2\mathrm{Re}\,\dot{P}^{RL} / \left(1+\left|\dot{P}^{RL}\right|^2\right),\quad Y = 2\mathrm{Im}\,\dot{P}^{RL} / \left(1+\left|\dot{P}^{RL}\right|^2\right),\quad Z = 2\left\{\left|\dot{P}^{RL}\right|^2 / \left(1+\left|\dot{P}^{RL}\right|^2\right) - 0{,}5\right\}.$$

Using these equations we can connect the complex $\dot{\mu}$ -plane of radar objects with the sphere of unit radius(fig. 6).

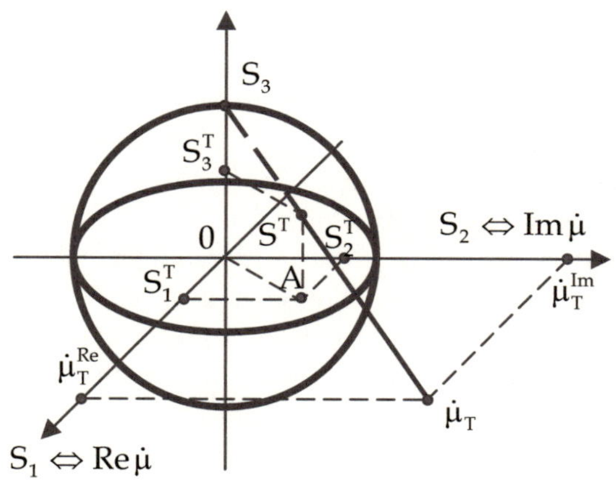

Fig. 6. Polarization sphere of radar objects

We will assume that the axes $\vec{S}_1, \vec{S}_2$ of the three-dimensional space S_1, S_2, S_3 are coinciding with real and imaginary axes $\mathrm{Re}\{\dot{\mu}\}$, $\mathrm{Im}\{\dot{\mu}\}$ of the radar objects complex plane $\dot{\mu} = \mathrm{Re}\,\dot{\mu} + j\,\mathrm{Im}\,\dot{\mu}$ respectively. In accordance with stated above, all point of this S_T -sphere will be connected one-to-one with corresponding points of radar objects complex $\dot{\mu}$ -plane.

Let us consider now that a radar object is defined on the $\dot{\mu}$ -plane by the point $\dot{\mu}_T = \mu_R + j\mu_I$. Then we will connect the point $\dot{\mu}_T$ with the sphere north pole by the line, which crosses the sphere surface at point S^T . The projections S_1^T, S_2^T, S_3^T of the point S_T to the axes S_1, S_2, S_3 will be defined to modified stereographic projection equation. It is not difficult to see that these values are satisfying to the unit sphere equation $\left(S_1^T\right)^2 + \left(S_2^T\right)^2 + \left(S_3^T\right)^2 = 1$.

Thus, all points of the complex plane of radar objects are corresponding one-to-one to points of the sphere S_T . We will name this sphere as the unit sphere of radar object .

2.3 Scattering operator of distributed radar object and its factorization

Let us to write now the Jones vector of the field scattered by the RDRO in the form

$$\left\|\bar{E}_{S\Sigma}(k,\ \delta\varphi)\right\| = \frac{\exp\{-j2kR_0\}}{R_0\sqrt{4\pi}} \left\| \begin{matrix} \sum\limits_{M=1}^{N} \dot{S}_{11}^{M} \exp\left\{j2k\eta^{M}\right\} & \sum\limits_{M=1}^{N} \dot{S}_{12}^{M} \exp\left\{j2k\eta^{M}\right\} \\ \sum\limits_{M=1}^{N} \dot{S}_{21}^{M} \exp\left\{j2k\eta^{M}\right\} & \sum\limits_{M=1}^{N} \dot{S}_{22}^{M} \exp\left\{j2k\eta^{M}\right\} \end{matrix} \right\| \left\| \begin{matrix} \dot{E}_1^0 \\ \dot{E}_2^0 \end{matrix} \right\|, \tag{18}$$

where $\eta^m = x'_m \delta\varphi + z'_m$ and matrix

$$\left\|\dot{S}_{jl}^{\Sigma}(k,\ \delta\varphi)\right\| = \frac{\exp\{-j2kR_0\}}{R_0\sqrt{4\pi}}\left\|\begin{matrix}\sum\limits_{M=1}^{N}\dot{S}_{11}^{M}\exp\left\{j2k\eta^{M}\right\} & \sum\limits_{M=1}^{N}\dot{S}_{12}^{M}\exp\left\{j2k\eta^{M}\right\} \\ \sum\limits_{M=1}^{N}\dot{S}_{21}^{M}\exp\left\{j2k\eta^{M}\right\} & \sum\limits_{M=1}^{N}\dot{S}_{22}^{M}\exp\left\{j2k\eta^{M}\right\}\end{matrix}\right\| \tag{19}$$

is scattering operator, which includes space, polarization and frequency property of random distributed radar object. It follows from the expression (19) that all elements of the RDRO scattering operator $\left\|S_{jl}^{\Sigma}(k,\ \delta\varphi)\right\|$ are the function of two variables. These variables are as the wave vector absolutely value and positional angle φ. A dependence from the wave vector absolutely value is the frequency dependence, as far as for a media having the refraction parameter $n=1$ the wave vector has the form $k=2\pi/\lambda=\omega/c$, where c is light velocity and $\omega=2\pi f$. It is necessary to note here that scattered field polarization parameters at the scattering by one-point radar object are independent both from positional angle and frequency. For analysis of polarization-angular and polarization-frequency dependences of the field at the scattering by the RDRO we write the exponential function $\exp\left\{-j\left[2kx'_M\delta\varphi+2kz'_M\right]\right\}$ that has been included into the operator (19) elements. The index of this function is originated by the existence both angular and frequency dependences of the field scattered by the RDRO. We will rewrite this index for its analysis:

$$f_M(k,\delta\varphi)=2kx'_M\delta\varphi+2kz'_M. \tag{20}$$

Let's us assume that the initial wave is quasimonochromatic ($\Delta\omega/\omega_0 << 1$) and that radar radiation frequency arbitrary changes are not disturbing this condition. We can write the wave vector k absolutely value in the form

$$k=(\omega/c)=(\omega_0+\Delta\omega)/c, \tag{21}$$

where ω_0 is a mean constant frequency of radar radiation, and $\Delta\omega$ is a variable part originated by radar radiation frequency change or frequency modulation. The substitution of the expression (21) in the expression (20) give us

$$\begin{aligned} f_M(k,\ \delta\varphi) &= 2\left[(\omega_0+\Delta\omega)x'_m\delta\varphi/c\right]+2\left[(\omega_0+\Delta\omega)z'_M/c\right]= \\ &= 2(\omega_0x'_M\delta\varphi/c)+2(\omega_0z'_M/c)+2(x'_M\delta\varphi\Delta\omega+z'_M\Delta\omega)/c \end{aligned}. \tag{22}$$

As far as the value ω_0 is constant, then from all items of the expression (22) only the value $2x'_m\delta\varphi\Delta\omega/c$ is depending simultaneously both on variable positional angle $\delta\varphi$ and on frequency variable $\Delta\omega$. However, it is not difficult to see that the inequality

$$2(x'_M\delta\varphi\Delta\omega/c) << 2(z'_M\Delta\omega/c) \tag{23}$$

is correct under the condition $\delta\varphi << 1\ Rad$ (i.e. $\delta\varphi \le 10^{\circ}$). Taking into account this inequality, we can neglect by the value $2x'_M\delta\varphi\Delta\omega/c$ in the equation (23) and then we can rewrite it in the form

$$f_M(k_0,\ \delta\varphi,\ \Delta\omega)\approx 2k_0x'_M\delta\varphi+t'_M\omega, \tag{24}$$

where $\omega = \omega_0 + \Delta\omega$, $t'_M = 2z'_M / c$. Thus, the angular and frequency variables in the expression (24) are separated. It is so-called factorization operation. The value t_M is a doubled time interval, which is necessary for initial wave passage of a distance, which is a projection of segment z_M on the OZ' axis, i.e. on the propagation direction of radar initial wave. This analysis shows that the scattered field polarization parameters frequency dependence at the scattering by the RCRO is defined by the projections of the scattering centers co-ordinates on the OZ' axis, which is coinciding with the radar initial wave propagation direction. In other words, a frequency dependence is defined by the RCRO extension along the initial wave propagation direction. It follows simultaneously from the equation (33) that the scattered field polarization-angular dependence on the mean frequency ω_0 is defined by the values x'_m collection. These values are projections of scattering centers positions on the OX' axis that is perpendicular to radar initial wave propagation direction. So, an extension of the RCRO along the OX' axis is originated a polarization-angular dependence of field polarization parameters at the scattering by a RCRO.

3. Angular response function of a distributed object and its basic forms

Taking into account the results of subsection 2.3 we can now consider separately the polarization-angular and polarization-frequency forms of a distributed radar object responses on unit action, having circular polarization.

In accordance with the mentioned results the polarization-angular response of a complex object at mean frequency ω_0 is determined by extension of the object along the axis OX', that is perpendicular to direction of incident wave propagation's. Taking into account the expression (18) we can write the scattering operator (28) of the distributed radar object for the circular polarization basis in the form

$$\left\| \dot{S}_{jl}^{\Sigma,RL}(k_0,\ \delta\varphi) \right\| = \frac{\exp\{-j2k_0R_0\}}{R_0\sqrt{4\pi}} \begin{Vmatrix} \dot{S}_{11}^{\Sigma,RL}(k_0,\ \delta\varphi) & \dot{S}_{12}^{\Sigma,RL}(k_0,\ \delta\varphi) \\ \dot{S}_{21}^{\Sigma,RL}(k_0,\ \delta\varphi) & \dot{S}_{22}^{\Sigma,RL}(k_0,\ \delta\varphi) \end{Vmatrix}, \tag{25}$$

Where

$$\dot{S}_{11}^{\Sigma,RL}(k_0,\delta\varphi) = \sum_{M=1}^{N} \left|\dot{\Delta}^M\right| \exp\{j[2k_0x'_M\delta\varphi - \beta_{3M}]\};\ \dot{S}_{22}^{\Sigma,rl}(k_0,\delta\varphi) = -\sum_{m=1}^{N} \left|\dot{\Delta}^m\right| \exp\{j[2k_0x'_m - \beta_{1m}]\};$$

$$\dot{S}_{12}^{\Sigma,rl}(k_0,\delta\varphi) = \dot{S}_{21}^{\Sigma,rl}(k_0,\delta\varphi) = j\sum_{m=1}^{N} \left|\dot{\Sigma}^m\right| \exp\{j[2k_0x'_m\delta\varphi - \beta_{2m}]\}.$$

Here $\beta_{2M} = \arg\dot{\Sigma}^M - 2kz'_M$; $\beta_{1M} = \arg\dot{\Delta}^M - 2\theta_M - 2kz'_M$; $\beta_{2M} = \arg\dot{\Sigma}^M - 2kz'_M$; $\beta_{3M} = \arg\dot{\Delta}^M + 2\theta_M - 2kz'_M$ and values $\dot{\Delta}^M = \dot{\lambda}_1^M - \dot{\lambda}_2^M$, $\dot{\Sigma}^M = \dot{\lambda}_1^M + \dot{\lambda}_2^M$ are the difference and union of $M-th$ elementary scatterers eigen values . If the Jones vector of the incident wave is right circular polarized, we can write for the circular Jones vector of the wave, scattered by a distributed radar object in the form

$$\vec{E}_S^{\Sigma,RL}(k_0,\ \delta\varphi)=\frac{\exp\{-j2k_0R_0\}}{R_0\sqrt{4\pi}}\left\|\begin{matrix}-j\sum\limits_{M=1}^{N}\left|\dot{\Sigma}^M\right|\exp\{j[\Omega_M\delta\varphi-\beta_{2M}]\}\\ \sum\limits_{M=1}^{N}\left|\dot{\Delta}^M\right|\exp\{j[\Omega_M\delta\varphi-\beta_{1M}]\}\end{matrix}\right\|. \tag{26}$$

We are using here the notion of spatial frequencies $\Omega_M = 2k_0x'_M$ (Kobak, 1975), (Tatarinov et al, 2006) that allows us to consider the elements of the Jones vector (26) as the sum of a large number harmonic oscillations. The moving coordinate of these oscillation is the variable positional angle $\delta\varphi$. The frequencies of these oscillations are determined by projections of the coordinates of scattering centers T_M on the axis OX'.
Amplitudes of oscillations are the values $\left|\dot{\Sigma}^M\right|$, $\dot{\Delta}^M$ can be characterized by the Rayleigh distribution (Potekchin et al, 1966) and the random initial phases β_{1M}, β_{2m} may have the uniform distribution into the interval $0\div 2\pi$. Stochastic values of the spatial frequencies Ω_M may have the uniform distribution in the interval $\Omega_{MIN}\div\Omega_{MAX}$. This interval correspond to domain of definition $x'_{MIN}\div x'_{MAX}$ along the OX' axis. Thus, we can consider the sum (26) as a complex stochastic function of the moving coordinate $\delta\varphi$.
The circular polarization ratio for mean frequency ω_0 we can write using elements of Jones vector (26) in this case will have the form

$$\dot{P}_S^{RL}(k_0,\ \delta\varphi)=j\sum_{M=1}^{N}\left|\dot{\Delta}^M\right|\exp\{j[\Omega_M\delta\varphi-\beta_{1M}]\}\Big/\sum_{M=1}^{N}\left|\dot{\Sigma}^M\right|\exp\{j[\Omega_M\delta\varphi-\beta_{2M}]\}. \tag{27}$$

This ratio represents an angular distribution of the polarization parameters of an RCRO and it is the polarization-angular response function of a random distributed radar object on the unit action, having the form of a circular polarized wave.
Polarization-angular response function (27) is a generalization of the point object response (16a) on the unit action, having the form of a circular polarized wave. Both the polarization properties of scatterers, and geometrical parameters of a random distributed radar object are represented into the polarization-angular response (27). We will transform every item of the numerator of (27) in the following form

$$\begin{aligned}\left|\dot{\Delta}^M\right|\exp\{j[\Omega_M\delta\varphi-\beta_{1M}]\}&=\left|\dot{\Sigma}^M\right|\left(\left|\dot{\Delta}^M\right|/\left|\dot{\Sigma}^M\right|\right)\exp\{j[\Omega_M\delta\varphi-\beta_{1M}]\}=\\&=\left|\dot{\Sigma}^M\right|\left|\dot{\mu}^M\right|\exp\{j[\Omega_M\delta\varphi-\beta_{1M}]\}\end{aligned}$$

Here the values $\left|\dot{\mu}^M\right|$ are determined by expression (10) and represent modules of elementary scatterer's T_M complex degree polarization anisotropy. The values $\left|\dot{\mu}^M\right|$, that are describing the polarization properties of elementary reflectors of an RDRO, make up a general expression by using the weight factors $\left|\dot{\Sigma}^M\right|=\left[\left(\lambda_1^M\right)^2+\left(\lambda_2^M\right)^2+2\lambda_1^M\lambda_2^M\cos\Delta\varphi_M\right]^{0,5}$.
Taking into account this fact, we can find

$$\dot{P}_S^{RL}(k_0,\ \delta\varphi) = j\sum_{M=1}^{N}\left|\dot{\Sigma}^M\right|\left|\dot{\mu}^M\right|\exp\{j(\Omega_M\delta\varphi - \beta_{1M})\} \Big/ \sum_{M=1}^{N}\left|\dot{\Sigma}^M\right|\exp\{j(\Omega_M\delta\varphi - \beta_{2M})\}. \quad (28)$$

The weight factors $\left|\dot{\Sigma}^m\right|$ are connected with the radar cross sections of elementary scatterers. The angular distribution of the polarization ratio (28) completely describes the polarization structure of the field, scattered by a complex object

$$\dot{P}_S^{RL}(k_0,\ \delta\varphi) = \tan\left[\alpha(k_0,\ \delta\varphi) + \pi/4\right]\exp\{j2\beta(k_0,\ \delta\varphi)\}. \quad (29)$$

Here values $\alpha(k_0,\ \delta\varphi)$ and $\beta(k_0,\ \delta\varphi)$ are angular distributions both of the ellipticity angle and the orientation angle of the polarization ellipse of the scattered field.
Existing measurement methods allow us to carry out direct measurements of the module of a polarization ratio. Thus, we have the possibility for the direct measurements of ellipticity angle of the scattered wave. The measurement of the orientation needs indirect methods.
First of all we shall consider the opportunity of the characteristics of an ellipticity angle in the analysis of wave polarization, scattered by random distributed objects. We will use all forms of complex radar object polarization-angular response, which are different functions of an ellipticity angle. The following parameters are connected with an ellipticity angle value:

- The value $\tan(\alpha + \pi/4) = \left|\dot{P}^{RL}(k_0,\ \delta\varphi)\right|$, determined in the interval $0 \le \tan(\alpha + \pi/4) \le \infty$;
- The coefficient of ellipticity $K(k_0,\ \delta\varphi) = \tan\alpha$, determined in the interval $-1 \le \tan\alpha \le 1$. This coefficient is connected with the module of circular polarization ratio $\left|\dot{P}^{RL}\right|$ by

$$K(k_0,\ \delta\varphi) = \tan\alpha = \left[\left(\left|\dot{P}^{RL}\right| - 1\right)/\left(\left|\dot{P}^{RL}\right| + 1\right)\right] = \left[\tan\left(\alpha + \pi/4\right) - 1\right]/\left[\tan\left(\alpha + \pi/4\right) + 1\right]; \quad (30)$$

- -The third normalized Stokes parameter $S_3 = \sin 2\alpha$, determined in the interval $-1 \le S_3 \le 1$.

This parameter is connected with the square of the circular polarization ratio module as

$$S_3(k_0,\ \delta\varphi) = \sin 2\alpha(k_0,\ \delta\varphi) = \left[\left|\dot{P}_S^{RL}(k_0,\ \delta\varphi)\right|^2 - 1\right]/\left[\left|\dot{P}_S^{RL}(k_0,\ \delta\varphi)\right|^2 + 1\right]. \quad (31a)$$

The Stokes parameter S_3 and the ellipticity coefficient K are connected by the expression

$$S_3 = \sin 2\alpha = \sin(2\arctan K) = 2K/\left(1 + K^2\right). \quad (31b)$$

The inverse function $K(S_3)$ is the solution of the equation $S_3K^2 - 2K + S_3 = 0$. We will choose the solution $K(S_3) = \left(1 - \sqrt{1 - S_3^2}\right)/S_3$ from two versions $K_{1/2} = \left(1 \pm \sqrt{1 - S_3^2}\right)/S_3$. It follows from conformity $K = -1$, $S_3 = -1$; $K = 0$, $S_3 = 0$; $K = 1$, $S_3 = 1$ that only solution

(4c) remains. Thus, we can use the initial polarization-angular response $\left|\dot{P}_S^{rl}(k_0, \delta\varphi)\right| = \tan\left[\alpha(k_0, \delta\varphi) + \pi / 4\right]$ and two other forms of responses - $K(k_0, \delta\varphi) = \tan\left[\alpha(k_0, \delta\varphi)\right]$ and $S_3(k_0, \delta\varphi) = \sin 2\alpha(k_0, \delta\varphi)$.

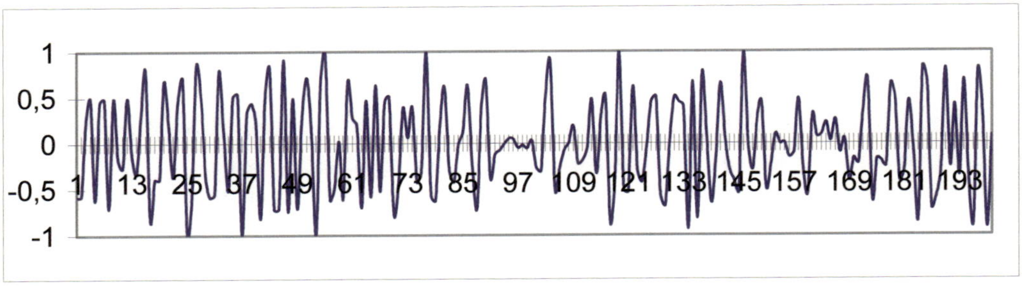

Fig. 7. The experimental realization of polarization- angular response function $S_3(\delta\varphi)$

For example, the experimental realization, having the form of narrow-band angular dependence $S_3(\delta\varphi)$ has shown on the Fig. 7. The angular extension of this experimental realization is $\pm 20^0$ at the observation to radar object board . The samples of polarization-angular response function are following with the angular interval $0{,}2^0$.

4. An emergence principle and polarization coherence notion

The analysis of an electromagnetic field polarization properties at the scattering by space distributed radar object is closely connected with two key problems. The first problem is the influence of separated scatterers space diversity on scattered field polarization. The second key problem of polarization properties investigation at the scattering by distributed radar object is connected with scattered field polarization properties definition on the base of the emergence principle with the use of possible relations between complex radar object parts polarization properties.

4.1 An emergence principle and space frequency notion for a simplest distributed object. polarization proximity and polarization distance

Let us to define a field, scattered by RDO using the Stratton-Chu integral (1), which allows us to represent this field as the union of waves scattered by elementary scatterers ("bright" or "brilliant" points), forming complex object. For the case when every elementary scatterer is characterizing by its scattering matrix $\left\|S_{jl}^M\right\|$; $(i, l = 1, 2)$ then the scattered field complex vector can be defined in the form

$$\left\|\dot{E}_S^{\Sigma}(k_0, \delta\varphi)\right\| = \frac{\exp\{-j2k_0R_0\}}{R_0\sqrt{4\pi}} \sum_{M=1}^{N} \left\|S_{jl}^M\right\| \left\|\dot{E}_0\right\|, \tag{32}$$

where R_0 is a distance between the radar and object gravity center, $\delta\varphi$ is a positional angle of the object and $\left\|\dot{E}_0\right\|$ is the complex vector of initial wave. It is necessary to indicate here that the expression (32) has been represented only individual polarization properties of every from scatterers, which are forming a large distributed radar object. Unfortunately, a large system property in principle can not be bringing together to an union of this system elements properties. The conditionality of integral system properties appear by means of its elements relations. These relations lead to the "emergence" of new properties which could not exist for every element separately. The emergence notion is one from main definitions of the systems analysis (Peregudov & Tarasenko, 2001). Let us consider the simplest distributed radar object in the form of two closely connected scatterers A and B (reflecting elliptical polarizers), which can not be resolved by the radar. These scatterers are distributed in the space on the distance l and are characterizing by the scattering matrices in the Cartesian polarization basis:

$$\|S_1\| = \begin{Vmatrix} \dot{a}_1 & 0 \\ 0 & \dot{a}_2 \end{Vmatrix}, \|S_2\| = \begin{Vmatrix} \dot{b}_1 & 0 \\ 0 & \dot{b}_2 \end{Vmatrix}. \tag{33}$$

It will be the case of coherent scattering and its geometry is shown on the fig. 8.

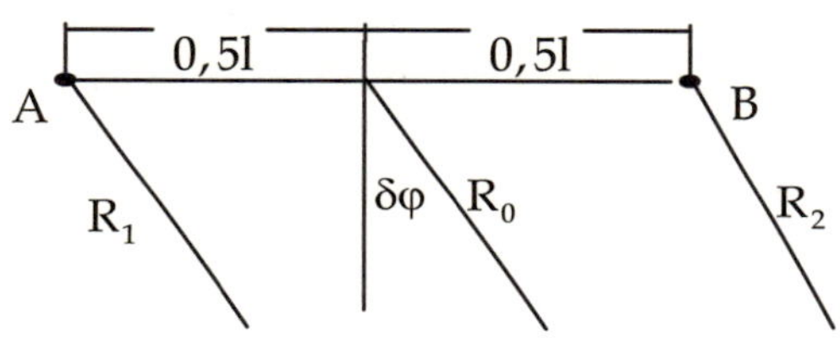

Fig. 8. The scattering by two-point radar object

Here the distances R_1, R_2 between the scatterers and arbitrary point Q in far zone can be written in the form $R_{1,2} \approx R_0 \pm 0,5l \sin\delta\varphi \approx R_0 \pm 0,5l\delta\varphi$ under the condition $0,5l << R_0$. Using these expressions, we can find the Jones vector of the scattered field for the case when radiated signal has linear polarization 45^0

$$\dot{\vec{E}}_S(\delta\varphi) = \frac{\sqrt{2}}{2} \begin{Vmatrix} \dot{a}_1 \exp(j\xi) + \dot{b}_1 \exp(-j\xi) \\ \dot{a}_2 \exp(j\xi) + \dot{b}_2 \exp(-j\xi) \end{Vmatrix}, \tag{34}$$

where $\xi = kl\theta$. Let us to define now a polarization- energetical response functions in the form of Stokes momentary parameters S_0, S_3 angular dependences

$$S_0(\delta\varphi) = \dot{E}_X(\delta\varphi)\dot{E}_X^*(\delta\varphi) + \dot{E}_Y(\delta\varphi)\dot{E}_Y^*(\delta\varphi); \;\; S_3(\delta\varphi) = i[\dot{E}_X(\delta\varphi)\dot{E}_Y^*(\delta\varphi) - \dot{E}_Y(\delta\varphi)\dot{E}_X^*(\delta\varphi)].$$

The expanded form of the energetically response function $S_0(\delta\varphi)$ can be found as

$$S_0(\delta\varphi) = 0,5\left[S_0^A + S_0^B\right] + \sqrt{a_1^2 b_1^2 + a_2^2 b_2^2 + \dot{a}_1^* \dot{a}_2 \dot{b}_1 \dot{b}_2^* + \dot{a}_1 \dot{a}_2^* \dot{b}_1^* \dot{b}_2} \cos(2\xi + \eta_1), \tag{35a}$$

where $\eta_1 = \arctan\left\{\left[\operatorname{Im}\left(\dot{a}_1 \dot{b}_1^* + \dot{a}_2 \dot{b}_2^*\right) / \operatorname{Re}\left(\dot{a}_1 \dot{b}_1^* + \dot{a}_2 \dot{b}_2^*\right)\right]\right\}$ and $S_0^A = a_1^2 + a_2^2$, $S_0^B = b_1^2 + b_2^2$. The values S_0^A, S_0^B are the Stokes zero-parameters of elementary scatterers A and B. The polarization-angular response function $S_3(\delta\varphi)$ has the form

$$S_3(\delta\varphi) = 0,5\left[S_3^A + S_3^B\right] + 2\sqrt{a_1^2 b_1^2 + a_2^2 b_2^2 - (\dot{a}_1^* \dot{a}_2 \dot{b}_1 \dot{b}_2^* + \dot{a}_1 \dot{a}_2^* \dot{b}_1^* \dot{b}_2)}\sin(2\xi + \eta_2), \quad (35b)$$

where $\eta_2 = \arctan\left\{\left[\operatorname{Im}\left(\dot{a}_1\dot{b}_2^* - \dot{a}_2\dot{b}_1^*\right) / \operatorname{Re}\left(\dot{a}_1\dot{b}_2^* - \dot{a}_2\dot{b}_1^*\right)\right]\right\}$ and $S_3^A = -05j\left(\dot{a}_1\dot{a}_2^* - \dot{a}_1^*\dot{a}_2\right)$, $S_3^B = -0,5j\left(\dot{b}_1\dot{b}_2^* - \dot{b}_1^*\dot{b}_2\right)$ are the 3-rd Stokes parameters of elementary scatterers A and B.

The angular harmonic functions $\cos[2kl\delta\varphi + \eta_1]$, $\sin[2kl\theta + \eta_2]$ in the expressions (35a,b), are representing the influence of scatterers A and B space diversity to the scattered field polarization-energetically parameters distribution in far zone. The derivative from angular harmonic functions full phases $\psi(\delta\varphi) = 2kl\delta\varphi + \eta_k$ ($k = 1, 2$) along the angular variable is the space frequency $f_{SP} = (1/2\pi)(d/d\delta\varphi)[2kl\delta\varphi + \eta_k] = 2l/\lambda$.
Now we will analyze the amplitudes of angular harmonic functions $\cos[2kl\delta\varphi + \eta_1]$, $\sin[2kl\delta\varphi + \eta_2]$. Let us write first of all the polarization rations $\dot{P}_A = \dot{a}_2/\dot{a}_1$ and $\dot{P}_B = \dot{b}_2/\dot{b}_1$ which are characterizing the point radar objects A and B on the complex plane of radar objects . We can find the spherical distance between the points S_A, S_B, laying on the surface of the Riemann sphere having unit diameter, which are connected with points $\dot{P}_A$, $\dot{P}_B$ of radar objects complex plane. The coordinates of the points S_A, S_B on the sphere surface are $X_1 = \operatorname{Re}\dot{P}/(1+|\dot{P}|^2)$; $X_2 = \operatorname{Im}\dot{P}/(1+|\dot{P}|^2)$; $X_3 = |\dot{P}|^2/(1+|\dot{P}|^2)$ and a spherical distance between these points can be found in the form (Tatarinov et al, 2006)
$\rho_S(S_A, S_B) = |\dot{P}_A - \dot{P}_B| / \sqrt{1+|\dot{P}_A|^2}\sqrt{1+|\dot{P}_B|^2}$, (36) where $|\dot{P}_A - \dot{P}_B|$ is the Euclidian metric on the complex plane of radar objects. After substitution of the polarization ratios $\dot{P}_A = \dot{a}_2/\dot{a}_1$ and $\dot{P}_B = \dot{b}_2/\dot{b}_1$ into the expressions (46) we can write

$$\rho_S(S_A, S_B) = \frac{\sqrt{|\dot{P}_A|^2 + |\dot{P}_B|^2 - \left(\dot{P}_A\dot{P}_B^* + \dot{P}_A^*\dot{P}_B\right)}}{\sqrt{1+|\dot{P}_A|^2}\sqrt{1+|\dot{P}_B|^2}} = \sqrt{\frac{a_1^2 b_1^2 + a_2^2 b_2^2 - (\dot{a}_1^*\dot{a}_2\dot{b}_1\dot{b}_2^* + \dot{a}_1\dot{a}_2^*\dot{b}_1^*\dot{b}_2)}{\left(a_1^2 + a_2^2\right)\left(b_1^2 + b_2^2\right)}}, \quad (37)$$

where the value

$$D = \left[a_1^2 b_1^2 + a_2^2 b_2^2 - (\dot{a}_1^*\dot{a}_2\dot{b}_1\dot{b}_2^* + \dot{a}_1\dot{a}_2^*\dot{b}_1^*\dot{b}_2)\right] / \left(a_1^2 + a_2^2\right)\left(b_1^2 + b_2^2\right) \quad (38)$$

is so-called polarization distance between two waves (or radar objects polarization states), having different polarizations (Azzam & Bashara, 1980), (Tatarinov et al, 2006). It is not difficult to demonstrate that the waves having coinciding polarizations ($\dot{P}_A = \dot{P}_B$) are having the polarization distance value $D = 0$ and the waves having orthogonal polarizations ($\dot{P}_B = -1/\dot{P}_A^*$) have the polarization distance value $D = 1$. Thus, it follows from (37) and (38) that $a_1^2 b_1^2 + a_2^2 b_2^2 - (\dot{a}_1^*\dot{a}_2\dot{b}_1\dot{b}_2^* + \dot{a}_1\dot{a}_2^*\dot{b}_1^*\dot{b}_2) = D\left(a_1^2 + a_2^2\right)\left(b_1^2 + b_2^2\right)$.
We can use also so-called polarization proximity value $N = 1 - D$. Using values N, D we can rewrite the expressions (35a,b) in the form

$$S_0(\delta\varphi) = 0,5\left[S_0^A + S_0^B + 2\sqrt{S_0^A}\sqrt{S_0^B}\sqrt{N}\cos(2\xi + \eta_1)\right]. \quad (39)$$

$$S_3(\delta\varphi)=0,5\left[S_3^A+S_3^B+2\sqrt{S_0^A}\sqrt{S_0^B}\sqrt{D}\sin(2\xi+\eta_2)\right]. \quad (40)$$

We can consider these expressions as generalized interference laws as far as these expression are the generalization of Fresnel-Arago interference laws (Tatarinov et al, 2007). It follows from the expression (39) that the orthogonal polarized waves can not give an interference picture, as far as for the polarization proximity value $N=0$. However, the expression (40) demonstrates that in this case we will have the maximal value of this interference picture visibility. It follows from expressions (40) that for every Stokes parameters have the place some constant component, which is defined by the according Stokes parameters of both objects (A and B), and space harmonics function $\cos[2kl\delta\varphi+\eta_1]$, $\sin[2kl\delta\varphi+\eta_2]$, having amplitudes $2\sqrt{S_0^A}\sqrt{S_0^B}\sqrt{N}$, $2\sqrt{S_0^A}\sqrt{S_0^B}\sqrt{D}$ and space initial phase η_k. So, the polarization-energetically properties of complex radar object can not be found only with the use of its elements properties. The conditionality of integral system properties appear by means of its elements relations. These relations in our case are polarization distance and polarization proximity. The use of these values leads to the "emergence" of new properties which did not exist for every element separately.

4.2 A polarization coherence notion and its definition as the correlation moment of the forth order

Let us to define a momentary visibility of generalized interference law (39) in the form

$$W=\left[S_0^{MAX}(\theta)-S_0^{MIN}(\theta)\right]/\left[S_0^{MAX}(\theta)+S_0^{MIN}(\theta)\right]=2\sqrt{S_0^A}\sqrt{S_0^B}\sqrt{N}/\left(S_0^A+S_0^B\right). \quad (41)$$

The equation (41) is coinciding with well known expression for partial coherent field interference law visibility (Born & Wolf, 1965), (Potekchin & Tatarinov, 1978)

$$W=\left[I^{MAX}(\theta)-I^{MIN}(\theta)\right]/\left[I^{MAX}(\theta)+I^{MIN}(\theta)\right]=2\sqrt{I_1}\sqrt{I_2}\gamma_{12}/[I_1+I_2],$$

where I_1, I_2 are power of waves summarized and γ_{12} is a coherence degree. If $I_1=I_2$ then an interference law visibility is defined by coherence degree having the second order.

So, we can claim, that from physical point of view the parameter N can be considered as polarization coherence parameter, which defines a proximity of elementary scatterers polarization states, analogously coherence degree of stochastic waves summarized. In this case we have "momentary" value of polarization coherence, at the some time a coherence degree γ_{12} is the correlation value. In this connection it is necessary to analyze statistical effects and polarization coherence mean value.

If we will consider the interference law (39) visibility, then we can see that it is defined by a value $\sqrt{N}$, which is a magnitude of space harmonic function $\cos[2kl\delta\varphi+\eta_1]$. It is necessary to point out that a value $\sqrt{N}$ is corresponding to polarization coherence of the second order. However, it is clear that value N is corresponding to polarization coherence of the forth order. On the fig. 9 the interference law (39) is presented for the case $S_0^A=S_0^B$. In this case the interference law visibility is defined by value $\sqrt{N}$.

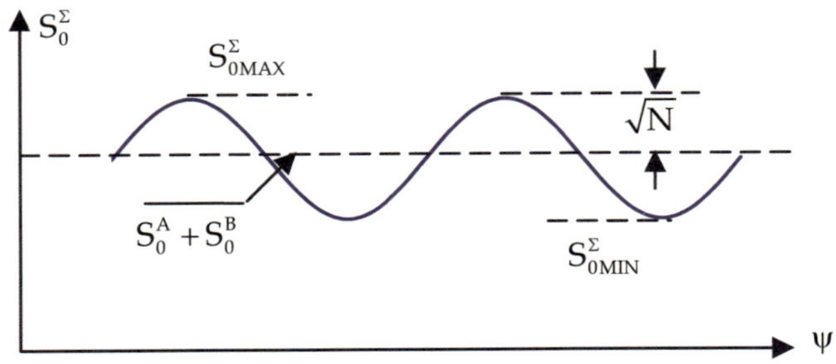

Fig. 9. To polarization coherence definition

A magnitude of space harmonic function $\sin[2kl\delta\varphi+\eta_2]$ into the interference law (40) for the third Stokes parameter is defined by elementary scatterers A and B polarization states distance $\sqrt{D}$. As far as $\sqrt{D}=\sqrt{1-N}$, then a value $\sqrt{D}$ can be considered also as polarization coherence of the second order.

It follows from the expressions (39, 40) that polarization states proximity and distance are included into the interference laws in the form $\sqrt{N}$ and $\sqrt{D}$. It provides power dimension for these laws. Let us to find now an autocovariance function of the interference law (39) for polarization coherence mean value definition. We will assume here that space harmonics amplitudes $\sqrt{N}$ and space initial phase η are random independent variables. For this case their two-dimensional probability distribution can be presented as two one-dimensional distributions densities product $W_2\left(\sqrt{N},\ \eta\right)=W_1\left(\sqrt{N}\right)W_1(\eta)$. We will assume also that $W(\eta)=1/2\pi$). We can presuppose also that $S_0^A=S_0^B=S_0$. At that time auto covariance function can be defined in the form of the mean value

$$K_S^0(\Delta\varphi)=S_0^2\left\{1+\overline{N\cos[2kl\delta\varphi+\eta_1]\cos\left[2kl(\delta\varphi+\Delta\varphi)+\eta_1\right]}\right\}=S_0^2\left[1+B_S^0(\Delta\varphi)\right] \tag{42}$$

that is statistical moment of the forth order. Here the function $B_S^0(\Delta\theta)$ is the autocorrelation function of scattered field intensity

$$B_S^0(\Delta\varphi)=\int_0^{\infty}\int_{-\infty}^{\infty}\left(\sqrt{N}\right)^2\cos[2kl\delta\varphi+\eta_1]\cos\left[2kl(\delta\varphi+\Delta\varphi)+\eta_1\right]W_1\left(\sqrt{N}\right)W_1(\eta)d\left(\sqrt{N}\right)d\eta\,. \tag{43}$$

The integration of this expression gives us

$$B_S^0(\Delta\varphi)=\frac{0{,}5}{2\pi}\int_0^{\infty}\int_{-\pi}^{\pi}\left(\sqrt{N}\right)^2\cos(2kl\Delta\varphi)W_1\left(\sqrt{N}\right)d\left(\sqrt{N}\right)d\eta=0{,}5\overline{N}\cos(2kl\Delta\varphi), \tag{44}$$

where $\overline{N}$ is the mean value of elementary scatterers A and B polarization states proximity. It is defined amplitudes of space harmonics collection having $f_{SP}=2l/\lambda$.

Thus, both autocovariance function and autocorrelation function are correlation function of the forth order and they are describing the intensity correlation for interference law (39). In this connection autocovariance function (42) is the interference law of the forth order. A visibility of this law is defined by polarization coherence degree $\overline{N}$.

For the interference law (40) under the condition $S_0^A = S_0^B = S_0$ autocovariance function has the form

$$K_S^3(\Delta\varphi) = 0,25S_0^2\left\{\left[\left(S_3^{\Sigma}\right)^2 / S_0^2\right] + 2B_S^3(\Delta\varphi)\right\} \tag{45}$$

and it is (how earlier) statistical moment of the forth order. Here $S_3^{\Sigma} = S_3^A + S_3^B$, and function $B_S^3(\Delta\varphi)$ is autocorrelation function of the third Stokes parameter angular distribution. Using the assumption how earlier, we can write

$$B_S^3(\Delta\varphi) = \frac{0,5}{2\pi}\int_0^{\infty}\int_{-\pi}^{\pi}\left(\sqrt{D}\right)^2\cos(2kl\Delta\varphi)W_1\left(\sqrt{D}\right)d\left(\sqrt{D}\right)d\eta = 0,5\overline{D}\cos(2kl\Delta\varphi), \tag{46}$$

where $\overline{D}$ is the mean value of elementary scatterers A and B polarization states distance, which was defined by the average of random values D statistical set. The autocovariance function (45) is the interference law of the forth order . A visibility of interference law (45) is defined by polarization coherence degree $\overline{D} = 1 - \overline{N}$ by virtue of the result (46).

The joint experimental investigation of generalized Fresnel - Arago interference laws in conformity to polarization-energetically properties of two-elements man-made radar objects were realized in the International Research Centre for Telecommunication-Transmission and Radar of TU Delft (Tatarinov et al, 2004). In this subsection we present an insignificant part of these results for the following objects: 1). Two trihedral, where the first was empty and the second was arranged by the elliptic polarizer in the form of special polarization grid. The transmission coefficients along the OX and OY axes are $b_Y = 0,5b_X$ and mutual phase shift between polarizer eigen axes is $\varphi_{XY} = \pi/2$ ($P_A = 1$; $\dot{P}_B = j0,5$; $N = 0,5$; $D = 0,5$); This object is presented on the fig. 10. 2).Two trihedral, where the first was empty and the second was arranged by the linear polarizer in the form of the special polarization grid. ($N = 0,5; D = 0,5$);

Fig. 10. Two-point radar object N1

The phase centers of the trihedral were distributed in the space on the distance 100 cm, the wave length of the radar was 3 cm. For these parameters the space frequency and space period are $f_{SP} = 2l/\lambda(Rad)^{-1}$, $T_{SP} = 0,015Rad$ (or 0.855^0). The construction, where the trihedral were placed, has rotated with the angular step $0,25^0$.

When the object includes the trihedral arranged by the elliptic polarizer and empty trihedral (combination N1), the polarization proximity and distance theoretical estimation is $N = D = 0,5$. On the Fig.11a,b the experimental angular harmonics functions (generalized

interference pictures) $S_0(\theta)$, $S_3(\theta)$ are shown. It follows from these figures that the visibility for interference picture $S_0(\theta)$ is $W_0 \approx 0{,}3$ that corresponding to polarization proximity $N_0 = 0.54$ (theoretical estimation is N=0.5). The visibility for $S_3(\theta)$ is $W_3 = 1$ that corresponding to polarization distance $D = 0{,}5$.
For the system including the trihedral arranged by the linear polarizer and empty trihedral (object N2), we can find the theoretical estimation visibility values $W_0 = 0{,}66;\ W_3 = 1$ that correspond to polarization proximity values $N_0 = \sqrt{W_0} = 0{,}82;\ N_3 = \sqrt{W_3} = 1$. On the Fig.12a,b the angular harmonics functions $S_0(\theta)$, $S_3(\theta)$ for this situation are shown.

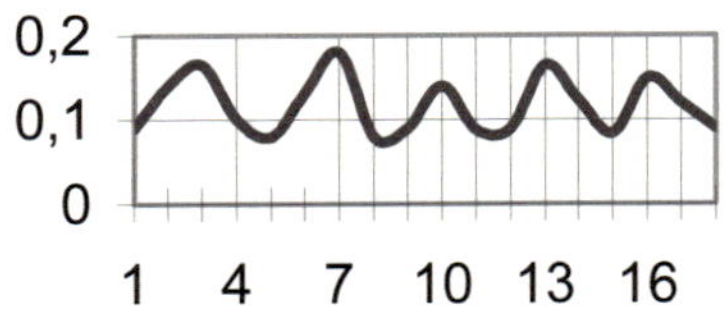

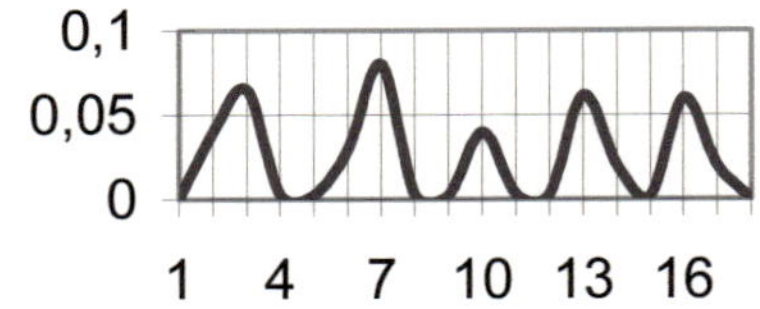

Fig. 11a. Generalized interference law for the parameter $S_0(\theta)$ (object N1)

Fig. 11.b. Generalized interference law for the parameter $S_3(\theta)$ (object N1)

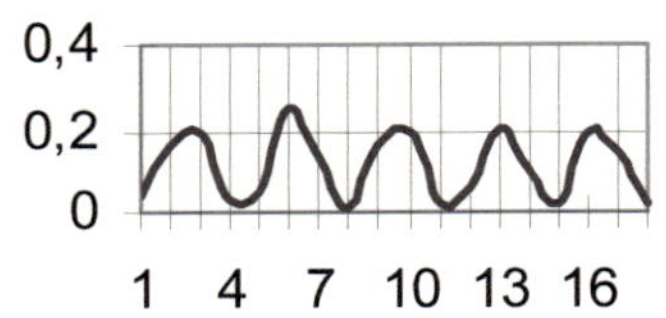

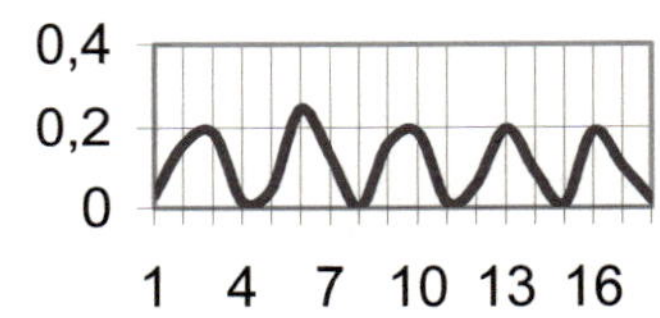

Fig. 12a. Generalized interference law for the parameter $S_0(\theta)$ (object N2)

Fig. 12.b. Generalized interference law for the parameter $S_3(\theta)$ (object N2)

The experimental estimation with the use of Fig.12a,b gives us $N_0 \approx 0{,}85;\ N_3 = 1$ what is the satisfactory coinciding with the theoretical estimation.

5. Polarization – energetic parameters of complex radar object coherent image formation as the interference process. Polarization speckles Statistical analysis

It is demonstrated in the given subsection that the scattered field polarization-energetically speckles formation at the scattering by multi-point random distributed radar object (RDRO) is the interference process. In this case the polarization-energetic response function of a RDRO can be considered as space harmonics collection. Every space harmonic of this collection will be initiated by one from a great many scattered interference pair, which can be formed by multi-point RCRO scatterers. In this connection every space harmonic will have an amplitude, which will be defined by a value of this pair scatterers polarization states proximity (or distance). As far as the RCRO elementary scatterers positions are stochastic, at the positional angle change and a random number of interference pairs, having the same space diversity under the condition of these pair scatterers polarization states proximity stochastic difference, we have the classical stochastic problem. This setting of a problem has been formulated in the first time.

Let's to consider the scattering by a multi-point (complex) radar object (see Fig. 13). For the case of coinciding linear polarization both for transmission and receiving we can write the field scattered by a point X_I (RCS of this scatterer is σ_I) for some point Q in far zone

$$\dot{E}_S(\theta) = -\frac{\exp(j2kR_0)}{R_0\sqrt{4\pi}}\sqrt{\sigma_I}\dot{E}_0 \exp(-j2kX_I\theta),$$

where $R_I \approx R_0 - X_I\theta$ is the distance between the scatterers X_I and X; $\dot{E}_0$ and $\dot{E}_S$ are initial and scattered field electrical vectors respectively. For the case when a scatterers are characterizing by the scattering matrix $\left\|\dot{S}_I^{ik}\right\|;(i,k=1,2)$ then the scattered field complex vector will be connected with initial field complex vector as

$$\dot{\vec{E}}_S(\theta) = -\frac{\exp(j2kR_0)}{R_0\sqrt{4\pi}}\left\|\dot{S}_I^{ik}\right\|\dot{\vec{E}}_0 \exp(-j2kX_I\theta). \tag{47}$$

Let us consider now the electromagnetic field polarization-energetic parameters distribution formation as the interference process at the scattering by multi-point RDRO. For the example we will find that the electrical vector of the field, scattered by 4-points complex object for the case of coinciding linear polarization both for transmission and receiving:

$$\dot{E}_S(\theta) = -\frac{\exp(j2kR_0)\dot{E}_0}{R_0\sqrt{4\pi}}\sum_{I=1}^{4}\sqrt{\sigma_I}\exp(-j2kX_I\theta).$$

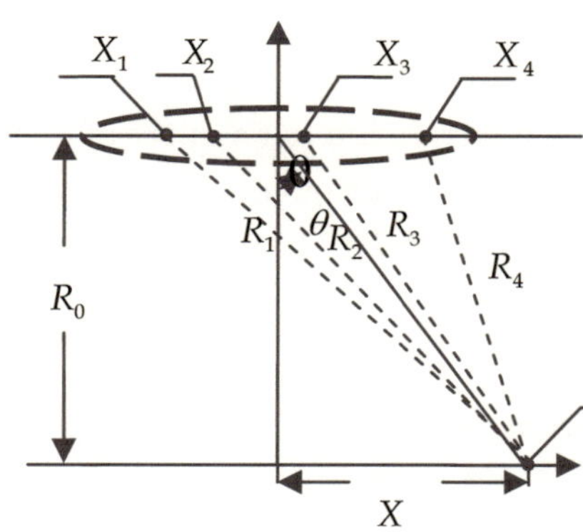

Fig. 13. Waves scattering by multi-point RDRO

Now we can define the instantaneous distribution of scattered field power in the space as the function of the positional angle θ:

$$P(\theta) = \dot{E}_S(\theta)\dot{E}_S^*(\theta) = \sigma_1 + \sigma_2 + \sigma_3 + \sigma_4 + 2\sqrt{\sigma_1\sigma_2}\cos(2kd_{12}\theta) + 2\sqrt{\sigma_1\sigma_3}\cos(2kd_{13}\theta) +$$

$$+2\sqrt{\sigma_1\sigma_4}\cos(2kd_{14}\theta) + 2\sqrt{\sigma_2\sigma_3}\cos(2kd_{23}\theta) + 2\sqrt{\sigma_2\sigma_4}\cos(2kd_{24}\theta) + 2\sqrt{\sigma_3\sigma_4}\cos(2kd_{34}\theta). \tag{48}$$

So, the instantaneous distribution of scattered field power in the space as the function of the positional angle θ is formed by the union of elementary scatterer radar cross section (4 items) plus 6 cosine oscillations. It is not difficult to see that every cosine functions are caused by the interference effect between the fields scattered by a pair of elementary scatterers forming the RCRO. The number of this pairs can be found with the use binomial coefficient

$$C_M^N = M!/\left[N!(M-N)!\right],$$

where M is a number of values, N is a number elements in the combination. In the case when $M=4$, $N=2$, we have $C_4^2=6$. So, the angular response function of the complex radar object considered will include 6 space harmonic functions as the interference result summarize how it follows from the expression (48) where the values $d_{12}=X_1-X_2$; $d_{13}=X_1-X_3; d_{14}=X_1-X_4; d_{23}=X_2-X_3; d_{24}=X_2-X_4; d_{34}=X_3-X_4$ are the space diversity of scattered elements for every interference pair. The space harmonic function $\sqrt{\sigma_i\sigma_k}\cos(2kd_{ik}\theta)$ corresponds to the definition that was done in (Kobak, 1975), (Tatarinov et al, 2007) . In accordance with this definition, the harmonic oscillation in the space having the type $\cos(2kd\theta)$ is defined by the full phase $\psi(\theta)=2kd\theta=(2\pi/\lambda)2d\theta$, the derivative from which is the space frequency $f_{SP}=2d/\lambda$ having the dimension Rad^{-1}. The period $T_{SP}=1/f_{SP}=\lambda/2d$ has the dimension Rad, which corresponds to this frequency.
So, a full power distribution of the field, scattered by complex radar object, is an union of the interference pictures, which are formed by a collection of elementary two-points interferometers.
Thus, we can write a scattered power random angular representation, depending on the positional angle, in the form

$$P(\theta)=\sum_{m=1}^{M}\sigma_m^2+2\sum_{1}^{C}\sqrt{\sigma_i\sigma_k}\cos(2kd_{ik}\theta),$$

where $C=C_M^2$ is combinations number, M is a full number of RCRO elementary scatterers.
It was demonstrated above that the electromagnetic field Stokes parameter S_0, S_3 angular distribution at the scattering by two-point distributed object has the form

$$S_0(\theta)=S_0^a+S_0^b+2\sqrt{S_0^a}\sqrt{S_0^b}\sqrt{N_{ab}}\cos(\xi+0{,}5\varphi);\; S_3(\theta)=S_3^a+S_3^b+2\sqrt{S_0^a}\sqrt{S_0^b}\sqrt{D_{ab}}\cos(\xi-0{,}5\varphi),$$

where $\xi=2kl\theta$. It follows from this expression that the space harmonics functions $\cos(2kl\theta\pm\eta)$ are having amplitudes $\sqrt{S_0^a}\sqrt{S_0^b}\sqrt{N_{ab}}$ or $\sqrt{S_0^a}\sqrt{S_0^b}\sqrt{D_{ab}}$. Here the values N_{ab}, D_{ab} are a proximity (distance) of distributed object elementary scatterers polarization states respectively.
Taking into account above mentioned, we can write the Stokes parameters angular distribution for the field, scattered by random complex radar object as an union of the generalized interference pictures, which are formed by a collection of elementary two-points interferometers (see Fig.13):

$$S_0(\theta)=\sum_{m=1}^{M}S_0^m+2\sum_{1}^{C}\sqrt{S_{0i}S_{0k}}\sqrt{N_{ik}}\cos(\xi_{ik}+\eta_{ik}),$$

$$S_3(\theta)=\sum_{m=1}^{M}S_3^m+2\sum_{1}^{C}\sqrt{S_{0i}S_{0k}}\sqrt{D_{ik}}\cos(\xi_{ik}+\eta_{ik}),$$

where $C=C_M^2$ is combinations number. An amplitude of every space harmonics and initial space phases of these harmonics will be stochastic values and the further analysis must be statistical. First of all we will find a theoretical form of scattered field Stokes parameter S_3 angular distribution autocorrelation function. As far as we would like to find the autocorrelation function (not covariance function!), we must eliminate a random constant item $\sum_{m=1}^{M}S_3^m$ from the stochastic function $S_3(\theta)$ for the guarantee of zero mean value. Taking into account that the value $\sum_{m=1}^{M}S_3^m$ can be as no stationary stochastic function, the average must be made using a sliding window. After a mean value elimination and normalization we can write stochastic stationary function $S_3(\theta)$ in the form

$$S_3(\theta)=\sum_{1}^{C}\sqrt{D_{ik}}\cos(2kd_{ik}\theta+\eta_{ik})$$

Its autocorrelation function can be found as

$$B_S(\Delta\theta)=\sum_{N=1}^{C}\int_{-\infty}^{\infty}\int_{-\infty}^{\infty}\left(\sqrt{D_N}\right)^2\cos[2kd_N\theta+\eta]\cos[2kd_N(\theta+\Delta\theta)+\eta]W_2\left(\sqrt{D},\ \eta\right)d\left(\sqrt{D}\right)d\eta\,. \quad (49)$$

Here amplitudes $\sqrt{D}$ and space initial phase η of space harmonics are random values, which can be characterized by two-dimensional probability distribution density $W_2\left(\sqrt{D},\eta\right)$, and $\Delta\theta=\theta_1-\theta_2$. We will suppose that random amplitudes and phases are independent variables. For this case two-dimensional probability distribution can be presented as two one-dimensional distributions densities product

$$W_2\left(\sqrt{D},\ \eta\right)=W_1\left(\sqrt{D}\right)W_1(\eta)\cdot$$

Let's suppose also that random phase has the uniform probability distribution density on the interval $(-\pi,\ \pi)$ i.e. $W(\eta)=1/2\pi$. A probability distribution density for the random amplitude $\sqrt{D}$ can be preassigned, however for all cases it will be one-sided. After the integration we obtain the value of double integral in the form

$$I=\frac{0{,}5}{2\pi}\int_{0}^{\infty}\int_{-\pi}^{\pi}\left(\sqrt{D_N}\right)^2\cos(2kd_N\Delta\theta)W_1\left(\sqrt{D}\right)d\left(\sqrt{D}\right)d\eta=0{,}5<D_N>\cos(2kd_N\Delta\theta), \quad (50)$$

where $<D_N>$ is the polarization distance mean value, which was found by the average along the statistical ensemble of random values D_N for all space harmonics having the space frequency $f_{SP}^N=2d_N/\lambda$. Thus, we can write the theoretical form of scattered field Stokes parameter angular distribution autocorrelation function in the form

$$B_S(\Delta\theta)=0{,}5\sum_{N=1}^{C}<D_N>\cos(2kd_N\Delta\theta)\cdot \quad (51)$$

Taking into account that the every item of the union (51) is the autocorrelation function for an isolated space harmonic oscillation $S_N(\theta) = \sqrt{D_N}\cos(2kd_N\theta + \eta_N)$ having random amplitude $\sqrt{D_N}$ and random initial space phase η_N, i.e.

$$B_{SN}(\Delta\theta) = 0{,}5 < D_N > \cos(2kd_N\Delta\theta) \tag{52}$$

it is not difficult to see that the autocorrelation function of the Stokes parameter stochastic realization is the union of individual autocorrelation functions of all space harmonics:

$$B_S(\Delta\theta) = \sum_{N=1}^{C} B_{SN}(\Delta\theta). \tag{53}$$

Let's now to find a complex radar object averaged space spectra using the expressions (8) for polarization-angular response autocorrelation function. The power spectra for the case of isolated space harmonic can be found as the Fourier transformation above the autocorrelation function (52):

$$P(\Omega_{SP}) = \int_{-\infty}^{\infty} B_{SN}(\Delta\theta)\exp(-j\Omega_{SP}\Delta\theta)d(\Delta\theta) = 0.5 < D_N > [\delta(\Omega_{SP} - \Omega_{SP}^N) + \delta(\Omega_{SP} + \Omega_{SP}^N)], \tag{54}$$

where $\Omega_{SP} = 2\pi f_{SP} = 2\pi(2d/\lambda)$ is a space frequency. The spectra lines are placed on the distances $\pm\Omega_{SP}^N$ from the co-ordinates system origin and their positions are defined by the space frequency $f_{SP}^N = 2d_N/\lambda$ of two-point radar object. This space frequency is corresponding to space diversity of two reflectors distributed in the space. The intensity of power spectra lines is determined by polarization distance between polarization states of two scatterers forming the radar object.

The full space spectra of stochastic polarization-angular response, i.e. Fourier transformation of the autocorrelation function (53) is:

$$P(\Omega_{SP}) = 0{,}5\sum_{N=1}^{C} < D_N > [\delta(-\Omega_{SP}^N) + \delta(+\Omega_{SP}^N)]. \tag{55}$$

It is necessary to indicate here that a connection between scattered (diffracted) field polarization parameters and polarization parameters distribution along a scattering (diffracting) object in the form of Fourier transformation pair is established in the first time. However, this connection is correct for fourth statistical moments: scattered field intensity correlations (include mutual intensity) and polarization proximity (distance) distribution along a scattering (diffracting) object.

In the conclusion we consider some results of scattered field polarization parameters investigation at the scattering by random distributed object having a lot of scattering centers - "bright" points. It follows also both from theoretical and experimental investigations results that polarization-angular response function of a RCRO in the form of the 3-rd Stokes parameter angular dependence corresponds to a narrow-band random process. The experimental realization of this parameter has shown on the fig.7. The angular interval for this dependence is $\pm 20^0$. The rotated caterpillar vehicle (the sizes 5,5x2,5x1,5 m) placed on the distance 2 km was used as complex radar object. The autocorrelation functions (ACF) of this object response $S_3(\Delta\theta)$ are shown on the fig.14. The ACF on the angular interval $\pm 20^0$ concerning the direction to the object board is designated by dotted line and the ACF into the same interval in direction to the stern of the object is continue line. The measurements in these directions allow us to take into account the difference in the radar object space spectra band at its observation in areas of perpendiculars to the board and to the stern of the object. On the fig. 15 RDRO mean power spectra are shown. Dotted line is corresponding to direction to the object board and continue line corresponds to object stern.

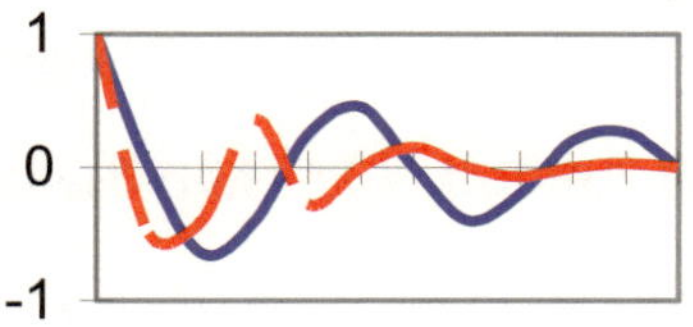

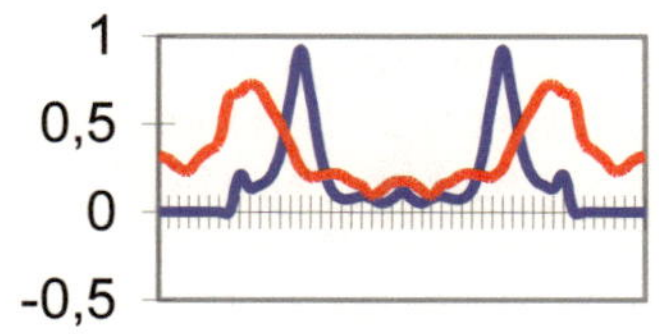

Fig. 14. Autocorrelation functions of RDRO Fig.15. Mean power space spectra of RDRO stochastic polarization-angular response

6. Conclusion

In the conclusion we can to indicate that in the Chapter proposed a new statistical theory of distributed object polarization speckles (coherent images) has been developed. The use of fourth statistical moments and emergence principle allow us to find the answers for a series of problems which are having the place at the electromagnetic waves coherent scattering by distributed (complex) radar objects.

7. References

Proceedings of the IEEE. (1965). Special issue. Vol. 53., No.8, (August 1965)

Ufimtsev, P. (1963). *A Method of Edge Waves in Physical Diffraction Theory,* Soviet Radio Pub. House, Moscow, Russia

Proceedings of the IEEE. (1989). Special issue. Vol. 77., No.5, (May 1965)

IEEE Transaction on Antennas and Propagation . (1989). Special issue. No.5, (May 1965)

Ostrovitjanov, R. & Basalov F. (1982). *A Statistical Theory of Distributed Objects Radar,* Radio and Communication Pub. House, Moscow, Russia

Shtager, E. (1986). *Waves Scattering by Complicated Radar Objects,* Radio and Communication Pub. House, Moscow, Russia

Kell, R. (1965). On the derivation of bistatic RCS from monostatic measurements. *Proceedings of the IEEE,* Vol. 53, No. 5, (May 1965), pp 983-988

Stratton, J. & Chu, L. (1939). Diffraction theory of electromagnetic waves. *Phys. Rev.,* Vol. 56, pp 308-316

Tatarinov, V. ; Tatarinov S. & Ligthart L. (2006). *An Introduction to Radar Signals Polarization Modern Theory (Vol. 1 : Plane Electromagnetic Waves Polarization and its Transformations),* Tomsk State University Publ. House, ISBN 5-7511-1995-5, Tomsk, Russia

Shtager, E. (1994). Radar objects characteristics calculation at random earth and sea surface. *Foreign Radioelectronics,* No. 4-5, (May 1994), pp 22-40, Russia

Steinberg, B. (1989). Experimental localized radar cross section of aircraft. *Proceedings of the IEEE,* Vol. 77, No. 5, (May 1989), pp 663-669

Kobak, V. (1975). *Radar Reflectors,* Soviet Radio Pub. House, Moscow, Russia

Kanareikin, D.; Pavlov, N. & Potekchin V. (1966). *Radar Signals Polarization,* Soviet Radio Pub. House, Moscow, Russia

Pozdniak, S. & Melititsky V. (1974). *An Introduction to Radio Waves Polarization Statistical Theory,* Soviet Radio Pub. House, Moscow, Russia

Franson, M. (1980). *Optic of Speckles.* Nauka Pub. House, Moscow, Russia

Peregudov, F. & Tarasenko, F. (2001). *The Principles of Systems Analysis,* Tomsk State University Publ. House, Tomsk, Russia

Azzam, R. & Bashara, N. (1977). *The Ellipsometry and Polarized Light,* North Holland Pub. House, New York-Toronto-London

Tatarinov, V. ; Tatarinov, S. & van Genderen P. (2004). A Generalized Theory on Radar Signals Polarization in Space, Frequency and Time Domains for Scattering by Random Complex Objects. *Report of IRCTR-S-004-04,* Delft Technology University, the Netherlands

Born, M. & Wolf, E. (1959). *Principles of Optics.* Pergamon Press, New-York-Toronto-London

Potekchin, V. & Tatarinov, V. (1978). *The Coherence Theory of Electromagnetic Fielg,* Svjaz Pub. House, Moskow, Russia

Tatarinov, V. ; Tatarinov S. & Kozlov, A. (2007). *An Introduction to Radar Signals Polarization Modern Theory (Vol. 2: A Statistical Theory of Electromagnetic Field),* Tomsk State University Publ. House, ISBN 978-5-86889-476-3, Tomsk, Russia

25

Radar Meteor Detection: Concept, Data Acquisition and Online Triggering

Eric V. C. Leite[1], Gustavo de O. e Alves[1], José M. de Seixas[1], Fernando Marroquim[2], Cristina S. Vianna[2] and Helio Takai[3]
[1]*Federal University of Rio de Janeiro/Signal Processing Laboratory/COPPE-Poli*
[2]*Federal University of Rio de Janeiro/Physics Institute*
[3]*Brookhaven National Labaratory*
[1,2]*Brazil*
[3]*USA*

1. Introduction

In the solar system, debris whose mass ranges from a few micrograms to kilograms are called meteoroids. By penetrating into the atmosphere, a meteoroid gives rise to a meteor, which vaporizes by sputtering, causing a bright and ionized trail that is able to scatter forward Very High Frequency (VHF) electromagnetic waves. This fact inspired the Radio Meteor Scatter (RMS) technique (McKinley, 1961). This technique has many advantages over other meteor detection methods (see Section 2.1): it works also during the day, regardless of weather conditions, covers large areas at low cost, is able to detect small meteors (starting from micrograms) and can acquire data continuously. Not only meteors trails, but also many other atmospheric phenomena can scatter VHF waves and may be detected, such as lightning and e-clouds.

The principle of RMS detection consists in using analog TV stations, which are constantly switched on and broadcasting VHF radio waves, as transmitters of opportunity in order to build a passive bistatic radar system (Willis, 2008). The receiver station is positioned far away from the transmitter, sufficiently to be bellow the horizon line, so that signal cannot be directly detected as the ionosphere does not usually reflect electromagnetic waves in VHF range (30 - 300 MHz)(Damazio & Takai, 2004). The penetration of a meteor on Earth's atmosphere creates this ionized trail, which is able to produce the forward scattering of the radio waves and the scattered signals eventually reach the receiver station.

Due to continuous acquisition, a great amount of data is generated (about 7.5 GB, each day). In order to reduce the storage requirement, algorithms for online filtering are proposed in both time and frequency domains. In time-domain the matched filter is applied, which is optimal in the sense of the signal-to-noise ratio when the additive noise that corrupts the received signal is white. In frequency-domain, an analysis of the power spectrum is applied.

The chapter is organized as it follows. The next section presents the meteor characteristics, and briefly introduces the several detection techniques. Section 3 describes the meteor radar detection and the experimental setup. Section 4 shows the online triggering algorithm performance for real data. Finally, conclusions and perspectives are addressed in Section 5.

2. Meteors

Meteoroids are mostly debris in the Solar System. The visible path of a meteoroid that enters Earth's (or another body's) atmosphere is called a meteor (see Fig. ??). If a meteor reaches the ground and survives impact, then it is called a meteorite. Many meteors appearing seconds or minutes apart are called a meteor shower. The root word meteor comes from the Greek $\mu\epsilon\tau\epsilon\omega\rho o\nu$, meaning "high in the air". Very small meteoroids are known as micrometeoroids, 1g or less.

Many of meteoroid characteristics can be determined as they pass through Earth's atmosphere from their trajectories, position, mass loss, deceleration, the light spectra, etc of the resulting meteor. Their effects on radio signals also give information, especially useful for daytime meteor, cloudy days and full moon nights, which are otherwise very difficult to observe. From these trajectory measurements, meteoroids have been found to have many different orbits, some clustering in streams often associated with a parent comet, others apparently sporadic. Debris from meteoroid streams may eventually be scattered into other orbits. The light spectra, combined with trajectory and light curve measurements, have yielded various meteoroid compositions and densities. Some meteoroids are fragments from extraterrestrial bodies. These meteoroids are produced when these are hit by meteoroids and there is material ejected from these bodies.

Most meteoroids are bound to the Sun in a variety of orbits and at various velocities. The fastest ones move at about 42 km/s with respect to the Sun since this is the escape velocity for the solar system. The Earth travels at about 30 km/s with respect to the Sun. Thus, when meteoroids meet the Earth's atmosphere head-on, the combined speed may reach about 72 km/s.

A meteor is the visible streak of light that occurs when a meteoroid enters the Earth's atmosphere. Meteors typically occur in the mesosphere, and most range in altitude from 75 to

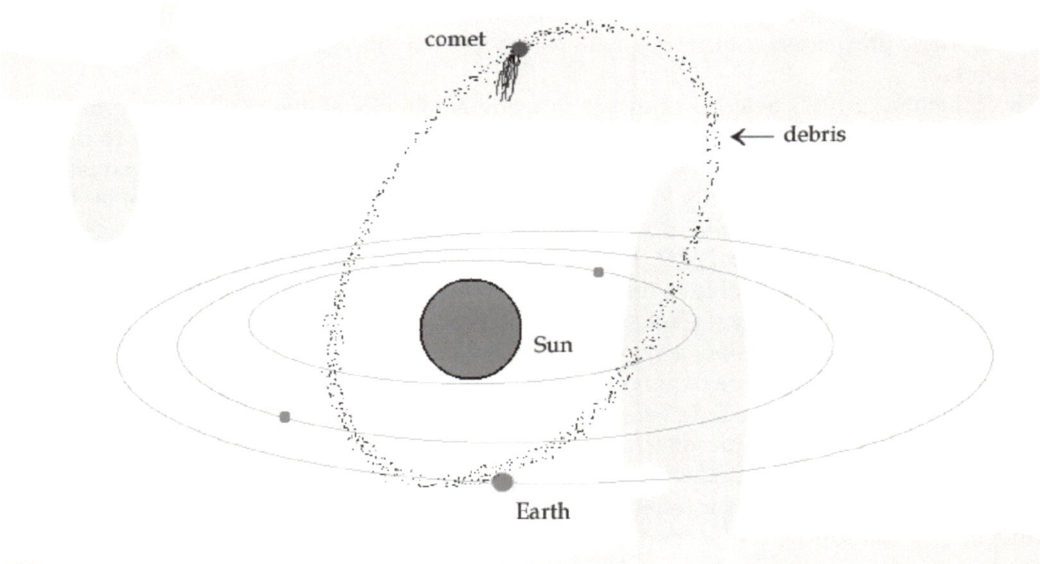

Fig. 1. Debris left by a comet may enter on Earth's atmosphere and give rise to a meteor.

100 km. Millions of meteors occur in the Earth's atmosphere every day. Most meteoroids that cause meteors are about the size of a pebble. They become visible in a range about 65 and 120 km above the Earth. They disintegrate at altitudes of 50 to 95 km. Most meteors are, however, observed at night as low light conditions allow fainter meteors to be observed.
During the entry of a meteoroid or asteroid into the upper atmosphere, an ionization trail is created, where the molecules in the upper atmosphere are ionized by the passage of the meteor (Int. Meteor Org., 2010). Such ionization trails can last up to 45 minutes at a time. Small, sand-grain sized meteoroids are entering the atmosphere constantly, essentially every few seconds in any given region of the atmosphere, and thus ionization trails can be found in the upper atmosphere more or less continuously.
Radio waves are bounced off these trails. Meteor radars can measure also atmospheric density, ozone density and winds at very high altitudes by measuring the decay rate and Doppler shift of a meteor trail. The great advantage of the meteor radar is that it takes data continuously, day and night, without weather restrictions. The visible light produced by a meteor may take on various hues, depending on the chemical composition of the meteoroid, and its speed through the atmosphere. This is possible to determine all important meteor parameters such as time, position, brightness, light spectra and velocity. Furthermore it is possible also to obtain light curves, meteor spectra and other special features.The radiant and velocity of a meteoroid yield its heliocentric orbit. This allows to associate meteoroid streams with parent comets. The deceleration gives information regarding the composition of the meteoroids. From statistical samples of meteor heights several distinct groups with different genetic origins have been deduced.

2.1 Meteor observation methods

There are many ways to observe meteors:

- **Visual Meteor Observation -** Monitoring meteor activity by the naked eye. Least accurate method but easy to carry out in special by amateur astronomers. Large numbers of observations allow statistically significant results. Visual observations are used to monitor major meteor showers, sporadic activity and minor showers down to a zenithal hourly rate (ZHR) of 2. The observer can count and estimate the meteor magnitude using a tape recorder for later to plot a frequency histogram. The visual method is very limited since the observer cannot work during the day or cloudy nights. Such an observation can be quite unreliable when the total meteor activity is high e.g. more than 50 meteors per hour. The naked eye is able to detect meteors down to approximately +7mag under excellent circumstances in the vicinity of the center of the field of view (absolute magnitude - mag - is the stellar magnitude any meteor would have if placed in the observer's zenith at a height of 100 km. A 5th magnitude meteor is on the limit of naked eye visibility. The higher the positive magnitude, the fainter the meteor, and the lower the positive or negative number, the brighter the meteor).
- **Photographic Observations -** The meteors are captured on a photographic film or plate (Hirose & Tomita, 1950). The accuracy of the derived meteor coordinates is very high. Normal-lens photography is restricted to meteors brighter than about +1mag. Multiple-station photography allows the determination of precise meteoroid orbits.

  Photographic methods can hardly compete with video advanced techniques. The effort to be spent for the observation equipment is much lower than for video systems. For this reason photographic observations is widely used by amateur astronomers. On the other hand, the photograph methods allow to obtain very important meteor parameters: accurate position, height, velocity, etc. The sensitivity of the films must be considered. There is now very sensitive digital cameras with high resolution for affordable prices, which produce a great impact to this technique. This method is restricted also to clear nights.

- **Video Observations -** This technique uses a video camera coupled with an image intensifier to record meteors (Guang-jie & Zhou-sheng, 2004). The positional accuracy is almost as high as that of photographic observations and the faintest meteor magnitudes are comparable to visual or telescopic observations depending on the used lens. Meteor shower activity as well as radiant positions can be determined. Multiple-station video observations allow the determination of meteoroid orbits.

  Advanced video techniques permit detection of meteors up to +8mag. Video observation is the youngest and one of the most advanced observing techniques for meteor detection. Professional astronomers started to use video equipment at the beginning of the seventies of the last century. Currently the major disadvantage is the considerable price of a video system.
- **Telescopic Observations -** This comprises monitoring meteor activity by a telescope, preferably binoculars. This technique is used to determine radiant positions of major and minor showers, to study meteors much fainter than those seen in visual observations ones, which may reach +11mag. Although the narrower field, the measurements are more precise.
- **Radio Observations -** Two main methods are used, forward scatter observations and radar observations. The first method is easy to carry out, but delivers only data on the general meteor activity. The last is carried out by professional astronomers. Meteor radiants and meteoroid orbits can be determined. Radar meteors as well as telescopic ones may be as faint as +11mag.

Radio meteor scatter is an ideal technique for observing meteors continuously, day and night and even in cloudy days. Meteor trails can reflect radio waves from distant transmitters back to Earth, so that when a meteor appears one can sometimes receive small portions of broadcasts from radio stations up to 2,000 km away from the observing site.
The technique is strongly growing in popularity amongst meteor amateur astronomers. In the recent years, some groups started automating the radio observations by monitoring the signal from the radio receiver with a computer and even in cloudy days (see Fig. 2). Even for such high performance, the interpretation of the observations is difficult. A good understanding of the phenomenon is mandatory.

3. Meteor radio detection

Measurements performed by Lovell in 1947 using radar technology of the time showed that some returned signals were from meteor trails. This was the start of a technique known today as RMS, which was intensely developed in the 50's and 60's. Both experimental and theoretical work have been developed. Today, radio meteor scatter can be easily implemented having in hands an antenna, a good radio receiver and a personal computer.
There are two basic radar arrangements: backscattering and forward scattering. Back scattering is the traditional radar, where the transmitting station is near the receiving antenna. Forward scattering is used when the transmitter is located far from the receiver. Both arrangements are used in the detection of meteors. Back scatter radar tends to be pulsed and forward scatter continuous wave (CW). Forward scatter radar shows an increase in sensitivity

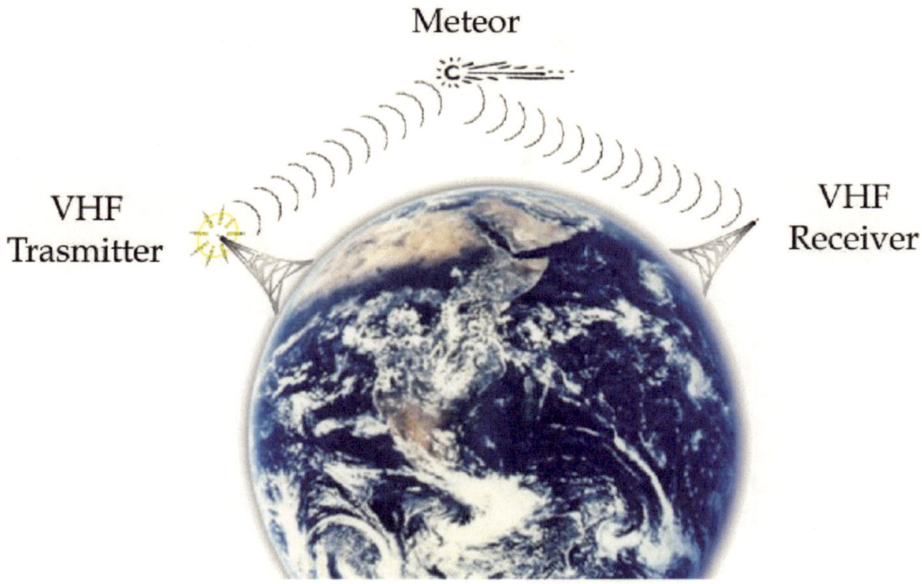

Fig. 2. RMS detection principle.

for the detection of small trails when compared to a same power backscatter due to the differences in aspect ratios. Forward scattering also avoids possible confusion of echos by the ionosphere as discussed by (Matano et al., 1968).
One of the main challenges to estimate the signal return power and its duration lies in a better understanding of the lower atmosphere chemical properties. At higher altitudes where meteors produce ionization trails, 80 to 120 km, the return signal duration only depends on the hot plasma diffusion rate. At lower altitudes, electron attachment to molecular oxygen limits the signal duration for their detection. In addition, the shorter mean free path causes the electron to scatter while radiating and therefore dampening the return power. An energy of $1eV$ electron will roughly undergo 10^9 collisions per second, or 10 collisions in a one wavelength at 100 MHz. The formalism to evaluate both signal duration and reflected power is well understood for meteor trails.
A specular reflection from an electron cloud only happens when a minimum free electron density is reached. This is known from plasma physics and is given by:

$$\nu_p = \sqrt{\frac{n_e e^2}{\pi m_e}} \tag{1}$$

where n_e, e and m_e are the electron density, charge and mass, respectively, which takes a value of $ne = 3.8 \times 10^{13} m^{-1}$ for $f = 55.24MHz$ (channel 2) and $ne = 5.6 \times 10^{13} m^{-1}$ for $f = 67.26MHz$ (channel 4). Below this critical density the reflection is partial and decreases with decreasing electron density. A total reflection happens because the electron density is high enough so that electrons reradiate energy from its neighbors. This happens in meteor trails that are usually called to be in an overdense scattering regime. The converse is the underdense, for which the density is lower and there is no re-radiation by electrons in the cloud. Both regimes are well

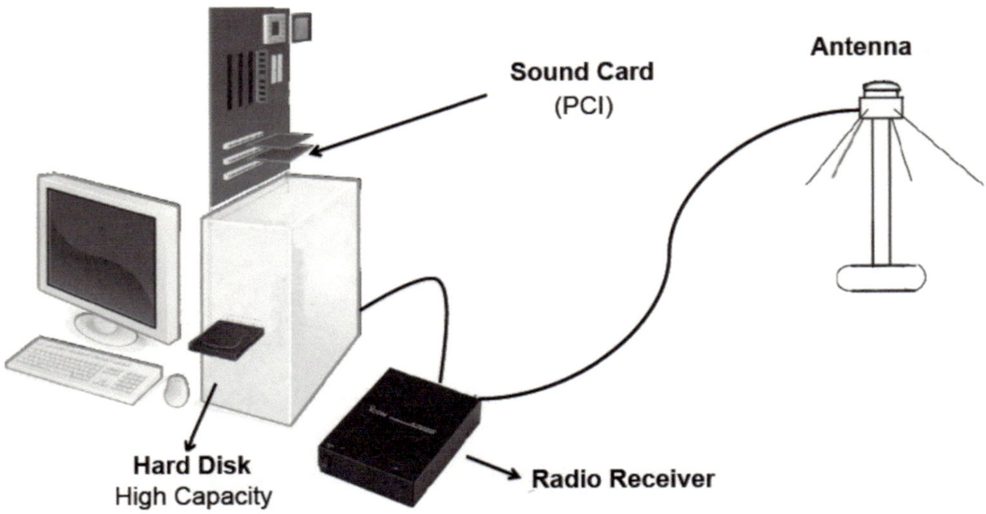

Fig. 3. Experimental setup for radar signal reception.

known from the radio meteor scatter science. Meteor ionization is produced at altitudes above 80 km where the atmosphere is rarefied and gases are from the meteor elements itself. The lifetime of the ionization trail produced by a meteor is a function of diffusion that cools the hot trail and recombination of electrons to the positive ions. Because of the elevated temperature the ionization lasts typically from 0.2 to 0.5 seconds.

The formalism to calculate the reflected power by a meteor trail is well developed both as a model and numerical integration. Models provide good means to understand the underlying processes and for the case of meteors they have been perfected over decades to provide reliable values for power at the receiver. Development of these models is driven by the application known as meteor burst communication where the ionization trails are used to bounce VHF for distances over 2,000 km.

3.1 Experimental setup

As an example, the setup for experimental data acquisition used here to quote the performance of the online detection algorithms is shown in Fig. 3. It includes a double dipole (in "V", inverted) antenna (Damazio & Takai, 2004) for a nearly vertical detection, a computer controlled radio receiver tuned to video carrier of an analog TV channel and a personal computer equipped with an off-board sound card, able to perform sampling rates up to 96 kHz. Due to continuous operation, a hard disk of high capacity is required.

4. Online triggering

The continuous acquisition is an inherent characteristic to radar technique. Acquiring data continuously means generating a great amount of data, which must be stored for a posterior analysis, or processed online for the extraction of the relevant information. Moreover, most of the data are from background noise events, which makes it difficult the detection of interesting events due to the data volume. If the online trigger is not implemented, the full data storage requires a more complex storage system, which increases the final cost of the experimental

setup. In other hand, if the data are processed online, only what is judged interesting will be retained, which translates into a significant reduction on the data volume to be stored.
In order to obtain an efficient detection and classification of received signals, online algorithms are designed in both time and frequency domains. In time-domain, the matched filter is applied. In frequency-domain, an analysis of the cumulative spectral power is applied. The next subsections provide a brief description of such techniques.

4.1 Signal detection

When the RMS (Radio Meteor Scatter) technique is considered, the signal detection problem can be formulated as the observation of a block of received data for decision among two hypotheses (Shamugan & Breipohl, 1998): H_0, also called the null hypothesis, which states that only noise is present, and H_1, also called the alternate hypothesis, which states that the block contains meteor signal masked by additive noise. In a simpler case, the signal to be detected may be known a priori (deterministic signal detection), and samples are corrupted by noise. Due to natural randomness of the occurrence of meteor events, the signal generated by them is considered as a stochastic process (Papoulis, 1965). Thus, from an observation Y of the incoming signal, $P(H_i|y)$ with $i = 0,1$ represents the probability, given a particular value $Y = y$, that H_i is true. The decision in favor of each hypothesis considers the largest probability: if $P(H_1|y) > P(H_0|y)$ choose H_1, or if $P(H_0|y) > P(H_1|y)$ choose H_0:

$$\frac{P(H_1|y)}{P(H_0|y)} \underset{H_0}{\overset{H_1}{\gtrless}} 1 \tag{2}$$

Through the Bayes' rule for conditional probabilities (Papoulis, 1965), we can write $P(H_i|y)$ as

$$P(H_i|y) = \frac{f_{Y|H_i}(y|H_i)P(H_i)}{f_Y(y)} \tag{3}$$

and the ratio in equation 2 becomes

$$\frac{f_{Y|H_1}(y|H_1)P(H_1)}{f_{Y|H_0}(y|H_0)P(H_0)} \underset{H_0}{\overset{H_1}{\gtrless}} 1 \tag{4}$$

or

$$\frac{f_{Y|H_1}(y|H_1)}{f_{Y|H_0}(y|H_0)} \underset{H_0}{\overset{H_1}{\gtrless}} \frac{P(H_0)}{P(H_1)} = \gamma. \tag{5}$$

The ratio at the left in Equation 5 is called the likelihood ratio, and the constant $P(H_0)/P(H_1) = \gamma$ is the decision threshold.
Due to noise interference and other practical issues, the detection system may commit mistakes. The meteor signal detection system performs a binary detection, so that two types of errors may occur:

- Type-I: Accept H_1 when H_0 is true (which means taking noise as a meteor signal and produce a false alarm).
- Type-II: Accept H_0 when H_1 is true (which means to miss a target signal).

The probability to commit a type-I error is called *false alarm probability*, denoted as P_F, and the probability of type-II error is called *probability of a miss* (P_M) (Shamugan & Breipohl, 1998). In addition, we can define $P_D = 1 - P_M$, which is called the detection probability. The decision threshold can be handled to produce acceptable values for both detection and false alarm probabilities. If the decision threshold is varied, the Receiver Operating Characteristics (ROC) curve can be constructed (Fawcett, 2006). This means to plot P_D versus P_F. As the signal-to-noise ratio (SNR) decreases, detection efficiency deteriorates, which translates into ROC curves near the diagonal and for a given fixed P_D, the false alarm probability increases. Therefore, the detection system can be designed by establishing the desired P_D and minimizing P_F, which is known as the Neyman-Parson detector (Trees - Part I, 2001). Another useful performance index is the sum-product (Anjos, 2006), which is defined as

$$SP = \frac{(P_D + 1 - P_F)}{2}(P_D(1 - P_F)). \tag{6}$$

By maximizing the SP index, a balanced detection efficiency is achieved for both hypotheses H_0 and H_1.

4.2 The matched filter

In the case the signal to be detected is known (deterministic), from Equation 5, considering that the block of received data $s[n]$ comprises N samples and the additive noise is white (Trees - Part I, 2001), we have:

$$\frac{\prod_{i=1}^{N} f_{Y_i|H_1}(y_i|H_1)}{\prod_{i=1}^{N} f_{Y_i|H_0}(y_i|H_0)} \underset{H_0}{\overset{H_1}{\gtrless}} \gamma. \tag{7}$$

Now if the noise is Gaussian with zero mean and variance σ^2, the Equation 7 becomes

$$\frac{\prod_{i=1}^{N} \frac{1}{\sqrt{(2\pi\sigma)}} exp(-\frac{(y_i - s[i])^2}{2\sigma^2})}{\prod_{i=1}^{N} \frac{1}{\sqrt{(2\pi\sigma)}} exp(-\frac{y_i^2}{2\sigma^2})} \underset{H_0}{\overset{H_1}{\gtrless}} \gamma. \tag{8}$$

Taking the natural logarithm and rearranging the terms, we have

$$\mathbf{y}^T\mathbf{s} \underset{H_0}{\overset{H_1}{\gtrless}} \sigma^2 ln(\gamma) + \frac{1}{2}(\mathbf{s}^T\mathbf{s}) \tag{9}$$

or

$$\mathbf{y}^T\mathbf{s} \underset{H_0}{\overset{H_1}{\gtrless}} \gamma'. \tag{10}$$

Thus, in the presence of additive white Gaussian noise, the decision between the two hypotheses is given by the inner product between the received signal and a copy of the target signal. This approach is known as the matched filter, which is proved to be optimal in the sense of the signal-to-noise ratio (Trees - Part I, 2001).

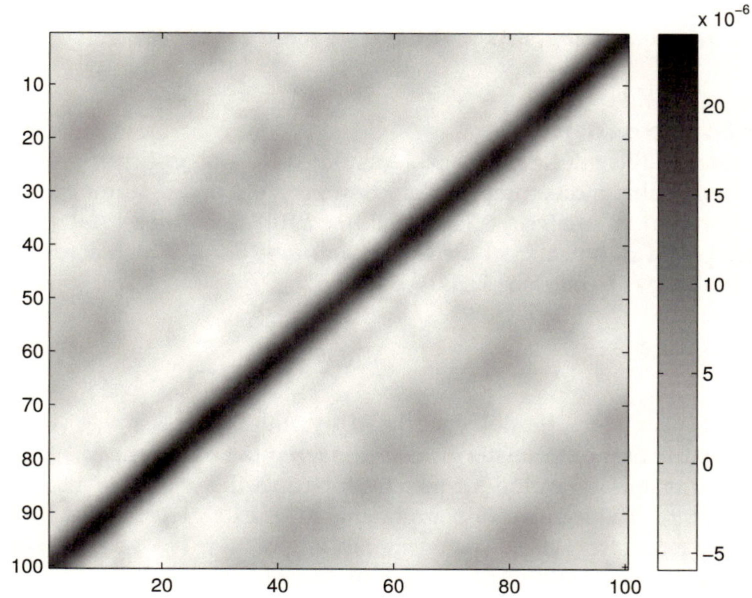

Fig. 4. Covariance matrix for the noise process (development set).

4.2.1 Noise whitening

If the additive noise is a colored noise, is desirable to apply a whitening filter (Whalen, 1995) on a preprocessing phase. When a given zero-mean signal $\mathbf{y}$ is said white, its samples are uncorrelated and the corresponding variance is unitary (Hyvärinen, 2001). As a consequence, its covariance matrix equals de identity matrix:

$$E[\mathbf{y}\mathbf{y}^T] = \mathbf{I}. \tag{11}$$

It is possible to obtain a linear transformation that applied to a process $\mathbf{y}$ produces a new signal process $\mathbf{v}$ that is white. A common way to obtain the whitening transformation is through the decomposition of the covariance matrix into its eigenvalues and eigenvectors (Hyvärinen, 2001):

$$E[\mathbf{y}\mathbf{y}^T] = \mathbf{E}\mathbf{D}\mathbf{E}^T \tag{12}$$

where $\mathbf{E}$ is the orthogonal matrix of eigenvectors and $\mathbf{D}$ is the diagonal matrix of the eigenvalues. The whitening matrix is then obtained through (Hyvärinen, 2001):

$$\mathbf{W} = \mathbf{E}\mathbf{D}^{-1/2}\mathbf{E}^T. \tag{13}$$

And the signal transformation

$$\mathbf{v} = \mathbf{W}\mathbf{y} \tag{14}$$

obtains the white signal process $\mathbf{v}$.

The covariance matrix for the raw data (see Section 4.2.3 next) is shown in Fig. 4. The covariance matrix exhibits crosstalks, which point out a deviation from a fully white noise process. The whitening transformation applied to the development set (see next section) results on a perfect diagonal matrix and the results are well generalized for the testing set (see Fig. 5).

4.2.2 Stochastic process detection

In this case, matched filter design may be generalized for stochastic process detection (Trees - Part III, 2001). For this more complex detection problem and assuming Gaussian process, the principal component analysis (PCA) (Jolliffe, 2010) is applied, and the meteor signal becomes decomposed into principal (deterministic) directions, which are obtained from the Karhunen-Loève series (Trees - Part I, 2001)

$$Y = \sum_{i=1}^{N} c_i \phi_i \tag{15}$$

The series coefficient c_i describes all the stochasticity of the process and ϕ_i is an eigenvector of the covariance matrix of the stochastic process (assumed to be zero-mean). Associated to each eigenvector, there is an eigenvalue λ_i, which represents the energy of the process retained in the direction of ϕ_i. The number of components to be extracted may be limited to a given amount of energy reconstruction, allowing signal compaction. The discarded components are typically associated to noise and do not help in the signal detection task. After decomposition, a filter is matched to each component, resulting in the detection system shown in Fig. 6.

In Fig. 6, h_i provides the weighting of each matched filter in the overall decision, corresponding to the energy fraction of each principal component (when the noise is with spectral height $N_0/2$):

$$h_i = \frac{\lambda_i}{N_0/2 + \lambda_i} \tag{16}$$

Typically, the masking noise process is not white and the detection system can be implemented as shown schematically in Fig.7.

Due to the fact that the detection must run online, the speed and complexity of the applied technique must be considered for the implementation. For meteor signals, the stochastic detection may roughly be approximated by considering the process represented by a deterministic target signal, which can be a specific event, considered the most representative of the process, or the process mean (see Fig. 8). This simplification has been successful in high-energy physics for particle detection (Ramos, 2004).

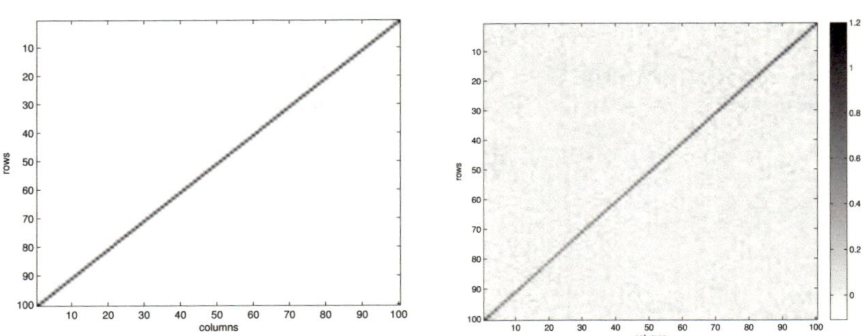

Fig. 5. Covariance matrix after whitening for both (a) development and (b) testing sets.

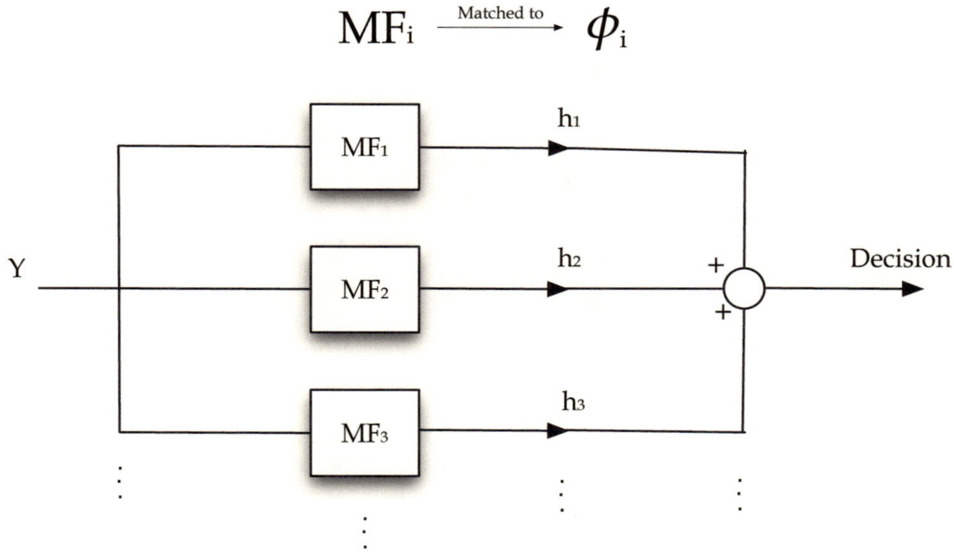

Fig. 6. Block diagram of the matched filter for stochastic signals.

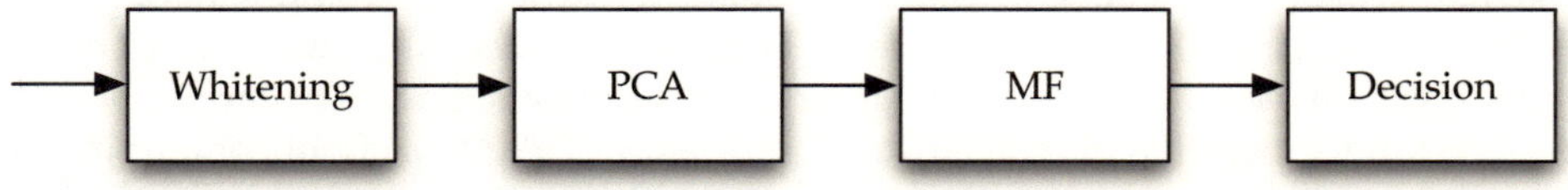

Fig. 7. Block diagram of the matched filter for stochastic signals considering colored noise.

4.2.3 Experimental data

Experimental meteor signals were selected through visual inspection and split into both development and test sets. The first set was used to design the filter, and the second to evaluate the generalization capability of the design and quotes the performance efficiency. Due to variations on signal width, a fixed time window of 1 second was chosen, which is large enough to accommodate most interesting events. Signals were synchronized by their peak values in the acquisition time windows. Noise data from different days of acquisition were also split into development and testing sets. Each noise data set comprised 500 signals, which were obtained through visual inspection. Fig. 9 shows a Gaussian fit applied to the noise histogram, which reveals that the noise may roughly be considered a zero-mean Gaussian process. This approximation facilitates the matched filter design, as shown above.

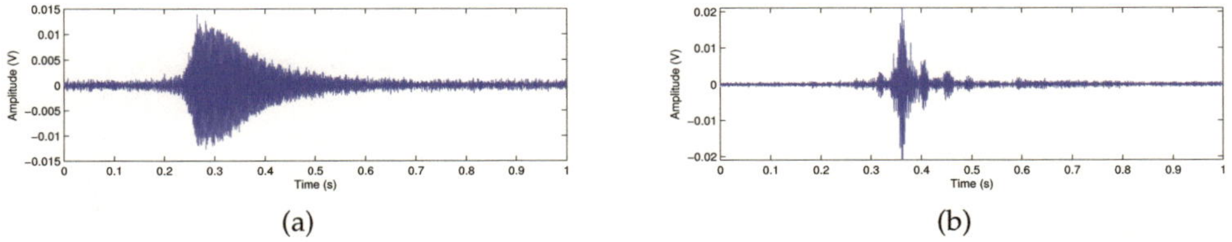

Fig. 8. Underdense trails signals: (a) typical event and (b) the process mean.

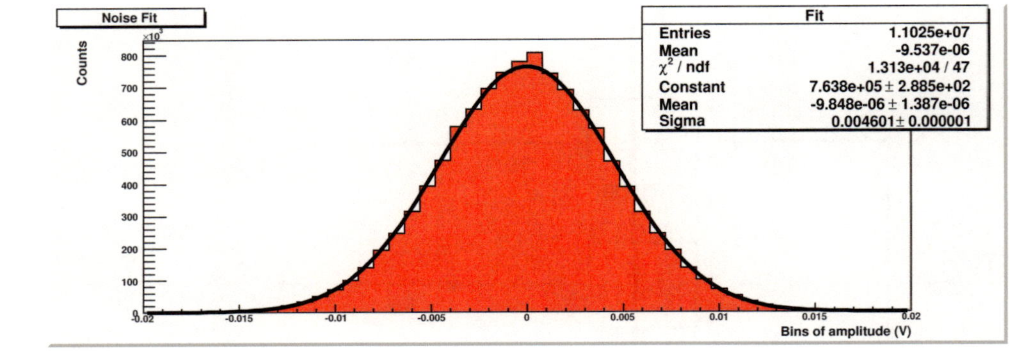

Fig. 9. Gaussian fit applied to noise histogram.

Some steps of the data acquisition, such as low-pass filtering, produced a colored noise, so that the whitening preprocessing should be applied in order to design the detection system based on a matched filter.

4.2.4 Filter performance

Considering the matched filter design approximation discussed in section 4.2.2, the best results for the detection based on matched filter were obtained using as a target signal the mean signal, which was obtained from underdense trails of the development set. Fig. 10 shows the ROC curves for the testing set for both the matched filter and simple threshold detection. It can be seen that the matched filter system achieves a much better performance. For the development set, the filter achieves an efficiency of 99.3% without committing errors of type-I. For an efficiency of 100%, the false alarm probability reaches 0.2%. For the test set, the numbers are 98.8% of detection efficiency for a P_F equal to zero and 1.5% of false alarm probability for 100% detection efficiency (see Fig. 10). Choosing the decision threshold by maximizing the SP index (see Fig. 11 (a)) leads us to an efficiency of 99.4%, for 0.2% of false alarm and considering the testing set.

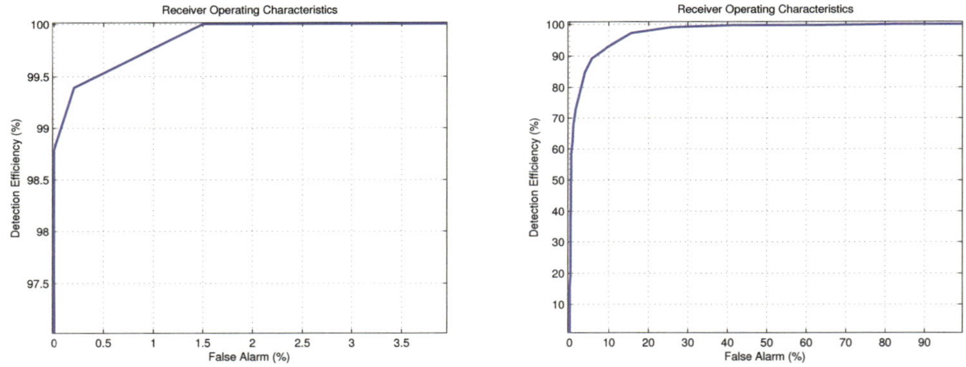

(a) (b)

Fig. 10. ROC curves for: (a) matched filter (deterministic approach) and (b) for threshold detection. Both for testing set.

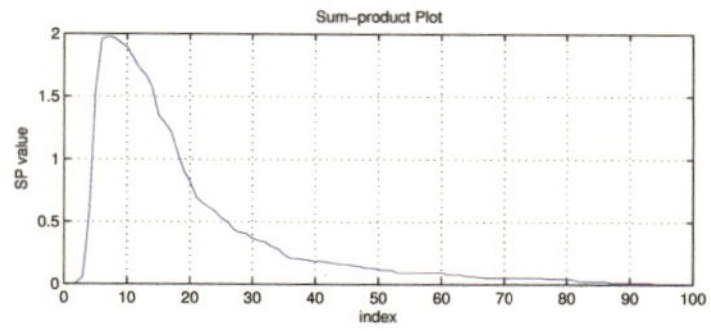

(a) Sum-product plot.

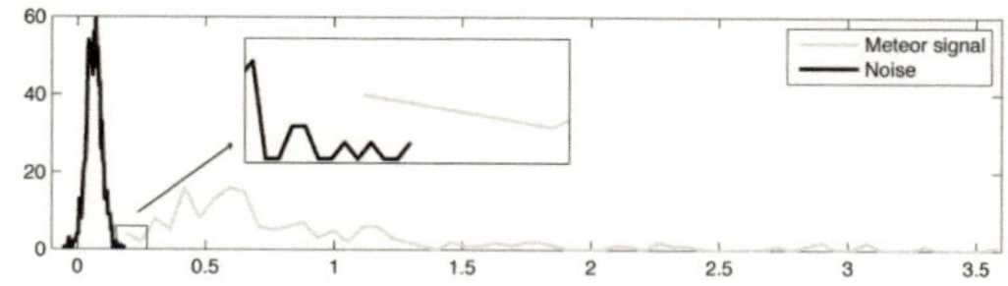

(b)

Fig. 11. (a) SP plot for choosing the decision threshold and (b) Filter output distributions (in the detail, the superposition of the curves for noise and meteor signal).

4.3 Frequency-domain analysis

Radar detection of meteors can also be performed in frequency-domain. The spectral information is obtained by applying the Fourier transform or its variants.

4.3.1 Cumulative spectral power analysis

In frequency-domain, the information about the meteor events is concentrated within a narrow band of the spectrum, corresponding to the demodulated video carrier.
For a WSS random process, the power spectrum is defined as the Fourier transform of the autocorrelation sequence

$$S_y(\omega) = \sum_{k=-\infty}^{\infty} R_y(k)e^{-jk\omega}. \tag{17}$$

If only a segment of the signal of length N is available, the autocorrelation can be estimated through (Hayes, 1996)

$$\hat{R}_y(k) = \frac{1}{N}\sum_{n=0}^{N-1} y(n+k)y^*(n). \tag{18}$$

Changing the upper limit to $N-1-k$, we guarantee that only values of $y(n)$ in the range $[0, N-1]$ will contribute to the sum. Now considering the signal $y_N(n)$, which results from the product of $y(n)$ with a rectangular window of length N samples,

$$y_N(n) = \begin{cases} y(n) & ; \quad 0 \leq n < N-1 \\ 0 & ; \quad otherwise \end{cases}, \tag{19}$$

the estimate of the autocorrelation sequence becomes

$$\hat{R}_y(k) = \frac{1}{N}\sum_{n=-\infty}^{\infty} y_N(n+k)y_N^*(n). \tag{20}$$

Taking the Fourier transform and using the convolution theorem (Hayes, 1996),

$$\hat{S}_y(\omega) = \frac{1}{N}Y_N(\omega)Y_N^*(\omega) = \frac{1}{N}|Y_N(\omega)|^2 \tag{21}$$

where $Y_N(\omega)$ is the Fourier transform of $y_N(n)$. Thus, the power spectral density (PSD) is proportional to the squared magnitude of the Fourier transform. This estimate is known as periodogram (Hayes, 1996).
Blocks of 30s of acquired signal were windowed using non overlapping rectangular windows and the short-time Fourier transform (STFT) was applied (Oppenheim, 1989). Then, the PSD

is estimated via periodogram. Considering the sampling frequency of 22,050 Hz, windows of 256 samples correspond to a time length of approximately 11 ms, which provides a good resolution for the detection and allows wide-sense stationarity (Papoulis, 1965). The 30 s of acquired data were split into 2,584 windowed segments. For the data windows, the peak values are stored. Fig. 12 shows a spectrogram for a 30 s data block.

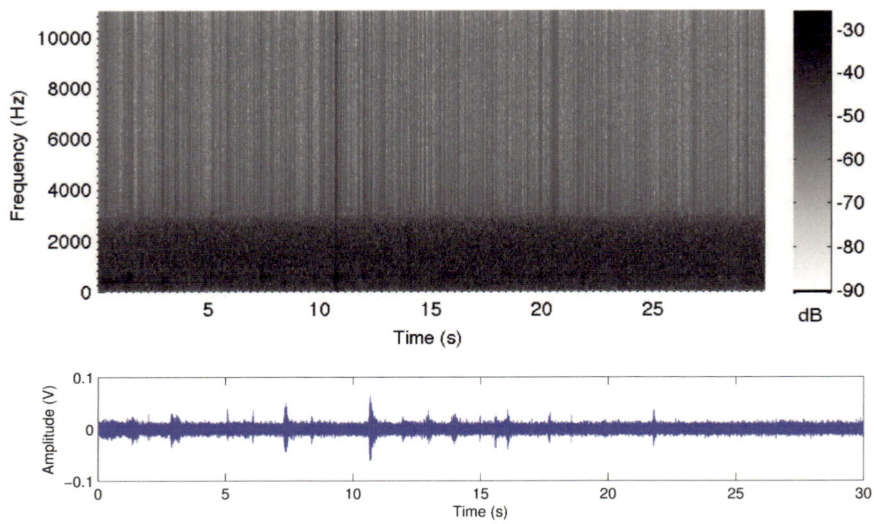

Fig. 12. Spectrogram for a 30*s* data block.

The obtained curve is then accumulated over all time windows, producing the curve shown in Fig.13. The accumulating process produces a curve that exhibits small fluctuation and monotonically increases. The slope in the cumulative power is due to the background noise and it is estimated through a straight-line fit (Fig. 13, dashed line), which is then subtracted from the curve. The resulting accumulated curve is also shown (thicker curve).

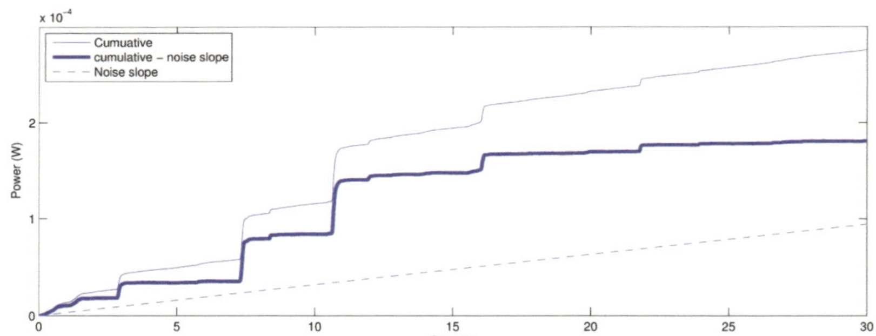

Fig. 13. Curves generated in the cumulative spectral power analysis.

Fluctuations in the slope threshold values are used to define the starting and ending samples for triggering. For the algorithm design, 40 blocks of acquired data were used, which comprised 233 meteor events. For evaluating the performance and generalization, 50 blocks comprising 261 meteor events were selected. Accumulating windows peak values, the cumulative power spectrum algorithm detected 17 fake events over 1,500 s of data (6.7 %

of false alarm), which means approximately 1 fake event per each 100 seconds. Therefore, from this test sample, the algorithm avoided 220 fake events to be recorded (280 MB less per hour). In a full day, the online filter would avoid 6.7 GB of noise to be recorded.

5. Summary and perspetives

Meteor signal detection has been addressed by different techniques. A new detection technique based on radar has advantages, as simplicity of the detection stations, coverture and capacity to be extended for other detection tasks, such as cosmic rays, lightning, among others. Due to its continuous acquisition characteristic, online triggering is mandatory for avoiding the storage of an enormous amount of background data and allow focusing on the interesting events in offline analysis. Both time and frequency domain techniques allow efficient meteor signal detection. The matched filter based system achieves the best performance, and has good advantages, such as it is easy to implemented and has fast processing speed. In frequency domain, a power spectrum analysis also achieves good results. This approach may also be further developed to include a narrowband demodulation in the preprocessing phase. As phase delays are produced by the different paths the traveling wave finds between the transmition, oscillations can be observed (see Fig. 14) mainly in underdense trails. These reflections can be seen as an amplitude modulation, similar to the modulation on sonar noise caused by cavitation propellers (Moura et al., 2009).

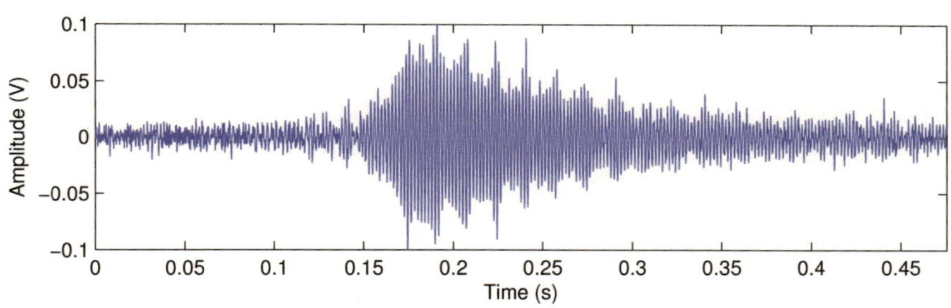

Fig. 14. Amplitude modulation on a underdense signal.

Therefore, a DEMON (Demodulation of Envelope Modulation On Noise) analysis may be applied. The acquired signal is filtered by a lowpass filter, to select the band of interest for the meteor signals. Then the signal is squared in a traditional amplitude demodulation, for the extraction of the envelope. Due to the low frequencies of the oscillations (typically tens of Hz), resampling is performed, after the anti-alias filtering. Finally a FFT is applied, and the frequency the envelope is identified. Other possible approach is to apply computational intelligence methods.

6. References

Anjos, A. R. dos; Torres, R. C.; Seixas, J. M. de; Ferreira, B.C.; Xavier, T.C., *Neural Triggering System Operating on High Resolution Calorimetry Information*, Nuclear Instruments and Methods in Physics Research (A), v. 559, p. 134-138, 2006.

Damazio, D. O., Takai, H.,*The cosmic ray radio detector data acquisition system*, Nuclear Science Symposium Conference Record, 2004 IEEE, On page(s): 1205-1211 Vol. 2.

Fawcett, T. *An introduction to ROC analysis*. Pattern Recognition Letters, 27, 861-874, 2006.

Guang-jie, W., Zhou-sheng, Z.*Video observation of meteors at Yunnan Observatory*. Chinese Astronomy and Astrophysics, Volume 28, Issue 4, October-December 2004, Pages 422-431.

Hayes, M.H. *Statistical Digital Signal Processing and Modeling*, ISBN: 0-471-59431-8, John Wiley and Sons Inc., New York, 1996.

Hirose H., Tomita, K., *Photographic Observation of Meteors*. Proceedings of the Japan Academy, Vol.26 , No.6(1950)pp.23-28.

Hyvärinen, A., Karhunen, J. and Oja, E. (2001). *Independent Component Analysis*, ISBN: 0-471-40540-X, John Wiley & Sons, .inc. 2001.

International Meteor Organization, *www.imo.net*, access September, 2010.

Jolliffe, I.T., *Principal Components Analysis* , ISBN: 0-387-95442-2, second edition, Springer New York, 2010.

McKinley, D.W.R., *Meteor Science and Engineering*, Ed. McGraw-Hill Book Company, New York 1961.

Matano, M. Nagano, K. Suga and G. Tanahashi. Can J. Phys. 46 (1968), p. S255.

Moura, M. M., Filho, E. S., Seixas, J .M., 'Independent Component Analysis for Passive Sonar Signal Processing', chapter 5 in *Advances in Sonar Technology*, ISBN:978-3-902613-48-6, In-Teh, 2009.

Oppenheim, A.V., and R.W. Schafer, *Discrete-Time Signal Processing*, Prentice-Hall, 1989, pp.730-742.

Papoulis, A., *Probability, Random Variables, and Stochastic Processes*, ISBN: 0-07-048448-1, McGraw-Hill Book Company Inc., New York, 1965.

Ramos, R. R., Seixas, J. M., *A Matched Filter System for Muon Detection with Tilecal*. Nuclear Instruments & Methods in Physics Research, v. 534, n. 1-2, p. 165-169. 2004.

Shamugan, K.S., Breipohl, A.M. *Random Signals - detection, estimation and data analysis*, John Wiley & Sons, New York, 1998.

Trees, H.L.Van. *Detection, Estimation, and Modulation Theory, Part I*, ISBN: 0-471-09517-6, John Wiley & Sons, New York, 2001.

Trees, H.L.Van. *Detection, Estimation, and Modulation Theory, Part III*, ISBN: 0-471-10793-X, John Wiley & Sons, New York, 2001.

Whalen A. D. *Detection of Signals in Noise*. Second Edition. ISBN: 978-0127448527, Academic Press, 1995.

Willis, N.C., 'Bistatic Radar', chapter 23 in *Radar Handbook*, third edition, (M.I. Skolnik ed.), ISBN 978-0-07-148547-0, McGraw-Hill, New York, 2008.

Wislez, J. M. *Forward scattering of radio waves of meteor trails*, Proceedings of the International Meteor Conference, 83-98, September 1995.

26

Electromagnetic Waves Propagating Around Buildings

Mayumi Matsunaga[1] and Toshiaki Matsunaga[2]
[1]*Ehime University*
[2]*Fukuoka Institute of Technology*
Japan

1. Introduction

It is a matter of great concern that places where no electromagnetic waves are reached are seen even nowadays when various types of wireless equipment are available anywhere without any concern. That is to say, the fact that places where no electromagnetic waves are reached are found is a problem bringing about unpleasantness to the users, and concurrently is a problem to be solved for those engaged in communication business. Here arises a skepticism why places where no electromagnetic waves are reached are in existence. The matter believed to be the greatest cause of the above is attenuation and interference generated by encounter of the electromagnetic waves with their obstacles. For example, almost all of the base stations (base exchanges) of cellular phones are established outdoors. To accomplish indoor-use of cellular phones, the electromagnetic waves should be aligned so that it will enter the spot deep enough from the entrance in the inside of the buildings by overcoming the obstructing walls. However the electromagnetic waves are liable to be attenuated when they go through the walls, and the waves reflected by the walls interferes with the ones that are going to reach the walls. As a result, the electromagnetic waves are made weak in the vicinity of the buildings or inside of them. This is believed to consequently be linked with creation of difficulty in achieving wireless communications.

It remains to be seen in what a manner the number of the places where no electromagnetic waves are reached is being reduced. As a matter of the fact however, none of easy method to solve this problem is available, and the number of such places has to be reduced one by one every day by repeating such strenuous operations as allowing the places where no electromagnetic waves are reached to be identified and by permitting the waves to be reached on such places with the change of the spots where base stations are settled together with adjustment of the output. Such strenuous trials are put to action by the hands of various researchers with a view to eliminating the troublesomeness of the work. At this stage, let the research that has been made to now be reviewed. With the operations to identify the places where none of electromagnetic waves are reached, two methods, i.e. the one to measure the waves and the other one in accordance with simulation are available. Despite the above, it might be next to impossible to recognize the field strength of the electromagnetic waves in the whole area where wireless communication is utilized. Therefore proposals for a simulation method that can adjust the settling position of the base station or output have been made by many researchers with respect to several methods such as the one to estimate attenuation loss of the electromagnetic on a propagation route utilizing the building height in the communication area obtained by residence maps and its distribution (Kita et al., 2007; Kitao & Ichitsubo, 2008; Xia, 1997) together with the state of the roads (Ikegami et al., 1984; Walfisch & Bertoni, 1988) or the one to estimate the propagation route in accordance with the Ray Tracing method (Lim et al., 2008). However with these methods, difficulties are pointed out in purport that they just enables the attenuation amount of the electric field strength outdoors to be estimated roughly along the propagation route, and real values of the electric field strength are greatly different from the estimated values. In addition the electric field strength distribution in the inside of the building cannot be estimated. Studies to enhance the estimation accuracy by solving these problems are also under way. Landron and Lim (Landron et al., 1996; Lim, 2008) release reports stating to the effect that consideration of the outside wall shape of the building enhances estimation accuracy. In the meanwhile, proposal is by Axiotis (Axiotis & Theologou, 2003) with an estimation method of the electric field strength extended into the inside of the building. However no observation is made to ascertain how electromagnetic waves propagating into the inside of the building are changed according to the shape of the outside wall and structure of it. Such being the case, we, the authors of this paper, have made research to explain how the electromagnetic waves propagating not only in the vicinity of the building but also through the inside are changed (Matsunaga et al., 2009; Matsuoka et al., 2008a; Matsuoka et al., 2008b). Special importance is attached to the detection by measurement, and studies are being made to comprehend whether estimation by means of simulation will make it possible to obtain the electric field strength distribution explaining to what extent the detection will be close enough to the measurement (Matsunaga et al., 1988; Matsunaga et at., 1996).

In this chapter, details are described with the method to measure the change to explain in what a manner the electromagnetic waves propagating in the vicinity of the building will change according to the difference in wall shape or the building or structure of it. In addition comparison is shown between the results from the measurement and the result obtained by the simulation in accordance with the FVTD, a kind of time domain difference method. Furthermore it is shown that as a result of such studies, 2 types of epoch-making effective discoveries as shown below are claimed. The first thing is that with many of the conventional methods, on the supposition of the building being dealt with just as a concrete square pillar the whole of which was filled with concrete to the extent of its pivotal point. However it is understood that great difference in the electric field strength is in existence between the building supposed to be comprised of the wall and inside space and that of the electromagnetic waves propagating in the vicinity of the building and through the inside of it. The second thing is that it is also understood that the amount of the reflection is greater with the concrete wall having round convexities on the outside wall and the amount of the invasion is smaller than with the reinforced-concrete wall.

However it is regrettable to state that the authors of this paper themselves are never free from defects in the research. That is to say, although it is understood that conduction simulation in consideration of the shape of the wall or structure of it makes it possible to estimate accurately the electric field distribution in the vicinity of it or inside of it, the authors are not aggressive enough to grapple with the problem of the simulation for improvement so as to allow the electromagnetic waves to reach the place where no electromagnetic waves generated in the vicinity of a specific building are reached. Nothing has been obtained with the result explaining that it is possible to shut the electromagnetic waves intruding from the outside or to allow the electromagnetic waves propagating through the room to be made homogeneous on supposition of, e.g., a tile as convexities on the wall surface by adequately adjusting the size of the tile, raw material, attaching position, etc. this might be called a future assignment.

2. A way of measurement

In this section, a way of measurement of the electromagnetic waves propagating in the vicinity of the building and inside of it is described. It is explained what kind of influence will be exercised on the electromagnetic waves propagating around the building by the shape or structure of the wall of the building. Details of the way of measurement are provided with: (1) Explanation of measurement methods. (2) Composition of the measurement systems such as measurement units and equipment. (3) Measurement procedure.

2.1 A method of measurement

Around a scaled-down model building which is settled in a radio-frequency anechoic chamber, a virtual 2-dimensional space is furnished, and measurement of the electromagnetic waves is made in the inside of the space. With the role of the measurement at this stage, it is necessary to use a measurement method from which the influence exercised by the factors except for the shape of the building or structure of it is removed as far as possible deducing from the fact that the shape of the wall of the building or change of the structure of it is the influence to be exercised on the electromagnetic waves propagating around the building. For such a reason, measurement is made in a virtual 2-dimensional space composed in a radio-frequency anechoic chamber.

First of all, let it be understood that measurement is made by using a scaled-down model regarding it as the building utilized for the experiment, because it is difficult to settle a real building in the radio-frequency anechoic chamber owing to its size. At this stage, a scaled-down model is a model building taken up based on the idea that the size of the building is made smaller by shortening the wavelength of the wave source used for the measurement, keeping constant the ratio of the size of the real building complying with the wavelength of the electromagnetic waves used for general mobile wireless devices such as mobile telephones, in-room wireless LAN, RFID, etc. Incidentally in this chapter, the scaled-down model building used for measurement shall be called a building model in this chapter since now on.

Secondly the virtual 2-dimensional space that has been referred to before is a space obtained by actually composing the 2-dimensional space used, for example, in the 2-dimensional simulation in the radio-frequency anechoic chamber. As illustrated in Figure 1, the said virtual 2-dimensional space is made real by putting the building model having conductor plates wide enough to be equivalent to the electric wall between the upper and the lower sides as seen above. Thus making measurement in the 2-dimensional space makes it possible to remove the influence exercised by the change of the building in a height direction, and it becomes possible to consider the influence of shape or structure change exclusively in a lateral direction. When the electromagnetic waves propagating in the vicinity of a building that is exceedingly great in comparison with the wavelength is measured or in case in-room propagation on a spot where a base station is located in the building is measure, it is easy to comprehend what shape of structure of the wall in the inside or outside of the building will exercise influence on the electromagnetic waves propagating around the wall so long as measurement is made in a 2-dimensional space rather than in a 3-dimensional space.

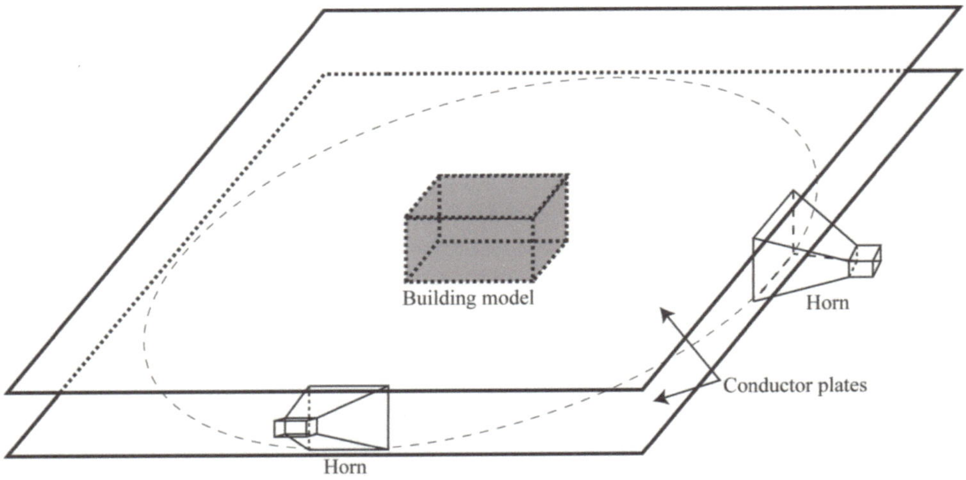

Fig. 1. A measurement unit comprised of a virtual two dimensional measurement space

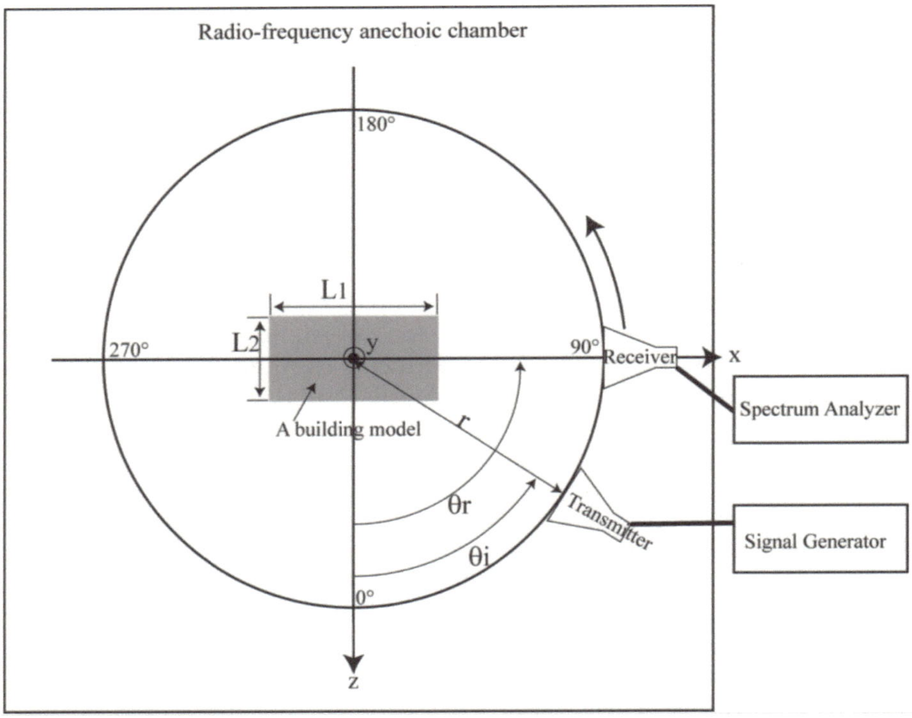

Fig. 2. A schematic diagram of the measurement system

Fig. 3. A photograph of measurement system inside our radio-frequency anechoic chamber

2.2 Composition of measurement systems

Description is hereunder made with the measurement system such as the building model or measurement units composed in the radio-frequency anechoic chamber. Illustrated in Figure 2 is a schematic diagram of the measurement system. In the left side of the figure, a top view of the measurement unit composed in the radio-frequency anechoic chamber is provided, and furthermore how the said unit is linked with the measurement equipment settled outside the radio-frequency anechoic chamber is shown. Illustrated in Figure 3 is a photograph in the radio-frequency anechoic chamber. It is noticed that virtual 2-dimensional space is composed in the center of the photograph. The building model is as a matter of reality settled in the inside of the space although no trace of the model is to be seen in the photograph. Now that, explanation is hereby made with measurement unit composed in the virtual 2-dimensional space by using Figure 2. First of all, coordinate axis is established as seen in Figure 2. And the building model whose configurational dimension is $L_1 \times L_2$ is placed at its center. Around the building model, both transmitting and receiving antennae placed on a circle whose radius is r is settled. Thereafter an incident wave is provided from the transmitting antenna fixed at an angle, and the electric field strength distribution around the building model is measured rotating the receiving antenna around the building. A horn antenna was utilized as the transmitting/receiving antenna, and the source wave is provided by the transmitting antenna using a signal generator. Meanwhile measurements of the electric field strength are made by means of a spectrum analyzer connected to the receiving antenna. In this connection, the settlement angles of the transmitting and receiving antennae are defined as θ_i and θ_r as the angles from the z axis.

2.3 Measurement procedure

At the final stage, actual measurement procedure is described. First a single piece of the building model is placed on a pivotal point of the measurement unit. Secondly the transmitting antenna is fixed on an angle θ_i on the circle whose radius is r from the pivotal point. Thirdly the receiving antennal, which is settled on the circle whose radius is r from the pivotal point, is placed on a lateral side of the transmitting antenna close enough to the right side. Thus the angle θ_i on that position and the electric field strength are measured. Fourthly the receiving antenna is moved by $\Delta\theta$ in a counterclockwise direction, and the electric field strength is measure. Thereafter measurement is continuously made as far as the receiving antenna comes immediately to the side of the left of the transmitting antenna in accordance with the fourth procedure.

3. Measurement results

In this section, the results are shown with the electric field strength distribution in the vicinity of the building model having various types of shapes of the wall and structure of it obtained in accordance with the measurement methods referred to above. In advance of exhibiting the measurement results, description of the building model used before the measurement is at first made, and secondly description is again made with the measurement conditions regarding the size of the building model and detailed dimensions of the shapes or structure of the wall together with the positions of the transmitting/receiving antennae are made. Thereafter with the measurement results of the electric field strength distribution brought about by using the building model are shown, observing the individual factors in comparison with them.

3.1 Individual types of the building models

First of all description is made with the building models used for the measurement. There are 4 types illustrated in Figure 4, and each of them is: (a) A square pillar model where the building is regarded as a concrete square pillar. (b) A building model with flat walls where the building is dealt with as the one comprised of a flat wall and inside space. (c) A building model with reinforced-concrete walls where the building is dealt with as being comprised from a flat wall and inside space. (d) A building model with walls having round convexities that are dealt with as being comprised of a wall having periodic convexities on the outside of it and inside space. In this connection, details of the part of the round convexities of the wall model having round convexities are defined as illustrated in Figure 5.

3.2 Measurement conditions

At the next stage, measurement conditions are described. In Table 1, the measurement systems and detailed dimensions of the building model defined in Figures 2 and 4 are concisely listed. First with respect to the measurement systems, the electric field strength distribution around the building model was measured allowing an electromagnetic wave with frequency f = 9.35 GHz to be radiated from the transmitting antenna fixed on a position whose angle θ_i = 0°on a circle whose radiation r = 1000 mm, rotating the receiving antennal by individually $\Delta\theta$ = 1°. Furthermore, detailed dimensions of the individual portions of the building model in Figure 4 are described. The description is made on the assumption that the configurational dimensions of the building model are as L_1 = 700 mm and L_2 =350 mm throughout the whole models. Meanwhile with a model having a wall, description is likewise made on the assumption that its wall thickness is T = 45 mm. The reinforced-concrete wall model was composed by inserting metal bars with a diameter w = 2 mm into the concrete wall in a series at an interval p = 10 mm. Both the tips of these bars are connected with the conductor plates used for composing a 2-dimensional space.

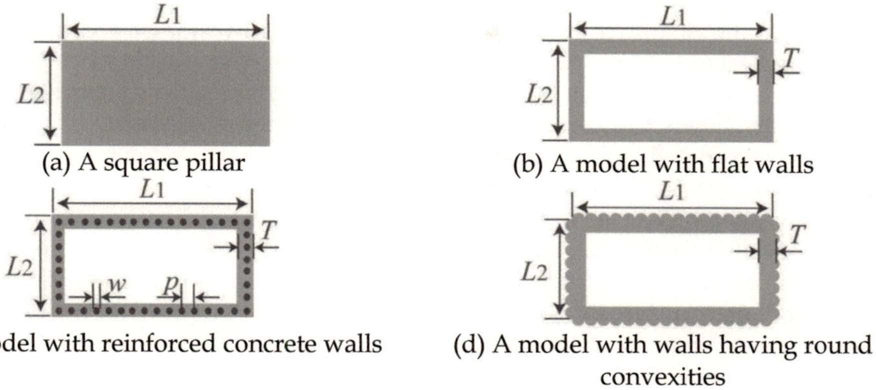

(a) A square pillar

(b) A model with flat walls

(c) A model with reinforced concrete walls

(d) A model with walls having round convexities

Fig. 4. The plane figures of building models

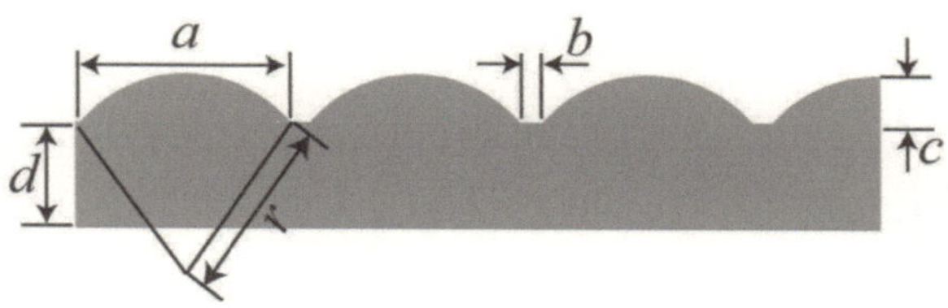

Fig. 5. Detailed figure of the round convexities in Figure 4(d)

f	r	θ_i	$\Delta\theta$	L_1	L_2	T	w	p
9.35 GHz (λ=32.0 mm)	1000 mm (31.25λ)	0 °	1 °	700 mm (21.88λ)	350 mm (10.94λ)	45 mm (1.41λ)	2 mm (0.06λ)	10 mm (0.31λ)

Table 1. Detailed measurements of the measurement system and building models in Figures 2 and 4

With respect to the wall model having round convexities, measurement is made by using 2 types of building models, that is to say, in case of the model whose round convexities are slightly greater and in case of the model whose round convexities are slightly smaller than the wavelength of the source wave. Listed in Table 2 are the detailed dimensions of the portion of the round convexities of the 2 types of building model in accordance with the definition in Figure 5.

	r	a	b	c	d
Big	60.0 mm (1.88λ)	77.5 mm (2.42λ)	10.0 mm (0.31λ)	14.2 mm (0.44λ)	36.4 mm (1.14λ)
Small	30.0 mm (0.94λ)	38.7 mm (1.21λ)	5.0 mm (0.16λ)	7.1 mm (0.22λ)	40.1 mm (1.25λ)

Table 2. Detailed measurements of the round convexities defined in Figure 5

3.3 Comparison among the measurement results obtained by using the individual building models

By comparing the experimented value obtained in response to the allusion referred to above with the measurement by using the individual building models, it is observed what influence the difference of the wall structure will exercise on the electric field distribution propagating in the vicinity of the building. First, illustrated in Figure 6 are the measurement values of the electric field distribution around the square pillar and the ones of the flat wall model comprised of the wall closer to the structure of the real building and the inside space simultaneously shown. Comparison of the 2 types of the measurement values reveals that great difference is noted in the electromagnetic waves in the rear side of the building whose receiving angle ranges close enough from 50 degrees to 220 degrees owing to the existence of the inside space. That is to say, it is understood that the penetrating wave directed rearward exhibits increase ranging from 20 dB to 30 dB exclusively with respect to the flat wall model in the inside of which space is in existence. It can be understood from the result that with the electromagnetic waves propagating in a direction of the other side of the building viewed from the transmitting point, almost all of them have been successful enough to reach there by penetrating the building. Contrarily, it can safely be said that just a slight amount of the waves have been successful in reaching there by diffraction. It is therefore suggested, it can be said, that whether the building should be a square pillar model or a building model with flat walls is a very important point in heightening the simulation value.

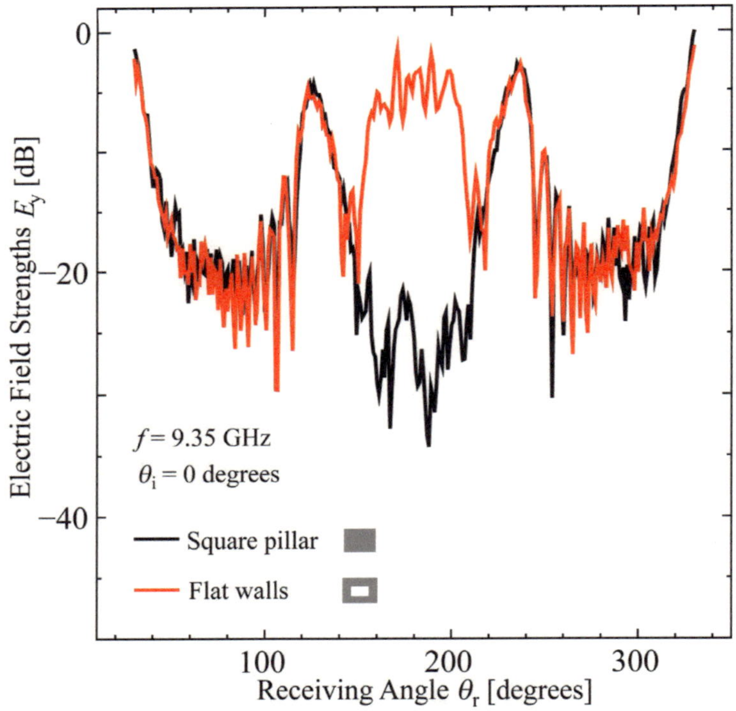

Fig. 6. A comparison of measurement results of electric field strength around the square pillar model and around the flat wall model comprised of the wall and inside space

Illustrated in Figure 7 are the measurement values of the electromagnetic waves around the reinforced-concrete wall model obtained by allowing the building to come closer to the real building and the ones around the flat wall model having no metal skeletons in it simultaneously shown. Comparison of these measurement values reveals that the electric field whose receiving angle ranges from 150 degrees to 220 degrees strength in the rear side of the building is decreased less than approximately 10dB. In addition in the vicinity of 90 degrees and 270 degrees as well, it is understood that the electric field strength is rather than decreased when metal skeletons are available. From the fact that such change cannot be witnessed in the result in Figure 6, it can safely be affirmed that existence of the metal skeletons in the inside of the concrete wall results in not only the penetration of the electromagnetic wave in a rear side of the building but also the propagation of the electric wave in a lateral side of the building is decreased.

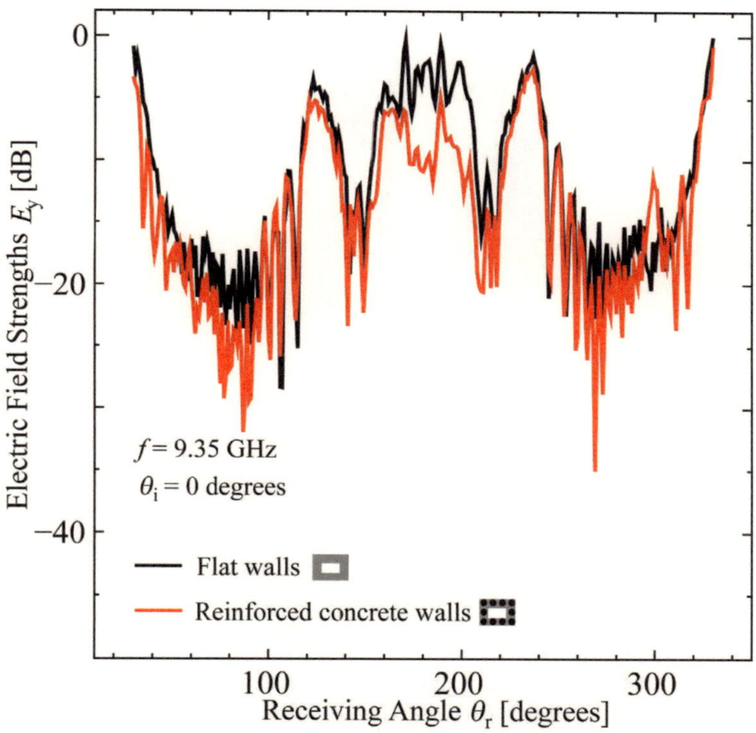

Fig. 7. A comparison of measurement results of electric field strength around the flat wall model and around the reinforced-concrete wall model

Illustrated in Figure 8 are distributions of the electric field strength around a building model with walls where round convexities are periodically in existence on the outside wall and around the building model with flat walls are simultaneously shown. By comparing these two types of measurement values, it is explained that existence of round convexities on the outside wall decreases the penetrated wave approximately 10 dB to 20dB in a rear direction of the building in the vicinities ranging from 150 degrees to 220 degrees. The matter to which attention should be arrested is that despite the fact that the two results are almost the same in the right and left square portions in the rear side of the building in the vicinities ranging from 110 degrees to 140 degrees and from 220 degrees to 250 degrees, in the closer vicinities ranging to the incidence side than these angles the electric field strength in case the round convexities are in existence is higher than in case they are not in existence. That is to say, it is imagined that although reflection in an incident side is increased and penetration in the rear side is decreased, the diffraction in a rear side is not so changed.

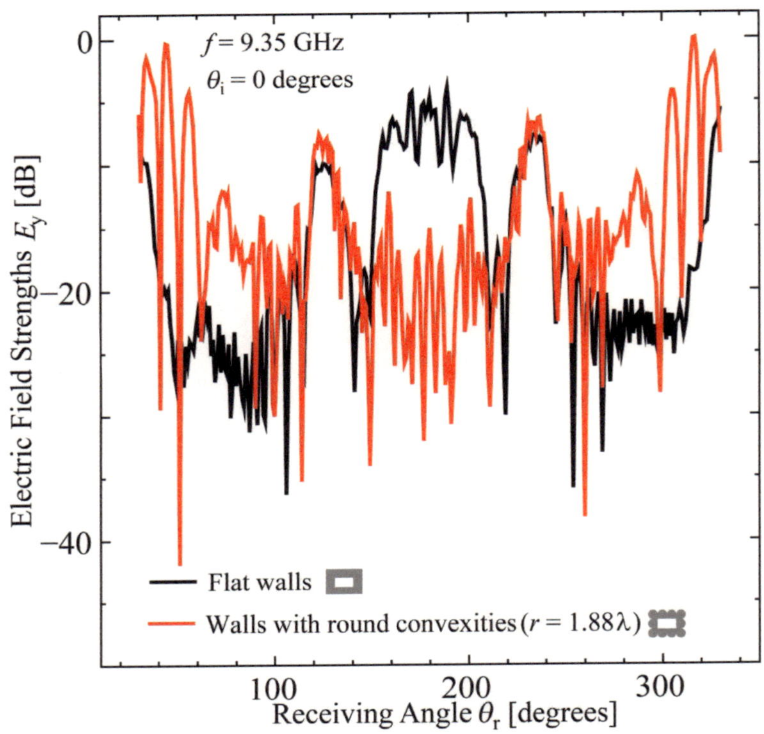

Fig. 8. A comparison of measurement results of electric field strength around the flat wall model and around the building model with walls where round convexities are periodically in existence on the outside wall

Illustrated in Figure 9 are the measurement results simultaneously shown with the electric field strength distributions in case the radius of the round convexity is slightly greater than the wavelength and in case the radius is slightly smaller than the length in relation to a building model with walls having round convexities. Comparison of these two measurement results reveals that penetrating wave is increased with the model whose radius of the round convexities is smaller, and the reflection wave is decreased. However from the fact that considerable change is noted with the electric field strength on the square portion in the diagonally rear side of the building as is known from the result in Figure 8, it is imagined that almost none of influence by diffraction is to be seen.

At the final stage, the whole of the results illustrated in Figure 6 through Figure 8 are shown simultaneously in Figure 10. By so doing, it is explained that the electromagnetic waves propagating in the vicinity of the building are subject to change depending upon the thickness of the wall of the building, presence or absence of metal skeletons, and shape of the surface of the outside wall. Especially the fact that existence of the round convexities on the outside wall encourages the reflection waves propagating from the lateral side in a direction of the incident side to be increased rather than the fact that reinforcement bar is in existence in the inside of the wall. This evidently brings about decrease of the penetrating waves in the rear side of the building. Accordingly for adjustment of the electric field strength distribution in the inside and outside of the building with which construction is already finalized, it is suggested that it is effective to paste convexities e.g. tiles onto the wall of the building after finalization of the construction.

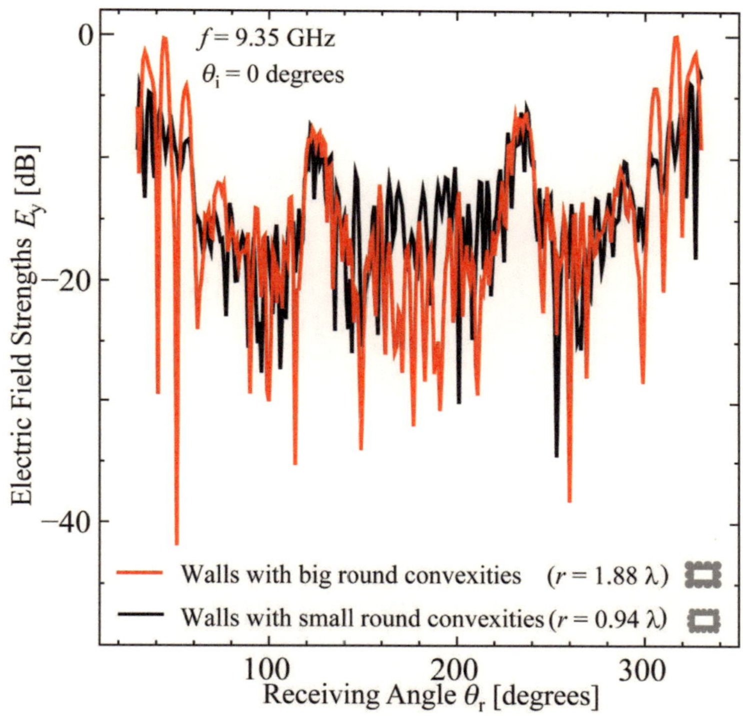

Fig. 9. A comparison of measurement results of electric fields around two building models which have different sizes of round convexities each other on their walls

4. Application of the measurement results to simulation

Making use of the measurement utilizing a building model, observation has been made to now in consideration of the influence of the shape of the wall or structure of the building that exercises influence on the electromagnetic waves propagating around the wall. From now on, investigation is made to determine whether these observation results should be applied to the simulation or not. Incidentally as a way to conduct the simulation, Finite Volume Time Domain method (hereinafter FVTD) method was taken up. The said method is a method similar to FDTD method widely applied to various types of electromagnetic field analysis, by which Maxwell's equation is discretized based on volume integral. For this reason several features are pointed out: The lag equivalent to half an amount of space difference quantity is produced when FDTD method is utilized, whereas the fact that the electric field and magnetic field are placed in the same position when the FVTD method is dealt with. With the electromagnetic field analysis having a boundary that is hard to be dealt with on an orthogonal coordinate as seen for example in a complicated surface area boundary, the FVTD method is preferable because the method very widely simplified rather than the FDTD method. Thus analysis time is also shortened. This is the reason why the FVTD method is adopted in this paper as a simulation method by which comparison of measurement values is made. Meanwhile with the FVTD method, readers are advised to refer to the details offered by Uchida et al., (Uchida et al., 1996a; Uchida et al., 1996b) and Yee and Chen (Yee & Chen, 1994).

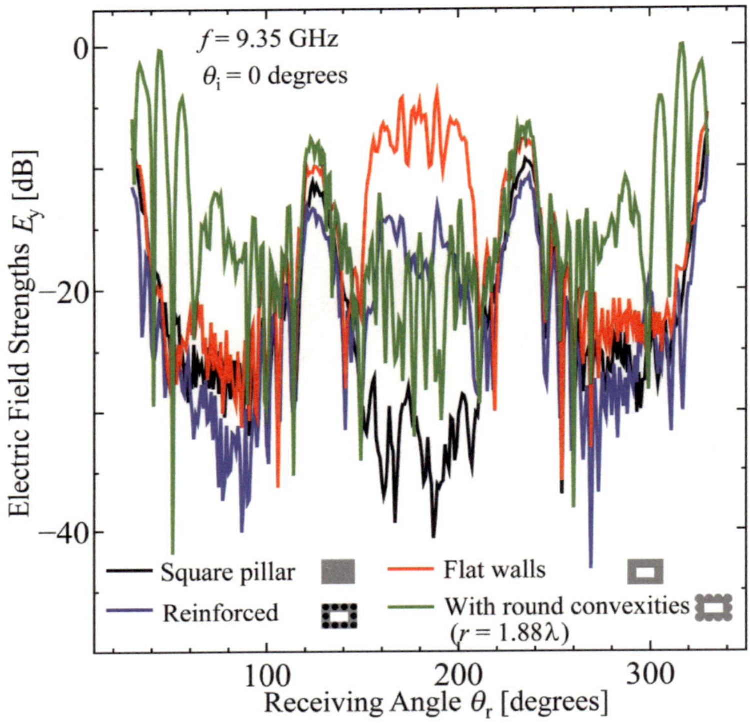

Fig. 10. The whole of the results illustrated in Figure 6 through Figure 8 are shown simultaneously

In this connection, an analytical model in accordance with FVTD method is described. Illustrated in Figure 11 is an analytical compositional diagram to make FVTD analysis with respect to the 2-dimensional virtual space portion on the measurement system referred to in Figure 2. First the analytical space is divided into $N \times N$ cells, and PML adsorption layer having 20 cells is located on the outer side of the analytical model with a view to suppressing the reflection from the surrounding boundary of the analytical mode. At this stage, the analysis is made by dividing the time difference quantity into 64 per cycle.

4.1 Comparison between the measurement result and analytical result in accordance with FVTD method

First of all, the following are determined in the simulation using FVTD method by comparing the measurement result with the simulation result to which the FVTD method is applied. (1) Electrical constant of the concrete. (2) Partitioning number *N*. With the dielectric constant of the concrete, the value calculated in accordance with the dielectric constant calculation method (Matsuoka et al., 2009) based on the measurement is, first of all, applied to FVTD method, and furthermore fine adjustment is made by comparing it with the measurement results. On that occasion, almost none of adjustment is required with the specific relative permittivity, and slight adjustment was just necessary with the conductivity. As a result, the dielectric constant of the concrete is obtained as specific relative permittivity $\varepsilon_r = 6.0$ and conductivity $\sigma = 0.1$ S/m. With regard to the partition cell number, it is likewise decided by comparing the measurement result with the simulation result that $N = 4460$. Illustrated in Figure 12, the results obtained by conducting the simulation in accordance with FVTD method using the dielectric constant and partition cell number determined as above together with the measurement value are simultaneously shown. From these results, it is explained that the electric field strength distribution changes in a complicated manner depending on the places around the building, but the simulation value exhibits its manner of the change faithfully almost to the measurement value. From the above, it is construed that the simulation close enough for the actual situation has been possible with the use of the measurement value.

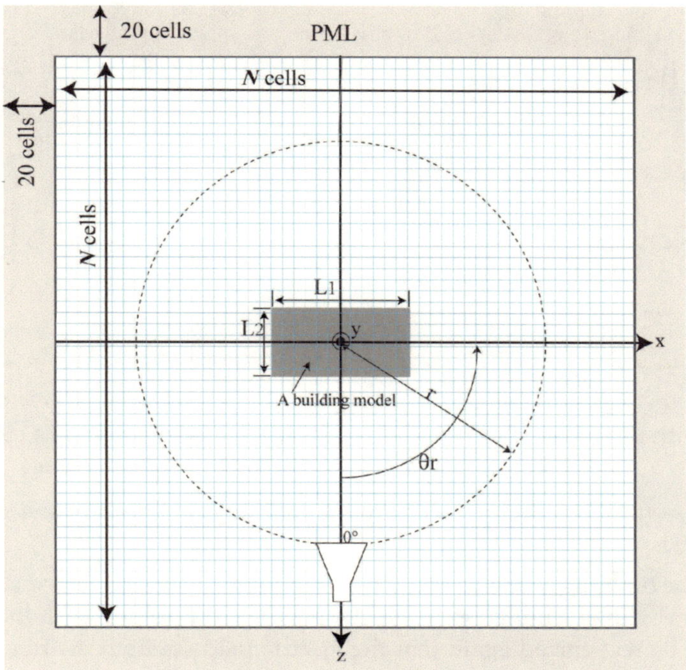

Fig. 11. A plane figure of the analytical compositional diagram for FVTD analysis

4.2 Field strength distribution simulation in the analytical model

As the final trial, the result obtained by conducting simulation with the electric field strength distribution of the whole of the analytical space utilizing FVTD is shown. Illustrated in Figure 13 are the results obtained by simulating the electric field strength distribution, using FVTD method. It is evident from Figure 11 that in what place of the

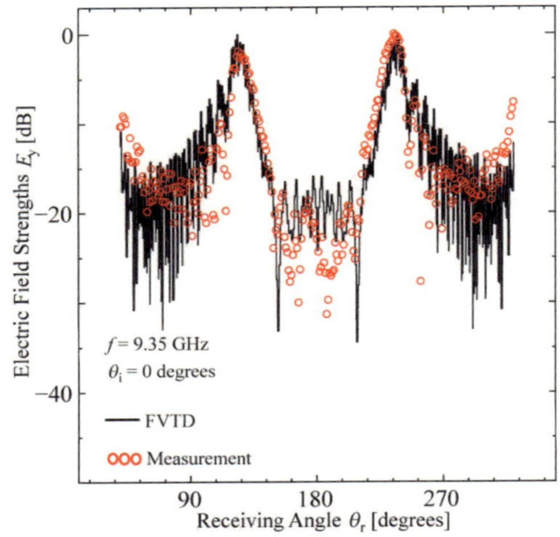

(a) A square pillar

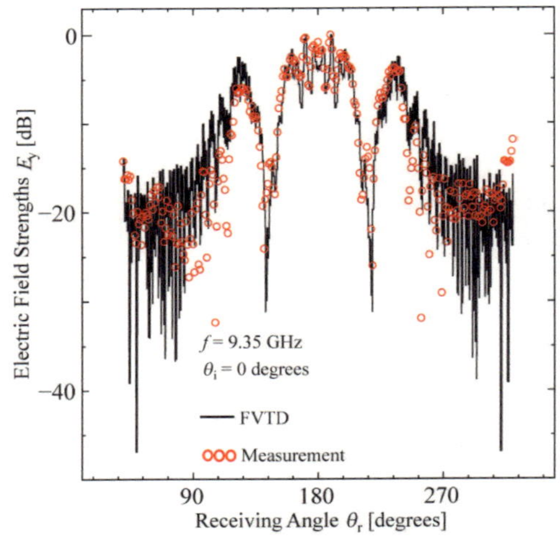

(b) A model with flat walls

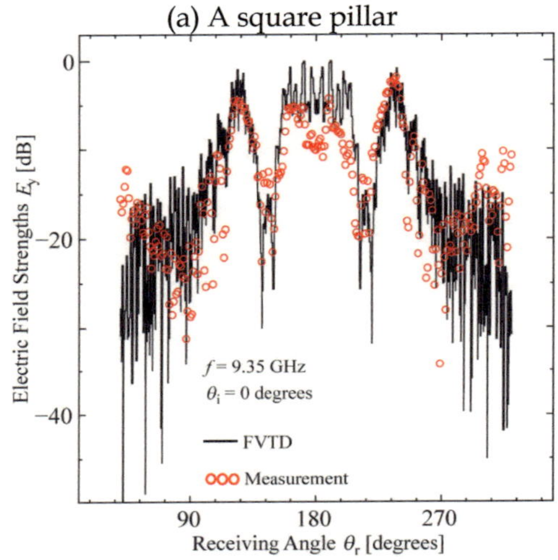

(c) A model with reinforced concrete walls

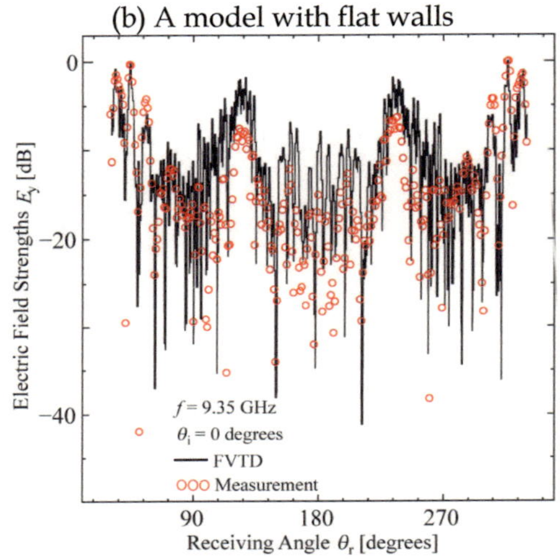

(d) A model with walls having round convexities

Fig. 12. Comparisons of the electric fields around buildings obtained by measurements and the FVTD method

analysis space the building model is and where the source wave is. Thus readers are advised to observe the individual simulation results by making comparison with the said figure. By so doing, it can be recognized again that the electric field strength distribution around the building is widely subject to change. First it is explained that the square pillar model is widely different from the other models in the electric field strength distribution. It is likewise explained that to faithfully simulate the electric field strength distribution, it is necessary for the building to be actually approached to some extent. Secondly none of great difference can be seen with either of the model devoid of reinforcement bar in the wall and the one in possession of such reinforcement bar. From this it can safely be said that in the simulation, it is not so much necessary to keep in mind the presence and absence of the reinforcement bar. Finally it is made known that existence of round convexities on the outside wall brings about great change to the electric field strength distribution in the inside and outside of the building. From this it can safely be affirmed that with the shape of the wall surface, conducting simulation considering as profoundly as possible encourages the electric field strength distribution close enough to reality to be obtained very easily.

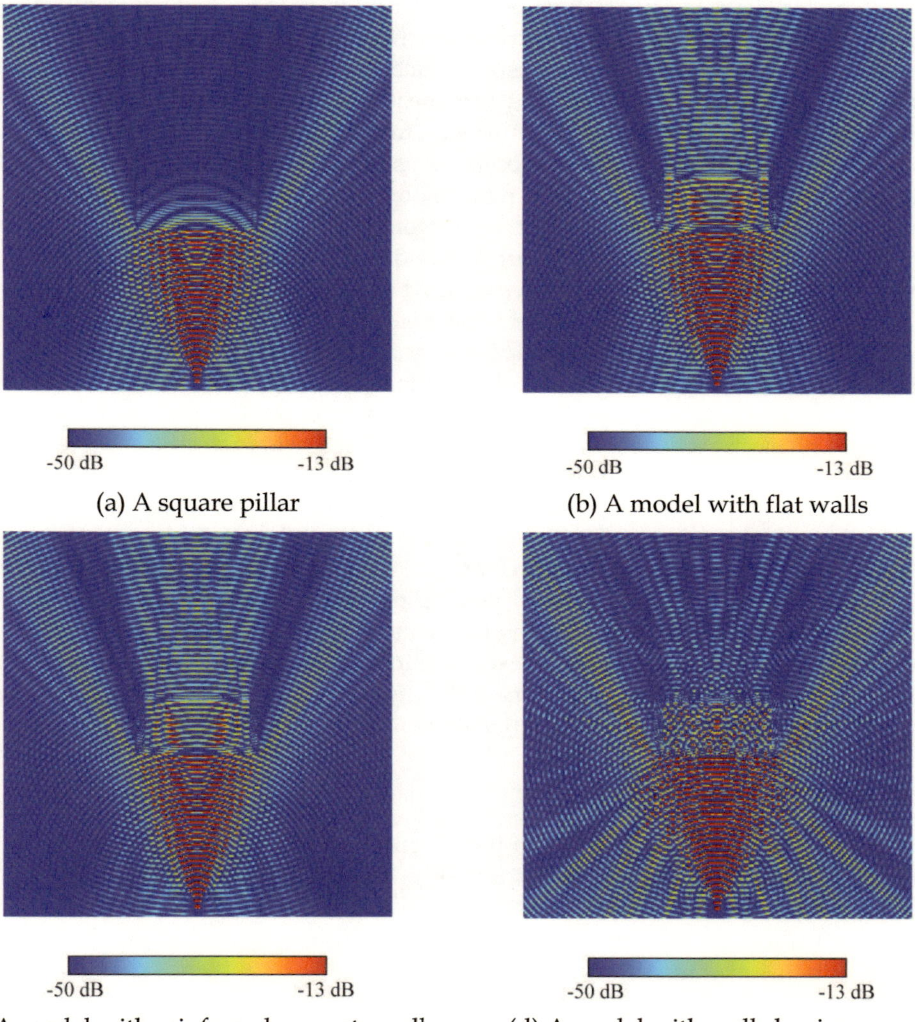

(a) A square pillar

(b) A model with flat walls

(c) A model with reinforced concrete walls

(d) A model with walls having round convexities

Fig. 13. Distribution of electric field strength obtained by the FVTD method

5. Concluding remarks

The electromagnetic waves propagating in the vicinity of the building or through the inside of it are influenced by the shape of the wall of the building or structure of it, and therefore the propagation becomes a complicated one. This brings about a portion where sufficient

electric field strength can be obtained and a portion where such a factor cannot be obtained. This is a matter of seriousness with wireless communication. To solve this problem, it is necessary to be acquainted with first of all the electric field strength distribution in the vicinity of the building. Electromagnetic wave propagation simulation is quite an effective measure to be very easily acquainted with such electric field strength distribution in the vicinity of the building, but simulation results different from the actual value are liable to be obtained depending on the occasion in a manner how the structure or shape of the wall of the building should be considered. Keeping such a situation in mind, the electric field strength distribution around the building was measured using a scaled-down model of the building having shapes of various walls or structure of them. As a results, it is explained that: (1) It is advisable to deal with the building as the one comprised of a wall and inside structure rather than the one as a square pillar model filled with concrete to the extent of the inside. (2) It is unnecessary to consider the structure of the wall inside so much as with the case of the reinforced concrete, but it is understood that the surface shape such as of tiles is preferable to consider. Meanwhile by investigating the dielectric constants of the concrete or cell sizes based on the comparison between the measurement and the simulation results, it is shown that simulation close enough to actual measurement becomes possible.
Despite the above, no improvement has been attempted with the simulation to allow the electromagnetic waves to reach any place, taking account of the fact that there are not a few places where no electromagnetic waves are reached. Such being the case, a project is under way to develop new methods in future to decrease the regions where none of sufficient electric field strength is obtained due to change in the wall shape, taking up the examples actually full of problematic points. With the effectiveness of the method, recognition is to be made as a matter of course in accordance with the measurement method introduced in this chapter. When such research advances, possibilities are wide spread before us, availing ourselves of the technology with accomplishment of successful monitoring in a detailed manner, explaining what objects are in existence in what a place in what a situation in the building. This is connected to development of information communicating technology by which peoples' life is steered into a direction of a more abundant and wealthier state.

6. Acknowledgments

Part of this research has been conducted with a grant provided by the Ministry of Internal Affairs and Communications of Japan's Strategic Information and Communications R&D Promotion Programme (SCOPE) (082309008) and the Ministry of Education, Culture, Sports, Science and Technology of Japan's Grant-in-Aid for Young Scientists (A) (21686035), for which the authors including myself feel deeply grateful.

7. References

Axiotis, D. I & Theologou, M. E. (2003). An Empirical Model for Predicting Building Penetration Loss at 2GHz for High Elevation Angles. *IEEE Antennas and Wireless Propagation Letters,* Vol.2, p.234--237, 2003.

Ikegami, F.; Yoshida, S.; Takeuchi, T. & Umehira, M. (1984). Propagation Factors Controlling Mean Field Strength on Urban Streets. *IEEE Transactions on Vehicular Technology,* Vol. 32, No. 8, pp. 822–829.

Kita, N.; Yamada, W. & Sato, A. (2007). New Method for Evaluating Height Gain at Subscriber Station for Wireless Access Systems in Microwave Band. *IEICE Transactions on Communications,* Vol.E90-B, No.10, pp.2903-2914.

Kitao, K. & Ichitsubo, S. (2008). Path Loss Prediction Formula in Urban Area for the Fourth-Generation Mobile Communication Systems. *IEICE Transactions on Communications,* Vol.E91-B, No.6, pp.1999-2009.

Landron, O.; Feuerstein, M. J. & Rappaport, T. S. (1996). A Comparison of Theoretical and Empirical Reflection Coefficients for Typical Exterior Wall Surfaces in a Mobile Radio Environment. *IEEE Transactions on Antennas and Propagation,* Vol.44, No.3, pp.341-351.

Lim, J.; Koh, I.; Park, Y.; Moon, H.; Jo, H.; Yook, J. & Yoon, Y. (2008). Improving the Accuracy of Ray Tracing Estimation Considering Inhomogeneous Building Surfaces in Urban Environments. *IEICE Transactions on Communications,* Vol.E91-B, No.12, pp.4067-4070.

Matsunaga, M.; Matsunaga, T. & Sueyoshi, T. (2009). An analysis of the effects of wall shapes on electromagnetic waves propagating around buildings. *Proceedings of the 39th Microwave Conference,* pp.990- 993, Rome, Italy.

Matsunaga, T.; Uchida, K. & Noda, T. (1988). Progation of Electromagnetic Waves in a Concrete Tunnel with a Step-Junction. *IEICE Transactions on Communications Japanese Edition,* Vol.J71-B, No.2, pp.309-311.

Matsunaga, T.; Uchida, K. & Kim, K. (1996). Electromagnetic Wave Propagations in Two Dimensional Tunnels with Fundamental Junctions. *IEICE Transactions on Communications Japanese Edition,* Vol.J79-B2, No.7, pp.399-406.

Matsuoka, T.; Matsunaga, M. & Matsunaga, T. (2008a). Analysis of Wave Propagation in a Concrete Building Model by the CIP Method. *Proceedings of the 2008 International Symposium on Antennas and Propagation,* pp.798-801, Taipei, Taiwan.

Matsuoka, T.; Matsunaga, M. & Matsunaga, T. (2008b). An Analysis of the Electromagnetic Waves Radiated from a Line Source which is Close to a Concrete Wall by using the CIP Method. *IEEJ Transactions on Fundamentals and Material,* Vol. 128, No. 2, pp.53-58.

Matsuoka, T.; Oshinomi, T.; Matsunaga, M. & Matsunaga, T. (2009). A Measurement Method of Electrical Parameters of Dielectric Materials by Using Cylindrical Standing Waves. *Proceedings of the 2009 International Symposium on Antennas and Propagation,* pp. 584-587, Bangkok, Thailand.

Uchida, K.; Matsunaga, T.; Kim, K. & Han, K. (1996a). FVTD Analysis of Electromagnetic Wave Propagation in Two Dimensional Tunnels with Fundamental Junctions. *IEICE Transactions on Electronics Japanese Edition,* Vol.J79-C1, No.7, pp.210-216.

Uchida, K.; Han, K.; Ishii, K.; Matsunaga T. & Kim, G. (1996b). An FVTD Version of Berenger Absorbing Boundary Condition for a Lossy Medium. *IEICE Transactions on Electronics,* Vol.E79-C, No.11, pp.1625-1627.

Walfisch, J. & Bertoni, H. L. (1988). A Theoretical Model of UHF Propagation in Urban Environments," *IEEE Transactions on Antennas and Propagation,* Vol. 36, No. 12, pp.1788–1796.

Xia, H. H. (1997). A Simplified Analytical Model for Predicting Path Loss in Urban and Suburban Environments. *IEEE Transactions on Vehicular Technology,* Vol.46, No.4, pp.1040-1046.

Yee, K. S. & Chen, J. S. (1994). Conformal Hybrid Finite Difference Time Domain and Finite Volume Time Domain. *IEEE Transactions on Antennas and Propagation,* Vol.42, No.10, pp.1450-1455.